OXFORD
UNIVERSITY PRESS

ASPIRE
SUCCEED
PROGRESS

GEOGRAPHY

for Cambridge International AS & A Level

Revision Guide

David Davies

Oxford excellence for Cambridge AS & A Level

OXFORD

OXFORD
UNIVERSITY PRESS

Great Clarendon Street, Oxford, OX2 6DP, United Kingdom

Oxford University Press is a department of the University of Oxford. It furthers the University's objective of excellence in research, scholarship, and education by publishing worldwide. Oxford is a registered trade mark of Oxford University Press in the UK and in certain other countries

First published in 2016

British Library Cataloguing in Publication Data
Data available

978-0-19-830703-7

14

Paper used in the production of this book is a natural, recyclable product made from wood grown in sustainable forests. The manufacturing process conforms to the environmental regulations of the country of origin.

Printed in Great Britain by Ashford Colour Press Ltd., Gosport

The questions, marks awarded, sample answers and comments that appear in this book were written by the author. In examination, the way marks would be awarded to answers like these may be different.

Figure 10.4 (p. 142) is reproduced by permission of Cambridge International Examinations.

Acknowledgements
The publishers would like to thank the following for permissions to use their photographs:

Cover image: Shutterstock; p42: © (2016) United Nations. Reprinted with the permission of the United Nations.; p84: © Petegar/Istock; p85: Land Transport Authority of Singapore; p86: Aran Ho Yeow Yong/Wikipedia; p91: © Image Source Plus / Alamy Stock Photo; p119: cccarto.com.

Artwork by GreenGate Publishing Services and OUP.

We are grateful to Carlos Vargas-Silvas of University of Oxford for permission to use the material on page 65.

This Revision Guide refers to the Cambridge International AS & A Level Geography (9696) Syllabus published by Cambridge Assessment International Education.

This work has been developed independently from and is not endorsed by or otherwise connected with Cambridge Assessment International Education.

Contents

AS Level (Core)

A Level (Advanced Options)

1 Hydrology and fluvial geomorphology

1.1 The drainage basin system

A drainage basin is the area drained by a river and its tributaries. A drainage basin system is an open system as water can be added as an input or lost as an output. Within the system water can flow/be transferred in a number of ways from a variety of stores.

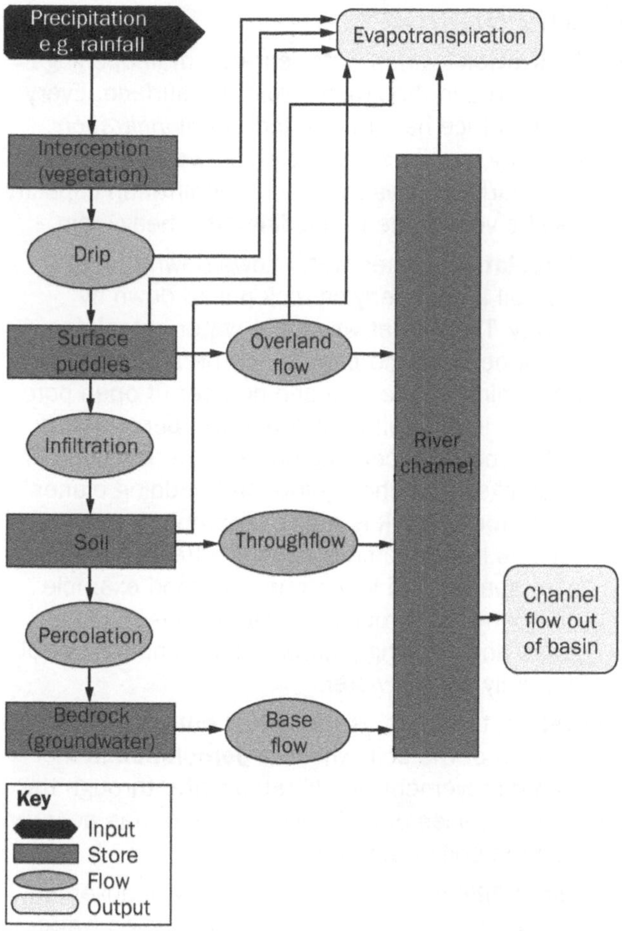

Fig 1.1 The processes operating within drainage basin systems

Inputs and outputs

Inputs are the addition of water to a drainage basin in the form of precipitation. The inputs can occur in a number of forms (rain, snow, hail, etc.) and at different times, intensities and frequencies throughout the year. In some parts of the world, such as high mountainous areas and in polar and temperate climates, snow and glacial meltwater is an important input, especially in spring and summer when it can produce floods.

Outputs are the losses of water from a drainage basin in terms of evaporation, evapotranspiration and river/channel flow.

- **Evaporation** is the loss of water from the land surface and bodies of water as it transfers from a liquid to a gaseous state (water vapour) by application of heat. The rate at which it will take place will depend on the temperature; the higher the temperature the higher the rate of evaporation.
- **Evapotranspiration** refers to the combined loss from both evaporation and transpiration. **Transpiration** is the loss of water vapour from the stomata in the leaves of plants and trees. Rates of transpiration will depend on the amount and type of vegetation. The typical tropical forest ecosystem has very high rates of transpiration. More precipitation than evaporation or evapotranspiration occurs over the land but most of the Earth's evaporation (86 per cent) and precipitation (78 per cent) take place over the oceans.
- **River discharge** is the volume of water being discharged by a river. It is normally expressed as the volume of water passing a point in the river channel in a given unit of time. This is commonly the number of cubic metres per second – abbreviated as "cumecs".

Stores

The stores in a drainage basin are where water is held in some part of the drainage basin for any length of time from seconds – on a leaf, to hours and days in depression stores such as puddles and lakes or in the soil, to several thousand years in an underground aquifer. Stores include soil stores – water retained within the pore spaces of the soil (called **interstices)** and groundwater stores – water that has percolated and is held in rocks below the top of the water table.

- **Interception** and **interception stores** – this is precipitation which is intercepted on its way to the land surface by leaves, plants and trees. It slows up the arrival of rainwater at the surface and reduces the amount that reaches the ground as some or all will evaporate as it lies and flows over the leaves, stems and trunks of the trees and plants. It is important in that it reduces the amount of water available for overland flow/ surface runoff and therefore helps reduce the possibility of soil erosion and flooding.
- **Surface water or depression stores** – this is water that is stored on the surface in the form of puddles, ponds, streams and lakes – often above an impermeable surface or where the ground or soil below is fully saturated.

- **Soil** or **soil moisture stores** – this is water retained within the gaps and pore spaces in the soil. *Antecedent moisture* is the moisture retained in the soil *before* a rainfall event.
- **Groundwater stores** – this is water that has percolated downwards and is held in gaps such as cracks, joints, bedding planes, fault lines and pore spaces in rocks in the underground **aquifer**. The top of the aquifer is called the **water table**. The factors that influence the amount of water that reaches the groundwater store are precipitation amount and intensity, surface flow and throughflow. The time it takes is controlled mainly by the speed of infiltration and percolation which, in turn, are controlled by porosity and permeability of both the soil and the underlying bedrock.
- **Channel store** – this is the volume of water contained in the river channel. Once in the river channel water will flow to the sea or lake and be lost from the drainage basin system.

Flows

The main flows found within a drainage basin system can be either above ground or below ground:

Above ground

- **Throughfall** – this is precipitation that makes it directly to the land surface without being intercepted by the plant canopy. Some of this throughfall may be intercepted by leaves and this water may then flow off the leaves and drip to the ground as **dripflow** – some plants have developed *drip tips* and waxy shiny surfaces on their leaf surface to get water off the leaf surface quickly.
- **Stemflow** – is the flow of water from precipitation down the stems of plants. If they reach the trunks of larger trees the water will flow down the trunks of trees, termed **trunk flow**, to reach the land surface. The **interception** of precipitation by trees and other vegetation means that the soil may be protected from rainwater impact and the water that is intercepted is then released slowly to the land surface allowing it to infiltrate more easily. In areas which experience high, intense periods of rainfall this slow release of water will prevent excess overland flow.
- **Overland flow** – when water flows over the land surface. There are two types of overland flow – **channel flow** and **sheet flow**. Channel flow is when the water is flowing in small channels, **rills**, which are less than 30 cm in width/depth, in a defined stream or in a river channel. Sheet flow is normally a relatively rare event and takes place when there is a layer/sheet of water on the ground surface. It may occur in two ways. Firstly, when there is either **excess overland flow**, when rainfall or water arrives too quickly on the land surface and does not have enough time to infiltrate the soil, i.e. when rainfall intensity exceeds the infiltration capacity of the surface. Secondly, it may occur when water flows onto a relatively impermeable surface, such as a clay soil, which has a very slow infiltration capacity as it contains very tiny pore spaces (while a sandy soil may exhibit a fast infiltration capacity as it has large pore spaces). A clay soil may therefore have water quickly building up on its surface and this may then start to flow over the surface, as overland flow, possibly causing soil erosion. Without a protective cover of trees and vegetation soil erosion is a distinct possibility. The occurrence of overland flow will be increased on slopes, or when there is **saturation overland flow** – when all the open/pore spaces in the underlying soil and rock are filled with water, which means that water is forced to flow over the land surface.

Below ground

- **Infiltration** – when water enters small openings and pores in the ground from the surface. Every land surface has its own individual *infiltration capacity* i.e. the speed at which water enters that land surface. Areas with a low infiltration capacity can be very prone to flooding after heavy rain.
- **Percolation** – when water flows down through the soil and underlying rock pulled down by gravity. The rate at which the water percolates will depend on the ***porosity*** of the soil or rock – depending on the size and number of open pore spaces in the soil or rock and the ***permeability*** of the rock – depending on the size and number of cracks, fault lines, joints and bedding planes in the rock. Chalk is a good example of a porous rock as it may actually absorb water in its many pore spaces and limestone is a good example of a permeable rock as it usually has many joints and bedding planes within it, but does not normally absorb water.

Infiltration, therefore, is the actual entry of water into the surface of the soil, whereas ***percolation*** is the downward movement of infiltrated water through the pores and spaces of soil once the water has actually entered the soil or surface.

- **Throughflow** is the lateral (sideways) movement in soil of infiltrated water. It occurs when water that has infiltrated the surface is retained in the soil. The water then moves horizontally (parallel to surface) through the soil, down a slope towards a river channel, usually along well-defined lines of seepage (called **percolines**) that have been formed in the soil or above an impermeable layer (for example, when there is a clay layer in the soil called a **clay pan**).
- **Baseflow** (**groundwater flow**) is water that has infiltrated and percolated into the bed rock below the soil that then moves laterally under gravity or hydrostatic pressure in a downslope direction to feed springs and river channels. Baseflow will normally increase where conditions encourage infiltration and percolation such as during periods of steady rainfall or where the soils and/or underlying rocks/geology are permeable and porous.

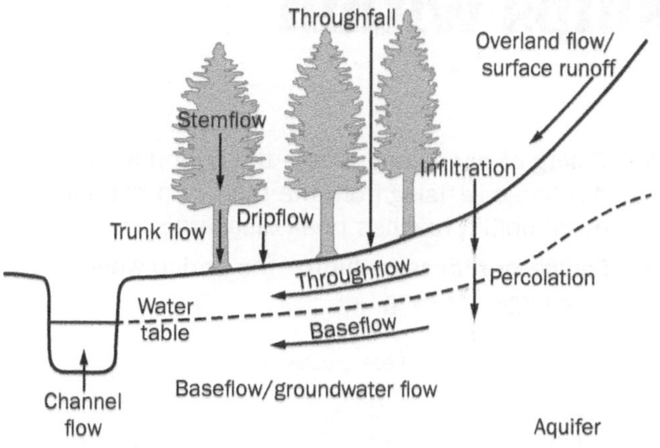

Fig 1.2 Flows

Underground water

Water tables

When water infiltrates the land surface, it becomes **groundwater**. It may then percolate, under the influence of gravity, through pores, cracks, joints and bedding planes and reach an area of saturation where all pores, joints, etc are full of water. The top of this saturated zone is called the **water table**. It may be variable in height depending upon the nature of the rock and the level of precipitation input and evapotranspiration output. The water table will generally mirror/follow the shape of the surface topography and water will flow, under the influence of gravity and by the hydraulic gradient, to a point in the river basin where it will appear either as a **spring** or by contributing to river discharge as baseflow. It may also be abstracted by humans in wells or boreholes.

The height of the water table will vary according to the season – winter or summer, wet or dry, and the amount of precipitation input and evapotranspiration output. Within an aquifer there will be a zone of permanent saturation, called the **phreatic zone**.

Recharge

Recharge of the groundwater takes place when water is added to the aquifer. Recharge takes place when precipitation on the land surface exceeds evapotranspiration and water then infiltrates the ground and percolates down to the aquifer. How long the groundwater takes to recharge will be controlled by the speed of infiltration and percolation.

Exam-style questions

1. Define the term drainage basin system as it applies to a river basin. [2]
2. Describe what is meant by the term interception storage. [2]
3. What are the zones found in an aquifer and its water table? [2]
4. Define the terms interception and stemflow.[4]
5. Define the terms throughfall and throughflow. [4]
6. Define the terms water table and springs. [4]
7. Identify and briefly describe two stores found in a river basin. [4]
8. Describe how groundwater recharge occurs. [3]
9. Describe how and when overland flow may occur. [3]
10. Explain how throughflow and groundwater flow (baseflow) occur. [4]
11. Describe the difference between infiltration and percolation. [4]
12. Briefly indicate how rates of infiltration might vary with the intensity of rainfall. [4]
13. Explain how precipitation received by a river basin may reach the river channel. [6]
14. Explain how water reaches, is stored in and removed from an aquifer. [6]
15. What is meant by inputs and outputs within a drainage basin system? [6]
16. With the help of a labelled diagram, show how water makes its way through a drainage basin system. [6]
17. Explain how water from surface storage reaches groundwater storage. [6]
18. Define the terms precipitation intensity and infiltration capacity. [4]
19. Describe how precipitation intensity might affect the surface flow of water in a river basin. [4]

1.2 Discharge relationships within drainage basins

The components of hydrographs (storm and annual)

There are two sorts of hydrographs – **storm** and **annual**. The term **river regime** may be seen in some textbooks as an alternative to the term annual hydrograph.

An **annual hydrograph** displays the pattern of seasonal variation that takes place to a river's discharge in a typical year. It is shown by graphs like the one in fig 1.3, where the peak in the summer months is explained by snow melt or a summer monsoon.

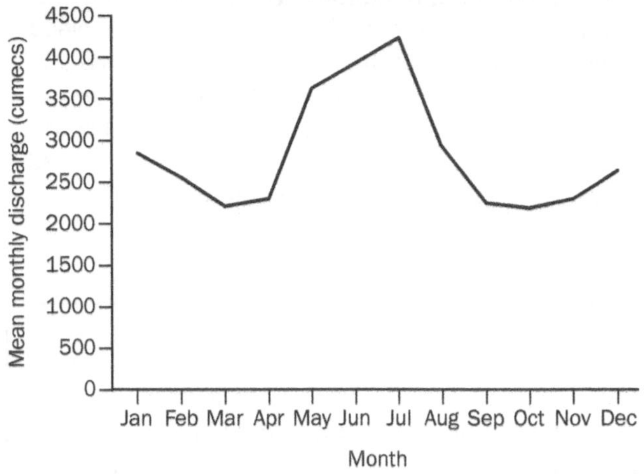

Fig 1.3 A typical annual hydrograph

Storm hydrographs show how river discharge responds to a rainfall event. It has a number of important components. The **lag time** between **rainfall peak** and **flood discharge peak** indicates how quickly the precipitation is reaching the river channel after a rainfall event.

Storm hydrographs allow an estimate to be made of the relative importance of the **quick flows** (mainly overland flow/surface runoff) and the **baseflow/ groundwater flow**.

The steepness of the **rising** and **falling/recession limbs** also indicates how quickly the precipitation input is reaching the river channel and being taken away by it. A storm hydrograph plots two variables – the rainfall received during a rainfall event (in millimetres.) and the river discharge (measured in cubic metres per second – shortened to cumecs).

A typical storm hydrograph (fig 1.4) will have several key components:

- **Rainfall peak** – when the highest amount of rainfall occurs during a rainfall event. If it is the same each hour the median time is taken – i.e. the middle of the event.
- **Peak discharge or peak flow** – when the highest amount of discharge occurs.
- **Lag time** – the time between the rainfall peak and peak discharge.
- **Rising or ascending limb** – the period when discharge is rising from the start of a rainfall event until it reaches peak discharge.
- **Falling or recession limb** – the period when discharge is falling.

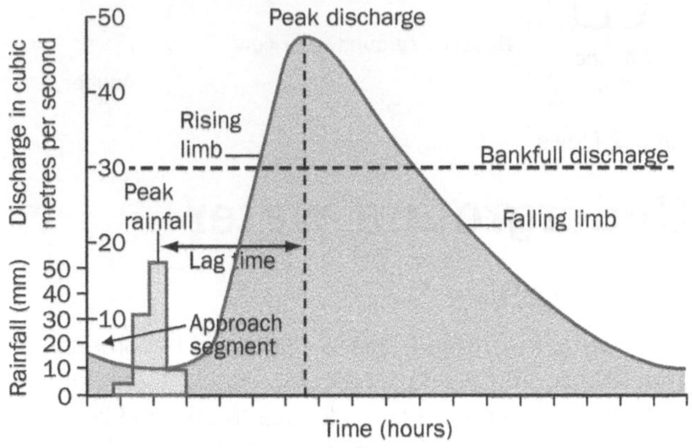

Fig 1.4 A typical storm hydrograph

In addition to these key components, several other pieces of information may be provided, such as:

- **Approach segment** – the period of time before water from the rainfall event gets into the river channel.
- **Bankfull discharge** – when the discharge of the river is at the top of the river banks – any further added discharge will cause the river to flood the surrounding land and flow over its floodplain.
- The different **flows** that make up the total discharge. Each of the three flows will arrive, and peak, at different times – the quickest to arrive and peak will be the overland flow/surface runoff, followed by the next fastest – throughflow and, lastly, baseflow/groundwater flow.

Influences on hydrographs

The shape of a hydrograph may be influenced by several climate factors, such as how the precipitation amount, intensity and type, temperatures and evapotranspiration vary over the year.

Climate

Precipitation type and intensity

Precipitation type is the form in which precipitation is received by the drainage basin system. Rain will be available to the system very quickly, whereas snow will delay the impact on the system; but it may then have a dramatic effect on the hydrograph as it may be released quickly as meltwater.

Precipitation intensity is the rate at which precipitation is received at the ground surface – it is the amount of precipitation in millimetres divided by the time.

altering river channels can also reduce the potential flood impact upon people living in a drainage basin. Extreme rainfall events, however, are still likely to result in flooding and so it is often the case of flood mitigation rather than flood prevention.

The soft engineering methods used in a river catchment can include:

- **Afforestation/Reafforestation** to increase interception and to protect and stabilise the soil.
- **Changing farming practices** in the catchment. This may include the use of terracing, contour ploughing, cover crops, seed drilling and strip crops to protect the soil from erosion and stop eroded sediment filling up river channels and reducing their capacity.
- **Preventing infrastructure development** on threatened floodplains.
- **Floodplain retreat** – some government agencies, along the Mississippi for example, are actually buying floodplain land from landowners and allowing these to flood during flood events as they reduce flood peaks further down the river.

CS Flooding in Pakistan – the Indus River Basin (2010)

Pakistan has a long history of flooding. Between 1950 and 2010, floods in the Indus river basin have killed a total of 8887 people, affected 109822 villages, and caused a cumulative **direct economic loss** of about $19 billion.

Impacts of the 2010 flood

- The 2010 flood, which affected all the provinces and regions of Pakistan, killed 1600 people, caused damage totalling over $10 billion, and inundated an area of about 38600 km². This flood was Pakistan's most damaging on record.
- In the country as a whole, the floods damaged nearly 2 million houses and displaced a population of over 20 million.
- Flood damage occurred mainly in the agriculture and livestock sector (50 per cent lost), followed by housing (16 per cent) and transport and communications (13 per cent). The prolonged inundation of large areas of cultivated land resulted in massive losses in the agriculture sector. Forecasts estimated that Pakistan's GDP growth rate of 4 per cent prior to the floods could turn to minus 2 per cent to minus 5 per cent followed by several additional years of below-trend growth.
- Crop losses were not just in vegetables and cereals but in crops such as cotton that supplied textile manufacturing, Pakistan's largest export sector. There was also the loss of over 10 million heads of livestock which, along with the loss of arable crops, reduced agricultural production by more than 15 per cent. As milk supplies fell by 15 per cent, the price of milk increased by 5 US cents per litre.
- Manufacturing companies, such as Toyota and Unilever Pakistan, had to make cuts in production, as infrastructure such as roads and bridges were destroyed and damaged.
- In September 2010, the International Labour Organisation reported that the floods had cost more than 5.3 million jobs. The province of Sindh in the south of Pakistan was most badly affected – about 1 million houses were damaged and of these, 66 per cent were completely demolished, while the remaining damaged houses were often no longer habitable. There was also extensive damage to schools, roads and bridges, telephone lines and electric supply lines.

What causes the flooding?

Flooding in Pakistan has generally been caused by the heavy concentrated rain received in the monsoon season in the months of July and August. However, the 2010 floods were a combination of natural and human factors that produced flood peaks that were far in excess of the river channels capacity to hold them.

The **natural factors** included:

- The **monsoon rainfall** in July 2010 was exceptional – some areas in Northern Pakistan received more than three times their annual rainfall in a matter of 36 hours. These abnormal monsoon rains amounted to double the 50-year average annual rainfall.
- A **steep topography** in the upper basin produced rapid overland flow/surface runoff

The **human factors** included:

- The Indus basin **lacks an appropriate flood policy** and has an **inadequate flood-control infrastructure**.
- **Lack of protective vegetation** due to river basin developments increased runoff into river channels. Deforestation in the upper catchment was a major factor.
- **Poor anticipation** of the scale of the flood – it was known in advance by a couple of days that around 4–10 km³ of water would pass through the Taunsa, Guddu and Sukkur dams, but no measures were taken to mitigate the effects of these huge volumes of floodwater downstream from the dams.

Flood mitigation measures

- The Government of Pakistan has been relying on a traditional flood control approach based on structural, hard engineering measures, through the building of dams and building of artificial levees. Between 1950–2010 the government spent $1.2 billion repairing flood damage, developing a flood-forecast system and building new levées. However the 2010 flood exposed weaknesses in these systems.
- The Pakistani government does not have an approved water policy, but there is a draft **National Water Policy** that recognises the need for appropriate flood management. This includes:
  - the continued construction of flood-protection facilities and the maintenance of existing infrastructure
  - reviewing the design and maintenance standards of existing flood protection structures
  - the establishment and promotion of **flood zoning**, and the enforcement of appropriate land uses in flood threatened areas
  - strict reservoir operating rules
  - the effective use of non-structural measures
  - the creation of flood response plans.

Structural measures

- 6 000 km of **levées** provide the bulk of flood protection. The levées now cover most of the critical stretches along the river channels and are the main flood-protection infrastructure. The height of these levées remains arbitrarily fixed at 1.8 meters (6 feet), which is higher than the previously observed peak flood levels in the basin. However, due to changes in channel shape the channel capacity at any location may change and require higher levées. Channel shape monitoring is needed to accurately determine the optimal levée height at different locations.
  The remote location and the inadequate maintenance of many levées has proved to be a major challenge.
- 1410 **spurs** (these are stone walls constructed transversely or obliquely to the flow direction to divert flooding at critical locations) that have been built since 1960 to protect the main towns and important infrastructure, such as major roads and railways.
- The 2010 flood demonstrated the effectiveness of the country's two main **reservoirs** at Mangla and Tarbela. However, sedimentation in both has significantly reduced their storage capacities. Pakistan currently has a water storage capacity equal to about 30 days of mean annual discharge, and this water is mostly used for irrigation and energy generation.

Non-structural measures

Pakistan is looking to improve its flood forecasting and early flood warning systems. At present it does not cover the whole Indus river basin. The Indus is a trans-boundary river – parts of its catchment are in neighbouring countries. Currently, cooperation with these neighbouring countries is limited so that key information regarding precipitation input and river discharge from the important upper areas of the catchment is both insufficient and delayed.

Exam-style questions

1. Briefly explain what is meant by the term recurrence interval in flood prediction. [3]
2. Describe and explain the effects on a drainage basin of building dams for water storage.[6]
3. Using examples, explain how human activities in a drainage basin may affect channel flow. [8]
4. Explain how urban growth could affect the flows within a river channel. [6]
5. How can changes in land-use affect flows and stores in a drainage basin? [6]
6. How can the abstraction (removal) and the storage of water by humans affect flows and stores within a drainage basin? [6]
7. Explain how river floods might be predicted. Giving examples, describe the methods that could be used to reduce the effects of flooding within a drainage basin. [8]
8. What are the main causes of river floods and to what extent can their effects be limited by human intervention? [8]
9. Describe the extent to which it may be possible to predict and prevent the flooding of rivers? [8]
10. Explain the causes of river floods. Describe the effects such conditions may have on the river channel and the landforms found in river valleys. [8]
11. Outline the causes of floods in a river catchment and explain how such floods may be either prevented or ameliorated (reduced). [8]

2 Atmosphere and weather

2.1 Diurnal energy budgets

The diurnal (daily) energy budget accounts for how much energy is received by the Earth and its atmosphere from the Sun each day, how much energy is then lost to space, as well as how much energy is retained by the Earth and its atmosphere. The main parts of the diurnal energy budget are summarised in fig 2.1.

Quantifying changes in the incoming, absorbed and outgoing radiation is required to accurately model the Earth's climate and predict climate change.

During the hours of daylight, the Earth receives incoming solar radiation, some of which is then reflected by clouds and the Earth's surface; some is absorbed into the surface and sub-surface, which then transfers this heat to the overlying atmosphere – an example of sensible heat transfer. This energy may then be lost to space by long-wave earth radiation. Latent heat transfer may take place when some of this energy is absorbed by, or released from a substance during a phase change, for example from a gas to a liquid or a solid or vice versa, for example, when water changes to water vapour by evaporation.

The process of evaporation may then be reversed when the water vapour is changed back to water by the process of condensation.

During the night, the Earth continues to emit long-wave earth radiation and latent heat transfer can take place resulting in the formation of dew. Dew is droplets of water that appear on exposed objects in the early morning or evening due to condensation. As the exposed surface cools by radiating its heat, water vapour condenses at a rate greater than it can evaporate, resulting in the formation of water droplets. Sensible heat transfer may again take place and absorbed energy is returned to the Earth.

Incoming (shortwave) solar radiation

The Earth's main source of energy is **incoming short wave solar radiation** from the Sun. This energy fuels the Earth's weather system.

How much incoming solar radiation is received by the Earth is controlled by **four** factors:

1. The **solar constant** – the energy released by the sun. This does vary in amount as it is linked to the amount of sunspot activity on the Sun.
2. The **distance of the Earth from the Sun** – our distance is not constant due to orbital rotation; this can cause a 6 per cent variance in the amount of solar energy being received.
3. The **altitude of the Sun** in the sky – as the Earth is a sphere, the amount of incoming solar radiation being received varies greatly depending on the angle of the Earth's surface that it is falling on. The same amount of solar radiation above 60° north and south of the Equator has to cover twice the land surface compared to the Equator.

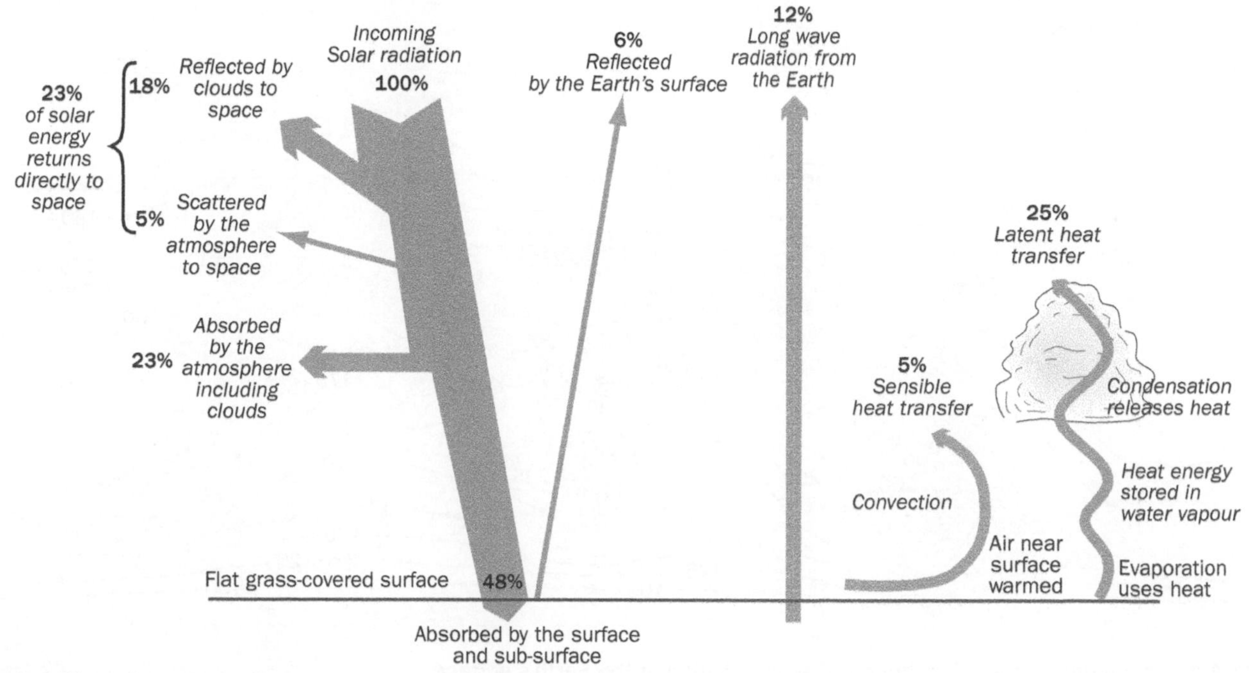

Fig 2.1 Diurnal energy budget

4. The **length of night and day** experienced – as the Earth is tilted at an angle of 23½° there is a long period of time during the year when areas north of the Arctic Circle (66½° north of the Equator) and south of the Antarctic Circle (66½° south of the Equator) do not receive any incoming solar radiation; the areas between the two Tropics of Cancer and Capricorn (at 23½° north and south of the Equator, respectively) receive high amounts all year round (fig 2.2).

Notice that the solar radiation received at the Equator is lower than that at the Tropics. This is because of cloud cover over the Equator which is produced by the uplift of the warm and moist air.

The **absorption** is mainly by the gases ozone, water vapour and carbon dioxide and minute particles of ice and dust in the atmosphere.

Reflected solar radiation

As the radiation from the Sun passes through the atmosphere, some is absorbed by liquids, gases and solids. Some is reflected and scattered, especially by the tops of clouds. The amount of energy that is reflected by a surface is determined by the reflectivity of that surface, called the **albedo**. Albedo is expressed as a percentage. A high albedo means the surface reflects the majority of the radiation that hits it and absorbs the rest. A low albedo means a surface reflects a small amount of the incoming radiation and absorbs the rest. For instance, fresh snow reflects up to 95 per cent of the incoming radiation. Generally, dark surfaces have a low albedo and light surfaces have a high albedo.

Thin clouds reflect 30–40 per cent, thicker clouds 50–70 per cent, while towering cumulonimbus clouds can reflect up to 90 per cent.

Energy absorbed into the surface and subsurface

This incoming short wave solar radiation is converted into heat energy when it reaches the surface of the Earth. Incoming solar radiation exceeds outgoing heat energy for many hours after noon and equilibrium is usually reached in mid afternoon, from 3–5 p.m.

The amount of energy absorbed by the surface and sub-surface during daylight hours can be affected by a variety of factors, such as the presence of large bodies of water and snow cover. These can have a high albedo and reflect as much as 80–90 per cent of the incoming radiation.

Some of the incoming energy will be transferred from the surface into the sub-surface soil and rocks by conduction. A light-coloured soil or rock, like chalk, is a poor conductor, so heating will mainly be confined to the surface; this explains the high temperatures of 50–60 °C recorded in hot deserts in daytime. In contrast, a dark volcanic soil or dark rocks like basalt and slate, with low albedos of 5–10 per cent, will absorb heat well.

The moisture content of a sub-surface soil will also affect its ability to conduct heat. A coarse sandy soil that has large pore spaces will be a poor conductor of heat, so the heat will concentrate on the surface, whereas a soil with a high water content will conduct heat down into the sub-surface and so the soil surface will be cooler.

Wind can remove heat quicky from a land surface, while the amount of cloud cover and the amount of water vapour in the atmosphere will affect the amount of reflection of incoming radiation and therefore the amount of incoming radiation that will reach the surface and sub-surface.

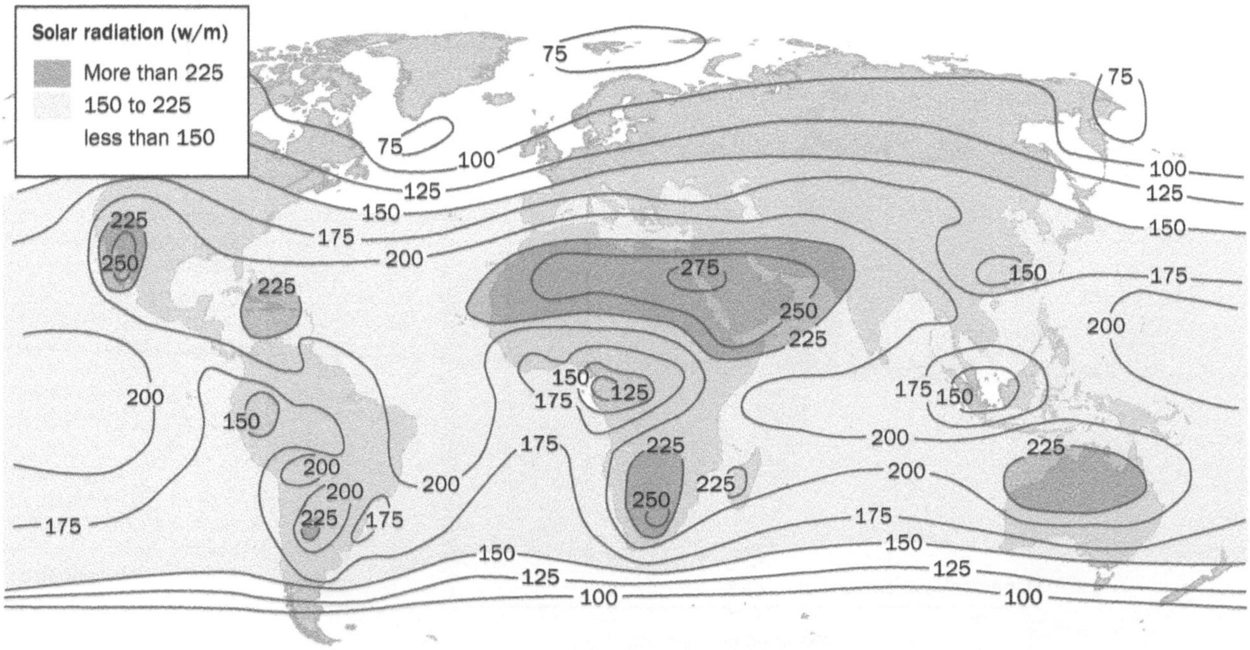

Fig 2.2 The annual average distribution of solar radiation at the Earth's surface

Long-wave radiation

As the Earth's surface warms up, it then radiates energy, at a longer wavelength, back to the atmosphere as Earth or terrestrial **long-wave radiation**. Of this, 94 per cent is absorbed by the greenhouse gases in the atmosphere such as carbon dioxide, water vapour and methane, warming the atmosphere and producing the natural **greenhouse effect**. The remaining 6 per cent is lost to space.

Without the greenhouse effect the Earth's average temperature would be about 33 °C colder than it is at present and life as we know it would not be possible.

Sensible heat transfers

During the day, when incoming short-wave solar radiation enters the atmosphere it is absorbed by the land surface before being re-radiated as long-wave Earth radiation which then heats the air above it. This is an example of a **sensible heat transfer**. **Sensible heat** is the energy required to change the temperature of a substance **with no phase change**. The temperature change can come from the absorption of sunlight by the soil or the air itself. Or it can come from contact with the warmer air caused by release of latent heat (by direct conduction).

Latent heat transfers

Latent heat is the energy absorbed by or released from a substance during a phase change from a gas to a liquid or a solid or vice versa, for example, when water changes to water vapour by **evaporation**. When heat is taken from the atmosphere to help with this process it will result in the atmosphere being cooled.

When the process of evaporation is reversed, for example when water vapour is changed to water by **condensation**, heat energy is released into the atmosphere which will heat up as a result. The main processes that do this type of transfer are radiation, conduction and convection.

During the night, there is no incoming solar radiation, which means that the only source of energy is the radiation that is being held/retained within the atmosphere. The main energy flow is therefore a net loss of heat from the land, which cools the air from the surface upwards.

Exam-style questions

1. Define the terms sensible heat transfer and latent heat transfer. [4]
2. Define the terms solar radiation and earth (terrestrial) radiation. [4]
3. Explain one way in which solar radiation is reflected. [3]

2.4 The human impact

The enhanced greenhouse effect and global warming

Evidence for the greenhouse effect

The **greenhouse effect** is the process by which the Earth's atmosphere is warmed as certain gases within the atmosphere absorb some of the long-wave earth radiation being emitted by the Earth and re-radiate some of it back. The most common greenhouse gases are Water Vapour, Carbon Dioxide, Methane, Chlorofluorocarbons (CFCs) and Nitrous Oxide.

The greenhouse gases allow the incoming short-wave solar radiation to pass through them, but they are then very effective in trapping the outgoing long-wave terrestrial radiation. Without these greenhouse gases life on the Earth as we know it could not exist. The greenhouse gases combine to raise the average temperature of the Earth by 33 °C.

Before the rapid growth of the human population and its various activities in the last 200 years, which have altered the amount of greenhouse gases, the atmosphere was fairly balanced with the carbon dioxide being produced by animals and humans equalling the amount being taken up in plants as part of the process of photosynthesis.

The **enhanced greenhouse effect (EGE)** and **global warming** are the terms used to describe the build-up of greenhouse gases and the impact it has had in the last 200 years by human actions and activities.

One of the most significant changes in the last 60 years has been the rise in carbon dioxide from 315 ppm to 392 ppm today (fig 2.11). Increasing amounts of heat are now being retained in the Earth's atmosphere leading to a global rise in temperatures.

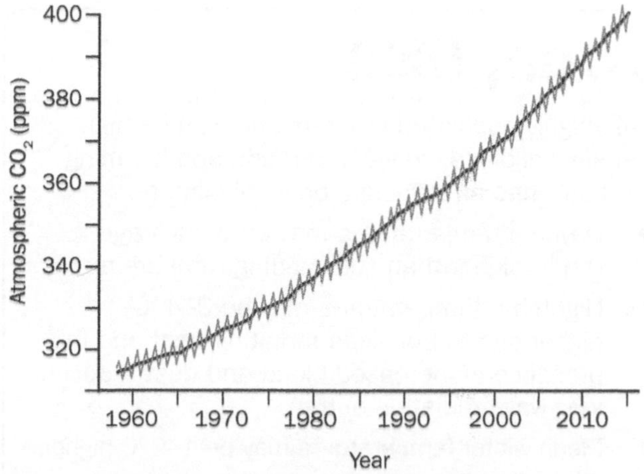

Fig 2.11 Changes in atmospheric CO_2 registered at the Mauna Loa Observatory (Hawaii, 1960–2015). In February 2015, the level of CO_2 reached 400 ppm.

Possible causes

The rise in **carbon dioxide** started with:

- clearing of forests by burning – this has the double effect of increasing carbon dioxide levels but also of removing trees which convert carbon dioxide to oxygen – and land cultivation
- industrialisation since the nineteenth century has put large amounts of carbon dioxide into the atmosphere from the burning fossil fuels.
- emissions from internal combustion engines and jet engines.

Methane is the second biggest contributor to global warming. It is increasing by 1 per cent per year, but it absorbs 25 times more heat than the equivalent amount of carbon dioxide. Methane may be produced from:

- cattle, which emit 100 million tonnes per year
- wet rice fields
- natural wetlands – both of which release methane by decomposing organic matter
- melting of permafrost in Arctic areas which releases methane from the organic matter that was previously frozen and is now melting as a result of global warming.

CFCs are human produced synthetic chemicals used as propellants in spray cans, coolants in fridges, freezers and air conditioning systems. Though alternatives to this chemical exist, they are not being fully used and the amount of CFCs is increasing by 6 per cent a year. The problem with CFCs is that they are 10 000 times more efficient at trapping long-wave earth radiation than carbon dioxide.

There are also concerns about the possible contribution from **urban heat islands (UHIs)** to global warming. An urban heat island is a city or urban area that is significantly warmer than its surrounding rural areas, due to human activities. The temperature difference is usually larger at night than during the day, and is most apparent when winds are weak. The main cause of the urban heat island effect is from the modification of land surfaces, though waste heat generated by energy usage is a secondary contributor. As a population centre grows, it tends to expand its area and increase its average temperature. Monthly rainfall may also be greater downwind of cities, partially due to the UHI. Increases in temperature within urban centres may also increase the length of growing season, and decreases the occurrence of weak tornadoes. The UHI can also decrease air quality by increasing the production of pollutants such as ozone, and decrease water quality as warmer water flows into local rivers and streams and puts stress on their ecosystems.

Impacts of global warming

The effect of global warming has been to increase the Earth's average temperature by about 0.5 °C since 1950 and to potentially produce a further increase of 2 °C by 2100.

This will have several impacts (fig 2.12) including:

- polar ice sheets and glacier ice melting
- rising sea levels – threatening low lying countries such as the Maldives, Bangladesh, several small Pacific Island nations such as Kiribati, Tuvalu, Marshall Islands, Vanuatu, and the Netherlands; the level of economic development (MICs/LICs) in providing coastal defences will vary enormously
- changing climatic patterns – a pole-ward shift of climatic belts
- an increase in dynamism of climatic systems and more extreme weather events such as storms, hurricanes, floods and droughts; the greater amount of heat producing more dynamic systems, when linked with rises in sea level, could lead to increased evaporation and rainfall and more frequent and larger storms
- conversely, the greater amount of heat in the atmosphere means that more water vapour can be held in the atmosphere, therefore producing less rainfall, extinction of plant and animal species.

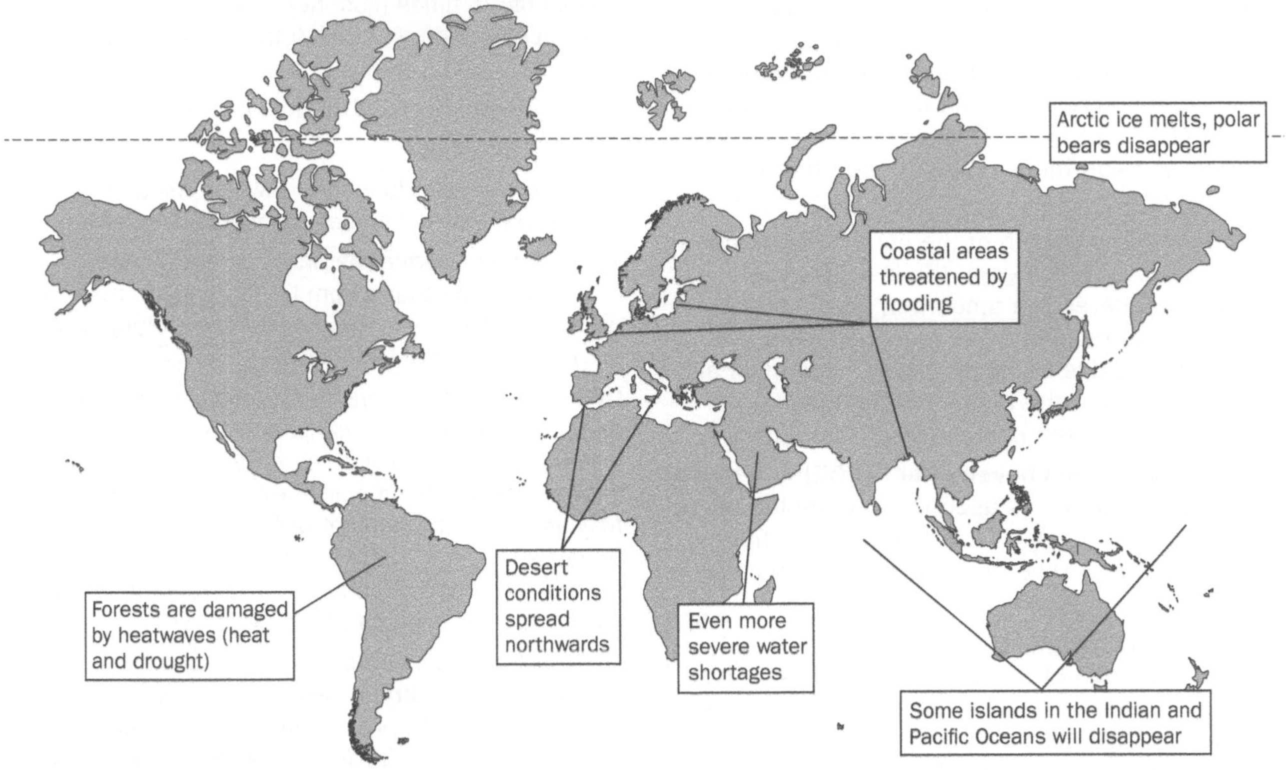
Arctic ice melts, polar bears disappear
Coastal areas threatened by flooding
Forests are damaged by heatwaves (heat and drought)
Desert conditions spread northwards
Even more severe water shortages
Some islands in the Indian and Pacific Oceans will disappear

Fig 2.12 Worldwide impacts of global warming

CS Los Angeles, California, USA

The Los Angeles metropolitan area is the second-largest metropolitan area, after New York, in the USA with a population of 18.2 million people. Its land area is 12 562 km², making it the largest in the United States.

In the 1930s, Los Angeles was nothing like it is today. It was covered by irrigated orchards with a high temperature of around 36 °C. However, as urbanisation took place, vegetation and trees were replaced with concrete and metal and the average temperature of Los Angeles increased steadily and reached 41 °C, and higher, in the 1990s. Fig 2.13 shows how average annual temperatures have increased in Los Angeles from 1878–2008.

This highly populated urban region with its high density buildings, congested roads and booming economy has an influence on local climate:

- Daytime temperatures may be on average 0.6 °C higher than surrounding rural areas.
- Nighttime temperatures may be 3–4 °C higher due to buildings radiating heat and the presence of increased cloud and dust reducing long-wave radiation output.
- Mean winter temperatures may be 1–2 °C higher.
- Mean summer temperatures may be 5 °C higher.
- Mean annual temperatures may be 0.5 to 1.5 °C higher than the surrounding areas.

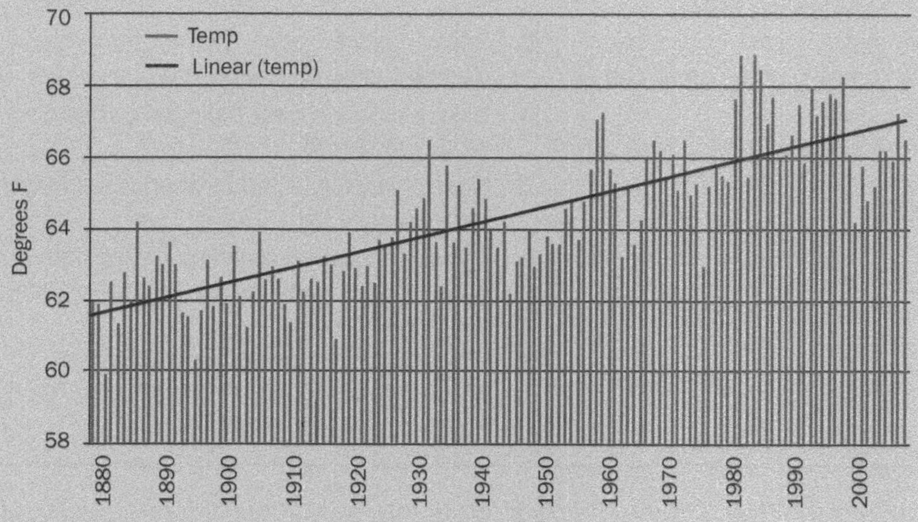

Fig 2.13 Change in annual temperatures in Los Angeles, 1987–2008

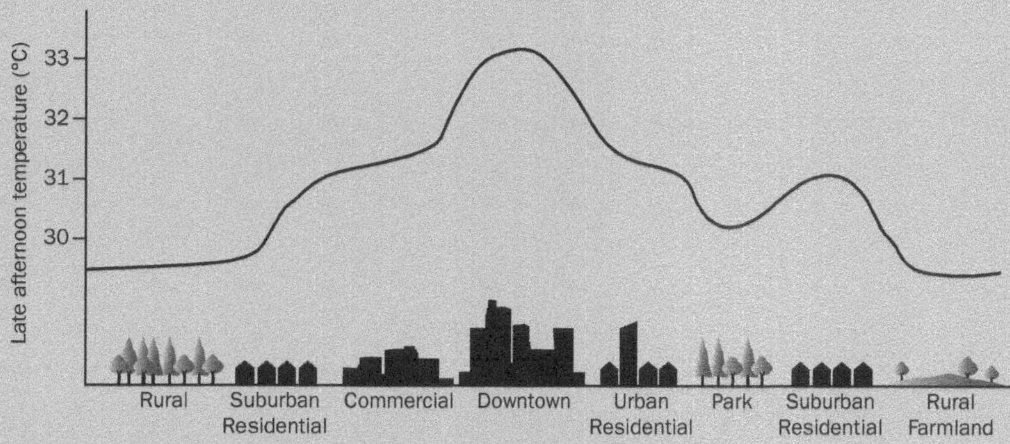

Fig 2.14 A typical thermal profile of an urban area in the United States

The reasons for these differences include:

- **Urban heat island**: buildings in urban areas absorb more radiation than vegetation. This is then re-radiated back to the atmosphere at night, raising temperatures and giving the urban heat island effect, which has an effect on the thermal profile of an urban area (fig 2.14).

In this typical thermal profile of the UHI (fig 2.14), the rural thermal field is interrupted by a steep temperature gradient at the rural/urban boundaries. There is then an increasing temperature until the highest temperature point is reached in the urban core or city centre. The island-shaped pattern is not quite uniform with a few peaks and depressions due to the presence of particularly hot points (i.e. micro urban heat islands) associated with features such as car parks/parking lots, shopping malls, industrial facilities, etc, and cold points due to features such as parks, fields, water bodies such as lakes, ponds and rivers, etc.

Temperature is not the only change – London receives approximately 270 fewer hours of sunlight than the surrounding countryside due to clouds and smog.

Building materials are usually very good at insulating, or holding in heat. For example, concrete can hold roughly 2000 times as much heat as an equivalent volume of air. This insulation makes the areas around buildings warmer. Therefore, buildings in cities tend not to reflect heat but absorb it. The albedo of urban areas is also lower, allowing for the greater absorption of heat energy. The buildings become stores of heat, which can then be released during the night. In addition, there is less evaporation so less energy is needed for the evaporation process, making it more available in the form of heat.

- **Convectional activity**: the higher temperatures and convectional heating may lead to an increased likelihood of thunder storms and hail in urban areas.
- **Wind**: buildings also provide increased friction with the wind which produces lower wind speeds – up to 30 per cent lower – than in rural areas. However, high-rise buildings and skyscrapers may channel winds and so increase wind speeds in the gaps between them.

Exam-style questions

1. Explain how an increase in greenhouse gases may cause changes to both temperature and precipitation. [5]
2. Explain the possible causes of present day global warming and describe the possible climatic effects. [10]
3. Explain how human activities have contributed to global warming and why the consequences of sea-level rise may be more severe in some areas than others. [10]
4. What is meant by the term urban heat island? [4]
5. Briefly describe one effect that atmospheric pollution may have upon urban climates. [3]
6. Give reasons why air pollution is higher in urban areas. [3]
7. Explain why nighttime temperatures may vary across a city. [5]
8. Explain the extent to which the climate in urban areas differs from that in the surrounding countryside. [10]
9. Why do urban areas often experience warmer, wetter conditions and more fog than surrounding rural areas? [10]

3 Rocks and weathering

3.1 Plate tectonics

The subject of **plate tectonics** brings together several theories which attempt to describe and explain the global distribution of earthquakes, volcanoes, fold mountains and the theory of continental drift.

The general consensus, at the present time, is that on top of the Earth's **inner and outer core** is the **mantle**, made up of semi-molten magma on top of which the Earth's **lithosphere** moves around and is split into several parts – **plates** – of varying size (fig 3.1).

The mechanism for this movement is made of huge **convection currents** which are created by the radioactive decay of the Earth's core. As these convection currents rise up and spread out as they reach the underside of the Earth's **lithosphere**, their friction with the crust drags the plates apart and they move across the **asthenosphere** and this causes some of them to collide with each other. Fig 3.1 shows the distribution of the main plates.

The Earth's crust is called the **lithosphere**. It is a relatively inflexible layer of solid rocks. Its depth varies from 6–70 km, with a density of about 3 grams/cm^3 and it is split into plates. It floats on a much denser layer beneath it, called the **asthenosphere**. There are two types of plates – the **continental** plates and the **oceanic** plates.

Continental plates carry the land continents and are generally older, thicker (more than 33 km) and lighter than the oceanic plates. The continental crust is mainly composed of an igneous rock called granite. The continental crust is sometimes called **sial**, as it is mainly made up of silica and aluminium.

Oceanic plates are underneath the ocean basins. They are made up of younger sediments lying on top of denser, heavier basaltic lavas. They are younger, thinner (10–16 km) and more dense than the continental plates as they are made up of **sima**, which is a mixture of silica and magnesium. As the oceanic plates are subducted beneath the continental plates, melting occurs in the **Benioff zone** and the molten magma that is produced forces its way through faults to the surface to form volcanoes, such as Mount St. Helens in the north west of the USA. Volcanoes can also form in weaknesses in a plate called **hot spots**.

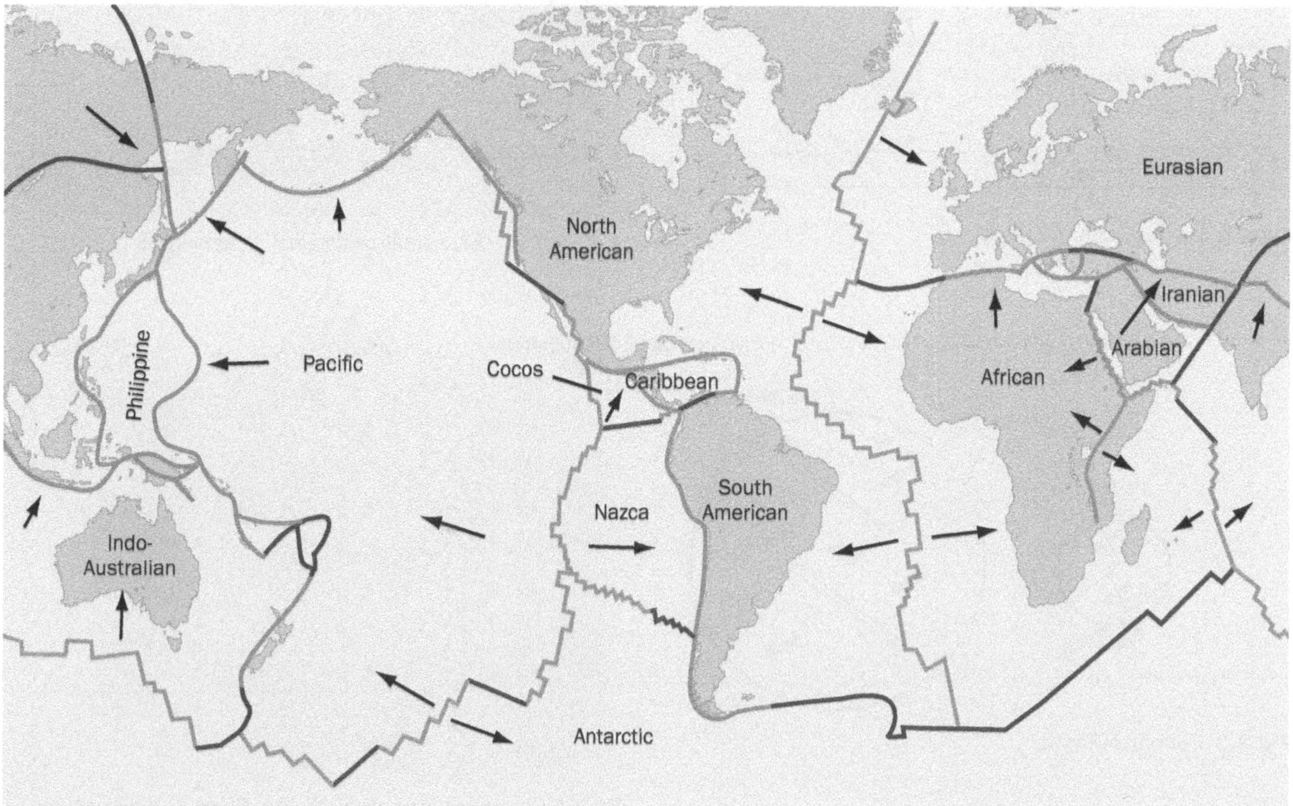

Fig 3.1 Plates, plate boundaries and plate movements

Types of plate boundaries

Divergent (constructive) plate margins

These are found where two plates are moving apart. **Sea floor spreading** is the sideways/lateral expansion/extension of some ocean floors as the oceanic plates move apart. The **Mid-Atlantic ridge** is a very long, high submarine mountain chain. It stretches through the centre of the Atlantic between Africa and Europe to the east and North and South America to the west (fig 3.2). It is composed of volcanic lava and it varies in both width and height. High volcanic peaks appear above the ocean surface forming the islands of the Azores, Ascension Island, Tristan da Cunha and Iceland. The Mid-Atlantic ridge is very irregular, with huge pieces of the plates offset at an angle sliding past each other to form **transform faults**, creating a series of short zigzags to accommodate the movement of the plates (fig 3.3).

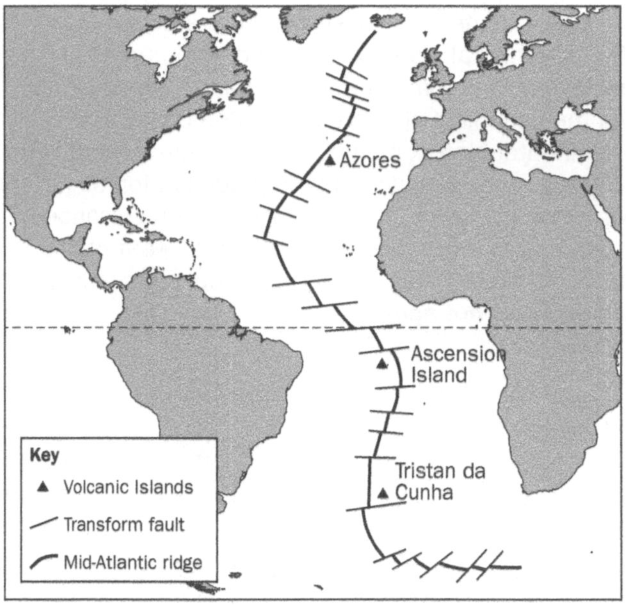

Fig 3.2 The Mid-Atlantic ridge

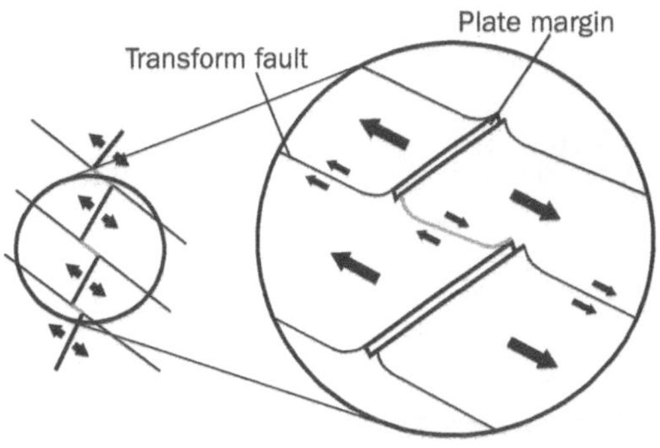

Fig 3.3 Transform faults

Conservative (transform) plate margins

These are found where plates slide past each other and commonly produce earthquakes, but land is neither destroyed nor created along these margins – for example, along the west coast of North America where the North American and Pacific plates move past each other. As no subduction occurs, there is no melting of the crust and so there is no volcanic activity associated with this type of plate margin.

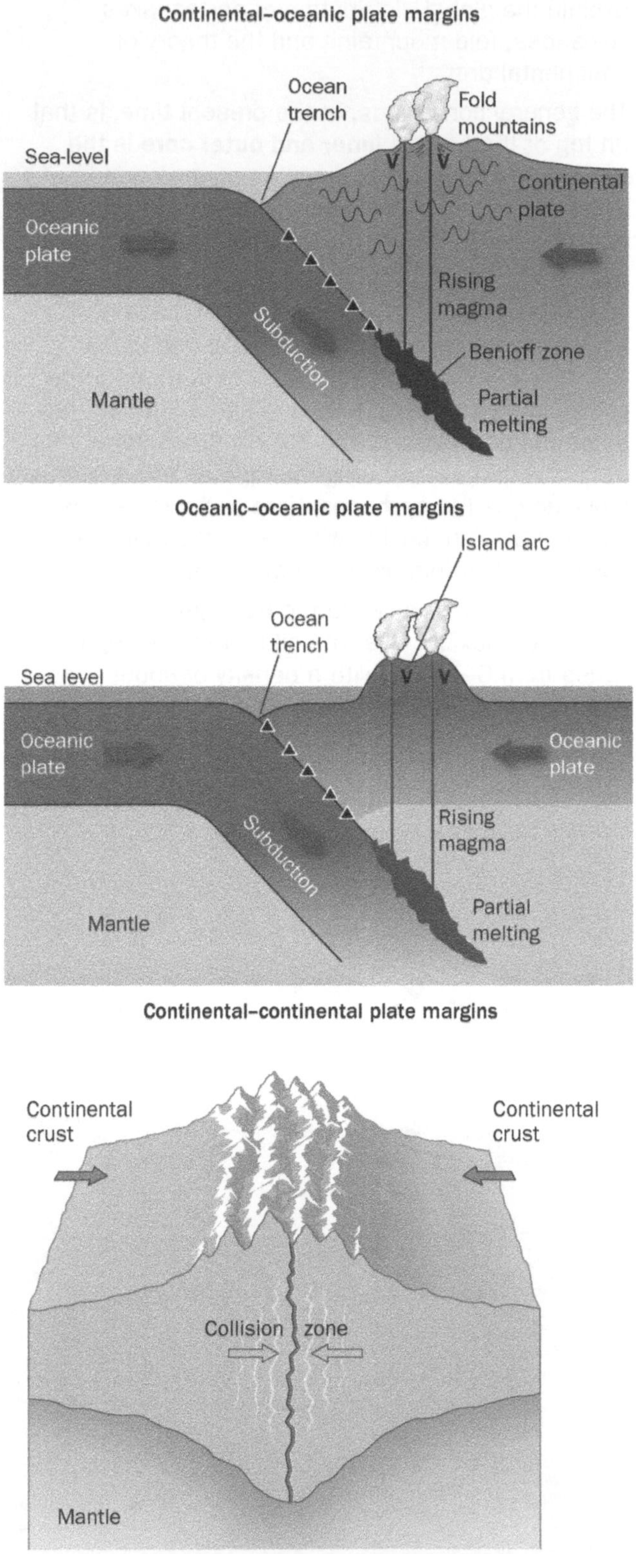

Fig 3.4 Cross-sections through the different types of convergent plate margins

Convergent (destructive) plate margins

They are found where two plates are moving towards each other and are classified by referring to the type of the plates colliding (fig 3.4).

- **Continental – oceanic margins** are seen when the denser oceanic plate subducts a continental plate, triggering earthquakes and forming volcanoes. This leads to the formation of **ocean trenches**. An example of an ocean trench is the Mariana Trench, which at 10 994 metres deep is the deepest part of the world's oceans. Located in the western Pacific Ocean and to the east of the Philippines, it is about 2550 kilometres long and only 69 kilometres wide (on average).
- **Oceanic – oceanic margins** are found where two convergent oceanic plates meet. At the point of impact or subduction a deep ocean trench may be formed where the oceanic crust moves downwards. The heavier, melting subducting oceanic plate rises through the thinner, lighter oceanic plate above it, upwelling magma and forming **island arcs** which are chains of volcanic islands, as in the Caribbean islands.
- **Continental – continental margins (collision plate margins)** are found where two lighter, continental plates collide with each other. This leads to formation of **fold mountains** as neither plate can sink into the denser rocks below. Instead, they are crushed, crumpled and forced upwards, usually folding in the process.

Exam-style questions

1. Define the term tectonic plate and give two differences between oceanic and continental tectonic plates. [4]
2. Define the terms island arc and ocean trench. [4]
3. Briefly describe sea floor spreading. [3]
4. Describe one landform that may develop at a convergent plate boundary. [3]
5. With the aid of diagrams describe the features that may be found at a divergent plate boundary. [8]
6. How does an understanding of plate tectonics help to explain the development of the Earth's crustal features, such as the development of mountains? [10]
7. How can the theory of plate tectonics be used to explain the formation and distribution of volcanoes, ocean trenches and island arcs? [10]

3.2 Weathering

There is often a lot of confusion over the terms **weathering** and **erosion**, many people think that they are the same process. They are, in fact, very different.

Weathering is the gradual breakdown or decay of rocks *in situ* (in their original place) at or close to the ground surface. There are two major types of weathering:

1. **Physical weathering** is the mechanical breakdown of rocks largely due to temperature change. The chemical composition of the rocks remains unaltered. Physical weathering includes: freeze–thaw weathering; heating and cooling – sometimes called exfoliation (or onion skin weathering); salt crystal growth/crystallisation and the growth of vegetation roots.
2. **Chemical weathering** is the breakdown of rocks through chemical changes to the composition of the rock. It includes the processes of: solution, carbonation, hydration, hydrolysis and oxidation.

Physical (mechanical) weathering processes

Freeze–thaw weathering

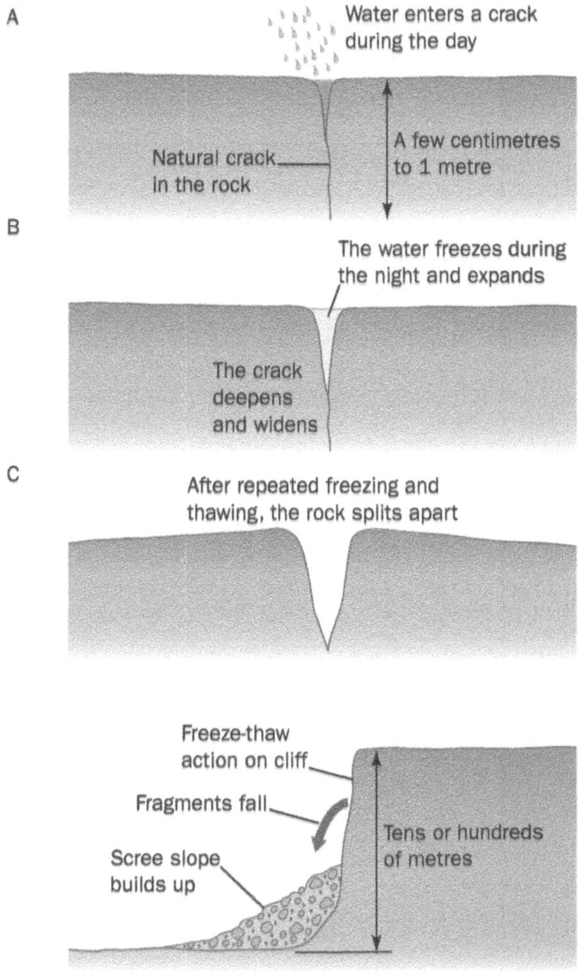
A
Water enters a crack during the day
Natural crack in the rock
A few centimetres to 1 metre
B
The water freezes during the night and expands
The crack deepens and widens
C
After repeated freezing and thawing, the rock splits apart
Freeze-thaw action on cliff
Fragments fall
Scree slope builds up
Tens or hundreds of metres

Fig 3.5 Freeze–thaw weathering

When water freezes, it expands its volume by 9 per cent. If it freezes within a rock, freeze–thawing can take place. **Freeze–thaw weathering** (fig 3.5) occurs in rocks which have cracks in them, are jointed, or have bedding planes that allow water to enter the rock. The climate must allow temperatures to fluctuate below and above freezing.

A long series of cycles of water freezing and thawing in the cracks can lead to the separation of blocks of rock producing landscapes filled with loose blocks of rocks called **block fields** (sometimes called felsenmeer). It takes place when water gets into small spaces and cracks in a rock. If it then freezes and expands it can put enormous pressures of up to 2 100 kg/cm^2 on the rock!

Most rocks can only take pressures of about 500 kg/cm^2 before they split apart. Where bare rock is exposed on a cliff or slope, fragments of rock may be forced away from the face. Freeze-thaw is most effective where temperatures frequently fall below freezing (i.e. daily/diurnally rather than seasonally) and where there is an availability of water, therefore temperatures should not generally be below –5 °C to –15 °C; if it does, there is unlikely to be any water present as it will be frozen permanently. Therefore this process tends to be most active in polar periglacial areas and in mountainous alpine areas.

The fragments of rock that fall to the bottom of a cliff or slope form a large pile of rocks called **scree**.

Frost shattering is another type of freeze–thaw weathering which takes place in porous rocks, such as chalk. These rocks absorb water, but can also break off in sheets. When the water freezes within these rocks the process is so powerful that the whole rock can fall apart and shatter.

Heating/cooling

Heating/cooling weathering is most effective in dark crystalline rocks, which absorb heat. This process requires a wide diurnal temperature range, ideally from greater than 25 °C down to 0 °C. Therefore, the clear skies and high temperature range of tropical hot deserts are prime locations for this process.

As rocks are poor conductors of heat, this leads to the surface of the rock expanding, causing **granular disintegration** as the surface breaks up into small grains of rock or **block disintegration** as larger blocks or plates of rock are detached from the main body of rock (fig 3.6).

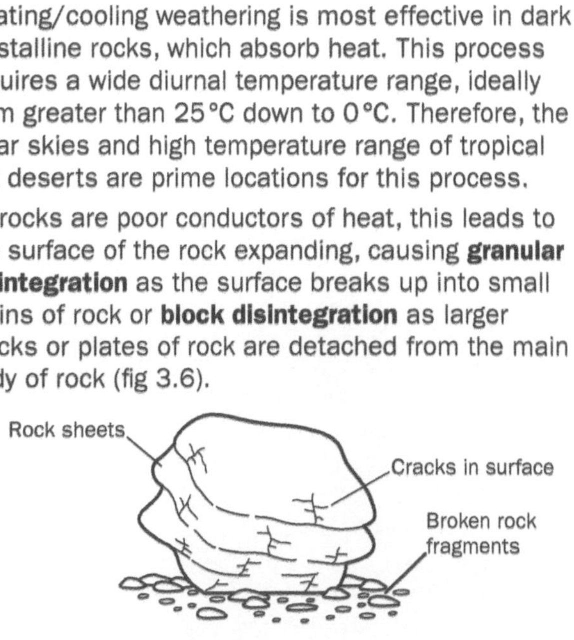
Rock sheets
Cracks in surface
Broken rock fragments

Fig 3.6 Block disintegration

Salt crystal growth

This process takes place in areas with temperatures of around 26–28 °C, when two types of commonly occurring salt – sodium sulphate and sodium carbonate – expand in size by up to 300 per cent. When salt is dissolved in water (brine), evaporation will leave behind salt crystals which attack and disintegrate the surface of the rock. This explains why houses close to the sea need repairing frequently, as salt weathering causes paint to flake.

Pressure release (dilatation)

This process takes place when heavy overlying rocks are removed by erosion. The release of this pressure and weight causes the exposed rocks to expand which may cause them to split apart parallel to the rock surface. The removal of melting ice sheets and glaciers from an area is a primary cause of pressure release.

Vegetation roots

Roots can grow into joints and bedding planes where there may be moisture and washed in soil. Seedlings may grow in such areas. In both cases extreme pressure can be exerted by the growth of roots leading to, or adding to, mechanical weathering break up.

Chemical weathering processes

- **Hydrolysis** is a complex chemical process involving water that affects the minerals in igneous and metamorphic rocks. It is particularly active in weathering the feldspar found in granite whereby material is removed in solution leaving a clay, called **kaolinite/kaolin** or China clay.
- **Hydration** affects minerals that have the capacity to take up water. As they absorb/soak up the water they expand and change their chemical composition which often makes them weaker and less resistant to erosion. For example, gypsum changing into anhydrite.
- **Carbonation** is particularly associated with limestone and chalk. Precipitation falling through the atmosphere absorbs carbon dioxide to form a weak carbonic acid. This acidulated rainwater will react with any rock containing calcium carbonate (mainly the many different types of limestone) converting it to calcium bicarbonate which then dissolves. Carbonation is more active in cold conditions, as cold water can hold more carbon dioxide.

General factors affecting the type and rate of weathering

Climate acts as the main overall influence on weathering, affecting the rate (speed) and the type of weathering. This will in turn influence the thickness of the weathered material found on a slope – the **regolith**.

It also affects slope transportational processes – the processes that move weathered material down a slope – including rainwash and a wide range of mass movements. It plays an important role in both physical and chemical weathering and it also supplies water for chemical processes and influences the speed of chemical reactions.

Temperature and **precipitation** control the type and speed of weathering processes (fig 3.7). For example, in physical weathering, water is required for freeze-thaw, as well as temperatures that move above and below freezing. In exfoliation, high temperatures are needed for the rock surfaces to expand and these need to fall for the surface to contract.

Chemical weathering is most effective in areas with both high precipitation and high temperatures which speed up most of the process. The active retreat or lowering of a slope will be more pronounced in areas where chemical weathering is active, such as the humid tropics. Areas with little precipitation, such as the arctic tundra and deserts will have chemical weathering processes that are slow and relatively ineffective.

Rock type refers to the composition of the rock and it influences both the rate and the type of weathering. This is due to three factors:

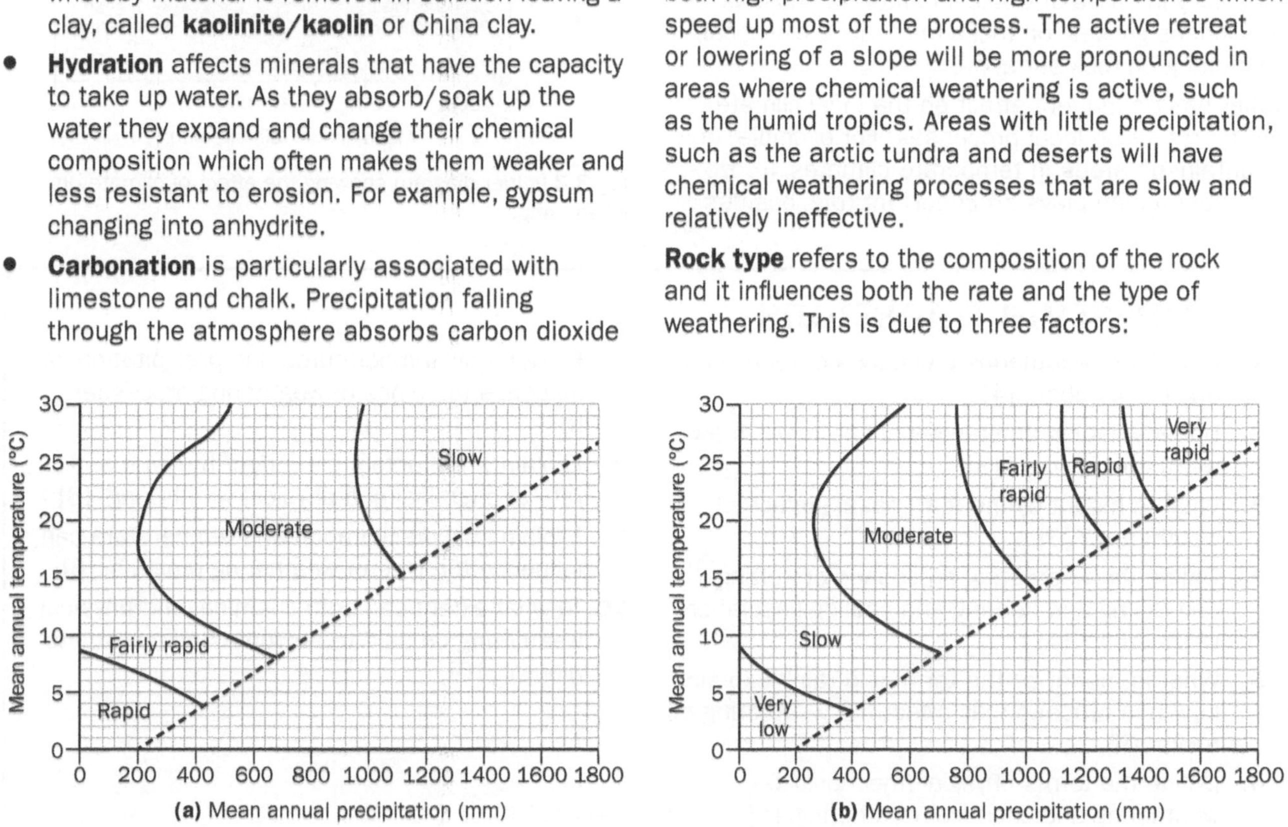

Fig 3.7 The influences of temperature and precipitation on (a) physical and (b) chemical weathering

1. Differences in the chemical composition of rocks. For example, limestone is composed of calcium carbonate which makes it very susceptible to chemical weathering through solution and carbonation.
2. Differences in the cements that hold together sedimentary rocks. For example, some cements are iron oxide-based and are very susceptible to oxidation, whereas silica-based cements made of quartz are very resistant to oxidation.
3. The presence or absence of pores (holes) will affect its vulnerability to weathering processes. Porous rocks (e.g. chalk) have a high proportion of cracks, while non-porous rocks (such as slate) do not.

Rock structures, such as bedding planes, joints, fault lines and cracks that are present in rocks will often have a major influence on the angle of a slope. Earth movements can tilt these structures. The angle of a tilted rock is called its **dip**. This can also affect the susceptibility of a slope to mass movement. Where the angle of the dip is inwards, towards the slope, the slope will tend to be stable; if the dip is down a slope, the slope may fail and rock slides can occur.

Vegetation intercepts precipitation and so it can protect a slope from weathering and erosion, especially in wet tropical areas. When vegetation is sparse, the transport of material and runoff from a slope will be greater.

The breakdown and decomposition of vegetation can produce humic acids which are especially effective in adding to the amount of chemical weathering, especially in areas of tropical rainforest. Mosses and lichens growing on rocks may cause biological weathering on rock surfaces.

Relief has an indirect effect on the local climate, and therefore on local processes. For example, in mountainous areas in temperate climates, freeze–thaw action also plays an important role, besides the dominant chemical weathering processes. In these areas, the height of the land may cause freezing temperatures and freeze–thaw action on exposed rocks.

In areas of steep relief, slope processes, such as slides and flows, can expose rock to various processes of weathering. In contrast, in lowland areas, rock may be protected by thick layers of soil and weathered material. The accumulation of water at the base of slopes may also provide more water for chemical processes to take place.

The **aspect** of a slope may also affect rates of weathering. In the Northern Hemisphere, rates of physical weathering are greater on north-facing slopes (sometimes called **ubac** slopes), which experience longer periods of freezing temperatures having longer periods of time in the shadow and a lack of direct sunlight compared to south-facing slopes (sometimes called **adret** slopes). The opposite is the case in the Southern Hemisphere.

Peltier diagrams produced in the 1950s show how rates of weathering are related to the availability of water and average annual temperatures (fig 3.8).

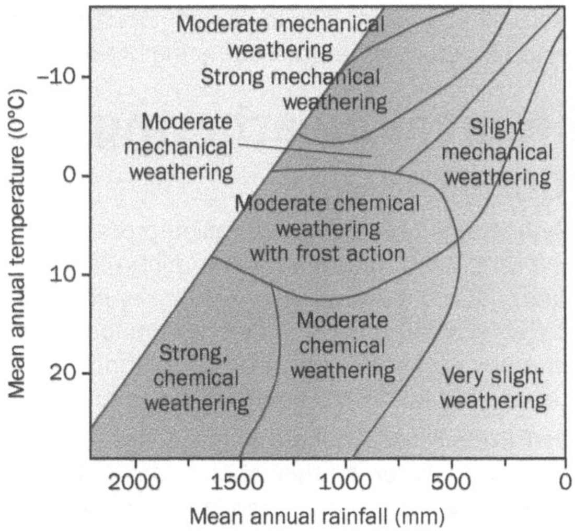

Fig 3.8 Peltier diagram showing the effect of climate on weathering

Exam-style questions

1. Define the weathering processes of hydrolysis and carbonation. [4]
2. Under what circumstances might freeze–thaw weathering occur? [3]
3. What is meant by the term basal surface of weathering? [3]
4. Define the weathering processes of wetting and drying and heating and cooling (insolation weathering). [4]
5. What is meant by the term acid rain? Explain how this process may affect the weathering of rocks. [5]
6. Define the terms physical (mechanical) weathering and chemical weathering. [4]
7. Explain how temperatures and precipitation can influence the types of weathering processes. [6]
8. Explain what factors can affect the type and rate of chemical and physical weathering. [8]
9. Explain how precipitation and temperature can influence the type of weathering process. [6]
10. To what extent can both chemical and physical weathering be said to be controlled by climate alone? [10]

3.3 Slope processes

Mass movement

A **mass movement** is the movement down a slope of weathered rock and soil, responding to the pull of gravity. When the pull of gravity is greater than the force of friction and resistance, a slope will fail and material will start to move downwards.

It is the primary control of slope steepness, because the angle of a slope will depend upon its shear strength and shear resistance.

Slope failure will depend on two factors:

1. **Shear strength** – how much internal resistance there is within a rock, or slope, to stop part of the slope falling, slipping or sliding down the slope.
2. **Shear stress** – the forces that are trying to pull part of the slope down the slope. These include extra weight being added by water being absorbed into the slope or weight of added material being put on the slope, or the weight of extra vegetation growing on a slope.

Therefore resistant, hard rocks, such as granite, will be able to create steeper slopes than weaker, less resistant rocks such as sandstones and clays. Where a permeable rock lies on top of an impermeable rock it can lead to slope failure as the permeable rock may slide and move over the impermeable rock. Softer, less resistant rocks, such as clays and mudstones, are usually more affected by mudflows and, sometimes, rotational slides.

There is a wide variety of mass movements taking place on slopes, with heave, flow, slide and fall being the four main ones.

Heave (soil creep and solifluction)

Soil creep is probably the most common and widespread of mass movements. It is a slow and almost imperceptible movement of particles down a slope under the influence of gravity. As the movement is very slow, it has limited impact on overall shape. Its main impact is to smooth and round the slope. It leads to the accumulation of soil on the upslope side of fences, walls and hedges and causes trees and telegraph poles to become out of vertical alignment. Soil creep may result in the formation of small pressure ridges (**terracettes**) on a hill side. Soil creep involves the mechanism of **heave**, where soil particles rise towards the surface due to wetting or freezing, only to drop back vertically to the slope when drying or thawing occurs. In this way, over many cycles, particles slowly move downhill.

Solifluction is very similar to soil creep, but is normally a slightly faster down slope movement (5 cm to 1 metre per year) of materials that have a high water content. It occurs in the cold periglacial regions of the world and in cold, high mountainous areas. Solifluction takes place in the summer when the surface ice melts to form a saturated active layer. This slowly slips and flows downhill on top of the frozen permafrost below.

Flows

Flows are much faster forms of mass movement. These may take place as **earthflows** at relatively slow speeds of 1 to 15 km per year when material is transported on slopes of 5–15° with a high water content. On steeper slopes **mudflows** may take place where speeds increase to between 1 and 40 km per hour, especially after heavy rainfall, adding both volume and weight to the soil. The heavy rain (or excess supplies of water from springs, etc.) increases the pore water pressure which forces the particles into a rapidly flowing mass of material.

Slides

Slides can be extremely rapid processes. They occur where a complete mass of material detaches itself from a slope and slides downhill. There are two types of slides – **planar** – where the mass movement leaves behind a flat slide plane, such as along bedding planes or fault lines, and **rotational** slides or slumps – where the material slides out from a slope in a curved motion.

Landslides take place at speeds of between 1 and 100 metres per second on slopes that are often greater than 40° and have a low water content. Here the material slides down a steep slope and forms a mass of broken fragments at the bottom of a slope.

Landslides are most active in areas of high relief and unstable slopes. In rock slides, the effects of bedding and joint planes are important in allowing the rock to fragment and they also provide slide planes. Landslides are the result of sudden and massive slope failure.

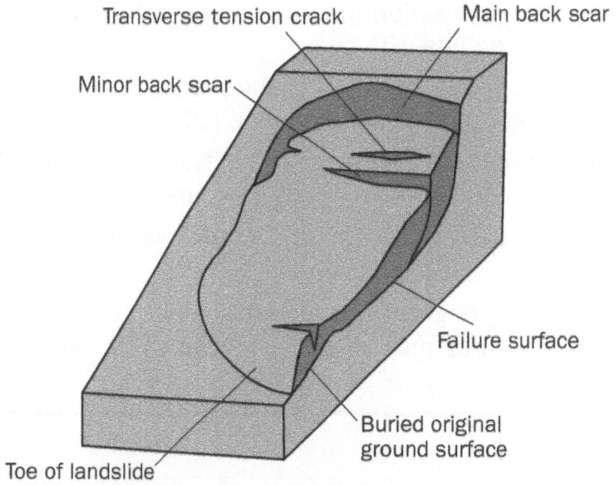

Fig 3.9 Features of a rotational slide

Landslides occur where the shear stress exceeds the shear strength of a material and this allows the material to slide. They are particularly common in tectonically unstable mountainous areas (areas prone to lots of earthquakes).

People can play a part in producing instability in slopes and landslides through increases in weight (by constructing buildings, reservoir, etc.), by undercutting a slope – for example in building roads across slopes

– and by diverting water onto slopes. The result of a landslide is to produce a shallower slope where the angle the landslide comes to rest has been reduced and the length of the slope has been increased. Landslides often occur along slide planes that are often influenced by bedding and joint planes.

Falls

Rockfalls occur on steep, often almost vertical jointed and fractured rock faces where the cohesion between masses of rock is overcome and the shear strength of the material is exceeded. They often result after several freeze–thaw cycles, or by repeated wetting and drying and other weathering processes.

Water and sediment movement on hillslopes

Water and sediment movement on hillslopes will take place as a result of rainsplash and surface runoff.

Rainsplash

Rainsplash occurs when rain falls with sufficient intensity. If it does, then as the raindrops hit bare soil, it is able to detach and move soil particles a short distance by the impact of the falling raindrop. Over two hundred tonnes of soil particles per hectare may be dislodged during a single rainfall event and then moved downslope! However, as the soil particles can only be moved a few centimetres at most by this process, it is all merely redistributed back over the surface of the soil. As rainsplash requires high rainfall intensities, it tends to be most effective under heavy convective rainstorms in the world's tropical regions.

Surface runoff

- **Sheetwash**: as surface runoff moves downslope as a thin sheet of water, it will move only slowly, have low energy, and will be generally incapable of detaching or transporting soil particles. However, on steeper slopes it will have more energy and the loose dislodged soil particles may be moved downslope by water flowing overland as a sheet, **sheetwash**. A more or less uniform layer of fine soil particles may be removed from the entire surface of an area, sometimes resulting in the extensive loss of fertile topsoil from a field. Sheetwash commonly occurs on recently ploughed fields or in areas with poorly consolidated soil material with little or no protective vegetative cover.
- **Rills**: where sheetwash takes place the soil's surface will be lowered slightly. In time, these preferential flow paths will be eroded to form small, well-defined channels, called **rills**. These may quickly develop and enlarge into gullies. Rills form efficient pathways for the removal of both water and sediment from hillslopes.

Exam-style questions

1. Define the terms rock slide and heave. [4]
2. Briefly describe the differences between a flow and a slide. [3]
3. Briefly describe how these processes can affect the shape of slopes. [3]
4. Describe the process of solifluction and explain the conditions in which it occurs. [4]
5. Describe soil creep and explain why it occurs at such low velocities. [4]
6. Briefly describe the effects that the process of soil creep may have on the shape of a slope. [3]
7. Describe the conditions under which a rock fall may take place. [3]
8. Explain how landslides occur and describe the effects they have upon slopes. [6]
9. Explain how rock type and structure, climate and vegetation can affect the form and development of slopes. [10]
10. How and to what extent can human activities affect the shape and form of slopes? [10]

3.4 The human impact

Increasing and decreasing stability of slopes

Human activities may either increase or decrease the stability of slopes. Human activities may decrease slope stability by mining, quarrying or undercutting a slope through road or railway construction. Also, large amounts of weight may be added to a slope by depositing the waste material from mining and quarrying.

Once the stability of a slope has been decreased it can lead to landslips, landslides and mudflows (e.g. in Aberfan in 1966 in South Wales, UK, when the collapse of coal waste on a hillside after heavy rain caused an estimated 100 000 m^3 of coal waste to destroy part of the town, including a local junior school, killing 147 people, including 116 children and five of their teachers) which attempt to re-establish the stability of the slope profile.

The rapid urbanisation of unsuitable sites in fast growing cities has led to some catastrophic landslides, for example in the favelas of Rio de Janeiro and São Paulo in Brazil, and in Hong Kong, where landslips have killed 480 people since 1948.

There are activities with less obvious effects, such as deforestation and the removal of vegetation from a slope or the diversion of drainage channels off roads and urban areas onto slopes and damming, which can affect the movement of both water and materials down slopes. All these can also have catastrophic local effects.

Strategies to modify slopes to reduce mass movements

Strategies to modify slopes to reduce mass movements include:

- **Grading**, where the geometry of a slope may be changed by removing material from the slope and the slope may be re-graded to a lower angle through a combination of slope reduction and infilling at the foot of the slope.
- **Hydrogeological**, where the groundwater level is lowered or the water content of the slope material is reduced. Shallow drainage trenches may be cut into the slope when the potential slope movement is a shallow landslide affecting the ground to a depth of only 5–6 metres. When there is deeper slope movement, deep drainage has to be introduced, often in addition to the shallow drainage trenches.
- **Mechanical**, where attempts are made to increase the shear strength of the slope by using **rock anchors**, **rock or ground nailing/ pinning** to counter the destabilising forces acting on a slope. Steelwire mesh **netting** can also be used for slope stabilisation, with the slope surface being covered by a steel-wire mesh, which is fastened to the slope and tensioned. It is a cost-effective approach.

CS Southern Brazil (January 2011)

From 1900–2015 there were 22 major landslide events in Brazil, killing 1 641 people, affecting 4.24 million people and causing $86 million worth of damage to property and infrastructure.

The majority of these events were in the coastal mountains of mid-southern Brazil. During the Southern Hemisphere summer season of December to March, the combination of steep slopes, heavy rainfall, residual soils, and weathered rocks have made the coastal mountains of mid-southern Brazil particularly susceptible to major, catastrophic landslide activity.

In January 2011 a massive landslide in the mountainous Serrana region of the state of Rio de Janeiro left over 900 people dead and 18 000 homeless. 300 mm of rainfall within 24 hours triggered many landslides. In the worst incident, more than 200 people were buried alive when the Morro do Bumba favela collapsed.

The region has become more vulnerable due to unchecked deforestation turning steep slopes into dangerous landslide-prone areas. Illegally occupied hillsides, with poor foundation quality and the physical expansion of urban areas by rapid population growth has also contributed to the increased vulnerability of the residents in the region. This disaster caused millions of dollars in damage. Apart from the destruction of infrastructure, local authorities also had to spend millions on the provision of temporary housing.

Responses to the landslides

The Brazilian national and urban authorities have designed short and long-terms solutions to deal with the disaster.

In the **short-term**, the urban authority gave affected families shelter for up to six months and the federal government offered to pay the rent of 2 500 families for an indeterminate period, set up a centre to register missing persons, and relocated families to better organise shelters in churches, warehouses, and stadiums.

In the **medium and long-terms**, the federal government allocated $460 million for reconstruction funds and promised to remove bureaucratic bottlenecks, register people to help with long-term housing needs, map out and evacuate risk-areas where residents are holding on to their homes, and review the current procedure of cleaning up disasters rather than stopping them from happening – being pro-active rather than re-active, and to prioritise prevention by controlling urban expansion pressures and natural resource exploitation, especially in areas characterised by mountainous topography.

Rio de Janeiro has put into operation a landslide early warning system "**Alerta-Rio**" dividing the urban area, for warning purposes, into four alert zones. Alerta-Rio has two different alerts: rainfall warnings and landslide warnings.

Other responses by the government include mapping geological risk areas, identification of support facilities (places to serve as temporary shelter during heavy rains, usually churches, schools, kindergartens, etc.), and of safer routes towards them, as well as points for the installation of sound alarms (horns or sirens).

Exam-style questions

1. How and to what extent can human activities affect the shape and form of slopes? [10]
2. With the use of one or more case studies explain how human activity may decrease the stability of a slope. [6]
3. What strategies can be used to modify slopes so they are less prone to mass movement? [10]

4 Population

4.1 Natural increase as a component of population change

Definitions

TIP

The terminology relating to this topic must be correctly understood and used. Be careful in giving these definitions and make sure you give all the parts of the definition. For example, the crude death/mortality rate is: the number of *deaths per 1 000 people in a year* – for the definition to be correct, all three terms in italics need to be mentioned.

- **Natural increase rate** – is the **birth rate** minus the **death rate**. It is normally expressed as a percentage per year. The natural increase rate is the rate at which the population is growing naturally, excluding gains from migration. Sometimes a country may experience a negative natural increase (called natural decrease). A natural decrease may occur in some of the fluctuations in stages 1 and 4 of the demographic transition model or in stage 5. Stage 5 examples include Japan, Germany, Russia and many countries in Eastern Europe.
- **Birth rate** (sometimes called the **crude birth rate**) – is the average number of children born (live) per 1 000 people per year. In 2011 the highest rate was in Niger in the Sahel region of Africa with 46/1 000 and the lowest was Germany and Japan with 8/1 000.The use of the word "crude" is because the rate applies to the whole population and does not take into account the age and sex of the population.
- **Death/mortality rate** (sometimes called the **crude death rate**) – is the number of deaths per 1 000 people per year.
- **Fertility rate** – may be expressed as: (i) an individual rate – the average number of children each woman in a country will give birth to or (ii) a general rate – the number of births in a year per 1 000 women or as a per cent of childbearing age, 15–49/50 or a similar age.
- **Infant mortality rate** – is the number of deaths of babies/children under one year of age or before their first birthday per thousand live births per year.
- **Life expectancy** is the average number of years from birth that a person is expected to live.

Levels of fertility around the world

The birth rate is sometimes used to give an idea of the fertility of a country, but a more precise measure is to use the **fertility rate**. The **total fertility rate** is sometimes used. This is defined as the average number of live births in a year per 1 000 women of childbearing age (15–49 years).

Niger in West Africa had the highest total fertility rate in 2014 (7.3) and Macao and Hong Kong in China the lowest (1.0). The world average total fertility rate between 2010 and 2015 was 2.5. The HICs have the lowest total fertility rate for a group of countries (1.7) and Africa, the highest (4.7); a major factor in this is the relative use of contraception by women – 71 per cent in the HICs and 29 per cent in Africa. Fig 4.1 illustrates how the total fertility rate has fallen globally and by region since 1950 and the predicted trend to 2050, while fig 4.2 shows the total fertility rate in 2010 for a few selected countries.

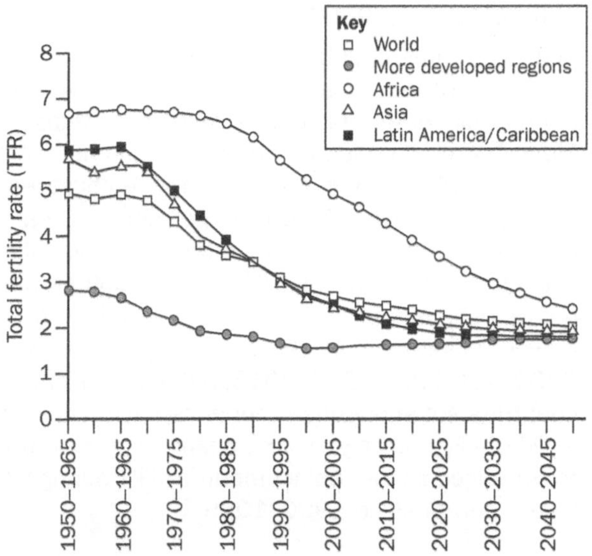

Fig 4.1 Total fertility rate – trends by region, 1950–2050

A fertility rate of 2.1 children per woman is the **replacement level fertility**. Below this, a country's population will fall – unless it is kept high or even rises due to migration into that country. In 2010, 87 countries were either at or below this level, reflecting one of the most fundamental and dramatic social changes in human history which allowed more women to work and more children to be educated.

factions steal food aid, block commercial food deliveries and systematically wreck local markets. Fields are often mined and water wells contaminated, forcing farmers to abandon their land. Fighting also forces millions of people to flee their homes, leading to hunger emergencies as the displaced find themselves without the means to feed themselves. The conflicts in Afghanistan, Syria and Libya are recent examples. In the newly independent country of South Sudan, internal tribal conflicts remain and up to 2 million people have had to leave their homes and farms and it has often been impossible for farmers to grow crops and rear animals. Those that do may find their crops and animals destroyed or stolen. More than 70 000 people have died so far from a mixture of starvation and diseases often caused or made worse by malnutrition.

Increasing food production

Increasing food production may involve developing skills and technology suited to the level of wealth, knowledge and skills of local people and is developed to meet their specific needs. This may include several approaches.

Use of appropriate technology

- The building of small **earth dams** and digging **wells and boreholes** to provide water for basic irrigation projects.
- Methods of **soil conservation**, such as planting trees to make **shelter belts** to protect soil from wind erosion in dry periods. Low stone walls can be built along the contours of a slope to stop runoff and allow it time to enter the soil. This helps to prevent soil erosion and increases the amount of water in the soil, making it available for crops.
- **Tied ridging** – low walls of soil are built in a field to form a grid of small squares which stops runoff and again allows water to drain into the soil. Root crops such as potatoes and cassava are grown on the soil walls.
- **Strip/inter Cropping** involves alternate strips of crops at different stages of growth across a slope. This limits runoff as there is always a strip of crops to trap water and prevent soil moving down the slope.
- **Improved food storage** which allows food to be kept fresh and edible for longer periods of time and protected from being eaten by rats and insects and from diseases.

The Green Revolution

The Green Revolution started in the mid 1960s with the development of **high yielding varieties** (**HYV**s) of five of the world's major cereal crops – rice, wheat, maize, sorghum and millet. These new hybrid varieties of crops were:

- resistant to drought
- higher yielding, often by two to four times, than traditional species of these crops
- growing in less time (shorted growing season), allowing more crops to be grown in a year in some areas.

Several countries, such as India, Indonesia and the Philippines, set up research programmes to investigate how to increase rice yields. One result was a new variety of rice called IR8, produced by crossing a semi-dwarf variety from China with a stronger taller variety from Indonesia.

The result was a stronger, shorter variety of rice which could be planted closer together, had a shorter growing season (4 months instead of 5 months) and a much higher yield (5 tonnes instead of 1.5 tonnes per hectare) than traditional varieties of rice. However, this new variety needed expensive fertilisers and more water for irrigation, which smaller farmers could not afford to buy. IR8 also attracted more pests than the traditional varieties so expensive pesticides had to be used.

Generally, the Green Revolution had several positive results:

- Farm incomes increased, raising the standard of living of many people in rural area. Families had money to pay for the education of their children, giving them access to qualifications that could allow them to get a better job in future.
- Increasing yields meant that crops could be exported.
- Improved people's diets and fewer food shortages in some areas.

However, the Green Revolution had its problems and critics, such as:

- The HYV crops need bigger, more expensive inputs of fertilisers, pesticides and herbicides.
- The mechanisation of jobs on the farms triggers increased unemployment, increasing poverty for some and forcing some people to migrate to cities in search of jobs.
- Much of the farmland is now being used for growing one HYV crop. HYV crops are often lower in minerals and vitamins than the local varieties they have replaced, which means that they do not provide people with the same level of nutrition compared to their traditional crop varieties.

Irrigation

Irrigation is an artificial watering of crops. It is mainly used in dry areas and in periods of rainfall shortfalls, but also to protect plants against frost. Additionally irrigation helps to suppress weeds growing in rice fields.

Surface irrigation has been the most common method of irrigating agricultural land, where water moves over and across the land by simple gravity flow in order to wet it and infiltrate into the soil. Surface irrigation can be subdivided into furrow, border strip or basin irrigation. It is often called flood irrigation when it results in flooding or near flooding of the cultivated land.

Increasingly, **drip irrigation**, where water is delivered at or near the root zone of plants is used as it is a more water-efficient method of irrigation, since evaporation and runoff are minimised. In modern agriculture, drip irrigation and is also the means of delivery of fertiliser – a process known as **fertigation**.

Modern drip irrigation has arguably become the world's most valued innovation in agriculture since the invention of the impact sprinkler in the 1930s, which replaced flood irrigation. A further development is **sub-surface drip** irrigation (SDI) which uses a permanently, or temporarily buried, dripper line or a drip tape located at or below the plant roots. It is becoming popular for row crop irrigation, especially in areas where water supplies are limited or recycled water is used for irrigation.

Overuse of irrigation and poor irrigation practices have led to increased soil salt content, **salinity**, in some areas, reducing the productivity of the land. Irrigation salinity is caused by water soaking through the soil level adding to the groundwater below. This causes the water table to rise, bringing dissolved salts to the surface. As the irrigated area dries, the salt remains and can bring crop production to a halt.

In New South Wales, Australia irrigation salinity is solved through a **salt interception scheme** *that pumps saline groundwater into evaporation basins, protecting approximately 50 000 hectares of farmland in the area from high water tables and salinity. The subsequent salt has various uses, including as an animal feed supplement. The programme has returned to production over 2 000 hectares of previously barren farmland and encouraged the regeneration of native eucalypts.*

Other problems caused by irrigation include:

- increased competition for water, from individual farmers, communities and even countries
- over-extraction of water can lead to the depletion of underground aquifers
- ground subsidence, e.g. New Orleans, Louisiana, USA, may occur as water is removed for irrigation from the underground aquifer.

Selective breeding

Farm animals have been undergoing human-managed selection ever since their original domestication. Initially, selection was probably limited to docility and manageability, but in the last 60 years breeding programmes have focused on the genetic improvement of production traits, such as milk yield, growth rates and weight gain.

The most widely used **dairy** production systems now use high-nutrient-input high-milk-output systems. In many countries yield per cow has more than doubled in the last 40 years. The average milk yield for dairy cows has increased from 4 200 to up to 12 000–17 000 kg of milk per lactation between 1950 and 2015. Genetic engineering and selective breeding in the **beef cattle** industry has brought tremendous increases in productivity. Between 1980 and 2010 the mature size of cows has increased by over 135 kg. Today the average cow in the U.S. weighs 610 kg. Herd size too is increasing. Cattle ranches in Brazil can have over 125 000 head of cattle and the country as a whole has 210 million cattle reared for both domestic consumption and for export, to supply the ever increasing demand for beef, especially from China.

Carrying capacity

Carrying capacity means the maximum number of people that can be supported by the resources of an area.

The definition of carrying capacity is sometimes seen as very subjective and statistically complex to actually measure. The carrying capacity of an area may not be about absolute/real numbers of people, but of the living standards that people have and the differences/contrasts/disparities that may occur – for example between rural and urban areas.

The discovery of new resources, such as an oil field or other mineral resources may change the "ceiling". Also, most areas and environments are not closed systems and emigration may occur as a response to potential overpopulation, before disaster/famine or starvation is likely to occur.

Innovations such as improvements in agriculture and water management can lead to an increase in carrying capacity. However, there are certain aspects of a resource base which are effectively fixed, for example, thin or stony soils, mountainous landscapes or a lack of land area.

The term **population ceiling** is sometimes used in describing the carrying capacity as a barrier, or cap, to future population growth.

Overpopulation, optimum population and underpopulation

- **Overpopulation** occurs where a country or region has too many people overusing the resources at a given level of technology.
- **Optimum population** is where a balance between population and resources exists. There is a debate as to whether optimum population actually exists in the classic manner as a point in time/absolute number of people. It is also seen as a very subjective term, as a government's point of view of what is an optimum population may not be that of the people's.
- **Underpopulation** occurs in a country or region where there are too few/not enough people to make full use of the resources at a given level of technology. Underpopulation is said to

occur where there are too few people to exploit the resources of an area fully and therefore to attain the highest standard of living for all the people, at any given level of technology. It is usually calculated and expressed at national or sometimes at a regional scale. Most areas considered as underpopulated today occupy a large territory and are resource rich, such as Australia, Canada, Kazakhstan and Mongolia.

- **Population density** is the average number of people per unit area, such as per square kilometre, in a region or country. There may be a situation where two countries have the same population density, but one may be viewed as overpopulated and one as underpopulated.

These two terms are relative, and they express population in relation to the resources in a country or region at a given level of technology. So a resource rich HIC with sophisticated technology may be said to be underpopulated, but an LIC with few resources (soil, climate, minerals, etc.) and traditional technology may be overpopulated.

Areas may become underpopulated when there has been a depopulation of rural areas in LICs where agricultural output has fallen or land has been abandoned due to rural-urban migration, the impact of natural catastrophe – such as drought or floods, war, or the impact of HIV/AIDs.

Exam-style questions

1. What is meant by the term carrying capacity in connection with population? [3]
2. Describe the constraints that may limit the capability of resources to feed a country's population. [4]
3. Why are some areas of the world considered underpopulated? [8]
4. Explain why the concept of a population ceiling may be of limited use in reality. [8]
5. What is meant by the term underpopulation? [4]
6. Assess the success of attempts to sustain an increasing population using technology and innovation. [10]
7. Explain the reasons why family planning services may not be available to everyone in LICs. [5]
8. Explain why birth rates may vary over time. [5]
9. Using one or more examples, describe and explain the problems that may be caused by a falling birth rate. [8]
10. Why is population growth unsustainable, given current population growth rates and existing resources? [10]
11. What is meant by the term fertility rate? [3]
12. Explain two factors that may influence levels of fertility. [4]
13. How far do you agree with the view that population change is predictable? [10]
14. What is meant by the term overpopulation? [3]
15. What methods can be used to reduce the problem of overpopulation? [4]
16. Explain the causes and the consequences of food shortages. [8]
17. Explain why the theory of optimum population may not exist in real life. [5]

4.4 The management of natural increase

CS Population policy in Singapore

Singapore is a modern city-state and island country in south-east Asia. It lies off the southern tip of the Malay Peninsula and is comprised of the main island – linked by a causeway and a bridge to the southern tip of Malaysia – and about 50 smaller islands.

Once a colonial outpost of the UK, Singapore has become one of the world's most prosperous places – with glittering skyscrapers and a thriving port.

Most of its population lives in public housing tower blocks. They enjoy one of the world's highest standards of living. Singapore has a highly developed trade-oriented market economy and the third highest per-person GDP in the world; its Purchasing Power Parity (PPP): $85 427 in 2015.

Population control policies in Singapore have gone through two distinct phases.

- The first phase attempted to slow and reverse the boom in births that started after Second World War.
- The second phase, from the 1980s onwards, attempted to encourage parents to have more children because birth rates had fallen below replacement levels.

As fig 4.11 shows, Singapore experienced a long period of rapid population growth from the 1960s as a result of falling death rates, high birth rates and high immigration, producing an annual growth rate of 4.4 per cent. During this time, the crude birth rate peaked at 42.7 per 1 000.

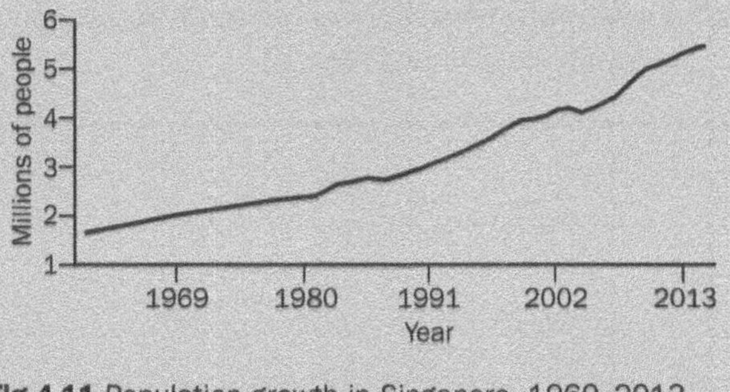

Fig 4.11 Population growth in Singapore, 1969–2013

The Singapore government saw rapid population growth as a potential threat to the living standards, education and health services and the political stability of the country. The government thought that this could result in:

- widespread unemployment
- a shortage of housing
- insufficient educational and healthcare services for the people
- increasing pressure on the limited resources of the country

Family planning

Beginning in 1949, family planning services, including clinical services and public education on family planning, were offered by the private **Singapore Family Planning Association**.

The policies were accompanied by publicity campaigns urging parents to "**Stop at Two**" and arguing that large families threatened parents' present livelihood and future security.

The government introduced a set of policies known as **population disincentives** to raise the costs of families having a third, fourth and subsequent children and encourage small families. These disincentives included:

- maternity hospitals charged progressively higher fees for each additional birth
- large families received no extra consideration in public housing assignments
- top priority in the competition for enrolment in the most desirable primary schools was given to only children and to children whose parents had been sterilised before the age of forty
- workers in the public sector would not receive maternity leave for their third child or any subsequent children.

The Singapore government launched the National Family Planning Programme in 1966 using the slogan "**Girl or Boy – Two is Enough**". Subsidised family planning services were provided through maternal and child health clinics. Women who had given birth were advised to go to these clinics or their family doctors for family planning services and they were able to gain advice and consultation on contraception and purchase contraceptives at reduced prices.

In 1970 abortion and voluntary sterilisation were legalised and incentives such as tax relief, priority housing and paid maternity leave were further implemented to encourage smaller family sizes.

Outcomes of family planning

These measures proved to be successful. Singapore had developed into a well-educated, highly urbanised society. Also, an increasing numbers of women were entering the workforce or continuing into higher education and the trend of later marriages was becoming more accepted.

A downward trend started in the total fertility rate from 2.1 babies per woman in 1975 to 1.4 in 1986.

By 1965 the crude birth rate was 29.5 per 1000 and the annual rate of natural increase had been reduced to 2.5 per cent.

However, the campaign was so successful that certain unforeseen problems began to surface. With the birth rates falling and death rates remaining low, Singapore's elderly population increased in proportion to the population that was working.

The government recognised the importance of having enough young people to replace the ageing workforce so that the economy would remain competitive with other countries. There was also a concern that a reduction in the number of young male adults enlisting in the armed forces would have serious consequences for national security.

Pro-natal policies

In 1986 the government decided to revamp its population policy to reflect its identification of the low birth rate as one of the country's most serious problems. The old family planning slogan of "**Stop at Two**" was replaced by "**Have Three or More, If You Can Afford It**". A new package of incentives for large families reversed the earlier incentives for small families. It included:

- tax rebates for third children
- up to four years' unpaid maternity leave for civil servants
- pregnant women were to be offered increased counselling to discourage "abortions of convenience" or sterilisation after the birth of one or two children
- a public relations campaign to promote the joys of marriage and parenthood
- in March 1989, the government announced a S$20000 tax rebate for fourth children born after January 1, 1988
- for children attending government-approved childcare centres, parents were given a $100 subsidy per month regardless of their income
- third child families were given priority over small families for school registration.

The initial response to the new policy was positive as total fertility rate rose from 1.4 in 1986 to 1.96 per woman, but still far short of the replacement level of 2.1. The government reacted in 1987 by urging Singaporeans not to "passively watch ourselves going extinct" but the rate dropped to 1.9 in 1990 and continued to fall to 1.6 in 1999.

The continuing decline in replacement level in Singapore was due to increasing numbers of Singaporeans not getting married.

Women were having children at an older age because couples were marrying later, resulting in families having fewer children. In 2001, the government offered the **Children Development Co-Savings Scheme** (or **Baby Bonus Scheme**). The aim of the **Baby Bonus Scheme** was to remove the financial obstacles associated with having more children. When a family had more children, a **Children Development Account** was set up by the government for the second child of the family.

The government put $500 into the account annually and matched, up to another $1000, every dollar deposited into the account by the family.

For the third child, the government contributed $1000 to the account annually and matched, up to another $2000, in contributions by the family. This incentive was valid until the children reached six years of age and could be used for the education and development of any child in the family.

Outcome of pro-natalist policies

Despite all these pro-natal policies Singapore's population, at 5.47 million, saw its slowest growth in 10 years in 2014. The total fertility rate was 1.19 in 2014, compared to 1.29 in 2013, again, well below the replacement rate of 2.1 per cent.

The 65 and above age group now forms 12.4 per cent of the population in 2014, up from 11.7 per cent in 2013. The total number of marriages and births also fell.

Due to the continued low total fertility rate, the Singapore government has varied its immigration policy over the years. As the demand for labour grew with industrialisation, foreign talent with professional qualifications, as well as less-skilled foreign workers, has made up a significant and increasing proportion of Singapore's total population since 2000.

CS The "one-child" policy in China

China is the only country in the world to introduce quotas to control the number of children people can have. In 1970 the Chinese government issued three policies in an attempt to reduce the birth rate:

1. Late marriage – men were encouraged to marry no earlier than 28 years old (25, in rural areas) and women no earlier than 25 years old (23, in rural areas).
2. Longer spacing between births – couples were encouraged to allow at least a four-year gap after the first child before having another baby.
3. Fewer children – it was suggested that urban families should be limited to two children, and rural families to three children.

In 1979, the Chinese authorities further tightened their control and limited all households to only one child. The goal of this policy was to limit China's population to 1.2 billion by the year 2000.

The policy proved extremely difficult for many families, with a complicated system of rewards, incentives, fines and punishments, which included:

- massive advertising campaigns to explain the reasons behind the policy
- tax incentives for those who had just one child
- women being forced or coerced into having abortions
- free health and education services for single children, but heavy fees for additional children
- neighbours being encouraged to "inform" on their family or friends who were expecting or considering having more than one child.

The policy has been heavily criticised for taking away the rights of the individual to choose the size of their family, but was it successful in slowing China's population growth. At least 300–400 million births were prevented, easing the pressure on China's resources.

Consequences of the policy

During the 1970s the average number of children per woman in China dropped from 6 to 2.5 and between 1950 and 2005 the crude birth rate dropped from 44/1000 to 14/1000 – a figure comparable to many HICs. The policy was harder to enforce in some remote rural areas where there was great pressure on women in rural farming areas to produce sons to work on the land. Several unexpected consequences have come to light since 1979:

- There is a new generation of indulged and spoiled children – China's "**little emperors**". With one child in a family being the centre of attention of two parents and four grandparents, they can end up overweight, arrogant and lacking in social skills.
- This **four-two-one** problem will also mean that one child will have to look after possibly six elderly people in the future. There is now evidence to show that the standard of care for the elderly in China has already declined sharply.
- Currently, an average gender ratio at birth of 119 boys to 100 girls exists, but in some rural areas it can be as high as 140:100. Selective abortion is practised where mothers can have a scan to determine the sex of their child and abort a female foetus, or practise infanticide or deliberately neglect a female child. As a result of this, a high percentage of unmarried men now live in China. Millions of extra boys have been born and now 41 million male bachelors in China will not have women to marry.
- China now has a shrinking population of economically active, working aged adults.
- Wealthy couples can buy their way around the system – there is a $30000 fine for having a second child which many couples can afford.

CS India

In August 1999, India's population passed the 1 billion mark and in 2015 had reached 1.28 billion, increasing by 47 141 people daily. With 2.4 per cent of the world's land area, India is now home to 17.5 per cent of the world's population, with 15.5 million babies being born in India each year. When India gained independence in 1947 it had a crude birth rate of 45/1 000. It was one of the first countries in the world to adopt population control policies.

The government set a target to bring the crude birth rate down to 21/1 000 by the year 2000. Fifty years later, the crude birth rate had dropped to 27/1 000 and in 2013 it was 23/1 000 – a significant decrease.

What are India's population policies?

Throughout the implementation of India's family planning programme since the 1950s, the main emphasis has been on taking drastic measures to reduce overpopulation. The most severe of these was undoubtedly the policy of encouraged sterilisation. Government agencies were given sterilisation quotas to achieve among employees. Workers were often rewarded with a radio or television if they successfully convinced enough people to have the operation.

This policy reached its most extreme in 1976 when India declared a state of emergency and began "forced sterilisation" in poor neighbourhoods. This policy led to international criticism and did not result in the dramatic reduction in the crude birth rate that had been hoped for.

A number of factors have undermined the effect of India's policies:

- the sterilisation programme ignored other influences on the birth rate, including issues of poverty and inequality
- early marriage – one in two girls marry before they are 18 and many start having children straight away
- people are living longer, largely because of an improved diet, and the death rate declined faster than the birth rate
- a built-in momentum – 36 per cent of the population is in the reproductive age group. Even if these men and women have small families, the sheer numbers of the next generation will lead to a further increase in population before it starts to shrink in future generations.

In spite of these limitations, India's rate of population growth is decreasing. In 1991, India's annual population growth rate was 2.15 per cent. By 1997 this figure had dropped to 1.7 per cent, which indicates that India was making some progress through its population policies. However, in real terms, between 2010 and 2015, India's yearly population growth was still over 16 million people – the equivalent of adding the population of the Netherlands each year! Yet India's population pyramid for 2010–2015 still shows the classic pyramid shape of an LIC with a wide base of children and young adults, tapering to a relatively small elderly population. Despite some small successes, if India is to be successful in slowing population growth, more has to be done.

Exam-style question

1. To what extent have attempts to reduce birth rates been successful in one country that you have studied? [10]

5 Migration

5.1 Migration as a component of population change

Migration is the movement of people to live or to work. It can be either **internal** (movement within a country) or **external** (movement to another country). People who leave a country are called **emigrants** while people who arrive in a country are called **immigrants**.

Migration can be **permanent**, when the migrant moves away forever, **temporary**, when the migrant returns to their home country at some time in the future or **daily**, when the migrant returns to their own country after each day's work. The reasons for migration can be **forced**, where there is no choice for the migrant, or **voluntary**, where it is the own choice of the migrant.

Causes of migration

Push and pull factors

The causes why people migrate may be one or more of two sets of factors, called **push and pull factors** as explained by Everett S. Lee in his 1966 model (fig 5.1).

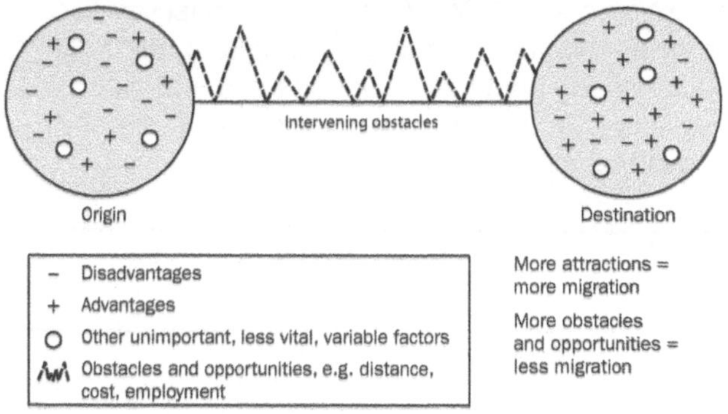

Fig 5.1 A model of migration proposed by Lee, 1966

Push factors operate in the **source area** (origin) and promote emigration, causing people to move away. Pull factors operate in the **receiving area** (destination) promoting immigration, attracting people to move there. Push factors include:

- natural disasters and events – such as volcanic eruptions, earthquakes, tsunamis, tropical storms/hurricanes/cyclones/typhoons, floods, droughts and rising sea levels
- unemployment
- lack of work opportunities
- escape from poverty and low incomes
- war
- racial, political or religious intolerance
- high crime rates
- housing shortages
- land shortages
- famine or lack of food.

Pull factors include:

- employment
- higher wages
- availability of food supplies
- better housing and education opportunities
- higher standard of living
- greater racial, political and religious tolerance
- more attractive living environment
- "bright lights" syndrome
- less crime.

A general view taken of many migrants is that most are "young, male and jobless". However, the type of migrant varies according to the type of migration. In international migration, the "young, male and jobless" label is valid, but migration flows are quite diverse. The contrast of migrants in and between LICs and HICs illustrates how diverse migrants can be.

There are several ways in which potential migrants receive information about possible destinations, including:

- through government agencies or advertising
- through media reports in newspapers, TV, the Internet and the radio
- by taking holidays to the destination/tourism
- from returning migrants
- as a result of hearsay and rumour.

Chain migration

Chain migration is where migrants from a particular area may follow others from that area to a particular city or neighbourhood. A chain migration may begin with migrants learning of opportunities. They may then be provided with transportation, and have their initial accommodation and employment arranged.

A chain migration may result in either a temporary or a permanent move. Italian immigration in the late nineteenth and early twentieth century relied on a system of both chain and return migration. Chain migration helped Italian men emigrate to the United States for work as migrant labourers. They generally left Italy due to poor economic conditions and a lack of job opportunities and they then returned to Italy as relatively wealthy people by Italian standards after working in the United States for a number of years.

Patterns of migration

Movements may appear to be random or individual, but distinct patterns of migration can often be identified. There may be the urban-rural movement from towns and cities; stepped migrations or, for example, to a capital city or to the coast or the sunbelt for retirement such as to the south of Spain in Europe or Florida in the USA.

There are several connections between **a person's age and migration**. Commonly, in voluntary migrations, younger adults are more likely to migrate than older adults, children migrate with parents and many movements relate to reaching certain stages in the life cycle, such as in making a career, or retiring. For example, only a small percentage of people aged over 60 years migrate. Possible reasons for this include:

- retirement is a settled phase of life
- much migration relates to employment and in this age group most are retired
- many older people have limited means (pensions) and moving can be costly
- older people may want to avoid the hassle and hard work of moving
- older people have less energy and may be in poorer health
- seniors may have less appetite for change, so ignore push/pull factors
- the familiar may be more appealing to them than the new
- habit patterns and networks of family and friends are established.

However, forced (involuntary) migrations affect everybody and are not linked to age.

Migration may bring negative social, economic and environmental impacts for the areas undergoing out-migration and may be linked to the type of migrant who leaves: young, often male, better educated and more dynamic. However, there can be some positives for the donor areas, such as lower unemployment, cheaper housing, less pressure on resources such as food supply and less pressure on the environment. This could well depend on the nature of the donor area and the volume of migration as both areas and migrants are far from uniform in terms of their resources and their population.

Constraints, obstacles and barriers to migration

Several constraints, obstacles and barriers may need to be overcome in order for migration to take place and migrants may be put off, deterred or delayed. These constraints may also restrict the size of migration flows. The constraints may differ according to the social context of the migrants and the development level of a country. The most frequently recognised constraints are:

- the distance to the destination
- the cost of getting there
- the need for a permit
- possible civil unrest
- the availability of imperfect information for the migrants to make their decision.

Migration decision-making is often complex and differs between individual people. A combination of pull factors and constraints may be involved, including:

- betterment e.g. maximising economic opportunity and the economic returns/remittances
- distance to migrate e.g. in relation to the cost of travel and the possibility of visits back home
- information available e.g. in the media, from relatives or friends, or rumour
- the contacts a person has e.g. at work, or with family members and friends
- the opportunities available e.g. possible job openings, what transport is available
- the barriers to migration e.g. political, economic, personal, physical
- the character of the individual migrant e.g. are they a risk-taker, are they optimistic or pessimistic, what fears do they have
- the impact of a new culture and if there is a similar group from your area at the destination to share experiences and speak your native language with.

Exam-style questions

1. Describe one way that the government of a receiving country can affect immigration? [2]
2. Describe the ways in which potential migrants may receive information about possible destinations. [6]
3. Describe and explain the possible impacts that migration may have on receiving countries. [5]
4. Explain the role that push and pull factors may have in migration decision-making. [6]
5. "Most migrants are economic migrants." How far do you agree with this statement? [10]
6. "Migration decision-making is often complex and differs between individuals." How far do you agree with this statement? [10]

6.2 Urban trends and issues of urbanisation

Urban growth and the process of urbanisation: causes and consequences

The process of **urbanisation** can be defined as the progressive concentration of population into urban areas over time.

Urbanisation is caused by a combination of high natural increase of population in cities and the process of rural–urban migration. The process of urbanisation is closely associated with economic development. Urbanisation took place earlier in the HICs of Europe and North America and in some MICs, where often more than 90 per cent of the population lives in towns and cities. The process of urbanisation started to accelerate over 200 years ago, as these countries went through the Industrial Revolution. People left their jobs in agriculture in rural areas and migrated to the growing towns and cities to work in factories. In these countries, urbanisation is now either very slow or has stopped.

In 2007, the majority of people worldwide lived in urban areas, for the first time in history; this is referred to as the arrival of the "Urban Millennium" or the "tipping point". In the future, it is estimated that 93 per cent of urban growth will occur in developing nations, with 80 per cent occurring in Asia and Africa. The LICs and MICs in Africa and south-east Asia have much lower levels of urbanisation at the moment, because industrialisation took place later.

Cities offer a larger variety of services than rural areas, including specialist services not found in rural areas. These services require workers, resulting in more numerous and varied job opportunities. Elderly people may be forced to move to cities where there are doctors and hospitals that can cater for their health needs. Varied and high-quality educational opportunities are another factor in urban migration.

Urbanisation also creates opportunities for women, creating a gender-related transformation, where women are engaged in paid employment and have access to education. This, in turn, may cause fertility to decline. However, women are sometimes still at a disadvantage due to their unequal position in the labour market, their inability to secure assets independently from male relatives and exposure to violence.

Rural migrants can be attracted by the possibilities that cities can offer, but often settle in shanty towns and experience extreme poverty. The inability of countries to provide adequate housing for these rural migrants is related to overurbanisation, a phenomenon in which the rate of urbanisation grows more rapidly that the rate of economic development, leading to high unemployment and high demand for resources.

- The main trends shown in fig 6.1 are: all world regions show a predicted increase in urban population by 2030

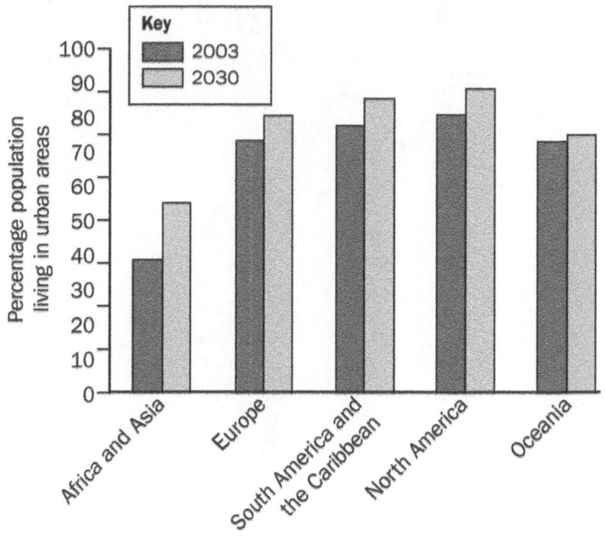

Fig 6.1 Predicted urban population increase (2003–2030)

- the increase is highly variable, from Oceania's 2 per cent to a massive 17 per cent for Africa and Asia
- all four world regions had urban population of 73 per cent or higher in 2003
- only Africa and Asia had a much lower level in 2003 (38 per cent) so the majority of the population still lived in rural areas. This predicted to increase significantly, by about 17 per cent by 2030.

The two cities where the increase in total population is expected to be the greatest between 2000 and 2015 are Lagos – 10.9 million and Bombay (Mumbai) – 9.3 million (fig 6.2) The large population increase expected in many cities can be explained by the combination of high rates of natural increase and of rural–urban migration.

Urbanisation is part of a sequence of processes that may also involve the processes of **suburbanisation, counterurbanisation** and **re-urbanisation**. This sequence has occurred in many HICs, including the UK and Germany.

Suburbanisation (1860s–1960s)

The main reason for the sudden surge in **suburbanisation** involves the outward growth of a city by the construction of suburban housing. In the UK, the main reason for this process was the construction of the urban railway networks. This triggered a huge increase in house building as where there was increased competition for available land, people could now leave the crowded inner city areas and live along the railway networks. This was firstly only the richer middle classes who could afford to move. But, after the First World War (1914–18), there was a massive move to build public housing for the working classes.

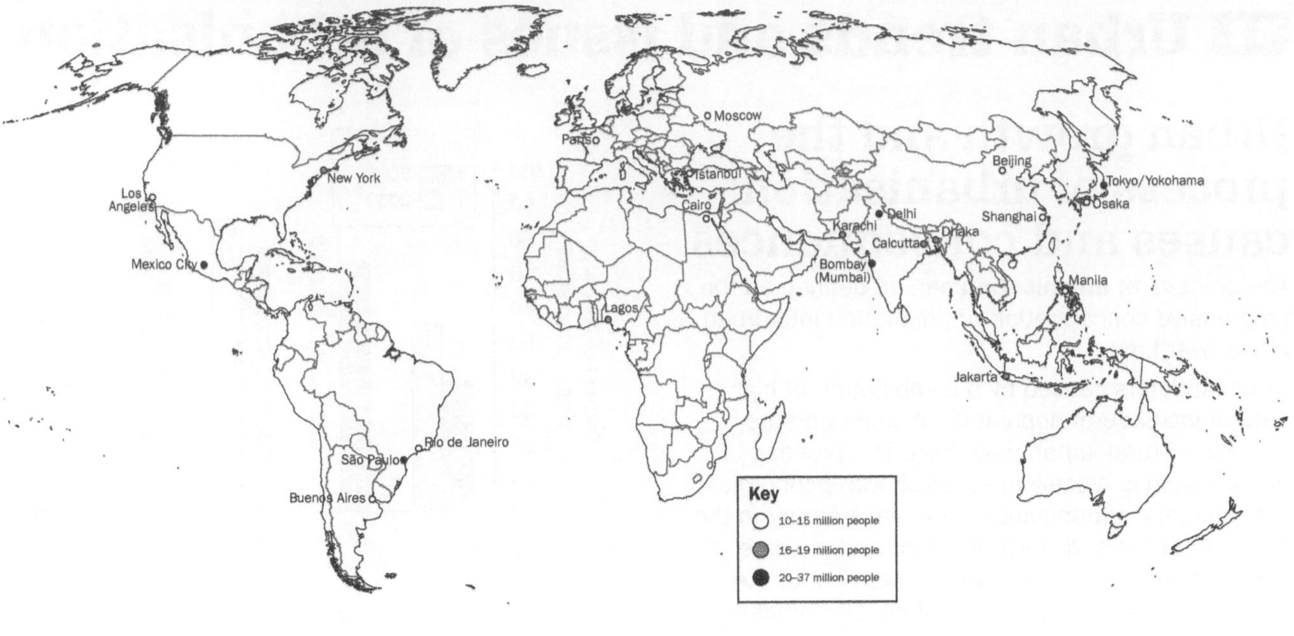

Fig 6.2 World's largest cities in 2015

This resulted in the building of 4.3 million houses in the period up to the Second World War (1939–45). Of this total, over 30 per cent was built by local authorities with the financial support of the UK government and was often referred to as "council housing".

This housing remained in the ownership of the councils and the people who inhabited them paid a small rent to the council. Having the backing of the government finance meant that the essential, and expensive infrastructure – such as sealed roads, piped clean water, sewage systems , electricity, gas, telephone and street lighting – could be put into place.

Counterurbanisation (1960s–1990s)

Counterurbanisation involves the movement of people out of urban areas into smaller towns and villages. For example, from London into the smaller towns of south-east England, such as Reading. Many cities in HICs, such as Detroit in Michigan, USA, have been losing population from their central areas.

Urban renewal

Urban renewal has taken place in many cities and involves redevelopment in areas of moderate to high density urban land-use. Urban renewal has had both successes and failures. It began in the late 19th century in HICs and experienced an intense phase in the late 1940s. The process has had a major impact on many urban landscapes, and has played an important role in the development of cities around the world.

Urban renewal involves the relocation of businesses, the demolition of structures, the relocation of people, and the government purchase of property for city-initiated development projects. In some cases, renewal may result in urban sprawl and less congestion when some areas of cities witness the construction of new motorways, freeways and expressways.

Urban renewal has been seen as a means of economic development. In some cases it is seen as enhancing existing communities and in other cases as the destruction and demolition of neighbourhood communities.

Many cities link the revitalisation of the central business district and the gentrification of residential neighbourhoods as examples of urban renewal. Over time, urban renewal has evolved in many cities as a policy based less on destruction and more on renovation and investment, and today is an integral part of many local governments, often combined with small and big business incentives.

Re-urbanisation (1990s onwards)

Re-urbanisation is the movement back to the city at a later stage in the urbanisation process. Many HIC cities have attempted to reverse the population decline taking place in the central areas of their major cities. In the UK millions of pounds of government money has been spent on rejuvenating inner city areas to make them more attractive to live and work. There is also pressure on the government to develop older **brownfield** sites and relieve the pressure on the surrounding countryside areas – the **greenfield** sites.

CS London

Inner London had a population of 5 million in 1900 but this had halved to 2.5 million by 1983. Since 1983 the loss of population has been reversed, due to reurbanisation, and the population of Inner London has risen to 3.2 million by 2011. Many of the people who have moved into London are young adults. The newly developed inner city areas in London provide several benefits:

- people have easy access to shops and services, including entertainment
- people can save money by living close to work
- people have easy access to a vibrant, active, cultured urban environment, especially appealing to young adults
- there may be refurbished heritage properties or new build housing to live in.

There are several examples of re-urbanisation in London, including the London Docklands area. This was re-developed from the mid 1970s by the **LDDC** – the **London Docklands Development Corporation** and lies to the east of the city along the River Thames.

Causes of urbanisation

HICs

HICs are characterised by having low rates of urban growth. Cities in HICs grew many years ago due to rural–urban migration so that a majority of the total population of many HICs are now living in urban areas. Added to this is the fall in birth and death rates and the consequent natural increase in population. This migration to the cities of many HICs has now all but ceased.

In many HICs there is actually a decline in population in urban areas rather than an increase which can be explained by counter-urbanisation and industrial decline.

LICs

LICs are characterised by rapid growth rates of urban growth. There are two main causes:

1. Rural–urban migration due to perceived opportunities/changing agricultural economies, etc.
2. Population growth of the existing urban population due to lack of family planning, need for children, cultural desire for large families, etc.

This rapid growth has led to a number of problems, including the uncontrollable growth of shanty towns, unemployment, pollution and social unrest.

World cities

World cities are of **global significance**. They tend to be large (both in population, in area and in their global reach). They are very powerful economically, and they are familiar to a large number of the Earth's population from the media, travel and world events.

They are often major financial hubs containing the headquarters of major global banks and businesses and they are at the centre of regional transport networks. Politically, many world cities are capitals or regional capitals and homes to governmental institutions such as the UN and the WTO. They provide the venues for major summit meetings.

Socially, they are multi-ethnic, diverse and they have a vibrant sports, arts and cultural life.

Causes of the growth of world cities

Possible factors contributing to the growth of the world cities include the emergence of:

- a fast growing, increasingly integrated, world economy, based in world cities
- the transport (especially air) and telecommunications revolutions
- urbanisation and the preferences of people for urban living and working
- urban agglomeration i.e. the growth of conurbations and megalopolises
- facilitation e.g. architecture, planning, finance, infrastructure, stability, etc.

These factors combine to give the following characteristic features of world cities:

- a variety of international financial services, notably in finance, insurance, real estate, banking, accountancy, and marketing along with the existence of financial headquarters, a stock exchange and major financial institutions
- the headquarters of several multinational corporations
- domination of the trade and economy of a large, often national, surrounding region
- major manufacturing centres with port and container facilities
- considerable decision-making power on a daily basis and at a global level
- centres of new ideas and innovation in business, economics, culture and politics
- centres of media and communications for global networks
- a high percentage of residents employed in the services sector and information sector
- high-quality educational institutions, including renowned universities, international student attendance and research facilities
- multi-functional infrastructure offering some of the best legal, medical and entertainment facilities in the country.

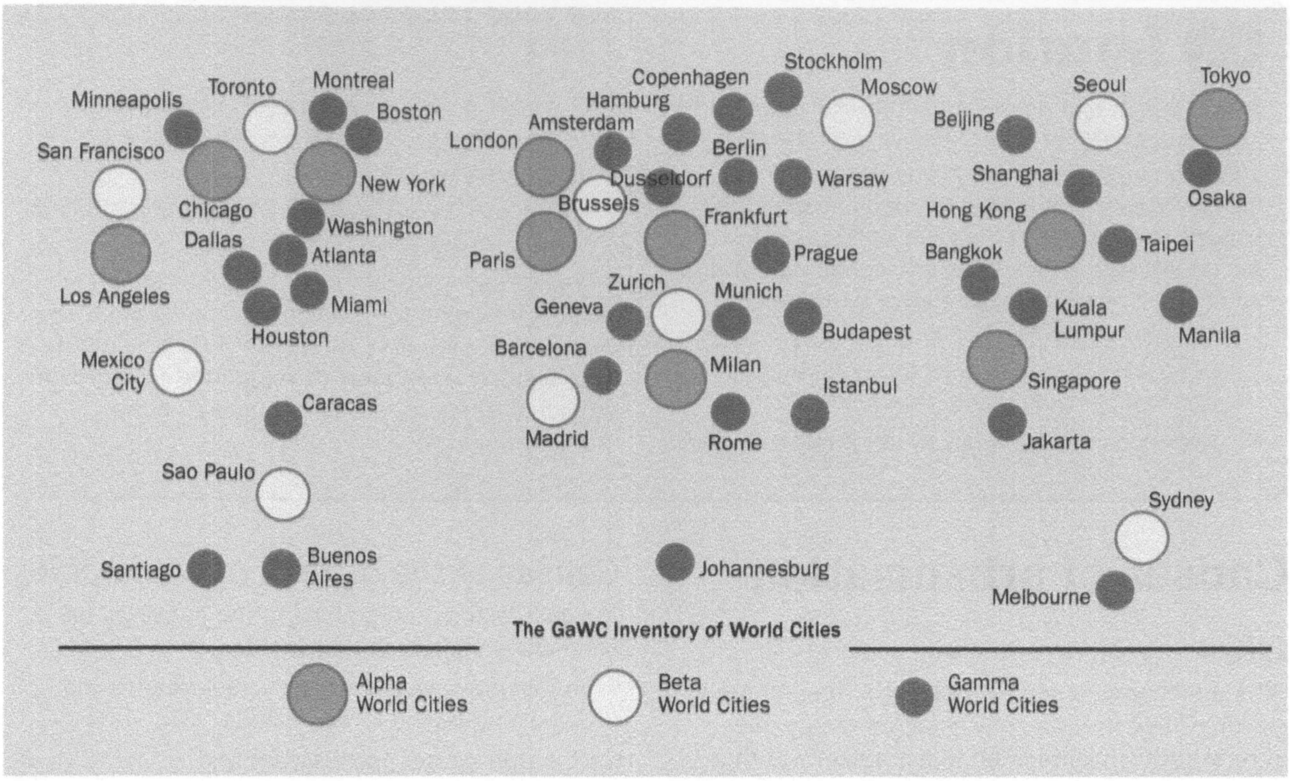

Fig 6.3 World cities based upon their level of advanced producer services. Global service centres are identified and graded for accountancy, advertising, banking/finance and law. Aggregating these results produces a list of 55 world cities at three levels: 10 alpha world cities, 10 beta world cities and 35 gamma world cities. These are mainly concentrated in three areas: North America, Western Europe and Pacific Asia (Source: GaWC Research Network)

Development of a hierarchy of world cities

It is possible to identify a hierarchy of world cities (fig 6.3). The GaWC (Globalization and World Cities Research Network) identified the position of a city based on the **service value of a city** to a particular company, meaning the importance of the company's office from that city within the global network of the company's offices.

Its disadvantages though are that it does not offer any basis to work out the differences between the different cities – differentiation especially between the top cities of London, Paris and New York; and it does not indicate which cities are the growing, emergent cities. There have been many attempts by governments to discourage further development of the largest urban centres in some countries. Two broad government approaches exist:

1. Restricting development of the largest centres. This can be done by:
   - **decentralisation**, such as moving government departments to other centres, as in some Caribbean islands
   - by withholding permission for development, such as TNC investment or housing
   - by protective measures such as a Greenbelt policy.
2. Encouraging the development of other urban areas. This can be done by:
   - Developing the concept of the **growth pole**, such as Zimbabwe's growth points, which have proved quite successful. This diverts new growth to other cities and towns.
   - Developing a **new town** or **expanding town policy** – London has eight new towns, such as Bracknell and Crawley, which are now well established outside of its Green Belt. Milton Keynes took this a step further when it was established as a new city to the north of London. Expanded towns have the advantage of already having set up much of the expensive basic physical, economic and social infrastructure in place to which new development can be added such as Reading to the west of London.

6.3 The changing structure of urban settlements

Factors affecting the location of activities in cities

There are several social, economic, environmental and political factors that usually combine to determine the location of activities within urban areas. These factors are dynamic and change over time with the result that the urban locations for retailing, services and manufacturing change and can be grouped in social, economic, environmental and political factors. The development of these factors over time is reflected and explained in the evolution of urban land-use models.

Urban land-use models

Several theoretical urban models have been put forward over the last 80 years to try to describe and explain patterns and zones you may expect to see in a "typical" city or large town. These result from social, economic, environmental and political factors

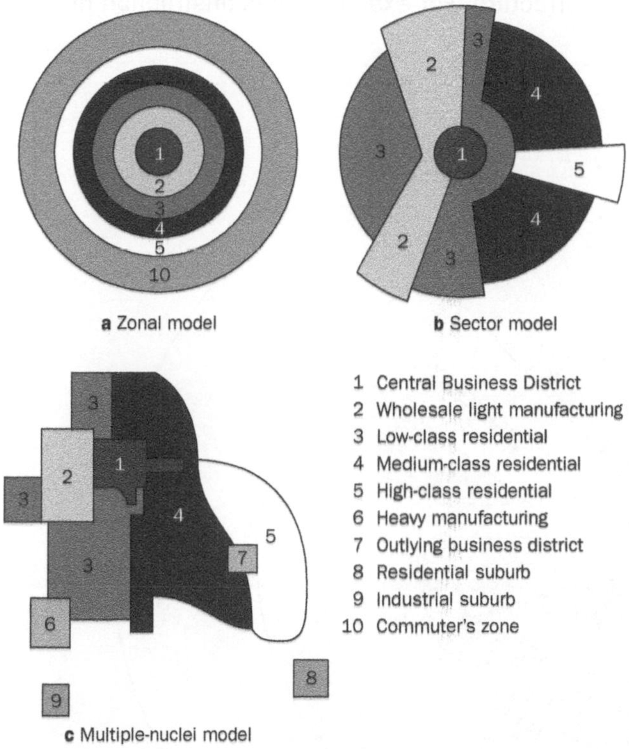

Fig 6.4 Urban settlement models

- **The concentric model (Burgess)**

  The concentric model was devised in 1925 by **Burgess**, based on his work on the cities of mid-west USA, in particular Chicago. He related a city's growth to a natural ecological succession where growth and change took place by a sequence of invasion and succession. Although outdated now in many ways, it was important in providing a theoretical foundation on which other researchers could investigate further. It had 4 main assumptions:

  1. A uniform, sometimes called an isotropic, land surface.
  2. Free competition for the space in the city.
  3. Easy access to the centre of the urban area from all directions.
  4. Continual migration into the city with new development expanding out from the centre.

  The end result of this was that the city developed as a series of concentric rings, as seen in the zonal model (fig 6.4a).

- **The sector model (Hoyt)**

  The sector model developed by **Homer Hoyt in 1939** was based on a larger sample of 142 cities, but only in the USA. Again, he placed the Central Business District (CBD) in a central location as it had maximum accessibility (fig 6.4b).

  However, he found that once land uses established themselves near to the city centre they expanded out from the centre in a distinct wedge/sector shape as the city grew. This expansion was normally along major transport routes.

  Low income housing was found in the least favourable locations, while high income housing was in the most favourable, with a middle income group in an intermediate position between the two.

  The original high and middle class areas, in the oldest housing near the city centre, was abandoned in time and became occupied by low income groups, who often subdivided the larger houses.

- **The multiple nuclei model (Harris and Ullman)**

  This was developed by Harris and Ullman in 1945 who put forward the idea that urban areas developed around several centres or nuclei (fig 6.4c). Some of these centres would have been the centres of villages and small towns swallowed into the city area as it expanded, while others would have been newly created.

  They put forward the ideas of attraction and repulsion – similar activities tend to group together, others repulse each other.

  For example, the residential segregation of ethnic groups may develop within an urban area. This can be due to the location of a person's family, community, village, people, tribe, and language group as well as religion.

Factors affecting residential patterns

- **Forces of attraction:** like attracts like, so new arrivals have natural bonds with and links to others from the same ethnic group through shared language, tradition, religion, family or friendship ties. They also may have similar needs both social: for interaction, marriage partners, schooling, worship etc; and economic, e.g. to shop for food items or clothing specific to their ethnic needs or working for each other. Beyond this, residential segregation enhances well-being through feelings of security and security in numbers, the maintenance of traditions and identification with "home".

- **Forces of repulsion:** unliked ethnic groups may repel, keeping others out. At worst, this relates to ethnic distrust, rivalry, disputes and violence, some of which may be longstanding. It may be an expression of the needs of minorities to survive in urban areas and to "defend" themselves socially. It may be fostered by the working of the property market e.g. landlords, by discriminatory behaviour and by planning decisions.
- **Urban model in LICs (Griffin and Ford)**

  Several urban models have been based on LIC cities. The one developed by Griffin and Ford (fig 6.5) is based on studies carried out on Latin American cities. As with the cities of HICs, the CBD is centrally located and has the same characteristics (referring to offices, government buildings, retailing etc). Many develop a commercial spine which extends out from the CBD and an elite residential sector on either side of this.

Surrounding the central area is a **zone of maturity** which contains a mixture of older traditional housing. This used to house the elite high income group who now lives in the elite sector. These buildings are in relative decline and have often been subdivided. Some have been replaced by self-built housing.

Outside of this is a **zone of "in situ" accretion** which has a very wide variety of housing, both in type and quality. The provision of services is variable in having sealed road surfaces, electricity, water and sewerage. New government housing projects are often focused in this zone.

Industries tend to be located near the centre because this area has both power and water supplies.

Surrounding the periphery of the city is a **zone of squatter settlements** which is home to new migrants.

The main features of the model's residential pattern are that the urban area is divided by income (an economic factor) and geographical location.

There are several **economic factors** which help to create residential districts with different characteristics within many large urban areas. One of the most important factors is the level/amount of personal or household income. This affects the ability of residents to pay for housing. However, the reasons may also include investment by local authorities in public sector housing or in infrastructure. For example, the construction of

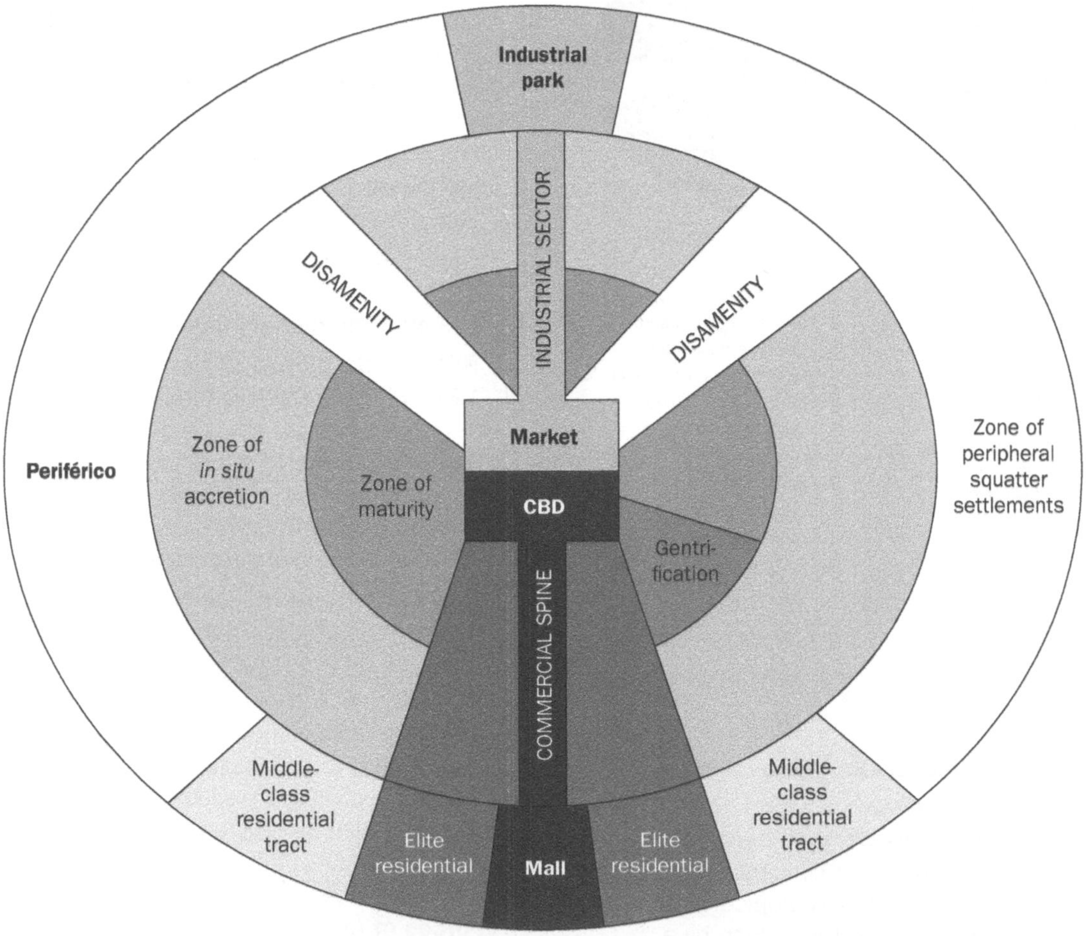

Fig 6.5 Urban model of a Latin American city (Griffin and Ford)

transport networks which may then affect an area's desirability if access is good and commuting to work is possible.

Many factors promote ethnic segregation in cities, such as language, kinship, religion, culture, support, business opportunities and networks. This shows the importance of social factors in determining residential patterns.

Many businesses today might choose to locate in the **outlying business district (OBD)** of a town or city rather than in its central business district. This is because high costs, congestion, lack of space for expansion, unpleasant environment in the CBD, whereas the OBD provides better communications, cheaper land costs and a more pleasant working environment.

Changes in location of manufacturing industry

During the Industrial Revolution of the nineteenth century, industry tended to be located in the inner cities. By the twentieth century though, an inner city location started to become very disadvantageous. In most cases this was due to increasingly poor accessibility to central locations for workers, raw materials and for the distribution of finished products. However there were many other factors, including:

- Many of the early factories, such as textile mills were several storeys high, unsuitable for new manufacturing methods which needed single-storey, low-rise factories and space for their modern production methods.
- The old inner city locations left no space for industries to expand or reorganise into new, large industrial estates.
- The old industrial sites were often polluted, very costly to clean up and decontaminate and there were also increasing concerns over air and noise pollution and public safety in inner city areas.
- Inner city areas faced increasing demands from other land users who wanted a central location, particularly for retailing and commercial activities.

Changes in location of retailing

There have been enormous changes in retailing habits in the last 45 years, with people having increased access to private motor cars. In the UK there were 2.5 million vehicles in 1954, in 2015 there were 35.6 million! This has led to several important new developments:

- **Suburban CBDs** – as urban areas grew, people started to find themselves at considerable distances both in space and time from the central CBDs, so suburban retail and commercial centres expanded
- **Retail parks** with large areas of free car parking and stores with very large floor space grew along ring roads and major intersections.
- **Urban superstores** and **hypermarkets** were built at points of high accessibility and high consumer demand – led by large supermarket corporations, such as Walmart, Carrefour and Tesco.
- **Out of town shopping centres** with one stop shopping, in a controlled environment, with eating and leisure facilities included.
- **Internet shopping** and **home delivery** from very large retail outlets threaten to irreversibly change traditional shopping habits.

Peripheral locations are increasingly popular in many urban areas because they offer several advantages, including more space, better access and parking (often for free), cheap(er) land, linkage for deliveries and a better shopping environment.

Changes in location of services

Apart from retailing, other services have moved to either the suburbs or to the rural–urban fringe. These include new sport complexes and stadia, including gold courses, large secondary schools, further education colleges and universities and regional health centres and hospitals. As with retailing, the reasons for their location include cheaper available land, more space for expansion, good access to transport lines and a pleasant environment.

Where the service sector has expanded in the suburbs and rural–urban fringe, there has been a decline in service provision in many inner city areas.

Many **key workers**, who usually have average or lower than average salaries, have found themselves in inner city locations where they are unable to buy or rent accommodation or afford the cost of transport. They include teachers, nurses, paramedics, police and fire officers. All of this has led to further decline in the inner city areas.

The changing Central Business District (CBD)

The Central Business District (CBD; fig 6.6) is the central part of a city, characterised by a high concentration of retailing, banks, office and administration. It is often one of the most dynamic and rapidly changing areas/zones in a city.

Whilst the four main functions found within a CBD are usually retailing, offices, banks and administration, other functions may include professional services such as lawyers, transport such as the bus or railway stations and public buildings such as courts, libraries and museums. Some CBDs may also have parks, manufacturing and residential (usually apartments) buildings.

Fig 7.6 Corestones in Namibia

Fig 7.6, shows an example of **corestones** resulting from **weathering** on granite in a tropical environment in Namibia.

The rounded core stone rocks, as well as the smaller angular rocks, known as **clasts**, are the result of **exfoliation**.

Limestone

Karst is a landscape formed from the dissolution of soluble calcareous rocks such as limestone, dolomite, and gypsum. It is characterised by underground drainage systems, often with well-developed underground caves in which the surface water enters through sinkholes and dolines. Fig 7.7 shows some landforms found in limestone (karst) regions of the tropics.

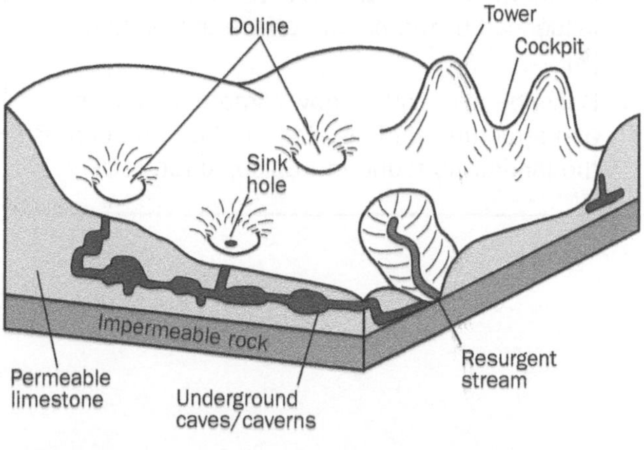

Fig 7.7 Typical landforms found in karst regions

In all karst areas, the key weathering process is **carbonation** – the action of acidulated rainwater on calcium carbonate. Although water is less able to dissolve carbon dioxide in tropical areas owing to higher temperatures, the presence of large amounts of organic matter produces high amounts of carbon dioxide in the soilwater. The joints and bedding planes found in limestones greatly increases the surface area which can be subjected to carbonation.

In the humid tropics, due to the high temperatures and abundant precipitation, the process is considerably enhanced, leading to large scale karst features. These can include all the generally found landforms of **underground caves**, **limestone pavements**, **clints**, **grykes**, **dolines** and **poljes**.

Tower karst and cone karst

Tower karst is characterised by upstanding rounded blocks set in a region of low relief. The towers can vary in size. They have steep sides, with cliffs and overhangs, and caves and solution notches at their base. The steepest towers are found in areas of **cone karst**, where residual cones of limestone are on massive, gently tilted limestone. The towers originate as residual cones of limestone and are then steepened by water table undercutting from surrounding alluviated plains. The towers are often found standing above large, flat flood plains and swamps and may show undercutting from rivers and lakes. Tower karsts are thought to represent the last remnants of limestone outcrops.

The limestone towers, which can be up to 200 metres in height, are the result of differential weathering and erosion along lines of structural weakness. The towers have nearly vertical walls and gently domed or serrated summits.

Tower karst occurs throughout south-east Asia. By far the most extensive and best developed tower karst is in the Guangxi province of southern China. Other areas of tower karst are found in Vietnam, Thailand and Malaysia, and in the Chinese Sea, where it is submerged. Cone karst occurs in Cuba, Madagascar and Puerto Rico.

Cockpit karst

Cockpit karst is a landscape pitted with smooth-sided soil-covered depressions and cone-like hills. Cockpit karst is characterised by groups of hills, fairly uniform in height. The **cockpits** are formed after underground limestone cave systems grow and then collapse leaving large depressions in between limestone cones.

- Cockpit karst is the beginning of the development. After cave systems developed, grew and collapsed, the former caves form huge valleys and the limestone in between remains as hills.
- Cone karst is the more common and less spectacular form of this landscape with limestone hills, residual cones, typically covered by rain forest.
- Tower karst is the spectacular form with 30–300 m high towers with vertical or overhanging sides. The walls are typically bare rock, as the walls are too steep for vegetation.

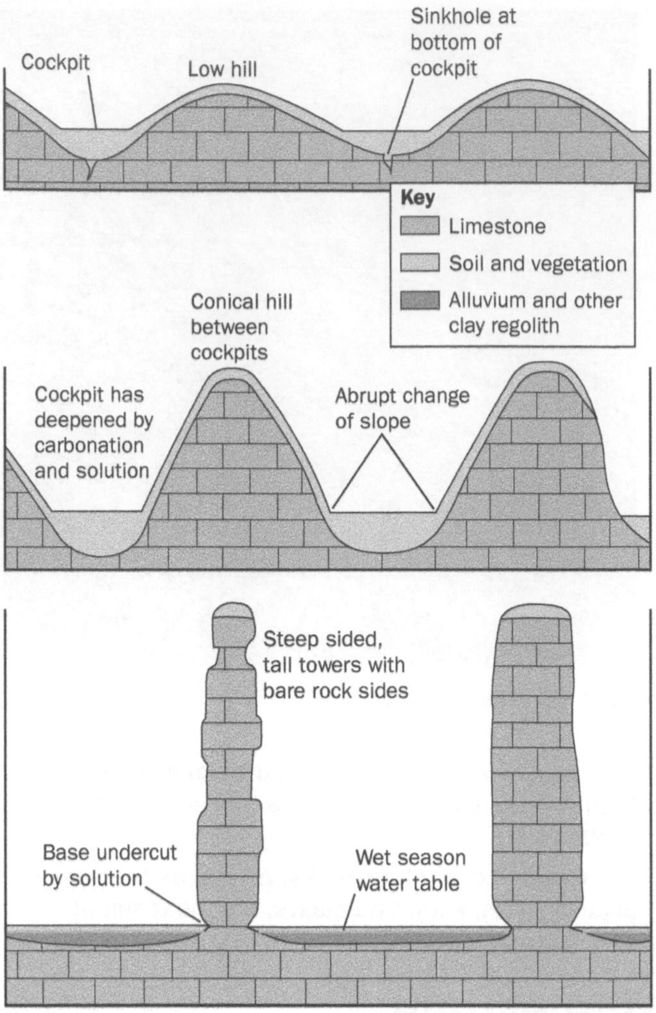

Fig 7.8 Cockpit karst, cone karst and tower karst

Exam-style questions

1. Describe the main type of weathering that takes place in tropical environments. [6]
2. Explain how weathering contributes to the formation of landforms in areas composed of either limestone or granite. [8]
3. Contrast the formation of landforms in granite and landforms in limestone in tropical climates. [10]
4. Explain what is meant by the term basal surface of weathering. [2]
5. Describe how the development of granite landforms in tropical climates may be influenced by the basal surface of weathering? [8]
6. Describe and explain how climate, rock type and structure are important in the formation of the landforms found in tropical karst. [10]

7.3 Humid tropical (rainforest) ecosystems and seasonally humid tropical (savanna) ecosystems

Plant communities

Climax vegetation

Climax vegetation is a plant community that has become adapted to environmental conditions, reaching a stable state. Theoretically, a climatic climax community will only change if the climate changes. Generally, the climax community contains a few dominant species, which have beaten other species in the competition for light, nutrients and space.

Sub-climax vegetation

If vegetation does not reach its climax as a result of interruptions by local factors, such as changes in soil type or differences in the parent rock, a sub-climax vegetation develops. Such interruptions are known as **arresting factors**.

Plagioclimax

A **plagioclimax** is a plant community kept stable by continuous human activity You could contrast the climax vegetation of a tropical rainforest (TRF) with a secondary plagioclimax forest which has undergone slash and burn, or between a wooded climax savanna and a grassland savanna, where human activities have removed the trees. The plagioclimax represents the intervention of human activities, or other circumstances, that bring about an alteration to the natural vegetation.

Tropical rainforests

Distribution

Fig 7.9 shows the distribution of the **main** areas of tropical rainforest.

The distribution of tropical rainforest is contained within a fairly narrow band of 20° north and south of the Equator. The distribution extends into the continental interiors of both South America and Africa but otherwise it is closer to coastal locations. It is clear to see that this distribution is climatically controlled. It is found in the humid tropical areas where there is equatorial uplift of air due to intense solar radiation and convergence at the Earth's surface. This produces high temperatures and year-round precipitation. The close proximity to the oceans adds maritime humidity.

Vegetation

Vegetation in a TRF is characterised by distinct layers. This structure is primarily influenced by sunlight.

Tropical rainforests are closed canopy forests with at least three vegetation layers/strata.

1. The **upper canopy layer** can reach a height of 30–60 metres with trees that emerge (the emergent layer) from the canopy, sometimes reaching 80 metres. Emergent trees are spaced wide apart, and have umbrella-shaped canopies. Exposed to drying winds, emergent trees tend to have small, pointed leaves that are dark green, small and leathery to reduce water loss in the strong sunlight. The leaves also have to be able to withstand high levels of UV radiation in their exposed position. These very tall trees have straight, smooth trunks with few branches. The trunks of some trees, such as the cocoa and bread fruit trees, often sprout flowers and fruits. Their root system is very shallow as the nutrients and water they require are found in the top few

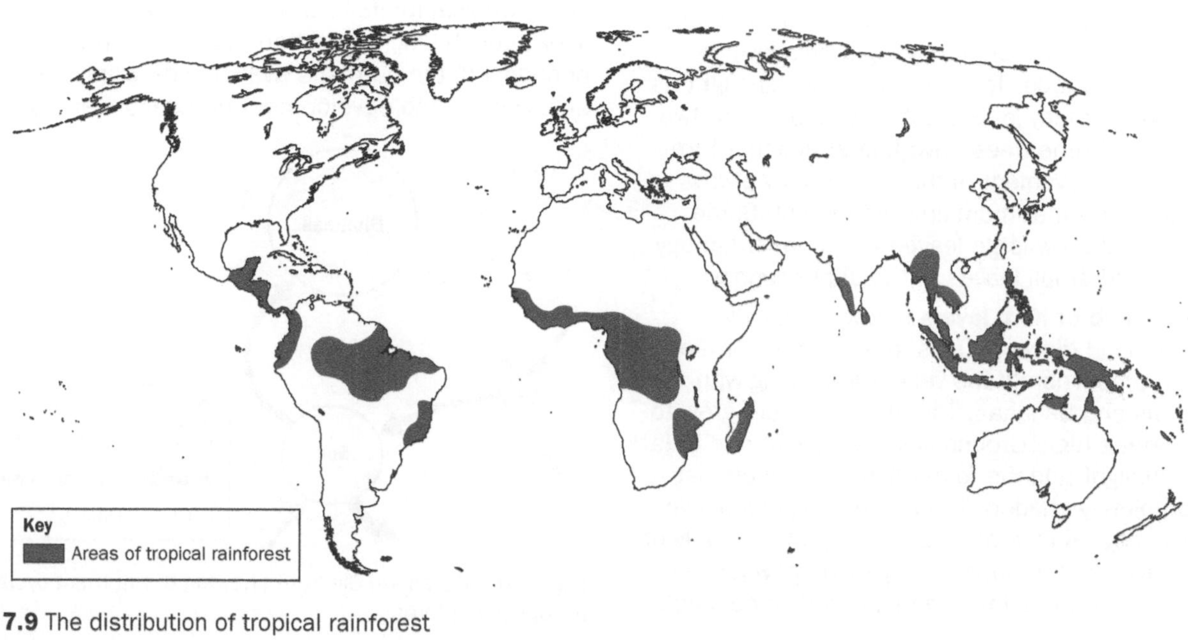

Fig 7.9 The distribution of tropical rainforest

centimeters of the soil. To support their size they grow buttress roots. These layers have a large proportion of broad-leaved trees in all stages of growth, so that leaves, flowers and seeds are constantly being grown and shed all year round.

Within the tree canopy there are thick woody climbers called **lianas**. A liana is a long-stemmed, woody vine that is rooted in the soil at ground level and may grow up to 60 cm in diameter and 100 metres in length. These structural parasites climb up the trunks and limbs of canopy trees to get access to sunlight. Lianas compete intensely with the canopy trees, and can greatly reduce tree growth. Lianas also provide access routes in the forest canopy for many arboreal animals, including ants and many other invertebrates, lizards, rodents, sloths, monkeys, and lemurs.

Most tropical rainforests contain **stranglers**, the common name for a number of tropical and subtropical plant species, including some banyans and unrelated vines. They all share a common "strangling" growth habit that is found in many tropical forest species, particularly of the genus Ficus, the fig trees. These plants begin life when their seeds, often bird-dispersed, germinate in crevices of canopy trees. The seedlings grow their roots downward and envelop the host tree while also growing upward to reach into the sunlight zone above the canopy. The original support tree can sometimes die, so that the strangler fig becomes a "columnar tree" with a hollow central core.

With the high rainfall, many of the trees and plants have made adaptations that help them shed water off their leaves quickly; many plants have **drip tips** that allow rain to run off and some leaves have waxy or oily coatings to help them shed water. This keeps them dry and prevents moulds and mosses from forming on them.

2. The next layer, the **sub**, or **lower canopy** is usually around 20 metres high. It is made up of the trunks of canopy trees, shrubs, plants and small trees. There is little air movement in this sheltered environment and, as a result, the humidity is constantly high. This level is in constant shade. To absorb as much sunlight as possible, many leaves are very large, up to two metres. Some trees have leaf stalks that turn with the movement of the sun so they always absorb the maximum amount of light. Some trees will grow large leaves at the lower canopy level and small leaves in the upper canopy.
3. The **shrub or herb layers** below are poorly developed discontinuous layers. Ferns form a high proportion of the vegetation, along with palms and bananas. The shrubs usually grow to 5 metres high. Ground cover is sparse due to lack of sunlight and the forest floor may be almost completely shaded. The shrub and ground layer receiving as little as 2–3 per cent of the sunlight received by the upper canopy, except where a canopy tree has fallen and created an opening.

The nutrient cycle

In 1976, P. F. Gersmehl developed a model to illustrate **nutrient cycles** within ecosystems. Circles of proportionate size are drawn to represent the stores of nutrients within the ecosystem – biomass, litter and soil. The nutrient transfers and inputs/outputs are represented by arrows of varying thickness, to show the relative rates of transfer between the stores.

The **inputs** include nutrients such as carbon and nitrogen dissolved in the precipitation and the minerals from the weathered parent rock.

The **outputs** include the loss of nutrients from the soil by leaching and by surface runoff.

The **flows** or **transfers** include leaf fall from the biomass to the litter, the decomposition of litter transfers nutrients to the soil, and then the uptake of nutrients from the soil by trees and other plants in the ecosystem.

Tropical rainforests have extremely rapid rates of nutrient transfer, due to their high temperatures, rainfall and humidity. The biomass (the living vegetation including the roots) is the largest store of nutrients. The litter or decaying matter is the smallest store because the nutrients are processed very efficiently by the very abundant decomposers including bacteria, fungi, and termites which are helped by the high availability of nutrients and high temperatures in the forest. Nutrients are transferred rapidly from the litter to the soil and they are almost immediately absorbed by the tropical vegetation. Nutrients are not stored in the soil for long; however they can be quickly lost by the process of leaching if the forest is cleared.

The natural tropical rainforest system is self-perpetuating while it is undisturbed. Any human intervention in the form of shifting, cultivation or commercial clearance can have a huge impact. The replacement of trees by crops means that a proportion of the nutrient stores will be lost due to the biomass being removed as crops are removed. This affects the flow between the litter and the biomass. Similarly, the exposure of the soil to heavy tropical rainfall can result in heavy nutrient loss through leaching. When the cleared area is abandoned, the nutrients available may only be sufficient for the development of a secondary forest.

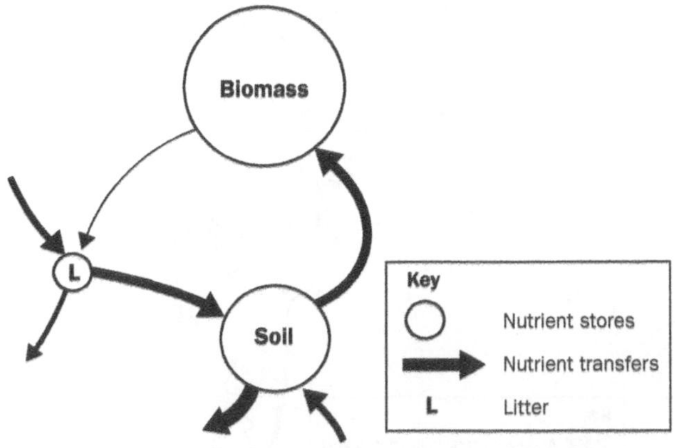

Fig 7.10 A Gersmehl diagram showing the nutrient cycle in a tropical rainforest.

Energy flow and trophic levels

Energy transfer within an ecosystem can be shown by a pyramid diagram showing the four main trophic levels. At each trophic level, some of the energy contained is available as food for the next level in the pyramid. Each layer in the pyramid from the bottom up decreases in size as around 90 per cent of the energy contained within each level is lost through the living processes, such as respiration, movement and excretion. As only 10 per cent of the energy is available for the next trophic level as food, the number of living organisms within an ecosystem decreases as the trophic level increases. Starting from the base of the pyramid, the four levels are:

1. The **Primary Producers** or **autotrophs** which are the green plants who produce their own food through photosynthesis using energy from sunlight.
2. The **Primary Consumers** or **herbivores** are found in the second trophic level. These are insects, fish, birds and mammals which consume the primary producers.
3. The **Secondary Consumers** or **carnivores which** are meat-eaters and they consume the herbivores.
4. The **Tertiary Consumers** are found at the top of the trophic pyramid. They are the top predators and eat the secondary consumers. They may be **omnivores** feeding on both the herbivores and the plants.

At all levels in the trophic pyramid **detritivores** and **decomposers** may be operating. A detritivore is an animal that feeds on dead material or waste products. A decomposer is an organism that breaks down the dead plants, animals and waste matter. Fungi and bacteria are decomposers.

Food chains and food webs

Within the trophic pyramids there are **food chains** which also illustrate the flow of energy through an ecosystem. As with the four levels in a trophic pyramid, there are usually four links in a food chain. Each link in the chain feeds on and therefore obtains energy from the preceding link and each link is then consumed by, and provides energy for, the following link.

Within the tropical rainforest there are an enormous number of individual food chains and the overall picture can be shown by means of putting all the food chains together as a **food web**. Many of the animals, especially the omnivores, have a very varied diet and any one species of animal or plant is likely to be the food for a large number of different consumers. As a result, there are extremely complex feeding interactions within the tropical rainforest ecosystem.

Soil formation

Soils in tropical rainforest areas develop on deep accumulations of weathered rock, regolith, and therefore have deep B and C horizons composed of slightly modified parent material. The A horizon on top is relatively thin, low in nutrients and poorly sorted. They are often referred to as **oxisols**, **latosols** or **ferralsols**.

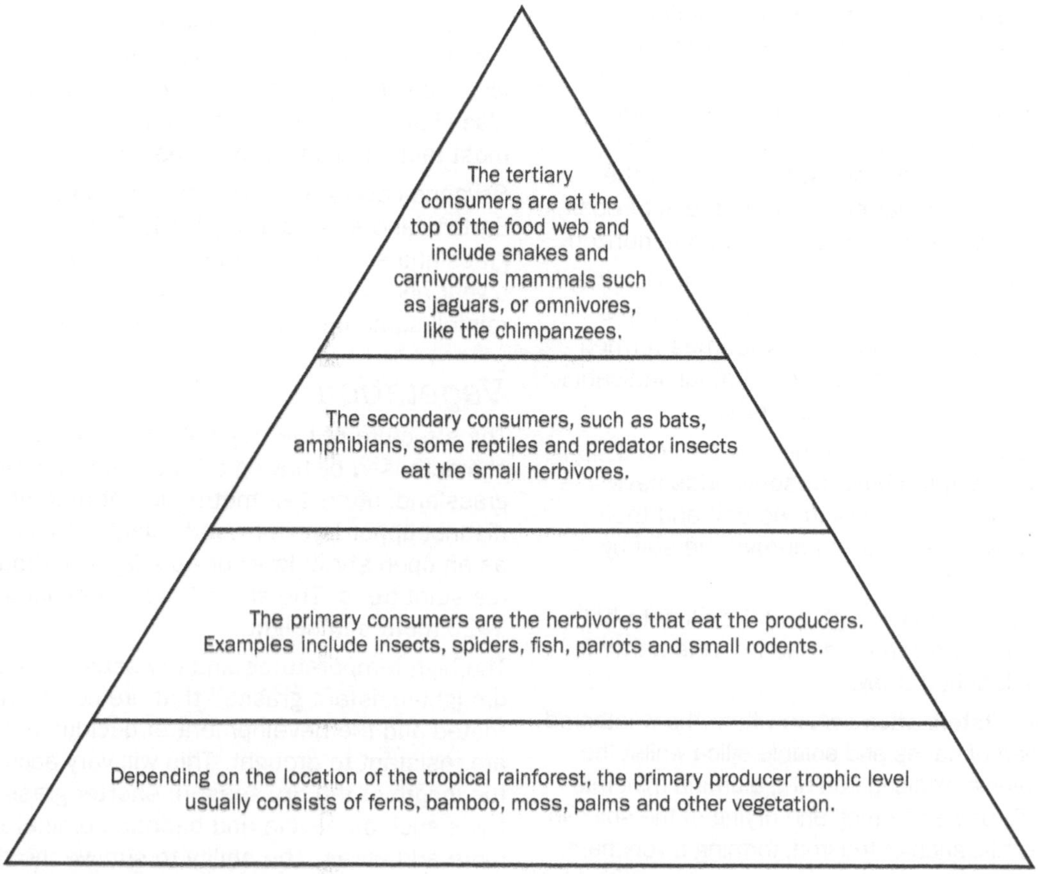
The tertiary consumers are at the top of the food web and include snakes and carnivorous mammals such as jaguars, or omnivores, like the chimpanzees.
The secondary consumers, such as bats, amphibians, some reptiles and predator insects eat the small herbivores.
The primary consumers are the herbivores that eat the producers. Examples include insects, spiders, fish, parrots and small rodents.
Depending on the location of the tropical rainforest, the primary producer trophic level usually consists of ferns, bamboo, moss, palms and other vegetation.

Fig 7.11 A typical rainforest energy pyramid

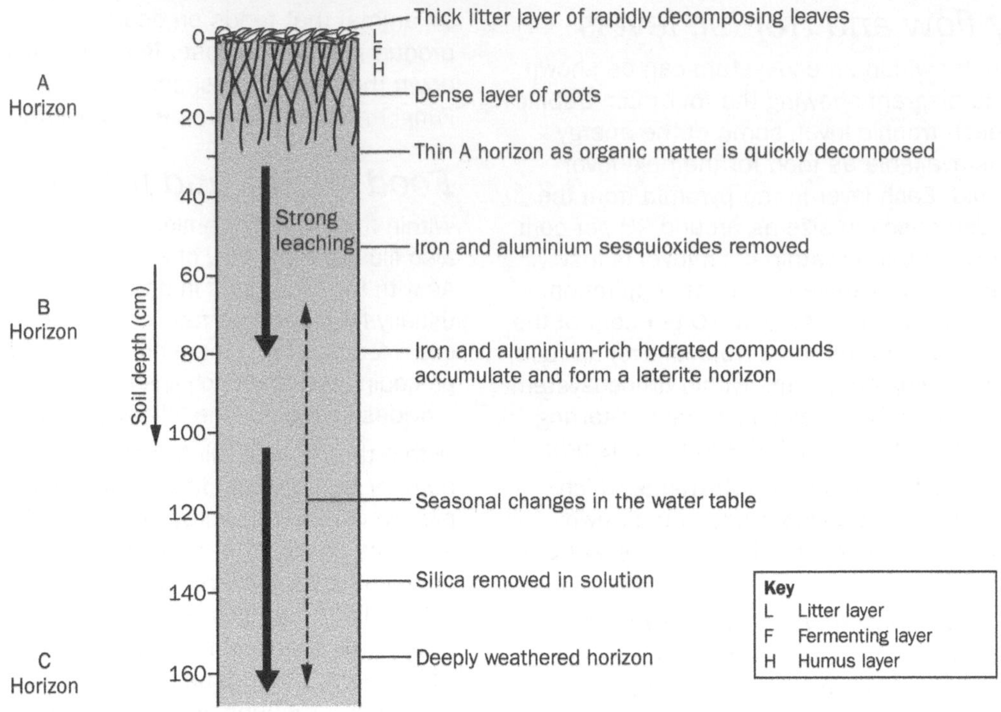

Fig 7.12 Typical tropical rainforest soil profile

Fig 7.12 shows a typical tropical latosol soil profile with its various horizons.

The high annual temperature and high annual rainfall of tropical rainforests leads to very rapid chemical weathering of the underlying bedrock. The high rainfall encourages rapid leaching in which all unstable elements are washed from the soil, leading to laterisation. This leads to a deep soil profile, up to 30 m deep.

The continuous leaf fall in the tropical rainforest provides a substantial litter layer. However, as decomposition is rapid, the humus layer below the leaf litter is thin, and this is quickly incorporated into the soil. There is a high level of fauna activity within the top layer of the soil which leads to the mixing of the organic matter. There is a lack of easily recognisable soil horizons below the humus horizon.

Tropical forest soils are acidic because the dominant movement of water is downward, leading to rapid leaching. These soils (latosols, sometimes termed ferralsols) are often red or yellow in colour, indicating the presence of ferric iron compounds.

The acidic water breaks down the clay particles into silica and **sesquioxides**. Sesquioxides have little ability to store nutrients in the soil and their presence indicates a need to improve the soil by adding organic matter.

Soils where the rainfall is high and the climate hot are often intensely weathered and infertile as their nutrients are leached away.

Ferrallitisation/laterisation occurs where the weathered layer is leached of bases and soluble silica whilst the relatively insoluble oxidised iron and alumina minerals accumulate. Repeated wetting and drying of the soil can lead to a crystallisation of the iron, forming a very hard, virtually impenetrable **duricrust** several metres thick.

Humid tropical (savanna) ecosystems

The change from a tropical rainforest to tropical savanna grassland is gradual (see fig 7.13).

Distribution

A **savanna** is a grassland ecosystem. Trees are sufficiently widely spaced so that the canopy does not close. This open canopy allows sufficient light to reach the ground to support an unbroken herbaceous layer consisting primarily of grasses. Savannas are also characterised by seasonal water availability, with most rainfall falling in one season.

Savanna covers approximately 20 per cent of the Earth's land area (see fig 7.14). Rainfall in tropical grassland savannas is between 500 and 1300 mm a year and can be highly seasonal. The entire year's rainfall sometimes falls within a couple of weeks.

Vegetation

The structure of the tropical savannas is characterised by having a lower layer of perennial grassland, often 1–2 metres tall at maturity and a distinct upper layer of woody plants. This can exist as an open shrub layer or as a layer of drought resistant trees. The shrub/tree layer varies from 1–20 metres in height.

The high temperatures and dry season encourage drought-resistant grasses that are coarse and long rooted and the development of deciduous trees that are resistant to drought. This will vary according to the length of the dry season; shorter grasses and trees such as acacia and baobab dominate in the more arid areas. The ability to survive the dry season and germinate when the rains arrive is crucial to

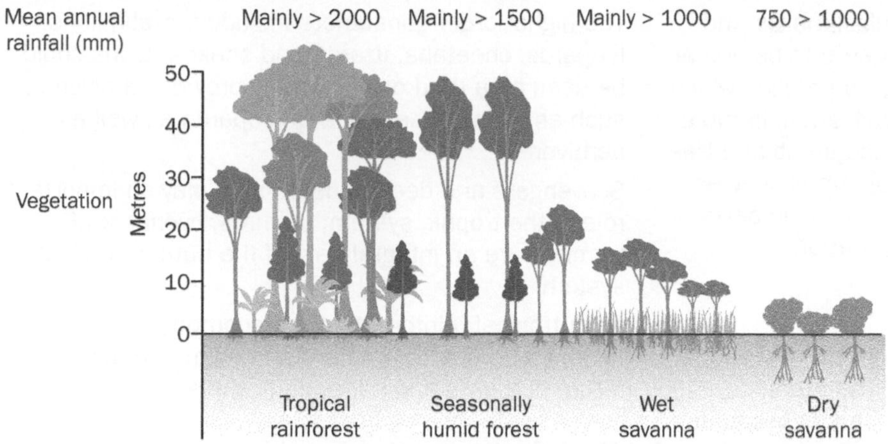

Fig 7.13 Differences in vegetation between tropical rainforest and dry savanna areas

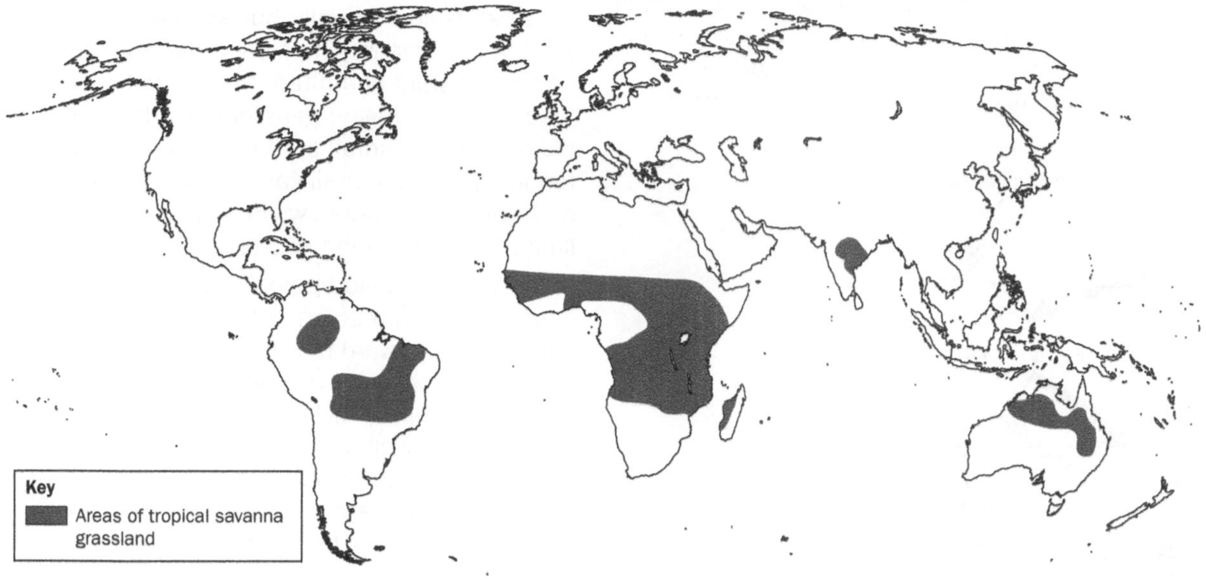

Fig 7.14 Distribution of tropical savanna grassland

many of the ground cover plants. The role of fire (either human or natural) also influences the types of plants found in grassland areas.

1. **Grasses**, such as elephant grass are tall (up to 3.5 m) and are prone to dry season fires. Fire is an important factor in the development of savannas and African herdsmen and agriculturalists use fire to encourage perennial grasslands with underground stems. The grasses die off in the dry season leaving only their roots. They grow back quickly and in clumps.
2. **Trees**. Savannas are seasonally humid and as such can vary from the semi-arid through to the reasonably well watered. In well-watered savannas, low woodlands develop such as acacia and baobab trees which exhibit **xerophytic** tendencies. A **xerophyte** is a species of plant that has adapted to survive in an environment with little water and these may develop along water courses. Some plants can store water in roots, trunks, stems, and leaves. Water storage in swollen parts of the plant is known as **succulence**.

   Baobabs and Australian bottle trees have large swollen trunks. They also have a thick bark to protect them from fires in the dry season, long tap roots to reach underground moisture and few leaves to reduce water loss through transpiration. The baobab also has a shiny and slick outer bark. This unique adaptation allows the baobab tree to reflect light and heat, keeping it cool in the intense savanna sun. It is also possible that the reflective nature of the bark may aid in protecting the tree from the effects of wildfires.

   Another adaptation that the baobab tree has developed to help it conserve water is a spongy bark. The bark of the baobab is more porous than regular wood, making it able to absorb moisture like a sponge. This allows the tree to absorb as much water as possible in times of rain and store it for use during times of scarcity or drought. The baobab tree has adapted its stems to catch every bit of water it can, from morning dew to summer downpours. Its stems form u-shaped funnels, allowing water to channel into holding canals, so the plant has time to absorb the water.
3. The **fauna** has a great range of hoofed mammals (ungulates) as well as a diverse set of carnivores, which vary according to the type of savanna.

Savannas are often regarded as sub-climaxes and those in south-east Asia are considered to be human generated. Overgrazing produces bare patches which can encourage soil loss and desertification. In more humid areas overgrazing could induce shrub and tree growth. The introduction of non-indigenous species (such as grasses in Australia and acacias in parts of Africa) has also impacted upon vegetation.

The nutrient cycle

The soil store of a savanna contains few nutrients, with the majority remaining in the litter store throughout the year (see fig 7.15). This is because the rate of decomposition of organic matter is slow during the dry season. The low levels of moisture and high temperatures greatly restrict bacterial decomposition. During the wet season the process of leaching removes nutrients from the soil and results in a hard, lateritic, crust forming that hinders plant growth. The biomass store varies enormously with the seasons.

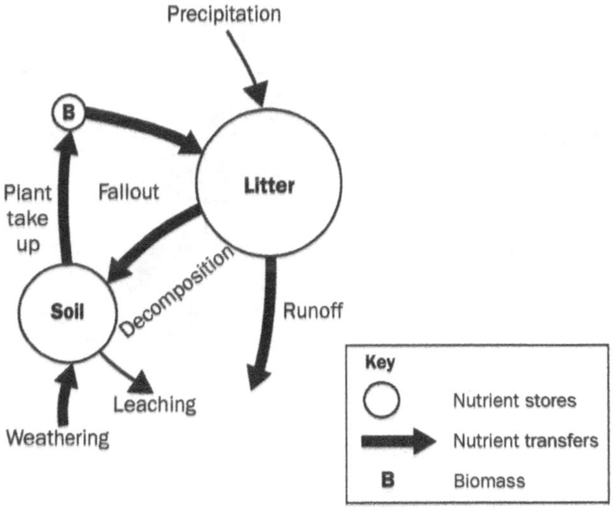

Fig 7.15 A Gersmehl diagram showing the nutrient cycle in a savanna

Food webs

In tropical savannas, tall grasses are the primary producers in the trophic pyramid, along with shrubs and sparse trees.

The primary consumers are zebras, giraffes, wildebeest, and elephants.

The higher order consumers include jackals, lions, leopards, cheetahs, lizards and snakes. Lions could be seen as a third order as they prey on carnivores, such as young cheetahs and leopards as well as herbivores.

Scavengers and decomposers also play an important role in the trophic system; vultures, hyenas and termites are an integral part of the nutrient cycling system.

As in tropical rainforests, the decomposers, bacteria, fungi, etc. break down plant and animal remains and return nutrients and minerals to the soil.

Soil formation

The soils in the savannas are **lateritic** and when baked hard they can inhibit tree and shrub growth. Fires can occur naturally, but are also sometimes started by humans. The plants therefore need to be fire resistant and often rely on fire to spread their seed and trigger germination. The amount and distribution of rainfall are important factors in soil formation. High rainfall can lead to base leaching and clay translocation, while seasonal rainfall may limit movement to the wet season.

In seasonally humid tropical savanna areas, soil type can vary in relation to the length of the dry season. The upward movement of water in the dry season may lead to **calcification**, producing a calcium rich upper layer and a higher pH as a result. The seasonally alternating upward and downward movements of water and minerals may produce upper cemented layers of laterite. The savanna soils often show merging horizons but generally they are red coloured as in tropical rainforests. The grassland vegetation can produce a thin dark brown humus layer of tropical brown earths. The soil depth tends to be around 1 to 2 metres – much less than in tropical rainforests.

Fig 7.16 shows the location of two soil types found in a tropical savanna. The position of the two soils on the slope is a result of water movement through soil – leaching occurs at the top of the slope and the accumulation of finer clay material and water at foot of slope.

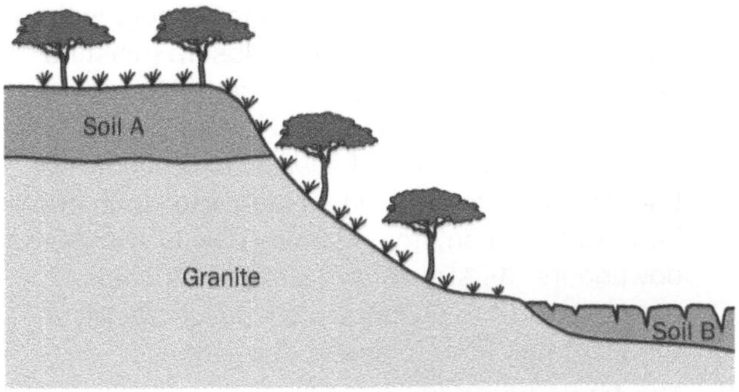

Fig 7.16 Soil types in a tropical savanna grassland

Tropical savanna grassland with widely spaced acacia trees

- Soil A is an oxisol which has a clayey or loamy texture and a low base exchange capacity. It is often characterised by extreme weathering of its minerals to form free oxides and kaolin clay. Heavily leached and infertile, it can be described as a latosol.
- Soil B is a vertisol with a high base exchange capacity and weakly developed soil horizons. Vertisol soils have a high clay content. The repeated wetting and drying of the soil can cause vertical swelling, shrinking and cracking. Vertisols generally form in tropical grasslands with a marked dry season.

Exam-style questions

1. With the use of a diagram, explain the characteristics features and processes operating in a soil found in either humid tropical (rainforest) ecosystems or seasonally humid tropical (savanna) ecosystems. [6]
2. Describe the vegetation structure and soils that are found in either humid tropical ecosystems or seasonally humid tropical ecosystems. [8]
3. Explain the process of laterisation and describe how it influences the development of soils in tropical climates. [8]
4. Using examples, describe how the climate, vegetation and soil forming processes develop a typical profile characteristics. [8]
5. Explain the ways in which characteristics of soils in tropical rainforests differ from the soils in the savanna. [6]
6. Describe the differences between the climax and plagioclimax vegetation found in humid tropical climates. [8]
7. Describe the patterns of nutrient cycling found in either tropical rain forests or savannas ecosystems. [6]
8. Describe and explain the energy flows and trophic levels for one tropical ecosystem. [6]

7.4 Sustainable management of tropical environments

Sustainable management is management that allows human activities to continue while causing as little harm to the ecosystem as possible. Therefore the continued existence of the ecosystem is to some extent ensured. A sustainable approach to management should attempt to ensure that the ecosystem is able to replace itself at a greater rate than it is being destroyed.

CS The Alto Juruá Extractive Reserve (AJER), Brazil

Threats

The reasons for tropical forest clearance in Brazil are varied, from commercial logging to small-scale farming and vegetable production.

Impacts

Soil leaching and degradation occur where there has been the greatest loss of biomass, for example through clear felling.

Similarly, there is a negative impact on trophic levels and animal populations in the Brazilian rainforests. Burning, however, may lead to short-term increases in soil nutrients and the longer term impact will depend upon the subsequent use of the area, the input of fertilisers, the nature of ground cover and any secondary re-growth of forest cover that takes place.

Management

The management of sustainable tropical rainforest ecosystem has proved difficult because of the nature of the ecosystem. With so much of the nutrient store in the biomass, it means that any action or management that alters or destroys the biomass can result in the destabilisation of the ecosystem. Clearly some activities are easier to manage than others. Therefore, low level shifting cultivation or selected felling with government licensing allow for the recovery and a general maintenance of the system. However, clear felling, large scale cultivation and grazing deplete the nutrient stores, and can be characterised by diminishing returns, loss of biodiversity and global implications for climate.

Brazil has established **extractive reserves** as a way to improve the rights of forest-dwelling populations, while also providing protection measures for the Amazon forest. The first extractive reserve, the 506 200 hectare Alto Juruá Extractive Reserve (AJER), is located in the westernmost part of the Amazon, and was created in 1990.

The AJER area is sparsely populated with approximately 4 600 people, who are primarily rubber tappers and subsistence farmers. Local communities now have more secure land rights which have led to diversification of the local economy. Beans have replaced rubber as the primary commodity, and are grown mainly on riverbanks. The population has grown in these areas, while declining in remote forest areas.

The AJER has also seen an increase in livestock raising, which has become the second largest source of income. Subsistence uses of forest resources in the AJER (for food production, home construction, etc.) have changed little since the establishment of the reserve, and have been estimated at 65 per cent of total extractive production.

An analysis of forest cover changes during the first decade of AJER's establishment (1989–2000) indicates that deforestation has only occurred in 1 per cent of the area. AJER continued to support a primarily mature forest cover over 99 per cent of its area during this time period.

There have been indications of a recovery of threatened species such as jaguar, tapir, peccaries, and several species of primates, assumed to be linked to the depopulation of remote forest areas.

However, the commercial shift to beans and cattle-raising competes with the forest and this could create conflicts between the goals of providing livelihood opportunities and conserving forests. At present, however, the AJER reserve "has been reasonably able to match the conservation, social, and development targets envisioned".

CS The Masai tribe in southern Kenya and northern Tanzania

The Masai practise nomadic farming, a traditional method of farming that allows vegetation to recover from animal grazing whenever the farmers move on to another area. To find enough grazing, the Masai move their cattle to higher land – mostly above 2000 metres – throughout the dry season. In the wet season they return to below 2000 metres.

However, in the last 40–50 years the Masai's nomadic way of life and farming have been disrupted due to commercial pressures and government policies. The ecosystem has also started to suffer.

Commercial farmers, encouraged by government policies, have moved into the best dry-season land and converted it for commercial agriculture. As savanna is converted into cropland, the natural vegetation and its biomass store are removed and the soil's nutrients are rapidly used up.

The Masai have become dependent on food produced in other areas such as maize, rice, potatoes, cabbage etc. Traditionally, the Masai believe that cultivating the land for crop farming is a crime against nature. Once the land is cultivated, it is no longer suitable for grazing. The concept of private ownership was, until recently, a foreign concept to the Masai. Their land has now been subdivided into group and individual ranches. In other parts of Masai land, people subdivided their individual ranches into small plots, which are sold to private developers.

The new land management system of individual ranches has economically polarised the Masai; some have substantially increased their wealth at the expense of others. The largest loss of land, however, has been to national parks and reserves, in which the Masai people are restricted from accessing critical water sources and pasture. The subdivision of Masai land reduced land size for cattle herding, reduced the number of cows per household, and reduced food production. As a result, Masai society, which once was a self-sufficient society, is now facing many social-economic and political challenges. The level of poverty among the Masai people is increasing.

The establishment, since the 1950s, of national parks, such as the Serengeti, Amboseli and Masai Mara National Park, to conserve wildlife and encourage tourism, has restricted human access to the parks including the Masai and their grazing herds. Though the Masai do not traditionally grow crops, they often rent out their land to farmers from other regions of Kenya, who fence it off. Some Masai pastoralists are finding that they no longer have enough grazing space; they are caught between large-scale farming on one side and the national parks, like the Masai Mara, on the other, all being developed on their traditional land.

These interventions have forced the nomadic Masai farmers onto marginal land. Their traditional pastoral migration patterns have been disrupted and they have been compelled to use smaller areas of land for their cattle. Overgrazing has been the inevitable result.

The population of the area has also expanded rapidly over the past 30 years. This has resulted in larger herds grazing the savanna grassland and more trees being cut down for fuel. As vegetation is removed, the risk and incidence of soil erosion has increased and lead to the deterioration of the latosol soils. Desertification occurs in extreme cases. Water is also becoming more scarce. Diverting water for tourist use has put pressure on local water reserves, leaving local people, plants and animals short of water. Tourist hotels also sometimes dump waste into rivers.

The solutions to these problems are difficult to put into operation. They include the controlled management of grazing over designated areas and involving the Masai more in the tourism industry, allowing them to benefit from tourist income. At the present time, most of the money generated by tourism and national parks leaks out of the area to the national government and tourist companies.

Exam-style questions

1. Describe the climate, vegetation, soils and nutrient flows found in either the tropical rainforest ecosystem or the savanna ecosystem. How have human activities influenced the development of this ecosystem? [8]
2. With reference to either the tropical rainforest ecosystem or the savanna ecosystem, explain what is meant by sustainable management and suggest ways by which it might be achieved. [8]
3. Describe how human activities affected the natural processes found in either tropical rainforest ecosystem or the savanna ecosystem. [8]
4. Using an example or examples, discuss the extent to which sustainable development is possible in either the tropical rainforest or the savanna ecosystems. [8]

8 Coastal environments

8.1 Coastal processes

Waves

Waves are orbital oscillations of water particles generated by the frictional drag of the wind moving over the sea surface. Wave size and amplitude of waves depends on several factors, including **fetch** (the distance over which the wind has been blowing over the sea) wind strength, duration of wind, depth of water and shape of the coastline.

Wave refraction

Wave energy is not evenly distributed along a coastline, as coastlines vary in shape and in their offshore profile. A wave approaching from deep water can be divided into equal sections of energy by lines/wave rays drawn at right angles to the wave crest. These lines are called wave **orthogonals**. A wave approaching a coastline will firstly encounter shallower water in front of headlands. This shallower water will slow the wave down, as friction takes place between the wave and the sea bed. The section of the wave in front of adjacent bays will still be in deeper water and continue to travel, at a faster speed than the section encountering the headland. The result is that the **wave crest line** will start to bend or "refract". This is **wave refraction**. When looking at this pattern it can be seen that the wave energy becomes concentrated around headlands and dispersed in the bays. This means that erosion will be concentrated on the headlands and deposition will take place in the bays. Fig 8.1 illustrates this process.

Wave types

Waves can be of various types, amplitudes and lengths. They can be classified in various ways (for example by their height, length, and wave period – the time taken for a complete wave to pass a fixed point) or in terms of being either high or low energy waves or as constructive and destructive waves. The terms **swash** and **backwash** describe the movement of a wave as it breaks up the beach (swash) and back down the beach (backwash).

- **Constructive (depositional) waves** (also called **spilling waves**) have a strong swash and weak backwash. This will result in sand and shingle being moved onshore, building up the beach profile with a steep gradient. Constructive waves tend to form over a long distance/fetch.They have an elliptical orbit and are usually small, low, flat waves with a long wave length – up to 100 metres. They have a low frequency – there will be about 6–8 every minute.
- **Destructive (erosional) waves** (also called **surging** or **plunging waves**), in contrast, have a steep, plunging break, producing a relatively feeble swash compared to their backwash – which will mean that sand and shingle will be moved offshore, therefore reducing and lowering the beach profile. Destructive waves tend to form over a short distance – they have a relatively short fetch. They have a circular orbit and are usually large, high, steep waves with a short wave length – 20 metres and they have a high frequency – there will be about 10–12 every minute.

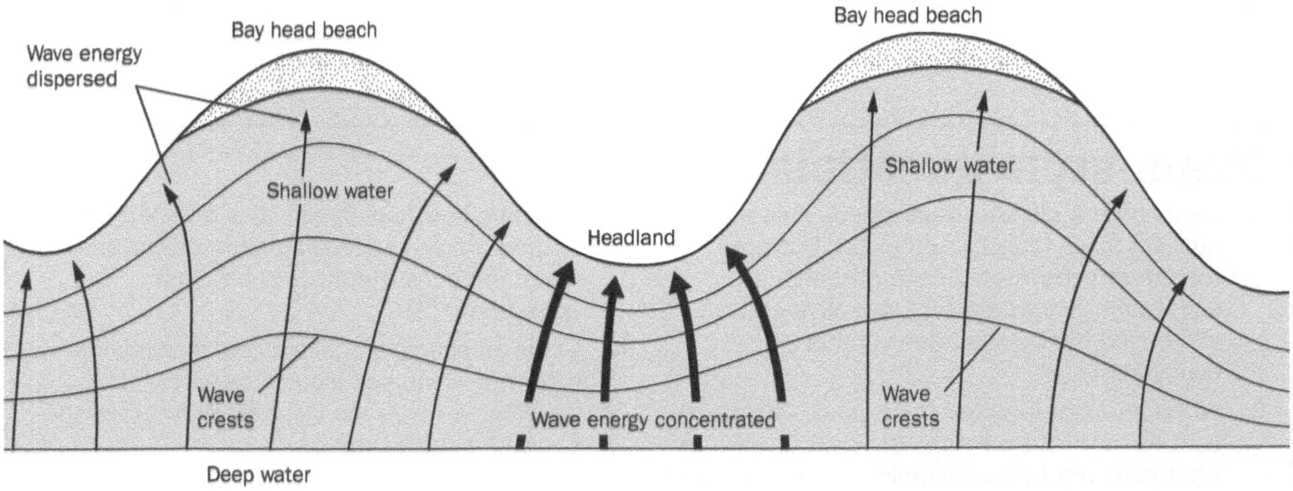

Fig 8.1 Wave energy concentrated on headlands

Constructive waves are usually low energy waves which push material up the beach. Destructive waves are usually high energy waves which pull material down the beach depositing it at the base of the beach profile.

Marine erosion

- The erosion of a coast is a function of sub-aerial weathering and movement, marine erosion, rock type and structure. The processes of sub-aerial weathering contribute to slope instability, which can be further aided by marine action in removing and transporting material. The geology of a coast has a strong influence, both in terms of the rock type as well as of the structures found in rocks (joints, bedding planes, faults, angle of dip, etc.). **Hydraulic action** is the effect of the wave hitting, colliding and collapsing onto the base of a cliff. The sudden release of energy helps weaken and break down rocks in the cliff and weaken any structures. Large waves in the Atlantic and Pacific Oceans can average 11 tonnes per m^2 and reach 30 tonnes per m^2 on exposed coastlines, causing considerable damage in a short period of time.
- **Abrasion** is the sand-papering effect of rocks being dragged over rocky platforms by waves and currents. It often occurs at the same time as corrasion.
- **Cavitation** is the trapping and compression of air in cracks, joints and fault lines, etc. in the rocks of the cliff, followed by the decompression of the air as the water retreats. The pressure this puts on the rocks of the cliff will weaken them and allow them to be removed by hydraulic action.
- **Corrasion** is the result of materials, from sand to boulder sized, being carried by the waves striking against the cliff base causing the erosion and undercutting of the cliff, often forming a cliff notch at the base of the cliff.
- **Solution** (sometimes called **corrosion**) involves the dissolving of chemicals in rocks. Limestone is particularly vulnerable to solution.
- **Attrition** is the process that causes the break up and rounding of rock particles as they knock against each other. Rock particles moved by the swash and backwash of waves continuously roll and grind against each other and are made smaller and rounder as a result.

Sub-aerial weathering

The types of sub-aerial weathering will vary according to climate and rock-type.

Physical

- **Freeze-thaw weathering** is most common in coastal areas in polar and temperate climates where freezing temperatures cause water to freeze in rocks exposed on the coast. When water freezes, it expands its volume by 9 per cent. It takes place when water gets into small pore spaces and cracks the rock. If it then freezes and expands, it can put enormous pressures of up to 2 000 kilograms per cm^2 on the rock and split it. Where bare rock is exposed on a coastal cliff, face fragments of rock may be forced away from the face. These fragments of rock may then fall to the bottom of the cliff where they form a large pile of rocks called scree, which may then be broken up by wave attrition and removed by wave action.
- **Exfoliation** (or **onion skin weathering**) is most common in hot tropical climates where the surface temperatures of coastal rocks exposed to the sun can reach over 90 °C during the day and then drop below 0 °C during the night! Bare rock surfaces will expand and contract each day as temperatures rise and fall. These daily stresses can cause the surface layer of the rock to separate and peel away from the rock.
- **Salt weathering** takes place when salt water dries out causing the formation of salt crystals. As they grow in openings and cracks in the surface of the rocks they can disintegrate the surface of the rock.

Chemical

- **Carbonation** occurs in those rocks exposed on cliffs with a high content of calcium carbonate such as limestone and chalk. Carbonic acid (formed by CO_2 being dissolved in rainwater) reacts with the calcium carbonate to form calcium bicarbonate. Calcium bicarbonate is soluble, so the rock is dissolved and carried away.
- **Oxidation** involves chemical reactions with oxygen, often dissolved in water. Iron minerals are especially susceptible to this process.
- **Hydration** involves absorption of water leading to swelling within rocks. Some rocks, especially those rich in clay minerals, may absorb water, putting stress on the rocks and often leading to their disintegration. The constant wetting and drying at the coast due to rising and falling tides results in expansion and contraction, causing rocks to break apart.
- **Biological weathering** may include both physical and chemical processes. The physical processes can include animals such as limpets, clams and other molluscs, marine worms and fish grazing on and burrowing into rocks. Burrowing clams on the English coast have lowered chalk wave-cut platforms by as much as 2.3 centimetres a year, while 100 grazing periwinkles can remove up to 86 cubic centimetres a year! In addition, plant roots can prise jointed rocks.

Mass movements

Mass movement is a common feature of most cliffs. It involves downhill movement of material due to gravity.

Mass movements can be particularly active on cliffed coasts because undercutting of the cliff by wave action makes them unstable. **Rockfalls** and **landslides** are common mass movements:

- **Rockfalls** may occur when the waves undercut the cliffs and sub-aerial weathering loosens fragments of rocks on the cliff face.
- **Landslides** may occur when the rocks in the cliff become saturated with water from percolating rainfall. They may be triggered by the undercutting of the cliff by the waves and/or by increasing the weight of the cliff by adding rainwater. The saturated material may then collapse to form a flow of earth and mud. The processes of marine erosion and longshore transportation then act upon the toe of the collapsed material.

Marine transportation and deposition

Longshore drift

Sediment is either dissolved, carried in suspension or rolled along the sea bed. Most sediment is transported along a coastline by **longshore drift** (fig 8.2). The **sources** of sediment include material supplied to the coast by rivers, the erosion of cliffs and coral reefs and from the offshore sea floor. Beaches themselves also supply sediment from destructive waves and rip currents.

Sediment may also be transported in a suspended form by waves and currents. Heavier material, such as pebbles and cobbles, can be rolled by the process of traction, or bounced through the process of saltation, along the sea floor by storms and in high energy environments. The most common means is by longshore currents giving rise to beach and longshore drift.

Deposition occurs where the current slackens and the load can no longer be suspended. This occurs at changes in coastal alignment or where the current slackens due to estuaries or shelving, leading to the formation of spits, beaches, bars, tombolos and cuspate forelands, etc. (which will be explained further later in the chapter).

Sediment cells

Coastlines can be divided up into distinct sediment cells (sometimes called **littoral cells**):

- A **coastal sediment cell** is a system by which sediments are sourced, transported and deposited within a part of a coast and offshore area. They appear as a closed coastal system with relatively little transfer of sediment between cells.
- Their sediment is derived from estuaries and coastal erosion, transported by processes such as longshore drift and deposited in the form of beaches, bars, spits etc. Thus sediment can be seen in terms of **sources** (rivers, estuaries, cliff erosion, etc.) **transportation** (longshore drift and tidal movements) and **deposition** (in the form of beaches, bars, spits salt marshes, dunes, etc.)
- Sediment cells are very vulnerable to human activities in terms of disrupting the supplies or movement of sediment. Therefore any protection of a coastline from erosion can disrupt sediment supplies. Features such as groynes and breakwaters can trap sediment and disrupt its movement. Dredging and sediment removal through blow outs in sand dunes, etc. can also cause cumulative changes to the operation of the cells and be the cause of several coastal management problems.
- Eleven sediment cells have now been identified around the coastline of England and Wales. To further study those in more detail, these sediment cells can be sub divided into **sub-cells**. From the information gathered **shoreline management plans** have been drawn up. These have to be used by all County and Local Authorities when they plan to do any work on or near the coastline. There is very little movement of sediment between these cells, so if sediment is removed it cannot be replaced quickly. Any beach which loses or is starved of sediment is much less effective in protecting the coast and so rates of erosion and flooding will increase. All of this activity has major cost implications for local and national government.

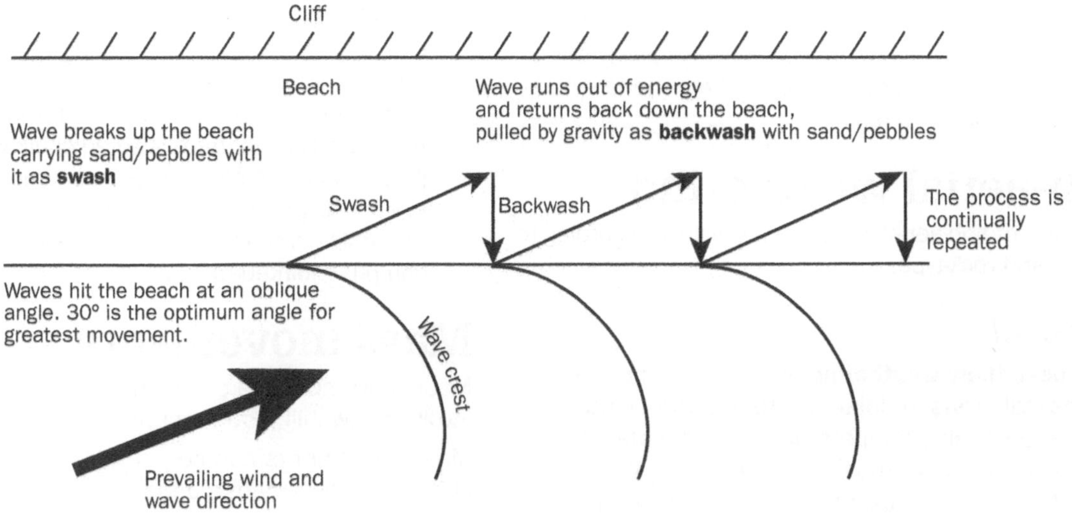

Fig 8.2 The process of longshore drift

5. Water quality and turbidity – reefs will not recover unless the sediment input stops. Areas open to waves and currents will get flushed out quicker than sheltered reefs.
6. The proximity (nearness) of nearby healthy reefs – reefs recover more quickly if they are near healthy reefs, which supply them with young larvae.
7. The measures taken to deal with the cause of the problem and measures taken to help recovery – this includes regeneration by transplanting corals or using artificial structures to rehabilitate corals.

The management of coral reefs to ensure their longevity very much depends on managing and regulating their use and exploitation and the human threats explained previously. Banning, limiting, regulating and managing sustainably certain activities on and near coral reefs would help their future longevity.

The "**Reefs at Risk**" programme uses an indicator, based on desk studies of data and maps, designed to highlight where coral reef degradation can be expected and predicted. The indicator measures the potential risks associated with four categories of threats posed by human activity around the coast:

- Coastal development, assessed by the number of cities of various sizes, settlements, airports and military bases, mines and tourist resorts at certain distances from the coast.
- Marine-based pollution, assessed by looking at the location of ports, oil related installations, shipping lanes, etc.
- Over-exploitation, assessed by estimating overfishing and the incidence of destructive fishing practices.
- Inland pollution and erosion (siltation), assessed by looking at the erosion potential of the land surface within river catchments flowing into coral reef areas.

Depending on the assessed severity of the threats, coral reefs are assigned as:

- high risk – if at least one of the threats above is a high severity threat
- medium risk – if at least one of the threats above is a medium severity threat
- low risk – if all the four threats have a low severity.

This system does not take into account natural disasters such as hurricanes.

The main global threats are a rise in sea temperature causing the deaths of corals, often evident as coral bleaching. Global warming and rising sea level may lead to long-term destruction of reefs as the growth of corals may not be fast enough to compensate for the relatively rapid rise in sea levels. Rising temperatures and carbon emissions may also affect the chemical balance of the oceans again acting as threat.

Sea level change can be both **isostatic** and **eustatic** (including the thermal expansion of the oceans due to the current rise in ocean temperatures). A reduction in sea level will expose coral and lead to its dying, a rapid rise in sea level may prevent coral growth from keeping pace and die at deeper water depths. Upward movements of the coastline of western Sumatra as a result of the earthquake that also generated the Asian tsunami in December 2004 left large sections of coral reef exposed to the air for longer periods than normal and led to their death.

Exam-style questions

1. What are the main conditions required for the formation of coral reefs? [6]
2. Describe the main threats to coral reefs. Explain how these threats may be managed. [8]

8.4 Sustainable management of coasts

CS Coastal management and defence – Unawatuna beach, southern Sri Lanka

People will normally defend a coastline when something of economic value is threatened by coastal erosion. This can be a settlement, an industrial area, a port or an important transport link such as a road or railway.

People can then do one of three things – they can **hold the line**, by building coastal defences, **retreat the line**, and abandon the coast or, advance the line by building up the coastline and extending it out to sea.

Unawatuna beach was once named "The Best Beach in the World" by Discovery Television Channel. In international travel guides it is described as one of the top ten beaches in the world. After the Asian tsunami devastated large sections of the eastern and southern coastline of Sri Lanka on 26 December 2004, a decision was made to provide long-term protection for the shallow bay in which the small tourist town of Unawatuna is located. Unfortunately, the building of the breakwater and possible changes to the offshore topography during the tsunami caused the beach to begin re-aligning its plan shape.

Between 2008 and 2014, sand began moving from the eastern end of the beach to accumulate at the western. As a result, the eastern end saw the almost complete removal of the beach in front of several properties, mainly hotels, restaurants and cafes, see fig 8.12. Waves were able to reach these properties and started to undermine their foundations.

A variety of schemes and methods were then used by property owners and the government in an attempt to defend these properties – a mixture of **hard engineering** and **soft engineering**.

Hard engineering

The hard engineering involved building permanent, rigid structures – small, 4 metre high, vertical **sea walls**. These have been the traditional high cost solution to protect valuable stretches of coast and property. They proved an unsuccessful solution

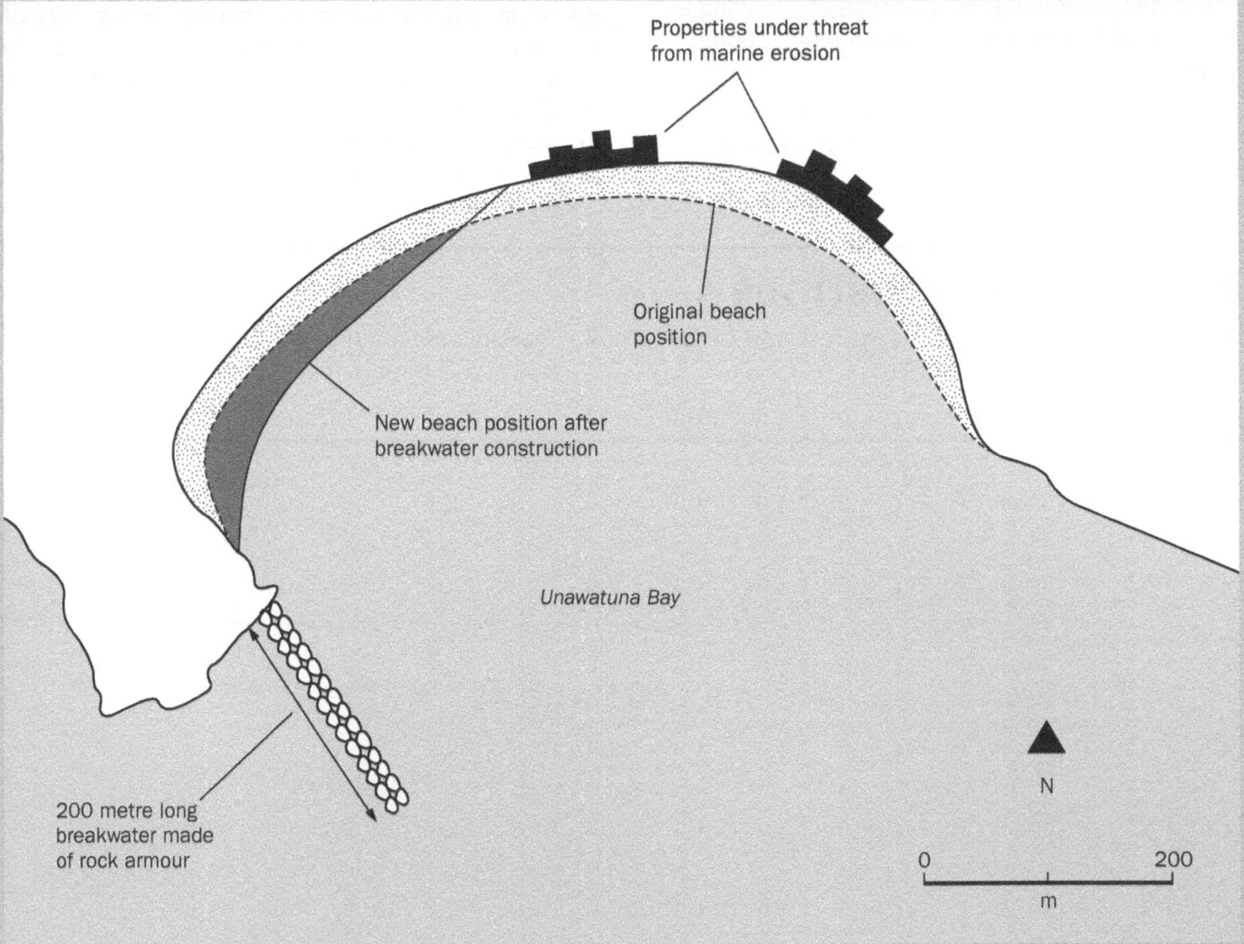

Fig 8.14 Unawatuna beach in southern Sri Lanka

due to **wave reflection**. The vertical sea walls reflected the waves and caused the beach to be further scoured/eroded, resulting in the sand being carried offshore. With no beach, the sea walls started to be undercut.

The sea walls should have been **inclined**, not vertical, to **deflect** rather than **reflect** waves. There should also have been a curved wave return wall, at the top of the wall, to further deflect the waves.

The portion of the rock armour breakwater above sea level was removed in the mistaken belief that this would allow sea and wave circulation and movement to return to normal. As these movements take place at or below sea level the impact of this was negligible.

Other hard engineering methods involving rock armour and gabions also failed as wave energy at the location had not been measured and exceeded the capacity of all these methods to stay in place and survive the wave erosion.

The rock armour was made up of rocks too small for the levels of wave energy. It had been deposited randomly and was unstable and easily moved by larger waves. The geotextile bags were placed vertically and reflected the waves, increasing beach scour and the gabions were broken up by waves too large for their location.

Soft engineering

Also, a new artificial beach was added to further protect the wall. This is called **beach nourishment**. Several hundred tonnes of sand were brought by truck from harbour dredging at another location on the coast. Unfortunately, the sand was too fine for the level of wave energy on the beach and was quickly removed by wave action.

In 2015, an international coastal management company was engaged to pump several hundred thousand tonnes of sand from offshore sources onto Unawatuna beach and built the beach up both in height and width. However, this solution, though currently protecting the hotels and other properties, may not be sustainable as the breakwater remains in place and the reason for the original movement of the beach has not been ascertained. As is the case globally, expensive coastal protection methods are often used with little knowledge or study of the coastal processes operating at specific locations.

Exam-style questions

1. With reference to a section of a coastline that you have studied, explain how the management of a coastline may create more problems than are solved. [8]
2. Describe and evaluate the level of success of coastal protection methods. [8]
3. Using an example, describe the threats that may affect coastlines and evaluate the possible ways that can be taken to protect coasts from such threats. [8]

9 Hazardous environments

9.1 Hazards resulting from tectonic processes

The theory of **plate tectonics** puts forward the idea that the **lithosphere** – the outer layer of the Earth consisting of the **crust** and **upper mantle** – is divided into a number of **tectonic plates**. These tectonic plates are segments/sections of the Earth's crust, much like the fragmented shell of a cracked hard-boiled egg. The edges of these plates – the **plate margins/boundaries** – are where the **oceanic** and/or **continental crustal plates** come into contact with one and another. They are zones where the plates can undergo great stress and change.

There are two types of crust:

- **Continental Crust** – older, lighter, cannot sink and is permanent. It is typically 200 kilometres thick.
- **Oceanic Crust** – younger, heavier, can sink and is continually being formed and destroyed. It is typically 100 kilometres thick.

There are four types of plate margin/boundary:

1. **Divergent** (or **constructive) margins/boundaries** – here two plates move **away** from each other. Any gap that appears between these plates fills with molten magma and forms/constructs new crust. One example is the Mid-Atlantic Ridge which stretches down the middle of the North and South Atlantic Oceans. Another is the Great African Rift valley. Plate movement ranges from a typical 10–40 mm on the Mid-Atlantic Ridge (about as fast as fingernails grow), to about 160 mm a year on the Nazca Plate (about as fast as hair grows)
2. **Convergent** (or **destructive) margins/boundaries** – these are found where plates made of heavier oceanic crust move **towards** plates made of lighter continental crust. Where they meet, the heavier oceanic crust is forced down under the lighter continental crust and forms a **subduction zone**. As the oceanic crust sinks deeper, it enters the **Benioff Zone** where it melts and forms magma. This magma may rise to the surface and form volcanoes, as in the Andes mountains in South America. Where this happens in the ocean, the volcanoes may form **island arcs**, as in the islands of the Caribbean and in the Aleutian Islands south-west of Alaska.
3. **Collision margins/boundaries** – these are found where two plates made of lighter continental crust collide with each other. As this lighter type of crust cannot sink, or be destroyed, the rocks between the colliding plates crumple and fold in to large mountain ranges such as the Himalayas.
4. **Transform** (or **conservative**) **margins/** boundaries – These are found where two plates are **sliding** past each other. The most famous example is the San Andreas Fault Line in California, USA. As the two plates are not colliding or tearing apart, there is no new land formed and there are no volcanoes being formed. However, they can produce violent earthquakes when they "stick" together.

The Earth's crust is not uniform in its thickness and in some thinner areas **hotspots** are found where convection currents bring plumes of **magma** – hot melted rock below the surface of the Earth – towards the surface, resulting in isolated volcanic areas usually far from plate margins/boundaries. When magma extrudes/appears onto the surface of the earth, it is referred to as **lava**.

Although the Earth's plates may move very slowly in human terms, it is this movement that is responsible for some of the most spectacular landscapes, processes and hazards on Earth. It is at the edges of the plates, where two or more meet, that these processes and hazards take place.

Distribution

Where do earthquakes occur?

To a considerable extent earthquake epicentres mirror the distribution of tectonic plate edges/margins/boundaries. Most of the seismic activity takes place at the boundaries of the tectonic plates and this activity can be associated with volcanic eruptions. The margins of the Pacific plate, the Pacific Ring of Fire, is prominent, and mid-ocean boundaries are particularly prone to earthquakes.

Therefore, the west coasts of North, Central and South America, the Aleutian Islands off south-west Alaska, Japan, the Philippines, SE Asia and New Zealand are all prone to earthquake activity as are the northern and eastern edges of the Mediterranean Sea and Iran and Iraq in the Middle East. Over 80 per cent of earthquake epicentres are found along the western coasts of the Americas, the Pacific islands and the Aleutian Island arc.

Minor earthquakes may also be triggered by human structures such as dams/reservoirs. Lake Mead, formed by the construction of the Hoover Dam on the Colorado River in the western USA, triggered 6000 minor earthquakes after its creation due to the weight of water on the Earth's crust

Where and how volcanic eruptions occur?

There are three common locations for volcanoes:

1. **Divergent (mid-ocean) margins** allow the extrusion of magma as part of sea floor spreading (e.g. Iceland).
2. **Convergent (destructive) margins** give rise to volcanoes as the magma from crustal melting in the **Benioff zone** makes its way upwards through cracks and faults in the crust to the surface.
3. The distribution of those volcanoes formed at **hot spots** differs from that of other types of volcano. The hot spots, for example Hawaii in the Pacific Ocean and Reunion in the Indian Ocean, allow magma to rise to the surface through faults and weaknesses within the plates as the plates pass over the hot spot. Hot spots are located above permanent and deep seated sources of magma plumes. Faults in the Earth's crust allow magma

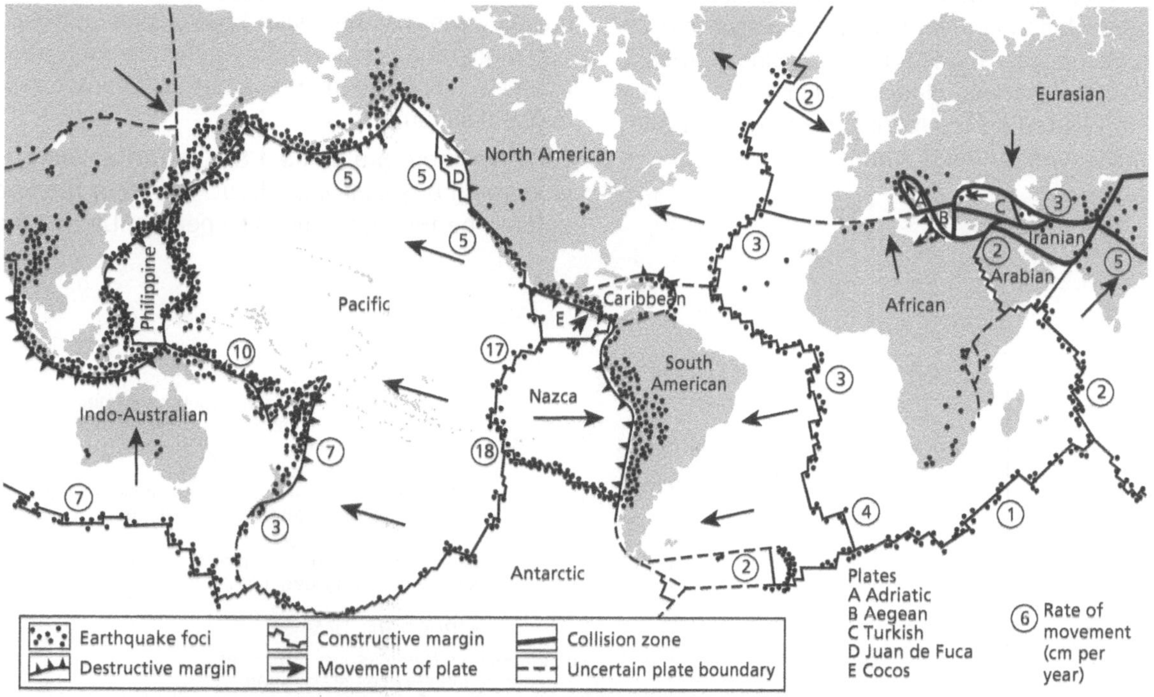

Fig 9.1 Global distribution of earthquakes

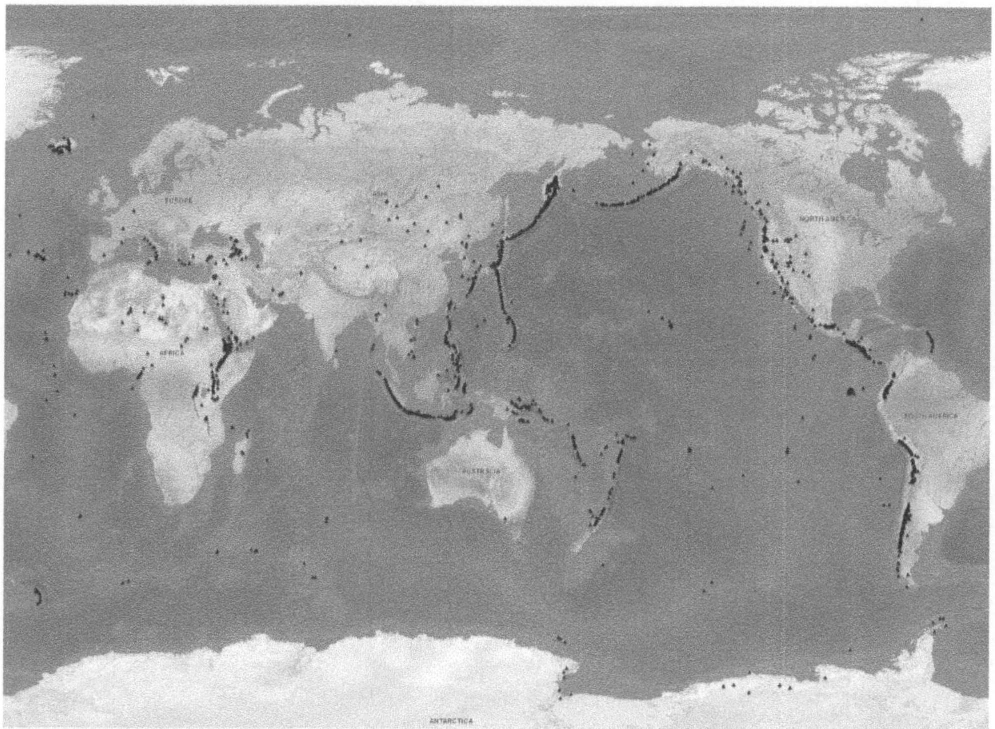

Fig 9.2 Global distribution of volcanoes

to erupt at the surface to produce basaltic types of volcanoes, e.g. in Hawaii in the Pacific Ocean and Reunion in the Indian Ocean. As the plates are moving so the volcanoes will move over time, producing a chain of extinct volcanic cones, craters, etc., many of which are located under the ocean. The distribution of hot spots tends to be in the central parts of plates, away from the margins. Japan has over 100 active volcanoes, more than almost any other country and accounts alone for about 10 per cent of all active volcanoes in the world. The volcanoes belong to the Pacific Ring of Fire, caused by **subduction zones** of the Pacific plate beneath continental and other oceanic plates along its margins. Japan is located at the junction of 4 tectonic plates – the Pacific, Philippine, Eurasian and North American plates, and its volcanoes are mainly located on 5 subduction-zones related to the movement of these plates.

Earthquakes and resultant hazards

An **earthquake** is a sudden and violent shaking of the ground. It results from a sudden release of energy in the form of seismic waves in the Earth's crust due to a sudden slippage or "snapping" along a fault, often at a plate margin.

The point within the crust where the earthquake happens is called the **focus**. The point on the ground surface immediately above the focus is called the **epicentre**.

Earthquakes are mainly caused by displacement or movement along faults due to the build-up of tension, but they can also be caused by volcanic activity and large landslides. Human activities such as coal mining or constructing reservoirs can sometimes cause earthquakes too.

Each day there are at least 8 000 earthquakes globally. In a typical year about 49 000 of these can actually be felt and noticed by people. Up to 18 of

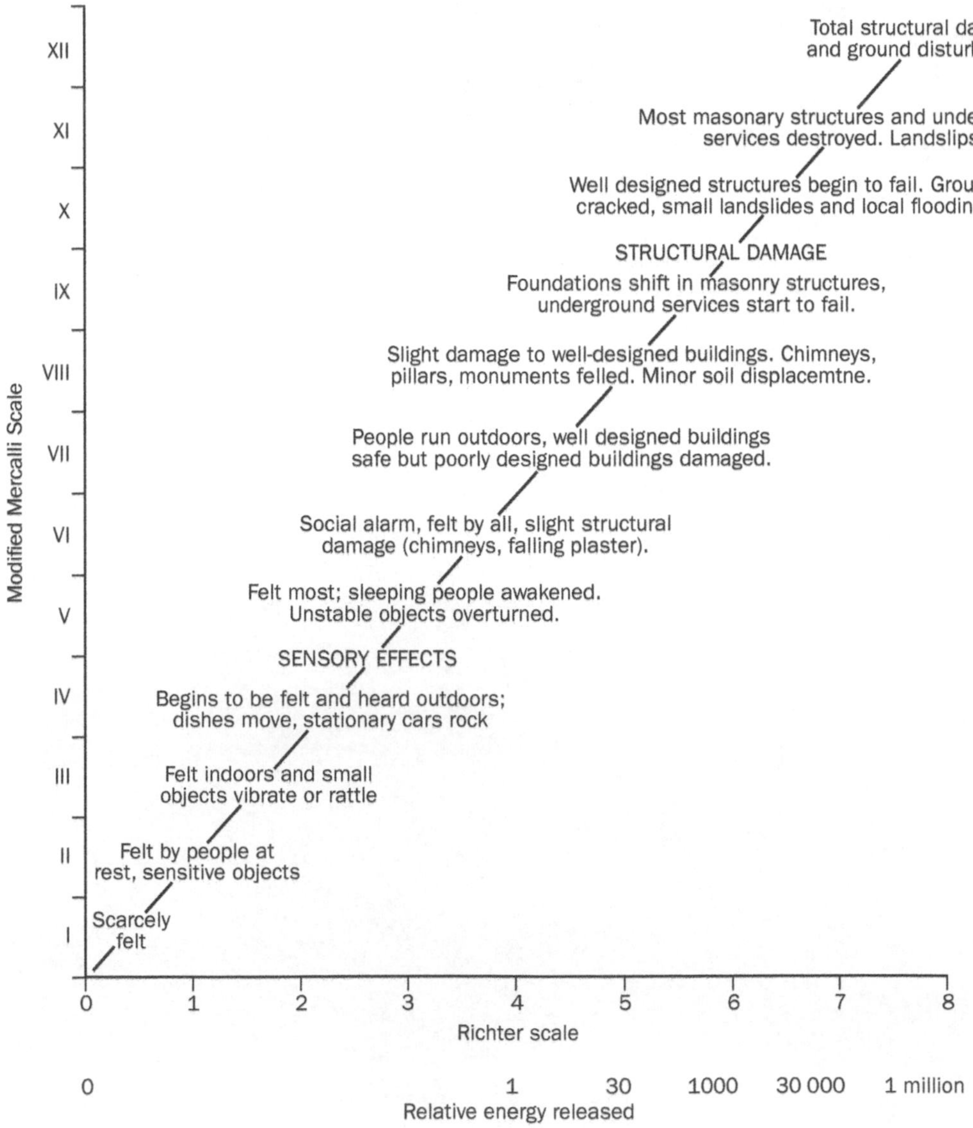

Fig 9.3 Measures of earthquake intensity

these can cause serious damage to buildings and possibly injure and kill people.

The **frequency of earthquakes** is closely related to the rates of plate movement and they are often most intense when associated with **subduction zones** (e.g. through the north and eastern Mediterranean, around the Aegean Sea and off the coast of Chile).

Earthquakes are recorded using a **seismometer**. Seismometers are instruments that measure and record motions of the ground, including those of the seismic waves generated by earthquakes. The trace (graph) showing the earthquake is known as a seismogram.

There are two common scales used to measure the size/magnitude of an earthquake – the Richter and Mercalli Scales:

1. The **size** of an earthquake is reported by the **Richter Scale** – on this scale an earthquake with a magnitude of 3 or lower is almost imperceptible, while one with a magnitude of 7 or higher can cause serious damage over large areas.
2. The **intensity of shaking** is measured on the **Mercalli Scale** – this considers the impacts of an earthquake.

The following diagram, taken from November 2005 Paper 2 Question 6, shows the effects associated with different levels of earthquake magnitude on the modified Mercalli and Richter Scales. Most hazardous impacts begin to occur after 4 on the Richter scale and the effects escalate as the energy release is increased logarithmically.

Earthquakes cause damage in five main ways:

1. **Ground movement** can cut or break anything that crosses a fault line: e.g. tunnels, highways, railroads, electricity lines, water and gas mains.
2. **Shaking** is the greatest threat. Modern buildings in earthquake threatened areas can be designed to survive earthquakes, but older buildings are much more likely to be damaged. People can be killed or injured by the collapsing buildings.
3. **Liquefaction** occurs when the shaking turns solid ground into mud. Buildings may then sink into this mud. It takes place in unconsolidated sediment containing groundwater. Liquefaction can cause the loss of soil from hill slopes or the collapse of earth walls that support dams.
4. **Subsidence** can badly damage tunnels, roads, railways, electricity and gas lines. By lowering coastal areas they can allow the sea to flood the land.

The extent to which the hazardous impact of an earthquake is related to its strength will vary as the impact is related to many other factors: the nature of the bedrock; the nearness of the area to the epicentre; the depth of the earthquake; the density of the population in the affected area; the level of preparedness of the area – which is often related to levels of economic development; the nature of the landscape and therefore the liability of **landslides**; the potential creation of "**quake lakes**", as landslides block and dam rivers which may then collapse and cause devastating floods further down the river valley and even the time of day.

5. **Tsunamis** are waves generated by the violent displacement of the ocean bed. The long, large waves can be generated by underwater earthquakes, submarine volcanic eruptions or landslides. At sea they are barely noticeable, but their long wave length gives them remarkable speed and energy. When they hit shallow coastlines they produce towering waves that retreat, sucking water back before hitting the coast with massive strength and height. They can be up to 30 metres in height. In estuaries and narrow bays their power is funnelled producing destructive waves that sweep all before them and extend some considerable distance inland. Buildings and settlements may be swept away and thousands of people drown.

Earthquakes may result in:

- **Loss of life due to building and bridge collapses (or drowning through tsunamis)**. Most people agree though that human death is the most significant human impact of earthquakes.
- **Collapse of buildings or the destabilisation of the base of buildings** which may lead to their future collapse in future earthquakes.
- **Damage** to houses, buildings, roads and bridges.
- **A lack of basic human necessities** such as power, water, sewage.
- **Disease**. The damage caused to water pipes can allow them to be contaminated with sewage and this triggers cholera epidemics.
- **Higher insurance premiums**. After earthquakes insurance companies will charge a lot more for insurance cover.

Earthquakes can also lead to volcanic eruptions, which cause further damage such as substantial crop damage – where crops are covered in ash or destroyed by lava flows.

Impacts of earthquakes: Chile 2014 and Nepal 2015

In April 2014, a magnitude 8.2 earthquake took place in northern Chile. Six people died, 2500 homes were damaged and 80000 people were displaced. Just over one year later, in April 2015 a 7.8 magnitude earthquake struck Nepal. Over 8800 people were killed, entire towns and villages flattened and thousands of people left homeless. This raises the question as to how two such similar earthquakes have such disparate effects?

Much of the answer lies in levels of wealth and building standards. After Chile experienced the world's largest earthquake, 9.5 on the Richter scale in 1960, where over 5500 people died, the country has focused on modernising its buildings and designing them to withstand the shaking produced by large earthquakes.

Unfortunately, in Nepal, large numbers of buildings followed any sort of **building code or regulations**, and many collapsed when the earthquake struck. Many of the people live in houses highly vulnerable to earthquake shaking – brittle unreinforced brick masonry.

The geology of both areas though is also different. Nepal lies on a continental collision zone between the two lighter continental plates of India and Eurasia and its actual earthquake fault is buried deep underground and any evidence of surface ruptures is quickly covered by sediment transported down by heavy monsoon rain from the surrounding Himalayas and then covered in dense vegetation.

Also, the speed of this continental collision of about 4.5 centimetres every year, meaning that major earthquakes only hit Nepal every few decades. Whereas Chile's destructive plate margin is more obvious – an oceanic trench where the oceanic Pacific plate subducts underneath the continental South American plate at a rate of nearly 10 centimetres per year – with major earthquakes occurring every year, making earthquake-resilience a priority for Chile.

As an example of a continental collision margin, Nepal's is relatively simple and has been relatively well studied. Geologists had identified and reported Nepal's most vulnerable section of the fault just three weeks before the earthquake occurred. Other examples of continental collision margins have widely dispersed faults spread over thousands of kilometres.

Responses

In many earthquakes, like Nepal in 2015, logistics, coordinating the efforts of governments and aid organisations big and small, quickly become the focus for those overseeing disaster relief operations. Delays in distributing aid become a growing problem and often lead to protests, as in Haiti in 2010 and Nepal.

The **physical geography** of mountainous areas, such as in Nepal, also mean that relief efforts to reach remote rural settlements are made much more difficult by landslides that block and destroy roads. In Nepal 80 per cent of the buildings in some villages were destroyed. An estimated 800000 houses, schools and government buildings collapsed or were badly damaged. The UK's Department for International Development (DFID) spent more than $30 million, between 2011–15, on an earthquake resilience programme in Nepal.

A major focus has been the improvement of building regulations and enforcing them. The Asian Development Bank, one of the biggest investors in Nepal, acknowledges that building codes are ignored and that enforcement of regulations is very poor. The approval of design documents for buildings can be bought through corruption and bribery.

One positive effect of this long-term aid is that most districts around Nepal now have earthquake committees. People have been educated on where to shelter during an earthquake, where to congregate afterwards and how to give basic first aid. Hospitals and schools have been retrofitted to withstand tremors. The police, army and local Red Cross teams have practiced rescue drills.

Volcanoes and resultant hazards

Volcanoes form where there is an opening in the Earth's crust called a **vent**. **Magma** from deep below the crust will then be forced to the surface and may come out in three different ways:

1. It may flow out of a vent as molten **lava**.
2. It may explode out as **volcanic bombs**.
3. It may appear as **ash**.

Over time, this material may build up to form a volcano. The central vent of the volcano is connected to a store of magma below the surface, which is known as the **magma chamber**. The extrusion of lava from the vent leads the creation of a volcanic cone. With each new eruption, new layers are added to the cone, in the first instance by the lava and pyroclastic materials, which are ejected from the volcano, and then by the volcanic ash which later settles from the air.

Types of eruption

Lava is molten rock (magma) that has reached and then flowed over the ground surface. Depending on its composition and temperature, lava can be very fluid or very sticky (**viscous**).

Some eruptions are **violent** and spectacular others are **very gentle**. **Volcanologists** classify eruptions into several different types. Some are named for particular volcanoes where the type of eruption is common; others concern the resulting shape of the eruptive products or the place where the eruptions occur

There are two types of lava that cause different disruptions:

1. **Basaltic lava**.
2. **Acidic lava**.

A key factor is the major difference between **basaltic** and **acidic** lava. **Basaltic** lava is more effusive with quieter eruptions and mostly lava flows, the temperature of the lava being quite high – about 1000 °C. It is usually associated with spreading centres and hot spots. **Acidic lava** is generally more explosive with ash clouds, pyroclastic flows,volcanic bombs, etc. The lava is cooler, 700–800 °C, and is mostly associated with subduction zones.

Eruptions can be classified according to the violence of their eruption:

- **Plinian** eruptions occur where magma is **rhyolitic** in its composition. The high viscosity of this type of magma prevents the escape of volcanic gases, leading to highly explosive eruptions.

- **Vulcanian** eruptions tend to be short lived, lasting only a few hours. Found in areas where lava is highly viscous, a build-up of pressure within the volcano means that these eruptions are relatively intense. The Indonesian island of **Krakatoa** was the most violent in recent history and took place in 1883. The noise could be heard 4 700 kilometres (3 000 miles) away and it completely blew away the top of the volcano and produced a massive **tsunami**.
- **Hawaiian** eruptions are different; fluid basaltic lava is thrown into the air in jets from a vent or line of vents (a fissure) at the summit or on the flank of a volcano. The jets can last for hours or even days, a phenomenon known as **fire fountaining**. Because these flows are very fluid, they can travel miles from their source before they cool and harden. **Mauna Loa** in **Hawaii**, the largest volcano on Earth, has been gently pouring out lava for at least 700 000 years and may have emerged above sea level about 400 000 years ago. It has poured out about 75 000 cubic kilometres of lava in that time!

Other factors such as the penetration of water and the vent being blocked by a hardened block of lava – a volcanic plug – may be important in determining the type of eruption.

Measuring explosive eruptions

The **Volcanic Explosivity Index (VEI)** is commonly used and is a relative scale that enables explosive volcanic eruptions to be compared with one another. It is very valuable because it can be used for both recent eruptions that scientists have witnessed and historic eruptions that happened thousands to millions of years ago.

Most lava flows can be easily avoided by a person as they do not move much faster than walking speed. Because lava flows are extremely hot – between 1000–2 000 °C – they can cause severe burns and often burn down vegetation and structures such as buildings and roads. They also create enormous amounts of weight and pressure, which can crush or bury property.

Fluid flows of lava are hotter and move the fastest. They can form streams of lava, or spread out across the landscape in lobes. **Viscous flows** are cooler and travel shorter distances, and can sometimes build up into lava domes or plugs which may collapse to form **pyroclastic flows**. **Lahars (mudflows)** are formed by debris caused by the volcano.

- **Pyroclastic flows (nuées ardentes)**

Pyroclastic flows are an explosive eruptive phenomenon. They are a mixture of pulverised rock, ash, and hot gases, and can move at speeds of hundreds of kilometres per hour. They are gravity-driven, which means that they flow down the slopes of volcanoes. They account for more deaths, 40 000 in recorded history, than any other volcanic hazard. Gases within the flow might occasionally ignite to form a fireball called a nuée ardente.

Pyroclastic flows of any kind are deadly. They can be extremely fast-moving and are also extremely hot – up to 400 °C. The speed and force of a pyroclastic flow, combined with its heat, mean that they can often destroy anything in their path, either by burning or crushing or both. An example of the destruction caused by pyroclastic flow is the abandoned city of Plymouth on the Caribbean island of **Montserrat**. When the Soufrière Hills volcano on the island began erupting violently in 1996, pyroclastic flows travelled down valleys in which many people had their homes, and covered the capital city of Plymouth. This part of the island is a no-entry zone and has been evacuated, all that remains are the buildings which have been destroyed, buried and melted by the heat of the pyroclastic density currents. Highly productive farmland has now been completely lost and 60 per cent of the island's population have permanently emigrated having lost their farms, houses and businesses.

- **Lahars (mudflows)**

Lahars are mudflows made up of volcanic debris. They commonly form when water (rainfall or snowmelt) mixes with ash on the slopes of a volcano.

Lahars flow like a liquid, but because they contain suspended volcanic material, they usually have a consistency similar to wet concrete. They flow downhill and will follow depressions and valleys. Lahars can travel at speeds of over 80 kph and reach distances tens of kilometres from their source volcano. If they are formed during a volcanic eruption, they may be at boiling point.

Lahars are extremely destructive. They will either bulldoze or bury anything in their path, sometimes in deposits many metres thick. Lahars can be detected in advance by **acoustic (sound) monitors**, which gives people time to reach high ground. They can also sometimes be channeled away from buildings and people by concrete barriers, although it is impossible to stop them completely.

The Nevado del Ruiz eruption in Colombia in 1985 produced a particularly devastating lahar that killed almost 25000 people in the town of Armero, marking one of the worst volcanic disasters in history. This tragedy could have been easily avoided if clear warnings by volcanologists had been taken seriously.

- **Volcanic landslides**

Volcanic landslides are large masses of rock and soil that fall, slide, or flow very rapidly under the force of gravity down the sides of volcanoes. The debris they produce may move in a wet or dry state. They commonly originate as massive rockslides or avalanches which disintegrate during movement into fragments ranging in size from small particles to enormous boulders.

Volcano landslides range in size from less than 1 km^3 to more than 100 km^3. Their high velocity and momentum allows them to travel uphill and to cross valley divides up to several hundred metres high. For example, the landslide at Mount St. Helens on May 18, 1980, had a volume of 2.5 km^3, reached speeds of 50–80 m/s (180–288 km/hr), and surged up and over a 400 metre tall ridge located about 5 km from the volcano.

Landslides are common on volcanoes because their massive cones typically rise hundreds to thousands of metres above the surrounding terrain and are often weakened by the very process that created them – the rise and eruption of molten rock. Each time magma moves toward the surface, the surrounding rocks are moved aside by the molten rock, often creating internal shear zones or over-steepening one or more sides of the cone. Magma that remains within the cone releases volcanic gases that partially dissolve in groundwater, resulting in a hot acidic hydrothermal system that weakens rock by altering rock minerals to clay. Furthermore, the tremendous mass of thousands of layers of lava and loose fragmented rock debris can lead to internal faults that move frequently as the cone "settles" under the downward pull of gravity.

A number of factors can trigger a landslide:

- the intrusion of magma into a volcano
- explosive eruptions (magmatic or phreatic–steam-driven explosions)
- a large earthquake directly beneath a volcano or nearby
- intense rainfall that saturates a volcano or adjacent ash covered hillslopes with water, especially before or during a large earthquake.

A landslide typically destroys everything in its path and may generate a variety of related activity. Historically, landslides have caused explosive eruptions, buried river valleys with tens of metres of rock debris, generated lahars, triggered waves and tsunamis, and created deep horseshoe-shaped craters.

Volcanic ash

Volcanic ash consists of powder-size to sand-size particles of igneous rock material that have been blown into the air by an erupting volcano. Volcanic ash is composed of irregularly-shaped particles with sharp, jagged edges and can be an abrasive material. Volcanic ash particles are very small in size and this allows it to be carried high into the atmosphere by an eruption and carried long distances by the wind. Volcanic ash particles are insoluble in water. When they become wet they form a mud that can make highways and runways very slippery or they can form a deadly lahar.

Ash can have a number of different impacts:

1. **People** exposed to falling ash can suffer from respiratory problems including nose and throat irritation, coughing, bronchitis-like illness and discomfort while breathing.
2. **Livestock** suffer the same eye and respiratory problems as humans. Animals may be unable to eat if the ash covers their food source, such as grassland. Those who do eat from an ash-covered food source often suffer from a number of illnesses and need to be evacuated.
3. **Buildings** can collapse and be damaged under the weight of ash (dry ash weighs about ten times the density of fresh snow). When mixed with water, ash can be corrosive to metal roofing materials. Air conditioners can be damaged if their filters are clogged or their vents are covered by volcanic ash. Appliances such as vacuum cleaners and computer systems are especially vulnerable because they process lots of air.
4. **Vehicles** also process enormous amounts of air and ash can damage engine parts and filters. If the wipers are used the abrasive ash can scratch the window, sometimes producing a frosted surface that is impossible to see through.
5. **Aircraft** engines also process enormous amounts of air. If volcanic ash is pulled into a jet engine melt in the engine and stick to the inside of the engine. This restricts airflow through the engine and can stop the engine with catastrophic results.
6. **Water supply systems** such as a river, reservoir or lake, can fill with suspended ash which must be filtered out before use. Processing this water containing abrasive ash can be damaging to pumps and filtration equipment.

The 2010 Eyjafjallajökull eruption in south-east Iceland, although relatively small, had a huge impact on air travel across northern Europe due to the ash plume that drifted over NW Europe's busiest airports. About 7 million passengers worldwide were stranded as the result of airport closures. Many of these were on holiday and unable to return to work, which left businesses without employees. One estimate put the loss in productivity at $600 million a day. The airline industry suffered large losses, estimated by the International Air Transport Association at $1.8 billion worldwide over six days.

The lifespan of a volcano

Volcanoes can be **active, dormant** or **extinct**. An **active** volcano is one that has recently erupted and is likely to erupt again. There are about 1700 active volcanoes in the world today – Indonesia has over 140 of them!

A **dormant** volcano is one that has erupted in the last 2000 years and may possibly erupt again – these can be dangerous as they are difficult to predict – the one on the Caribbean island of Montserrat last erupted 500 years ago but has made up for this with its massive eruptions in the last few years.

An **extinct** volcano has long since finished erupting – the UK's volcanoes last erupted over 50 million years ago.

Predicting volcanoes and earthquakes

Figure 9.4 shows some of the methods that can be used to predict and monitor both earthquakes and volcanoes.

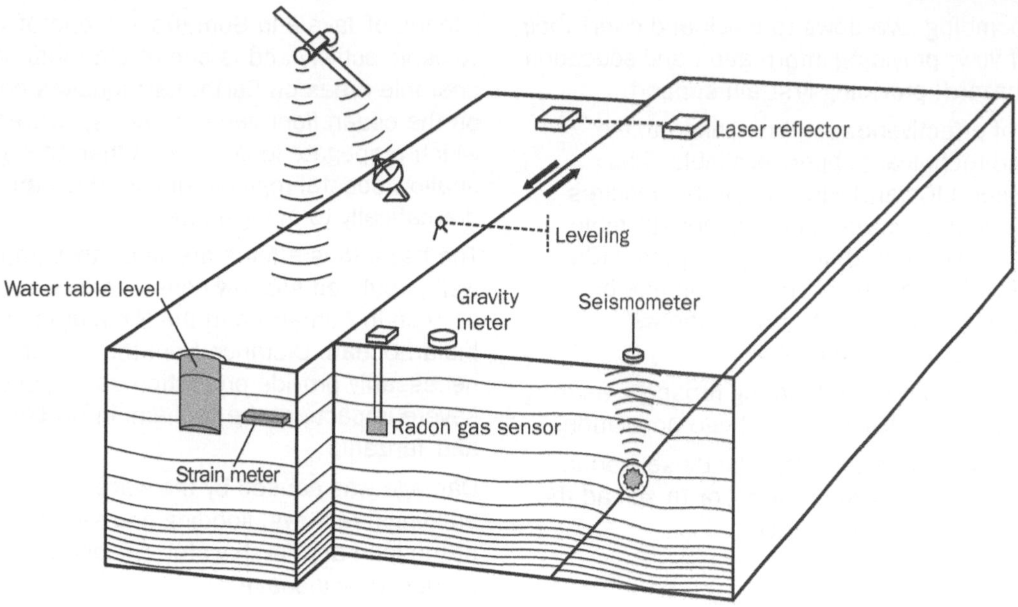

Fig 9.4 Predicting and monitoring earthquakes

Volcanoes

Prediction has improved enormously in recent years and this involves closely monitoring volcanoes for earthquake activity, ground swelling/ground deformation, gas discharges. All of this requires considerable human resources and expensive equipment. There are instances of successful prediction, such as Mt. St Helens in the USA and Mt. Pinatubo in the Philippines, but, as there are potentially 2000 destructive volcanoes around the world, monitoring all of them is impossible.

There are a number of ways we can monitor volcanoes:

- **Satellite imagery and data** and **air photographs** can measure extremely small **changes in the topography/ground deformation** of the landscape and identify any bulging of the land surrounding a volcano as magma approaches the surface under a volcano. They can also monitor **electrical and magnetic fields** which indicate the ascent of magma towards the surface.
- **Heat sensitive probes** can be used to indicate rising magma and to gauge the size and growth of the underground magma chambers, although this may only give a limited amount of time before the actual eruption. The **temperature** of the ground, both at the surface and subsurface, and the groundwater as well as the **level of groundwater** can also indicate imminent volcanic activity.
- **Gravity meters** record changes to gravity in rocks as they become stressed and detect rising dense plumes of iron-rich magma.
- **Strain meters** measure the stretching and compression of the crust.
- **Tilt meters detect** bulging of the land surface caused by raising magma.
- **Radon gas emissions** and the analysis of gases released from fumeroles can indicate that there is disturbance to underlying rock minerals and the existence of other gases can indicate the filling of the magma chamber.
- The study of **volcanic periodicity** from **historic records** can also allow some estimate of the probability of an eruption.

Earthquakes

Earthquakes cannot be predicted with any sense of certainty or accuracy. Methods of earthquake prediction largely depend upon the study of the **locations** and **frequencies** of past earthquakes. It is then possible to calculate the **probability** of one occurring in the future at particular locations. Research undertaken on the North Anatolian fault line in Turkey has been used to predict a possible future earthquake under the capital city, Istanbul, in the future, as earthquakes seem to be moving west along the fault line.

There are two main ways we can monitor earthquakes:

Seismic monitoring can indicate a series of minor earthquakes that may precede a major earthquake.

Monitoring seismic gaps can identify stretches of faults that have not had an earthquake for a long time (this is a seismic gap). In theory, with the stresses building up in this zone, an earthquake is likely to occur.

Reducing the impact of volcanic eruptions and earthquakes

Volcanic eruptions

Measures to reduce volcanic impacts include: prediction the eruption and ensuring the evacuation of people before the eruption; diverting the lava flows in artificial channels or by cooling the front of the lava flows by pumping water on to the margins of

Hot arid areas

- Hot arid (desert) climates have a marked water deficit, where the annual loss of water through evapotranspiration is greater than the gain through precipitation.
- Low annual rainfall (less than 250 mm and as low as 0.5 mm) which can be variable. Rainfall is scarce due to the existence of sub-tropical highs producing sinking stable air. The timing and location of rainfall is difficult to predict.
- There can be individual storms of considerable magnitude giving rise to severe flooding. Most storms, however, are of low intensity and are very localised. On coastal fringes, fogs occur (e.g. Namibia).
- Clear skies give rise to large diurnal ranges of temperature, which are often greater than seasonal temperature ranges. Daytime temperatures are often high, over 40 °C and often exceeding 50 °C, but can fall below freezing at night.
- High seasonal temperatures and limited vegetation give rise to low humidity levels.
- Some variation in temperatures is evident, for example coastal deserts are cooler than those in continental interiors where ground surface temperatures are even higher.

Semi-arid climates

- Semi-arid climates may be defined as having an average rainfall of 250–600 mm.
- Rainfall patterns are unpredictable and subject to great fluctuations. Semi-arid areas have low and irregular patterns of precipitation that are seasonal, with long dry seasons. Sometimes, rainfall can be relatively high, but is poorly distributed and subject to rapid loss through high rates of runoff or evapotranspiration.
- Temperatures are generally high throughout the year with a mean annual temperature above 18 °C.
- As these climates tend to be located around the hot arid climatic areas, the basic causes of their aridity are similar, i.e. in areas of sub-tropical high pressure, in continental interiors, having rainshadow effects.
- Rainfall is often convectional and therefore unpredictable.
- Some areas, such as the Sahel, in Sub-Saharan Africa, can be affected by the movement of the Inter Tropical Convergence Zone (ITCZ). The ITCZ affects the pattern of rainfall in many semi-arid areas, and the length of wet and dry seasons. Longer-term changes in the ITCZ can result in severe droughts or flooding in semi-arid areas.
- All areas are subject to periodic prolonged drought brought about by failure of the monsoons or movement of the ITCZ.

Exam-style questions

1. Compare and contrast the characteristics of the climates found in hot arid and semi-arid areas. [8]
2. Describe the distribution of the large hot desert areas in the world and explain the factors that account for that distribution. [8]

10.2 Landforms of hot arid and semi-arid climates

Weathering processes

Physical

Exfoliation (insolation weathering)

- The considerable **diurnal range** of temperatures in deserts of up to 30°C is thought to lead to powerful insolation weathering in the form of **exfoliation** (also called **thermal fracturing** or **onion skin weathering**).
- The very high daytime temperatures (up to 50°C in hot arid climates), cause the expansion of the surface layers, followed by cooling in the nighttime and contraction. Over time, this will lead to the fracturing of the surface layers of rocks resulting in layers peeling away, hence the term onion-skin weathering.
- The differential expansion of the various minerals found in exposed rocks also leads to the **granular disintegration** of exposed rock surfaces. Physical weathering processes result in the build-up of piles of angular rocks – **scree** – at the base of steep slopes and mountainsides.

Freeze–thaw weathering

- This can occur in areas of high altitude deserts such as the Colorado Plateau in the USA. Water enters cracks in rocks and in cold conditions, freezes and expands.
- This expansion causes the cracks to widen and can lead to disintegration, especially after repeated freezing and thawing.

Salt weathering

- Salt-crystal growth causes the disintegration of rocks when saline solutions seep into cracks and joints in the rocks and then evaporate, leaving salt crystals behind. These salt crystals expand in the sun's heat, exerting pressure on the rock. The most powerful salts are sodium sulphate, magnesium sulphate, and calcium chloride. Some of these salts can expand their volume by up to three times or more.

Chemical

Chemical weathering processes such as **hydration**, **hydrolysis** and **oxidation** are also common processes in deserts, suggesting that the presence of water is of critical importance.

Tafoni

- These are small honeycomb-like features found in exposed granular rocks such as sandstone, granite, and sandy-limestone.
- They have rounded entrances and smooth concave walls.
- Scientists believe that this weathering could be made possible through the action of dew collecting on the undersides of rocks.

The abundant deposits on the margins of limestone **wadis** are likely to have resulted from the collapse of jointed rocks after selected chemical attack along joint lines and bedding planes. Rounded boulders in granite areas and the existence of weathering caverns also indicate the importance of chemical weathering rather than of thermal fracture.

Erosion

The present day erosive processes include both wind and water and can be explained in terms of **stream floods** and **sheet floods**, as well as wind erosion. The problem in using these processes to explain the major landforms we find today lies in the small scale effects of wind erosion and the episodic nature of water action. In neither case is it likely to occur frequently, or intensively, enough to produce the sizeable and extensive landforms of the desert piedmont or extensive wadi systems found in many arid and semi-arid areas. Current processes tend to mould and modify existing landforms, although episodic events such as sudden floods can have some impact in transporting and depositing alluvial materials.

The formation of wadis and their alluvial infill can be caused by stream floods, as can the development of extensive drainage systems and residual playa lakes. The enormous sand seas can be attributed to the product of vast inland oceans in the past. Clearly much of this activity took place in periods of wetter climates in the past and much of what we see in present day desert landscapes represents relict landforms that are only episodically affected by storm events or by the ongoing action of the wind which mainly transports already eroded material around.

Transport and deposition

Wind

In the case of wind, the loose friable surfaces can be removed by the strong desert winds. Very small particles, measuring less than 0.2mm in diameter, can be picked up by desert winds and carried for hundreds, or even thousands, of kilometres. Suspended on air currents, dust from Africa's Sahara Desert sometimes crosses the Atlantic Ocean before landing in the west Atlantic and Caribbean Sea.

On the other hand, sand particles, which typically measure 0.2 to 6.0mms in diameter, can be carried only by extremely strong winds. Silt and other very small-sized particles fill the air during dust storms, but these and most other wind-borne grains are too small to cause erosion or sandblasting of major landforms that stand high above the desert floor and so this wind-blown sediment tends to cause the most erosion at a height of no more than 25 centimetres.

Lighter material is transported as dust storms, whilst heavier material is pushed along the surface by saltation, traction – rolling or creep. Material that is airborne can be used as an abrasive agent upon rock surfaces due to the high wind speeds found in many deserts, although usually only at lower levels below 2 metres.

- **Wind/Aeolian (**from the name Aeolus, the god of the winds, in Greek mythology) **erosion** can be very effective due to high pressure gradients, a lack of vegetation and the aridity of the desert surface. Therefore much loose material can be moved by the wind. Suspension in the wind allows the formation of large dust clouds and these may well have an impact in removing material from **deflation** (the lowering of the land surface due to the removal of particles by the wind) and other hollows, many of which can be of considerable size.
- **Saltation (bouncing/hopping)** is probably a more important mechanism in terms of the abrasion of landforms, but is unlikely to occur at heights above 2 metres and it is concentrated up to 0.5m above the surface. It mainly polishes hard rocks and undercuts weaker ones. The **surface creep** of sand sized material will also contribute to deflation.

Dunes

Sand is deposited when the wind encounters some type of an obstacle: rocks, vegetation, or a man-made structure. As wind passes over an obstacle, the wind's velocity increases. Once on the other side of the obstacle, its velocity decreases and any sand or particles it was transporting, drop out to begin forming a mound. When there is a steady supply of sand carried by a steady wind that comes in contact with an obstacle or irregularity in the flat ground surface, a sand dune forms.

The type of dune formed is influenced by the direction and strength of the wind, the amount of sand it carries, and the shape of the land.

All dunes have a gently sloping windward face and a steeply sloping leeward or slip face. The slope of the windward face is usually between 10° and 20°, while the slip face has a slope of a much greater angle, up to 32°, at which angle sand will start to slip down the face of the dune. The windward face is usually hard packed and smooth, but occasionally cut by minor grooves. The slip face is soft and unstable.

As wind passes over the windward face of a dune, it moves sand along the surface through traction-rolling and sliding movements, and saltation – the bouncing movement of sand caused by the wind. Once over the crest of the dune, sand flows down the slip face. This action – the eroding of sand on the windward face and the deposition of sand on the slip face – causes the dune to move or migrate with the wind like a slow moving wave.

It is rare that a single dune forms in an area. Most often, dunes form in groups called dune fields on broad flat land where winds blow steadily and sand is plentiful. Large dune fields, such as those in the Arabian Desert, are called sand seas or ergs (Arabic for ocean). Although dunes make up only 20 per cent of the total desert landscape, these landforms may cover thousands of square kilometres and reach heights of up to 500 metres.

There are five types of common dune forms: barchan, parabolic, linear, transverse, and star.

- A **barchan** is a crescent or U-shaped dune that has its "horns" or tips pointing downwind or away from the wind. Barchans arise where sand supply is limited, where the ground is hard, and where wind direction is fairly constant. They form around shrubs or larger rocks, which act as anchors to hold the main part of the dune in place while the tips migrate with the wind. They are mobile features and can change form if wind direction changes i.e. from a barchan into linear/seif dune.
- A **parabolic** dune is similar in shape to a barchan, but its tips point into the wind. Its formation is also influenced by the presence of some type of obstruction, such as a plant or a rock. Just the opposite of a barchan, a parabolic is anchored at its tips by the obstruction, which acts to block the wind, while its main body migrates with the wind, forming a depression between the tips. Because of this formation, parabolic dunes are also known as blowout dunes.
- A linear, or longitudinal, **seif** dune is one that forms where sand is abundant and cross winds converge, often along seacoasts where the winds from the sea and winds from the land meet and push the sand into long lines. These high, parallel dunes can be quite large reaching 200 metres in height and over 100 kilometres in length. The crests or summits of linear dunes are often straight or slightly wavy. Linear dunes form with a bimodal wind system.
- A **transverse** dune also forms where sand supply is great. This dune is a ridge of sand that forms perpendicularly to the direction of the wind. The slip face of a transverse dune is often very steep. A group of transverse dunes resembles sand ripples on a large scale. Transverse dunes typically form with a unimodal wind system.
- A **star** dune forms where there is plentiful sand and many dominant winds come from various directions – a multi modal wind system. As its name implies, a star dune resembles a star with its many arms pointing out in different directions. The crests on the arms slope upward, meeting to form a point in the middle of the dune similar to that of a pyramid. The largest and highest dunes are often star dunes.

All the five major dunes can be further categorised into simple, compound, and complex types:

- When they occur in their original states, all dunes are simple.
- When a smaller dune forms on top of a larger dune of a similar type and orientation to the wind, the entire structure is known as a compound dune.
- When a smaller dune forms on top of a larger dune of a different type, it is known as a complex dune.

Water

Water erosion occurs after sudden heavy downpours of rain and is more effective in enclosed channels such as wadis where rock debris and sediment will also be transported in violent flash floods. Erosion can be significant, if short-lived, as the materials abraid the sides and bottom of the channel. Where wadis emerge onto the desert plain, the flow spreads out in a series of **anastomosing channels**, where the velocity is checked and slowed and only the lightest material travels further away.

Water, however, may be a powerful agent; though its infrequency makes it less of a landform-altering agent as compared to past pluvial periods the effects of its erosion, transportation and deposition is evident in the features found in many arid and semi-arid areas such as **wadis/arroyos/canyons, mesas/buttes, alluvial fans/bahadas, pediments and playas**.

The action of stream/channel floods and sheet floods produce many of the features found in desert **piedmonts** as well as the extensive wadi systems of many desert highland areas. As mentioned, in the current climatic conditions that we have now, such limited activity is unlikely to be able to produce such extensive landforms nor could it account for the retreat of mountain fronts, **pedimentation** and the relict features of mesas and buttes.

A few deserts are crossed by "exotic" rivers (such as the Nile, the Colorado, and the Yellow Rivers) that derive their water from outside the desert. Such rivers infiltrate soils and evaporate large amounts of water on their journeys through the deserts.

- **Rainfall in semi-arid climates**

Rainfall in semi-arid areas is **episodic** and intense, which leads to low rates of interception due to the lack of vegetation. Due to precipitation intensity exceeding infiltration capacity, infiltration rates are low as is percolation. Runoff is generally rapid, although temporary surface stores can be created in flat areas. Runoff is also increased due to the emulsification/hardening of the dry sandy surfaces and the formation of a **duricrust**. As a result, **throughflow** and groundwater stores are low as is baseflow.

This is reflected in the channel flow in many rivers in arid areas having a **flashy response** in their hydrographs. The hydrological regimes (water flows and stores) in arid and semi-arid environments are limited due to the arid conditions but are still important. Rivers are either **exotic** (Nile) or **intermittent** and they are contained within **closed basins** as the water in them fails to reach the sea. Desert streams (wadis) characteristically have high sediment loads when they flow.

- **Rainfall in hot arid climates**

The seasons in hot arid areas are generally warm throughout the year and very hot in the summer. The winters usually bring little rainfall. Rainfall is very low and/or concentrated in short bursts between long rainless periods and falls in the form of sudden, violent thunderstorms. Also, evaporation rates regularly exceed rainfall rates.

There may be several storms in a year, or none for several years: average rainfall is, therefore, deceptive. Deserts receive runoff from **ephemeral**, or short-lived streams fed by rain and snow from adjacent highlands.

Yardangs

A yardang is a wind-sculpted, streamlined ridge that can stretch for over one kilometre in length and 30 meters in height. It forms when strong winds blowing primarily in one direction remove all the sand in an area down to the bedrock. If the bedrock is slightly soft or porous, winds will erode the bedrock, sand blasting hollows out of the soft parts of its surface. Over time, wind erosion removes enough material to leave a sleek-shaped ridge, similar in shape to the bottom of an overturned boat that runs parallel to the wind.

Zeugens

Zeugens are more rare and form isolated table-like masses of resistant rock when weaker underlying rock has been eroded. Like yardangs, they are sculpted by the wind. They are formed by wind abrasion where a surface layer of hard rock is underlain by a layer of soft rock into a ridge and furrow landscape. The undercutting effect is concentrated near ground level, where sand movement is greatest, and is enhanced in areas of near-horizontal strata when the lowest bed is relatively weak. The ridges, called zeugens, may be as high as 30 metres. Ultimately they are undercut and gradually worn away.

Wadis and arroyos

A wadi is a steep-sided valley with a flat valley floor. It is formed by intense stream flows following a storm. An arroyo is a stream bed which is usually dry, except during flash floods.

Fig 10.2 shows how these features are caused by erosion from running water, in river channels, in a wetter past and/or stream flood episodes; their steep sides result from lateral erosion and lack of weathering that would be needed to develop a "normal" valley side, v-shaped, slope profile. When an arroyo finally opens onto a flat, broad plain, the rushing water flows out and drops its load of sediment, forming an alluvial fan.

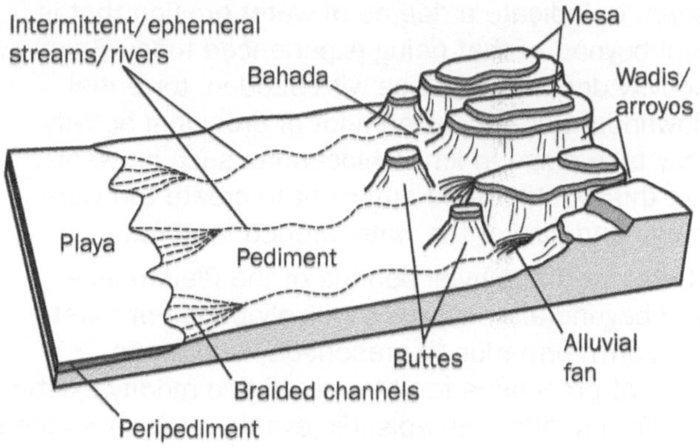

Fig 10.2 Formation of wadis/arroyos

Wadis formed due to water action produce steep sided valleys that meander through upland desert areas. They are often choked with alluvium giving rise to a flat bottom. They are a response to **stream floods**. Where the water is released onto the pediment they can cause **sheet flooding** that then removes material from the pediment and **peripediment**. The deposition of material at the mouth of the wadi can lead to the development of the new landforms. The stream channels can erode into the mountain front, leading to its retreat and the development of a **knick** as the mountain front retreats.

Some wadis may have been formed in the past, during the Pleistocene's pluvial periods when climatic belts shifted southwards, giving rise to a wetter climate than the one we experience today.

Alluvial fan

An **alluvial fan** is deposited as a fan-shaped mass where a rapidly flowing stream leaves a steep and narrow valley and enters an area of gentle gradient. Reduction in gradient, and thus stream velocity, occurs, causing overloading and deposition of alluvium. The convergence of adjacent alluvial fans into a single apron of deposits against a slope is called a **bahada** (Spanish for "slope") or a compound alluvial fan.

Inselberg

An **inselberg** is a prominent steep-sided hill composed of strong, solid rock that sits isolated in a desert plain. Ayers Rock in central Australia, known as Uluru to the Anangu Aboriginals, the native people of the region, is an inselberg. Like other inselbergs, Ayers Rock's rounded appearance is caused by weathering in which its durable surface has been worn away in successive layers.

Pediment

A **pediment** is essentially an erosion surface even though it may be covered with a layer of deposition. Pediments could be an active basal slope left by recession of the mountain front or they may result from lateral planation. Streams emerge from the wadis, depositing their sediment with the loss of velocity on the pediment. The pediment is developed at a constant angle, between 1° and 7°, due to sheet floods and eventually the water enters a playa lake, which is usually ephemeral as the water evaporates after rainfall, leaving behind salt deposits.

Playa

A **playa** is a lake developed in an internal drainage basin from intermittent streams from the occasional rainfall. However, the lake that forms is rapidly evaporated and its site becomes occupied by a saline or alkaline crust or mudflat. Permanent water bodies are rare in deserts. When precipitation does occur in a desert, it often runs down steep hills to form temporary surging streams in low-lying areas before evaporating or sinking into the ground.

When the water falls on fairly flat areas, it may collect in these internal drainage basins, forming a small lake. It may last for a while before the water evaporates or is absorbed. What remains after the water is gone are the sediments it collected as it flowed along the desert surface. Those sediments, mostly clay, silt, and various salts, form a level, broad, cracked surface – a playa. When water is still present, these bodies are called **playa lakes**.

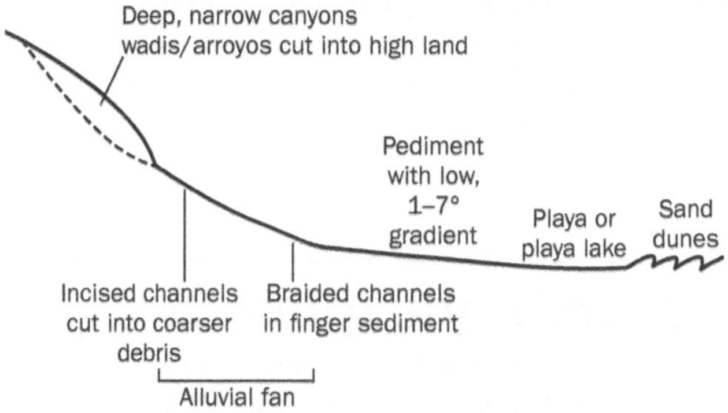

Fig 10.3 Landforms in a hot arid environment

Pleistocene pluvials

The Pleistocene period, lasting from about 2 588 000 to 11 700 years ago, was characterised by the movement equatorward of wetter, cooler conditions. This produced **pluvials**, giving the desert areas that we have currently more temperate and humid conditions. As a result, major depressions and extensive plains could have been formed by fluvial erosion (**peneplanation**) and rivers could have eroded flat areas between present day **mesas** and **buttes** as in Arizona, south-west USA.

Wetter conditions found around 6 000 years ago in the Sahara allowed human occupation with cultivation and hunting, in savanna-like conditions (seasonally humid) which would not be possible today. Similarly, the developed water course systems (wadis) of the Hoggar, Tassili and Tibesti mountains in the Sahara suggest that far more rainfall would have been needed to bring about such high levels of erosive activity. This is confirmed by the greater size of a mega Chad lake compared to the size of the Lake Chad we see in the southern Sahara today.

Peneplains

A **peneplain** is a low-relief plain representing the final stage of fluvial erosion during times of extended tectonic stability. The existence of peneplains, and peneplanation as a geomorphological process, are not without controversy, due to the fact that there is a lack of contemporary examples and that there is uncertainty in identifying relic examples.

After the rivers and streams in an area have reached "base level", lateral erosion is dominant as the higher areas between the streams are eroded. Finally, when areas of upland are almost fully eroded away the floodplains of the rivers and streams merge together in an area of very low to no topographic relief.

Such large amounts of fluvial deposition can therefore be used to explain the existence of the very large sand seas that we see today and the processes of fluvial erosion and stream and surface wash can account for inselbergs (pediplanation), mountain fronts, deeply dissected upland areas, wadis, etc. These processes result from wetter climates and they had a major impact upon the larger environmental processes (for example, weathering, availability of sand, pediplanation etc.) as well as their erosive impact on individual landforms.

The low rates of current processes have led to the dominance of these relict features in the present day arid landscapes.

Other landforms

The existence of large scale drainage networks, wadis, alluvial fans, as well as former lake sites, all seem to indicate a degree of water erosion that is well beyond of that being experienced today. Episodic activity does takes place with sudden, torrential downpours of rain, when a lot of erosional activity may take place in limited locations such as wadis, but this would not be sufficient to create the wide range and size of the water-eroded features.

Therefore, the pluvial periods of the Pleistocene and beyond are seen as the likely period of most landform formation in present-day arid areas. Current processes tend to mould and modify existing landforms, although episodic events such as sudden floods can have some impact in transporting and depositing alluvial materials. For example, current wind erosion and deposition are still active, giving rise to depressions, yardangs, zeugens and dune systems. It seems unlikely that the present-day episodic nature of water action or the small-scale effects of wind erosion could produce the sizeable and extensive landforms of the desert piedmont or extensive wadi systems found in many arid and semi-arid areas. Clearly much of this landscape-forming activity took place in past pluvial periods. Much of what we see in present-day desert landscapes represents relict landforms that are only episodically affected by storm events or the ongoing action of the wind, which mainly transports already eroded material around.

Exam-style questions

1. Explain the processes of weathering, erosion and transportation in hot arid areas and describe their effects upon the landscapes and rocks found in such areas. [8]
2. Describe and explain the effects that wind erosion may have on the development of desert landforms. [8]
3. With the use of diagrams, describe the different types of sand dunes that may be found in hot arid environments and explain how they may have been developed. [8]
4. Describe and evaluate the relative importance of water erosion in producing landforms found in hot arid areas. [8]
5. Describe and explain the evidence for past changes in climate since the Pleistocene. [6]
6. Describe the extent to which landforms in hot arid areas may be the product of past erosional and depositional processes. [8]
7. Describe and explain the formation of the different shapes of desert sand dunes. [8]

10.3 Soils and vegetation

Soil processes

Hot arid

In **hot arid climates** soils are generally infertile, alkaline and saline. They are characterised by a lack of organic material, a thin, shallow **regolith**, a scarcity of micro-organic action and a lack of water. There is little sorting of surface material and in character, colour and mineral content they reflect parent material rather than any soil-forming processes. As precipitation is less than PET (Potential evapotranspiration), the dominant movement is upwards, which can increase salt in the upper layers, giving very alkaline soils.

- **Capillary movement** – Soils are often sandy, grey in colour and salt and calcium rich. Many will have accumulations of mineral salts at or near the surface due to capillary movement, producing a salt crust (**duricrust**) on their surface. Arid soils can be subject to leaching, whereby soluble salts accumulate at a depth related to the water table, providing one of few distinct horizons.
- **Salinisation** – Salt accumulation (**salinisation**) occurs more often close to, or on, the surface. This can prove toxic to plant life. Vegetation cover is low and with this lack of protection they are prone to soil erosion. Water controls both the amount and distribution of plant matter. Soils are thin and lacking in horizons and structure; arid desert soils are lacking in humus and are low in inorganic matter – less than 1 per cent.

Semi-arid

In semi-arid climates soils are also alkaline, shallow and lacking in clear horizons. Organic material is usually only in the top 25 centimetres and is developed under a discontinuous vegetation cover. Salts are often concentrated in the upper horizons. The increased organic matter may be enough to colour the top soil to create chestnut-brown soils.

Such soils are difficult to manage because of a lack of structure and they are prone to wind erosion if disturbed. They can respond to irrigation initially, but there are often long-term problems of salt accumulation, especially where they have been irrigated. In semi-arid soils the cultivation of drought-resistant crops may be possible as well as managed grazing but again, both pose difficulties.

Vegetation

Biomass productivity

- **Hot arid**

In hot arid climates the rate of nutrient cycling is low with a **Net Primary Productivity** of less than 0.003 kilos/square metre/annum (tropical rainforests in contrast have 2.2kgs). It is higher in semi-arid areas – up to 0.2kgs in more favourable areas.

In hot arid climates, the soil is the largest nutrient store. There is a very slow rate of decomposition of dead organic matter and a lack of leaching. The climate, biodiversity, nutrient cycling and soils limit the possibilities for plant development as the harsh climate of aridity with high temperatures and winds is not conducive to biodiversity, which remains very low. Plant densities remain very low which means a lack of litter stores and very limited biomass. With the soils also being structureless, lacking **humus** and characterised by upward capillarity and salt accumulation, these environmental conditions do not provide fertile ground for development.

- **Semi-arid**

In semi-arid climates the biomass is the largest nutrient store with sufficient seasonal water for the growth of vegetation. The stores and nutrient flows are much larger than in arid deserts. The transfer of nutrients to the biomass is limited due to the lack of water. Loss from the biomass is high, due to the loss of plant material in periods of drought. This gives a high degree of **fragility**.

Animals tend to be scarce. Those that do live in this climate are small and able to endure long periods without water or food. Adaptation leads to better survival prospects for plants and animals.

Plant adaptation

Two classes of vegetation are evident:

1. **Perennials** which are often succulent, dwarfed and woody.
2. **Ephemerals** (or annuals) which have a short life cycle after rain. Most are **xerophytes** and have mechanisms for gaining and storing water (deep roots, waxy leaves etc.).

- **Xerophytes** have adapted to resist drought and to some extent salinity. Cacti usually have special means of storing and conserving water. Transpiration is reduced by having spines and thorns for shade; few or no leaves to reduce their leaf surface, with dense hairs covering glossy, waxy leaf surfaces; the closure of stomata and some have a thick bark that reduces their loss of water through transpiration; shallow root systems to take advantage of short periods of rainfall and/or deep roots to reach a deepwater table. Others have the ability to store water in leaves, roots and bulbous stems (for example, succulents like cacti, prickly pear, euphorbias and the baobab tree). Many have long periods of seed storage before germination – several years in some cases. Germination is then often triggered by rare rains making the desert bloom.
- **Phreatophytes** are plants that have extremely long and large root networks roots, up to 45 metres long. This allows them to acquire moisture and maximise water gathering at or near the water table.

Pollution of water is caused in a number of ways:

- **Salinisation** – the groundwater in an aquifer may have a high salt content, especially in arid and semi-arid areas. Water added by irrigation may cause the aquifer to rise so that the top of the aquifer – the water table – comes into contact with the roots of the crop and may harm or kill the crop. This problem can be made worse by the build-up of salts in the groundwater and soil from the irrigation water itself. The groundwater may become up to 10 times more saline due to the salts added by irrigation. This is because the evapotranspiration of the irrigation water from the soil and plants leaves salts behind in the soil. Many crops will be unable to grow in such saline conditions so that farm land may have to be abandoned. Currently, about one million hectares of farmland worldwide go out of production every year because of salinity alone.
- **Waterlogging** – irrigated land is often not perfectly flat which leads to water accumulating in depressions and waterlogging the soil and killing some types of crop. Once again, costly drainage pipes and channels can be put in to get rid of this excess water. It is estimated that about 50 million hectares of irrigated land are affected by waterlogging problems resulting from poor irrigation management practices.
- **Pesticides and herbicides**

  If pesticides and herbicides are added to irrigated crops, these too can find their way into the aquifer and local rivers and, once again, affect the ecology.
- **Fertilisers**

Advantages and disadvantages aside, irrigation, through its impact on food production, has allowed the expansion of the world's population and it has increased the standard of living and quality of life of millions of people. However, the problems highlighted here will need to be addressed and quickly. At the very least, global food production will have to increase by about 50 per cent to feed possibly two billion more people by 2020, and a large part of that increase needs to come from irrigated agriculture.

What is the relationship between agricultural production and distance from markets where the products are sold?

Some of the basic principles in establishing the relationship between agricultural production and distance from markets were developed by farmer and amateur economist J.H. Von Thünen in 1826. His model was created before industrialisation and is based on the following limiting assumptions:

- The city is located centrally within an "Isolated State" which is self sufficient and has no external influences.
- The Isolated State is surrounded by an unoccupied wilderness.
- The land is an isotropic surface, it is completely flat and has no rivers or mountains to interrupt the terrain.
- The soil quality and climate are consistent.
- Farmers transport their own goods to market via oxcart, across land, directly to the central city. Therefore, there are no roads.
- Farmers act to maximise profits.

Von Thünen suggested that a pattern of four rings of agricultural activity would develop around the city:

1. Dairying and intensive farming would occur in the ring closest to the city. Since vegetables, fruit, milk and other dairy products must get to market quickly, they would be produced close to the city.
2. Timber and firewood would be produced for fuel and building materials in the second zone. Before industrialisation and the use of coal, wood was a very important fuel for heating and cooking. Wood is very heavy and difficult to transport, so it is located as close to the city as possible.
3. Extensive fields crops such as cereals for bread – as cereals last longer than dairy products and are much lighter than fuel, reducing transport costs, they can be located further from the city.
4. Ranching would be located in the final ring. Animals can be raised far from the city because they are self-transporting. Animals can walk to the central city for sale or for butchering.

Beyond the fourth ring lies the unoccupied wilderness, which is too great a distance from the central city for any type of agricultural product.

Even though the Von Thünen model was created in a time before factories, highways, and even railways, it is still regarded as an important model in geography, as it illustrates the balance between the cost of land and transportation costs. As one gets closer to a city, the price of land increases. The farmers balance the cost of transportation, land, and profit and produce the most cost-effective product for market. The relationships between agricultural land-use and market distance are very difficult to establish in the contemporary context. However, a strong relationship between the transport system and regional agricultural land-use patterns can be acknowledged at the continental level in North America.

The role of agricultural technology

The term agricultural technology may be interpreted to mean machines and other equipment. Traditional and modern agricultural technology aims to maximise production, in part by reducing variations in physical geography. For example:

- soil type by the use of fertilisers
- draining or irrigating unsuitable land

- chemicals to reduce pests
- the genetic modification of plants, etc.

Table 11.3 Advantages and disadvantages of agricultural technology

Advantages	Disadvantages
increased efficiency and productivity	increased rural unemployment
time-saving	pollution
the ability to cultivate larger areas	environmental degradation
better crop handling and crop quality	loss of tradition
	diminishing returns
	skills shortage
	increasing debt

It is estimated that between 10 and 40 per cent of food is "lost" after harvest in Africa and never gets eaten or sold. Consider how technology could address the following issues:

- Rodents and insects that get into sacks, stores, etc.
- Deterioration, for example, rotting, mildew, fungus
- Inefficient removal of crops from fields
- Intervening circumstances: personal, for example, illness; local, for example, political instability
- Severe weather events
- Transport problems
- Storage and distribution problems
- Handling problems, for example, the spillage of grain, errors
- Theft.

In any sustainable approach to agriculture, an important connection has to be made between the natural and socio-economic systems that influence agriculture and the farming methods that are used. There are many ways to improve the sustainability of a farming system. While these vary from country to country, region to region, farmers trying to take a more sustainable approach share some common practices. All these contribute to long-term farm profitability, environmental stewardship and to improving quality of rural life. These practices include:

- **Integrated Pest Management (IPM)** – managing pests by combining biological, cultural, physical and chemical tools in a way that minimises economic, health and environmental risks.
- **Rotational Grazing Management** uses intensive grazing systems that take animals out of barns to pastures to provide high-quality forage and reduced feed costs while avoiding manure build-up.
- **Soil Conservation** methods, including strip cropping, reduce tillage and "no-till", help prevent loss of soil due to wind and water erosion.
- **Water conservation and protection** are important parts of sustainable agriculture. Many practices improve quality of drinking and surface water, as well as protect wetlands. These play a key role in filtering nutrients and pesticides, in addition to providing wildlife habitat.
- Growing **Cover Crops** such as rye or clover in the off season, after harvesting a grain or vegetable crop, provides many benefits, including weed control, erosion control, and improved soil nutrients and soil quality.
- **Crop/Landscape Diversity** by growing a greater variety of crops on a farm reduces risks from extreme weather, market conditions or crop pests. It also contributes to soil conservation, wildlife habitat and increased populations of beneficial insects.
- **Nutrient Management**, the careful management of nitrogen and other plant nutrients improves the soil and protects the environment. Increased use of on-farm nutrient sources, such as manure and leguminous cover crops, reduces the need to buy fertiliser.
- **Agroforestry** covers a range of tree uses on farms, including interplanting trees with crops or pasture, better managing woodlots, and using trees and shrubs along streams as riparian buffer strips.
- **Marketing** – farmers often find that improved marketing provides a key way to enhance profitability. Direct marketing of agricultural goods to consumers, such as farmers' markets, roadside stands and community-supported agriculture, is becoming much more common.

Exam-style questions

1. Describe and explain the relationship between agricultural production and distance from market. [8]
2. Explain the possible advantages and disadvantages of extensive subsistence farming. [6]
3. Explain the advantages and disadvantages for farmers in using agricultural technology on their farms. [8]
4. What is meant by the term extensive farming? [4]

11.2 The management of agricultural change

CS Sheep farming in Australia

Australia is the world's biggest sheep producer. There are over 70 million sheep in Australia on 85 000 hectares of land. Sheep in Australia are raised either as lambs for meat or as older sheep for wool, with wool production historically forming the largest portion of industry revenue. Australia produces more than a quarter of the world's wool but demand is falling, influenced by new developments in synthetic fibres. Following years of losses, the sheep farming industry returned to growth for the majority of the past five years until 2015. The return of rain in 2009–10 marked the end of a long period of drought. Increased rainfall improved pasture feed, reducing the cost of keeping livestock and allowing farmers to expand production. However, industry revenue has remained volatile, moving in line with fluctuations in world commodity prices and rainfall. The industry revenue in 2014–15 reached $3.2 billion, employing 21 000 people.

Commercial sheep farming in Australia is found on very large farms in marginal areas – areas where other animals and crops would not be as successful or as profitable due to physical and human factors. Therefore, they are often found in areas of low rainfall, high temperatures and poor quality grazing where they are left to graze on grass or small bushes. Such land is also cheaper to buy. Per hectare, sheep farming has very low inputs of capital – much of the land that is used is of relatively small value as often it cannot be used for arable farming. Farms may need up to 25 hectares of grazing land per animal as grazing land is poor. Rearing animals in these areas produces the smallest profits per hectare of any type of commercial farming. However, Australia has very large areas of land available for this activity, so is suited to this type of farming.

Labour – it takes very few people to look after large numbers of sheep, as they can be left out in the fields all year round. Labour is also needed to gather the sheep together for shearing (the actual shearing is often done by groups of skilled shearers who move from farm to farm) and applying pesticides to their fleece and antibiotics to overcome any pests.

At both a national and local scale the industry faces several major challenges, most of them environmental, such as the following:

- Periodic droughts – some lasting several years have become increasingly common. A ten year drought in the south-east of Australia, between 2000–2010, has had a major impact on the industry. This makes the supply of water and food for the animals very difficult and increases costs to the farmer who may have to buy food for the sheep.
- Weed infestation – a variety of non-native plants have found their way into Australia and thrived, covering large areas of grazing land with plants inedible to sheep.
- Destruction of natural habitats and soil erosion – this is due to the grazing of sheep and, in the worst cases, sheep have overgrazed, destroying the natural protective vegetation cover for the soil which is then exposed and easily eroded by water and wind.
- Shortage of sheep shearers – a very tough, hard, manual job which has lost workers to other, easier jobs in the expanding Australian mining industry, for example.

The Australian government has a policy to support farm families and farm businesses in hardship, including assistance for drought-affected farmers. In 2014, a new national approach to drought programmes focusing on encouraging farmers to prepare for and manage the effects of drought and other challenges. The Farm Household Allowance provides assistance to farm families experiencing financial hardship, providing eligible farmers and their partners with up to three years of fortnightly income support. It is hoped that this will help sheep farmers overcome any difficult drought periods. The government actively supports industry change and adjustment through policy and programme development, and encourages the uptake of new and innovative approaches along the value chain that helps the sheep farming industry improve its responsiveness to the ever-changing global market.

Exam-style questions

1. Using a case study, explain the success of attempts to overcome the difficulties experienced in agricultural production. [8]
2. Using one or more examples, assess the role of the government in promoting change in agriculture [8]

11.3 Manufacturing and related service industry

Factors affecting physical location

Land

A large, flat site is easier to build on and for a factory to expand on in the future. A large site may be expensive to buy, so cheaper land is also an advantage. As a result, a large, flat floodplain or a coastal plain are popular sites for factories. The site does need to be well drained, though choosing either a river floodplain or a coastal plain needs careful thought.

Labour

Different industries need different types and numbers of workers. Some industries need large numbers of relatively **unskilled** workers; e.g. some types of farming, like market gardening and sorting and packing vegetables. These are called **labour intensive industries**. Other industries need relatively few workers but they must be **highly skilled,** such as the IT industry, often called **mechanised industries**.

Capital

Money and investment are needed to start an industrial enterprise. New factories are often highly expensive to build and establish. The new Volkswagen Phaeton car factory in Dresden, Germany cost 186 million Euros (roughly $208 million). Toyota plans to set up a small-car assembly factory in Bangalore, southern India, at a cost of more than $100 million. This means that, for both companies and governments, finding the optimum location for a factory is very important.

Markets

This is where the industrial products are sold. The size and the location of a market have now become more important than raw materials. Large MEDC cities like New York, Paris and London provide very large markets for many products. A prime location is near these cities or within reach of an easy, fast form of transport like a motorway. Also, locating near a factory that uses your products reduces transport costs, e.g. car component factories next to car assembly plants.

Materials

Raw materials are often very bulky, heavy and expensive to transport. An industry that uses large amounts of bulky raw materials will find it much easier to locate near the source of the raw materials or at a location where they can be cheaply transported to (such as a deepwater port), a **break of bulk** location, where bulky cargo is unloaded and then processed, so saving on very significant transport costs if the materials were to be transported inland by road or rail.

Access to cheap power sources

Many industries need large amounts of power and therefore a location beside a cheap source of power, such as fast flowing water for mills or a coalfield; this is both useful and often much cheaper.

In the past, one of the major sources of power was coal and therefore, a location on or near a major coalfield was a perfect location, e.g. South Wales, in the UK and the Ruhr, in Germany. Hydro Electric Power (HEP) is a cheap source of electricity and so is often used by power hungry industries, such as aluminium smelters.

Technology

The use of robotic machines that run with computer software to do repetitive jobs has transformed many factories, for example car assembly plants. The Internet, video conferencing and fax machines have released workers from their normal workplace. This means that many workers in tertiary industries can work from home. IT software companies, banks and insurance companies have been able to close offices in MEDCs where labour costs are high and open up in India where those are much lower. For example, Universal moved to New Delhi from London.

Economies of scale

A business with many small factories may not be very profitable compared to those which have just one large factory location. This is referred to as economics of scale, and is similar to buying in bulk to make a saving. Many businesses have closed their smaller plants and built larger ones to put all the industrial processes in one site. A fully integrated iron and steel works, such as the Llanwern iron and steel works in south Wales, is an example where all the iron and steel making processes are put on one site, from the blast furnace to the rolling mill. This has now become the normal practice for car assembly factories, such as the Nissan car factory in Washington, in north-east England.

Diseconomies of scale

Diseconomies of scale are the forces that cause large firms and governments to produce goods and services at increased per-unit-costs. Diseconomies of scale occur for different reasons, but all are as a result of the difficulties of managing a larger workforce. Poor communication and lack of motivation among employees may contribute to this situation. **Physical** diseconomies include a shortage of land for expansion, a shortage of labour, traffic congestion and urban decay. **Economic** diseconomies include rising rents and rates, rising labour costs, high levels of taxation. **Social** diseconomies include pollution and/or environmental problems, high crime rates, pressure from labour unions, pressure from "green" groups and government policy.

Industrial inertia

A stage may be reached where an industry prefers to continue to operate in its former location, although the main factors that caused it to initially locate there are no longer valid. For example, the raw material source is depleted, or an energy crisis has emerged. An industry may stay due to the level of its fixed costs (land, buildings and machinery), or if:

- There is linkage with other activities in the area.
- It is in a favourable location for transportation.
- There is a large pool of skilled labour in that area.

The Yamaguchisteel plant region, in the south-west of Honshu island, in Japan, is a good example of industrial inertia, because it has kept running even after the Chikulu coal mine (the primary easy access to raw materials factor) was depleted.

Transport

Transport costs can make up a larger proportion of production costs. Finding the cheapest forms of transport for moving raw materials and finished goods is very important. This has been made even more important with the global rise in oil prices. Bulky raw materials like crude oil, iron ore and wheat can be most cheaply moved by bulk carriers. Container ships can carry other goods relatively cheaply and efficiently as they are easy to transfer in comparison to road or rail. In 2014 it cost just $1 400 to transport a container from south-east Asia to the EU. A prime location could be where a good natural routeway (such as a river confluence or a port), provide the ways in which raw materials can be transported easily to and from factories. Valleys can accommodate roads, railways and canals so that where they meet in a confluence or where they link to the coast gives that location an advantage over others.

Government policies

Many national governments and the EU have a wide range of policies they can use to encourage industries to move to particular locations. A common policy is to decentralise their government departments. In the UK, the government has moved the Royal Mint and the Passport Office away from London to South Wales and Liverpool, parts of the Foreign Office to Milton Keynes and the Pensions Department to Newcastle. Governments can also provide incentives such as lower company taxes, subsidised wages, lower rents and improved infrastructures like better roads and railways. The role of government policies is sometimes seen as the key to the emergence and growth of the **Newly Industrialised Countries (NICs)**. Their policies can be important in several key areas:

- Socially, for example, investing in education
- Economically, for example, ensuring economic stability, offering incentives
- Environmentally, for example, developing infrastructure
- Politically, for example, entering into trade agreements, good governance
- Governments can encourage industries and businesses in many ways, for example, investing in an industrial estate, setting up an EPZ, education and skills training or putting in essential infrastructure such as efficient transport systems.

Industrial agglomeration; functional linkages; the industrial estate and the Export Processing Zone (EPZ)

The term **industrial agglomeration** refers to a high concentration of industrial activities in an area because industries may enjoy both **internal** and **external economies** when they cluster together (agglomerate).

There are two forms of agglomeration:

1. Where there is a concentration of related or well-linked factories that form a specialised industrial region.
2. Where there is a concentration of various kinds of factories in an industrial zone/estate in urban areas.

Over time, industrial agglomeration results in the growth of large industrial concentrations, producing different areal patterns of industrial land-use. They will have large numbers of associated and inter-dependent factories, surrounded and served by residential and commercial areas.

Industrial agglomerations are found in many countries, such as in Hong Kong in China – the Kwun Tong, Tai KokTsui and Tai Po Industrial Estates; in Sydney, Australia in the Paramatta and Alexandria areas; in major industrial cities, such as Shanghai (textiles), Nagoya/Toyota (car-making), Detroit (car-making) and in several industrial regions, such as Silicon Valley in California (electronics), around the Inland Sea of Japan (shipbuilding), the Pearl River Delta in

South China (toys). Industrial agglomerations have clear functional linkages between the individual industries.

There are five forms of functional/industrial linkages:

1. **Forward** linkage to the industry that consumes the industrial product.
2. **Backward** linkage to the industry that provides the raw material/component.
3. **Vertical** (one-to-one) linkage – a forward and backward linkage, as a raw material goes through several successive processes. For example, a chemical factory supplying the fibre to a textile factory supplying the cloth to a garment factory to make the finished garment.
4. **Horizontal** (many-to-one) linkage, in an industry relies on several/many others for supplies. For example, a car assembly factory being linked/ supplied by iron and steel, engine, tyre and glass factories.
5. **Diagonal** (one-to-many) linkage, in an industry that makes something that can be used in several linked industries. For example, screws, processing chips or a bottling or canning factory linked to several food and beverage factories.

Industries with simple vertical linkages have a very strong production relationship. They can obtain the greatest economic advantages (economies of scale), once they are grouped or agglomerated together in a small area, such as an industrial estate. The benefits from vertical linkages are a lower cost of transporting goods from factory to factory, for example the fully integrated plants of the iron and steel industry.

A broader view of linkages includes many services such as finance, sub-contracting, maintenance, advertising, packaging and transport.

Industrial agglomeration can bring savings in many of the production processes for a business:

- Energy savings
- The waste products or final products from one industry can be the raw materials of another
- Discounts can be obtained when several firms buy similar inputs in bulk
- Savings on advertising costs
- The presence of ancillary services
- Saving of storage
- The close relationships among factories makes it easy to solve the problems of similar nature and to maintain higher level of production skills
- A pool of skilled labour and managerial expertise
- Infrastructure savings on power, water and road networks.

Industrial estates

Manufacturing and related services can make further savings and have added advantages in locating in a purpose-built industrial area, such as science parks, industrial estates, Export Processing Zones (EPZs) and other purpose-built zones. The advantages may be many and are diverse, including:

- Financial incentives, e.g. relocation packages, subsidies, reduced taxation, preferential rates and rents
- An assured supply of utilities, e.g. electricity, water, gas
- Specialist disposal facilities for their waste products
- Good road access, near nodal points and highways
- On-site security
- Promotion and prestige
- Interaction with other businesses/functional linkages
- Agglomeration economies
- For existing businesses, overcoming problems of their current location, for example, traffic congestion, poor environmental quality, lack of space to expand, high rents and rates.

The concentration of industry can result in diseconomies of scale including pollution and other environmental impacts; a shortage of sites or room for expansion which can mean that land prices rise; congestion and transport or logistics issues such as delays if infrastructure is overwhelmed and not improved; an increase in rates and charges; over-stretched utilities and a growing undesirability of the area making it difficult to attract a quality workforce.

Export processing zones

An **Export Processing Zone (EPZ)** is a specific type of **Free Trade Zone**, set up generally in developing countries by their governments to promote industrial and commercial exports. It is a designated area, often near a sea port or airport location, where some of the normal trade barriers, such as tariffs and quotas, are suspended in the hope of attracting new business and foreign investment. They tend to have labour intensive manufacturing industries that import raw materials or component parts and export the finished products. These areas attract major TNCs due to their attractive tax breaks, infrastructure and linkages.

Most FTZs located in developing countries, such as Brazil, India, Indonesia, China, the Philippines, Malaysia, Bangladesh, Pakistan, Kenya, have EPZ programs. Initially, in 1997, 93 countries had set up export processing zones employing 22.5 million people, and five years later, in 2003, EPZs in 116 countries employed 43 million people.

CS Bangladesh EPZ

The **Bangladesh Export Processing Zone Authority (BEPZA)** is currently responsible for supervising the functions of the eight EPZs in the country. The objective was to attract both foreign and local investors to promote industrialisation in Bangladesh to boost economic growth. Factories have been set up in the Ready Made Garment (RMG) sector and in the technology sector to manufacture camera lenses for TNCs such as Fuji and Nikon, mobile parts for Sony and automobile parts for brands like Nissan and Mitsubishi. Investment has come from companies from 35 countries including the UK, USA, China, Germany, India, Malaysia and Australia.

There have been criticisms of EPZs including that sometimes the domestic government pays part of the initial cost of the factory setup, loosens environmental protections and promises not to ask payment of taxes for the next few years. When the taxation-free years are over, the corporation that set up the factory is often able to set up operations elsewhere for less expense.

Bangladesh's fast-growing garment industry – second only to China's in exports – has long provided jobs and revenue, while turning out the low-priced products for shoppers in the U.S., EU countries like the UK, and Australia. However, the industry has received heavy criticism for its low wages, workers being paid as little as $37 a month, and bad worker safety record. Nearly 800 people were injured in largely unreported fires in garment and textile factories in Bangladesh in 2013. Such factories, making products for a number of western HIC retailers, play a big part in Bangladesh's economy, with ready-made garments alone making up to 80 per cent of the country's $24 billion in annual exports. 112 people were killed in 2012 in a fire in a factory with 1 400 workers, making clothes for Wal-Mart, Disney, Sears and other major HIC retailers.

The formal and informal sector of manufacturing and services; causes, characteristics, location and impact

Employment can be divided into two sectors:

1. Formal employment.
2. Informal employoyent.

Formal jobs

These are "official" jobs where the worker is usually registered with the government and may be taxed, but, at the same time, will be eligible for paid holidays and health care benefits. These jobs usually provide better security of employment and are often better paid, with workers getting a regular weekly or monthly salary. The formal economy includes reported payroll items, income taxes, employee taxes and any other official economic factors. Jobs in the formal sector normally include those in government services, education and healthcare. Formal employment is most common in HICs.

Informal jobs

These jobs are often part time, cash-based and low paid. They are found in street markets as market traders, food stall workers, and shoe shiners or as farm workers such as fruit pickers. Informal employment is most common in LICs. The informal sector refers to parts of the economy that are not taxed, regulated, monitored or included in the gross national product.

The informal sector also includes the "black market", which involves the unregulated trade of goods and services.

Though difficult to measure or define, the informal sector is an important element in the functioning of any country. Many developing countries have economies much more heavily based in the informal sector simply due to the fact that regulations have not yet been established.

In developing countries, the largest part of informal work, around 70 per cent, is self-employed. The majority of informal economy workers are women. Above all, informal workers are among the most vulnerable. The informal economy is also characterised by the low productivity of enterprises due to low-skilled workers, outdated production systems and limited management capacity.

Informal employment is thought to reach 51 per cent in South America, 65 per cent in Asia, and 72 per cent in Sub-Saharan Africa.

Exam-style questions

1. Explain the term functional linkages. [2]
2. Describe the circumstances in which existing linkages may end. [6]
3. Explain how government policy can influence the location of manufacturing industry within one country you have studied. [8]
4. Explain the term industrial inertia. [2]
5. Describe the circumstances in which industrial inertia may occur. [6]
6. Explain the term diseconomies of scale. [2]
7. Using examples, describe how diseconomies of scale may develop in the manufacturing industry. [6]
8. Assess the success of Export Processing Zones (EPZs). [6]
9. Describe the main factors that can lead to the concentration of manufacturing industry in a particular location. [8]
10. Describe the diseconomies that may result from the concentration of industry. [6]

11.4 The management of change in manufacturing industry

CS India: A case study of a Newly Industrialised Country (NIC)

The following case study examines India's policies regarding its manufacturing industry and consequent changes in the character, location and organisation of its manufacturing, highlighting some of the issues faced and evaluating attempted solutions.

In 2013, the Indian economy was worth $1.842 trillion, the eleventh-largest global economy. With its average annual Gross Domestic Product (GDP) growth rate of 5.8 per cent over the past two decades, India is one of the world's fastest-growing economies. However, the country ranks 140th in the world in GDP per person.

Until 1991, all Indian governments followed protectionist policies that were influenced by socialist economics. Widespread state intervention and regulation largely excluded the economy from the outside world. However, an acute balance of payments crisis in 1991 forced the nation to liberalise its economy; since then it has slowly moved towards a free-market system by emphasising both foreign trade and encouraging FDI. The newly elected government in 2014 promised to speed up economic reforms to further encourage investment.

Having 487 million workers, the Indian labour force was the world's second-largest in 2011. The service sector made up 56 per cent of GDP, the industrial sector 26 per cent and the agricultural sector 18 per cent.

Until 1991 industry in India was heavily burdened by bureaucracy. Corruption was often involved in businesses obtaining compulsory licences to start up a company. This resulted in some powerful companies having near monopolies. The first major reforms in industrial regulation policy came in 1991 when India was on the verge of bankruptcy, due to heavy government borrowing in the 1980s. To obtain a loan from the International Monetary Fund (IMF) that was needed to avoid bankruptcy, the financial institution insisted that a condition of the loan was that India had to liberalise and open it up its economy to foreign competition.

In contrast, today there are thousands of small businesses established in India and it has several transnational companies (TNCs) that compete in the global market. These include **Tata**, who now own the Leyland, Jaguar and Land Rover vehicle companies, and **Arcelor Mittal**, the world's largest steel-making company. Four of India's most important manufacturing industries are the steel, pharmaceutical, automotive and IT industries. All have undergone changes in their character, location and organisation in recent years.

Steel

India is the fourth largest global producer of steel, with a production of 100 million tonnes for 2013 and is expected to reach 275 million tonnes by 2020. The Indian government has encouraged growth in the steel industry by reducing taxes on imported plant and machinery, customs deductions on steel processing, allowing 100 per cent Foreign Direct Investment (FDI) – currently worth $1 765 million, mainly from the USA, Japan and Australia – and using public-private partnerships developing the market for steel in large infrastructure projects and in the construction, automobiles and power sectors.

Pharmaceuticals

The Indian pharmaceutical industry is the third largest in the world by volume of sales and was worth $26 billion in 2013. With labour costs cheaper (research chemists' salaries are a fifth of those in the USA) and clinical trials a tenth the cost of those in the USA, it has had a growth rate of 14 per cent in global generic drugs. The government has encouraged FDI worth $9 776 million between 2000 to 2012 in both research and manufacturing. In 2013, there were 4 655 pharmaceutical manufacturing plants in India, employing over 345 000 workers.

Automotive

The Indian automotive industry has also undergone rapid growth in recent years and accounts for almost 7 per cent of GDP and employs about 19 million people both directly and indirectly. India is emerging as a global hub for auto component sourcing and is set to break into the top five vehicle producing nations worldwide. The country is also emerging as a sourcing hub for engine components. The Indian auto component sector covers a wide range of industries, including engine parts, drive transmissions and steering parts, body and chassis, suspension and braking parts, equipment and electrical parts.

From 2000 the government removed many trade restrictions and allows 100 per cent FDI in the automotive industry; between 2000 and 2014, and FDI into the Indian automobile industry was recorded at $9 344 million. There has been an influx of foreign auto manufacturers entering the Indian vehicle market, building cars, buses and trucks. Nissan India exports to about 101 countries worldwide,Volkswagen is looking at investing $249 million over the next five years and setting up a diesel engine manufacturing facility from 2014.

The foreign companies have located around Pune, Chennai, Bangalore and Gujurat. Mercedes-Benz was one of the first multinational auto companies to set up in India and located in Pune, near the existing motoring manufacturers. They were followed by the German Audi/Skoda/VW group and US General Motors, which have also recently set up plants in the Pune area.

The growth of the car industry seems set to continue its rise due to:

- The present low level of car ownership in India means that there is a large potential market. A car is still a luxury item in India, about 8 cars per 1 000 people, compared to 500 per 1 000 in Germany. However, a general rise in wealth of the Indian population has meant that domestic sales of automobiles during 2013 grew by 3.5 per cent over 2012.
- An increase in the number of skilledworkers, now 19 million.
- Easier access to finance through the Indian government, allowing 100 per cent FDI.
- An increase in the export of Indian manufactured cars; exports grew 7.2 per cent in 2013.

Information technology

India is the world's largest sourcing destination for the information technology (IT) industry, accounting for approximately 52 per cent of the global market. The industry employs about 10 million people and continues to contribute significantly to the social and economic transformation of the country.

The IT industry has not only transformed India's image on the global platform, but has also fuelled economic growth by energising the higher education sector, especially in engineering and computer science. India's cost competitiveness in providing IT services, approximately 3–4 times cheaper than the US and UK, continues to be its main advantage in the global sourcing market.

The Indian **IT industry's** contribution to India's GDP has been growing rapidly over recent years, from 1.2 per cent of GDP 1998 to 7.5 per cent in 2012. In 2013 software exports grew 12–14 per cent worth $84 billion and the domestic market grew 14 per cent to $185 billion. The main IT centres in India are found in Bangalore (often referred to as the "silicon valley" of India), Chennai, Hyderabad, Pune, Delhi and Kolkata.

These are the main cities for both Indian multinational companies like Infosys Technologies, Wipro and Tata Consultancy Services, along with overseas companies like HSBC, Dell, Microsoft, GE and Hewlett Packard. India exports IT services to 95 countries. In 2013 the industry added 188 000 jobs, taking the total number of direct jobs to three million.

Special Economic Zones (SEZ)

Throughout India there were about 143 SEZs operating in 2012 and by 2013 this had risen to 173. The Indian government's SEZ Act in 2006 aimed to set up SEZs. The objectives of India's SEZs are:

- Generation of additional economic activity
- Promotion of exports of goods and services
- Promotion of investment from domestic and foreign sources
- Creation of employment opportunities
- Development of infrastructure facilities.

SEZs are intended to increase industrial output and exports by:

- Tax incentives on the import and export of goods
- Exemption from various sales and services taxes, including full income tax exemption for a period of 5 years and an extra 50 per cent tax relief for additional two years
- Improving the supply of power, water, sewerage and sanitation
- Improving the infrastructure linking SEZs to non-SEZ areas through improved railways, roads and telecommunications
- Improved deepwater port and port handling facilities
- Improved safety and security measures
- Through liberalisation of laws regarding employment in Gujarat, workers in SEZs can have a one month notice of their jobs being terminated.

Exam-style questions

1. Describe how recent changes in manufacturing production in one named country have been successful. [8]
2. With reference to one country, describe and assess the success of attempts made to overcome issues affecting its manufacturing industry. [8]
3. Describe and explain the causes and consequences of industrial change, with reference to the manufacturing industry in a country that you have studied. [8]

12 Environmental management

12.1 Sustainable energy supplies

Non-renewable and renewable energy resources

Resources are features of the natural and human environment that can be used by people.

A **non-renewable energy resource** is one that is either finite or **non-sustainable**, as its continued use will eventually lead to its exhaustion, e.g. **fossil fuels** such as coal, oil, natural gas and peat.

A **renewable energy resource** is one that can be used continually without the fear of it running out – it is a **sustainable** resource, e.g. wind, water, geothermal, wave, tidal, biogas, biofuels (like ethanol) and solar energy.

The use of these energy resources is not evenly spread across the countries of the world – there is a very uneven distribution. Currently, the richest 25 per cent of the world's population in HICs use over 75 per cent of the world's available energy resources.

The term **sustainability** can be defined as **"development which meets present needs without compromising the ability to meet the needs of future generations"**.

Factors affecting energy demand and supply

Countries have a **total energy demand**. The term covers all requirements for energy, from firewood for heating and cooking to production of electricity and processed fuels. Demand covers domestic, industrial and service use.

Broadly speaking, energy consumption reflects levels of economic development, wealth and the structure of the economy. There is a high demand for energy in HICs and some MICs. There is also a link to climate, both cold areas (heating) and hot areas (air conditioning, etc.) Some parts of the world are energy-rich, for example, the Middle East and Russia for oil and gas, Norway for oil, gas and Hydro Electric Power (HEP) and Iceland for geothermal.

Energy policy

The **energy policy** of a national government can have a major influence. It explains the dominance of nuclear energy in France. In Japan, the use of nuclear energy is under debate following the earthquake and tsunami in 2011. Certain countries have committed themselves towards the Kyoto Protocol goals and the UN Climate Change Conference in Paris in 2015 and the EU has made commitments to increase its use of renewable forms of energy.

Resource endowment

Some areas, notably the Middle East countries such as Saudi Arabia and Kuwait, have a rich resource endowment and are rich in oil, so see little need to develop renewable forms of energy. For many LICs, though, there are real difficulties in developing their huge potential for renewable energy use. This includes:

- **A lack of finance as a result of national debt** – rural poverty/pricing structures and having other financial priorities, often in developing basic infrastructure such as road networks and healthcare.
- **A lack of technology** – **technology transfer** is needed for renewable forms of energy to be developed.
- **A lack of skills and technical expertise** – to both install and maintain.
- **Risk assessments** – for example, the potential for earthquake and cyclone/hurricane damage.
- Many LICs/MICs have an **established dependence on non-renewables** such as oil.
- **The higher relative costs for new renewable energy** compared to non-renewable energy.

Trends in energy consumption: non-renewable fossil fuels

Fossil fuels in 2012 accounted for 87 per cent of the world's power generation. It is forecast that this figure will fall to 55 per cent in 2040. In many countries, although the **proportion** of fossil fuels is decreasing, and of renewables increasing, this may hide an increase in the absolute use of fossil fuels.

Table 12.1 Proportions of energy supplied by different sources

Oil	Coal	Natural gas	Nuclear energy	Hydro Electric Power (HEP)
33%	30%	24%	4%	4%

Fossil fuels have enormous advantages in their relative abundance and ease of use with long established technology in obtaining, delivering and using them as an energy resource. They continue to be used and exploited at an ever-increasing rate despite the increasing use of renewable energy.

Oil and natural gas

These are the main sources of energy for many HICs and most have to import them. Their main advantages are that they are easy to transport and distribute by pipelines and tankers. They are less harmful to the environment than coal – gas is even cheaper and cleaner than oil. They can be used for generating electricity – gas is a very popular fuel for thermal power stations. Oil has extra benefits in that it provides the raw material for the petrochemical industry.

The disadvantages are that global reserves of oil may only last 45 years and gas 55 years at the rate they were being consumed in 2010. "**Peak Oil**' has been reached – the world now consumes more oil than it finds new oil fields.

There is the ever-present danger of pollution through oil spills – for example, during the 1990 Gulf War in Kuwait, several hundred oil wells were set alight causing massive air pollution; oil spillages at sea kill aquatic life and may have a massive impact on fishing industries as in the 2010 Deepwater Horizon disaster in the Gulf of Mexico. When burnt, gas and oil give off **nitrogen oxide** and **sulphur dioxide**, respectively, which contribute to **acid rain**.

Also, oil prices especially can fluctuate widely – for example from $150 to $40 a barrel between 2008–2009. In 2015, it stabilised at around $65 a barrel. Oil and gas pipelines can be targets for terrorism and political decisions can cause supply problems, for example, the turning off of gas supplies to Europe by Russia in 2008 and 2009.

Coal

Given that **coal** is heavy, wasteful and polluting, its continued importance globally as a source of energy would seem difficult to explain. However, there are many reasons for its continued dominance:

- It is **plentiful with very large reserves** spread throughout the world, which means that it is not vulnerable to geopolitical risks (unlike oil).
- It is **relatively cheap** to extract and can be stockpiled.
- It is a **very cost-effective** means of generating electricity and makes a moderate technological demand.
- It is found in many politically stable countries, so that **supplies are relatively safe** and guaranteed.
- Global **reserves of coal will last at least 115 years** at current rates of consumption.
- Improved mining technology have **improved the efficiency and the cleanliness of its emissions**.
- It can be **used for heating and making coking coal** (used in the iron and steel industry).

The use of coal increased by 48 per cent from 2000 to 2009, with most being used to produce electricity in thermal power stations. The world's two largest economies are also the largest users. China produces and consumes 40 per cent of the world's total; the USA produces 50 per cent of its electricity from coal.

However, coal has some major disadvantages:

- It causes air pollution through its production of **carbon dioxide** (a **greenhouse gas**) contributing to the **EGE – the Enhanced Greenhouse Effect**.
- This contributes to an increase in global warming, the melting of ice caps, a rise in sea level, which cause areas of coastal lowland to flood, and **sulphur dioxide** which produces **acid rain**, responsible for killing forests and aquatic life in rivers and lakes.
- It is often mined in open-cast mines which harm the natural environment, while deep coal mining is dangerous for miners, as seen in the Turkish mine disaster in May 2014 which killed over 200 miners.
- Coal is heavy and bulky to transport, so most thermal power stations have to be on, or beside, coal fields or near a deepwater port, as a **break of bulk** location, which allows the coal to be transported most cheaply in bulk ore carriers.

Nuclear energy

No other source of energy has caused more controversy. It is mostly used in countries which do not have their own, large supplies of fossil fuels – such as France, Japan, South Korea and Belgium. In 2014, 11 per cent of the world's electricity came from nuclear power with 30 countries operating 435 nuclear reactors for electricity generation. However, in 2012 Japan shut down 50 of its nuclear power plants indefinitely, following fears about their safety after the Japanese tsunami.

The USA produces the most nuclear energy, providing 19 per cent of the electricity it consumes, while France produces the highest percentage of its electrical energy from nuclear reactors – 74 per cent in 2013.

Nuclear energy policy differs between European Union countries as some, such as Austria, Estonia, and Ireland, have no nuclear power stations. In comparison, France has a large number of stations, with 16 in current use. Some countries, such as Germany and Sweden, are very dependent on the policies of their national governments towards nuclear power, and these have become more negative since the Fukishima nuclear incident following the Japanese Tsunami in 2011. Following the Fukushima incident, Germany decided to end the use of of nuclear energy by 2022.

Advantages of nuclear energy are:

- It is not a bulky fuel – 50 tonnes per year for a power station, compared to up to 540 tonnes of coal per hour for a large coal-fired thermal power station.
- Nuclear waste is very small in quantity and can be stored underground.
- It does not produce greenhouse gases or carbon emissions or contribute to acid rain and it takes away the dependency on imported oil, coal and gas.
- There are relatively large reserves of uranium in Kazakhstan, Australia, Canada and the USA and the stations have relatively low running costs.

Disadvantages of nuclear energy are:

- It can be very dangerous in the event of a nuclear accident and radioactive materials are released into the environment. In 2011, when the Fukishima reactor exploded in Japan, over 200 000 people had to be evacuated from the immediate area.
- Nuclear waste can remain dangerous for several thousand years and so there are expensive unknown storage problems.
- The cost of shutting down (decommissioning) nuclear reactors is very high and there is a constant debate as to who will pay for this – national governments or the electricity companies.

CS Chernobyl (Ukraine, 1986)

In 1986, a reactor at **Chernobyl** in the Ukraine exploded. The explosions and the resulting fire sent a plume of highly radioactive fallout into the atmosphere and over a large area. Four hundred times more fallout was released than by the atomic bombing of Hiroshima. The plume drifted over large parts of western Russia, Europe, and eastern North America, with light nuclear rain falling as far as Ireland.

Large areas in Ukraine, Belarus, and Russia were badly contaminated, resulting in the evacuation and resettlement of over 336 000 people. About 60 per cent of the radioactive fallout landed in Belarus. A 2005 report attributed 56 direct deaths (47 accident workers, and nine children with thyroid cancer), and estimated that there may be 4 000 extra cancer deaths among the approximately 600 000 most highly exposed people.

Trends in energy consumption: Renewable energy resources

In 2013, renewable forms of energy accounted for 5.1 (projected 10 per cent by 2030) per cent of global power generation, with the highest share (5.8 per cent) in Europe and Eurasia. The **growth** in renewable energy remains concentrated in the leading energy consuming countries in Europe and Eurasia, Asia Pacific, and North America.

Over 90 per cent of the population of many LICs (over two billion people) does not have access to electricity which most people in HICs take for granted. A similar number of people depend on fuels such as wood and charcoal which they have to cut and gather or use the dung of their animals to cook their daily meals. A growing population means that it is becoming increasingly difficult for many people to find sufficient and sustainable supplies of energy. The development of sustainable and renewable energy resources would greatly benefit these people and provide an alternative to the finite non-renewable fossil fuels that they depend on for their energy.

Hydro Electric Power (HEP)

HEP generates the highest proportion of renewable energy and 4 per cent of the world's total energy. In some countries, though, it is a very high proportion of their total energy use – in Norway, 96 per cent of electricity, Paraguay, 93 per cent, and Brazil, 86 per cent. HEP schemes also usually have multiple uses apart from generating electricity, including flood control, making water available for irrigation, creation of fisheries, improving river transportation and recreation.

Advantages of HEP are:

- It is a cheap power source (after the high initial costs of the dam).
- Dams can also help with flood control and provide water for the local population and for farming (irrigation) and industry.
- Dams can be stocked with fish and support a local fishery.
- They can be used for recreation and attract tourists and the new source of electricity may attract manufacturing industries and create new jobs.

Disadvantages of HEP are:

- Dams are expensive to build.
- The lakes they created may drown large areas of natural habitats and farmland.
- People may have to move (see the Three Gorges dam) and whole towns and communities may disappear along with historical and archaeological remains.
- Sediment may be trapped and carried by the river and gradually fill up within a few years.

Water quality in China

In China, research by the O.E.C.D. in 2007 found that:

- 300 million people use contaminated water daily.
- 190 million suffer from water-related illnesses annually.
- One third of all rivers, 75 per cent of major lakes and 25 per cent of coastal rivers are now classed as "highly polluted".

The Yellow River supplies water to over 150 million people and 15 per cent of China's agricultural land, but two-thirds of its water is now unsafe to drink, and 10 per cent is classified as pure sewage. Out of approximately 660 cities in China, only one, Lianyuan, with a population of about 200 000, is able to provide clean safe drinking water. In Beijing, residents have to boil their water or buy it in bottles.

Water pollution problems are also acute in rural areas with industries, such as pulp and paper, tanning, and chemical factories, attracted to river locations and discharging effluents.

Chinese coastal regions also have chronic environmental problems, as 60 per cent of China's marine pollution flows out of rivers.

Eutrophication results from the excessive concentrations of nutrients (nitrogen, phosphorus) and consequent phytoplankton or algae growth. This leads to algal blooms, where areas depleted of oxygen cause **hypoxia**. This affects the whole coastal ecosystem, with direct and indirect effects on human health, food supplies, and recreation.

Pollution has led to so-called "Red Tides". Some toxic forms of algae and phytoplankton thrive in nutrient-rich but oxygen-poor zones. Some phytoplankton contain reddish pigments and potent neurotoxins that paralyse fish and kill throughout the food chain. Humans may be affected if they eat shellfish. Skin irritation and burning eyes among swimmers also result.

By 2000, China was estimated to be losing over $100 million per year through red tide disruption to key inshore fisheries. The number of reported red tides escalated from 19 in 1999 to 77 in 2001.

Improving water quality

The main challenge is to balance long-term environmental quality with shorter-term goals of industrialisation and the demand for raising living standards.

The coastal pollution needs to be treated at its source by reducing the pollutants feeding the algae. From the mid-1990s, China's Ocean Agenda 21 developed an action framework for the protection of maritime resources, elimination of pollution, and the implementation of sustainable development, by an improved legal system and increased public participation. The polluter pays principle was invoked, with fines started to discourage misuse and abuse of sea areas.

An example of river management to improve water quality is the **Three Gorges Environmental Protection Programme** started in 2002, with a 10-year plan to build waste water treatment plants and waste disposal plants in the area, costing $4.8 million. It also includes:

- **Re-afforestation** in the surrounding catchment, to reduce sediment runoff
- **Replacement** of small polluting paper mills by larger more efficient plants
- **Control** of agricultural effluent.

In 2008, the Water Pollution Prevention and Control law came into effect. More supervision and accountability is to be created, the polluter pays principle being enforced at a more effective scale. Polluting industries may be ordered to shut down, and fined up to 50 per cent of their annual income from the previous year and 30 per cent of the direct damages caused by pollution.

Land degradation in rural areas

Types and causes of land or soil degradation

The degradation of land and soil is a complex issue as it is the result of the interaction of several natural physical processes and human activities. Its impact is both increasing and accelerating and in a world of 7.2 billion people, it is having a negative effect on food production.

Land degradation is a global problem and largely related to agricultural use. The major causes include:

Deforestation (when people cut forests, woodlands and shrublands) to obtain timber, fuelwood and other products—at a pace exceeding the rate of natural re-growth. This has become an increasing problem in semi-arid environments, where fuelwood shortages are often severe.

In rainforest areas, deforestation can be carried out for several reasons:

- **Logging for valuable timber**, like mahogany and teak, impacts the habitat for thousands of species of plants and animals and destroys delicate **food webs and food chains**. It also takes away habitat of indigenous peoples.
- **Plantation agriculture**, where the forest is cleared to create huge farms for growing plantation crops such as sugar cane and oil palms – both now in great demand as **biofuels**. Malaysia has cleared large areas of TRF and is now the world's biggest exporter of palm oil.
- **Cattle ranching** to meet growing demand for beef and burgers from HICs in particular.
- **New settlement** to provide land for small-scale farmers. The Brazilian government has used the rainforest to provide land for some of the country's 25 million landless people. Alongside

some stretches of the 12000 kilometres of new roads built through the rainforest, 10 kilometres-wide strips of land have been cleared to provide new settlers with farmland.

Overcultivation (peasants farming crops are being forced to increase the yield from their land) taking place in fallow periods – leaving the land bare to regenerate and regain nutrients – is being ignored and the soil is losing fertility. Rising populations are forcing farmers into cropping more marginal areas on desert fringes. This is fine in years of abundant rainfall but when the rains fail, the soils quickly degrade.

Overgrazing (grazing of natural pastures at stocking intensities above the livestock carrying capacity) is resulting in a decrease in the vegetation cover, a leading cause of wind and water erosion. As the number of people has increased, so has the worldwide animal population. Herds of cows, goats and sheep concentrate in certain areas, stripping the vegetation back and exposing the soil to erosion. Great pressure is put on cultivated areas around the boreholes and wells where the animals drink. The trampling of the ground by animals also leads to soil compaction, destroying the structure and leaving it open to erosion.

Population pressure (improper agricultural practices under population pressure) may lead to settlers moving to marginal land. It may lead people to plough land left fallow before it has recovered its fertility, or to attempt to obtain multiple crops by irrigating unsuitable soils.

High population density is not always related to land degradation. More often it is the practices of the human population that can cause a landscape to become degraded.

Approximately 20 million square kilometres of the world's land surface is in a state of degradation. Water and wind erosion account for more than 80 per cent of this degradation. Agricultural mismanagement, often in the form of **overgrazing**, has affected more than 12 million square kilometres worldwide. This means that 20 per cent of the world's pastures and rangelands have been damaged and the situation is most severe in Africa and Asia. In addition, huge areas of forest have been and are being cleared for logging, fuel wood, farming or other human uses.

Several processes contribute to soil degradation:

1. **Water erosion** accounts for nearly 60 per cent of soil degradation. There are many types of erosion including sheet erosion and rill and gully erosion.
2. **Wind erosion**, when dried soil particles are exposed to high winds.
3. The **over abstraction** of **groundwater,** which may lead to soils drying, leading to physical degradation.
4. **Salinisation** – is a common problem in hot arid areas where capillary action brings salts to the upper part of the soil profile. Soil salinity has been a major problem in parts of Australia following the removal of vegetation in dryland farming.
5. The **atmospheric deposition** of heavy metals and persistent organic pollutants may make soils less suitable to sustain the original land cover and land-use.
6. **Climate change** will probably intensify the problem of soil degradation.
   - Higher temperatures cause higher decomposition rates of organic matter. Soil organic matter is important as a source of nutrients and it improves moisture storage.
   - Increased number and severity of floods will cause more water erosion.
   - Increased number and severity of droughts will cause more wind erosion.

Managing soil degradation

The protection of rural environments is decided according to the needs of different locations and environments. The following management techniques could be used:

- **Soil conservation** often involves introducing **appropriate technology** to rural areas. This is technology which is suited to the level of wealth, knowledge and skills of local people and is developed to meet their specific needs.

  To reduce the risk of soil erosion, farmers can be encouraged to use more extensive management practices such as organic farming, afforestation, and pasture extension.

  Methods to reduce or prevent erosion can be mechanical, for example building physical barriers such as embankments and wind breaks, or they may focus on vegetation cover and soil husbandry. Overland flow can be reduced by increasing infiltration.

  The key is to **prevent or slow the movement of rainwater downslope**.
- **Afforestation** and **re-afforestation** can be used to combat soil erosion in many areas. Planting trees to make **shelter belts** protects soil from wind erosion in dry periods. In areas where wind erosion is a problem, shelter belts of trees or hedgerows act as a barrier to the wind and disturb its flow. Wind speeds are reduced, therefore having reduced ability to disturb the topsoil and erode particles.
- **Contour ploughing** takes advantage of the ridges formed at right angles to the slope to act to prevent or slow the downward accretion of soil and water. On steep slopes and in areas with heavy rainfall, such as the monsoon in South-East Asia, contour ploughing is insufficient and terracing is undertaken.
- **Bunding** involves building **low stone walls** along the **contours** of a slope to stop the runoff of rainwater allowing it time to enter the soil, helping to prevent soil erosion and increasing

the amount of water in the soil and making it available for crops. **Compartment building,** sometimes called **tied ridging**, is where low walls of soil are built in a grid of small squares, stopping rainfall runoff and allowing water to be drained into the soil. Crops such as potatoes and cassava can be grown on the soil walls.

- **Strip or inter cropping** which has alternate strips of crops being grown, at different stages of growth, across a slope to limit rainfall runoff, as there is always a strip of crop to trap water and soil moving down the slope.
- **Tier or layer cropping** where several types and sizes of crops are grown in one field to provide protection from rainfall and increase food and crop yields. For example, the top tier or layer, may be coconut trees, below this may be a tier of coffee or fruit trees, and, at ground level, vegetables or pineapples.
- **Terracing** is where the slope is broken up into a series of flat steps, with bunds (raised levees) at the edge. The use of terracing allows otherwise unsuitable areas to be cultivated.

Preventing erosion by different cropping techniques largely focuses on:

- maintaining a **cover crop** for as long as possible
- keeping in place the **stubble** and root structure of the crop after harvesting, to protect and stabilise the soil
- planting a **grass or alfalfa crop**. Grass and alfalfa roots bind the soil, minimising the action of the wind and rain on a bare soil surface.

Increasing the organic content of the soil by adding organic manure allows the soil to hold more water, preventing aerial erosion and stabilising the soil structure.

In addition, care is taken over the **use of heavy machinery** or keeping cattle on wet soils and ploughing on erosion-sensitive soil, to prevent damage to the soil structure.

Urban degradation

Urban degradation involves damage to the physical environment that threatens human welfare now and in the future. The increasing pace of urbanisation and the growing scale of urban industrial activity is exacerbating environmental degradation in many urban areas, and is increasing the vulnerability of the people who live in urban areas. The protection of urban environments requires several measures and in different urban areas the outcomes have been variable.

The degradation problems include air pollution, inadequate waste management and pollution of land, rivers and lakes and the coast. The most common air pollution sources are thermal power stations burning fossil fuels, vehicle emissions and various industrial sources.

Air pollution is also affected by meteorological conditions where calm air conditions can allow pollution levels to build up, as pollutants are not dispersed by wind. This can be made worse in some urban areas, as in Los Angeles and Mexico City where temperature inversions can trap air pollution at ground level for long periods. In Mexico City, every year, about four million tonnes of pollutants are released into the atmosphere from the metropolitan area. In an effort to counter this problem and to reduce pollution and the smog that goes with it, car tax discs in Mexico City have different colours, indicating the day of the week each car is not allowed to drive within the city. This policy is intended to force car drivers onto public transport for that day.

Land pollution includes the dumping of household, industrial and commercial wastes including toxic/hazardous waste in urban areas. Apart from contaminating the land surface, toxic chemicals can also leach from waste dumps and pollute groundwater and surface water runoff.

Legacy pollution is associated with many past industrial activities in older established urban areas. Most older urban areas have contaminated sites. The development of "Town gas" (coal gas) in many cities and towns in the UK in the 19th and 20th centuries left a legacy of pollution at former gas works where the toxic by-products of gas production had seeped into the soil, including heavy metals. These have to be removed before new developments can take place.

In Sydney, Australia, the site of the 2000 Olympics was a former landfill site. Approximately 160 hectares of the site were identified as containing wastes including power station ash, demolition rubble, asbestos, industrial hydrocarbons, domestic garbage, and dredging material from the Parramatta River. Between 1992 and 2000, the New South Wales state government allocated $137 million for remedial action to clean up polluted areas and this included the recovery, consolidation and containment of about 9 million cubic metres of waste. The remediation policy at the time was to safely contain and where possible treat, waste on site, rather than relocating it to other places.

The O2 site in Greenwich, in east London, built as the Millennium Dome in the late 1990s, had previously had various industrial uses. The land was previously derelict and contaminated by toxic sludge from East Greenwich Gas Works that operated from 1889 to 1985, as well as from tar distillation works and a benzene plant. The remedial actions needed for this development to take place included the removal of 7 million litres of tar, washing and cleaning 33 000 m^3 of soil, treating 66 000 m^3 of contaminated groundwater and effluent and recycling 245 000 m^3 of natural and engineering materials for backfill. The operation cost $33 million and took 14 months.

The UK government also spent $19 million cleaning up a part of the 2012 Olympic site in east London that was "grossly contaminated" with toxic waste left behind under a chemical storage facility that was bulldozed to make way for the main stadium.

In many US cities, open recreational areas including parks, gardens and school playing fields, are contaminated with industrial chemicals, in particular lead. This was associated with past industrial activities like lead milling and mining. Pittsburg, in the US state of Kansas, is one city which has expanded to cover areas of lead mining waste.

In 2010, in China, the major city of Wuhan, built 2400 apartments on the site of a former chemical plant. The construction was almost complete when it was discovered that the site was contaminated with antimony, a metallic element that can cause lung and heart problems. Plastic sheeting was spread over 21000 square metres to insulate the contaminated soil, and new soil was spread on the top of the plastic.

Water pollution is a major problem in many urban areas. In India, about 80 per cent of urban waste ends up in the country's rivers and poorly planned urban growth, poor management and a lack of government accountability all contribute to a major pollution problem. In Delhi, between 75 and 80 per cent of the pollution of the city's major river, the Yamuna River, is the result of raw sewage. This is combined with industrial runoff and rubbish thrown into the river, totalling over three billion litres of waste per day, a quantity well beyond the river's capacity to assimilate it.

Only 55 per cent of Delhi's 15 million residents are connected to the city's sewage system. An estimated 7 million people empty their wastewater and raw sewage into the Yamuna. Lack of sewer infrastructure means that 11 of the 17 sewage treatment plants in the city are under used, with a quarter of the plants running at less than 30 per cent of their capacity, as the sewage system is simply unable to deliver sewage to the plants. Also, the 1500 sprawling slums/bustees of New Delhi are not connected to the system and their wastewater and sewage drain into the river.

Exam-style questions

1. Explain using examples, different approaches to waste disposal. [8]
2. Using examples, explain how land pollution may be reduced. [8]
3. Describe and explain the main sources of air pollution. [8]
4. Using examples, describe and explain some of the potential links between pollution and ill health. [8]
5. Explain why it is difficult to solve the challenges posed by air pollution [6]
6. Explain how the burning of fossil fuels contributes to air pollution. [6]
7. Describe and explain the main causes of water pollution. [6]
8. Explain the ways in which water pollution may be reduced and water quality improved. [6]
9. Explain why water quality is an issue in both LICs/MICs and HICs. [8]
10. Using examples, explain the possible measures that can be taken to ensure that land does not become degraded. [8]
11. Describe and explain the possible causes of land degradation in rural environments. [8]
12. Using examples, explain how the different forms of land pollution may be reduced. [8]
13. Using examples, assess how incidents of accidental pollution may result in environmental degradation. [8]
14. Explain the causes of deforestation. [6]
15. Using one or more examples, describe and explain how levels of water quality may be improved. [8]
16. Explain why it is often very difficult to solve the problem of air pollution. [8]

12.4 The management of a degraded environment

CS Desertification in Burkina Faso, West Africa

Desertification can be defined, according to the Food and Agriculture Organization (FAO), as:

> 'The degradation of land in arid, semi-arid, and dry sub-humid areas. It is caused primarily by human activities and climatic variations. Desertification does not refer to the expansion of existing deserts. It occurs because dryland ecosystems, which cover over one-third of the world's land area, are extremely vulnerable to over-exploitation and inappropriate land-use.'

The country of Burkina Faso is located in the Western Sahel and lies within the transitional zone of desert and grassland. It receives more rainfall than countries to its north and people began to settle and farm intensively. This led to increasing desertification during the 1990s and resulted in increased hardship for rural communities and families. Many were forced to migrate further south.

In an attempt to remedy some of the problems, aid agencies such as the Eden Foundation (Oxfam) and USAID intervened both to supply the equipment needed for sustainable farming and to educate local people and communities on sustainable techniques.

Farmers were taught how to use drip irrigation techniques, using far less water than flood irrigation. This system drips water slowly to where each seed is planted. It reduces the amount of water lost to evaporation in normal flood irrigation, where much water is lost as it spreads in large quantities across the whole field.

Other techniques include:

- Placing large open containers at the foot of hundreds of rocky outcrops and hillsides. These large containers act as reservoirs and collect water, which would otherwise have been lost, running off the steeper ground. Pipelines linked these containers to local villages to supply fresh water, and to another pipeline network to supply their installed drip irrigation systems.
- The anchoring of thin arid soils by providing a permanent root system to provide protection for both the new crops and the soil from wind erosion.
- Creating stone walls (bunds) between rows of crops. This reduces ground level wind speeds and reduces soil erosion, allowing the seeds a better chance of taking root.
- The planting of perennial shrubs, and in the more arid areas, desert scrub plants such as creosote bushes. These plants survive year-round in arid conditions and contribute to reducing desertification in three main ways:
  1. They improve the thin soils by providing dead and decaying foliage to increase the depth and nutrient quality of the humus layer of soil.
  2. Their height provides some protection from wind erosion of the top soil.
  3. They have extensive root systems which help anchor the soil and further reduce wind erosion, allowing the humus layer to gain depth.

Microdosing involves the application of small, affordable quantities of fertiliser using a bottle cap, either during planting or as a top dressing, 3 to 4 weeks after germination. This technique maximises the use of fertiliser and improves productivity. Microdosing has increased sorghum and millet yields by up to 120 per cent and incomes by up to 50 per cent in more than 200000 households in Africa and has triggered the reintroduction of fertiliser use not only in Burkina Faso, but in Zimbabwe, Mozambique, South Africa, Niger and Mali.

In the continuous battle against desertification, the education and funding provided by NGOs and other aid agencies is important in helping growing populations in the Sahel to farm sustainably. Evaluating the success of management strategies is not always straightforward in the Sahel, as there are frequent outbreaks of civil war and insurgencies, such as in Mali, South Sudan and Somalia which make education and aid programmes difficult.

Exam-style questions

1. Using a named example, describe the attempts to improve the quality of environmental degradation. [8]
2. Using one named example, assess the effectiveness of attempts made to improve environmental quality. [8]
3. Using one named example of a degraded environment:
   (a) Describe and explain the factors which have contributed to the degradation of the environment at that location. [8]
   (b) Assess the level of success in upgrading the quality of your chosen environment. [6]

13 Global interdependence

13.1 Trade flows and trading patterns

Visible and invisible imports and exports

A **visible import** or **export** is a good or product that is traded and flows into, or out of, a country. It is visible in that it can be touched or seen, e.g. cereals such as wheat and maize, or manufactured products such as vehicles and machinery.

An **invisible import** or **export** is a product that is traded and flows in and out of a country. It is invisible in that it cannot be touched or seen physically, for example a service such as finance, information consultancy and tourism.

Global patterns of and inequalities in trade flows

The global economy has grown continuously since the end of the Second World War in 1945. Global growth has been accompanied by a change in the pattern of trade, which reflects ongoing changes in structure of the global economy. The main changes in the global economy are:

1. The emergence of regional trading blocs, where members freely trade with each other, but erect trade barriers for non-members, has had a significant impact on the pattern of global trade. While the formation of blocs, such as the European Union and NAFTA, has led to trade creation between members, countries outside the bloc have suffered from trade diversion.
2. In many HICs the trade in manufactured goods has fallen relative to its trade in commercial and financial services. Many of these advanced economies have experienced de-industrialisation, with less national output generated by their manufacturing sectors.
3. The collapse of communism led to the opening-up of many former-communist countries. These countries have increased their share of world trade by taking advantage of their low production costs, especially their low wage levels.
4. Newly industrialised countries (NICs) like India and China have dramatically increased their share of world trade and of manufacturing exports. China, in particular, has emerged as an economic super-power. China's share of world trade has increased in all areas, and not just in clothing and low-tech goods. For example, in 1995, the US had nearly 25 per cent of global trade in hi-tech goods, while China had only 3 per cent. By 2005, the US share had fallen to 15 per cent, while China's share had risen to 15 per cent.

Global trade and development

Trade is the exchange of goods and services. It is the major driving force for economic relations between countries and some countries benefit from trade more than others, due to having **comparative advantages**.

Global trade is a complex issue. National governments prioritise development through trade preserve and enhance the standard of living and quality of life of their people. HICs have great influence over international trade – see the following section on the World Trade Organization (WTO). Many HICs have transnational corporations (TNCs), many of which have more purchasing power than most LICs/MICs.

All countries need to be involved in trade with other countries, to trade their products for other goods or services that they need. All countries aim to have a trading surplus, where the value of the goods and services they sell or exports, exceeds the value of those goods and services they buy or import. Traditionally, HICs have had strong economies with positive trade balances, while many LICs/MICs have much weaker economies and negative trade balances. The balance **of trade** is calculated as exports minus imports (of visible goods) or the difference between them.

Factors affecting global trade

Resource endowment

The world's natural resources are unevenly distributed. Where countries and regions are endowed with **natural resources** they have a **locational advantage** and have the potential to establish industries and develop their economies. This locational advantage can be further enhanced if it can be combined with other locational advantages, such as access to labour, markets, capital, transport and favourable government policies and trade agreements.

Historical factors

Many of the poorest countries of the world are former colonies of European countries, such as the UK, France, and Spain. European countries have comparatively few resources compared to Africa and South America. They took over countries and established **colonies** to supply their raw materials as cheaply as possible. These raw materials were then used to develop their manufacturing industries. The colonies also provided a large captive market for these manufactured goods and products were sold back at high cost to the colony.

During their colonial periods, which for many lasted from the 1600s to the 1960s, their economies were geared to provide primary produce for their colonial rulers. For example, in the Caribbean Islands large plantations were established to grow sugar or bananas for sale in France or Britain. Ghana grew cocoa that was shipped to Britain to be manufactured into chocolate, Sri Lanka (then known as Ceylon), Kenya and India were used for supplying tea and Malaysia for rubber.

This pattern of world trade continues to the present day. Since independence, many LICs/MICs have remained primary producers as a result of this **colonial legacy**, growing crops or mining minerals for export; they are often tied to the same former colonial masters – the industrialised HICs. Transnational corporations (TNCs) based in the HICs have replaced the European nations to practice a form of **neo-colonialism**. In most cases, the prices of their commodities are still determined by the amount the buyers are prepared to pay rather than by a realistic consideration of the cost of production in the LIC/MIC.

Locational advantage

A locational advantage is the benefits or competitive edge derived from a place's unique nature and/or position.

The operation of locational advantage in international trade is seen in all economic sectors, for example, in the primary production of tropical crops; in the secondary sector in relation to mineral resource endowment; and in the tertiary sector in nearness to potential customers. Locational advantage affects competition and market share and it produces disparities within and between countries.

While location cannot be changed, it may be influenced by protective measures and developments such as EPZs and over time as global trading shifts.

Trade agreements

A trade agreement (also known as trade pact) is a tax, tariff and trade treaty between countries. The most common trade agreements are free trade agreements, which reduce or eliminate tariffs, quotas and other trade restrictions on items traded between the countries. Some trade agreements are quite complex (such as within the European Union), while others are less intensive (such as the North American Free Trade Agreement, NAFTA). The resulting level of economic integration depends on the specific type of trade pacts and policies adopted by the trade blocs. Typically, the benefits and obligations of the trade agreements apply only to their signatories.

The role and nature of Fairtrade

The definition of **Fairtrade** used by **FINE** (a grouping of four key fairtrade organisations), is: "a trading partnership, based on dialogue, transparency and respect that seeks greater equity in international trade".

It differs from standard trading practices in several ways and its aims to include improved trading for producers in LICs/MICs:

- It attempts to trade with poor and marginalised producer groups in such countries, to improve their trading conditions.
- A better financial outcome, for example, by paying a fairer price that covers the full cost of production and provides a living wage for the producers.
- Provides a loan/credit facility for producers when it is needed, so they have the funds to cover production costs.
- Developing knowledge and skills to improve lives more widely.
- Pays a premium for goods that can provide funding for social and infrastructural development work in communities where the producers live.
- Encourages the fair treatment of all workers and their families and ensures healthy and safe working conditions.
- Develops product certification and Fairtrade labeling.

The World Trade Organization (WTO)

The World Trade Organization (WTO) is an international body whose purpose is to promote **free trade** by persuading countries to abolish import tariffs and other barriers. As such, it has become closely associated with globalisation. It has 159 member states and the key players are USA, the EU, Japan.

It is based in Geneva and was set up in 1995 to replace another international organisation known as the **General Agreement on Tariffs and Trade** (GATT). It was formed in 1948, when 23 countries signed an agreement to reduce customs tariffs.

The WTO has a much broader scope than GATT. Whereas GATT regulated trade in merchandise goods, the WTO also covers **trade in services**, such as telecommunications and banking, and other issues such as **intellectual property rights**.

The WTO is the only international agency overseeing the rules of international trade. It has three basic roles:

1. To police free trade agreements
2. To settle trade disputes between governments
3. To organise trade negotiations.

In theory, WTO decisions are absolute and every member must abide by its rulings. So, when the US and the European Union are in dispute over trade in bananas or beef, it is the WTO which acts as both judge and jury. WTO members are empowered by the organisation to enforce its decisions by imposing trade sanctions against countries that have breached the rules.

The highest body of the WTO is the Ministerial Conference. This meets every two years and is the setting for negotiating global trade deals, known as "**trade rounds**" which are aimed at reducing barriers to free trade.

The WTO has been the focal point of criticism concerning the effects of free trade and economic globalisation. In particular, many countries have been waiting for the WTO to conclude a long-awaited global trade deal, intended to cut subsidies, reduce tariffs and give a fairer deal to developing countries.

Discussions on this (the so-called **Doha Round** of talks) began in 2001. But a breakthrough has proved elusive, with rows emerging among the WTO's key players over agricultural tariffs and subsidies.

Opposition to the WTO centres on four main points:

- That the WTO is too powerful, in that it can in effect compel sovereign states to change laws and regulations by declaring these to be in violation of free trade rules.
- That the WTO is run by the rich countries for the benefit of rich countries and large multinational corporations, harming smaller countries which have less negotiation power and that it does not give significant weight to the problems of developing countries. For example, rich countries have not fully opened their markets to products from poor countries.
- The WTO is indifferent to the impact of free trade on workers' rights, child labour, the environment and health.
- It lacks democratic accountability, in that its hearings on trade disputes are closed to the public and the media.

Supporters of the WTO argue that:

- It is democratic, in that its rules were written by its member states, many of whom are democracies, who also select its leadership.
- By expanding world trade, the WTO in fact helps to raise living standards globally.

The term **free trade** assumes there are no barriers/tariffs to trade between countries and that the cost of goods and services is determined by the balance between what the producer country wants for the goods and services and what the receiving country is prepared to pay for those goods and services. If a commodity or service is scarce and where there is competition for those goods or services, the producer country can exert influence on their cost. However, if there is an abundance or surplus supply of a good, then the consumer country can negotiate a lower cost.

Exam-style questions

1. To what extent may political factors cause large increases and large decreases in trade? [8]
2. Can global inequalities in trade flows be explained in terms of historical factors? [8]
3. How might poverty affect development? [8]
4. Describe the role of the World Trade Organization (WTO) and, using examples, evaluate its work. [8]
5. Explain why trade may advantage some people more than others. [8]
6. Explain the term Fairtrade and, using examples, explain what Fairtrade aims to achieve. [8]
7. Explain the meaning of the terms visible import and invisible export. [4]

13.2 International debt and international aid

Many countries are in debt. There are two types of debt:

1. **Public debt** is an internal debt, where money is owed by a government to financial organisations or individuals within a country.
2. **External, or foreign, debt** is money owed to creditors outside the country. These creditors can be **international organisations** such as the International Monetary Fund (IMF), the World Bank, the Asian Development Bank and the African Development Bank, other national **governments** and **TNCs**.

A country's debt is normally expressed in two ways: either as a ratio of debt to a country's GDP, or as a ratio of a country's external debt to the value of its exports.

Many countries have borrowed more money and accumulated more debt than it is feasible for them to pay back in the foreseeable future. Such debts are not exclusive to LICs/MICs. The global economic recession in 2008 saw the EU countries of Greece, Portugal, Ireland and Cyprus fall into heavy debt. Between 2008–2015, Greece took on multi-billion dollar loans from other Eurozone countries and the IMF, in an attempt to avoid economic collapse.

Causes of debt

There are several ways in which a country may find its way into debt. They include:

- Through a trade imbalance, where the value of imports is greater than the value of exports over a period of time.
- As a result of a currency devaluation or foreign exchange issues.
- Rising prices of key imported commodities. For example, the oil crisis of 1973 and subsequent rises in the price of oil since that time.
- An inability to make repayments on international loans, for example, Argentina being unable to repay the International Monetary Fund (IMF).
- Poor financial decision-making by a government due to inappropriate lending and borrowing in the 1960s and 1970s and excessive interest charges being imposed by creditors.
- Following independence, many former colonies were given large loans to develop their infrastructure and internal industries that would replace imports.
- Financial mismanagement of an economy, sometimes due to high levels of military spending.

Why do some countries find it difficult to get out of debt?

There are several reasons and in many countries these are multi-dimensional. **Economic and political factors** often dominate, for example, high interest rates on international loans, the demands of a growing population with increasing aspirations, the dominance of cheaper primary exports in their export portfolio, changes of government, political corruption and economic instability.

Some countries have also faced physical/environmental problems for example, the impact of **natural disasters** such as earthquakes, floods, droughts, hurricanes/cyclones and tsunamis, or **social issues** such as high levels of population growth, limited education provision and poor infrastructure.

The international debt crisis

The financial crisis of 2007–08 is considered by many economists to have been the worst financial crisis since the Great Depression of the 1930s. It threatened the collapse of large financial institutions, which was prevented by the bailout of banks by national governments, but stock markets still dropped worldwide. In many areas, the housing market also suffered, resulting in evictions, foreclosures and prolonged unemployment. The crisis played a significant role in the failure of key businesses, declines in consumer wealth estimated in trillions of U.S. dollars, and a downturn in economic activity leading to the 2008–2012 global recession and contributing to the European debt crisis.

The European debt crisis is a multi-year debt crisis that has been taking place in the European Union since the end of 2009. Several Eurozone member states (Greece, Portugal, Ireland, Spain and Cyprus) were unable to repay or refinance their government debt or to bail out over-indebted banks under their national supervision without the assistance of third parties like other Eurozone countries, the European Central Bank (ECB), or the International Monetary Fund (IMF).

Debt relief

Debt relief involves the partial or total forgiveness or cancelling of loans, in the recognition that LICs/MICs cannot repay them in full and that any future development will be badly impeded by indebtedness.

Recent initiatives to address debt relief include the World Bank and IMF **HIPC Initiative**. The **Heavily Indebted Poor Countries** (HIPC) are a group of 39 developing countries with high levels of poverty and

do not develop in such a linear fashion. Some skip steps, or take different paths. Rostow's theory can be classified as "**top-down**", or one that emphasises a "**trickle-down**" modernisation effect from urban industry and Western influence to develop a country as a whole. Later models have challenged this approach, emphasising a "**bottom-up**" development route, in which countries become self-sufficient through local efforts, and urban industry is not necessary.

The model also assumes that all countries have a desire to develop in the same way, with the end goal of high mass consumption, regardless of the impact on wider aspects of development. For example, while Singapore is one of the most economically prosperous countries, it also has one of the highest income disparities in the world.

Finally, the model disregards site and situation. It assumes that all countries have an equal chance to develop, without regard to population size, natural resources, or location. Singapore, for instance, has one of the world's busiest trading ports, but this would not be possible without its advantageous geography as an island nation situated between Indonesia and Malaysia.

Measuring global inequalities

There are various indices used to measure social and economic inequality.

Gross domestic product (GDP) and gross national product (GNP): measures of wealth

Gross Domestic Product (GDP) is the total value of good and services produced by a country in any one year. It is limited to goods and services produced within that country.

In the 1970s, development was seen almost exclusively as an **economic phenomenon**. It was hoped, at that time, that rapid overall growth and per person growth would "**trickle down**" to the majority of the population in the form of jobs and higher incomes. Therefore, **Gross National Product** (**GNP**) and **GNP per person** were seen as the key measures of development. GNP is the "total value of a country's economic production in one year" and is nearly always calculated on a **per person** basis so that differences in sizes of population are neutralised.

The **GNP** of a country is the GDP plus the income a country receives from abroad. This income from abroad includes dividends, interest and profit. GNP includes the value of all goods and services produced by nationals, whether in the country or not.

GNP remains the most commonly-used indicator and was used to divide the world's countries into HICs, MICs, NICs and LICs.

Disadvantages of measures

Although extremely widely used, the use of money to assess development has a number of disadvantages, including:

- The **real value** of the unit of currency for each country will change significantly over short periods of time, hence the use of US dollars as the means of comparison. Inevitably, the conversion process creates distortions because of different and changing inflation rates.
- International exchange rates do not necessarily reflect the relative purchasing power of one currency against another.
- A large part of the country's output does not enter international trade.
- It gives no indication of how national income is actually distributed in a country. Therefore, a rising level of both absolute and per person GNP can completely hide the fact that the poor are no better off in a country.

As a result of the limitations of using GNP, the World Bank devised the **Purchasing Power Parity per Person Index** (PPP), an indicator of buying power. The **PPP** attempted to take costs into account. It recognises that some currencies are weaker, that costs can vary and it tries to compare the real cost of living between countries.

For example, China's GDP is $1 400 but, when adjusted for PPP, it is $6 200 – more than four times greater. Japan's GDP is $37 600, but it is lowered when adjusted by the PPP to $31 400. By relating the average earnings to the ability to buy goods, the PPP therefore raises the GNP for developing countries and lowers it for developed countries. However, it still does not take into account regional variations or the social and environmental costs of development.

Multivariate analysis

Berry pioneered the concept of multivariate analysis (analysing several variables at once) and this led, in 1970, to the production of an Atlas of Economic Development which analysed 43 important development variables. They included:

- **Transportation –** such as kilometres of railways per unit area of a country.
- **Energy –** such as kilowatt hours of electricity per person.
- **Agricultural yields** such as rice and wheat yields.
- **Communications and other per person indices** such as newspaper and telephones per population unit (per 1 000).
- **GNP –** the national product per country or per person.
- **Trade –** the value of foreign trade, or exports and imports per person.
- **Demographic** – such as population density, crude birth and death rates, population growth rates and infant mortality rates.

One of the earliest attempts to use multivariate analysis was carried out by the United Nations Research Institute on Social Development (**UNRISD**) in 1970. A **Social Development Index** was developed using 16 core indicators (9 social and 7 economic). Interestingly, this social development index correlated less highly with per person GNP for HICs than LICs. In particular, social development was notably falling behind economic development in some oil producing and exporting countries (OPECs , which had very high GNP per person figures. The major criticisms of the Social Development Index are:

- Its concentration on measuring **inputs**, for example, the numbers of doctors or teachers per 1 000 population, or the number of school enrolments.
- Some indicators, such as animal products consumption were inappropriate for LICs.

Quality of life indices

Table 14.3 Quality of life indices

The Physical Quality of Life Index (PQLI)	The **PQLI** is an average of three key characteristics – **literacy, life expectancy** and **infant mortality**. Each is scaled from 0 to 100. For example, literacy rates of zero to 100 per cent would be scaled as 0 to 100 respectively, exactly as in the raw data. However, with life expectancy and infant mortality, the scaling is done in a different way. Each year, the world's shortest life expectancy and highest infant mortality is scaled as 0 and the opposite of each as 100. The composite index for each country is then calculated by averaging the three rates, giving equal weightings to each.
The Human Development Index (HDI)	The Human Development Index (HDI) measures a country's average achievements in three basic aspects of human development: **health, knowledge,** and **income**. From 1980, the United Nations has worked on the construction and refinement of the **Human Development Index,** which it uses in its annual Human Development Reports. HDI attempts to rank all countries based on three goals/outputs which result from overall development: **1.** Longevity (life expectancy at birth). **2.** Knowledge (measured by a weighted average of which 66.6 per cent is from adult literacy and 33.3 per cent is from mean years of schooling). **3.** Income as adjusted to measure real per person income, including purchasing power adjusted to local cost of living. From this information it is then possible to rank countries into groups. However, the concept of human development is much broader than the information that can be captured in the HDI, or any other of the composite indices in the Human Development Report (Inequality-adjusted HDI, Gender Inequality Index and Multidimensional Poverty Index). The HDI does not reflect political participation or gender inequalities.
The International Human Suffering Index (IHSI)	The **International Human Suffering Index (IHSI)** was developed in 1987 by the Population Crisis Committee in Washington USA. The index measures development based on the 10 variables. A country is ranked from 0 to 10 for each of the following indicators (0 is very good, 10 very bad): • Life expectancy • Daily calorie supply • Access to clean water • Per person income • Civil rights • Political freedom • Inflation • Communications • Percentage in secondary school • Immunisation of infants A low score (lowest of 4 achieved by Switzerland in 1991) indicates minimal human suffering. This Index highlighted that many areas of extreme human suffering are found in Africa (the poorest continent). The index is successful in that the 10 indicators selected have been chosen to genuinely reflect the overall quality of life. However, some of the points awarded do rely more on qualitative data than is normal for such development indicators.

Table 14.3 Quality of life indices (continued)

Multi-dimensional Poverty Index (MPI)	The **Multidimensional Poverty Index (MPI)** was developed by the UN in 2010. It replaced the earlier Human Poverty Index (HPI). The **MPI** uses different factors to determine poverty beyond income-based lists. It complements traditional income-based poverty measures by capturing the severe deprivations that each person faces at the same time with respect to education, health and living standards. The MPI assesses poverty at the individual level. If someone is deprived in a third or more of ten (weighted) indicators, the global index identifies them as "MPI poor", and the extent – or intensity – of their poverty is measured by the number of deprivations they are experiencing. The MPI can be used to create a comprehensive picture of people living in poverty, and allows comparisons both across countries, regions and the world, and within countries by ethnic group, urban/rural location, as well as other key household and community characteristics. This makes it invaluable as an analytical tool to identify the most vulnerable people – the poorest among the poor, revealing poverty patterns within countries and over time, enabling policy makers to target resources and design policies more effectively. The following ten indicators are used to calculate the MPI: **Education** (each indicator is weighted equally at 1/6) **1.** Years of schooling: deprived, if no household member has completed five years of schooling **2.** Child school attendance: deprived, if any school-aged child is not attending school up to class 8 **Health** (each indicator is weighted equally at 1/6) **3.** Child mortality: deprived, if any child has died in the family **4.** Nutrition: deprived, if any adult or child for whom there is nutritional information is malnourished **Standard of Living** (each indicator is weighted equally at 1/18) **5.** Electricity: deprived, if the household has no electricity **6.** Sanitation: deprived, if the household's sanitation facility is not improved (according to MDG guidelines), or it is improved, but shared with other households **7.** Drinking water: deprived, if the household does not have access to safe drinking water (according to MDG guidelines) or safe drinking water is more than a 30-minute walk from home roundtrip **8.** Floor: deprived, if the household has a dirt, sand or dung floor **9.** Cooking fuel: deprived, if the household cooks with dung, wood or charcoal **10.** Assets ownership: deprived if the household does not own more than one radio, TV, telephone, bike, motorbike or refrigerator and does not own a car or truck.

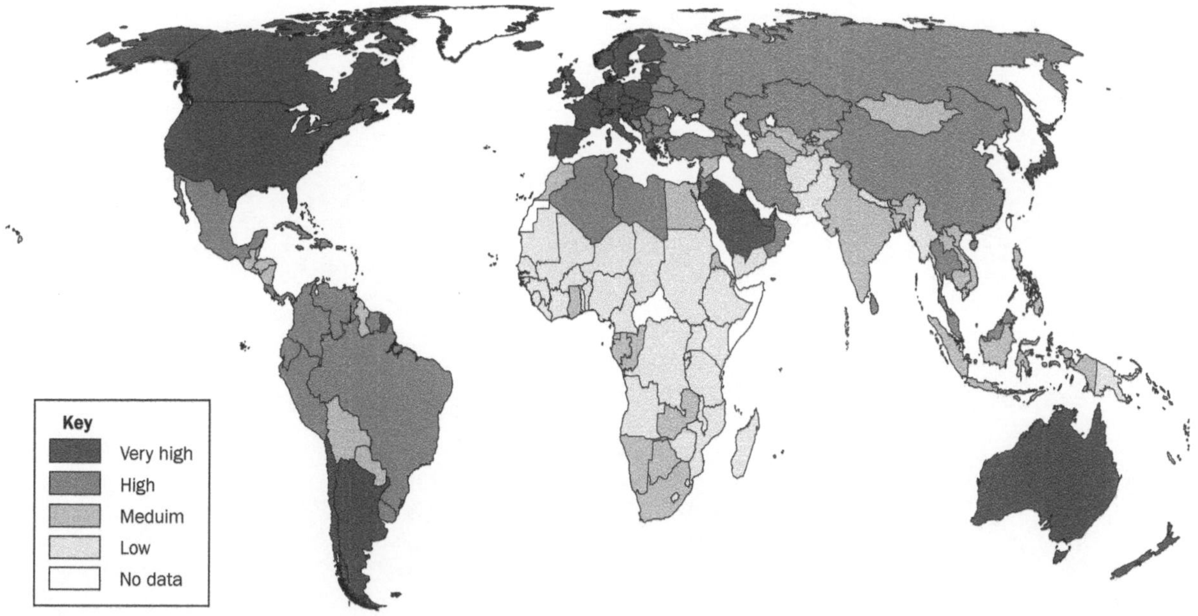

Fig 14.3 World map indicating the category of Human Development Index

Fig 14.3 is a world map indicating the category of Human Development Index by country, based on 2013 data. The darker the colour, the higher is the HDI for that country. Grey indicates no data available – includes Greenland, Somalia, North Korea and South Sudan.

The Happy Planet Index – a measure of well-being

The **Happy Planet Index (HPI)** shows how well countries use their natural resources to provide long and happy lives for their populations. It is made up of three indicators; **environmental impact, life expectancy** and **life-satisfaction**. There are strengths in using a multiple criteria index, such as HPI, compared to single criterion measures.

The strengths of the HPI include:

- Environmental impact – good to see included, pollution, degradation, etc.
- Life expectancy – relates to many elements, for example, housing, sanitation, health and education.
- Life satisfaction – relates to society, culture, "wealth does not necessarily make people happy", etc.

Its weaknesses include:

- The absence of a traditional economic measure.
- Difficulties in accessing such data in some countries.

The results for HPI show a dominance of South America, which is largely considered middle level development and "development" causing social and environmental problems in HICs, together with their higher expectations and greater inequality.

The advantages in using the HPI are that it is internationally recognised and accepted as a measure. The map also gives a useful overall impression. The disadvantages of the HPI are that it is only available at the national scale; therefore, it can hide any differences that exist between regions and gender. Also, the map has broad classes without numerical values, which limits its use.

Exam-style questions

1. Explain the advantages and limitations of using GDP per person as a measure of inequality. [6]
2. Using examples, describe and explain the part either the primary sector, or the tertiary sector can play in economic development. [8]
3. Explain why global inequalities in social and economic wellbeing are difficult to measure. [8]
4. Using examples, explain why some indicators of social and economic inequality are sometimes considered better than others. [8]

14.2 The globalisation of economic activity

What is meant by the term Foreign Direct Investment (FDI)?

Foreign Direct Investment (FDI) is an investment that is made to serve the business interests of an investor in a company in a different country from the investor's country. Normally, it involves a business and its foreign affiliate within a TNC and some element of interest and/or control. Different types of FDI may be identified, such as **greenfield** FDI, where there is an investment in new plant or facilities when starting up, or business and company **mergers**, which account for most FDI, enabling a TNC to expand overseas.

FDI can be **inward,** received, and **outward**, given/made. For example, the decision of Toyota, a Japanese TNC, the world's largest vehicle manufacturer, setting up a car assembly plant in Thailand, represents outward FDI from Japan, and inward FDI for Thailand.

There are certain political and economic circumstances which might discourage businesses putting FDI into a country. Business confidence is fundamental to FDI and this confidence may be affected by a number of circumstances:

- **Political** – for example, government instability, civil war, regime change, levels of government control, threats of terrorism, anti-globalisation and nationalism.
- **Economic** – for example, recession, debt, adverse exchange rates, loss of incentives, financial instability (which can be within a company or a country).

The New International Division of Labour

The **New International Division of Labour** (**NIDL**) is an outcome of globalisation. It is the spatial division of labour which occurs when the process of production is no longer confined to national economies. It is **new** in that it emerged recently associated with globalisation. It is **international** in that it takes place across countries in the global production network. The **division of labour** is due to the work being split up into different tasks/functions for efficiency. The term **NIDL** was put forward by theorists seeking to explain the spatial shift of manufacturing industries from advanced HIC countries to developing countries—an ongoing geographic reorganisation of production.

The "old" international division of labour, OIDL, reflected the colonial and immediate post-colonial realities that the industrialised societies of the West produced manufactured goods, while the rest of the world tended to produce one or two primary products per country. Under the OIDL, until around 1970, underdeveloped areas were incorporated into the world economy principally as suppliers of minerals and agricultural commodities. However, as developing economies are merged into the world economy, more production takes place in these economies.

The term New International Division of Labour (NIDL) may be referred to as the **international spatial division of labour** – with **international spatial** referring to between countries/across the world and the "**division of labour**" referring to the breakdown of production into jobs/tasks, to improve efficiency and therefore profitability.

The international spatial division of labour changes over time because of globalisation, changes in competitiveness between countries, changes in products and government policies. TNCs now look to LICs and NICs in order to cut their costs, with profitability being the key.

This division has led to a trend of transference, or what is also known as the **global industrial shift** or just **global shift**. This is where production processes are relocated from developed countries (the USA, Europe and Japan) to developing countries in Asia (for example China, Vietnam and India) and South America. This is because companies search for the cheapest locations to manufacture and assemble components, so low-cost labour-intensive parts of the manufacturing process are shifted to the developing world where costs are substantially lower.

NIDL occurs as a result of globalisation – the increasing interconnectedness of the world economy. As TNCs seek to remain competitive and to maximise their profits, they seek to minimising their costs.

NIDL changes over time because of globalisation, changes in competitiveness between countries, changes in products and changes in government policies.

NIDL has been driven by rising labour costs and high levels of industrial conflict in the West which reduced the profitability of **transnational corporations (TNCs)**.

Factors in the emergence and growth of newly industrialised countries (NICs)

Newly Industrialised Countries (NICs) are a varied group of countries, which have developed from being LICs, and have experienced the development of their secondary, tertiary (and quaternary) sectors.

Different generations of NICs have been identified, such as "the **Asian Tigers/Tiger Economies**" of some south-east Asian countries (like Singapore, Hong Kong, Taiwan and South Korea) Brazil, several of the East European countries and the **STIC's** or second tier income countries, such as Indonesia, Colombia and Turkey.

Factors that help to account for their emergence and growth include a combination of:

- **Social** factors, e.g. improved education and skills training, an existing work ethic.
- **Economic** factors, e.g. high levels of inward FDI, having stable currencies and the establishment of EPZs.
- **Environmental** factors, e.g. ease of accessibility and resources.
- **Political** factors, e.g. government planning, stability in a context of TNCs emerging and global production and markets.

The role of the governments in NICs has been put forward as the main factor in many NICs. It may be direct in terms of economic policy, budget priorities, offering financial incentives, membership of trade blocs, restricting the activities of trades unions and indirect, for example, in relation to investing in education and skills training, energy supply networks and infrastructure and investment in transport infrastructure. The government role may be observed spatially, for example in the existence of EPZs, SEZs, priority corridors and industrial estates.

Transnational Corporations (TNCs)

A global corporation, company or business with headquarters and research in one country and at least one, but often many more, branches and/or production centres in other countries.

Approximately 90 per cent of TNCs are based in HICs, especially the USA, the EU – mainly France, Germany, the UK, the Netherlands, Spain and Italy – Japan and Korea. The growth of Chinese companies, such as Huawei, will see China enter this group in the near future.

TNCs are very dominant players in the current global economy and are found in all sectors of industry. They directly employ around 45 million people and provide jobs indirectly for millions more workers and they currently control over 75 per cent of global trade, 40 per cent of which involves the movement of goods between units of the same TNC in different countries.

TNCs are very prominent in the primary industries where they:

- grow, process and distribute most of the world's food products
- harvest most of the world's timber and make most of its pulp and paper
- mine, refine and distribute most of the world's oil-based fuels
- extract most of the world's minerals
- and build most of the world's oil, gas, coal, HEP and nuclear power stations.

They are also responsible for the manufacturing of most of the world's motor vehicles, airplanes, chemicals, medicines, computers and home electronics, televisions, etc. They are also very prominent in the tertiary sector, supplying many of services linked with banking and finance, transport, tourism and, increasingly, in the education sector.

What are the reasons for the global spread of TNCs?

Over the past 35 years major technological advances in transport (including the development of containerisation, bulk carriers and air freight), along with developments in computerisation and telecommunications (satellites and Internet), have brought about the globalisation of the world's economy and the resultant growth in size and number of such TNCs.

What are the main factors which have encouraged the growth of TNCs?

There are a number of factors, in a context of globalisation, including:

- The stage of economic development and the emergence of new markets within an ever expanding global market.
- Changes in factors of production, for example,the ability to lower labour costs by moving some operations overseas.
- Improvements in transport technology which has led to a dramatic fall in the relative cost of transport.
- The relative ease and speed of transactions in global financial systems.
- Innovation in telecommunications and IT systems.
- Government role: reducing nationalism and protectionism and in giving incentives.
- A growing global consumer society and the role of media and advertising.

Why do TNCs operate in a wide variety of countries?

This can be explained in a number of ways:

- **Economic factors**, for example, comparative or competitive advantage and finding new markets.
- **Historical factors**, for example, colonial ties.
- **Political factors**, for example, economic colonialism.

Exam-style questions

1. Describe and explain the global organisation of one named TNC that you have studied. [8]
2. Explain the importance of an HIC location in the global organisation of one transnational corporation (TNC) that you have studied. [8]
3. Explain the term new international division of labour (NIDL). [6]
4. Explain the term foreign direct investment and describe why it occurs. [6]

CS Toyota

The Toyota Motor Company is part of a larger Toyota conglomerate. The automotive industry makes up over 90 per cent of the company's total sales. The remaining 10 per cent of its operations includes telecommunications, prefabricated housing (including earthquake resistant designs) and leisure boats. The location of its operations is a reflection of its corporate strategy plans and objectives with the aim of profit maximisation. Its business locations aim to increase its competitiveness and market penetration. Several of its manufacturing locations take advantage of government policies aimed at attracting the investment and jobs that a Toyota facility will bring to a country or region.

As can be seen by their operational locations, Toyota does not simply have high-end functions in MEDCs and low-end production functions in LICs.

The Toyota Motor Company was founded in 1937 (when it produced just 4 000 vehicles) by Sakichi Toyoda and by 1957 was exporting to the U.S. In 1984, Toyota began producing cars in the U.S. in a joint venture with General Motors.

Toyota was the largest automobile manufacturer in 2012 by production. In 2012, the company produced its 200 millionth vehicle. By 2013, Toyota was selling automobiles in over 160 countries and had 52 manufacturing companies in 27 countries on five continents, producing a full range of both parts and automobiles. Toyota also has Research and Development centres in the USA, Germany, France, UK, Spain, Belgium, Thailand, China and Australia, in addition to its Japanese facilities.

In 2014, Toyota had 333875 employees worldwide and was the fourteenth-largest company in the world by revenue.

Toyota, like many other TNCs, has realised the importance of creating a good public image and using environmentally-friendly practices and is famous for its **Toyota Production System**, with a main goal of eliminating waste. This has enabled Toyota to reduce pollution and production costs. Toyota's two factories in the USA have achieved "zero landfill status", as Toyota sells or gives away all waste products to companies that recycle the waste.

The Toyota Production System (TPS) is sometimes referred to as a **lean manufacturing** or **just-in-time** system, and is now used by many TNCs.

A Toyota employee, Taiichi Ohno, invented and developed the "just-in-time" philosophy. It allowed the company to reduce its parts inventory and efficiently produce only precise quantities of items, based on customer demand and with a minimum amount of waste.

The Toyota Product System relies on two basic concepts:

1. **Just-in-time (JIT) –** in which each part of the production process produces only what is needed by the next process in a continuous flow. When first introduced, this approach represented a radical departure from conventional manufacturing systems, which required large inventories in order to "push" as much product as possible through production lines, regardless of actual demand. The idea of JIT, on the contrary, was to produce only "what is needed, when it is needed, and in the exact amount needed" with the customer "pulling" production.
2. **Jidoka** (roughly translated as "automation with a human touch") – meaning when a problem occurs, the equipment stops immediately, preventing defective products from being produced. In addition, when a machine automatically stops, either because of a problem or because processing is completed, an alert is generated via display boards or other visual devices. Alternatively, operators who spot a problem are requested to pull the "and on", a cord hanging along the production line, to request immediate support. This allows all workers to easily identify the cause of the problem and prevent its recurrence.

The advantages of JIT and Jidoka is that they allow vehicles and products to be manufactured more efficiently and quickly.

By manufacturing products only as they are needed, and by keeping a tight control of quality, the Toyota Production System prevents waste and therefore reduces the amount of energy, raw materials and other resources used, making it a powerful asset in Toyota's approach towards sustainability.

It can lead to:

- A skilled and flexible workforce.
- A strong tradition of engineering and vehicle manufacturing and favourable working practices.
- A large domestic market for Toyota cars.
- Good transport links to customers and its 230 British and European supply partners.
- Ease of integration and communication, as English is very much the second language in Japan.
- Business and personnel support services to help the company and its people to integrate into the local communities.
- Local authorities in both locations assisted Toyota in providing an effective infrastructure, electricity, gas, water, telephones.
- A first class environment in which to live and work.
- A supportive positive attitude to inward FDI from the both the national and local government.

14.3 Regional development within countries

What is meant by the term core-periphery in relation to regional development?

J. Friedmann (1966) maintained that the world can be divided into four types of region:

1. **Core regions** are centres, usually metropolitan, with a high potential for innovation and growth, such as São Paulo in Brazil.
2. Beyond the cores are the **upward transition regions**, areas of growth spread over small centres rather than at a core. Development corridors are upward transition zones which link two core cities such as Belo Horizonte and Rio de Janeiro in Brazil.
3. The **resource-frontier regions** are peripheral zones of new settlement as in the Amazon Basin.
4. The **downward transition regions** are areas which are now declining because of exhaustion of resources or because of industrial change. Many "problem" regions of Europe are of this type.

This concept may be extended to continents. The capital-rich countries of Germany and France attract labour from peripheral countries like Spain, Greece, Turkey, and Algeria. Higher wages and prices are found at the core while the lack of employment in the periphery keeps wages low there. The result may well be a balance of payments crisis at the periphery, or the necessity of increased exports from the periphery to pay for imports. In either case, development of the periphery is retarded.

The model has been criticised in a number of ways. Most notably, it has been argued that uneven development is not the inevitable consequence of development, but of the particular mode of production being used to bring about that development, and that Friedmann's model represents the effects of the capitalist mode.

The **core** is the most developed region, both socially and economically, while the **periphery** is less developed and may be disadvantaged socio-economically.

The two are linked by flows of people, labour, materials and capital and a gradient may exist, as levels of development, economic activity and prosperity decrease from core to periphery.

How do "core" areas develop?

Several factors combine to lead to the development of core areas:

- Having environmental resources locally or within easy access, such as coal or iron ore.
- Historical initial development, possibly through a colonial legacy.
- The presence of elite groups and entrepreneurs.
- Multiplier effect, whereby wealth attracts development such as infrastructure and finance.
- Being the political seat of government and the prestige attached to this.

The flows into the growth core region can be explained as the growth core region is the first to develop and is therefore the most demanding in terms of labour, capital, and materials, for example, goods and commodities. It is also the region with the most opportunities, for example, jobs and potential for profit, so is the natural destination of flows. Therefore, the process of cumulative causation can be an important factor in developing a core region, with its initial advantage(s), leading to the attraction of labour, capital, innovation (and materials) with multiplier effects and spread effects occurring.

In looking at the development of one or more regions within a named country, it is useful to have an understanding of **cumulative causation**. The three key elements to look for are the initial advantages, spread (and backwash) effects and the cumulation, multiplier, up spiral or "snowball" phenomenon. It is possible to approach this using any region, core or peripheral.

A downward spiral (vicious circle) may occur in a peripheral region from which there is out-migration of labour. This process can take place between a poor rural area and an urban area, for example between parts of rural Kenya and Nairobi, or from parts of rural Tanzania to Dar es Salaam.

In contrast to this downward spiral witnessed in peripheral regions, a core region may experience an upward spiral (virtuous circle) from the inward migration of labour.

In the core region young and ambitious people arrive, boosting the labour force, making the core more attractive both for current economic activity and to new investment and (further) increasing core/periphery disparity. In association with this influx of new waged migrants, the local market and purchasing power increase which can lead to an increase in local services and to further attraction of activities.

Regional divergence and convergence

The increase in economic differences between regions is called **divergence**. A decrease in differences is called **convergence**.

The issue of convergence and divergence is of great importance in understanding economic development and economic growth.

What is meant by the terms spread effects and backwash effects?

- **Spread effects** are the movement of economic growth or initiatives out from the core, either spatially into the periphery or down the settlement hierarchy.
- **Backwash effects** are the negative impacts on or consequences for economy or economic growth in the periphery as a result of the tendency for resources, investment and labour being drawn into the core.

Spread effects may be a natural tendency, resulting from diseconomies in the core. Usually, spread has to be helped either by the government or authorities with the use of development projects and/or investment in the periphery. For example, initiating projects to establish growth poles, improving transport infrastructure, tor policies which will restrict growth in the core for example, through taxation and planning laws.

Changes may take place in the level of regional disparity within a country as it develops – sometimes referred to as the **core-periphery concept**. Initially, the country is economically undeveloped and the level of disparity is small. Theoretically, during economic development, regional disparities within a country may increase initially before decreasing. In the early stages of development there is a tendency, due to **cumulative causation**, for development to be concentrated in the core region(s) which benefits from **agglomeration economies** and the **multiplier effect**. Backwash effects, for example, selective migration, capital flows, etc., may enhance the level of disparity between core and periphery (indicated by the rising line). At the notional mid-point, disparities start to decline. As the country becomes more developed, government may have more capital (from greater tax yields) to spread development into peripheral regions. The core(s) may also suffer from overheating and diseconomies, which help to encourage businesses and entrepreneurs to seek better and more cost-effective locations in peripheral regions.

In reality, full convergence (to the right) may not be reached due to the disadvantages that the periphery may always have. Also, governments must balance the need to reduce disparities with attaining continued national economic growth and/or meeting other priorities.

Under what circumstances might it be possible to see capital, resources and labour move from the core to the periphery?

There are several circumstances in which this might take place:

- When **spread effects** operate. These can occur at an interim stage of development when, as the core expands, it reaches the point where diseconomies of scale set in and operating in the core becomes more difficult and costly. For example, as congestion increases, land prices may rise and businesses may decide to decentralise and find peripheral locations attractive.
- When **regional development initiatives** are put in place. Spread may be a natural tendency, as economies operate in the periphery against the diseconomies in the core, but in many cases, spread has to be helped and governments may attempt to reduce the congestion of the core and increase the development of the periphery or overcome regional disparities by using several types of schemes and initiatives, for example, in agriculture; mining; energy; dam building; manufacturing; improving transport infrastructure, encouraging tourism; establishing growth poles and by restricting growth in the core, for example, via taxation, planning laws, etc.
- When **urban-rural migration** occurs. For example, on retirement, especially of the more affluent who may innovate, or when family or community ties attract a young migrant to return home with their skills/savings to, for example, establish a new business.
- Through **remittances**. As part of the income of a rural-urban migrant is sent back to support the family. In the state of Kerala in southern India, 25 per cent of its GDP comes from remittances, much from the Gulf region and a great deal of this goes to their families in rural communities.

To what extent has it been possible to reduce regional disparities?

Some countries see regional disparity as the price that has to be paid for overall national development while other countries may be overwhelmed by the sheer scale of the backwash effects.

Others, possibly for political reasons, have little interest in reducing regional disparities. One indication of success may be seen by observing the tension between economic, social and political aspects which can result in differing priorities in reducing disparities.

Contents

AS Level

8 Waves

9 Interference

10 Particle physics

A Level

11 Motion in a circle

12 Oscillations

13 Gravitational fields

14 Electric fields

15 Capacitors

Introduction

Complete Physics aims to make your study of physics successful and interesting.

It has been written specifically to meet the requirements of the Cambridge International AS & A Level Physics syllabus (9702).

The book is divided into 24 chapters:

- Chapters 1–10 cover AS level.
- Chapters 11–24 cover A level.

New ideas are presented in the book in a careful step-by- step manner to allow you to develop a firm understanding of concepts and ideas. Key concepts that occur throughout the course are identified and discussed in detail at appropriate intervals in the book. The questions at the end of each chapter and after each key concept interval will help you to develop your grasp of the key concepts. These concepts are essential ideas, theories, principles or mental tools that help you to develop a deep understanding of their subject and make links between the different topics.

The key concepts in physics cover the following ideas.

- The development of **models of physical systems** is central to physics. Physics is the science that seeks to understand the behaviour of the Universe. Models simplify, explain and predict how physical systems behave.
- Physical models are usually based on prior observations or experiments, and their **predictions** are tested to check that they are consistent with **evidence** from further observations and experiments.
- **Mathematics** is used to express physical principles and models. It is also a **tool** to analyse theoretical models, solve quantitative problems and make predictions.
- Everything in the Universe comprises **matter** and/or **energy**. **Waves** transfer energy and are essential to many modern applications of physics.
- Matter and energy interact through **forces** and **fields**. The behaviour of the Universe is governed by fundamental forces that act over different length scales and magnitudes, ranging from the very small (particle physics) to the very large (cosmology).

Physics at AS and A Level will require you to describe and explain facts and processes in detail and with accuracy. The course is also about developing skills so that you can apply what you have learned.

The more basic topics are covered in the early chapters of the book and each chapter contains related topics. However, the topics do not have to be studied in strict numerical order, but can be followed in any sequence which suits you or your teacher.

The layout of the book is designed to cover information in a clear way that is easy to access. Its features include:

- **Accessible language** to improve your comprehension and understanding.
- **Purple type** to highlight words that can be found in the glossary. This enables you to easily access a full explanation of important terms used in the text. The comprehensive glossary provides you with clear and concise definitions of over 250 terms used throughout the book.
- **Extension material** helps to widen your horizons and stimulate an interest in broader aspects of physics. This extension material is **not** required for the Cambridge Assessment International Examinations (CAIE), but will serve you well in your further studies.
- **Notes** in boxes and the text to give you useful advice about certain aspects of a topic and so aid your learning. These may be helpful information or explanations to make concepts clearer.
- **Summary tests** to provide a quick check on how well you have learnt and understood the factual content of each topic. Answers are provided at the back of the book, so that the accurately completed test gives you a concise summary of the information in each topic.
- **End of chapter questions** to effectively consolidate the information learned across all of the topics in a chapter. Answers are provided at the back of the book.
- A **syllabus mapping grid** (p.vii) is provided showing the topic in the book that covers each syllabus topic. In addition, these pages also include an outline of the CAIE A level physics examinations structure.

In the Exam-style and Practice Questions sections at the end of each chapter, you will find the following icon:

In the Enhanced Online Student Book, this icon will launch additional digital resources to support your learning further. This content includes:

- **Worksheets** containing additional questions and practice
- **Interactive online quizzes**, including **multiple-choice** question practice
- Further support to improve your **mathematical skills**
- **Animations** illustrating key concepts

Visit **www.oxfordsecondary.com/bookshelf** to redeem your token code and access the Enhanced Online Student Book.

Answers to questions in this book and the syllabus matching grid are also available on the support website: **www.oxfordsecondary.com/caie-al-complete-science**

Specification grid

	Specification topic	Sub-topic	Book reference
1.1 1.2	Physical quantities SI units	1,2 1,2,3	***Part 1: Skills for starting physics***
1.3	Errors and uncertainties	1,2,3,4	***Part 2: Practical skills***
1.4	Scalars and vectors	1,2,3	1.1 Vectors and scalars
2.1	Equations of motion	1,4	1.2 Speed and velocity
		1,5,6	1.3 Acceleration
		5,6	1.4 Motion along a straight line at a constant acceleration
		7,8	1.5 Free fall
		3,4,5	1.6 Motion graphs
		7	1.7 More calculations on motion along a straight line
		9	1.8 Projectile motion 1
		9	1.9 Projectile motion 2
3.1	Momentum and Newton's laws of motion	1,2,5,6	2.1 Force and acceleration
		1,2,5	2.2 Using $F = ma$
		3,4,5	4.1 Momentum
3.2	Non-uniform motion	1,2	2.3 Terminal speed
3.3	Linear momentum and its conservation	2	4.2 Impact force
		1,2	4.3 Conservation of momentum
		1,2,3,4	4.4 Elastic and inelastic collisions
		1,2,4	4.5 Explosions
4.1	Turning effects of forces	1,2	3.2 The principle of moments
		3,4	3.3 More on moments
		1,2,3,4	3.6 Statics calculations
4.2	Equilibrium of forces	1	3.2 The principle of moments
		1	3.3 More on moments
		1,2,3,4	3.6 Statics calculations
		1,2,3	3.4 Stability
		1,2,3	3.5 Equilibrium rules
4.3	Density and pressure	1	5.1 Density
		2,3,4	5.2 Pressure
		5,6	5.3 Upthrust
5.1	Energy conservation	1,2	2.4 Work and energy
		5,6,7	2.6 Power
		3,4	2.7 Efficiency
		1–7	2.8 Renewable energy
5.2	Gravitational potential energy and kinetic energy	1–4	2.5 Kinetic energy and potential energy
		1,2,3,4	2.5 Kinetic energy and potential energy

Specification topic		Sub-topic	Book reference
6.1	Stress and strain	3,4,	5.4 Springs
		1,2,5,6	5.5 Deformation of solids
		1,2,5,6	5.6 More about stress and strain
6.2	Elastic and plastic behaviour	3,4	5.4 Springs
		1	5.5 Deformation of solids
		1,2,3	5.6 More about stress and strain
7.1		2,3,4	8.2 Measuring waves
7.2	Transverse and longitudinal waves	1,2	8.1 Waves and vibrations
7.3		1,2	8.9 The Doppler effect
7.4	Electromagnetic spectrum	1,2,3	8.1 Waves and vibrations 8.4 Electromagnetic waves and polarisation
7.5	Polarisation	2	8.4 Electromagnetic waves and polarisation
8.1	Stationary waves	1,2,3	8.5 Superposition
		2,3	8.6 Stationary and progressive waves
		2,3,4	8.7 More about stationary waves on strings
		2,3,4	8.8 Stationary waves in pipes
8.2	Diffraction	1,2	8.3 Wave properties
8.3	Interference	1, 3,4	9.1 Interference of light
8.4		1,2,3,4	9.2 More about interference
8.5	The diffraction grating	1,2	9.3 The diffraction grating
9.1	Electric current	1,2	6.1 Electric charge
		3,4	6.2 Current and charge
9.2	Potential difference and power	1,2,3	6.3 Potential difference and power
9.3	Resistance and resistivity	1,2,3,5,6	6.4 Resistance
		3,4,7,8	6.5 Components and their characteristics
10.1	Practical circuits	1,2	6.3 Potential difference and power
		3,4,5	7.3 E.m.f. and internal resistance
10.2	Kirchoff's laws	1,2 3,4,5,6,	7.1 Circuit rules 7.2 More about resistance
10.2		4,6,7	7.4 More circuit calculations
10.3	Potential dividers	1,2,3,4	7.5 The potential divider
11.1	Atoms,nuclei and radiation	1,2	10.1 The discovery of the nucleus
		2,3,4,5	10.2 Inside the atom
		7,9	10.3 The properties of α, β and γ radiation
		7,8,9,10,11	10.4 More about α, β and γ radiation
11.2	Fundamental particles	1,2,3,4,5,6	10.6 Fundamental particles
12.1	Kinematics of uniform circular motion	1,2,3	11.1 Uniform circular motion
	Centripetal acceleration	1,2	11.2 Centripetal acceleration
		1,,3,4	11.3 On the road
		1, 3,4	11.4 At the fairground
13.1	Gravitational field	1,2	13.1 Gravitational field strength

Specification topic		Sub-topic	Book reference
13.2	Gravitational force between point masses	3,4	13.3 Newton's law of gravitation 13.5 Satellite motion
13.3	Gravitational field of a point mass	1,2,3	13.4 Planetary fields
13.4	Gravitational potential	1,2,3	13.2 Gravitational potential 13.4 Planetary fields
14.1 14.2	Thermal equilibrium Temperature scales	1,2 1,2,3,4	18.2 Temperature scales
14.3	Specific heat capacity and change of state	1 2	18.3 Specific heat capacity 18.4 Change of state
15.1	The mole	1,2	19.2 The ideal gas law
15.2	Equation of state	1,2,3	19.1 Experiments on gases 19.2 The ideal gas law
15.3	Kinetic theory of gases	1,2,3,	19.3 The kinetic theory of gases
16.1	Internal energy	1,2	18.1 Internal energy and temperature 19.4 Thermodynamics of ideal gases
16.2	The first law of thermodynamics	1,2	18.1 Internal energy and temperature 19.4 Thermodynamics of ideal gases
17.1	Simple harmonic oscillations	1	12.1 Measuring oscillations
		1,2,5	12.2 The principles of simple harmonic motion
		1,2,3,4,5	12.3 More about sine waves
		1,2,3,4,5	12.4 Applications of simple harmonic motion
17.2	Energy in simple harmonic motion	1,2	12.5 Energy and simple harmonic motion
17.3	Damped and forced oscillations, resonance	3	12.6 Forced oscillations and resonance
18.1 18.2	Electric field and field lines Uniform electric fields	1,2,3 1	14.1 Electric field strength
18.3	Electric force between two point charges	1,2	14.1 Coulomb's law
18.4	Electric field of a point charge	1	14.3 Electric field strength of point charges
18.5	Electric potential	1,2, 3,4	14.2 Electric potential
19.1	Capacitors and capacitance	1,2	15.1 Capacitance
		3,4	15.2 Capacitors in series and parallel
19.2	Energy stored in a capacitor	1,2	15.3 Energy stored in a charged capacitor
19.3	Discharging a capacitor	1,2,3	15.4 Capacitor discharge
20.1	Concept of a magnetic field	1,2	16.1 Magnetic field patterns
20.2	Force on a current carrying conductor	1,2,3	16.2 The motor effect 16.3 Magnetic flux density
20.3	Force on a moving charge	1,2,3,4,5,6	16.4 Moving charges in a magnetic field 16.5 Charged particles in circular orbits
20.4	Magnetic fields due to currents		16.1 Magnetic field patterns
20.5	Electromagnetic induction	1,2,3,4,5	17.1 Generating electricity 17.2 The laws of electromagnetic induction
21.1	Characteristics of alternating currents	1,2,3,4	19.3 Alternating current

	Specification topic	Sub-topic	Book reference
21.2	Rectification and smoothing	1,2,3,4	19.3 Alternating current
22.1	Energy and momentum of the photon	1,2,3	20.1 Photoelectricity
		5	20.4 Wave–particle duality
22.2	Photoelectric effect	4	20.2 More about photoelectricity
22.3	Wave-particle duality	1,2,3,4,5	20.4 Wave–particle duality
22.4	Energy levels and line spectra	1,2,3	20.3 Energy levels and spectra
23.1	Mass defect and binding energy	1,2	21.4 Energy from the nucleus
		3,4,5,6	21.5 Binding energy
		5,6,7	21.6 Fission and fusion
23.2	Radioactive decay	1,2,3	21.1 Radioactive decay
		3,4,5,6	21.2 The theory of radioactive decay
		3,4,5,6	21.3 Radioactive isotopes in use
24.1	Production and uses of ultraound	1,2,3,4,5,6	22.1 Ultrasonic imaging
24.2	Production and use of X-rays	1,2,3,4	22.2 X-rays
24.3	PET scanning	1,2,3,4,5,6,	22.3 PET scanning
25.1	Standard candles	1,2,3,4	23.1 Astronomical distances
25.2	Stellar radii	1,2,3	23.2 Stellar radii
25.3	Hubble's theory and the Big Bang theory	1,2,3,4	23.3 Hubble's Law and the Big Bang

Summary of the CAIE AS and A Level physics examination assessment structure

The assessment objectives (AOs) are:

- **AO1** Knowledge and understanding
- **AO2** Handling, applying and evaluating information
- **AO3** Experimental skills and investigations

The assessment objectives are examined through the following papers.

Paper 1: Multiple choice Written paper, 1 hour 15 minutes, 40 marks	Questions are based on the AS Level syllabus content.	40 multiple-choice items of the four-choice type, testing assessment objectives AO1 and AO2
Paper 2: AS Level structured questions Written paper, 1 hour 15 minutes, 60 marks	Questions are based on the AS Level syllabus.	Structured questions testing assessment objectives AO1 and AO2 content
Paper 3: Advanced practical skills Practical test, 2 hours, 40 marks	This paper tests assessment objective AO3 in a practical context. Question 1 involves carrying out an experiment in which measurements are made, recorded and used in calculations and/or to plot graphs to draw conclusions. Question 2 involves carrying out an inaccurate experimental method, evaluating it and suggesting improvements.	Two questions assess the practical skills listed in the practical assessment section of the syllabus. The content of the questions may be outside the syllabus content.
Paper 4: A Level structured questions Written paper, 2 hours, 100 marks	Structured questions testing assessment objectives AO1 and AO2. Questions are based on the A Level syllabus.	Knowledge of material from the AS Level syllabus content will be required.
Paper 5: Planning, analysis and evaluation Written paper, 1 hour 15 minutes, 30 marks	Two questions testing assessment objective AO3. Questions are based on the A Level practical skills of planning, analysis and evaluation.	Knowledge of practical skills from the AS Level syllabus may be required. The content of the questions may be outside of the syllabus content.

Skills for starting physics Part 1

1. Using a calculator

Practice makes perfect when it comes to using a calculator. For A level Physics, you need no more than a scientific calculator. You need to make sure you master the technicalities of using a scientific calculator as early as possible in your A level Physics course. At this stage, you should be able to use a calculator to add, subtract, multiply, divide, find squares and square roots and calculate sines, cosines and tangents of angles. Further important calculator functions are described in Chapter 24 Mathematical skills.

2. Making measurements

You should know at this stage how to make measurements using basic equipment such as metre rules, protractors, stopwatches, thermometers, measuring cylinders, balances (for weighing an object) and ammeters and voltmeters. During the course, you will also be expected to use micrometers, (vernier) calipers, centre-reading meters (galvanometers), oscilloscopes and data-loggers, and Hall probes for measuring magnetic fields (A2 only). The use of these items is described for reference in part 2 of this introduction, on p.xvi–xxii (except for the Hall probe: see p.245).

Some useful reminders about making measurements:

- Check the **zero reading** when you use an instrument to make a measurement. For example, a metre ruler worn away at one end might give a zero error or a digital meter or analogue meter (i.e. a pointer with a scale) might not read zero when the input to the meter is zero.
- When a multi-range instrument is used, start with the **highest range** and switch to a lower range if the reading is too small to measure accurately.
- Make sure you record all your measurements in a **logical order**, stating the correct unit of each measured quantity.
- Don't pack equipment away until you are sure you have enough measurements or you have checked unexpected readings. See p.xxiv for **anomalous** measurements.

How to obtain accurate results from an experiment

- Make your measurements as precisely as possible and repeat them if possible to ensure they are reliable. See p.xv–xvi for more about **precision** and **reliability**.
- Reduce experimental errors in making measurements. Such errors are either:
  1. Random errors that cause repeat readings to differ at random and are caused by uncontrolled variables,
  2. Systematic errors such as zero errors that cause readings to be consistently higher or lower than they should be. See p.xvi for more about random and systematic errors.
- Where measurements are repeated, calculate and use the **mean values**. Assess the uncertainty in a measurement from the spread of the readings about the mean value. See p.xxiii for more about how to assess and use **uncertainties** in your measurements to find out how **accurate** your results are.

3. Using measurements in calculations

When you record a measurement, you must always note the correct unit as well as the numerical value of the measurement.

Learning outcomes

On these pages you will learn to:

- appreciate that all physical quantities consist of a numerical magnitude and a unit
- know the unit of each physical quantity listed in the syllabus
- make reasonable estimates of physical quantities in the syllabus
- recall the following SI base quantities and their units: mass (kg), length (m), time (s), current (A), temperature (K), amount of substance (mol) (Topic 19.2)
- express derived units in terms of SI base units
- use SI base units to check the homogeneity of physical equations
- use the following prefixes and their symbols to indicate decimal submultiples or multiples of both base and derived units: pico (p), nano (n), micro (μ), milli (m), centi (c), deci (d), kilo (k), mega (M), giga (G), tera (T)
- use the scientific conventions for labelling graph axes and table columns

The scientific system of units is called the SI system. The SI system is described in more detail later in this introduction, on p.xiv. The base units of the SI system you need to remember are listed below. All other units are derived from the SI base units. The symbol for each unit is shown in brackets. Later in your course you will also meet the kelvin (K), which is the SI unit of temperature (see p.272), and the mole (mol) which is the SI unit for the amount of a substance (see p.287).

- The metre (m) is the SI unit of length. Note also that 1 m = 100 cm = 1000 mm.
- The kilogram (kg) is the SI unit of mass. Note that 1 kg = 1000 g.
- The second (s) is the SI unit of time.
- The ampere (A) is the SI unit of current.

All other units are derived from the SI base units. For example,

- the SI unit of speed is the metre per second because speed = distance ÷ time. The unit is abbreviated as $m\,s^{-1}$ because it is the unit of distance (m) ÷ the unit of time (s)
- the SI unit of force is the newton (N). Because 'force = mass × acceleration' as explained in Topic 2.1, the newton expressed in terms of base units is the kilogram metre per second squared ($kg\,m\,s^{-2}$).

See p.xiii–xiv for more about derived units.

Powers of ten and numerical prefixes are used to avoid unwieldy numerical values. For example,

- $1\,000\,000 = 10^6$ which is 10 raised to the power 6 (usually stated as '10 to the 6')
- $0.000\,000\,1 = 10^{-7}$ which is 10 raised to the power −7 (usually stated as '10 to the −7').

Prefixes are used with units as abbreviations for powers of ten. For example, a distance of 1 kilometre may be written as 1000 m or 10^3 m or 1 km. The most common prefixes are shown in Table 1.

Table 1 *Prefixes*

Prefix	pico-	nano-	micro-	milli-	kilo-	mega-	giga-	tera-
Value	10^{-12}	10^{-9}	10^{-6}	10^{-3}	10^{3}	10^{6}	10^{9}	10^{12}
Prefix symbol	p	n	μ	m	k	M	G	T

Note that the prefix c for centi- stands for 10^{-2} or a hundredth (as in cm for centimetre) and prefix d for deci- stands for 10^{-1} or a tenth. For example, $1\,cm^3 = 1 \times (10^{-2})^3 = 10^{-6}\,m^3$.

Also, note that cubic centimetre (cm^3) and the gram (g) are in common use and are therefore allowed as exceptions to the prefixes in Table 1.

Standard form is usually used for numerical values smaller than 0.001 or larger than 1000.

- The numerical value is written as a number between 1 and 10 multiplied by the appropriate power of ten. For example:

$$64\,000\,m = 6.4 \times 10^4\,m$$
$$0.000\,005\,1\,s = 5.1 \times 10^{-6}\,s$$

- A prefix may be used instead of some or all of the powers of ten. For example,

$$35\,000\,m = 35 \times 10^3\,m = 35\,km$$
$$0.000\,000\,59\,m = 5.9 \times 10^{-7}\,m = 590\,nm$$

To convert a number to standard form, count how many places the decimal point must be moved to make the number between 1 and 10. The number of places moved is the power of ten that must accompany the number between 1 and 10. Moving the decimal place:

- to the left gives a positive power of ten (e.g. $64\,000 = 6.4 \times 10^4$)
- to the right gives a negative power of ten (e.g. $0.000\,005\,1 = 5.1 \times 10^{-6}$).

4. Using trigonometry

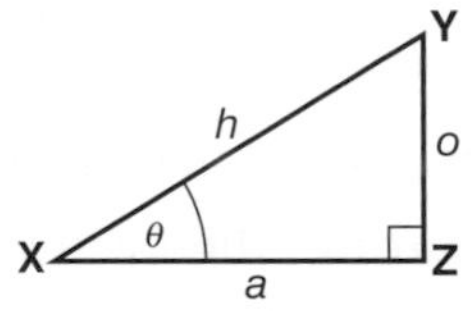

Figure 1 *A right-angled triangle*

The sine, cosine and tangent of an angle are defined from the right-angled triangle. Figure 1 shows a right-angled triangle XYZ, in which side XY is the **hypotenuse** (i.e. the side opposite the right angle), side YZ is **opposite** to angle θ and side XZ is **adjacent** to angle θ.

$$\sin\theta = \frac{YZ}{XY} = \frac{o}{h} \quad \text{where } o = YZ\text{, the side opposite angle } \theta$$

$$\cos\theta = \frac{XZ}{XY} = \frac{a}{h} \quad h = XY\text{, the hypotenuse}$$

$$\tan\theta = \frac{YZ}{XZ} = \frac{o}{a} \quad a = XZ\text{, the side adjacent to angle } \theta$$

Notice that when:

i $\theta = 0$, YZ = 0 so $\sin\theta = 0$ and XZ = XY so $\cos\theta = 1$

ii $\theta = 90°$, YZ = XY so $\sin\theta = 1$ and XZ = 0 so $\cos\theta = 0$

Pythagoras' theorem states that for any right-angled triangle:

The square of the hypotenuse = the sum of the squares of the other two sides

Applying Pythagoras' theorem to the right-angled triangle XYZ gives:

$$\mathbf{XY^2 = XZ^2 + YZ^2}$$

5. Symbols in science

Symbols are used in equations and formulae to represent physical quantities. In your previous course, you may have used equations with words instead of symbols to represent physical quantities. For example , you will have met the equation 'distance moved = speed × time'. Perhaps you were introduced to the same equation in the symbolic form '$s = vt$', where s is the symbol used to represent distance , v is the symbol used to represent speed and t is the symbol used to represent time. The equation in symbolic form is easier to use because the rules of algebra are more easily applied to it than to a word equation.

If you used symbols in your GCSE course, you might have met the use of s for distance and I for current. Maybe you wondered why we don't use d for distance instead of s; or C for current instead of I. The answer is that physics discoveries have taken place in many countries. The first person to discover the key ideas about speed was Galileo, the great Italian scientist; he used the word '*scale*' from his own language for distance, and therefore assigned the symbol s to distance. Important discoveries about electricity were made by Ampère, the great French scientist; he wrote about the intensity of an electric current, so he used the symbol I for electric current. The symbols we now use are used in all countries in association with the **SI system** of units.

Table 2

Physical quantity	Symbol	Unit	Unit symbol
Distance	s	metre	m
Speed or velocity	v	metre per second	$\mathrm{m\,s^{-1}}$
Acceleration	a	metre per second per second	$\mathrm{m\,s^{-2}}$
Mass	m	kilogram	kg
Force	F	newton	N
Energy or work	E	joule	J
Power	P	watt	W
Density	ρ	kilogram per cubic metre	$\mathrm{kg\,m^{-3}}$
Current	I	ampere	A
Potential difference or voltage	V	volt	V
Resistance	R	ohm	Ω

A full list of all the physical quantities, their symbols and units, that you are likely to meet in the AS or A level Physics course is given at the end of Topic 24.3 More about algebra.

6. Using equations

Equations often need to be rearranged. This can be confusing if you don't learn the following basic rules at an early stage:

- **Read an equation properly**. For example, the equation $v = 3t + 2$ is not the same as the equation $v = 3(t + 2)$. If you forget the brackets when you use the second equation to calculate v when $t = 1$, then you will get $v = 5$ instead of the correct answer $v = 9$. The first equation tells you to multiply t by 3 then add 2. The second equation tells you to add t and 2, then multiply the sum by 3.
- **Rearrange an equation properly**. In simple terms, always make the same change to both sides of an equation. For example, to make t the subject of the equation $v = 3t + 2$:

Step 1: Subtract 2 from both sides of the equation,

so $v - 2 = 3t + 2 - 2$
$= 3t$

Step 2: The equation is now $v - 2 = 3t$ and can be written

$3t = v - 2$

Step 3: Divide both sides of the equation by 3,

so $\frac{3t}{3} = \frac{v-2}{3}$

Step 4: Cancel 3 on the top and the bottom of the left-hand side, to finish with $t = \frac{v-2}{3}$

To use an equation as part of a calculation:

- Start by making the quantity to be calculated the **subject** of the equation.
- Write the equation out with the **numerical values** in place of the symbols.
- Carry out the calculation and make sure you give the answer with the **correct unit.**

Unless the equation is simple (e.g. $V = IR$), don't insert numerical values then rearrange the equation. Rearrange, then insert the numerical values; you are less likely to make an error if the numbers are inserted later in the process.

Estimating the result of a measurement or calculation is a useful technique to ensure big errors have not been made. For example, in a density measurement of a metal object, you would expect the result to be several times the density of water. A result of $800\,\mathrm{kg\,m^{-3}}$ compared with a value of $1000\,\mathrm{kg\,m^{-3}}$ for water means an error must have been made. Before carrying out an exact calculation of a physical quantity, a mental estimate of the approximate result of the calculation gives an 'order of magnitude' value. Combined with an awareness of the magnitudes of physical quantities, errors can thus be spotted and corrected before carrying out an accurate calculation. See Chapter 24 Mathematical skills for more about signs, equations and formulae.

7. More about SI units

In the SI system of units, every derived unit can be expressed in terms of the **base units** of the SI system. Table 3 at the end of Topic 24.3 More about algebra, shows how the derived units you meet at A level are related to the SI base units and, in some cases, how they relate to each other. For example, in terms of base units, the coulomb (C) is the ampere second (A s). Such knowledge can prove useful. For example, if you can't quite recall the formula for:

- Density, but you know that its unit is the kilogram per cubic metre ($kg\,m^{-3}$), you should be able to see that $kg\,m^{-3}$ is the unit for mass divided by volume.
- Centripetal acceleration, remembering that the unit of acceleration (i.e. $m\,s^{-2}$) is the same as the unit of speed^2 divided by the unit of distance (i.e. $m^2\,s^{-2}/m$) can lead you to the formula v^2/r for centripetal acceleration. See Topic 11.2 if necessary.

Derived units written in terms of their base units can be used to check equations. The physical quantities on each side of an equation must match in terms of base units. If they don't match, the equation cannot be correct. For example, consider:

- the equation $v = \sqrt{2gR}$, which is used to calculate the escape speed v of an object from the surface of a planet of radius R and surface gravitational field strength g
  Left-hand side base units = $m\,s^{-1}$
  Right-hand side base units = $\sqrt{ms^{-2} \times m} = m\,s^{-1}$
  The equation has the same combination of base units on each side, so it is correct; we say it is **homogeneous** in terms of the base units.
- the equation $W = QV$ is used to calculate the work done W to move a charge Q through a potential difference V
  Left-hand side base units (see Table 3) = $kg\,m^2\,s^{-2}$
  Right-hand side base units = $(A\,s) \times (kg\,m^2\,s^{-3}\,A^{-1})$
  $= kg\,m^2\,s^{-2}$
  The equation has the same combination of base units on each side, so it is homogeneous. Note that for simple equations such as this, homogeneity can be checked more quickly by recalling basic relationships between physical quantities. In this example, one volt is one joule per coulomb, so the unit of QV is the joule per coulomb × the coulomb, which is the joule.

The links between different units do not need to be made through the SI base units. For example, the volt (V) is the joule per coulomb ($J\,C^{-1}$), which is a useful link to remember as it helps you to develop your understanding of potential difference.

There are some units in the A level specification that are not SI units but they are used in specific situations for convenience. Those listed below are in common use.

1. The **atmosphere** (atm) is a unit of pressure equal to the mean pressure of the atmosphere at sea level and is equal to 101 kPa, or 1.01×10^5 Pa.
2. The **electron volt** (eV) is a unit of energy defined as the work done when an electron moves through a p.d. of 1 V.
   $1\,eV = 1.6 \times 10^{-19}\,J$.
   Note that $1\,MeV = 10^6\,eV = 1.6 \times 10^{-13}\,J$
3. The **kilowatt hour** (kW h) is a unit of energy equal to the energy supplied to a 1 kilowatt appliance in 1 hour which is 3.6 MJ.
4. The **light year** is the distance travelled in space by light in 1 year.
5. The **litre** is a unit of volume equal to $10^{-3}\,m^3$.

In the laboratory

The experimental skills you will develop during your course are part of the 'tools of the trade' of every physicist. Data loggers and computers are commonplace in modern physics laboratories, but awareness on the part of the user of precision, reliability, errors and accuracy are just as important as when measurements are made with much simpler equipment. The practical skills you develop in the laboratory will prepare you for the CIE Paper 3 practical tests and for the CIE Paper 5 written practical examination. Let's consider in more detail what you need to be aware of when you are working in the physics laboratory.

Safety and organisation

A science laboratory should be a safe place, but it can be dangerous if the safety rules and common sense are disregarded. Your teacher will give you a set of safety rules and should explain them to you. You must comply with them at all times. In addition, you must use your common sense and organise yourself so that you work safely. For example, if you set up an experiment with pulleys and weights, you need to ensure they are stable and will not topple over. Before you set up an electrical circuit, think about where to place the meters so you can read them easily.

Working with others

Most scientists work in teams, each person cooperating with the other team members to achieve specific objectives.

In your practical activities, you will often work in a small group of two or three people. In such group work, you need to cooperate with the others in the group so everyone understands the objectives of the practical activity and everyone participates in planning and carrying out the activity.

Planning

You will be asked to plan an experiment or investigation in Paper 5. The practical activities you carry out during your course should enable you to prepare a plan. Here are the key steps in drawing up a plan:

1. Decide in detail what you intend to investigate. Identify the variable you will alter (the **independent variable**) and the variable that will be altered as a result (the **dependent variable**). Note these variables and note the other variables that need to be controlled and kept constant. A **control variable** that can't be kept constant would cause the dependent variable to alter.
2. Select the equipment necessary for the measurements. Specify the range of any electrical meters you need.
3. List the key stages in the method you intend to follow and make some preliminary measurements to check your initial plans. Consider safety issues before you do any preliminary tests. If necessary, modify your plans as a result of your preliminary tests.
4. If the aim of your investigation is to test a hypothesis or theory or use the measurements to determine a physical quantity (e.g. resistivity), you need to describe how you will use your measurements to achieve the objectives of your plan.

Carrying out instructions and recording your measurements

In some investigations, and in the CIE Paper 3 tests, you will be expected to follow instructions supplied to you, either verbally or on a worksheet. You should be able to follow a sequence of instructions without guidance. However, always remember safety first and, if the instructions are not clear, ask your teacher to clarify them.

When you record your measurements, tabulate them with a column for the independent variable and one or more columns for the dependent variable. Remember to allow for repeat readings and average values, if appropriate. Include columns for values of quantities calculated from your measurements. The table should have a clear heading for each of the measured variables, with the unit symbol shown after the heading (e.g. current/A or I/A), as below.

Single measurements of other variables (e.g. control variables) should be recorded together, immediately before or after the table. Make sure you record the unit of each measurement. In addition, you should record the precision (i.e. the least detectable reading) of each measurement (which should be the same for each measurement of the same measured variable). This information is important when you come to analyse and evaluate your measurements.

Table 1 *Tabulating the measurements from an investigation of p.d. against current for a wire*

potential difference / V	current / A			average current / A
	1st set	2nd set	3rd set	

length of wire/m = ________
diameter of wire/mm = ________, ________, ________
average diameter of wire/mm = ________

Evaluating your results

You should be able to form a conclusion from the results of an investigation. This might be a final calculation of a physical quantity or property (e.g. resistivity) or a statement of the relationship established between two variables. As explained earlier, the degree of accuracy of the measurements could be used as a guide to the number of significant figures in a 'final result' conclusion. Mathematical links established or verified between quantities should be stated in a 'relationship' conclusion.

You always need to evaluate the conclusion(s) of an experiment or investigation to establish its validity. This evaluation could start with a discussion of the strength of the experimental evidence used to draw the conclusions:

- Discuss the reliability of the data and suggest improvements, where appropriate, that would improve the reliability. You may need to consider the effect of the control variables, if the experimental evidence is not as reliable as it should be.
- Discuss the methods taken (or proposed) to eliminate or reduce any random or systematic errors. Describe the steps taken to deal with anomalous results.
- Evaluate the accuracy of the results by considering the percentage uncertainty in the measurements. These can be compared to identify the most significant sources of uncertainty in the measurements, which can then lead to a discussion of how to reduce the most significant sources of uncertainty.

A Level

The practical skills assessed in Paper 5 involve planning an experiment, carrying out the plan and evaluating the results. Although planning skills are in Paper 5, such skills can be developed throughout the course. For this reason, information about planning an experiment is given in Practical Skills, p.xv. The section below concentrates on analysis, conclusions and evaluation skills beyond the Paper 3 requirements.

More about uncertainties

When a measurement in a calculation is raised to a power n, the percentage uncertainty is increased n times. For example, suppose you need to calculate the area A of cross-section of a wire that has a diameter of 0.34 ± 0.01 mm. You will need to use the formula $A = \pi d^2/4$. The calculation should give an answer of $9.08 \times 10^{-8}\,\text{m}^2$. The percentage uncertainty of d is $0.01/0.34 \times 100\% = 2.9\%$. So the percentage uncertainty of A is 5.8% ($= 2 \times 2.9\%$). The consequence of this rule is that in any calculation where a quantity is raised to a higher power, the uncertainty of that quantity becomes much more significant.

When two or more quantities are multiplied or divided by each other in a calculation, the overall percentage uncertainty in the result is the sum of the uncertainties of each quantity. For example, if the uncertainty in a resistance R is 5% and in a capacitance C is 4%, the uncertainty in RC is 9%.

Percentage uncertainties

To work out the percentage uncertainty of A, you could calculate:

- the area of cross-section A where $d = 0.34 - 0.01\,\text{mm} = 0.33\,\text{mm}$

  This should give an answer of $8.55 \times 10^{-8}\,\text{m}^2$.

- the area of cross-section A where $d = 0.34 + 0.01\,\text{mm} = 0.35\,\text{mm}$

  This should give an answer of $9.62 \times 10^{-8}\,\text{m}^2$.

Therefore, the area lies between $8.55 \times 10^{-8}\,\text{m}^2$ and $9.62 \times 10^{-8}\,\text{m}^2$.

In other words, the area is $(9.08 \pm 0.53) \times 10^{-8}\,\text{m}^2$

(as $9.08 - 0.53 = 8.55$ and $9.08 + 0.53 = 9.62$).

The percentage uncertainty of A is $0.53/9.08 \times 100\% = 5.8\%$.

This is twice the percentage uncertainty of d.

It can be shown as a general rule that for a measurement x:

the percentage uncertainty in x^n is n times the percentage uncertainty in x.

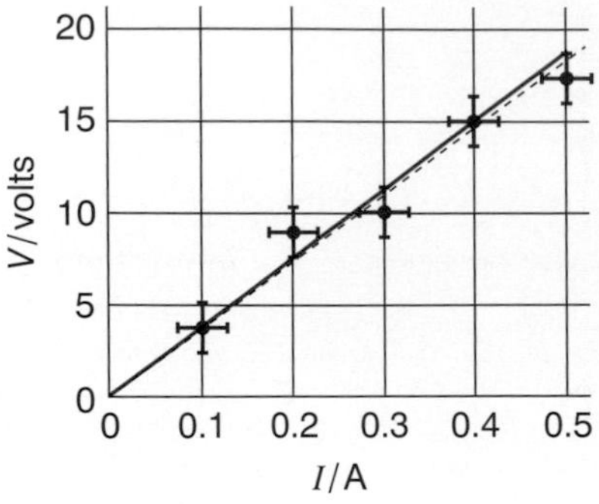

Figure 2 Error bars

Notes

1. An example of the worst acceptable line on a graph is shown in Figure 2 as the dashed line. In this case it still passes through the origin as the current was definitely zero when the voltage was zero. This might not always be the case. For example, in an investigation of a spring, there is an uncertainty in the unstretched length of a spring which gives an uncertainty at zero extension.
2. In the above example, check for yourself that the gradient of the dashed line is about 2.5% less than the gradient of the best-fit line. The gradient of the best-fit line is $37.5\,\Omega$ ($= 15.0\,V/0.40\,A$). So the uncertainty in the gradient is $\pm 0.9\,\Omega$ (0.9 rounded down from 0.9375 as the gradient value is given to one decimal place.)

More about Graphs

Error bars

The uncertainty of each measurement can be used to give a small range or **error bar** for each measurement. Figure 2 shows the idea. **A straight line of best fit** should pass through all the error bars such that there is an even distribution of points either side of the line. To estimate the uncertainty in the gradient, you could draw the steepest straight line of (or least steep straight line) passing through all the error bars. The uncertainty of the gradient is the difference between the gradient of this worst acceptable line and the 'even distribution' best fit line.

For graphs in which the points are clearly not in a straight line:

- The points may lie along a straight line over most of the range, but the points curve away from the straight line further along. If so, a straight-line relationship between the plotted quantities is valid only over the range of measurements which produced the straight part of the line.
- Two or three points might seem to lie on a straight line (see Figure 3). In this case, it cannot be concluded that there is a **linear** relationship between the plotted quantities.

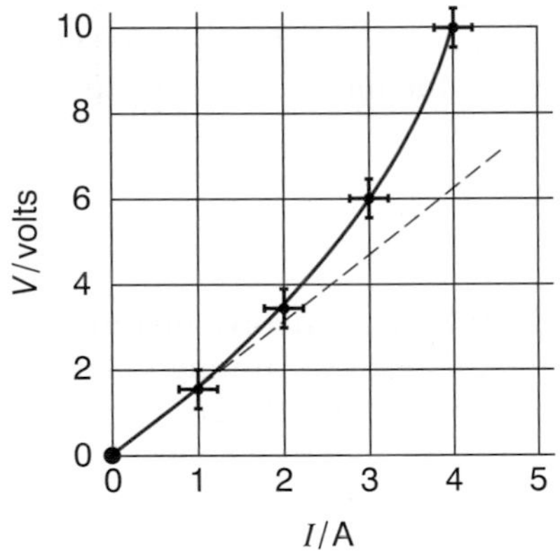

Figure 3 Curves

If there is a known mathematical equation between the two plotted quantities, this equation could be rearranged into a linear equation in order to produce a straight-line graph.

Some examples are:

1. In a falling ball experiment, the time t taken for a small ball to fall from rest through two light gates at different distances s apart is measured. The top light gate would need to be a fixed distance below the point of release of the ball. The kinematics equation $s = ut + ½\, at^2$ would give a curved graph if distance s is plotted on the y-axis and time t is plotted on the x-axis. If both sides of the equation are divided by t, the equation becomes $s/t = u + ½\, at$. This would give a straight-line graph of the form $y = mx + c$ if s/t is plotted on the y-axis and t on the x-axis. The gradient in this case is ½ a, where a is the acceleration, and the y-intercept is u, the speed of ball at the upper light gate.
2. In an investigation of the relationship between the current I and the p.d. V for a lamp, the relationship is non-linear as shown by Figure 2. The relationship could be of the form $V = kI^n$ where k and n are constants. This possible power relationship can be tested by taking base 10 logarithms (lg) of both sides to give $\lg V = \lg(kI^n) = \lg k + n \lg I$. Plotting $y = \lg V$ against $x = \lg I$ would give a straight line if the relationship is of the form $V = kI^n$. If so, the value of n can be determined as it is equal to the gradient of the line. You can learn more about logarithms in Topic 24.6.

3. In a capacitor discharge experiment, the capacitor is charged and then discharged through a resistor. The capacitor voltage V is measured at regular intervals as the capacitor discharged. A graph of V on the y-axis against time t (from the start of the discharge) on the x-axis gives an exponential decrease curve corresponding to the equation $V = V_0\, e^{-t/RC}$ where RC is the time constant of the discharge. The time constant can be determined by taking natural logarithms (ln) of both sides to give $\ln V = \ln (V_0\, e^{-t/RC}) = \ln V_0 - \left(\frac{1}{RC} \times t\right)$.

Comparing this with the straight-line equation $y = mx + c$ means that a graph of $\ln V$ against t should be a straight line with a negative gradient equal to $-\frac{1}{RC}$ and a y-intercept equal to $\ln V_0$. You will learn more about capacitor discharge and the exponential decrease in Topic 15.4 and in Topic 24.7.

Notes

1. A data analysis software package on a computer could be used to test different possible relationships, but you need to learn about the above use of logarithms in case they occur in your exams. See Topic 24.6 for more about logarithms.

2. Where logarithms are applied to data, units should be shown in a bracket with the quantity whose logarithm is being taken (e.g. $\ln (I/\mathrm{A})$ not $\ln I$). Also, for a logarithm of the value of a quantity, the number of decimal places should correspond to the number of significant figures in the value of the quantity.

More about evaluation

In Paper 5, your evaluation skills will be tested in a written paper in which you will be given an equation and some experimental data. From these you need to find the value of a constant to estimate the uncertainty in your answer. So, in addition to the practical skills you were tested on in Paper 3, apart from Paper 3 measurements, observation and data presentation skills, you need to be able to work out the uncertainty of a derived quantity from the uncertainties in the quantities within the derived quantity.

For example, in a magnetic field context, you may be given values and uncertainties for the force F on a current-carrying wire of length L in a uniform magnetic field, when the wire is at right angles to the field lines and there is a current I in the wire. You will be expected to calculate the magnetic flux density B of the field using an appropriate equation, in this case the equation $F = BIL$. To calculate B, you would be expected to rearrange the equation and use the rearranged equation $B = F/IL$ to calculate B. In addition, you would be expected to add together the percentage uncertainties in F, I and L to obtain the percentage uncertainty in B and then calculate the actual uncertainty in B.

If you are given data to plot a graph, you could be expected to estimate the uncertainty in the gradient of the line if the line is straight (or of a tangent to the line if it is curved) and use your estimate to find the uncertainty in a derived quantity. For example, in a capacitor discharge experiment, you could be expected to plot a straight-line graph of $\ln V$ against t in order to find the time constant and its uncertainty from the gradient of the line as explained on p.xxvi. You could then be asked to calculate the uncertainty of the capacitance C given the uncertainty in the resistance and your own calculated uncertainty in the time constant.

Uncertainty calculations are essential to give a proper evaluation of an investigation. By comparing percentage uncertainties in each measured quantity, you can pinpoint weaknesses in the method or procedure used, and you can also make suggestions about how to improve the methods or procedure. By combining percentage uncertainties, you can estimate the overall uncertainty in a derived quantity and thereby compare its value and uncertainty with a known value or with values obtained by other methods.

AS Level

This section of the book contains the material that you will cover in the first year of the Cambridge International AS & A Level Physics course.

The content builds on the physics you have studied earlier and is a foundation for the second year of your A Level studies.

The material is divided into two parts:

- Forces: Chapters 1–5
- Electricity, waves and radioactivity : Chapters 6–10

Each chapter is matched to the syllabus and is followed by practice questions that will test your understanding and give you practice at tackling examination-style questions.

1 Kinematics

1.1 Vectors and scalars

Learning outcomes

On these pages you will learn to:

- distinguish between scalar and vector quantities and give examples of each
- add and subtract coplanar vectors
- resolve a vector into two perpendicular components

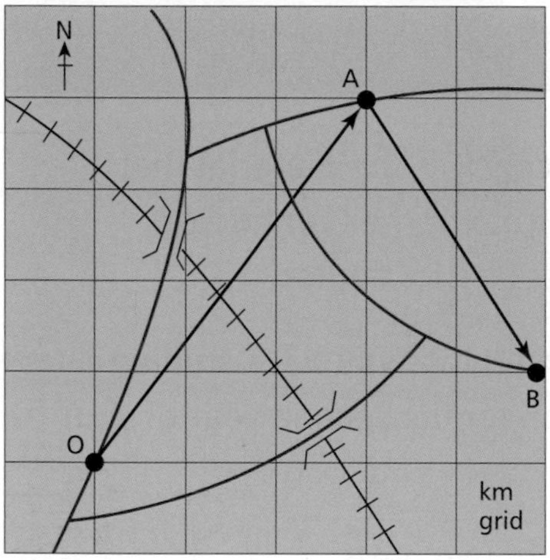

Figure 1 Map of locality

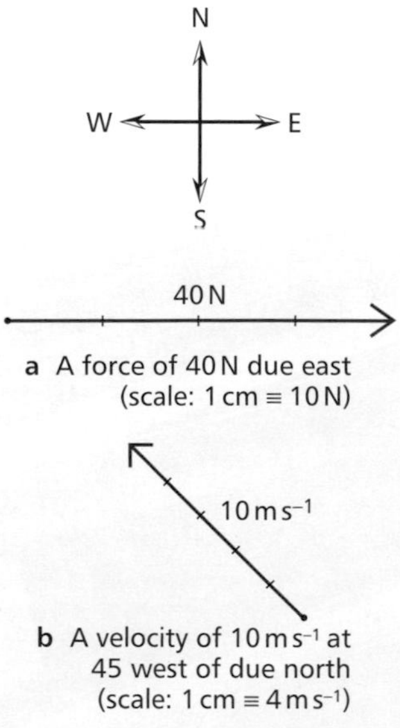

Figure 2 Representing a vector

Imagine you are planning to cycle to a friend's home several kilometres away from your home. The **distance** you travel depends on your route. However, the direct distance from your home to your friend's home is the same whichever route you choose. Distance in a given direction or **displacement** is an example of a **vector** quantity because it has magnitude and direction.

A vector is any physical quantity that has a direction as well as a magnitude.

Further examples of vectors include velocity, **acceleration**, force and **weight**.

A scalar is any physical quantity that has magnitude only and is not directional.

For example, distance is a scalar because it takes no account of direction. Further examples of scalars include mass, density, volume and energy.

Representing vectors

Any vector can be represented as an arrow. The length of the arrow represents the **magnitude** of the vector quantity. The direction of the arrow gives the **direction** of the vector.

- **Displacement** is distance in a given direction. The displacement from one point to another can be represented on a map or a scale diagram as an arrow from the first point to the second point. The length of the arrow must be in proportion to the least distance between the two points.
- **Velocity** is **speed** in a given direction. The velocity of an object can be represented by an arrow with length in proportion to the speed pointing in the direction of motion of the object.
- **Force and acceleration** are both vector quantities and therefore can each be represented by an arrow in the appropriate direction and of length in proportion to the magnitude of the quantity.

On a journey

Cyclists and hill walkers should always take a map and compass to make sure they do not get lost. A compass tells the user which direction is north. A map tells the user how far he or she has gone. Consider the map shown in Figure 1. Suppose your home is at O and your friend's home is at A. Your route-plan is to cycle along the road heading north, over the railway bridge, then turn east at the next road junction.

- The **distance** to be cycled can be estimated by measuring the length of the route on the map in centimetres, then using the map scale.
- The **displacement** or direct distance from O to A is marked on the map as an arrow OA.
- Your **velocity** at any point on your journey changes, because you change direction and because your speed changes. Suppose your speed as you pass over the railway bridge is 2.0 m s^{-1}. The direction in which you are travelling as you pass over the bridge is about 10° east of due north. You can check this on Figure 1 using a protractor. So your velocity at the railway bridge is 2.0 m s^{-1} in a direction which is 10° east of due north.

Vector components

The displacement vector **OA**, shown on Figure 1, can be represented:

- as an arrow of length in proportion to the direct distance of 5.0 km from O to A. The direction of the arrow would be 53° north of due east.
- as a map reference, one part stating how far A is east or west of O and the other stating how far A is north or south of O. The two parts of the map reference, referred to as **components**, may be written as (3.0 km, 4.0 km) where east/west is first. This is the same as writing the **coordinates** of a point on a graph as (x, y), where x is the distance from the origin along the x-axis and y is the distance from the origin along the y-axis.

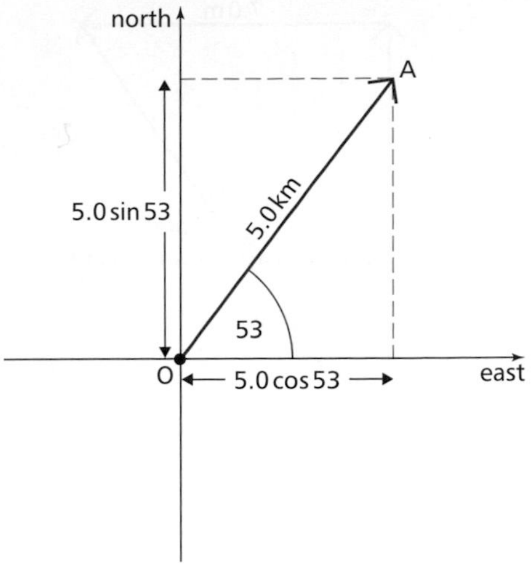

Figure 3 *Resolving a vector*

Resolving a vector into two perpendicular components

Resolving a vector is the process of working out the components of a vector in two perpendicular directions from the magnitude and direction of the vector. Figure 3 shows the displacement vector OA represented on a scale diagram that also shows lines due north and due east. The components of this vector along these two lines are 5.0 cos 53° km (= 3.0 km) along the line due east and 5.0 sin 53° km (= 4.0 km) along the line due north.

In general, to resolve any vector into two perpendicular components, draw a diagram showing the two perpendicular directions and an arrow to represent the vector. Figure 4 shows this diagram for a vector OP. The components are represented by the projection of the vector onto each line. If the angle θ between the vector OP and one of the lines is known:

- the component along that line = OP $\cos\theta$, and
- the component perpendicular to that line (i.e. along the other line) = OP $\sin\theta$.

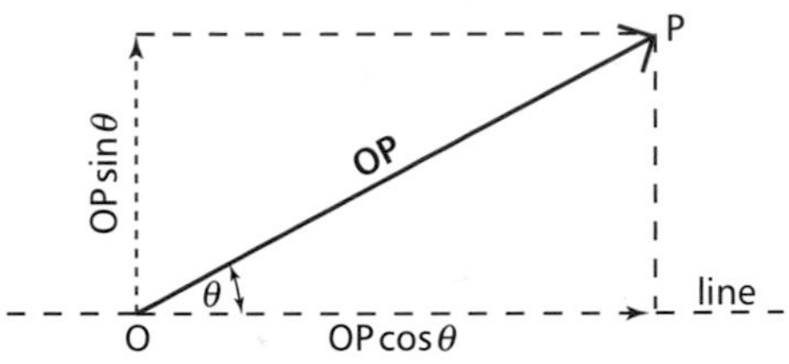

Figure 4 *The general rule for resolving a vector*

Worked example

An aircraft in level flight has a constant velocity of 50 m s^{-1} in a direction of 40° north of due east, as shown in Figure 5. The angle between the direction of its velocity and due east is therefore 40°.

The components of its velocity are, in m s^{-1},

- 50 cos 40° due east, and
- 50 sin 40° due north.

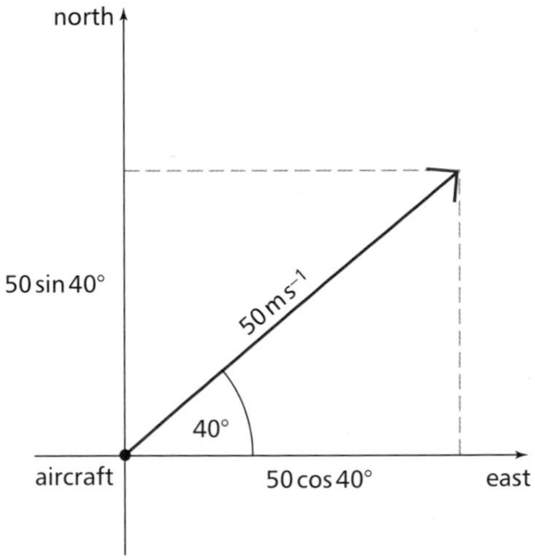

Figure 5 *Resolving velocity*

Addition of vectors

Using a scale diagram

Let's go back to the cycle journey in Figure 1. Suppose when you reach your friend's home at A, you then go on to another friend's home at B. Your journey is now a two-stage journey:

- **Stage 1, from O to A,** is represented by the displacement vector **OA**.
- **Stage 2, from A to B,** is represented by the displacement vector **AB**.

Figure 6 shows how the overall displacement from O to B, represented by vector **OB**, is the result of adding vector **AB** to vector **OA**. The **resultant** is the third side of a triangle, where OA and AB are the other two sides.

$$\mathbf{OB} = \mathbf{OA} + \mathbf{AB}$$

Use Figure 1 to show that the resultant displacement **OB** is 5.1 km in a direction 11° north of due east.

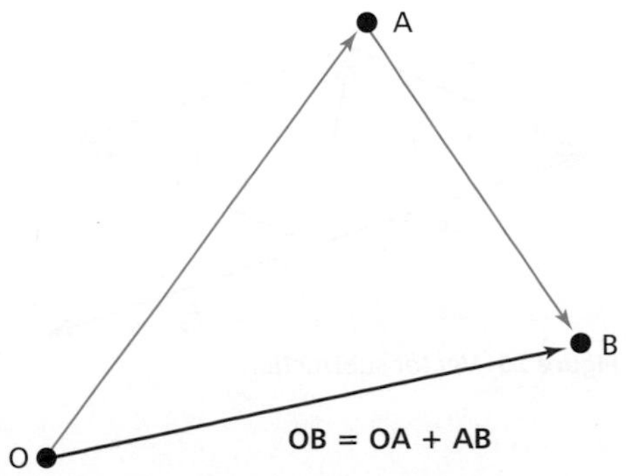

Figure 6 *Displacement from O to B*

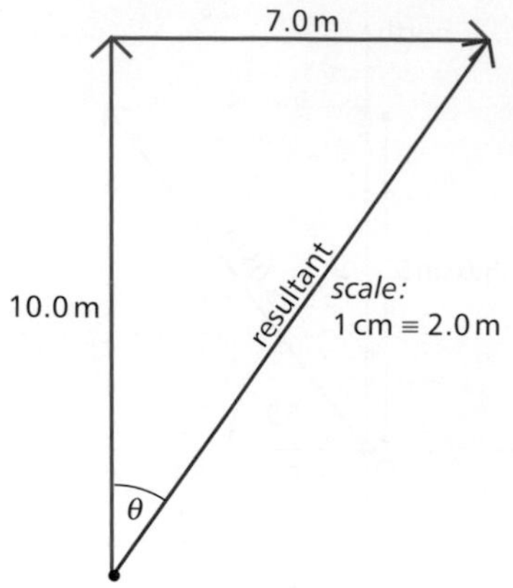

Figure 7 *Adding two displacements at right angles to each other*

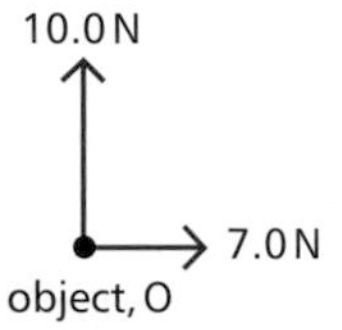

Figure 8 *Two forces acting at right angles to each other*

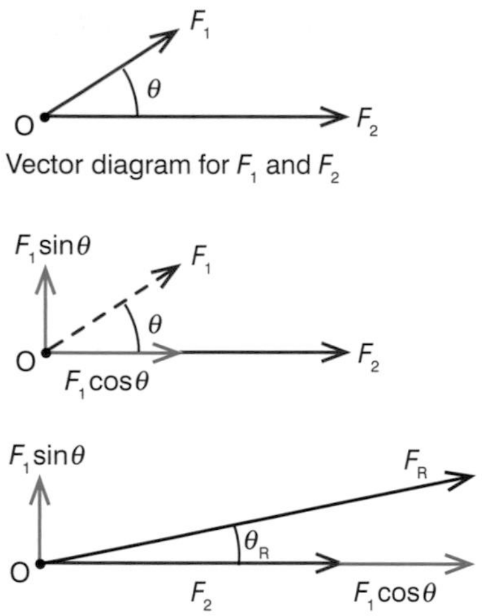

a Vector diagram for F_1 and F_2

b Using the components to find the resulting force F_r

Figure 9 *Using a calculator to find a resultant force*

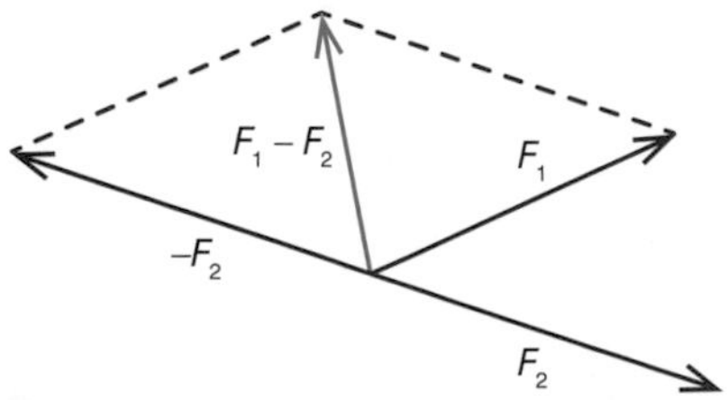

Figure 10 *Vector subtraction*

Using a calculator

1 Adding two perpendicular vectors

Suppose you walk 10.0 m forward then turn through exactly 90° and walk 7.0 m. At the end, how far will you be from your starting point? The vector diagram to add the two displacements is shown in Figure 7, drawn to a scale of 1 cm to 2.0 m. The two displacements form two sides of a right-angled triangle with the resultant as the hypotenuse. To find the resultant displacement:

- using Pythagoras' theorem gives 12.2 m ($= (10.0^2 + 7.0^2)^{1/2}$) for the magnitude of the resultant displacement (i.e. the distance), and
- using the trigonometry equation $\tan\theta = \dfrac{7.0}{10.0}$ gives 35° for the angle θ between the direction of the resultant displacement and the initial direction.

The method above can be applied to any two vectors at right angles to each other. For example:

Figure 8 shows an object, O, acted on by two forces, 7.0 N and 10.0 N perpendicular to each other. The vector diagram to add the two forces together would be as shown in Figure 7 if the labels were changed to 7.0 N and 10.0 N instead of 7.0 m and 10.0 m. As explained above, the resultant force is therefore 12.2 N at an angle of 35° to the 10 N force.

2 Adding two vectors that are at angle θ to each other

Consider an object, O, acted on by forces F_1 and F_2 at angle θ to each other, as shown in Figure 9a. The magnitude and direction of the resultant force F_R can be found by resolving one of the forces into components that are parallel and perpendicular to the other force, as shown in Figure 9b.

- Resolving F_1 parallel and perpendicular to F_2 gives $F_1\cos\theta$ for the parallel component and $F_1\sin\theta$ for the perpendicular component.
- Adding the components in each direction therefore gives the parallel component of F_R as $F_1\cos\theta + F_2$ and the perpendicular component as $F_1\sin\theta$.

Using Pythagoras' theorem to find the magnitude of the resultant force gives

$$F_R = [(F_1\cos\theta + F_2)^2 + (F_1\sin\theta)^2]^{1/2}$$

Because $\sin^2\theta + \cos^2\theta = 1$ for all angles of θ, it can be shown that

$$F_R^2 = F_1^2 + F_2^2 + 2F_1F_2\cos\theta.$$

Using the trigonometry rule for $\tan\theta$ to find θ_R, the angle between the resultant force and F_2 gives

$$\tan\theta_R = \frac{F_1\sin\theta}{(F_1\cos\theta + F_2)}.$$

Note

The resultant of two vectors that act along the same line has a magnitude that is:

- the **sum,** if the two vectors are in the *same* direction. For example, if an object is acted on by a force of 6.0 N and a force of 4.0 N, both acting in the same direction, the resultant force is 10.0 N.
- the **difference,** if the two vectors are in *opposite* directions. For example, if an object is acted on by a 6.0 N force and a 4.0 N force in opposite directions, the resultant force is 2.0 N in the direction of the 6.0 N force.

To subtract one vector from another, the above example illustrates a general method which is to reverse one vector and then add the reversed vector to the other. To reverse a vector, point it in the opposite direction to its original direction so the sign of each component is changed from + to – or – to +.

Worked example

$g = 9.81\,\text{m}\,\text{s}^{-2}$

A ball released from a height of 1.20 m above a concrete floor rebounds to a height of 0.82 m.

a Calculate: **i** its time of descent, **ii** the speed of the ball immediately before it hits the floor.

b Calculate: the speed of the ball immediately after it leaves the floor.

c Sketch a velocity–time graph for the ball. Assume the contact time is negligible compared with the time of descent or ascent.

Solution

a $u = 0$, $a = -9.81\,\text{m}\,\text{s}^{-2}$, $s = -1.2$ m

i To find t, use $s = ut + \frac{1}{2}at^2$

$$\therefore -1.2 = 0 + 0.5 \times -9.81 \times t^2$$
$$-1.2 = -4.905t^2$$
$$t^2 = \frac{-1.2}{-4.905} = 0.245$$
$$t = 0.49\,\text{s}$$

ii To find v, use $v = u + at$

$\therefore v = 0 + -9.81 \times 0.49 = -4.8\,\text{m}\,\text{s}^{-1}$ (– for downwards)

so the speed is $4.8\,\text{m}\,\text{s}^{-1}$

b $v = 0$, $a = -9.81\,\text{m}\,\text{s}^{-2}$, $s = +\,0.82$ m

To find u, use $v^2 = u^2 + 2as$

$$\therefore 0 = u^2 + 2 \times -9.81 \times 0.82$$
$$u^2 = 16.1\,\text{m}^2\,\text{s}^{-2}$$
$$u = +\,4.0\,\text{m}\,\text{s}^{-1}\ (+\text{ for upwards})$$

c See Figure 4. Note that the line has a constant gradient equal to the acceleration due to gravity, $-9.81\,\text{m}\,\text{s}^{-2}$, except on impact.

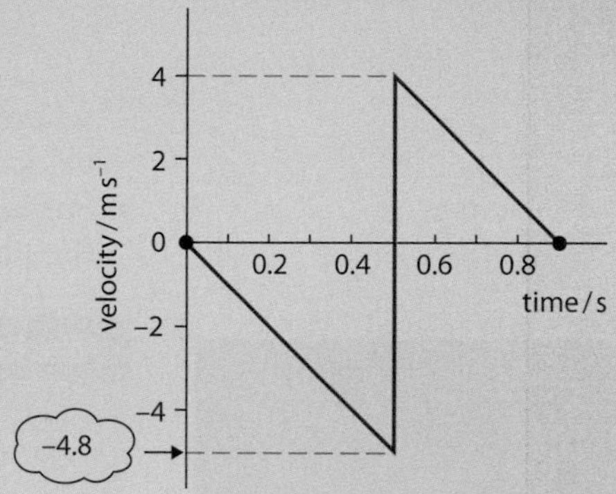

Figure 4

Summary test 1.6

$g = 9.81\,\text{m}\,\text{s}^{-2}$

1 A swimmer swims 100 m from one end of a swimming pool to the other end at a constant speed of $1.2\,\text{m}\,\text{s}^{-1}$; then swims back at constant speed, returning to the starting point 210 s after starting.

a Calculate how long the swimmer takes to swim from:
i the starting end to the other end,
ii back to the start from the other end.

b For the swim from start to finish, sketch:
i a displacement–time graph,
ii a distance–time graph.

c Sketch a velocity–time graph for the swim.

2 A motorcyclist travelling along a straight road at a constant speed of $8.8\,\text{m}\,\text{s}^{-1}$ passes a cyclist travelling in the same direction at a speed of $2.2\,\text{m}\,\text{s}^{-1}$. After 200 s, the motorcyclist stops.

a Calculate how long the motorcyclist has to wait before the cyclist catches up.

b On the same axes, sketch a velocity–time graph for: **i** the motorcyclist, **ii** the cyclist.

3 The graph (Figure 5) shows the velocity of a train on a straight track for 50 min after it left a station.

a i Describe how the displacement of the train from the station changed with time.
ii Sketch a graph to show how the displacement in part **i** varied with time.

b i Calculate how far from the station the train was after 50 min.
ii Calculate the total distance travelled by the train in this time.

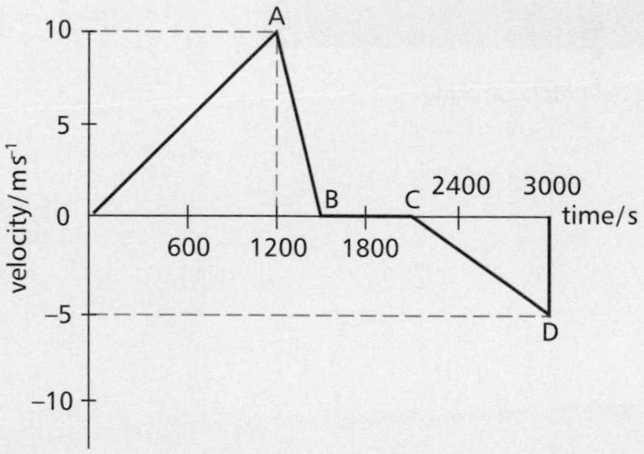

Figure 5

4 A ball is released from a height of 1.8 m above a level surface and rebounds to a height of 0.90 m.

a Given $g = 9.81\,\text{m}\,\text{s}^{-2}$, calculate: **i** the duration of its descent, **ii** its velocity just before impact, **iii** the duration of its ascent, **iv** its velocity just after impact.

b Sketch a graph to show how its velocity changes with time from release to rebound at maximum height.

c Sketch a further graph to show how the displacement of the object changes with time.

1.7 More calculations on motion along a straight line

Learning outcomes

On these pages you will learn to:

- solve problems including those with several stages using the equations for uniformly accelerated motion in a straight line

Motion along a straight line at constant acceleration

- The 'suvat' equations for motion at constant acceleration, a, are:

$$v = u + at \quad \text{(Equation 1)}$$

$$s = \frac{(u + v)t}{2} \quad \text{(Equation 2)}$$

$$s = ut + \tfrac{1}{2}at^2 \quad \text{(Equation 3)}$$

$$v^2 = u^2 + 2as \quad \text{(Equation 4)}$$

  where s is the displacement in time t, u is the initial velocity and v is the final velocity.
- For motion along a straight line at constant acceleration, one direction along the line is 'positive' and the other direction is negative.

Figure 1 *A space vehicle docking*

Worked example

A space vehicle moving towards a docking station, at a speed of $2.5\,\text{m s}^{-1}$, is 26 m from the docking station when its reverse thrust motors are switched on (to slow it down and stop it when it reaches the station). The vehicle decelerates uniformly until it comes to rest at the docking station, when its motors are switched off.

Calculate: **a** its deceleration, **b** how long it takes to stop, **c** its velocity, if its motors remained on for 5.0 s longer than necessary.

Solution
Let the + direction represent motion towards the docking station and – away from the station.

a and **b** Initial velocity, $u = +2.5\,\text{m s}^{-1}$, final velocity, $v = 0$, displacement, $s = +26\,\text{m}$.

To find its deceleration, a, use

$$v^2 = u^2 + 2as.$$
$$0 = 2.5^2 + 2a \times 26$$
$$-52a = 2.5^2$$
$$a = -\frac{2.5^2}{52} = -0.12\,\text{m s}^{-2}$$

To find the time taken, use

$$v = u + at.$$
$$0 = 2.5 - 0.12t$$
$$0.12t = 2.5$$
$$t = \frac{2.5}{0.12} = 21\,\text{s}$$

c Initial velocity, $u = 2.5\,\text{m s}^{-1}$, acceleration, $a = -\,0.12\,\text{m s}^{-2}$, time taken $t = 26\,\text{s}$

To calculate its velocity v after 26 s, use $v = u + at = 2.5 - 0.12 \times 26$

$$= -0.62\,\text{m s}^{-1}$$

The velocity of the space vehicle would be $0.62\,\text{m s}^{-1}$ away from the docking station after 26 s.

Two-stage problems

Consider an object released from rest, falling then hitting a bed of sand. The motion is in two stages:

1 **Falling** motion due to gravity; acceleration = g (downwards)
2 **Deceleration** in the sand; initial velocity = velocity of object just before impact.

The acceleration in each stage is **not** the same. The link between the two stages is that the velocity at the end of the first stage is the same as the velocity at the start of the second stage.

For example, consider a ball released from a height of 0.85 m above a bed of sand that creates an impression in the sand of depth 0.025 m (Figure 2). For directions, let + represent upwards and – represent downwards.

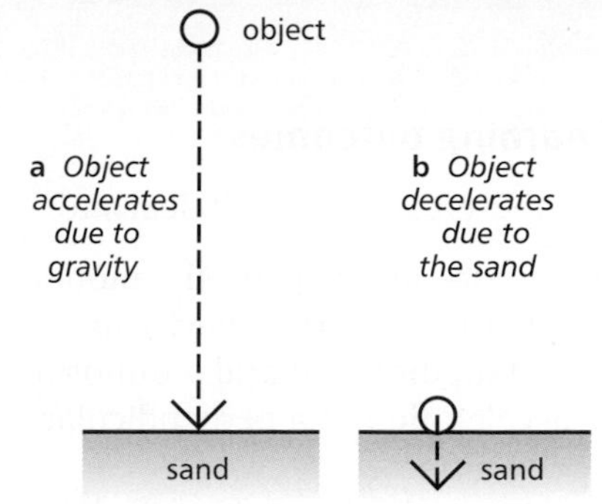

Figure 2 A two-stage problem

Stage 1

$u = 0$, $s = -0.85\,\text{m}$, $a = -9.81\,\text{m s}^{-2}$.

To calculate the speed of impact v, use $v^2 = u^2 + 2as$

$\therefore \quad v^2 = 0^2 + 2 \times -9.81 \times -0.85 = 16.7$

$\therefore \quad v = -4.1\,\text{m s}^{-1}$

Note

$v^2 = 16.7$, so $v = -4.1$ or $+\,4.1\,\text{m s}^{-1}$. The negative answer is chosen, as the ball is moving downwards.

Stage 2

$u = -4.1\,\text{m s}^{-1}$, $v = 0$ (as the ball comes to rest in the sand), $s = -0.025\,\text{m}$

To calculate the deceleration, a, use $v^2 = u^2 + 2as$.

$\therefore \quad 0^2 = (-4.1)^2 + 2a \times -0.025$

$\therefore \quad 2a \times 0.025 = 16.7$

$$a = \frac{16.7}{2 \times 0.025} = 334\,\text{m s}^{-2}$$

Note

$a > 0$ and therefore in the opposite direction to the direction of motion, which is downwards. Thus the ball slows down in the sand with a deceleration of $334\,\text{m s}^{-2}$.

Summary test 1.7

$g = 9.81\,\text{m s}^{-2}$

1 A vehicle on a straight downhill road accelerated uniformly from a speed of $4.0\,\text{m s}^{-1}$ to a speed of $29\,\text{m s}^{-1}$ over a distance of 850 m, when the driver braked and stopped the vehicle in 28 s.

a Calculate: **i** the time taken to reach $29\,\text{m s}^{-1}$ from $4\,\text{m s}^{-1}$, **ii** its acceleration during this time.

b Calculate: **i** the distance it travelled during deceleration, **ii** its average deceleration as it slowed down.

2 A rail wagon moving at a speed of $2.0\,\text{m s}^{-1}$ on a level track reached a steady incline, which slowed it down in 15.0 s and caused it to reverse. Calculate:

a the distance it moved up the incline,

b its acceleration on the incline,

c its velocity and position on the incline after 20.0 s.

3 A cyclist accelerated from rest at a constant acceleration of $0.4\,\text{m s}^{-2}$ for 20 s, then stopped pedalling and slowed to a standstill at a constant deceleration over a distance of 260 m.

a Calculate: **i** the distance travelled by the cyclist in the first 20 s, **ii** the speed of the cyclist at the end of this time.

b Calculate: **i** the time taken to cover the distance of 260 m after she stopped pedalling, **ii** her deceleration during this time.

4 A rocket was launched directly upwards from rest. Its motors operated for 30 s after it left the launch pad, providing it with a constant vertical acceleration of $6.0\,\text{m s}^{-2}$ during this time. Its motors then switched off (Figure 3).

a Calculate: **i** its initial velocity, **ii** its height above the launch pad when its motors switched off.

b Calculate its maximum height gain after its motors switched off.

c Calculate the velocity with which it would hit the ground, if it fell from maximum height without the support of a parachute.

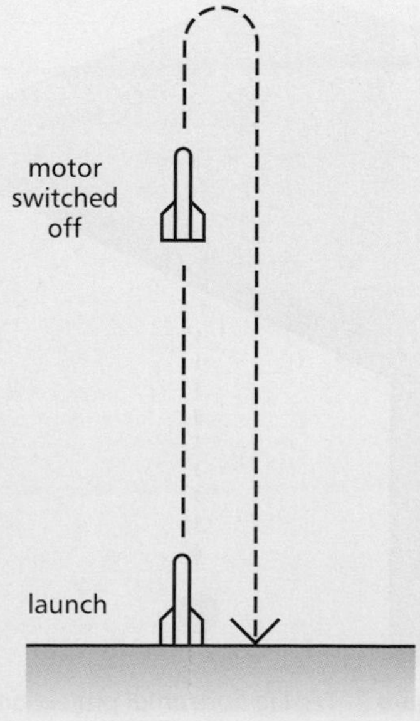

Figure 3

1.8 Projectile motion 1

Learning outcomes

On these pages you will learn to:

- describe and explain the motion due to a uniform velocity in a certain direction and a uniform acceleration in a perpendicular direction
- solve projectile problems where projection is horizontal using the equations for uniform acceleration in a vertical direction and for uniform velocity in a horizontal direction

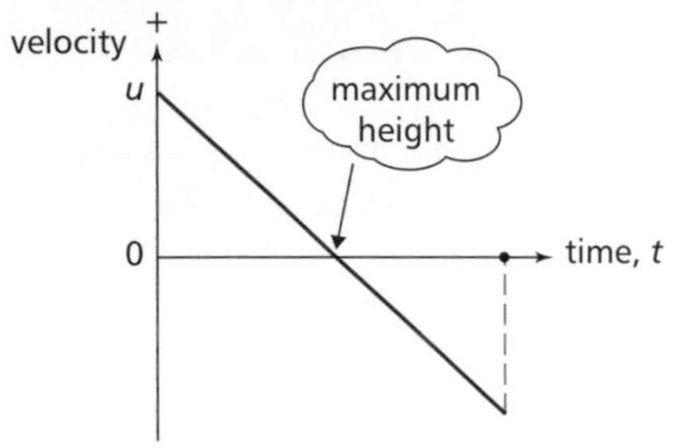

Figure 1 *Upward projection*

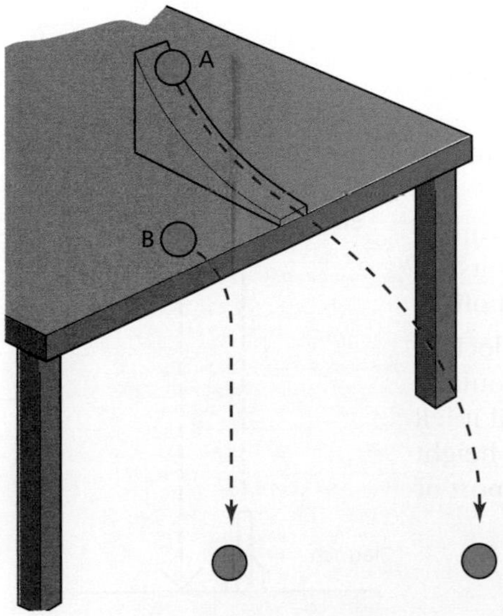

Figure 2 *Testing horizontal projection*

A **projectile** is any object acted upon only by the force of gravity. Air resistance is assumed to be negligible. Three key principles apply to all projectiles:

- The **acceleration** of the object is always equal to g and acts downwards, because the force of gravity acts downwards. The acceleration therefore only affects the vertical motion of the object.
- The **horizontal velocity** of the object is constant, because the acceleration of the object does not have a horizontal component.
- The motions in the horizontal and vertical directions are **independent** of each other.

Vertical projection

If an object is projected vertically, it moves vertically as it has no horizontal motion. Its acceleration is $9.81\,m\,s^{-2}$ downwards. Using the direction code '+ is upwards; – is downwards', its displacement, y, and velocity, v_y, after time t are given by:

$$v_y = u - gt$$

$$y = ut - \tfrac{1}{2}gt^2$$

where u is its initial velocity.

See Topic 1.7 for more about vertical projection.

Horizontal projection

A stone thrown from a cliff top follows a curved path downwards before it hits the water. If its initial projection was horizontal:

- Its path through the air becomes steeper and steeper as it drops.
- The faster it is projected, the further away it will fall into the sea.
- The time taken for it to fall into the sea does not depend on how fast it is projected.

Suppose two balls are released at the same time above a level floor, such that one ball drops vertically and the other is projected horizontally. Which one hits the floor first? In fact, they both hit the floor simultaneously. Try it! Why should the two balls hit the ground at the same time? They are both pulled to the ground by the force of gravity which gives each ball a downward acceleration, g. The ball that is projected horizontally experiences the same downward acceleration as the other ball. This downward acceleration does not affect the horizontal motion of the ball projected horizontally; only the vertical motion is affected.

Investigating horizontal projection

A stroboscope and a camera may be used to record the motion of a projectile. Figure 3 shows a multi-flash photograph of two balls, A and B, released at the same time. B was released from rest and dropped vertically; A was given an initial horizontal projection so it followed a curved path. The stroboscope flashed at a constant rate, so images of both balls were recorded at the same time.

- **The horizontal position** of A changes by equal distances between successive flashes. This shows that the horizontal component of A's velocity is constant.
- **The vertical position** of A and B changes at the same rate. At any instant, A is at the same level as B. This shows that A and B have the same vertical component of velocity at any instant.

The projectile path of a ball projected horizontally

An object projected horizontally falls in an **arc** towards the ground. If its initial velocity is U, then at time t after projection (Figure 4):

- The **horizontal component** of its displacement,

$$x = Ut$$

(because it moves horizontally at a constant speed)

- The **vertical component** of its displacement,

$$y = \frac{1}{2}gt^2$$

(because it has no vertical component of its initial velocity)

- Its velocity has:

  a horizontal component $v_x = U$, and

  a vertical component $v_y = -gt$

Note Its speed at time $t = \sqrt{(v_x^2 + v_y^2)}$

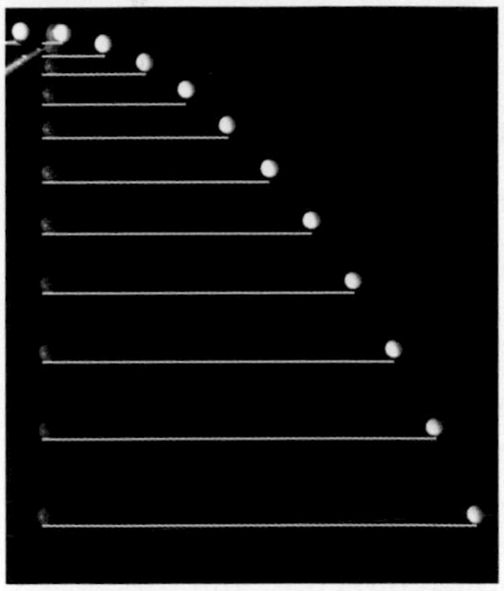

Figure 3 Multi-flash photo of two falling balls

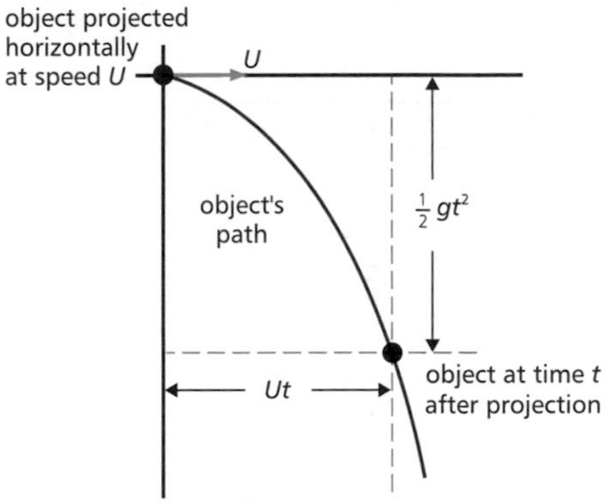

Figure 4 Horizontal projection

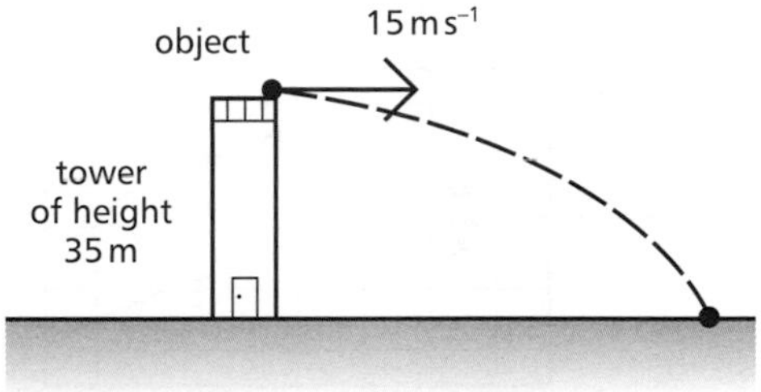

Figure 5

Worked example

$g = 9.81\,\text{m s}^{-2}$

An object is projected horizontally at a speed of $15\,\text{m s}^{-1}$ from the top of a tall tower (Figure 5) of height 35.0 m. Calculate:

a how long it takes to fall to the ground,

b how far it travels horizontally,

c its speed just before it hits the ground.

Solution

a $y = -35\,\text{m}$, $a = -9.81\,\text{m s}^{-2}$ (– for downwards)

$$y = \frac{1}{2}gt^2$$

$$\therefore t^2 = \frac{2y}{g} = \frac{2 \times -35}{-9.81} = 7.14\,\text{s}^2$$

$$\therefore t = 2.67\,\text{s}$$

b $U = 15\,\text{m s}^{-1}$, $t = 2.67\,\text{s}$

$x = Ut = 15 \times 2.67 = 40\,\text{m}$

c Just before impact, $v_x = U = 15\,\text{m s}^{-1}$ and $v_y = -gt = -9.81 \times 2.67 = 26.2\,\text{m s}^{-1}$

$\therefore$ speed just before impact, $v = (v_x^2 + v_y^2)^{½} = (15^2 + 26.2^2)^{½} = 30.2\,\text{m s}^{-1}$

Summary test 1.8

$g = 9.81\,\text{m s}^{-2}$

1 An object is released from a hot air balloon 50 m above the ground that is descending vertically at a speed of $4.0\,\text{m s}^{-1}$. Calculate:

a the velocity of the object at the ground,

b the duration of descent of the object,

c the height of the balloon above the ground when the object hits the ground.

2 An object is projected horizontally at a speed of $16\,\text{m s}^{-1}$ into the sea from a cliff top of height 45.0 m. Calculate:

a how long it takes to reach the sea,

b how far it travels horizontally,

c its impact velocity.

3 A dart is thrown horizontally along a line which passes through the centre of a dartboard 2.3 m away from the point at which the dart was released. The dart hits the dartboard at a point 0.19 m below the centre. Calculate:

a the time of flight of the dart,

b its horizontal speed of projection.

4 A parcel is released from an aircraft travelling horizontally at a speed of $120\,\text{m s}^{-1}$ above level ground. The parcel hits the ground 8.5 s later. Calculate:

a the height of the aircraft above the ground,

b the horizontal distance travelled in this time by **i** the parcel, **ii** the aircraft,

c the speed of impact of the parcel at the ground.

1.9 Projectile motion 2

Learning outcomes

On these pages you will learn to:

- compare projectile motion with the motion of electrons in a uniform electric field
- solve projectile problems where projection is not horizontal using the equations for uniform acceleration in a vertical direction and for uniform velocity in a horizontal direction

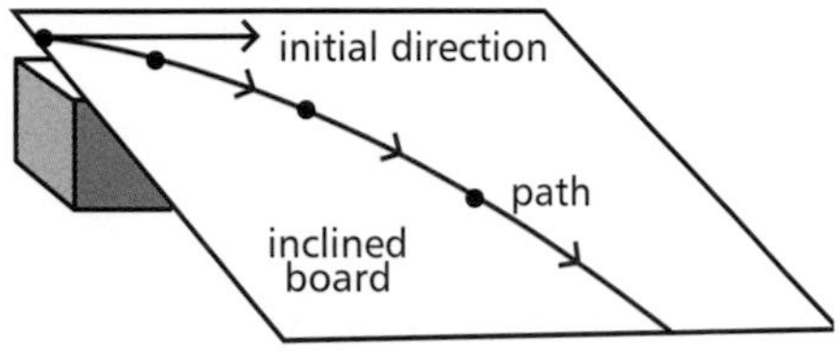

Figure 1 *Using an inclined board*

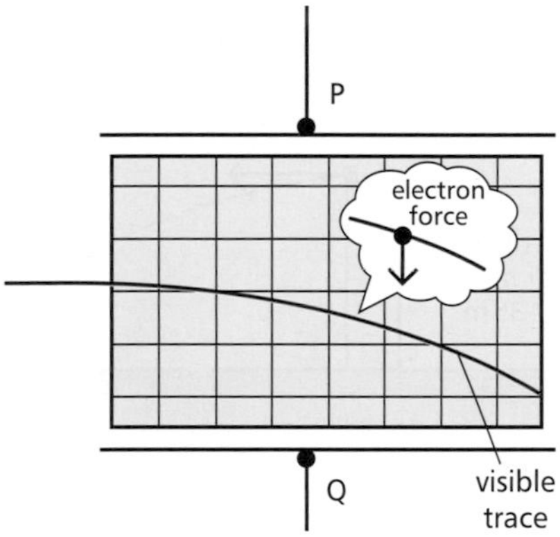

Figure 2 *An electron beam on a parabolic path*

(Vacuum tube used to contain apparatus not shown)

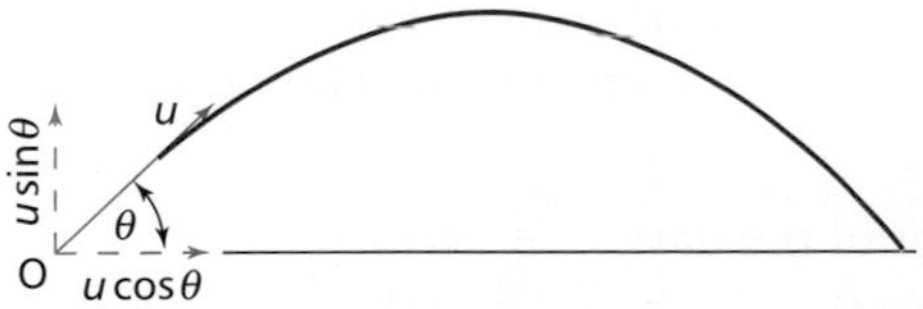

Figure 3 *Projectile motion*

Projectile-like motion

Any form of motion where an object experiences a constant acceleration in a different direction to its velocity will be like projectile motion. For example:

- The path of a ball rolling across an inclined board will be a projectile path. Figure 1 shows the idea. The object is projected across the top of the board from the side. Its path curves down the board and is **parabolic**. This is because the object is subjected to a constant acceleration acting down the board, and its initial velocity is across the board. The same equations as for projectile motion apply, with the motion down the board replacing the vertical motion.
- The path of a beam of **electrons** directed between two oppositely charged parallel plates is a **parabola**, as shown in Figure 2. Each electron in the beam is acted on by a constant force towards the positive plate, because the charge of an electron is negative. Therefore, each electron experiences a constant acceleration towards the positive plate. If its initial velocity is parallel to the plates, then its path is parabolic because its motion parallel to the plates is at **zero** acceleration; whereas its motion perpendicular to the plates is at **constant** (non-zero) acceleration.

Projection at angle θ and speed U above the horizontal

An object projected into the air follows a path that depends on its **speed** of projection and its **angle** of projection. In unit 1.8, we looked at the special cases of vertical projection and of horizontal projection. In this section, we will consider the general case of projection at angle θ to the horizontal. The analysis can be applied to the motion of any object acted on by a force that is constant in **magnitude** and **direction**. Air resistance is assumed to be negligible in this analysis.

For an object projected at initial speed U at an angle θ to the horizontal, as in Figure 3, its **initial velocity** has:

- a **horizontal** component $u_x = U\cos\theta$
- a **vertical** component $u_y = U\sin\theta$

Its **acceleration** has:

- a **horizontal** component $a_x = 0$
- a **vertical** component $a_y = -g$

For the **horizontal motion**, at time t after release:

- Its horizontal component of **velocity**, $v_x = U\cos\theta$ (unchanged), as its horizontal component of acceleration is zero.
- Its horizontal component of **displacement**, x = its horizontal component of velocity × time taken = $Ut\cos\theta$

For the vertical motion,

- Its vertical component of **velocity**, $v_y = U\sin\theta - gt$
  This is obtained by applying the equation '$v = u + at$' to the vertical motion with $U\sin\theta$ for the initial speed u and $-g$ for the acceleration a.
- Its vertical component of **displacement**, $y = Ut\sin\theta - \frac{1}{2}gt^2$
  This is obtained by applying the equation '$s = ut + \frac{1}{2}at^2$' to the vertical motion with $U\sin\theta$ for the initial speed u and $-g$ for the acceleration a.
- Its speed at time $t = \sqrt{(v_x^2 + v_y^2)}$, in accordance with Pythagoras' rule for adding the two perpendicular components of a vector.

A **positive value** of θ will signify projection at angle θ **above** the horizontal.

A **negative value** of θ will signify projection at angle θ **below** the horizontal.

Worked example

An arrow is fired at a speed of $48\,m\,s^{-1}$ at an angle of 50° above the horizontal. Calculate **a** how long it takes to reach maximum height; **b** its maximum height. $g = 9.81\,m\,s^{-2}$

Solution
$U = 48\,m\,s^{-1}$, $\theta = 50°$

a The vertical component of the initial velocity, $u_y = U\sin\theta = 48\sin 50° = 36.7\,m\,s^{-1}$

The vertical component of velocity at maximum height, $v_y = 0$

Using $v_y = U\sin\theta - gt$, $0 = 36.7 - 9.81t$

$$t = \frac{36.7}{9.81} = 3.74\,s = 3.7\,s \text{ to 2 s.f.}$$

b The maximum height can be calculated by using $y = Ut\sin\theta - \frac{1}{2}gt^2$ with $t = 3.74\,s$,

$\therefore$ maximum height $= 36.7 \times 3.74 - (0.5 \times 9.81 \times 3.74^2)$

$= 68.6\,m$

The effects of air resistance

A projectile moving through air experiences a force that drags on it because of the resistance of the air it passes through. This **drag force** is partly caused by friction between the layers of air near the projectile's surface where the air flows over its surface. The drag force:

- acts in the opposite direction to the direction of motion of the projectile, and it increases as the projectile's speed increases
- has a horizontal component that reduces its range
- reduces the maximum height of the projectile if its initial direction is above the horizontal and makes its descent steeper than its ascent.

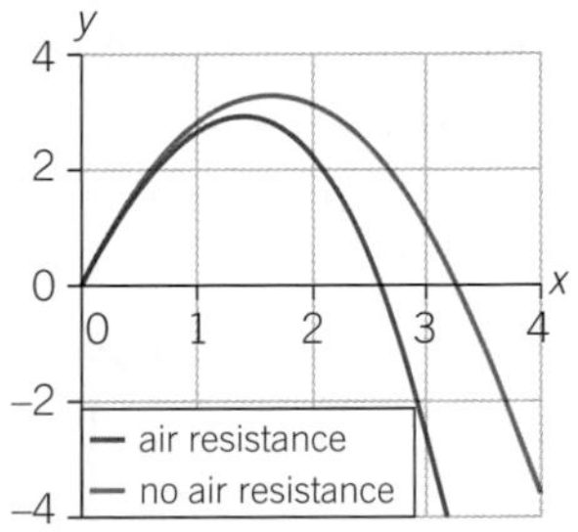

Summary test 1.9

$g = 9.81\,m\,s^{-2}$. For all questions, assume air resistance is negligible unless otherwise stated.

1 A ball was projected horizontally at a speed of $0.52\,m\,s^{-1}$ across the top of an inclined board of width 600 mm and length 1200 mm. It reached the bottom of the board 0.90 s later.

Calculate:

a the distance travelled by the ball across the board,

b its acceleration on the board,

c its speed at the bottom of the board.

2 An arrow is fired at a speed of $45\,m\,s^{-1}$ at an angle of 30° above the horizontal. Calculate:

a its maximum height,

b how long it takes to reach maximum height,

c how long it takes to return to the same horizontal level at which it started,

d the distance travelled horizontally to the point of return in **c**.

3 A cable car was travelling at a speed of $4.6\,m\,s^{-1}$ in a direction that was 40° above the horizontal when an object was released from the cable car (Figure 5).

a Calculate the horizontal and vertical components of velocity of the object at the instant it was released.

b The object took 5.8 s to fall to the ground below. Calculate: **i** the distance fallen by the object from the point of release, **ii** the horizontal distance travelled by the object from the point of release to where it hit the ground.

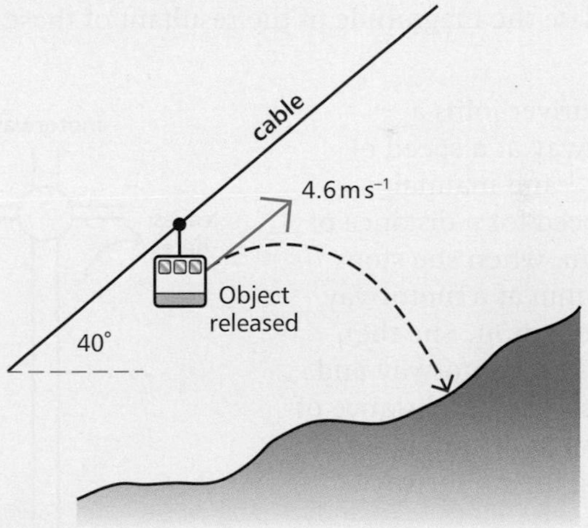

Figure 5

4 A cannon ball was fired from the top of a tower at an angle of elevation of 25°. The ball hit the ground 2.7 s later at a distance of 58 m away from the foot of the tower.

a Calculate the horizontal and vertical components of the velocity of the cannon ball at the instant it was fired.

b Calculate the height of the tower above the ground.

c Calculate the speed of impact of the cannon ball at the ground.

d Discuss the effect on the range of the cannon ball if air resistance had not been negligible.

Launch additional digital resources for the chapter

$g = 9.81\,\text{m s}^{-2}$

1 a i State the difference between a vector and a scalar quantity.

ii Give one example of a vector quantity other than force.

iii Give an example of a scalar quantity other than speed.

b A 6.0 N force and a 5.0 N force act at right angles to each other on an object of mass 3.0 kg.

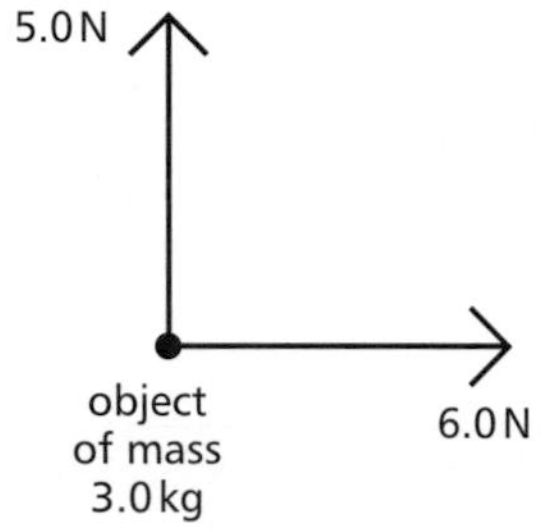

Figure 1.1

Calculate the magnitude of the resultant of these two forces.

2 A car driver joins a motorway at a speed of $22\,\text{m s}^{-1}$ and maintains this speed for a distance of 32.5 km, when she stops for 20 min at a motorway service station. She then rejoins the motorway and travels a further distance of 47.5 km in 40 min before leaving the motorway.

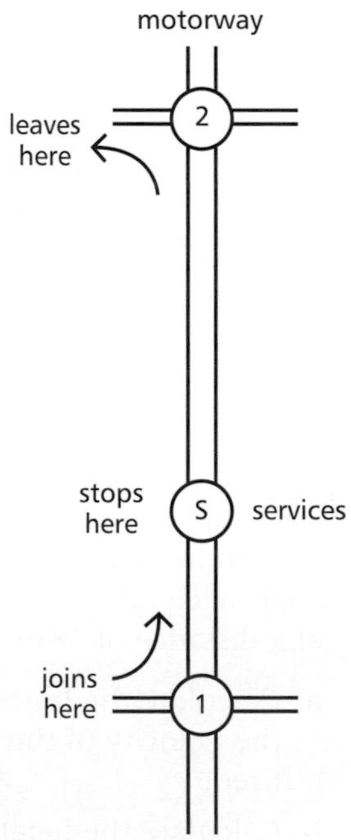

Figure 2.1

a Calculate the time taken for the first part of the journey.

b Sketch a speed–time graph for this journey on the motorway.

c Calculate the average speed of the car for the whole journey.

3 An aircraft taking off accelerated uniformly for 40 s from rest, before it left the ground after travelling a distance of 1600 m. Calculate:

a its average speed during take-off,

b its velocity at the instant it left the ground,

c its acceleration during take-off.

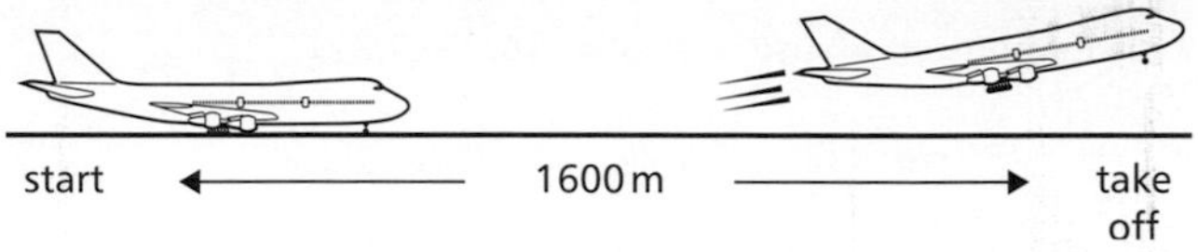

Figure 3.1

4 A train accelerated uniformly from rest for 40 s until its speed reached $15\,\text{m s}^{-1}$. It then travelled at a constant speed of $15\,\text{m s}^{-1}$ for a further 60 s, before slowing down at constant deceleration and stopping 20 s later.

a Sketch a speed–time graph to represent its motion.

b Calculate its acceleration and the distance it travelled in each of the three stages of its motion.

c Calculate its average speed.

5 An object released from rest at the side of a river bridge hit the water 2.2 s later. Calculate:

a the distance fallen in air by the object,

b the speed of the object just before it hit the water.

6 A stationary railway truck on an inclined track was struck by a shunting engine moving at a constant speed of $0.6\,\text{m s}^{-1}$. The impact caused the truck to move up the incline at an initial speed of $2.0\,\text{m s}^{-1}$. The truck slowed down and stopped 10 s later.

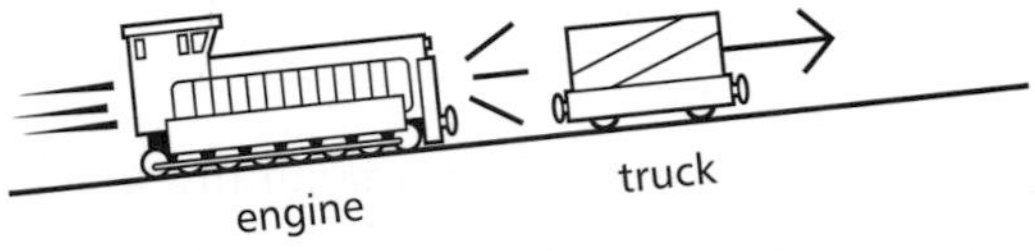

Figure 6.1

a Calculate: **i** the distance travelled by the truck before it stopped, **ii** the magnitude and direction of its acceleration.

b The truck stopped for an instant, then rolled back down the track and hit the engine again. The engine continued to move up the track at $0.6\,\text{m s}^{-1}$.

i Sketch a velocity–time graph on the same axes for the truck and the engine.

ii Use your graph to show that the truck hit the engine a second time 14 s after it was first hit.

iii Calculate the velocity of the truck immediately before this second impact.

7 A ball thrown vertically into the air left the thrower's hand when it was 1.6 m above the ground and hit the ground on return 3.1 s later.

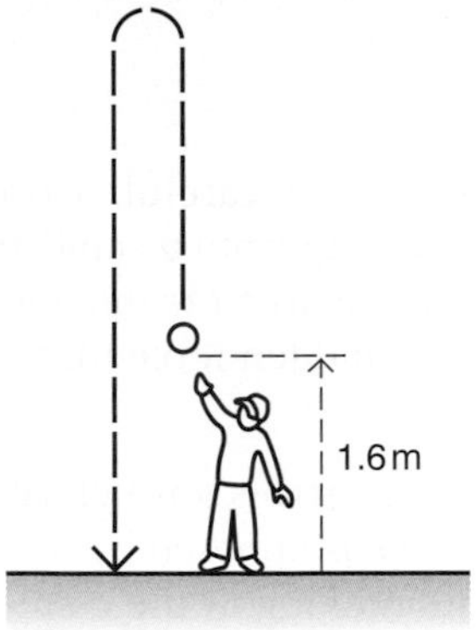

Figure 7.1

a Calculate: **i** the initial velocity of the ball, **ii** the maximum height of the ball above the ground.

b Calculate its velocity just before impact.

8 A cricketer on a level playing field threw a ball into the air at an angle of 30° above the horizontal. The ball was caught by another cricketer, 36 m away at the same level 1.8 s later.

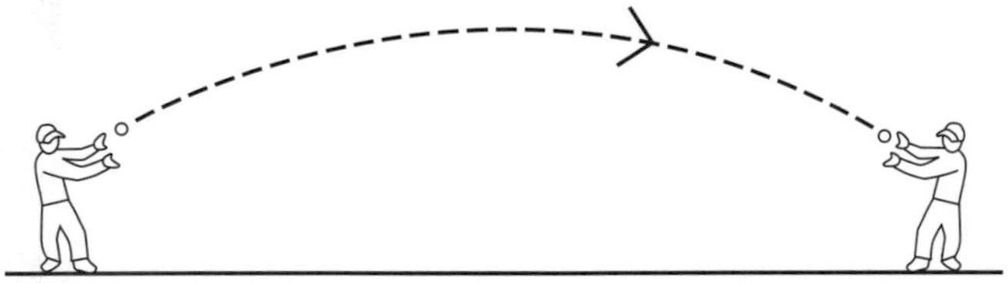

Figure 8.1

a Calculate the horizontal velocity of the ball.

b Show that the initial speed of the ball was $23\,\mathrm{m\,s^{-1}}$.

c Calculate the maximum height reached by the ball.

9 A rocket launched vertically accelerated at a constant acceleration of $6.0\,\mathrm{m\,s^{-2}}$ for 25 s before its fuel supply ran out. It then rose to maximum height and fell back to the ground.

a i Calculate its velocity and its height when its fuel ran out.

ii Show that it reached a maximum height of 3.0 km.

b i Calculate how long it took to fall from maximum height to the ground.

ii Calculate its velocity immediately before impact with the ground.

iii Sketch a velocity–time graph for the motion of the rocket from launch to impact.

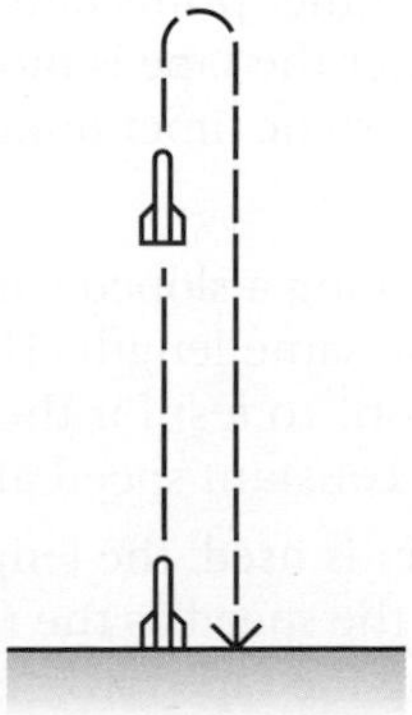

Figure 9.1

10 When a tennis player served a ball, the ball left the racquet horizontally at a speed of $23\,\mathrm{m\,s^{-1}}$ as shown in Figure 10.1.

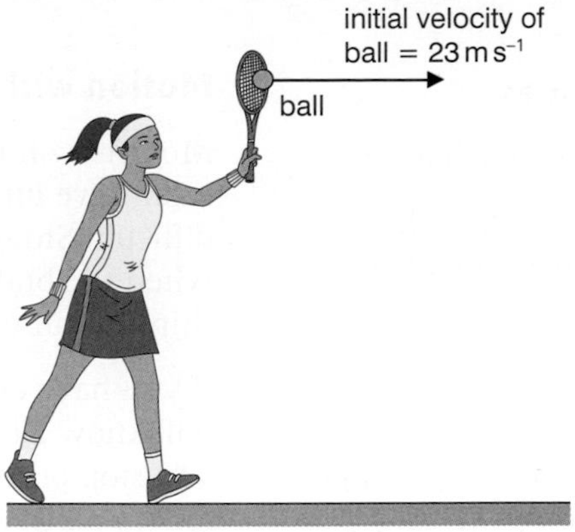

Figure 10.1

a After leaving the racquet, the ball just cleared the net after travelling a horizontal distance of 11.9 m to the net. Calculate the vertical displacement of the ball in this time. Assume air resistance is negligible.

b After clearing the net at a height of 0.92 m above the ground, the ball hit the ground, as shown in Figure 10.2.

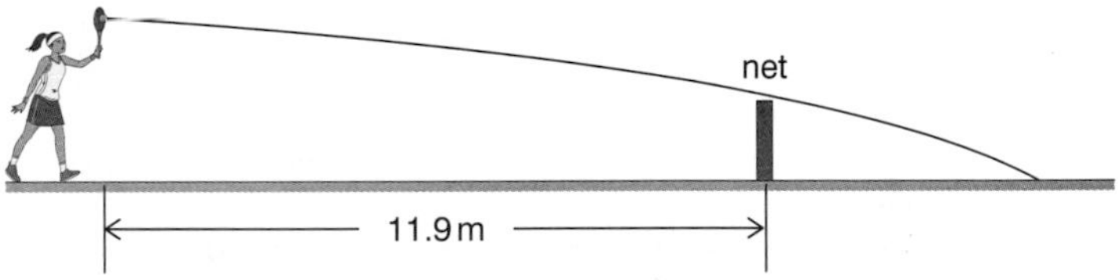

Figure 10.2

i Calculate the horizontal distance from the net to the point where the ball hit the ground.

ii Calculate the magnitude and direction of the ball's velocity immediately before it hit the ground.

c Discuss the effect on: **i** the magnitude, and **ii** the direction of the ball's velocity immediately before it hit the ground if the ball's initial direction had been above the horizontal.

2 Forces in action

2.1 Force and acceleration

Learning outcomes

On these pages you will learn to:

- show an understanding that mass is the property of a body that resists change in motion
- state Newton's first law of motion and give examples of where it applies
- recall and use Newton's second law of motion for constant mass in the form $F = ma$
- describe and use the concept of weight as the effect of a gravitational field on a mass
- recall that for a body of mass m, its weight $W = mg$, where g is the acceleration of free fall

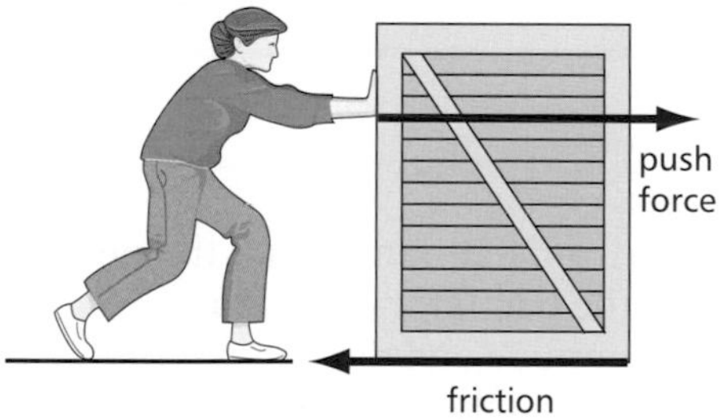

Figure 1 *Overcoming friction*

Motion without force

Motorists on icy roads in winter need to be very careful, because the tyres of a car have little or no 'grip' on the ice. Moving from a standstill on ice is very difficult. Stopping on ice is almost impossible, as a car moving on ice will slide when the brakes are applied. **Friction** is a hidden force that we don't usually think about until it is absent!

If you have ever tried to push a heavy crate across a rough concrete floor, you will know all about friction. The push force is opposed by friction and as soon as you stop pushing, friction stops the crate moving. If the crate had been pushed onto a patch of ice, it would have moved across the ice without any further push needed.

Figure 2 shows an air track which allows **motion** to be observed in the absence of friction. The glider on the air track floats on a cushion of air. Provided the track is level, the glider moves at a constant velocity along the track because friction is absent.

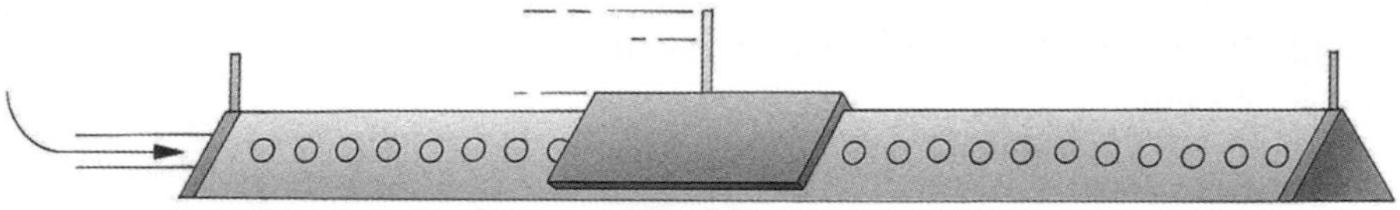

Figure 2 *The linear air track*

Newton's first law of motion

Objects either stay at rest, or remain in uniform motion (i.e. at constant velocity), unless acted on by a resultant force.

Sir Isaac Newton was the first person to realise that a moving object stays moving at constant velocity unless acted on by a force. He recognised that when an object is acted on by a **resultant** force, the result is to change the object's velocity. In other words, an object moving at **constant** velocity is either:

- acted on by **no forces**, or
- the forces acting on it are **balanced** (i.e. the **resultant force is zero**).

Investigating force and motion

How does the velocity of an object change if it is acted on by a constant force? Figure 3 shows how this can be investigated using a dynamics trolley and a ticker-tape timer. The timer prints dots on the tape at a constant rate of 50 dots per second; so the faster the tape is moving, the greater the spacing between adjacent dots. An electronic timer linked to a computer can be used in place of the ticker-tape timer.

The trolley is pulled along a sloped runway by means of one or more elastic bands stretched to the same length. The runway is sloped just enough to compensate for friction. To test for the correct slope, the trolley should move down the runway at constant speed after being given a brief push.

- If a ticker-tape timer is used, the length of each 'ten dot' section of the tape gives a measure of the speed as the middle of that section went through the ticker timer. Cutting the tape into 'ten dot' sections enables a speed–time chart to be made, as shown in Figure 4.

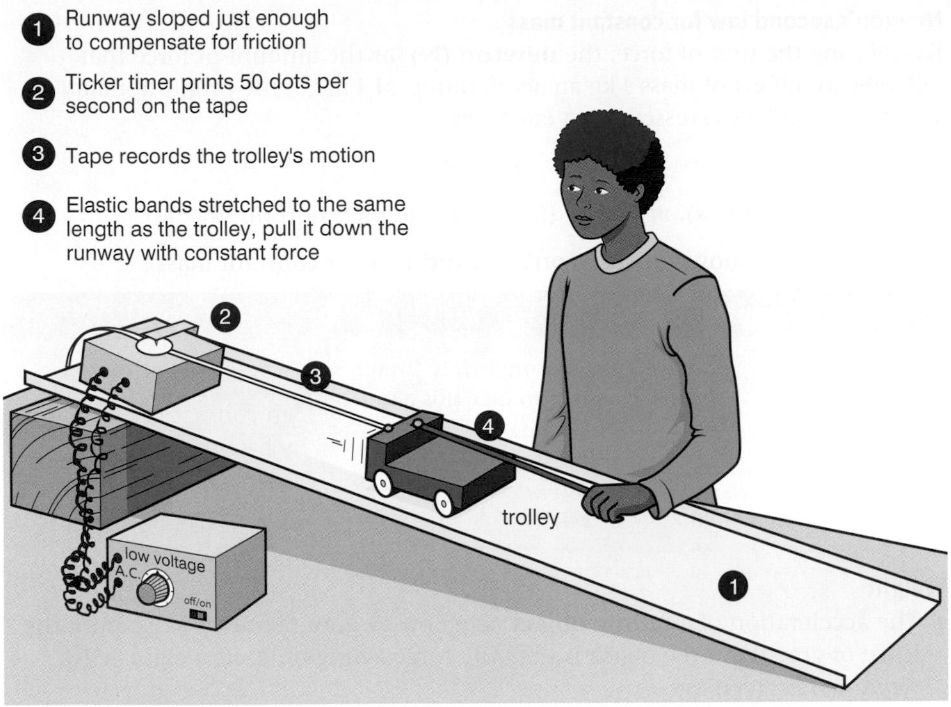

Figure 3 *Investigating force and motion*

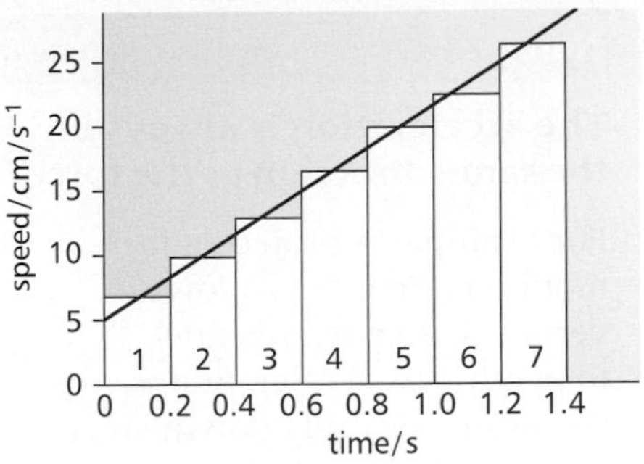

Figure 4 *Making a tape chart*

- If an electronic timer linked to a computer is used, the measurements should be displayed directly as a speed–time graph (Figure 5).

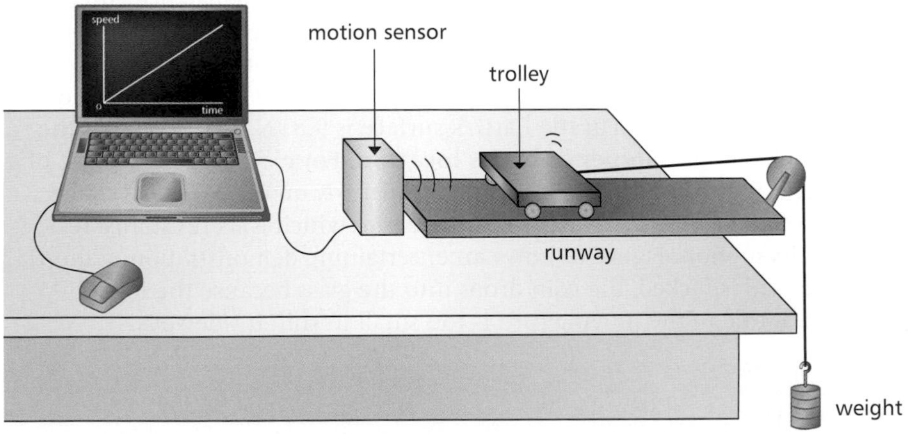

Figure 5 *Using a computer to measure acceleration*

The speed–time graph should show that the speed increased at a constant rate. The **acceleration** of the trolley is therefore constant, and can be measured from the speed–time graph. Table 1 shows typical measurements using different amounts of **force** (i.e. one, or two, or three elastic bands in 'parallel') and different amounts of **mass** (i.e. a single trolley, or a double trolley, or a triple trolley).

The results in Table 1 show that the force is proportional to the mass × the acceleration. In other words, if a force F acts on an object of mass m, the object undergoes an acceleration a such that:

F is proportional to ma

Table 1 *Varying force and mass*

Force (i.e. no. of elastic bands)	1	2	3	1	2	3
Mass (i.e. no of trolleys)	1	1	1	2	2	2
Acceleration/m s^{-2}	12	24	36	6	12	18
Mass × acceleration	12	24	36	12	24	36

Note

The acceleration is always in the same direction as the force

For example, a projectile in motion experiences a force vertically downwards due to gravity. Its acceleration is therefore vertically downwards, no matter what its direction of motion is.

Newton's second law for constant mass

By defining the unit of force, the **newton (N)**, as the amount of force that will give an object of mass 1 kg an acceleration of 1 m s^{-2}, the proportionality statement can be expressed as an equation:

$$F = ma$$

where F = force (in N), m = mass (in kg), a = acceleration (in m s^{-2}).

This equation is known as **Newton's second law** for constant mass.

Worked example

A vehicle of mass 600 kg accelerates uniformly from rest to a speed of 8.0 m s^{-2} in 20 s. Calculate the force needed to produce this acceleration.

Solution

Acceleration, $a = \frac{v-u}{t} = \frac{8.0-0}{20} = 0.4\,\text{m s}^{-2}$

Force, $F = ma = 600 \times 0.4 = 240\,\text{N}$

Weight

- The acceleration of a **falling object** acted on by gravity only is g. Because the force of gravity on the object is the only force acting on it, its weight W (in newtons) is given by:

  $$W = mg$$

  where m = the mass of the object (in kg).
- When an object is **in equilibrium**, the support force on it is equal and opposite to its weight. Therefore, an object placed on a weighing balance (e.g. a spring balance or a top-pan balance) exerts a force on the balance equal to the weight of the object. Thus the balance measures the weight of the object.
- g is also referred to as the **gravitational field strength** at a given position, as it is the force of gravity per unit mass on a small object at that position. So the gravitational field strength at the Earth's surface is 9.81 N kg^{-1}. Note that the weight of a fixed mass **depends on its location**. For example, the weight of a 1 kg object is 9.81 N on the Earth's surface and 1.6 N on the Moon's surface.
- The mass of an object is a measure of its **inertia**, which is its resistance to change of its motion. Figure 6 shows an entertaining demonstration of inertia. When the card is flicked, the coin drops into the glass because the force of friction on it due to the moving card is too small to shift it sideways.

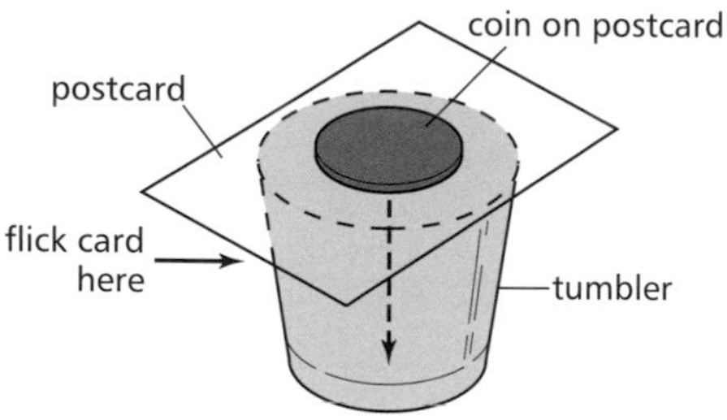

Figure 6 *An 'inertia' trick*

Summary test 2.1

$g = 9.81\,\text{N kg}^{-1}$

1 A car of mass 800 kg accelerates uniformly along a straight line from rest to a speed of 12 m s^{-1} in 50 s. Calculate:

a the acceleration of the car,

b the force on the car that produced this acceleration,

c the ratio of the accelerating force to the weight of the car.

2 An aeroplane, of mass 5000 kg, lands on a runway at a speed of 60 m s^{-1} and stops 25 s later. Calculate:

a the deceleration of the aeroplane,

b the braking force on the aeroplane.

3 a A vehicle, of mass 1200 kg, on a level road accelerates from rest to a speed of 6.0 m s^{-1} in 20 s, without change of direction. Calculate the force that accelerated the car.

b The vehicle in part **a** is fitted with a trailer of mass 200 kg. Calculate the time taken to reach a speed of 6.0 m s^{-1} from rest for the same force as in part **a**.

4 A bullet of mass 0.002 kg, travelling at a speed of 120 m s^{-1}, hit a tree and penetrated a distance of 55 mm into the tree. Calculate:

a the deceleration of the bullet,

b the impact force of the bullet on the tree.

Using F = ma

Two forces in opposite directions

- When an object is acted on by two **unequal** forces acting in **opposite** directions, the object accelerates in the direction of the larger force. If the forces are F_1 and F_2 where $F_1 > F_2$:

$$\textbf{Resultant force, } F_1 - F_2 = ma$$

where m is the mass of the object and a is its acceleration.

- If the object is on a horizontal surface and F_1 and F_2 are **horizontal** and in opposite directions, the above equation still applies. The support force on the object is equal and opposite to its weight.

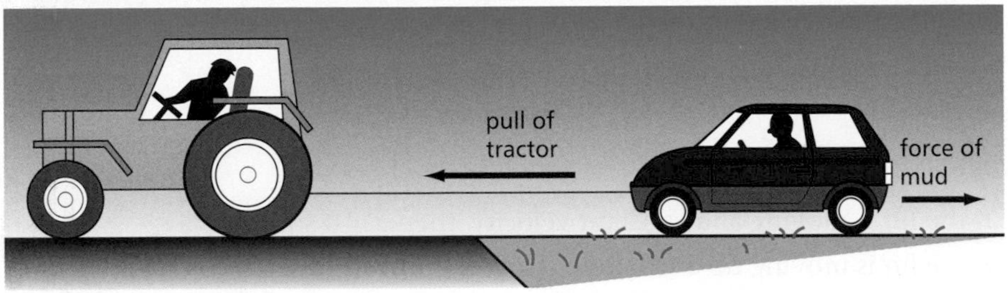

Figure 1 *Unbalanced forces*

Some examples are given below, where two forces act in different directions on an object.

Towing a trailer

Consider the example of a car of mass M fitted with a trailer of mass m on a level road. When the car and the trailer accelerate, the car pulls the trailer forward and the trailer holds the car back (Figure 2).

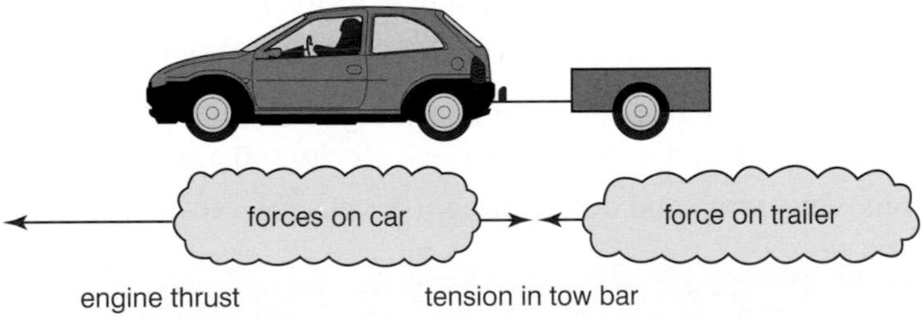

Figure 2 *Car and trailer*

- The car is subjected to a driving force F pushing it forwards (from its engine thrust) and the tension T in the tow bar holding it back. Therefore:

$$\textbf{Resultant force on car} = F - T = Ma$$

- The force on the trailer is due to the tension T in the tow bar pulling it forwards. Therefore:

$$T = ma$$

Combining the two equations gives:

$$F = Ma + ma = (M + m)a$$

Learning outcomes

On these pages you will learn to:

- recall the relationship $F = ma$ and solve problems using it, including where more than one force acts on object
- appreciate that acceleration and force are always in the same direction

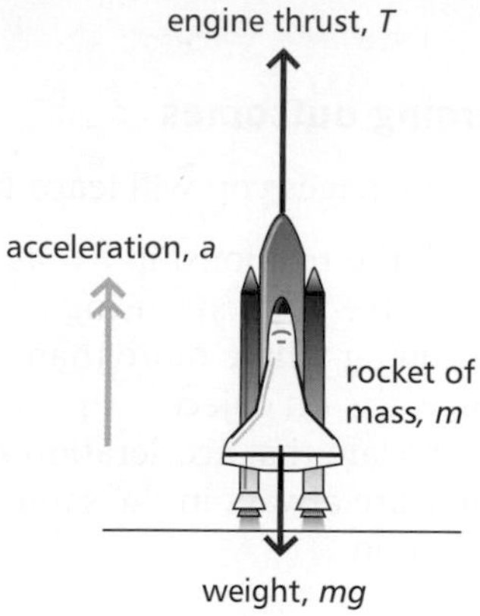

Figure 3 *Rocket launch*

Rocket problems

If T is the thrust of the rocket engine when its mass is m and the rocket is moving upwards, its acceleration a is given by:

$$T - mg = ma$$

Therefore: **thrust,** $\mathbf{T = mg + ma}$

The rocket thrust must therefore overcome the weight of the rocket for the rocket to take off (Figure 3).

Lift problems

Using 'upwards is positive' gives the resultant force on the lift as:

$$F = T - mg$$

where T is the tension in the lift cable and m is the total mass of the lift and occupants (Figure 4).

Therefore $$T - mg = ma$$

where a = acceleration.

- If the lift is moving at a constant velocity, then $a = 0$ so $T = mg$.
- If the lift is moving up and accelerating, then $a > 0$ so $T = mg + ma > mg$.
- If the lift is moving up and decelerating, then $a < 0$ so $T = mg + ma < mg$.
- If the lift is moving down and accelerating, then $a < 0$ (i.e. velocity and acceleration are both downwards and therefore negative) so $T = mg - ma < mg$.
- If the lift is moving down and decelerating, then $a > 0$ (i.e. velocity downwards and acceleration upwards and therefore positive) so $T = mg + ma > mg$.

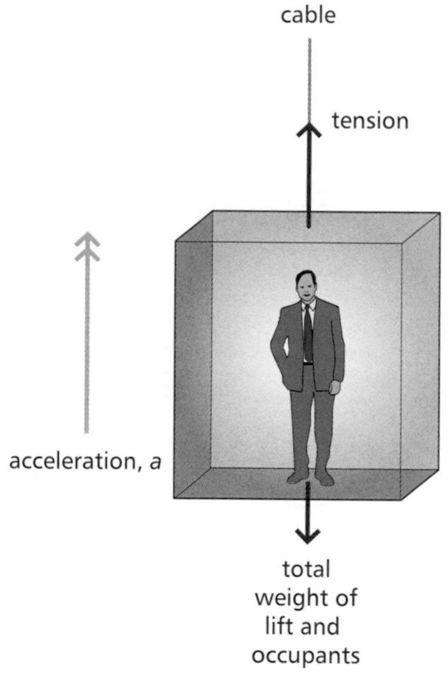

Figure 4 *In a lift*

Tension in the cable is less than the weight if:

- The lift is moving up and decelerating (i.e. velocity > 0 and acceleration < 0).
- The lift is moving down and accelerating (i.e. velocity < 0 and acceleration < 0).

Tension in the cable is greater than the weight if:

- The lift is moving up and accelerating (i.e. velocity > 0 and acceleration > 0).
- The lift is moving down and decelerating (i.e. velocity < 0 and acceleration > 0).

Worked example

$g = 9.81\,\mathrm{m\,s^{-2}}$

A lift of total mass 650 kg moving downwards decelerates at $1.5\,\mathrm{m\,s^{-2}}$ and stops. Calculate the tension in the lift cable during the deceleration.

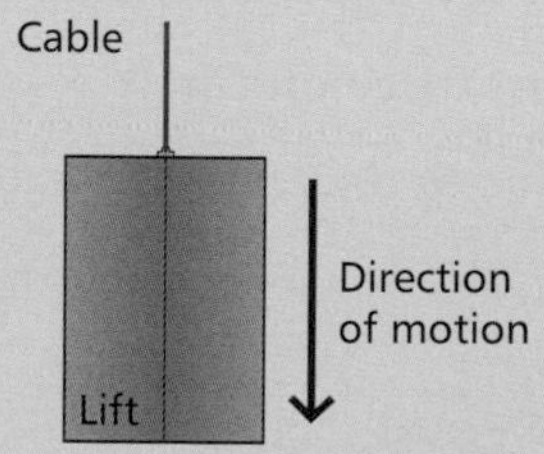

Figure 5

Solution

The lift is moving down so its velocity $v < 0$. Since it is decelerating, its acceleration a is in the opposite direction to its velocity, so $a > 0$. Therefore use $a = +1.5\,\mathrm{m\,s^{-2}}$ in the equation

$$T - mg = ma$$
$$T = mg + ma$$
$$= 650 \times 9.81 + 650 \times 1.5 = 7400\,\mathrm{N}$$

Further F = ma problems

Pulley problems

Consider two masses M and m (where $M > m$) attached to a thread hung over a frictionless pulley, as in Figure 6. When released, mass M accelerates downwards and mass m accelerates upwards. If a is the acceleration and T is the tension in the thread, then:

- On mass M, the resultant force $= Mg - T = Ma$.
- On mass m, the resultant force $= T - mg = ma$.

Therefore, adding the two equations gives:

$$Mg - mg = (M + m)a$$

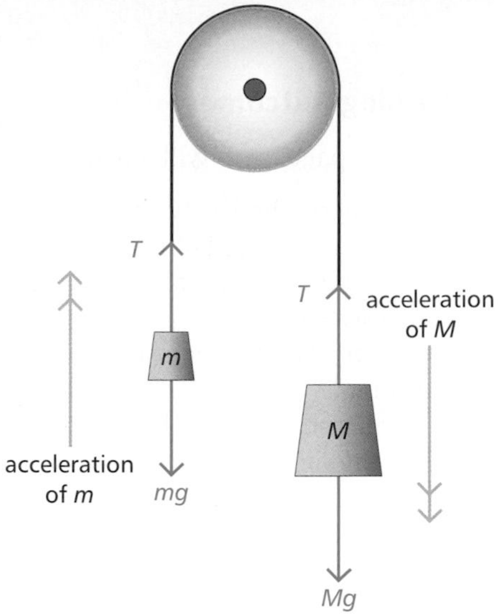

Figure 6 A pulley problem

Sliding down a slope

Consider a block of mass m sliding down a slope (Figure 7).
The component of the block's weight down the slope is $mg\sin\theta$.
If the force of friction on the block is F_0, then:

$$\text{Resultant force on the block} = mg\sin\theta - F_0$$

Therefore: $$mg\sin\theta - F_0 = ma$$

where a is the acceleration of the block.

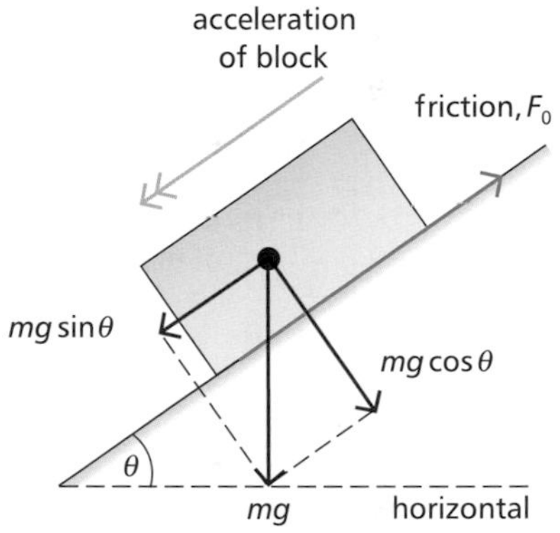

Figure 7

Note

With the addition of an engine force F_E, the above equation can be applied to a vehicle on a downhill slope of constant gradient. Thus:

$$F_E + mg\sin\theta - F_0 = ma$$

where F_0 is the combined sum of the force of friction and the braking force.

Summary test 2.2

$g = 9.81\,\text{m}\,\text{s}^{-2}$

1 A rocket of mass 550 kg blasts vertically from the launch pad at an acceleration of $4.2\,\text{m}\,\text{s}^{-2}$. Calculate:

  a the weight of the rocket,

  b the thrust of the rocket engines.

2 A car of mass 1400 kg, pulling a trailer of mass 400 kg, accelerates from rest to a speed of $9\,\text{m}\,\text{s}^{-1}$ in a time of 60 s on a level road. Assuming air resistance is negligible, calculate:

  a the tension in the tow bar,

  b the engine force.

3 A lift and its occupants have a total mass of 1200 kg. Calculate the tension in the lift cable when the lift is:

  a stationary,

  b ascending at constant speed,

  c ascending at a constant acceleration of $0.4\,\text{m}\,\text{s}^{-2}$,

  d descending at a constant deceleration of $0.4\,\text{m}\,\text{s}^{-2}$.

4 A brick, of mass 3.2 kg on a sloping flat roof, at 30° to the horizontal, slides at constant acceleration 2.0 m down the roof in 2.0 s from rest.
Calculate:

  a the acceleration of the brick,

  b the frictional force on the brick due to the roof.

Terminal velocity

Learning outcomes

On these pages you will learn to:

- describe qualitatively the motion of bodies falling in a uniform gravitational field with air resistance
- explain what is meant by terminal velocity
- explain why bodies falling in a fluid and powered vehicles reach a terminal speed and why their acceleration decreases to zero

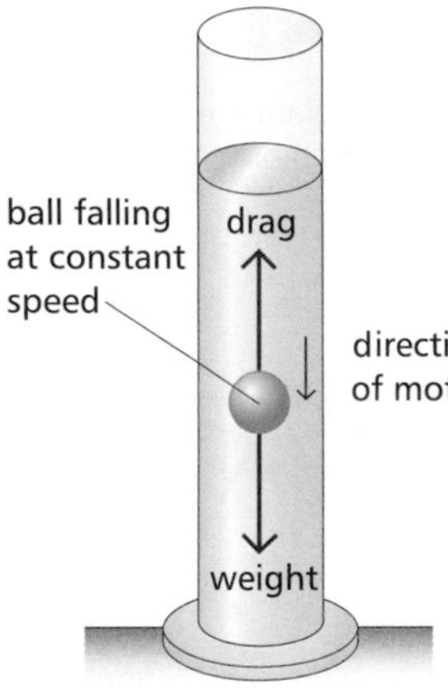

a Falling in a fluid

b Skydiving

Figure 1 *At terminal velocity*

> **Note**
>
> The acceleration at any instant is the gradient of the speed–time curve.

Drag forces

Any object moving through a fluid (something that can 'flow', i.e. a liquid or a gas) experiences a **drag force** due to the fluid. The drag force depends on:

- the shape and size of the object
- its speed
- the **viscosity** of the fluid (which is a measure of how easily the fluid flows past a surface).

The faster an object travels in a fluid, the greater the drag force on it.

Motion of an object falling in a fluid

- The speed of an object released from rest in a fluid increases as it falls, so the drag force on it due to the fluid increases. The resultant force on the object is the difference between the force of gravity on it (i.e. its weight) and the drag force. As the drag force increases, the resultant force decreases so the acceleration becomes less as it falls. If it continues falling, it attains **terminal velocity** when the drag force on it is equal and opposite to its weight. Its acceleration is then zero and its speed remains constant as it falls.
- Figure 2 shows how to investigate the motion of an object falling in a fluid. When the object is released, the thread attached to the object pulls a tape through a ticker timer which prints dots on the tape. The spacing between successive dots is a measure of the speed of the object, as the dots are printed on the tape at a constant rate. A tape chart can be made from the tape to show how the speed changes with time. The results show that:

The speed increases and reaches a constant value, which is the terminal velocity.

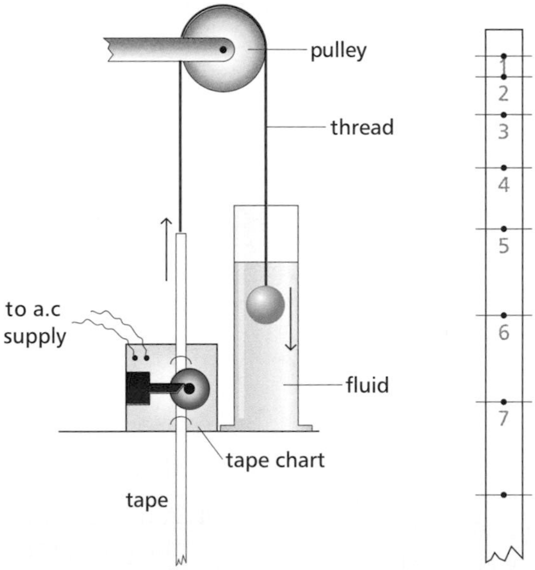

a Apparatus

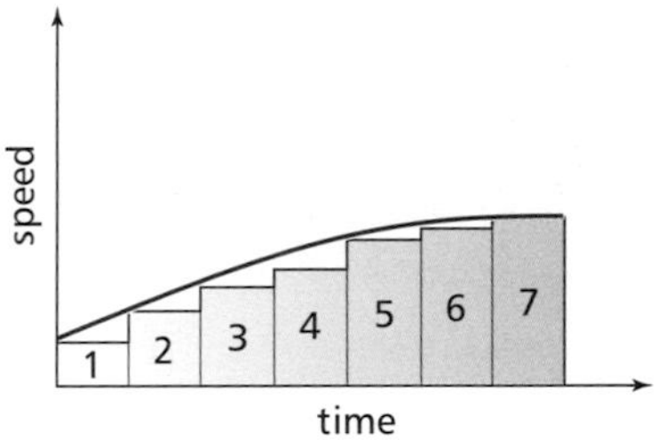

b Making a tape chart

Figure 2 *Investigating the motion of an object falling in a fluid*

- At any instant, the resultant force is $mg - F_D$, where m is the mass of the object and F_D is the drag force.

  Therefore: **Acceleration of the object** $= \dfrac{mg - F_D}{m} = g - \dfrac{F_D}{m}$

1 The **initial acceleration** is g, because the drag force is zero at the instant the object is released.
2 At the **terminal velocity**, the **potential energy** lost by the object as it falls is converted to **internal energy** of the fluid by the drag force.

Motion of a powered vehicle

The top speed of a road vehicle or an aircraft depends on its **engine power**, and its shape. A vehicle with a streamlined shape can reach a higher top speed than a vehicle with the same engine power that is not streamlined.

For a powered vehicle of mass m moving on a level surface, if F_E represents the **engine force** of the vehicle (i.e. the engine force driving the vehicle):

$$\textbf{Resultant force} = F_E - F_D$$

where F_D is the combined effect of friction and the drag force due to air resistance.

Therefore, its acceleration $a = \dfrac{(F_E - F_D)}{m}$

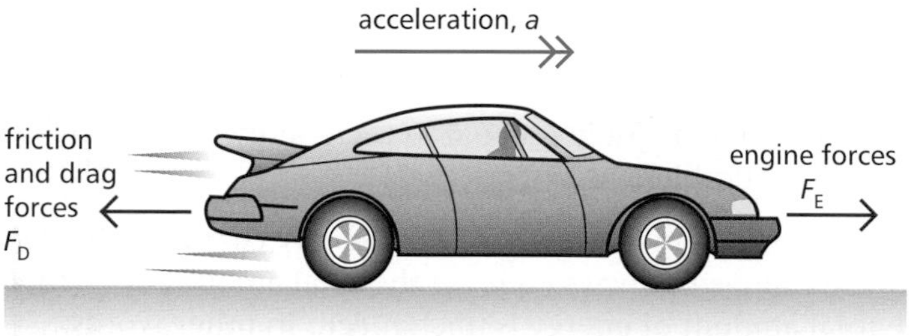

Figure 3

- Because the drag force increases with speed, the maximum speed (i.e. terminal velocity) of the vehicle, v_{max}, is attained when the drag force becomes equal and opposite to the motive force.
- At maximum speed, the **work done** by the engine is dissipated by the drag force and becomes internal energy of the surroundings. See Topic 2.6 for more about power.

Worked example

A car of mass 1200 kg has an engine which provides a motive force of 600 N. Calculate:

a its initial acceleration,

b its acceleration when the drag force is 400 N.

Solution

a The maximum acceleration is when the drag force is zero (i.e. when the car starts). The resultant force on the car is therefore 600 N at the start.

Therefore:

$$\text{initial acceleration} = \frac{\text{force}}{\text{mass}} = \frac{600\,\text{N}}{1200\,\text{kg}} = 0.5\,\text{m s}^{-2}$$

b When the drag force = 400 N,
the resultant force = motive force − drag force
= 600 − 400 N
= 200 N

$$\therefore \text{Acceleration} = \text{force/mass} = \frac{200\,\text{N}}{1200\,\text{kg}} = 0.16\,\text{m s}^{-2}$$

Hydrofoil physics

A hydrofoil boat travels much faster than an ordinary boat, because it has a powerful jet engine that enables it to 'ski' on its hydrofoils when the jet engine is switched on.

When the jet engine is switched on and takes over from the less-powerful propeller engine, the boat speeds up and the hydrofoils are extended. The boat rides on the hydrofoils so the drag force is reduced, as its hull is no longer in the water. At top speed, the motive force of the jet engine is equal to the drag force on the hydrofoils.

Figure 4 *A hydrofoil ferry*

Summary test 2.3

$g = 9.81\,\text{m s}^{-2}$.

1 a A steel ball of mass 0.15 kg released from rest in a liquid falls a distance of 0.20 m in 5.0 s. Assuming the ball reaches terminal velocity within a fraction of a second, calculate:
  i its terminal velocity,
  ii the drag force on it when it falls at terminal velocity.

b State and explain whether or not a smaller steel ball would fall at the same rate in the same liquid.

2 Explain why a cyclist can reach a higher top speed by crouching over the handlebars instead of sitting upright while pedalling.

3 A vehicle of mass 32 000 kg has an engine which has a maximum driving force of 4.4 kN and a top speed of $36\,\text{m s}^{-1}$ on a level road.
Calculate:

a its maximum acceleration from rest,

b the distance it would travel at maximum acceleration to reach a speed of $12\,\text{m s}^{-1}$ from rest.

4 Explain why a vehicle has a higher top speed on a downhill stretch of road than on a level road.

2.4 Work and energy

Learning outcomes

On these pages you will learn to:

- give examples of energy in different forms, its conversion and conservation, and apply the principle of conservation of energy to simple examples
- define work as the product of a force and the displacement in the direction of the force
- calculate the work done in a number of situations including the work done by a gas that is expanding against a constant external pressure: $W = p\Delta V$

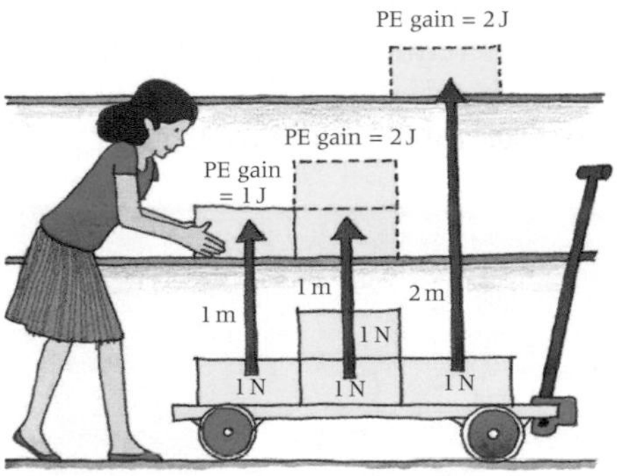

Figure 1 *Using joules*

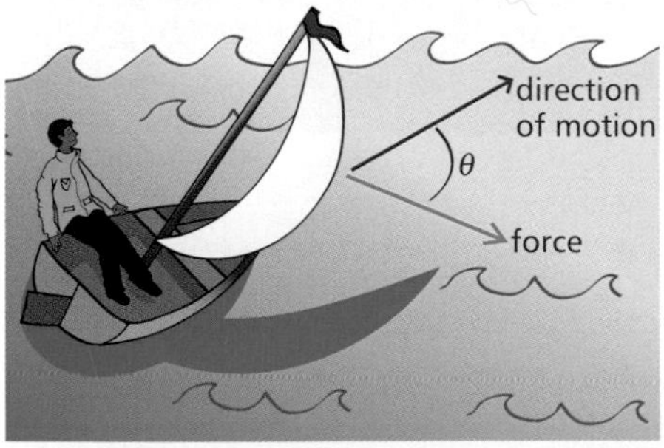

Figure 2 *Force and displacement*

Note

If $\theta = 90°$, which means that the force is perpendicular to the direction of motion, then, because $\cos 90° = 0$, the work done is zero.

Energy rules

Energy is needed to make stationary objects move, or to lift an object, or to change its shape, or to warm it up. When you lift an object, you **transfer energy** from your muscles to the object.

Objects can possess energy in **different forms**, including:

- **gravitational potential energy**, which is energy due to position in a gravitational field,
- **kinetic energy**, which is energy due to motion,
- **thermal energy**, which is energy due to the temperature of an object,
- **chemical** or **nuclear energy**, which is energy associated with chemical or nuclear reactions,
- **electrical potential energy**, which is energy of electrically charged objects,
- **elastic potential energy**, which is energy stored in an object when it is stretched or compressed.

Energy is measured in **joules (J)**. One joule is equal to the energy needed to raise a 1 N weight through a vertical height of 1 m.

Energy can be changed from one form into other forms. In any change, the total amount of energy after the change is always equal to the total amount of energy before the change. The total amount of energy is unchanged. In other words:

Energy cannot be created or destroyed.

This statement is known as the **principle of conservation of energy.**

Work

Work is done on an object when a force acting on it due to another object makes it move. As a result, energy is transferred to the object by the force from the other object. The amount of work done depends on the force and the displacement of the object. The greater the force or the further the distance, the greater the work done.

Work done = force × displacement in the direction of the force

The **unit of work** is the joule (J).

One joule is equal to the work done when a force of 1N moves its point of application by a displacement of 1m in the direction of the force.

Energy is also measured in joules, because the energy transferred to an object when a force acts on it is equal to the work done on it by the force. For example, a force of 2N needs to be applied to an object of weight 2N to raise the object steadily. If the object is raised by 1.5 m, the work done by the force is 3 J (= 2 N × 1.5 m). Therefore, the gain of potential energy of the object is 3 J.

Force and displacement

Imagine a yacht acted on by a wind force F at an angle θ to the direction in which the yacht moves, as in Figure 2. The wind force has a component $F\cos\theta$ in the direction of motion of the yacht, and a component $F\sin\theta$ at right angles to the direction of motion. If the yacht is moved a displacement s by the wind, the work done on it is equal to the component of force in the direction of motion × the displacement:

$$W = Fs\cos\theta$$

Force–distance graphs

- If a constant force F acts on an object and makes it move a displacement s in the direction of the force, the work done on the object $W = Fs$. Figure 3 shows a graph of force against distance in this situation. The area under the line is a rectangle, of height representing the force and of base length representing the distance moved. Therefore **the area represents the work done**.

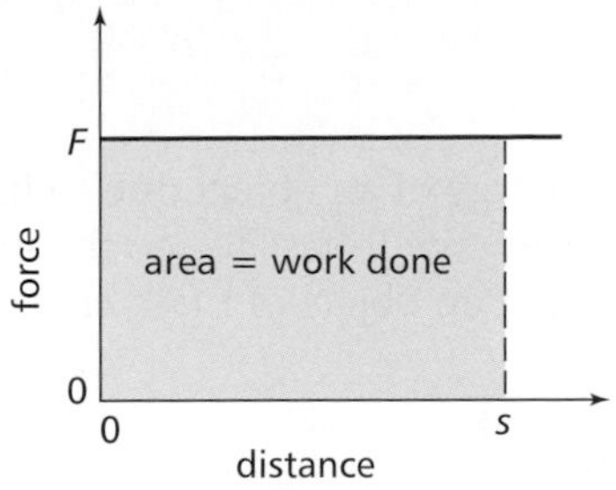

Figure 3 *A force–distance graph for a constant force*

- If a variable force F acts on an object and causes it to move in the direction of the force, for a small amount of distance Δs, the work done $\Delta W = F\Delta s$. This is represented on a graph of the force F against distance s by the area of a strip under the line of width Δs and height F. The total work done is therefore **the sum of the areas of all the strips** (i.e. the total area under the line). See Figure 4.

The area under the line of a force–distance graph represents the total work done.

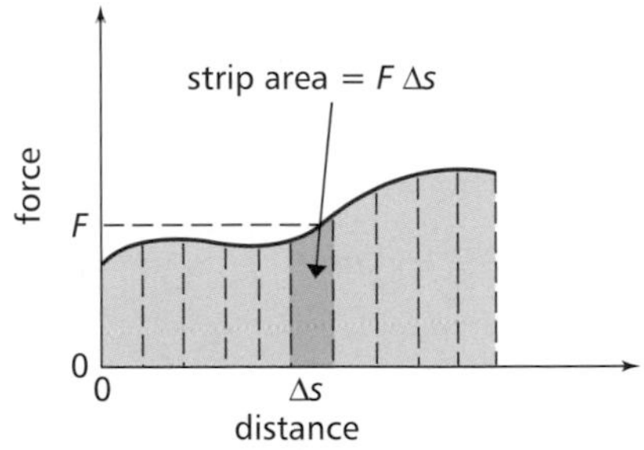

Figure 4 *A force–distance graph for a variable force*

Springs

For example, consider the force needed to stretch a spring. The greater the force, the more the spring is extended from its unstretched length. Figure 5 shows how the force needed to stretch a spring changes with the **extension** of the spring. The graph is a straight line through the origin. Therefore the force needed is proportional to the extension of the spring. This is known as **Hooke's law**. See Topic 5.3 for more about springs.

Figure 5 is a graph of force against distance; in this case, the distance the spring is extended. As explained above, the area under the line represents the work done to stretch the spring. Let F_0 represent the force needed to extend the spring to extension x_0. Therefore, because the area under the line from the origin to extension x_0 is a triangle of height F_0 and base length x_0:

Area under line = $\frac{1}{2}$ × height × base = $\frac{1}{2}F_0x_0$

Work done to stretch spring to extension e_0 = $\frac{1}{2}F_0x_0$

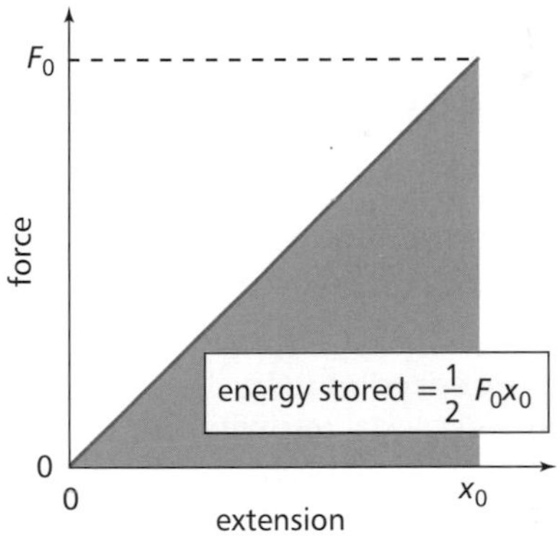

Figure 5 *Force against extension for a spring*

Summary test 2.4

1 Calculate the work done when:

a a weight of 40 N is raised by a height of 5.0 m,

b a spring is stretched to an extension of 0.45 m by a force of 20 N.

2 Calculate the energy transferred by a force of 12 N when it moves an object by a distance of 4.0 m:

a in the direction of the force,

b in a direction at 60° to the direction of the force,

c in a direction at right angles to the direction of the force.

3 A luggage trolley of total weight 400 N is pushed at a steady speed 20 m up a slope by a force of 50 N acting in the same direction as the object moves in. At the end of this distance, the trolley is 1.5 m higher than at the start. Calculate:

a the work done pushing the trolley up the slope,

b the gain of potential energy of the trolley,

c the energy wasted due to friction.

4 A spring that obeys Hooke's law requires a force of 1.2 N to extend it to an extension of 50 mm. Calculate:

a the force needed to extend it to an extension of 100 mm,

b the work done when the spring is stretched to an extension of 100 mm.

2.5 Kinetic energy and potential energy

Learning outcomes

On these pages you will learn to:

- derive, from the equations of motion, the formula for kinetic energy $E_k = \frac{1}{2}mv^2$ and recall and apply the formula
- distinguish between gravitational potential energy and elastic potential energy
- show an understanding of and use the relationship between force and potential energy in a uniform field to solve problems
- derive, from the defining equation $W = Fs$, the formula $\Delta E_p = mg\Delta h$ for potential energy changes near the Earth's surface
- recall and use the formula $\Delta E_p = mg\Delta h$ for potential energy changes near the Earth's surface

Note

The formula does not hold at speeds **approaching the speed of light**. Einstein's theory of special relativity tells us that the mass of an object increases with speed and that the energy of an object can be worked out from the equation $E = mc^2$, where c is the speed of light in free space and m is the mass of the object.

Note

The formula does not hold unless the change of height Δh is much smaller than the **Earth's radius**. If height Δh is not insignificant compared with the Earth's radius, the value of g is not the same over height Δh. The force of gravity on an object decreases with increased distance from the Earth.

Kinetic energy

Kinetic energy is the energy of an object due to its motion. The faster an object moves, the more kinetic energy it has. To see the exact link between kinetic energy and speed, consider an object of mass m, initially at rest, acted on by a constant force F for a time t.

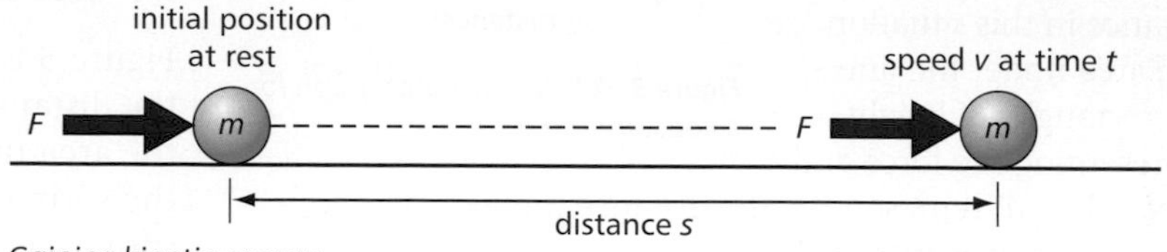

***Figure 1** Gaining kinetic energy*

Let the speed of the object at time t be v:

$\therefore$ Distance travelled, $s = \frac{1}{2}(u + v)t$

$= \frac{1}{2}vt$ because $u = 0$

Acceleration, $a = \frac{v - u}{t} = \frac{v}{t}$

Using Newton's second law:

$$F = ma = \frac{mv}{t}$$

$\therefore$ the work done, by force F, to move the object through distance s:

$$W = Fs = \frac{mv}{t} \times \frac{vt}{2} = \frac{1}{2}mv^2$$

Because the gain of kinetic energy is due to the work done:

$$\textbf{Kinetic energy, } E_K = \frac{1}{2}mv^2$$

Potential energy

Potential energy is the energy of an object due to its position.

If an object of mass m is raised through a vertical height Δh at steady speed, the force needed to raise it is equal and opposite to its weight mg. Therefore:

Work done to raise the object = force × distance moved = $mg\Delta h$

The work done on the object increases its **gravitational potential energy**:

Change of gravitational potential energy, $\Delta E_P = mg\Delta h$

At the Earth's surface, $g = 9.81\,\text{m}\,\text{s}^{-2}$.

Energy changes involving kinetic and potential energy

An object of mass m released above the ground

If air resistance is negligible, the object gains speed as it falls. Its potential energy therefore decreases and its kinetic energy increases.

After falling through a vertical height Δh, its kinetic energy is equal to its loss of potential energy.

In other words: $\frac{1}{2}mv^2 = mg\Delta h$

A pendulum bob

A pendulum bob is displaced from equilibrium and then released with the thread taut. The bob passes through the equilibrium position at maximum speed, and then slows down to reach maximum height on the other side of equilibrium. If its initial height above the equilibrium position is Δh_0, then whenever its height above the equilibrium position is Δh, its speed v at this height is such that:

Its kinetic energy = its loss of potential energy from maximum height

$$\frac{1}{2}mv^2 = mg\Delta h_0 - mg\Delta h$$

A fairground vehicle of mass m on a downward track

If a fairground vehicle was initially at rest at the top of the track, and its speed is v at the bottom of the track, then at the bottom of the track:

- its kinetic energy $= \frac{1}{2}mv^2$
- its loss of potential energy $= mg\Delta h$, where Δh is the vertical distance between the top and the bottom of the track
- the work done to overcome friction and air resistance $= mg\Delta h - \frac{1}{2}mv^2$.

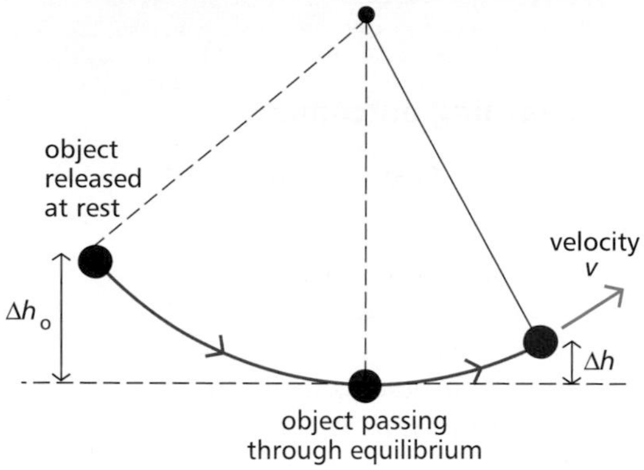

Figure 2 *A pendulum in motion*

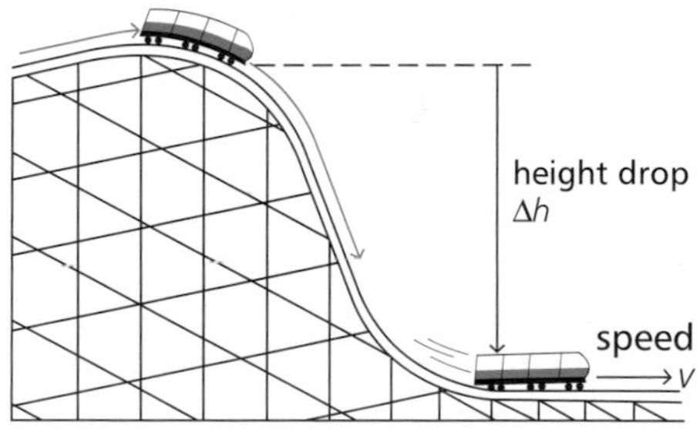

Figure 3

Worked example

$g = 9.81\,\text{m s}^{-2}$

On a fairground ride, the track descends by a vertical drop of 55 m over a distance of 120 m along the track. A train of mass 2500 kg on the track reaches a speed of $30\,\text{m s}^{-1}$ at the bottom of the descent, after being at rest at the top. Calculate:

a the loss of potential energy of the train,

b its gain of kinetic energy,

c the average frictional force on the train during the descent.

Solution

a Loss of potential energy $= mg\Delta h = 2500 \times 9.81 \times 55 = 1.35 \times 10^6$ J

b Its gain of kinetic energy $= \frac{1}{2}mv^2 = 0.5 \times 2500 \times 30^2 = 1.13 \times 10^6$ J

c Work done to overcome friction $= mg\Delta h - \frac{1}{2}mv^2 = 1.35 \times 10^6 - 1.13 \times 10^6 = 2.2 \times 10^5$ J

Because the work done to overcome friction = frictional force × distance moved along track,

$$\text{the frictional force} = \frac{\text{work done to overcome friction}}{\text{distance moved}} = \frac{2.2 \times 10^5\,\text{J}}{120\,\text{m}} = 1830\,\text{N}$$

Summary test 2.5

1 A ball of mass 0.50 kg was thrown directly up at a speed of $6.0\,\text{m s}^{-1}$. Calculate:

a its kinetic energy at $6\,\text{m s}^{-1}$,

b its maximum gain of potential energy,

c its maximum height gain.

2 A ball of mass 0.20 kg, at a height of 1.5 m above a table, is released from rest and it rebounds to a height of 1.2 m above the table. Calculate:

a i the loss of potential energy on descent,

ii the gain of potential energy at maximum rebound height,

b the loss of energy due to the impact.

3 A cyclist of mass 80 kg (including the bicycle) freewheels from rest 500 m down a hill. The foot of the hill is 20 m lower than the cyclist's starting point, and the cyclist reaches a speed of $12\,\text{m s}^{-1}$ at the foot of the hill.

Calculate:

a i the loss of potential energy,

ii the gain of kinetic energy of the cyclist and cycle,

b i the work done against friction and air resistance during the descent,

ii the average resistive force during the descent.

4 A fairground vehicle of total mass 1200 kg, moving at a speed of $2\,\text{m s}^{-1}$, descends through a height of 50 m to reach a speed of $28\,\text{m s}^{-1}$ after travelling a distance of 75 m along the track. Calculate:

a its loss of potential energy,

b its initial kinetic energy,

c its kinetic energy after the descent,

d the work done against friction,

e the average frictional force on it during the descent.

2.6 Power

Learning outcomes

On these pages you will learn to:

- define power as work done per unit time and derive power as the product of force and velocity
- solve problems using the relationships $W = Pt$ and $P = Fv$

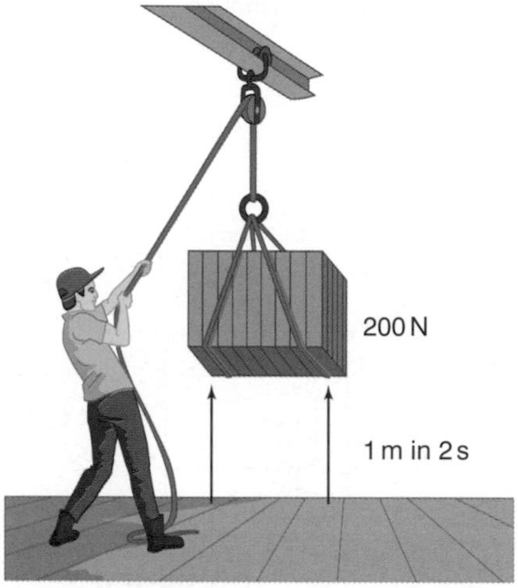

***Figure 1** A 100 watt worker*

Note

1 kilowatt (kW) = 1000 W, and 1 megawatt (MW) = 10^6 W

Power and energy

Energy transfers

Energy can be **transferred** from one object to another by means of:

- **Work done** by a force due to one object making the other object move.
- **Heat transfer** from a hot object to a cold object. Heat transfer can be due to conduction, convection or radiation.
- In addition, **electricity**, **sound waves and electromagnetic radiation**, such as light or radio waves, transfer energy.

In any energy transfer process, the more energy transferred per second, the greater the **power** of the transfer process. For example, in a tall building where there are two elevators of the same total weight, the more powerful elevator is the one that can reach the top floor faster. In other words, its motor transfers energy from electricity at a faster rate than the motor of the other elevator. The energy transferred per second is the **power** of the motor:

Power is defined as the rate of transfer of energy.

The unit of power is the watt (W), equal to an energy transfer rate of 1 joule per second.

If energy ΔE is transferred steadily in time t:

$$\textbf{Power, } P = \frac{\Delta E}{t}$$

Where energy is transferred by a force doing work, the energy transferred is equal to the work done by the force. Therefore, the **rate of transfer of energy** is equal to the work done per second. In other words, if the force does work W in time t:

$$\textbf{Power, } P = \frac{W}{t}$$

Power measurements

1 Muscle power

Test your own muscle power by timing how long it takes you to walk up a flight of steps. To calculate your muscle power, you will need to know your weight and the total height gain of the flight of steps:

- Your gain of potential energy = your weight × total height gain
- Your muscle power, $P = \dfrac{\text{energy transferred}}{\text{time taken}} = \dfrac{\text{weight} \times \text{height gain}}{\text{time taken}}$

A person of weight 480 N, who climbs a flight of stairs of height 10 m in 12 s, has leg muscles of power: $\dfrac{480\,\text{N} \times 10\,\text{m}}{12\,\text{s}} = 400\,\text{W}$

Since each leg does work while the other is being lifted, each leg would have an output power of 400 W.

2 Electrical power

The power of a 12 V light bulb can be measured using a joulemeter, as shown in Figure 2. The joulemeter is read before and after the light bulb is switched on. The difference between the readings is the energy supplied to the light bulb. If the light bulb is switched on for a measured time, the power of the light bulb can be calculated from the energy supplied to it / the time taken.

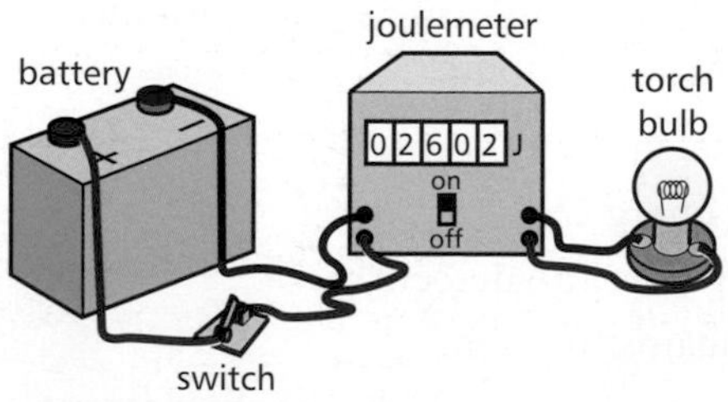

***Figure 2** Using a joulemeter*

Engine power

Vehicle engines, marine engines and aircraft engines are all designed to make objects move. The output power of an engine is called its **engine** power.

When a powered object moves at a constant velocity at a constant height, the resistive forces (e.g. friction, air resistance, drag) are equal and opposite to the engine force.

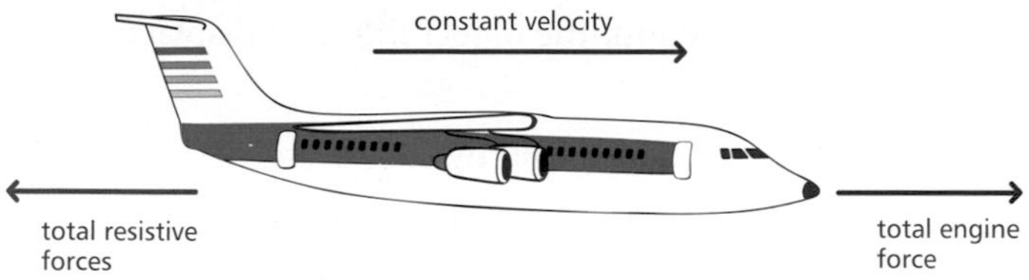

Figure 3 Engine power

The work done by the engine is converted to internal energy of the surroundings by the resistive forces.

For a powered vehicle driven by a constant force F moving at speed v:

Work done per second = force × distance moved per second

$\therefore$ Engine power of the engine, $P = Fv$

Worked example

An aircraft powered by engines that exert a force of 40 kN is in level flight at a constant velocity of $80\,m\,s^{-1}$. Calculate the engine power of the engine at this speed.

Solution

Power = force × velocity = $40\,000\,N \times 80\,m\,s^{-1} = 3.2 \times 10^6\,W$

When a powered object gains speed, the engine force exceeds the resistive forces on it.

Consider a vehicle that speeds up on a level road. The engine power of its engine is the work done by the engine per second. The work done by the engine increases the kinetic energy of the vehicle and enables the vehicle to overcome the resistive forces acting on it. Because the resistive forces increase the internal energy of the surroundings:

Engine power = energy per second wasted + gain of kinetic energy per second

Juggernaut physics

The maximum mass of a truck on UK roads must not exceed 44 tonnes, which corresponds to a total mass of 44 000 kg. This limit is set so as to prevent damage to roads and bridges. European Union regulations limit the engine power of a large truck to a maximum of 6 kW per tonne. Therefore, the maximum motive power of a 44 tonne truck is 264 kW. Prove for yourself that a truck with a engine power of 264 kW moving at a constant speed of $31\,m\,s^{-1}$ (= 70 miles per hour) along a level road experiences a drag force of 8.5 kN.

Figure 4 Heavy goods on the move

Summary test 2.6

$g = 9.81\,m\,s^{-2}$

1 A student of weight 450 N climbed 2.5 m up a rope in 18 s. Calculate:

a the gain of potential energy of the student,

b the energy transferred per second.

2 Calculate the power of the engines of an aircraft at a speed of $250\,m\,s^{-1}$, if the total engine thrust to maintain this speed is 2.0 MN.

3 A rocket of mass 5800 kg accelerates uniformly and vertically from rest to a speed of $220\,m\,s^{-1}$ in 25 s. Calculate:

a its gain of potential energy,

b its gain of kinetic energy,

c the power output of its engine, assuming no energy is wasted due to air resistance.

4 Calculate the height through which a 5 kg mass would need to be dropped to lose the same energy as a 100 W light bulb would use in 1 min.

2.7 Efficiency

Learning outcomes

On these pages you will learn to:

- recall and understand that the efficiency of a system is the ratio of useful work done by the system to the total energy input
- show an appreciation for the implications of energy losses in practical devices and use the concept of efficiency to solve problems

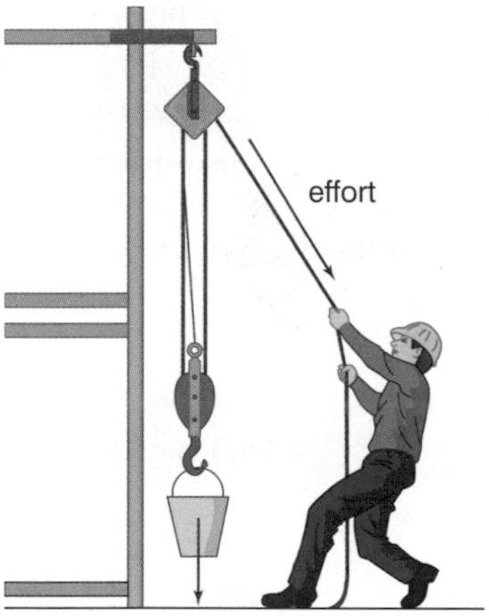

Figure 1 *Using pulleys*

Machine power

A machine that lifts or moves an object applies a force to the object to move it. If the machine exerts a force F on an object to make it move through a displacement s in the direction of the force, the work done W on the object by the machine can be calculated using the equation:

Work done, $W = Fs$

If the object moves at a constant velocity v due to this force being opposed by an equal and opposite force caused by friction, the object moves a displacement $s = vt$ in time t.

Therefore, the output power of the machine is given by

$$P_{OUT} = \frac{\text{work done by the machine}}{\text{time taken}} = \frac{Fvt}{t} = Fv$$

Output power, $P_{OUT} = Fv$, where F = 'output' force of the machine and v = speed of the object

Examples

1 An electric motor operating a sliding door exerts a force of 125 N on the door, causing it to open at a constant speed of $0.40\,m\,s^{-1}$. The output power is $125\,N \times 0.40\,m\,s^{-1} = 50\,W$. The motor must therefore transfer 50 J every second to the sliding door while the door is being opened.
Friction in the motor bearings and also electrical resistance of the motor wires means that some of the electrical energy supplied to the motor is wasted. For example, if the motor is supplied with electrical energy at a rate of $150\,J\,s^{-1}$ and it transfers $50\,J\,s^{-1}$ to the door, the difference of $100\,J\,s^{-1}$ is wasted as a result of friction and electrical resistance in the motor.

2 A pulley system is used to raise a load of 80 N at a speed of $0.15\,m\,s^{-1}$ by means of a constant **'effort'** of 30 N applied to the system. Figure 1 shows the arrangement. Note that for every metre the load rises, the effort needs to act over a distance of three metres because the load is supported by three sections of rope. The effort must therefore act at a speed of $0.45\,m\,s^{-1}$ ($= 3 \times 0.15\,m\,s^{-1}$).
   - The work done on the load each second = load × distance per second $= 80\,N \times 0.15\,m\,s^{-1} = 12\,J\,s^{-1}$
   - The work done by the effort each second = effort × distance each second $= 30\,N \times 0.45\,m\,s^{-1} = 13.5\,J\,s^{-1}$

   The difference of $1.5\,J\,s^{-1}$ is the energy wasted each second in the pulley system. This is due to friction in the bearings and also because energy must be supplied to raise the lower pulley. For example, if the weight of the lower pulley is 6 N, the potential energy gain each second by the lower pulley would be $0.9\,J\,s^{-1}$ ($= 6\,N \times 0.15\,m\,s^{-1}$) when the load is raised at a speed of $0.15\,m\,s^{-1}$. Thus the energy wasted each second due to friction would be $0.6\,J\,s^{-1}$ ($= 1.5 - 0.9\,J\,s^{-1}$).

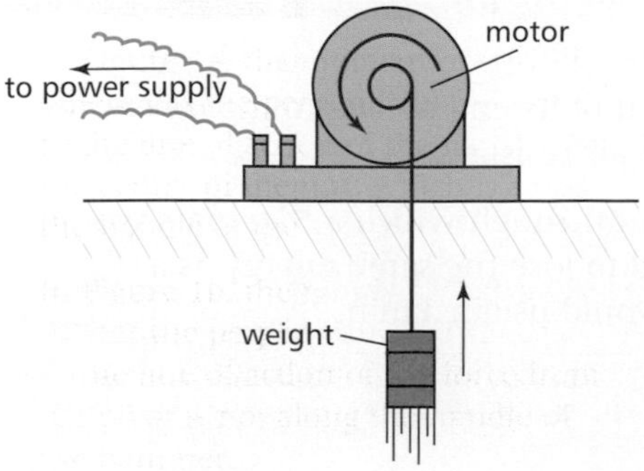

Figure 2 *Efficiency*

Efficiency measures

Useful energy is energy transferred for a purpose. In any machine, where friction is present, some of the energy transferred by the machine is wasted. In other words, not all the energy supplied to the machine is transferred for the intended purpose. For example, suppose a 500 W electric winch raises a weight of 150 N by 6.0 m in 10 s:

- The electrical energy supplied to the winch is $500\,W \times 10\,s = 5000\,J$
- The useful energy transferred by the machine is the potential energy gain of the load $= 150\,N \times 6\,m = 900\,J$.

Therefore, in this example, 4100 J of energy is wasted.

$$\text{The efficiency of a machine} = \frac{\text{useful energy transferred by the machine}}{\text{energy supplied to the machine}}$$

$$= \frac{\text{work done by the machine}}{\text{energy supplied to the machine}}$$

Note

- **Percentage efficiency = efficiency × 100%**

In the above example, the efficiency of the machine is therefore 0.18 or 18%.

- Also: $\textbf{Efficiency} = \dfrac{\textbf{output power of a machine}}{\textbf{input power to the machine}}$

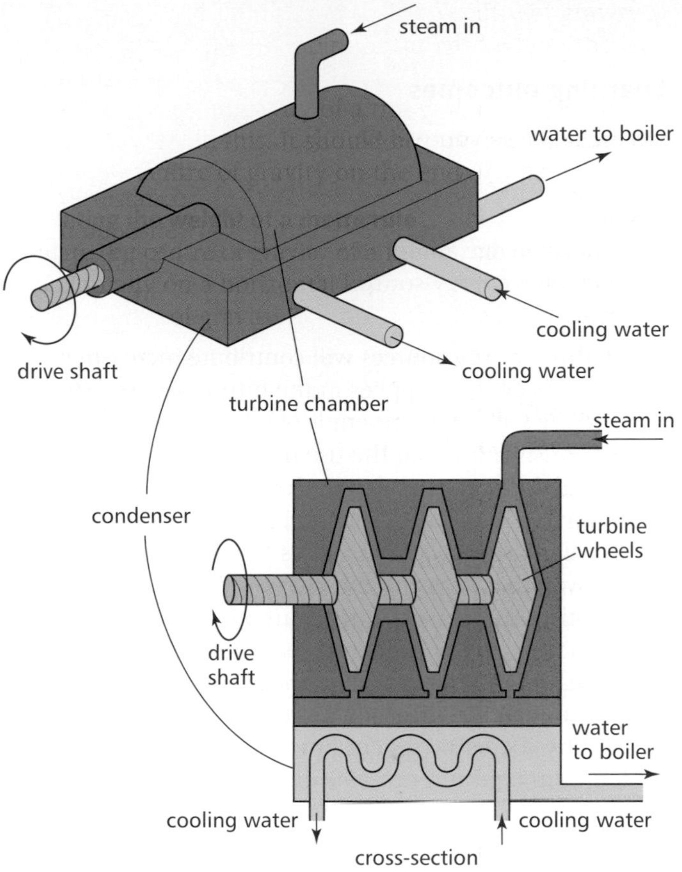

Figure 3 *A power station generator*

Wasting energy

In any process or device where energy is transferred for a purpose, the efficiency of the transfer process or the device is the fraction of the energy supplied which is used for the intended purpose. For example:

- A light bulb that is 10% efficient emits 10 J of energy as light for every 100 J of energy supplied to it by electricity. The rest of the energy is wasted as heat.
- An engine that is 45% efficient delivers 45 J of useful energy for every 100 J of energy supplied to it from its fuel. The rest of the energy is wasted as sound and heat.

Is it possible to stop energy being wasted as heat? In a power station, steam is used to drive turbines which turn the electricity generators. If the turbines were not kept cool, they would stop working because the **pressure** inside would build up and prevent steam entering. Stopping the heat transfer to the cooling water would stop the generators working. In general, energy tends to spread out when it is usefully used.

Summary test 2.7

1 In a test of muscle efficiency, an athlete on an exercise bicycle pedals against a braking force of 30 N at a speed of 15 m s^{-1}.

  a Calculate the useful energy supplied per second by the athlete's muscles.

  b If the efficiency of the muscles is 25%, calculate the energy per second supplied to the athlete's muscles.

2 A 60 W electric motor raises a weight of 20 N through a height of 2.5 m in 8.0 s. Calculate:

  a the electrical energy supplied to the motor,

  b the useful energy transferred by the motor,

  c the efficiency of the motor.

3 A power station has an overall efficiency of 35% and it produces 200 MW of electrical power. The fuel used in the power station releases 80 MJ kg^{-1} of fuel burned. Calculate:

  a the energy per second supplied by the fuel,

  b the mass of fuel burned per day.

4 A vehicle engine has a power output of 6.2 kW and uses fuel which releases 45 MJ kg^{-1} of energy when burned. At a speed of 30 m s^{-1} on a level road, the fuel usage of the vehicle is 18 km kg^{-1}. Calculate:

  a the time taken by the vehicle to travel 18 km at 30 m s^{-1},

  b the useful energy supplied by the engine in this time,

  c the overall efficiency of the engine.

wall X and 1.5 m from wall Y. Calculate the support force on the beam from each wall.

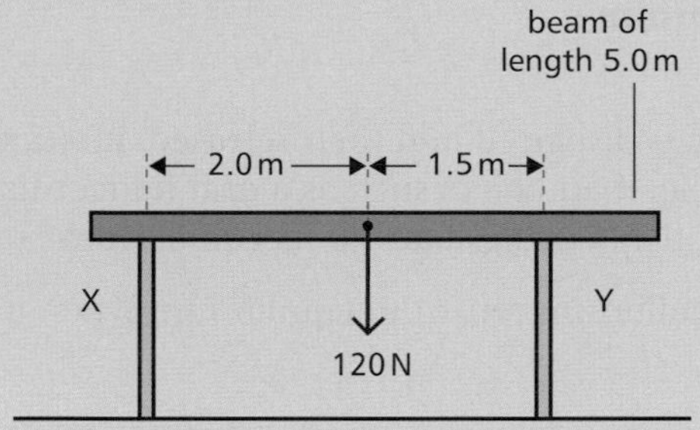

Figure 3

Solution

Let S_X and S_Y represent the support forces at X and Y.

$\therefore \quad S_X + S_Y = 120\,N$

Taking moments about X gives:

Sum of the clockwise moments

= weight of beam × distance from centre of gravity to X

= 120 N × 2.0 m = 240 N m

Sum of the anticlockwise moments

= support force S_Y × distance from X to Y

= S_Y × 3.5 m

∴ Applying the principle of moments gives:

$$3.5\,S_Y = 240$$

$$S_Y = \frac{240}{3.5} = 69\,N$$

Therefore $\quad S_X = W - S_Y = 120 - 69 = 51\,N$

Couples

A **couple** is a pair of equal and opposite forces acting on a body, but not along the same line. Figure 4 shows a couple acting on a coil. The couple turns, or tries to turn, the coil.

The moment of a couple is referred to as a **torque:**

Torque of a couple = one force × perpendicular distance between lines of action of the forces

The total moment is the same, regardless of the point about which the moments are taken.

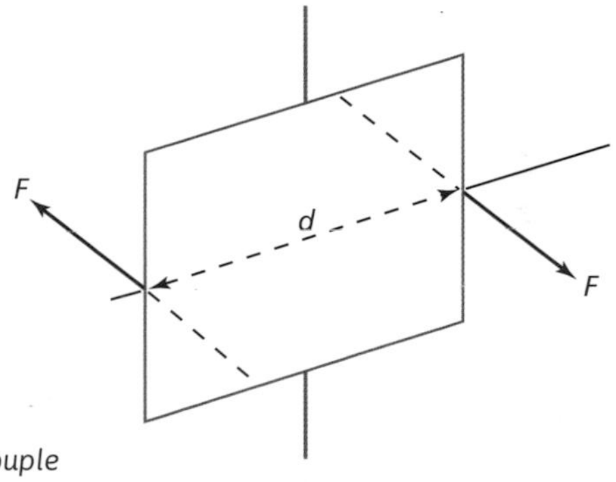

Figure 4 *A couple*

Summary test 3.3

1 A metre rule of weight 1.2 N rests horizontally on two knife edges at the 100 mm mark and the 800 mm mark. Sketch the arrangement and calculate the support force on the rule due to each knife edge.

2 A uniform beam of weight 230 N and of length 10 m rests horizontally on the tops of two brick walls 8.5 m apart, such that a length of 1.0 m projects beyond one wall and 0.5 m projects beyond the other wall (Figure 5).

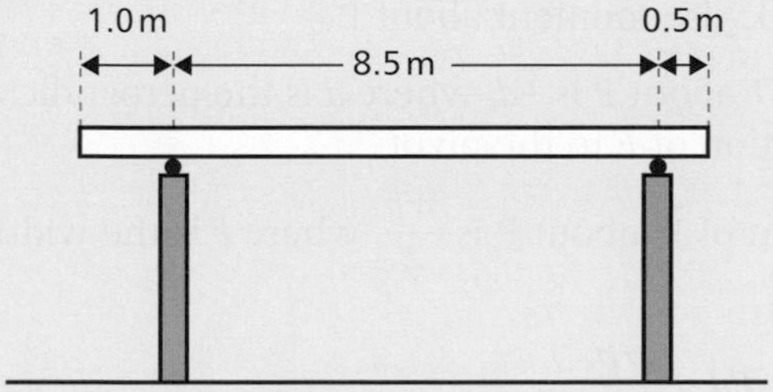

Figure 5

Calculate:

a the support force of each wall on the beam,

b the force of the beam on each wall.

3 A uniform bridge span of weight 1200 kN and of length 17.0 m rests on a support of width 1.0 m at either end. A stationary lorry of weight 60 kN is the only object on the bridge. Its centre of gravity is 3.0 m from the centre of the bridge (Figure 6). Calculate the support force on the bridge at each end.

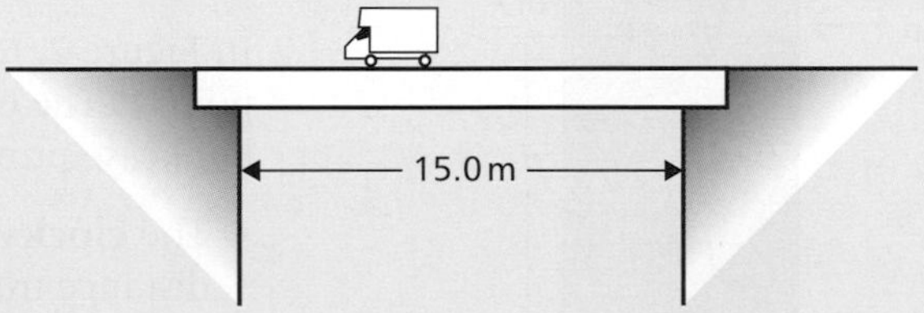

Figure 6

4 A uniform plank of weight 150 N and of length 4.0 m rests horizontally on two bricks. One of the bricks is at the end of the beam. The other brick is 1.0 m from the other end of the plank.

a Sketch the arrangement and calculate the support force on the plank from each brick.

b A child stands on the free end of the beam and just causes the other end to lift off its support.

Sketch this arrangement and calculate the weight of the child.

3.4 Stability

Learning outcomes

On these pages you will learn to:

- recognise stable and unstable equilibrium situations
- recognise how the stability of an object at rest on a flat surface is affected when the object or the surface is tilted

Stable and unstable equilibrium

Stable equilibrium

If a body in **stable equilibrium** is displaced and then released, it returns to its equilibrium position. For example, if an object such as a coat hanger hanging from a support is displaced slightly, it swings back to its equilibrium position.

Why does an object in stable equilibrium return to equilibrium when it is displaced and then released?

- The reason is that the centre of gravity of the object is directly below the point of support when the object is at rest. The support force and the weight are directly equal and opposite to each other when the object is **in equilibrium**.
- However, when it is displaced and then released, at the instant of release, the line of action of the weight no longer passes through the point of support so the weight **returns the object to equilibrium**.

Unstable equilibrium

A plank balanced on a drum is in **unstable equilibrium**. If it is displaced slightly from equilibrium then released, the plank will roll off the drum.

- The reason is that the centre of gravity of the plank is directly above the point of support when it is **in equilibrium**. The support force is exactly equal and opposite to the weight.
- If the plank is displaced slightly, the centre of gravity is no longer above the point of support. The weight therefore acts to turn the plank **further from the equilibrium position**.

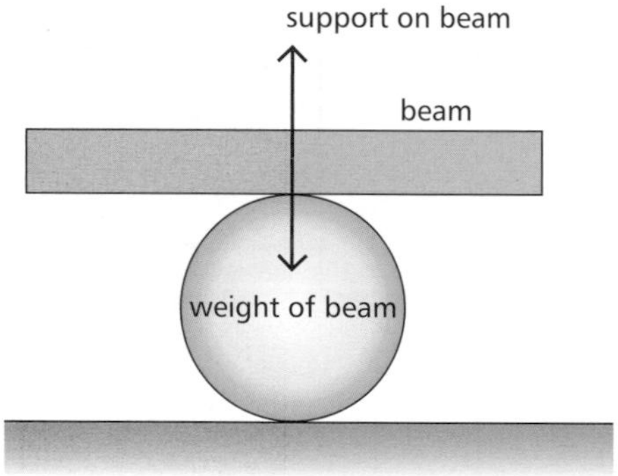

Figure 1 *Unstable equilibrium*

Tilting and toppling

Skittles at a bowling alley are easy to knock over because they are top-heavy. This means that the centre of gravity is too high and the base is too narrow. A slight nudge from a ball causes a skittle to tilt and then tip over.

Tilting

Tilting is where an object at rest on a surface is acted on by a force that raises it up on one side. For example, if a horizontal force is applied to the top of a tall free-standing bookcase, the force can make the bookcase tilt about its base along one edge.

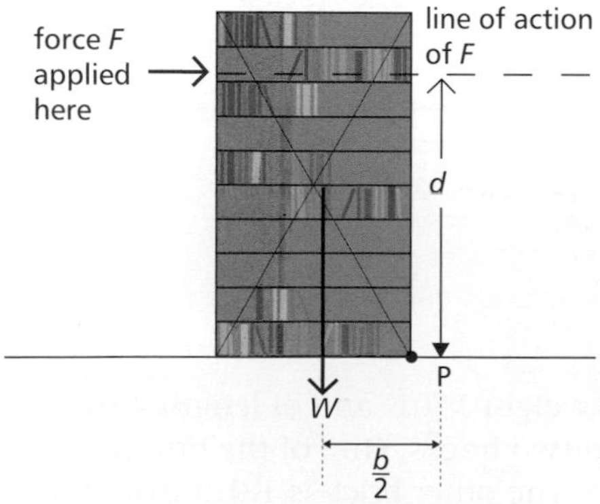

Figure 2 *Tilting*

In Figure 2, to make the bookcase tilt, the force must turn it clockwise about point P. The entire support from the floor acts at point P. The weight of the bookcase provides an anticlockwise moment about P.

- The **clockwise** moment of F about P is Fd, where d is the perpendicular distance from the line of action of F to the pivot.
- The **anticlockwise** moment of W about P is $\frac{Wb}{2}$, where b is the width of the base.

Therefore, for tilting to occur: $\boldsymbol{Fd > \frac{Wb}{2}}$

Toppling

A tilted object will **topple** over if it is tilted too far. For example, a tractor on a hill could topple over sideways if the hill is too steep. If an object on a flat surface is tilted more and more, the line of action of its weight (which is through its centre of gravity) passes closer and closer to the 'pivot'. If the object is tilted so much that the line of action of its weight passes beyond the pivot, the object will topple over if allowed to. The position where the line of action of the weight passes through the 'pivot' is the furthest it can be tilted without toppling. Beyond this position, it topples over if it is released. See Figure 3.

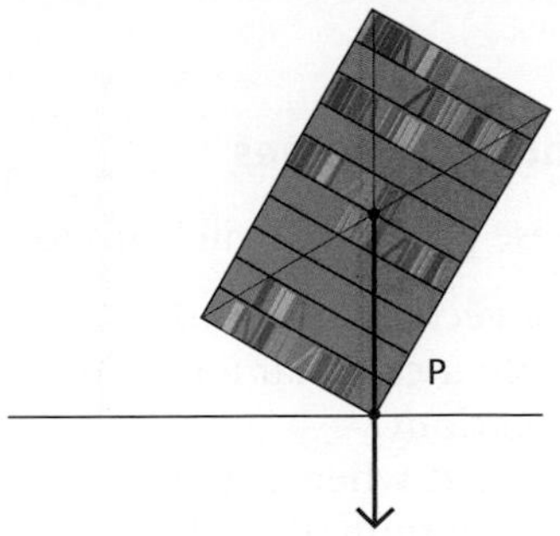

Figure 3 *Toppling over*

On a slope

A tall object on a slope will topple over if the slope is too great. For example, a high-sided vehicle on a road with a sideways slope will tilt over. If the slope is too great, the vehicle will topple over. This will happen if the line of action of the weight (passing through the centre of gravity of the object) lies outside the **wheel base** of the vehicle. In Figure 4, the vehicle will not topple over because the line of action of the weight lies within the wheel base.

Consider the forces acting on the vehicle on a slope when it is at rest. The sideways friction F, the support forces S_X and S_Y and the force of gravity on the vehicle (i.e. its weight) act as shown in Figure 4. For equilibrium, resolving the forces parallel and perpendicular to the slope gives:

- **Parallel** to the slope:

$$F = W \sin\theta$$

- **Perpendicular** to the slope:

$$S_X + S_Y = W \cos\theta$$

Note that S_X is greater than S_Y because X is lower than Y.

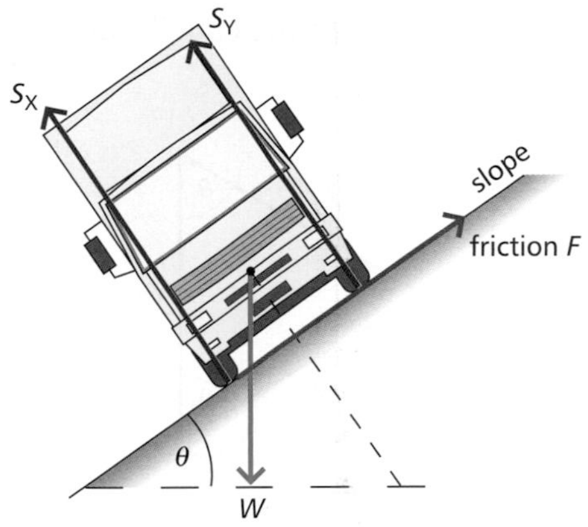

Figure 4 *On a slope*

Summary test 3.4

1 Explain why a bookcase with books on its top shelf only is less stable than if the books were on the bottom shelf.

2 An empty wardrobe of weight 400 N has a square base 0.8 m × 0.8 m and a height of 1.8 m. A horizontal force is applied to the top edge of the wardrobe to make it tilt. Calculate the force needed to lift the wardrobe base off the floor along one side (Figure 5).

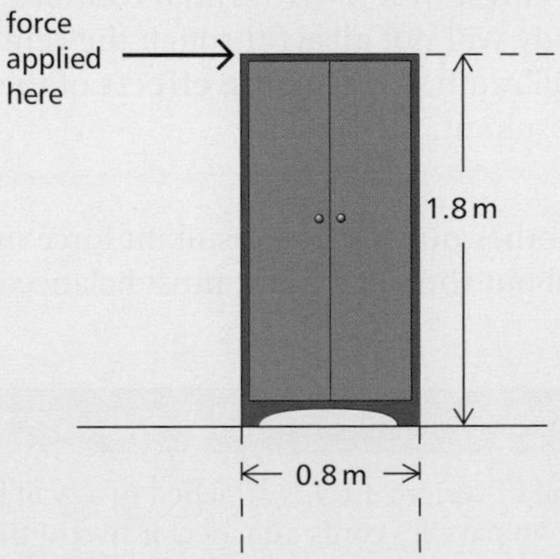

Figure 5

3 A vehicle has a wheel base of 1.8 m and a centre of gravity, when unloaded, which is 0.8 m from the ground (Figure 6).

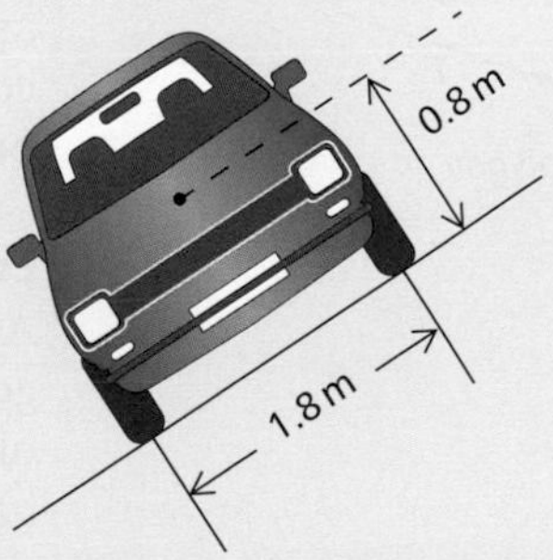

Figure 6

a The vehicle is tested for stability on an adjustable slope. Calculate the maximum angle of the slope to the horizontal if the vehicle is not to topple over.

b If the vehicle carries a full load of people, will it be more or less likely to topple over on a slope? Explain your answer.

4 Discuss whether or not a high-sided heavy goods lorry is more or less likely to be affected by strong side winds when it is fully loaded compared to when it is empty.

3.5 Equilibrium rules

Learning outcomes

On these pages you will learn to:

- use a vector triangle to represent coplanar forces in equilibrium
- recall that when a system is in equilibrium, there is no resultant force and no resultant torque acting on it
- apply the above general conditions for the equilibrium of a body

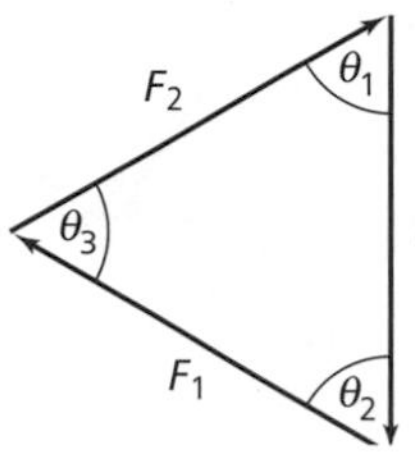

Figure 1 *Triangle of forces*

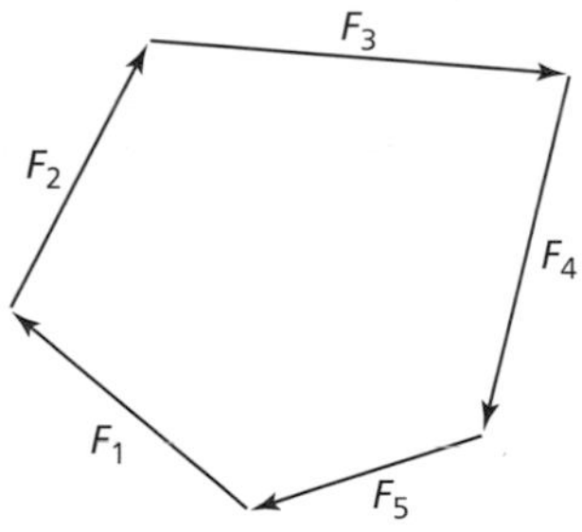

Figure 2 *The closed polygon*

The triangle of forces

For a point object acted on by **three forces** to be in equilibrium, the three forces must give an overall **resultant of zero**. The three forces as vectors should form a **triangle**. In other words, for three forces F_1, F_2 and F_3 to give zero resultant:

$$\text{Vector sum, } \boldsymbol{F_1 + F_2 + F_3 = 0}$$

Any two of the forces gives a resultant which is represented by the third side of the triangle. Therefore, for equilibrium, the third force must be represented by the third side of the triangle. For example, the resultant of $F_1 + F_2$ is equal and opposite to F_3:

i.e. $$\boldsymbol{F_1 + F_2 = -F_3}$$

The sine rule can be applied to the triangle of forces to find an unknown force or angle, given the other forces and angles in the triangle:

$$\frac{F_1}{\sin\theta_1} = \frac{F_2}{\sin\theta_2} = \frac{F_3}{\sin\theta_3}$$

where θ_1, θ_2 and θ_3 are the angles opposite sides F_1, F_2 and F_3 respectively.

The closed polygon

The triangle of forces rule can be extended for any number of forces acting on an object. If the object is in equilibrium, the force vectors drawn end-to-end must form a closed polygon. In other words, the tip of the last force vector must join the tail of the first force vector. Figure 2 shows the idea. Unfortunately, the sine rule can't be applied here; so the forces must be resolved in the same parallel and perpendicular directions to calculate an unknown force, given all the other forces.

The conditions for equilibrium of a body

Free-body force diagrams

When two objects interact, they always exert equal and opposite forces on one another. A diagram showing the forces acting on an object can become very complicated if it also shows the forces the object exerts on other objects. A **free-body force diagram** shows only the forces acting on the object.

Equilibrium

An object in equlibrium is either at rest or it moves with a constant velocity. In general, the forces acting on a body will not all act through the centre of gravity of the body. If the body is in equilibrium, the **turning effects** of the forces must balance out as well giving zero resultant.

For a body at rest:

- The **forces** must balance each other out (i.e. the resultant force must be zero).
- The **moments of the forces** about the same point must balance (i.e. the resultant torque must be zero).

Worked example

A uniform shelf of width 0.6 m and of weight 12 N is attached to a wall by hinges and is supported horizontally by two parallel cords attached at two of the corners of the shelf, as shown in Figure 3. The other end of each cord is fixed to the wall 0.4 m above the hinge. Calculate:

a the angle between each cord and the shelf,

b the tension in each cord.

Solution

a Let the angle between each cord and the shelf be θ.

From Figure 3, $\tan\theta = \dfrac{0.4}{0.6}$ so $\theta = 34°$

b Taking moments about the hinge gives:

- Sum of the clockwise moments
  = weight of shelf × distance from hinge to the centre of gravity of the shelf
  = 12 × 0.3 = 3.6 N m
- Sum of the anticlockwise moments = $2\,Td$, where T is the tension in each cord and d is the perpendicular distance from the hinge to either cord.

  From Figure 3, it can be seen that $d = 0.6\sin\theta = 0.6\sin 34 = 0.34$ m
  ∴ Applying the principle of moments gives:

$$2 \times 0.34 \times T = 3.6$$
$$T = 5.3\,\text{N}$$

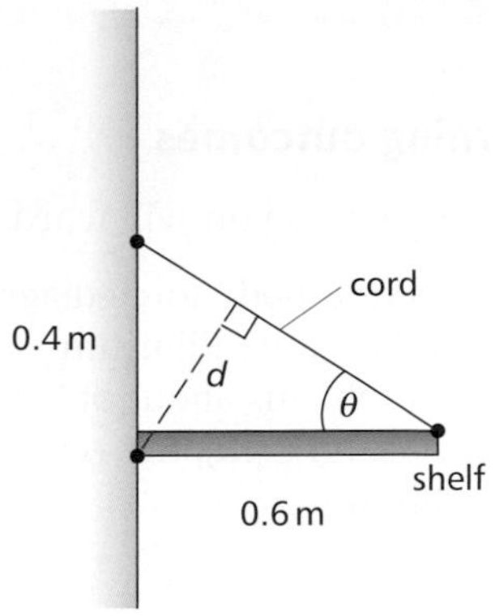

Figure 3

Summary test 3.5

1 A uniform plank of length 5.0 m rests horizontally on two bricks, which are 0.5 m from either end. A child of weight 200 N stands on one end of the plank and causes the other end to lift, so that it is no longer supported at that end.

Figure 4

Calculate:

a the weight of the plank,

b the support force acting on the plank from the supporting brick.

2 A security camera is supported by a frame which is fixed to a wall and ceiling (Figure 5). The supporting structure must be strong enough to withstand the effect of a downward force of 1500 N acting on the camera, in case the camera is gripped by someone below it.

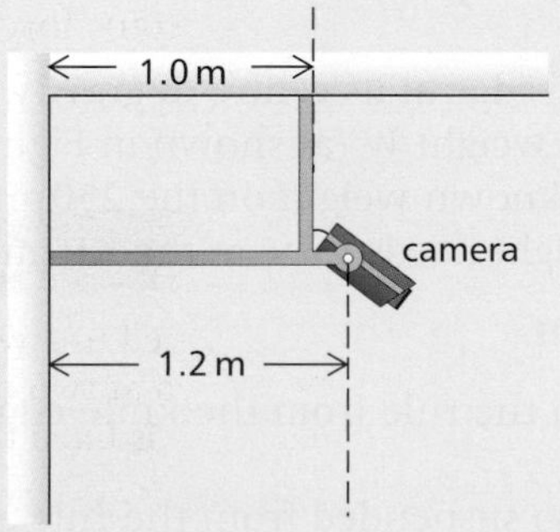

Figure 5

Calculate:

a the moment of a downward force of 1500 N on the camera about the point where the supporting structure is attached to the wall,

b the extra force of the vertical strut supporting the frame, when the camera is pulled with a downward force of 1500 N.

3 A crane is used to raise one end of a 15 kN girder of length 10.0 m off the ground. When the end of the girder is at rest 6.0 m off the ground, the crane cable is perpendicular to the girder (Figure 6). Calculate the tension in the cable.

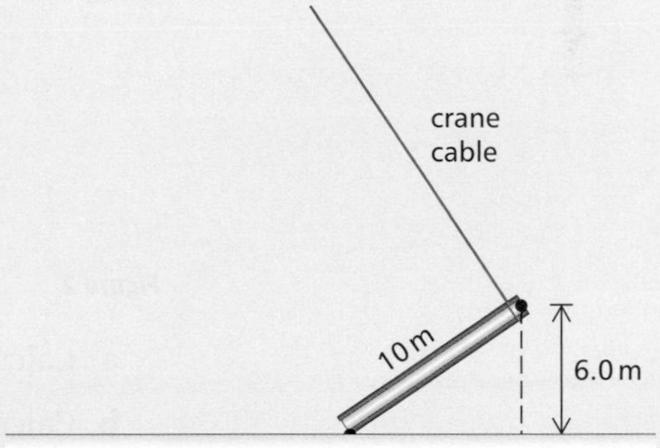

Figure 6

4 In question **3**, show that the support force on the girder from the ground has a horizontal component of 3.6 N and a vertical component of 10.2 kN. Hence calculate the magnitude of the support force.

3.6 Statics calculations

Learning outcomes

On these pages you will learn to:

- sketch free-body force diagrams for bodies in equilibrium
- solve problems about objects in equilibrium using the rules for equilibrium

1 Calculate the magnitude of the resultant of a 6.0 N force and a 9.0 N force acting on a point object when the two forces act:

a in the same direction,

b in opposite directions,

c at 90° to each other.

2 A point object in equilbrium is acted on by a 3 N force, a 6 N force and a 7 N force. What is the resultant force on the object if the 7 N force is removed?

3 A point object of weight 5.4 N in equilibrium is acted on by a horizontal force of 4.2 N and a second force F.

a By considering the triangle of forces rule, determine the magnitude of F.

b Calculate the angle between the direction of F and the horizontal.

4 An object of weight 7.5 N hangs on the end of a cord which is attached to the midpoint of a wire, stretched between two points on the same horizontal level. Each half of the wire is at 12° to the horizontal. Calculate the tension in each half of the wire.

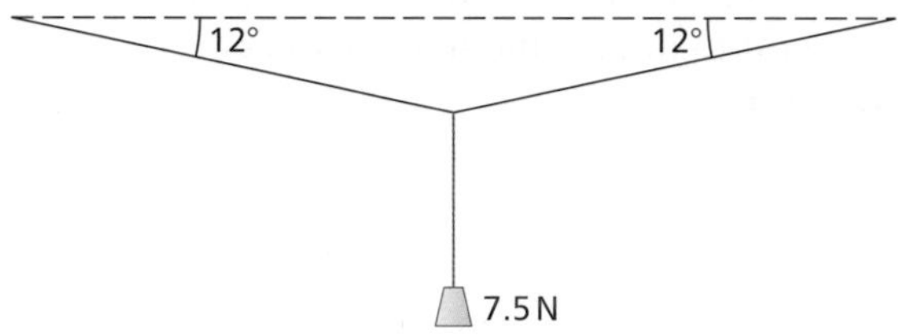

Figure 1

5 A ship is towed at constant speed by two tug boats, each pulling the ship with a force of 9 kN. The angle between the tug-boat cables is 40°.

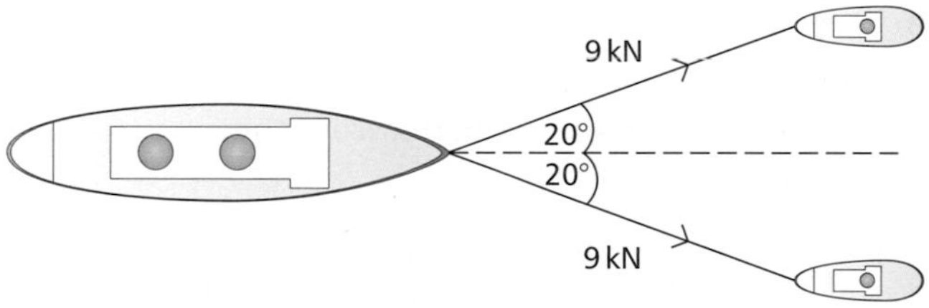

Figure 2

a Calculate the resultant force on the ship due to the two cables.

b Calculate the drag force on the ship.

6 A metre rule is pivoted on a knife edge at its centre of gravity, supporting a weight of 5.0 N and an unknown weight W (as shown in Figure 3). To balance the rule horizontally with the unknown weight on the 250 mm mark of the rule, the position of the 5.0 N weight needs to be at the 810 mm mark.

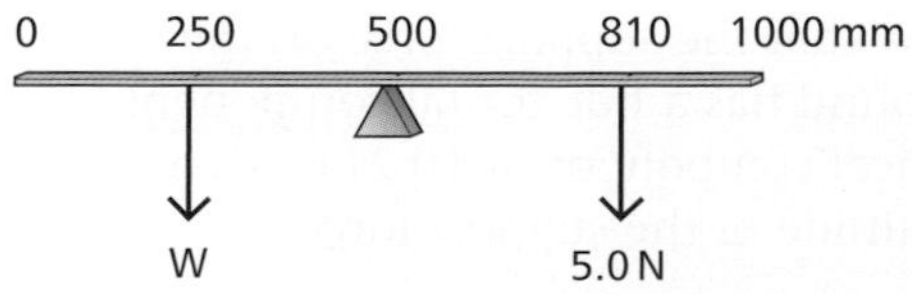

Figure 3

a Calculate the unknown weight.

b Calculate the support force on the rule from the knife edge.

7 In Figure 3, a 2.5 N weight is also suspended from the rule at its 400 mm mark. What adjustment needs to be made to the position of the 5.0 N weight to rebalance the rule?

8 A uniform metre rule is balanced horizontally on a knife edge at its 350 mm mark by placing a 3.0 N weight on the rule at its 10 mm mark.

a Sketch the arrangement and calculate the weight of the rule.

b Calculate the support force on the rule from the knife edge.

9 A diving board has a length 4.0 m and a weight of 250 N. It is bolted to the ground at one end and projects by a length of 3.0 m beyond the edge of the swimming pool (Figure 4). A person of weight 650 N stands on the free end of the diving board. Calculate:

a the force on the bolts,

b the force on the edge of the swimming pool.

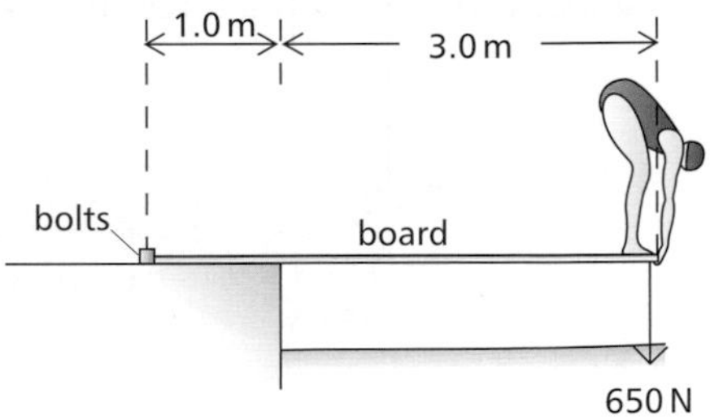

Figure 4

10 A uniform beam XY of weight 1200 N, and of length 5.0 m, is supported horizontally on a concrete pillar at each end. A person of weight 500 N sits on the beam, at a distance of 1.5 m from end X.

a Sketch a free-body force diagram of the beam.

b Calculate the support force on the beam from each pillar.

11 A bridge crane used at a freight depot consists of a horizontal span of length 12 m fixed at each end to a vertical pillar (Figure 5).

a When the bridge crane supports a load of 380 kN at its centre, a force of 1600 kN is exerted on each pillar. Calculate the weight of the horizontal span.

b The same load is moved across a distance of 2.0 m by the bridge crane. Sketch a free-body force diagram of the horizontal span and calculate the force exerted on each pillar.

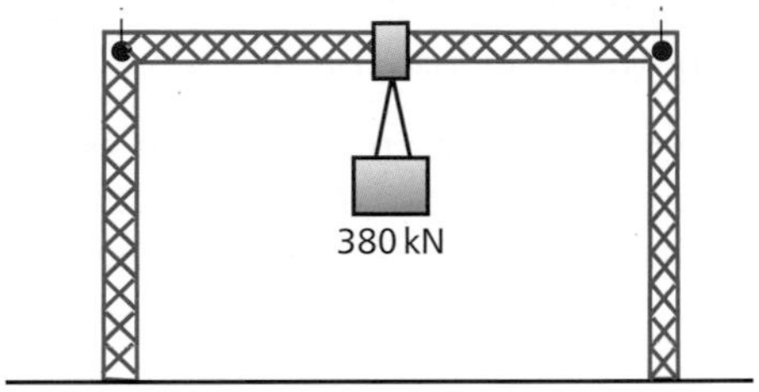

Figure 5

12 A curtain pole of weight 24 N and of length 3.2 m is supported horizontally by two wall-mounted supports X and Y, which are 0.8 m and 1.2 m from each end respectively.

a Sketch the free-body force diagram for this arrangement, and calculate the force on each support when there are no curtains on the pole.

b When the pole supports a pair of curtains of total weight 90 N drawn along the full length of the pole, what will be the force on each support?

13 A steel girder of weight 22 kN and of length 14 m is lifted off the ground at one end, by means of a crane. When the raised end is 2.0 m above the ground, the cable is vertical.

a Sketch a free-body force diagram of the girder in this position.

b Calculate the tension in the cable at this position and the force of the girder on the ground.

14 A rectangular picture of weight 24 N, hangs on a wall supported by a cord attached to the frame at each of the top corners (Figure 6). Each section of the cord makes an angle of 25° with the picture, which is horizontal along its width.

a Copy the diagram and mark the forces acting on the picture.

b Calculate the tension in each section of the cord.

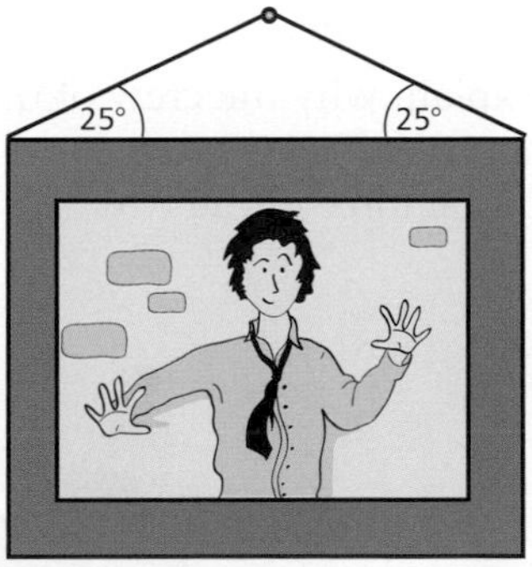

Figure 6

- The impact force F_1 on ball A changes the velocity of A from u_A to v_A in time t.

  Therefore, $F_1 = \dfrac{m_A v_A - m_A u_A}{t}$,

  where t = the time of contact between A and B,

  and m_A = the mass of ball A.
- The impact force F_2 on ball B changes the velocity of B from u_B to v_B in time t.

  Therefore, $F_2 = \dfrac{m_B v_B - m_B u_B}{t}$,

  where t = the time of contact between A and B,

  and m_B = the mass of ball B.

Because the two forces are equal and opposite to each other, $F_2 = -F_1$

Therefore,

$$\frac{m_B v_B - m_B u_B}{t} = -\frac{m_A v_A - m_A u_A}{t}$$

Cancelling t on both sides gives

$$m_B v_B - m_B u_B = - m_A v_A + m_A u_A$$

Rearranging this equation gives

$$m_B v_B + m_A v_A = m_A u_A + m_B u_B$$

Therefore,

the total final momentum = the total initial momentum

In other words, the total momentum is unchanged by this collision.

Note

If the colliding objects stick together as a result of the collision, they have the same final velocity. The above equation may therefore be written

$$(m_B + m_A)\,V = m_A u_A + m_B u_B$$

Testing conservation of momentum

Figure 3 shows an arrangement that can be used to test conservation of momentum using a motion sensor linked to a computer. The mass of each trolley is measured before the test. With trolley B at rest, trolley A is given a push so that it moves towards trolley B at constant velocity. The two trolleys stick together on impact. The computer records and displays the velocity of trolley A throughout this time. The computer display shows that the velocity of trolley A dropped suddenly when the impact took place. The velocity of trolley A immediately before the collision, u_A, and after the collision (V) can be measured. The measurements should show that the total momentum of both trolleys after the collision is equal to the momentum of trolley A before the collision. In other words,

$$(m_B + m_A)\,V = m_A u_A$$

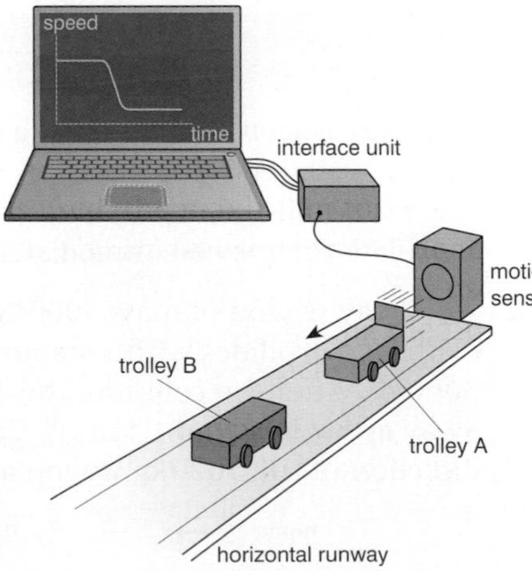

Figure 3 *Testing conservation of momentum*

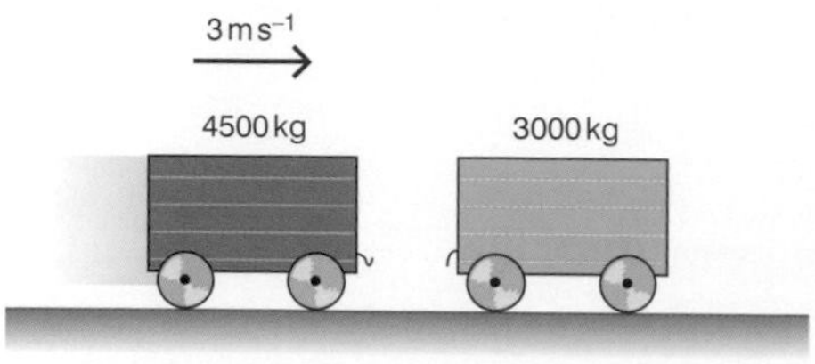

Figure 4 *Colliding wagons*

Worked example

A railway wagon of mass 4500 kg moving along a level track at a speed of $3.0\,\mathrm{m\,s^{-1}}$ collides with and couples to a second railway wagon of mass 3000 kg which is initially stationary. Calculate the speed of the two wagons immediately after the collision.

Solution

Total initial momentum

$= \text{initial momentum of A} + \text{initial momentum of B}$

$= (4500 \times 3.0) + (3000 \times 0) = 13\,500\,\mathrm{kg\,m\,s^{-1}}$

Total final momentum

$= \text{total mass of A and B} \times \text{velocity } V \text{ after the collision}$

$= (4500 + 3000)\,V = 7500V$

Using the principle of conservation of momentum,

$$7500V = 13\,500$$

$$V = \frac{13\,500}{7500} = 1.8\,\mathrm{m\,s^{-1}}$$

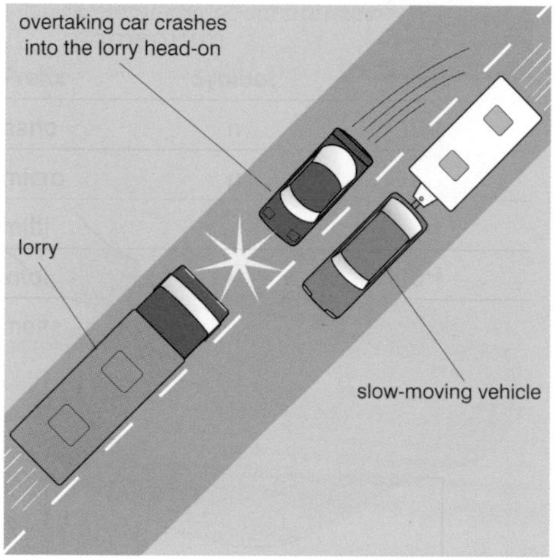

Figure 5 *A crash calculation*

Head-on collisions

Consider two objects moving in opposite directions that collide with each other. Depending on the masses and initial velocities of the two objects, the collision could cause them both to stop. The momentum of the two objects after the collision would then be zero. This could only happen if the initial momentum of one object was exactly equal and opposite to that of the other object. In general, if two objects move in opposite directions before a collision, then the vector nature of momentum needs to be taken into account by assigning numerical values of velocity + or − according to the direction.

For example, if a car of mass 600 kg travelling at a velocity of $25\,\mathrm{m\,s^{-1}}$ collides head-on with a lorry of mass 2400 kg travelling at a velocity of $10\,\mathrm{m\,s^{-1}}$ in the opposite direction the total momentum before the collision is $9000\,\mathrm{kg\,m\,s^{-1}}$ in the direction in which the lorry was moving. As momentum is conserved in a collision, the total momentum after the collision is the same as the total momentum before the collision. Prove for yourself that if the two vehicles stick together after the collision, they must have a velocity of $3.0\,\mathrm{m\,s^{-1}}$ ($= 9000\,\mathrm{kg\,m\,s^{-1}} / 3000\,\mathrm{kg}$) immediately after the impact in the direction in which the lorry was moving.

Summary test 4.3

1 A railway wagon of mass 3000 kg moving at a velocity of $1.2\,\mathrm{m\,s^{-1}}$ collides with a stationary wagon of mass 2000 kg. After the collision, the two wagons couple together. Calculate their speed immediately after the collision.

2 A railway wagon of mass 5000 kg moving at a velocity of $1.6\,\mathrm{m\,s^{-1}}$ collides with a stationary wagon of mass 3000 kg. After the collision, the 3000 kg wagon moves away at a velocity of $1.5\,\mathrm{m\,s^{-1}}$. Calculate the speed and direction of the 5000 kg wagon after the collision.

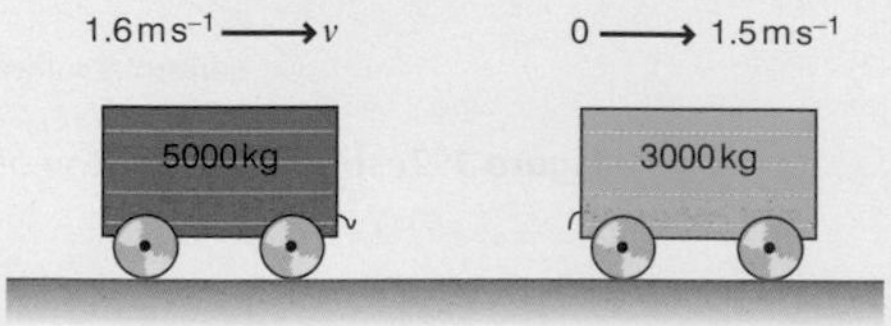

Figure 6

3 In a laboratory experiment, a trolley of mass 0.50 kg moving at a speed of $0.25\,\mathrm{m\,s^{-1}}$ collides with a trolley of mass 1.0 kg moving in the opposite direction at a speed of $0.20\,\mathrm{m\,s^{-1}}$. The two trolleys couple together on collision. Calculate their speed and direction immediately after the collision.

4 A ball of mass 0.80 kg moving at a speed of $2.5\,\mathrm{m\,s^{-1}}$ along a straight line collides with a ball of mass 2.5 kg which is initially stationary. As a result of the collision, the 2.5 kg ball is given a velocity of $1.0\,\mathrm{m\,s^{-1}}$ along the same line. Calculate the speed and direction of the 0.80 kg ball immediately after the collision.

Elastic and inelastic collisions

4.4

Learning outcomes

On these pages you will learn to:

- distinguish between an elastic and an inelastic collision
- recognise that momentum is always conserved in any collision, whereas kinetic energy is only conserved in an elastic collision
- apply the principle of conservation of momentum to elastic and inelastic interactions between bodies in both one and two dimensions
- recognise that, for a perfectly elastic collision, the relative speed of approach is equal to the relative speed of separation

Drop a bouncy rubber ball from a measured height onto a hard floor. The ball should bounce back almost to the same height. Try the same with a cricket ball and there will be very little bounce! A **perfectly elastic** ball would be one that bounces back to exactly the same height. Its kinetic energy just after impact must equal its kinetic energy just before impact. Otherwise, it cannot regain its initial height. There is no loss of kinetic energy in a perfectly elastic collision.

A perfectly elastic collision is one where there is no loss of kinetic energy.

- A squash ball hitting a hard surface bounces off the surface with little or no loss of speed. If the ball is perfectly elastic, there is no loss of speed on impact and no loss of kinetic energy.
- A very low speed impact between two cars is almost perfectly elastic, provided no damage is done. Some of the initial kinetic energy of the two vehicles may be converted to sound. However, if the collision causes damage to the vehicles, the kinetic energy after the collision is less than before so the collision is not elastic.

A totally inelastic collision is one where the colliding objects stick together.

- A railway wagon that collides with and couples to another wagon is an example of a totally inelastic collision. Some of the initial kinetic energy is converted to other forms of energy.
- A vehicle crash in which the colliding vehicles lock together is another example of a totally inelastic collision. The total kinetic energy after the collision is less than the total kinetic energy before the collision.

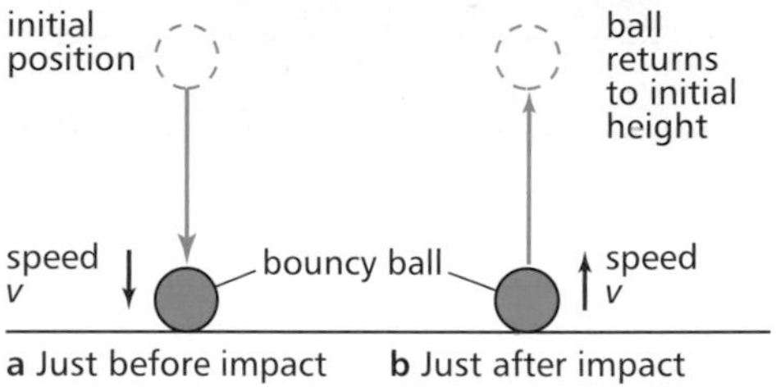

***Figure 1** An elastic impact*

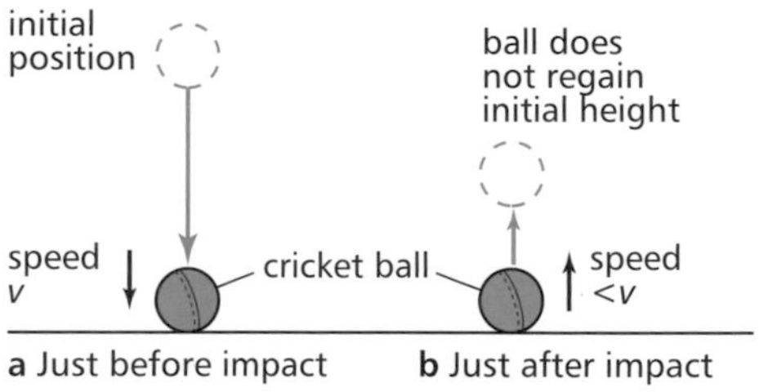

***Figure 2** A partially elastic impact*

A partially inelastic collision is where the colliding objects:

1 move apart, and

2 have less kinetic energy after the collision than before.

To work out if a collision is elastic or inelastic, the kinetic energy of each object before and after the collision must be worked out.

For a ball of mass m falling in air from a measured height H and rebounding to a height h, as shown in Figure 3:

- the kinetic energy immediately before impact = loss of potential energy through height $H = mgH$,
- the kinetic energy immediately after impact = gain of potential energy through height $h = mgh$.

So the height ratio $\frac{h}{H}$ gives the fraction of the initial kinetic energy that is recovered as kinetic energy after the collision.

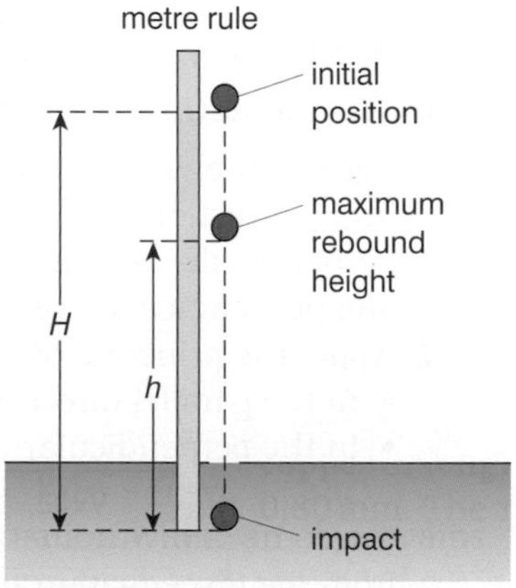

***Figure 3** Testing an impact*

Collisions between two objects

When two objects with known masses and initial velocities collide, if the final velocity of one of the objects is known, the final velocity of the other object can be determined using the principle of conservation of momentum. The kinetic energy of each object before or after the collision can then be calculated using the kinetic energy formula $E_k = \frac{1}{2}mv^2$. If the collision is elastic, the total kinetic energy of the objects before the collision is equal to their total kinetic energy after the collision.

For a perfectly elastic collison between two objects, using the principle of conservation of momentum and the rule that the total kinetic energy is unchanged in an elastic collision, it can be shown that **their relative speed of separation = their relative speed of approach.**

- **For a collision in which the objects move along the same straight line before and after the collision** (e.g. railway wagons on a straight track), velocities in one direction are assigned positive values and velocities in the opposite direction are assigned negative values, because velocity is a vector quantity.

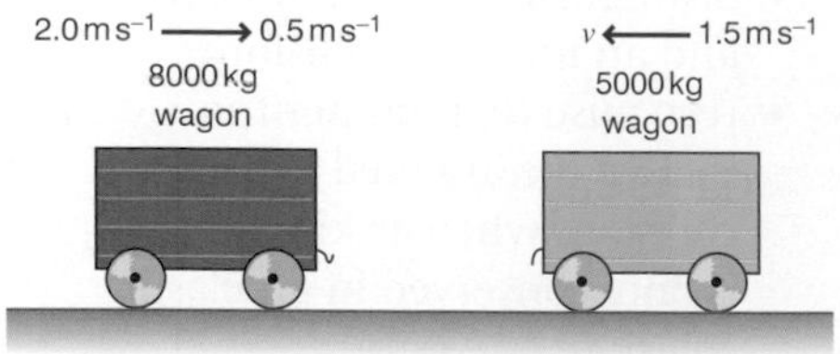

Figure 4

Worked example

A railway wagon of mass 8000 kg moving at a speed of 2.0 m s^{-1} collides with a wagon of mass 5000 kg moving at a speed of 1.5 m s^{-1} in the opposite direction. The collision reduces the speed of the 8000 kg wagon to 0.5 m s^{-1} without changing its direction. Calculate the speed V and direction of the second wagon after the collision.

Solution

The total initial momentum $= (8000 \times 2.0) + (5000 \times -1.5)$

$= 16\,000 - 7500 = 8500\,\text{kg m s}^{-1}$

(note the negative value of the initial velocity of the 5000 kg wagon)

The total final momentum $= (8000 \times 0.5) + (5000 \times V)$

Using the principle of conservation of momentum

$$(8000 \times 0.5) + (5000 \times V) = 8500$$

$$5000 \times V = 8500 - 4000 = 4500$$

$$V = \frac{4500}{5000} = +0.9\,\text{m s}^{-1}$$

Note that the relative speed of separation (0.4 m s^{-1} (= 0.9 m s^{-1} – 0.5 m s^{-1})) is not equal to their relative speed of approach (3.5 m s^{-1} (= 2.0 m s^{-1} + 1.5 m s^{-1})). So the collision is **inelastic**. Prove for yourself that 18.6 kJ of kinetic energy is transferred to the surroundings (as heat and sound).

__Figure 5__ This dramatic collision was used to demonstrate that special casks used to transport radioactive materials by rail could withstand high speed impacts. In this collision, a remotely controlled diesel locomotive was driven into a stationary cask. Even though the locomotive was destroyed, the cask itself was intact afterwards.

- **In an 'oblique' collision, two objects collide then move away in different directions from their initial directions.** Figure 6 shows an oblique collision between two spheres X and Y, where X is initially moving at velocity V towards Y, which is initially stationary. After the collision, the two objects move away at velocities v_X and v_Y in directions at angles θ_X and θ_Y to the initial direction of X. If the initial and final velocity of X is known and the masses of the two spheres are known, the momentum of Y after the collision can be determined using the principle of conservation of momentum:
  - **either by drawing a vector diagram**, as shown in Figure 7
  - **or by calculation,** as follows.
    1. Resolve the velocities v_X and v_Y parallel and perpendicular to the initial direction of X to give $v_X \cos\theta_X$ and $v_X \sin\theta_X$ for the parallel and perpendicular components of X and $v_Y \cos\theta_Y$ and $v_Y \sin\theta_Y$ for the parallel and perpendicular components of Y.
    2. Apply the principle of conservation of momentum in each direction:
       - In the parallel direction: $m_X U = m_X v_X \cos\theta_X + m_Y v_Y \cos\theta_Y$
       - In the perpendicular direction: $0 = m_X v_X \sin\theta_X + m_Y v_Y \sin\theta_Y$, where m_X and m_Y are the masses of X and Y, respectively.
    3. Insert the known values into the above equations to give a value of $v_Y \cos\theta_Y$ from the first equation and a value of $v_Y \sin\theta_Y$ from the second equation. Note that if the perpendicular component of X is assigned a positive value, then the perpendicular component of Y needs to be assigned a negative value (because they are in opposite directions).
    4. The values of v_Y and θ_Y can then be found using $v_Y{}^2 = (v_Y \cos\theta_Y)^2 + (v_Y \sin\theta_Y)^2$ and $\tan\theta_Y = v_Y \sin\theta_Y / v_Y \cos\theta_Y$.

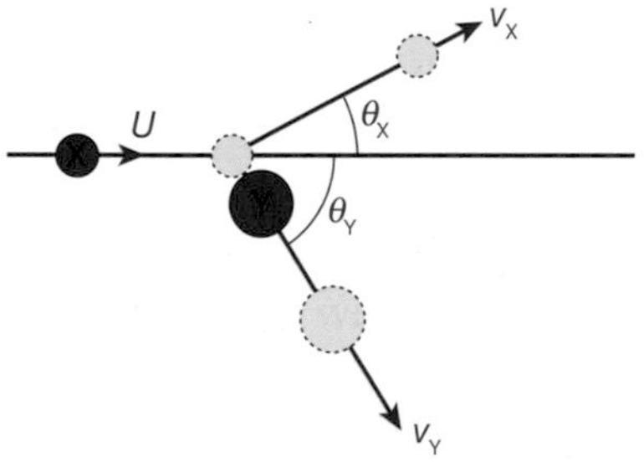

__Figure 6__ An oblique collision

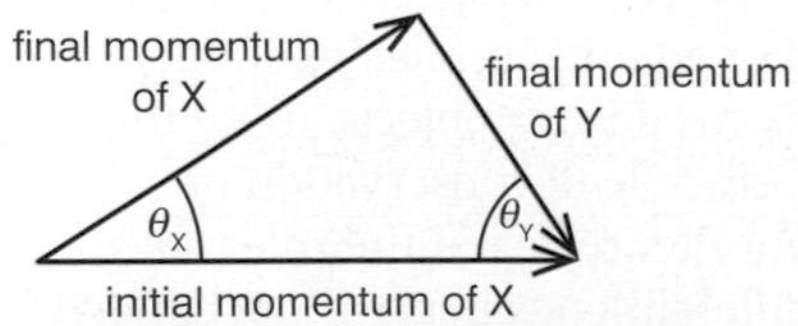

__Figure 7__ Conservation of momentum

Worked example

A sphere X of mass 0.10 kg moving at a speed of 2.0 m s^{-1} collides with a sphere Y of mass 0.20 kg, which is initially at rest. As a result, X is deflected by 30° at a speed of 1.6 m s^{-1}. Show that Y rebounds at a speed of 0.50 m s^{-1} in a direction at 53° to the initial direction of X.

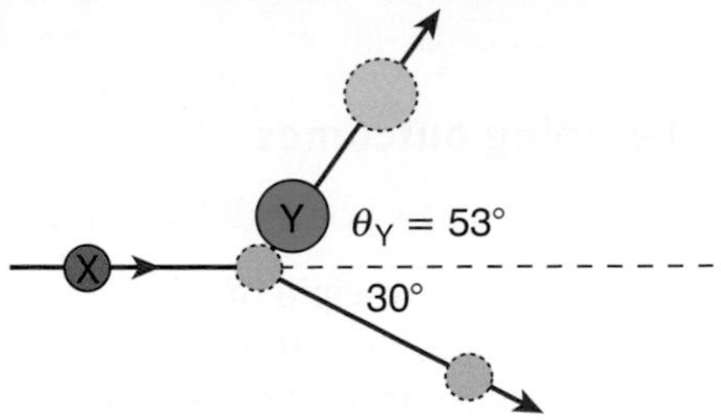

Figure 8

Solution 1: Vector diagram method

Initial momentum of X = 0.10 × 2.0 = 0.20 kg m s^{-1}

Final momentum of X = 0.10 × 1.6 = 0.16 kg m s^{-1} at 30° to the initial direction of X

Final momentum of Y = 0.20 × 0.50 = 0.10 kg m s^{-1} at cos 53° to the initial direction of X

Figure 9 shows the momentum vectors for the above three momentum values. The diagram shows that the vector sum of the final momentum vector for X and the final momentum vector for Y is equal to the initial momentum vector for X.

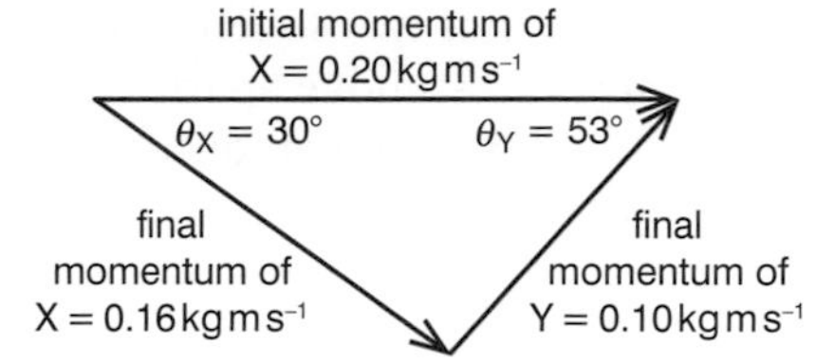

Figure 9

Solution 2: Calculation method

Apply the principle of conservation of momentum parallel and perpendicular to the initial direction of X.

In the parallel direction:

$$m_X U = m_X v_X \cos\theta_X + m_Y v_Y \cos\theta_Y$$

$$(0.10 \times 2.0) = (0.10 \times 1.6 \times \cos 30°) + (0.20 v_Y \times \cos\theta_Y)$$

$$(0.20 v_Y \times \cos\theta_Y) = (0.10 \times 2.0) - (0.10 \times 1.6 \times \cos 30°) = 0.20 - 0.139 = 0.061$$

$$v_Y \times \cos\theta_Y = 0.061 \div 0.20 = 0.305$$

In the perpendicular direction:

$$0 = m_X v_X \sin\theta_X + m_Y v_Y \sin\theta_Y$$

$$0 = (0.10 \times 1.6 \times \sin 30°) + (0.20\, v_Y \times \sin\theta_Y)$$

$$0.20 v_Y \times \sin\theta_Y = -0.10 \times 1.6 \times \sin 30° = -0.080$$

$$v_Y \times \sin\theta_Y = -0.080 \div 0.20 = -0.400$$

Therefore $v_Y = [(v_Y \cos\theta_Y)^2 + (v_Y \sin\theta_Y)^2]^{0.5} = [0.305^2 + 0.400^2]^{0.5}$

$= 0.50$ m s^{-1} to two significant figures

$$\tan\theta_Y = \frac{v_Y \sin\theta_Y}{v_Y \cos\theta_Y} = \frac{-0.400}{0.305} = -1.31$$

Therefore, $\theta_Y = 53°$

Summary test 4.4

1 **a** A squash ball is released from rest above a flat surface. Describe how its energy changes if:
i it rebounds to the same height,
ii it rebounds to a lesser height.

b In **a ii**, the ball is released from a height of 1.2 m above the surface and it rebounds to a height of 0.9 m above the surface. Show that 25% of its kinetic energy is lost in the impact.

2 A vehicle of mass 800 kg moving at a speed of 15.0 m s^{-1} collides with a vehicle of mass 1200 kg moving in the same direction at a speed of 5.0 m s^{-1}. The two vehicles lock together on impact. Calculate:

a the velocity of the two vehicles immediately after impact,

b the loss of kinetic energy due to the impact.

3 An ice puck of mass 1.5 kg moving at a speed of 4.2 m s^{-1} collides head on with a second ice puck of mass 1.0 kg moving in the opposite direction at a speed of 4.0 m s^{-1}. After the impact, the 1.5 kg ice puck continues in the same direction at a speed of 0.8 m s^{-1}. Calculate:

a the speed and direction of the 1.0 kg ice puck after the collision,

b the loss of kinetic energy due to the collision.

4 Bumper cars at fairgrounds are designed to withstand low-speed impacts without damage. A bumper car of mass 250 kg moving at a velocity of 0.9 m s^{-1} collides elastically with a stationary car of mass 200 kg. Immediately after impact, the 250 kg car has a velocity of 0.1 m s^{-1} in the same direction as it was initially moving.

a **i** Calculate the velocity of the 200 kg car immediately after the impact.
ii Show that the collision is elastic.

b **i** What is the relative velocity of approach?
ii What is the relative velocity of separation?

4.5 Explosions

Learning outcomes

On these pages you will learn to:

- apply the principle of conservation of momentum to explosions and other situations where two objects initially at rest push each other away
- recognise that two such objects initially at rest move apart with equal and opposite amounts of momentum

When two objects fly apart after being initially at rest, they recoil from each other with equal and opposite amounts of momentum.

Figure 1 *One false move and the rider and the skateboard will fly apart*

Consider Figure 2 where a trolley of mass m_A and a trolley of mass m_B, initially at rest, move away at speeds v_A and v_B respectively.

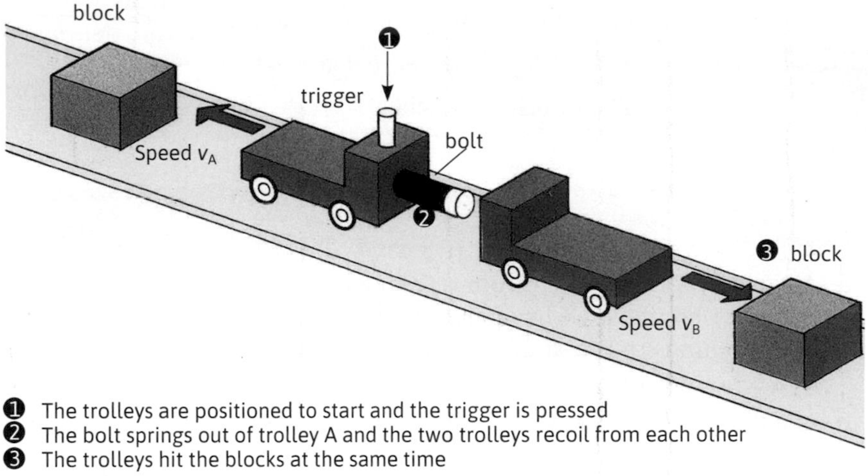
block
1
trigger
bolt
Speed v_A
2
3 block
Speed v_B
1 The trolleys are positioned to start and the trigger is pressed
2 The bolt springs out of trolley A and the two trolleys recoil from each other
3 The trolleys hit the blocks at the same time

Figure 2 *Flying apart*

The total initial momentum = 0

The total momentum immediately after the explosion = momentum of A + momentum of B

$$= m_A v_A + m_B v_B$$

Using the principle of conservation of momentum, $m_A v_A + m_B v_B = 0$

$$\therefore m_B v_B = -m_A v_A$$

The minus sign means that the two masses move away from each other in opposite directions. For example, if $m_A = 1.0\,\text{kg}$, $v_A = 2\,\text{m}\,\text{s}^{-1}$ and $m_B = 0.5\,\text{kg}$, then $v_B = \dfrac{-m_A v_A}{m_B} = -4.0\,\text{m}\,\text{s}^{-1}$. So A and B move away at speeds of $2\,\text{m}\,\text{s}^{-1}$ and $4\,\text{m}\,\text{s}^{-1}$ in opposite directions.

Testing a model explosion

In Figure 2, when the spring is released from one of the trolleys, the two trolleys, A and B, push each other apart. The bricks are positioned so that the trolleys hit the bricks at the same moment. The distance travelled by each trolley to the point of impact with the brick is equal to its speed multiplied by the time taken to travel that distance. Because the time taken is the same for the two trolleys, the distance ratio is the same as the speed ratio. Because the trolleys have equal (and opposite) amounts of momentum, the ratio of their speeds is the inverse of the mass ratio. The distance ratio should therefore be equal to the inverse of the mass ratio. In other words, if trolley A travels twice as far as trolley B, then the mass of A must be half the mass of B (so that they carry away equal amounts of momentum).

Summary test 4.5

1 A shell of mass 2.0 kg is fired at a speed of 140 m s^{-1} from an artillery gun of mass 800 kg. Calculate the recoil velocity of the shell.

2 In a laboratory experiment to measure the mass of an object X, two identical trolleys A and B, each of mass 0.50 kg, were initially stationary on a track. Object X was fixed to trolley A. When a trigger was pressed, the two trolleys moved apart in opposite directions at speeds of 0.30 m s^{-1} and 0.25 m s^{-1}.

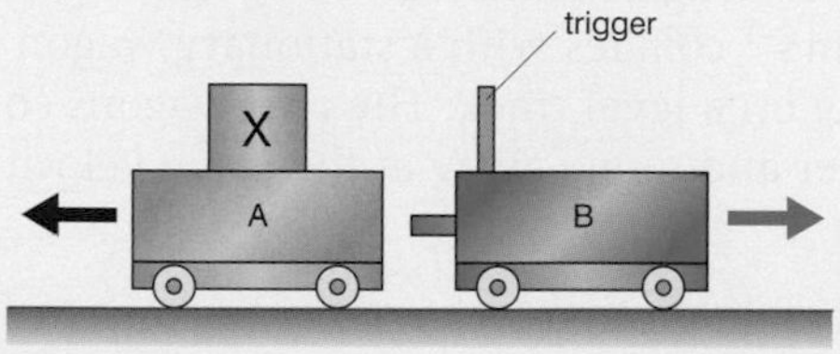

Figure 3

a Which of the two speeds given above was the speed of trolley A? Give a reason for your answer.

b Show that the mass of X must have been 0.10 kg.

3 Two trolleys, X of mass 1.2 kg and Y of mass 0.8 kg, are initially stationary on a level track.

a When a trigger is pressed on one of the trolleys, a spring pushes the two trolleys apart. Trolley Y moves away at a velocity of 0.15 m s^{-1}.

i Calculate the velocity of X.

ii Calculate the total kinetic energy of the two trolleys immediately after the impact.

b If the test had been carried out with trolley X held firmly, calculate the speed at which Y would have recoiled, assuming that the energy stored in the spring before release is equal to the total kinetic energy calculated in **a ii**.

4 A person in a stationary boat of total mass 150 kg throws a rock of mass 2.0 kg out of the boat. As a result, the boat recoils at a speed of 0.12 m s^{-1}. Calculate:

a the speed at which the rock was thrown from the boat,

b the kinetic energy gained by:
i the boat, **ii** the rock.

Chapter summary

- Momentum = mass × velocity
- $\text{Force} = \dfrac{\text{change of momentum}}{\text{time taken}}$
- The principle of conservation of momentum states that when two or more bodies interact, the total momentum is unchanged, provided no external forces act on the bodies.
- An elastic collision is one in which the total kinetic energy after the collision is the same as before the collision. For such a collision between two objects, the relative speed of approach is equal to the relative speed of separation.
- A totally inelastic collision is one in which the colliding objects stick together.
- In an explosion where two objects fly apart, the two objects carry away equal and opposite momentum.

Launch additional digital resources for the chapter

1 An object of mass 7.0 kg, initially at rest, is acted on by a force of 14 N for 10 s. Calculate:

a the gain of momentum of the object,

b the velocity of the object after 10 s.

2 A vehicle of mass 600 kg travelling at a velocity of 15 m s^{-1} is acted on by a braking force of 150 N. Calculate:

a the momentum of the object at 15 m s^{-1},

b the time taken to stop the object.

3 An object of mass 5.0 kg, initially at rest, is acted on by a force of 8.0 N for 12 s and is then brought to rest by a different force in 20 s.

a Show that the change of momentum of the object due to the 8.0 N force is 96 N s.

b Calculate the speed of the object after 12 s.

c Calculate the force needed to bring the object to rest in 20 s.

4 A stream of identical atoms of mass 1.0×10^{-25} kg hit a vertical surface normally at a speed of 750 m s^{-1} at a rate of 2000 atoms per second. The atoms stick to the surface on impact.

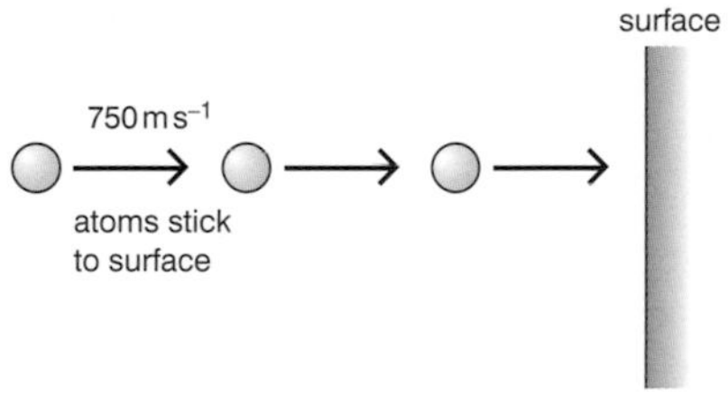

Figure 4.1

Calculate:

a the loss of momentum of a single atom,

b the average force of the atoms on the surface.

5 A molecule of mass 2.5×10^{-26} kg moving at a speed of 520 m s^{-1} collides normally with a wall and rebounds normally without loss of speed.

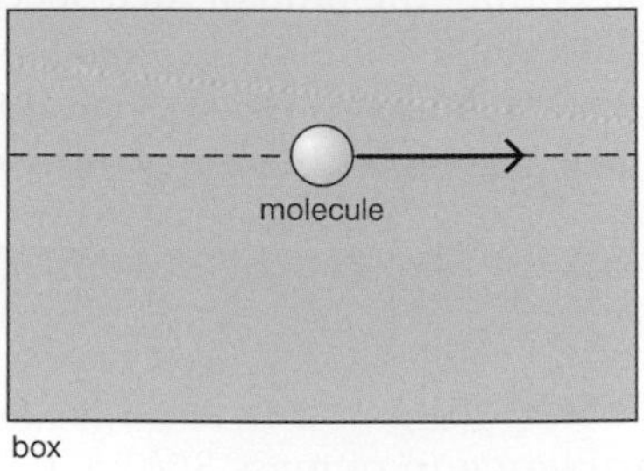

Figure 5.1

a Calculate the change of momentum of the molecule.

b The molecule is in a rectangular box and collides repeatedly with the same side of the box every 2.0 ms. Calculate the average force of impact of the molecule on the box.

6 A railway wagon of mass 2500 kg moving at a speed of 2.4 m s^{-1} collides with a stationary wagon of mass 1500 kg on a level track. The two wagons couple together and move away at the same velocity after the impact.

a Calculate their velocity after the impact.

b i Calculate the loss of kinetic energy due to the impact.

ii Discuss the energy changes that take place as a result of the impact.

7 In a road accident, a van of mass 1500 kg moving at a speed of 28 m s^{-1} runs into the back of a car of mass 900 kg moving in the same direction at a speed of 11 m s^{-1}. As a result of the impact, the car is pushed forward at a speed of 18 m s^{-1}.

a Calculate the velocity of the van immediately after the impact.

b Calculate:

i the loss of kinetic energy of the van,

ii the gain of kinetic energy of the car,

iii the total change of kinetic energy of the two vehicles.

c Discuss the effect of the impact on a person in the car.

8 In a radioactive decay, a nucleus of mass 4.0×10^{-25} kg, initially at rest, emitted an α-particle of mass 6.7×10^{-27} kg with a velocity of 1.5×10^{7} m s^{-1}.

a Calculate the velocity of recoil of the nucleus.

b Calculate the kinetic energy of:
i the recoil nucleus,
ii the α-particle.

9 A railway wagon P of mass 15 000 kg moving at a speed of 1.8 m s^{-1} collides on a level track with a 25 000 kg wagon Q, which then moves away from wagon P. Figure 9.1 shows how the velocity of Q changes during the impact.

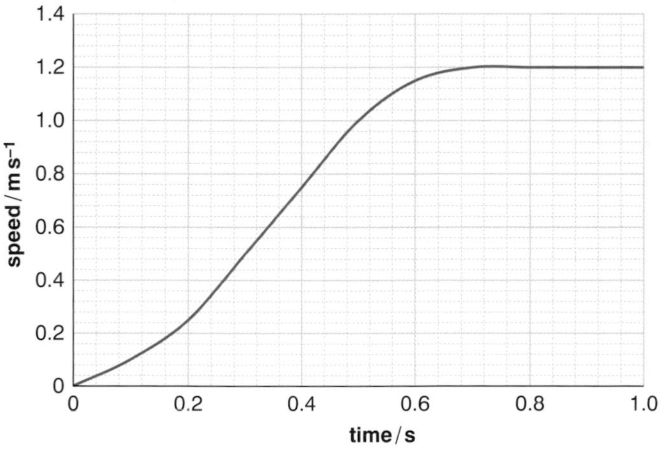

Figure 9.1

a Estimate the duration of the impact and hence calculate the force of the impact.

b **i** State the principle of conservation of momentum.
ii Determine the velocity of P immediately after the impact.

c **i** State what is meant by a *perfectly elastic collision*.
ii Calculate the change of kinetic energy of each truck due to the impact.
iii Discuss whether or not the collision is elastic.

10 A radioactive isotope emits α-particles which all have the same initial kinetic energy. When an α-particle is emitted by a stationary nucleus, the nucleus and the α-particle move away from each other in opposite directions at different speeds.

a Explain, in terms of momentum, why the nucleus and the α-particle move away from each other:
i in opposite directions,
ii at different speeds.

b The nuclei of the polonium isotope $^{210}_{94}Po$ emits α-particles of kinetic energy 8.5×10^{-13} J. Each such nucleus changes into a nucleus of a lead isotope when it emits an α-particle. The process is represented in Figure 10.1.

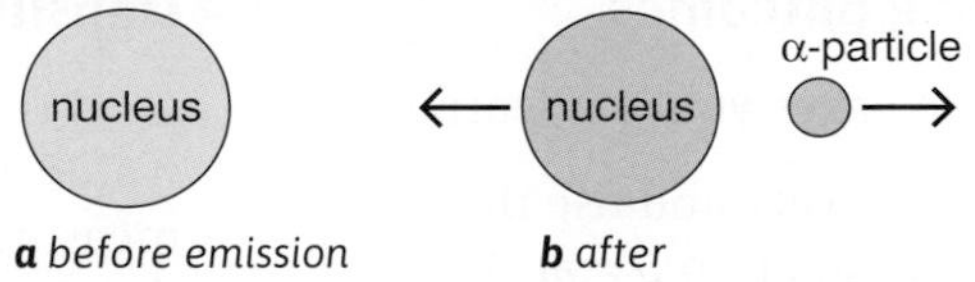

Figure 10.1

i Calculate the speed of an α-particle with this kinetic energy.
The mass of an α-particle = 6.6×10^{-27} kg.
ii The mass of the lead nucleus is 3.4×10^{-25} kg. Calculate the speed and the kinetic energy of the nucleus immediately after the α-particle has been emitted.

c The α-particles from this polonium isotope have a range in air of 39 mm at atmospheric pressure.
i Calculate the time taken for an α-particle to travel this distance.
ii Estimate the average deceleration of an α-particle over this distance.

11 A rubber band is stretched between two frictionless trolleys A and B in fixed positions on a horizontal surface, as shown in Figure 11.1. The trolleys, which have different masses, are then released at the same time so that they move towards each other until they collide. The mass of A is greater than the mass of B.

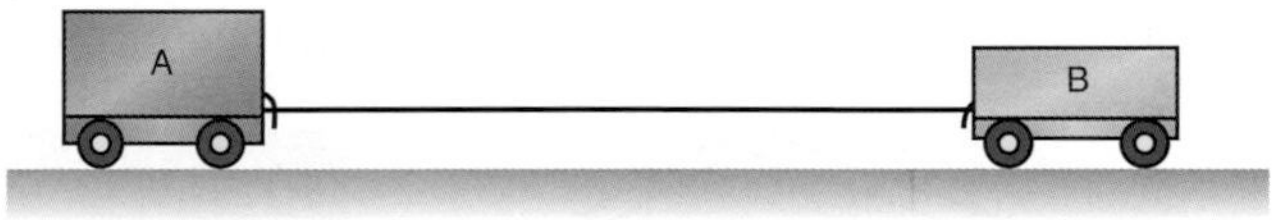

Figure 11.1

a **i** Immediately before the trolleys collide, explain why the two trolleys have equal and opposite momentum.
ii Explain why the two trolleys have different speeds immediately before they collide.

b State and explain which trolley has the greater kinetic energy immediately before the collision.

5 Forces and materials

5.1 Density

Learning outcomes

On these pages you will learn to:

- define density and use the density equation $\rho = m / V$ in calculations
- measure the density of liquids and regular and irregular solids

Density and its measurement

Lead is much more dense than aluminium. Sea water is more dense than tap water. To see how dense one substance is compared with another, we need to measure the mass of equal volumes of the two substances. The substance with the greater mass in the same volume is more dense. For example, a lead sphere of volume 1 cm^3 has a mass of 11.3 g; whereas an aluminium sphere of the same volume has a mass of 2.7 g, so lead is more dense than aluminium.

The density of a substance is defined as its mass per unit volume.

For a certain amount of a substance of mass m and volume V, its density, ρ (pronounced 'rho'), may be calculated using the equation

$$\textbf{density, } \rho = \frac{m}{V}$$

- The unit of density is the **kilogram per metre³ (kg m⁻³)**.
- Rearranging the above equation gives: $m = \rho V$

$$\text{or } V = \frac{m}{\rho}$$

More about units

Mass	1 kg = 1000 g
Length	1 m = 100 cm = 1000 mm
Volume	1 m^3 = 10^6 cm^3
Density	$1000\,\text{kg}\,\text{m}^{-3} = \dfrac{10^6\,\text{g}}{10^6\,\text{cm}^3} = 1\,\text{g}\,\text{cm}^{-3}$

Table 1 shows the density of some common substances in kg m^{-3}. You can see that gases are much less dense than solids or liquids.

Table 1 *Density*

Substance	Density / kg m^{-3}
Air	1.2
Aluminium	2700
Copper	8900
Gold	19 300
Hydrogen	0.083
Iron	7900
Lead	11 300
Oxygen	1.3
Silver	10 500
Water	1000

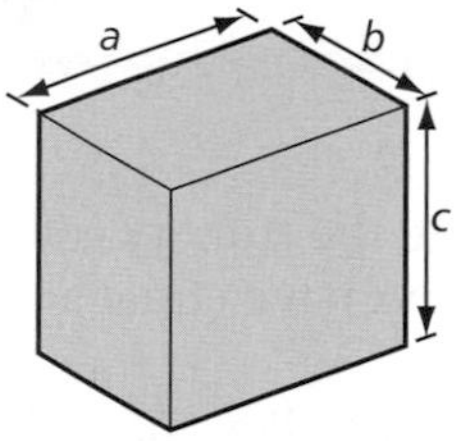

i Volume of cuboid = $a \times b \times c$

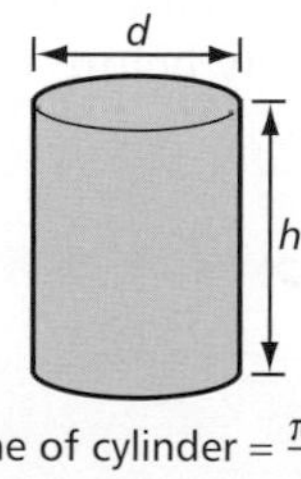

ii Volume of cylinder = $\dfrac{\pi d^2}{4} \times h$

Figure 1 *Volume formulae*

Worked example

Using the data above, calculate:

a the mass, in kilograms, of a piece of aluminium of volume 3.6×10^{-5} m^3,

b the volume, in m^3, of a mass of 0.50 kg of iron.

Solution

a $\rho = 2700\,\text{kg}\,\text{m}^{-3}$; mass $m = \rho V = 2700\,\text{kg}\,\text{m}^{-3} \times 3.6 \times 10^{-5}\,\text{m}^3 = 9.7 \times 10^{-2}\,\text{kg}$

b $\rho = 7900\,\text{kg}\,\text{m}^{-3}$; volume $= \dfrac{m}{\rho} = \dfrac{0.50\,\text{kg}}{7900\,\text{kg}\,\text{m}^{-3}} = 6.3 \times 10^{-5}\,\text{m}^3$

Density measurements

An unknown substance can often be identified if its density is measured and compared with the density of known substances. The following procedures may be used to measure the density of a substance.

- **A regular solid.** Measure its mass using a top-pan balance; measure its dimensions using Vernier calipers or a micrometer. Calculate its volume using the appropriate formula (e.g. volume of a sphere of radius r is $\frac{4}{3}\pi r^3$; volume of a cylinder of radius r and length L is $\pi r^2 L$). Calculate the density from $\dfrac{\text{mass}}{\text{volume}}$.

- **A liquid.** Measure the mass of an empty measuring cylinder. Fill the cylinder with the liquid and measure the volume of the liquid directly. Measure the mass of the cylinder and liquid to enable the mass of the liquid to be calculated. Calculate the density from $\frac{\text{mass of liquid}}{\text{volume}}$.

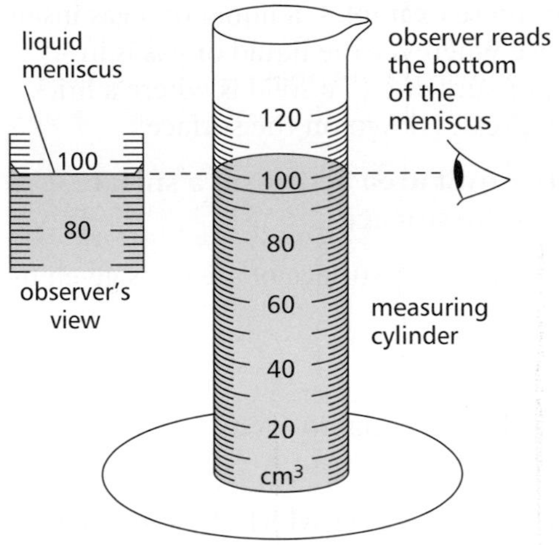
liquid meniscus
100
80
observer's view
120
100
80
60
40
20
cm³
observer reads the bottom of the meniscus
measuring cylinder

Figure 2 Using a measuring cylinder

- **An irregular solid.** Measure the mass of the object. Fill a displacement can with water up to the spout, as shown in Figure 3. Place a beaker of known mass under the spout. Immerse the object on a thread in the liquid and collect the overflow. Measure the mass of the beaker and overflow. Hence determine the mass of the overflow water and calculate its volume, given that the density of water is $1000\,\text{kg}\,\text{m}^{-3}$.
  Calculate the density of the object from $\frac{\text{its mass}}{\text{the overflow volume}}$.

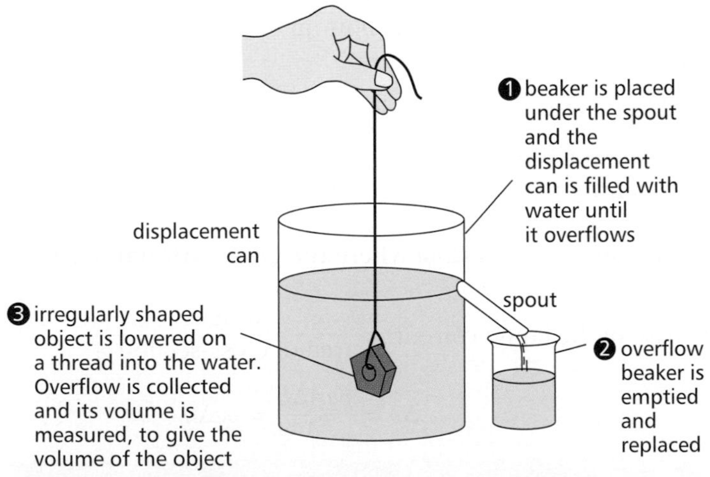
1 beaker is placed under the spout and the displacement can is filled with water until it overflows
displacement can
spout
3 irregularly shaped object is lowered on a thread into the water. Overflow is collected and its volume is measured, to give the volume of the object
2 overflow beaker is emptied and replaced

Figure 3 Measuring the volume of an irregularly shaped object

Density of alloys

An **alloy** is a solid mixture of two or more metals. For example, **brass** is an alloy of copper and zinc which has good resistance to corrosion and wear.

If an alloy consists of two metals A and B, then for volume V of the alloy:

- If the volume of metal A is v_A, the mass of metal A is $\rho_A v_A$, where ρ_A is the density of metal A.
- If the volume of metal B is v_B, the mass of metal B = $\rho_B v_B$, where ρ_B is the density of metal B.

$\therefore$ **Mass of the alloy,** $m = \rho_A v_A + \rho_B v_B$

Hence the density of the alloy, $$\rho = \frac{m}{V} = \frac{\rho_A v_A + \rho_B v_B}{V} = \frac{\rho_A v_A}{V} + \frac{\rho_B v_B}{V}$$

Worked example

A brass object consists of $3.3 \times 10^{-5}\,\text{m}^3$ of copper and $1.7 \times 10^{-5}\,\text{m}^3$ of zinc. Calculate the mass and the density of this object. The density of copper is $8900\,\text{kg}\,\text{m}^{-3}$; the density of zinc is $7100\,\text{kg}\,\text{m}^{-3}$.

Solution

Mass of copper = density of copper × volume of copper
$= 8900\,\text{kg}\,\text{m}^{-3} \times 3.3 \times 10^{-5}\,\text{m}^3 = 0.29\,\text{kg}$

Mass of zinc = density of zinc × volume of zinc
$= 7100\,\text{kg}\,\text{m}^{-3} \times 1.7 \times 10^{-5}\,\text{m}^3 = 0.12\,\text{kg}$

Total mass, $m = 0.29 + 0.12 = 0.41\,\text{kg}$

Total volume, $V = 5.0 \times 10^{-5}\,\text{m}^3$

Density of alloy, $\rho = \frac{m}{V} = \frac{0.41\,\text{kg}}{5.0 \times 10^{-5}\,\text{m}^{-3}} = 8200\,\text{kg}\,\text{m}^{-3}$

Summary test 5.1

1 A rectangular brick of dimensions $5.0\,\text{cm} \times 8.0\,\text{cm} \times 20.0\,\text{cm}$ has a mass of 2.5 kg. Calculate:
  a its volume,
  b its density.

2 An empty paint tin of diameter 0.150 m and of height 0.120 m has a mass of 0.22 kg. It is filled with paint to within 7 mm of the top. Its total mass is then 6.50 kg. Calculate, for the paint in the tin:
  a the mass,
  b the volume,
  c the density.

3 A solid steel cylinder has a diameter of 12 mm and a length of 85 mm. Calculate:
  a its volume (in m^3),
  b its mass (in kg); the density of steel is $7800\,\text{kg}\,\text{m}^{-3}$.

4 An alloy tube of volume $1.8 \times 10^{-4}\,\text{m}^3$ consists, by volume, of 60% aluminium and 40% magnesium.
  a Calculate the mass, in the tube, of:
    i aluminium,
    ii magnesium.
  b Calculate the density of the alloy; the density of aluminium is $2700\,\text{kg}\,\text{m}^{-3}$; the density of magnesium is $1700\,\text{kg}\,\text{m}^{-3}$.

5.2 Pressure

Learning outcomes

On these pages you will learn to:

- define and use the pressure equation $p = F / A$
- derive and use the equation $\Delta p = \rho g \Delta h$
- describe how to measure gas pressure
- use $\Delta p = \rho g \Delta h$ to compare the densities of two different liquids

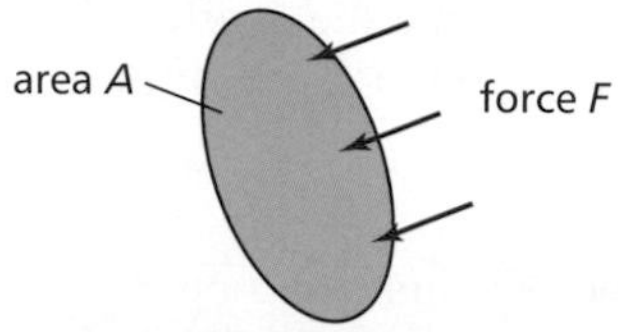

Figure 1 *Pressure*

Figure 2 *A snowmobile*

In climates where snow falls, caterpillar tracks on snow vehicles allow the vehicle to travel across snow without sinking into the snow. The tracks have a much greater contact area with the ground than ordinary tyres. Therefore, the pressure of the tracks on snow is much less than the pressure would be if the vehicle was fitted with tyres.

Pressure and force

Lots of people need to measure **pressure**. For example, nurses measure blood pressure, motor vehicle technicians measure tyre pressure and gas engineers measure the pressure of the gas supply. In these examples, a liquid or a gas inside a container presses on the container surface wherever the liquid or gas is in contact with the surface. An example of pressure due to a solid is where a brick rests on a surface. The weight of the brick presses down on the surface.

Pressure is defined as the force per unit area acting on a surface perpendicular to the surface.

- The pressure of a force F acting at right angles to a surface of area A is given by the equation

$$\textbf{pressure, } p = \frac{F}{A}$$

- The unit of pressure is the **pascal (Pa)**, which is equal to $1\,\mathrm{N\,m^{-2}}$.

Solid pressure

When a force acts on an object, the smaller the area over which the force acts, the greater the pressure of the force on a surface.

Liquid pressure

For any fluid at rest:

- the pressure at any point acts equally in all directions,
- the pressure increases with depth.

The **upthrust** on an object in a fluid is due to the increase of pressure with depth in the liquid. Imagine a ball in a liquid. The average force due to the liquid pressure on its lower half acts upwards. This is greater than the average force on its upper half which acts downwards. The ball therefore experiences an upward resultant force or upthrust due to the pressure of the liquid.

The greater the density of a liquid is, the greater its pressure is at any given depth. Consider the column of liquid in the container shown in Figure 3. The pressure caused by the liquid column on the bottom of the container is due to the weight of the liquid.

For a column of height Δh and area of cross-section A:

- the volume of liquid in the container = $A\Delta h$,
- the mass of liquid = its density × its volume = $\rho A \Delta h$ where ρ is the density of the liquid,
- the weight of the liquid = mass × g = $\rho A \Delta h g$ where the **gravitational field strength** $g = 9.81\,\mathrm{N\,kg^{-1}}$.

The pressure Δp at the base of the liquid column = $\dfrac{\text{weight}}{\text{area of cross-section } A}$

$$\Delta p = \frac{\rho g A \Delta h}{A} = \rho g \Delta h$$

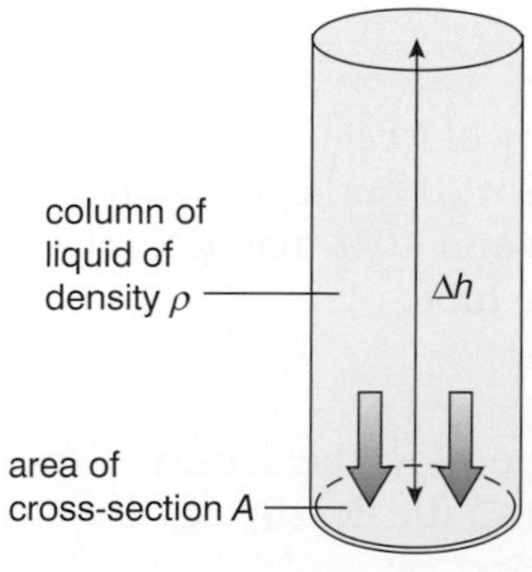

Figure 3 *Calculating liquid pressure*

Worked example

$g = 9.81\,\mathrm{N\,kg^{-1}}$

Calculate the pressure due to sea water of density $1050\,\mathrm{kg\,m^{-3}}$ at a depth of 200 m.

Solution

pressure $\Delta p = \rho g \Delta h = 1050\,\mathrm{kg\,m^{-3}} \times 9.81\,\mathrm{N\,kg^{-1}} \times 200\,\mathrm{m} = 2.1 \times 10^6\,\mathrm{Pa}$

Gas pressure

The pressure of a gas is due to molecules repeatedly undergoing elastic collisions with the surface of the container. The pressure of the impacts of the gas molecules on the surface is constant, assuming that the temperature of the gas is constant, and that there are a large number of gas molecules in the container.

The pressure of a gas in a container can be measured using an electronic pressure gauge or a U-tube manometer, as shown in Figure 4. The pressure of the gas forces the liquid in the manometer up the open side of the U-tube until it is at rest.

Because atmospheric pressure acts on the liquid on the 'open' side of the manometer, the gas pressure = the pressure due to the difference in liquid levels ($h\rho g$) + atmospheric pressure, where ρ is the liquid density.

- The 'excess' pressure of the gas (i.e. its pressure above atmospheric pressure) is equal to the pressure due to the difference in the level of liquid on each side.

Therefore, the pressure of the gas relative to atmospheric pressure = $h\rho g$

Atmospheric pressure varies slightly from day to day, changing with the local weather conditions. The mean pressure of the atmosphere at sea level is 101 kPa. Atmospheric pressure decreases with height. The Earth's atmosphere extends more than 100 km into space.

Comparison of liquid densities

The pressure of a liquid depends on its density. The densities of two immiscible liquids, such as oil and water, can be compared using the U-tube arrangement shown in Figure 5. The arrangement is set up by pouring water into one side of the U-tube and then when the water has settled carefully pouring oil into the other side. When the liquids settle, the oil level is higher than the water level on the other side. This is because the pressure is equal at the same level in both columns. So the pressure at the bottom of the column of oil (X in Figure 5) is equal to the pressure at the same level of the water column (Y in Figure 5).

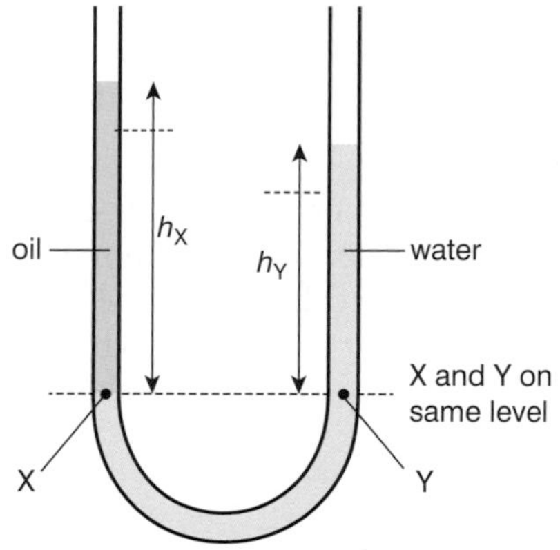

Figure 5 *Comparing densities*

To compare the densities of the two liquids, the length of each column is measured from the same level as the bottom of the oil column. Since the pressure of each measured column is the same, then for columns of height h_X for the oil and h_Y for the water,

$$\rho_X g h_X = \rho_Y g h_Y$$

Therefore:

$$\frac{\rho_X}{\rho_Y} = \frac{h_Y}{h_X}$$

Given that the density of water is $1000\,kg\,m^{-3}$ and knowing the measured values for h_X and h_Y, the density of the oil can be calculated.

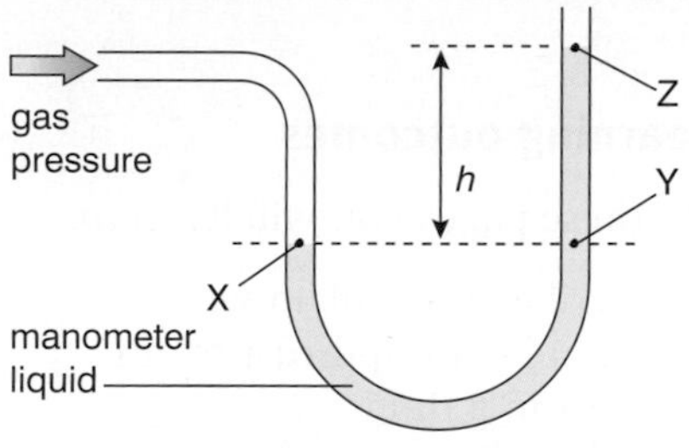

Figure 4 *The U-tube manometer*

Summary test 5.2

$g = 9.81\,m\,s^{-2}$

1 Calculate the pressure exerted by a paving stone of density $2500\,kg\,m^{-3}$ and of dimensions $0.80\,m \times 0.80\,m \times 0.05\,m$ when the stone rests flat on a smooth horizontal surface.

2 Calculate the force due to the atmosphere on the panes of a sealed double-glazed window of dimensions $1.5\,m \times 0.80\,m$ which has a vacuum between the two panes. The mean value of atmospheric pressure = 101 kPa.

3 The pressure in the tyres of a vehicle is 280 kPa. The contact area of each of the four tyres on the ground is $0.012\,m^2$. Calculate the weight of the vehicle.

4 The mean value of atmospheric pressure is 101 kPa.

a The density of water is $1000\,kg\,m^{-3}$. Calculate the depth of water that will give a pressure of 101 kPa.

b Domestic gas is supplied to homes at a pressure which is normally 2.5% above atmospheric pressure.

A gas engineer used a water-filled U-tube manometer to measure the gas pressure at a house and observed a difference of 0.15 m between its levels.

i Calculate the pressure of the gas supply in this house.

ii State and explain whether or not the gas pressure at the house is normal.

5.3 Upthrust

Learning outcomes

On these pages you will learn to:

- describe and explain what is meant by an upthrust on an object in a fluid
- explain the cause of an upthrust
- Calculate the upthrust on an object in a fluid
- explain whether or not an object in a fluid sinks or floats

When you go swimming, have you noticed that you feel lighter in the water? People with mobility problems often find it much easier to move in water than in air. Water exerts an upward force on the body. This force is called an **upthrust**.

Investigating upthrust

Use a newton meter to weigh a metal object in air.

- Repeat the test by weighing the same object when it is completely in the water.

You should find that the newton meter reading is less when the object is in water. This is because when the metal object is in the water, it experiences an upthrust. The difference between the two newton meter readings is equal to the upthrust on the object.

Repeat the test with the same object only partly immersed in the water. You should find that the newton meter reading is in between the two earlier readings. This is because the upthrust is less when the object is only partly immersed in the water.

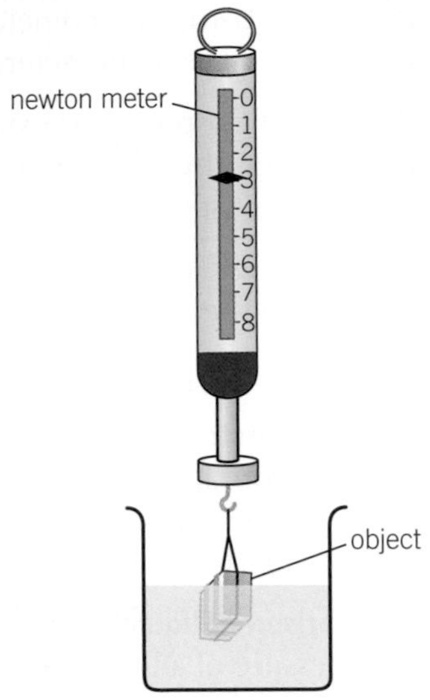

Figure 1 *Measuring an upthrust*

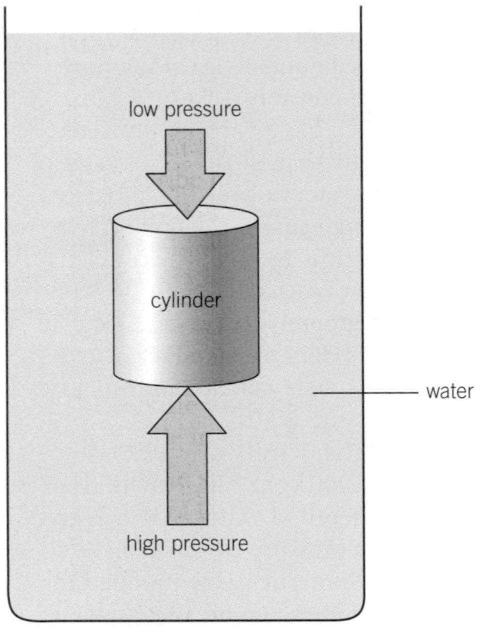

Figure 2 *Explaining upthrust*

Explaining upthrust

The water level in a water container rises when an object is lowered into the water. This is because the object **displaces** some of the water.

- The more the object is lowered into the water, the bigger the volume of water displaced, and the bigger the upthrust.
- When the object is fully immersed, the volume of water displaced is equal to the volume of the object.

Figure 2 shows a cylinder fully immersed in water. Because pressure increases with depth, the pressure of the water at the bottom of the cylinder is greater than the pressure on the top of the cylinder. So the upward force of the water on the bottom of the cylinder is greater than the downward force of the water on the top of the cylinder. The upthrust is the resultant of these two forces.

Float or sink?

A ship being loaded with cargo will float lower and lower in the water as the load is increased. The ship displaces more water when the load increases, so the upthrust increases. At any instant, the upthrust on the boat is equal to the weight of the ship and its cargo. If the ship is loaded too much, it sinks because it has displaced as much water as it possibly can, and because the upthrust cannot support the total weight.

An object floats when its weight is equal to the upthrust.

An object sinks when its weight is greater than the upthrust.

Figure 3 *A loaded ship*

The upthrust equation

In Figure 2, because the cylinder is vertical, the pressure difference Δp between the top and the bottom of the cylinder = $\rho g \Delta h$, where Δh is equal to the length of the cylinder, ρ is the density of the liquid, and $g = 9.81\,\mathrm{N\,kg^{-1}}$.

Therefore the upthrust on the cylinder is the pressure difference Δp × the cylinder's cross-sectional area:

$$(\rho g \Delta h) \times A = \rho g \Delta h A = \rho g V$$

where $V = \Delta h \times A$ = the cylinder volume.

For any object in a fluid:

the upthrust on the object = $\rho g V$

where ρ is the density of the fluid and V is the volume of fluid displaced by the object.

Notes

1 The upthrust equation applies to any object wholly or partially immersed in any fluid provided V is the volume of fluid displaced by the object.

2 Because mass m = density ρ × volume V, the upthrust = $\rho V g = mg$, where m is the mass of liquid displaced by the cylinder. Hence **the upthrust is equal to the weight of liquid displaced by the cylinder**. This statement is known as **Archimedes' principle**.

Worked example

A raft in water floats with its base horizontal at a depth d of 0.22 m below the water line. The raft has a base area A of $1.5\,\mathrm{m^2}$.

a Calculate the volume of water displaced by the raft.

b The density ρ of water is $1000\,\mathrm{kg\,m^{-3}}$. Calculate the upthrust on the raft and hence determine its weight.

Solution

a Volume of water displaced $V = dA = 0.22\,\mathrm{m} \times 1.5\,\mathrm{m^2} = 0.33\,\mathrm{m^3}$

b Upthrust = weight of water displaced

$$= \rho g V = 1000\,\mathrm{kg\,m^{-3}} \times 9.81\,\mathrm{N\,kg^{-1}} \times 0.33\,\mathrm{m^3} = 3240\,\mathrm{N}$$

Since the raft is floating, its weight is equal to the upthrust on it; hence its weight is 3230 N.

Density tests

Objects made of material such as cork or wood float in water, but metal objects sink. Objects that sink have a density greater than that of water. Objects that float have a density less than that of water. By observing if an object floats or sinks in water, you can tell if its density is less or greater than that of water.

An object that is more dense than water sinks because its weight is greater than the weight of water it displaces. So when it is fully immersed, it sinks because its weight is greater than the upthrust on it.

An object that is less dense than water floats because its weight is equal to the upthrust on it. This is because the density of the water is greater than that of the object so it displaces just enough water for the upthrust to be equal to the weight of the object.

Summary test 5.3

1 a Explain why it is difficult to hold an inflated plastic ball under water.

b Explain why cork is a suitable material for filling a life belt.

2 When an object is weighed using a newton meter, the reading on the newton meter is 5.20 N when the object is in air, and 3.3 N when the object is totally immersed in water.

a Explain why the reading on the newton meter is less when the object is in water.

b i Calculate the upthrust on the object and hence determine its volume.

ii Show that the density of the object is about $2700\,\mathrm{kg\,m^{-3}}$.

3 Three blocks A, B, and C are released in a bowl of a water.

Block A sinks to the bottom.

Block B floats with its top half above the water.

Block C floats with a small proportion above the water.

a Which object has the greatest density? Give a reason for your answer.

b Which object has the lowest density? Give a reason for your answer.

4 Figure 4 shows a weighted test tube floating vertically in water.

a Make a reasoned prediction about how the length of tube above the water depends on the total weight of the tube.

b Design an experiment to test your prediction, using the test tube and any other apparatus necessary.

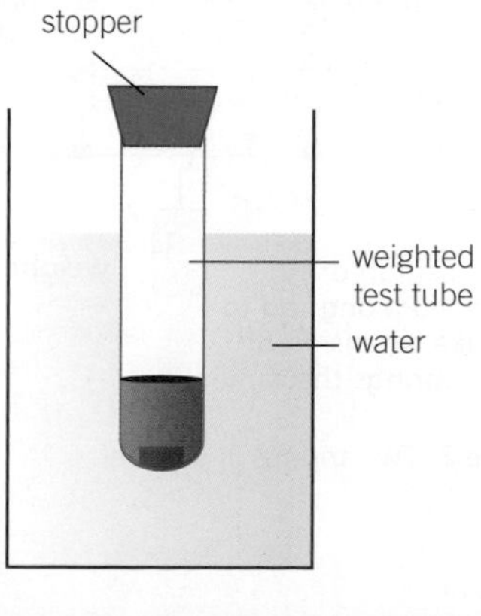

Figure 4

5.4 Springs

Learning outcomes

On these pages you will learn to:

- state Hooke's law for springs and explain what is meant by the elastic limit and the spring constant of a spring
- use Hooke's law to solve problems involving springs, including springs in parallel and in series
- calculate the energy of a stretched spring

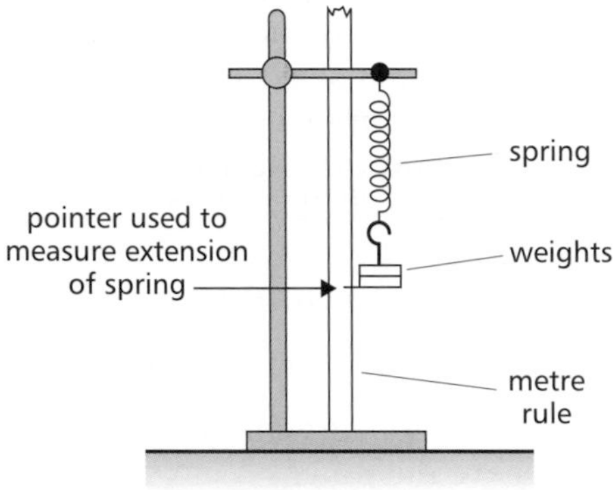

a *Testing the extension of a spring*

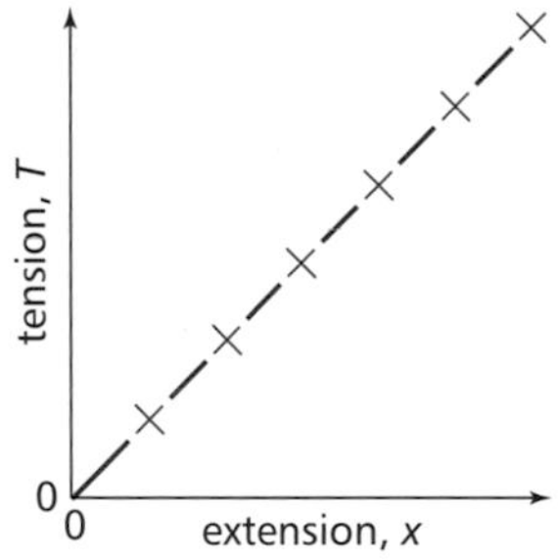

b *Hooke's law*

Figure 1

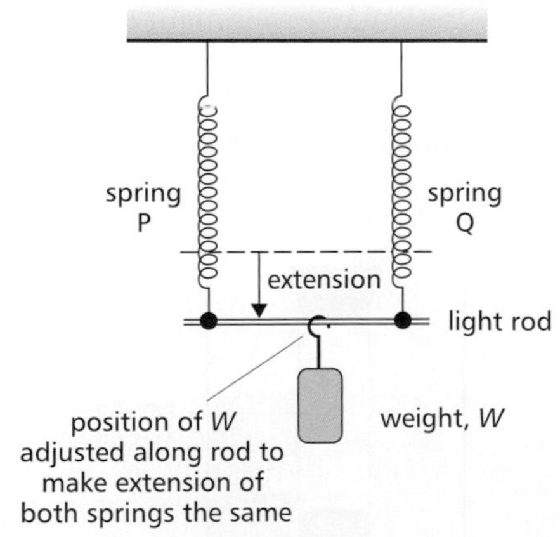

Figure 2 *Two springs in parallel*

Hooke's law

A stretched spring exerts a pull on the object holding each end of the spring. This pull, referred to as the **tension** in the spring, is equal and opposite to the force needed to stretch the spring. The more a spring is stretched, the greater the tension in it. Figure 1a shows a stretched spring at rest supporting a **load** consisting of some weights. This arrangement may be used to investigate how the tension in a spring depends on its extension from its unstretched length. The measurements may be plotted on a graph of tension v. extension (Figure 1b). The graph shows that the force needed to stretch a spring is proportional to the extension of the spring. This is known as **Hooke's law**, after Robert Hooke, a 17th-century scientist.

Hooke's law states that the force needed to stretch a spring is proportional to the extension of the spring from its natural length.

Hooke's law may be written as:

$$\textbf{Force, } F = kx$$

where k is the spring constant (sometimes referred to as the stiffness constant) and x is the extension.

- The greater the value of k, the stiffer the spring is. The unit of k is $\mathrm{N\,m^{-1}}$.
- The graph of F against x is a straight line of gradient k through the origin.
- If a spring is stretched beyond its **elastic limit**, it does not regain its initial length when the force applied to it is removed.
- A level maths students may meet Hooke's law in the form $F = \frac{\lambda x}{L}$ where L is the unstretched length of the spring and λ (= kL) is the spring modulus. λ is not in the specification for A level physics.

Worked example

A vertical steel spring fixed at its upper end has an unstretched length of 300 mm. Its length is increased to 385 mm when a 5.0 N weight attached to the lower end is at rest. Calculate:

a the spring constant,

b the length of the spring when it supports an 8.0 N weight at rest.

Solution

a Use $F = kx$ with $F = 5.0\,\mathrm{N}$ and $x = 385 - 300\,\mathrm{mm} = 85\,\mathrm{mm} = 0.085\,\mathrm{m}$.

Therefore $k = \frac{F}{x} = \frac{5.0\,\mathrm{N}}{0.085\,\mathrm{m}} = 59\,\mathrm{N\,m^{-1}}$

b Use $F = kx$ with $F = 8.0\,\mathrm{N}$ and $k = 59\,\mathrm{N\,m^{-1}}$ to calculate x:

$x = \frac{F}{k} = \frac{8.0\,\mathrm{N}}{59\,\mathrm{N\,m^{-1}}} = 0.136\,\mathrm{m}$

Therefore the length of the spring = 0.300 m + 0.136 m = 0.436 m

Springs in parallel

Figure 2 shows a weight W supported by means of two springs, P and Q, in parallel with each other. The extension, x, of each spring is the same. Therefore:

- The force needed to stretch P, $F_P = k_P x$
- The force needed to stretch Q, $F_Q = k_Q x$, where k_P and k_Q are the spring constants of P and Q respectively.

The weight W is supported by both springs, $W = F_P + F_Q = k_P x + k_Q x = k_{eff} x$ where the **effective spring constant,** $k_{eff} = k_P + k_Q$

Springs in series

Figure 3 shows a weight W supported by means of two springs joined end-on in 'series' with each other. The tension in each spring is the same and is equal to the weight W. Therefore:

- Extension of spring P, $x_P = \frac{W}{k_P}$
- Extension of spring Q, $x_Q = \frac{W}{k_Q}$, where k_P and k_Q are the spring constants of P and Q respectively.

$$\textbf{Total extension, } x = x_P + x_Q = \frac{W}{k_P} + \frac{W}{k_Q} = \frac{W}{k_{eff}}$$

where k_{eff}, the effective spring constant, is given by the equation $\frac{1}{k_{eff}} = \frac{1}{k_P} + \frac{1}{k_Q}$

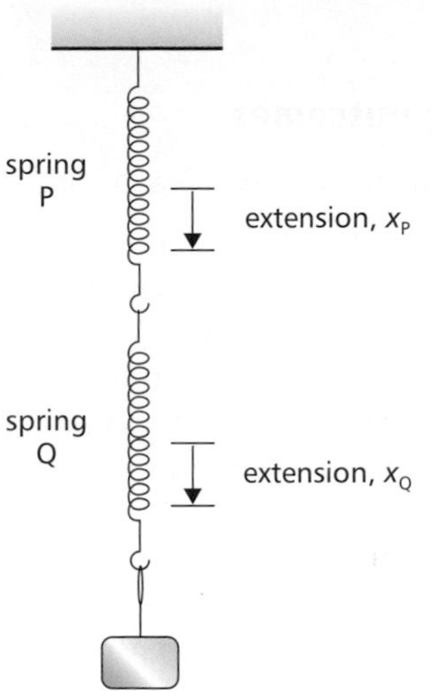

Figure 3 *Two springs in series*

The energy stored in a stretched spring

Elastic potential energy is stored in a stretched spring. If the spring is suddenly released, the elastic energy stored in it is suddenly converted to kinetic energy of the spring. As explained in Topic 2.4, the work done to stretch a spring by extension x_0 from its unstretched length is $\frac{1}{2}F_0x_0$, where F_0 is the force needed to stretch the spring to extension x_0. The work done on the spring is stored as elastic potential energy. Therefore the elastic potential energy E_p in the spring is $\frac{1}{2}F_0x_0$. Also, since $F_0 = kx_0$, where k is the spring constant, $E_p = \frac{1}{2}kx_0^2$.

Elastic potential energy stored in a stretched spring, $E_p = \frac{1}{2}F_0x_0 = \frac{1}{2}kx_0^2$

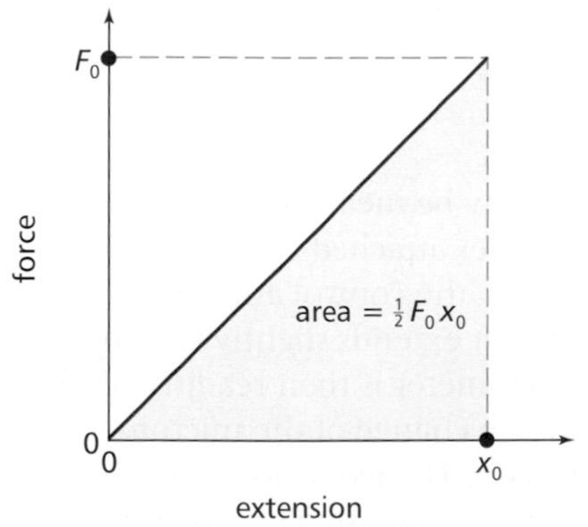

Figure 4 *Energy stored in a stretched spring*

Summary test 5.4

$g = 9.81\,\text{m}\,\text{s}^{-2}$

1 A steel spring has a spring constant of $25\,\text{N}\,\text{m}^{-1}$. Calculate:
  a the extension of the spring when the tension in it is equal to 10 N,
  b the tension in the spring when it is extended by 0.50 m from its unstretched length.

2 Two identical steel springs of length 250 mm are suspended vertically side-by-side from a fixed point. A 40 N weight is attached to the ends of the two springs. The length of each spring is then 350 mm. Calculate:
  a the tension in each spring,
  b the extension of each spring,
  c the spring constant of each spring.

3 Repeat questions **2a** and **b** for the two springs in 'series' and vertical.

4 An object of mass 0.150 kg is attached to the lower end of a vertical spring of unstretched length 300 mm, which is fixed at its upper end. With the object at rest, the length of the spring becomes 420 mm as a result. Calculate:
  a the spring constant,
  b the energy stored in the spring,
  c the weight that needs to be added to extend the spring to 600 mm.

5.5 Deformation of solids

Learning outcomes

On these pages you will learn to:

- appreciate that deformation is caused by a force and it can be tensile or compressive
- define stress, strain and the Young modulus and carry out calculations involving these quantities
- describe an experiment to determine the Young modulus of a metal in the form of a wire
- use a stress–strain graph to calculate the Young modulus of a material
- distinguish between elastic and plastic deformation of a material
- compare the properties of different materials in terms of their stress–strain curves

Force and solid materials

Look around at different materials and think about the effect of force on each material. To stretch, twist, or compress the material, a pair of forces is needed. For example, stretching a rubber band requires the rubber band to be pulled by a force at either end. Some materials, such as rubber, bend or stretch easily. The **elasticity** of a solid material is its ability to regain its shape after it has been deformed or distorted, and the forces that deformed it have been removed. An object that regains its shape after being deformed is said to have undergone **elastic deformation**. Deformation that stretches an object is **tensile**, whereas deformation that compresses an object is **compressive**.

The arrangement shown in Figure 1a in Topic 5.4 may be used to test different materials to see how easily they stretch. In each case, the material is held at its upper end and loaded by hanging weights at its lower end. The position of the pointer is measured as the weight of the load is increased in steps then decreased to zero. The extension of the strip of material at each step is its increase in length from its unloaded length. The tension in the material is equal to the weight. The measurements may be plotted as a graph of extension v. weight, as shown in Figure 1.

- A steel spring gives a straight line, in accordance with Hooke's law (see Topic 5.4).
- A rubber band at first extends easily when it is stretched. However, it becomes 'fully stretched' and very difficult to stretch further when it has been lengthened considerably.
- A polythene strip 'gives' and stretches easily after its initial stiffness is overcome. However, after 'giving', it extends little and becomes difficult to stretch.

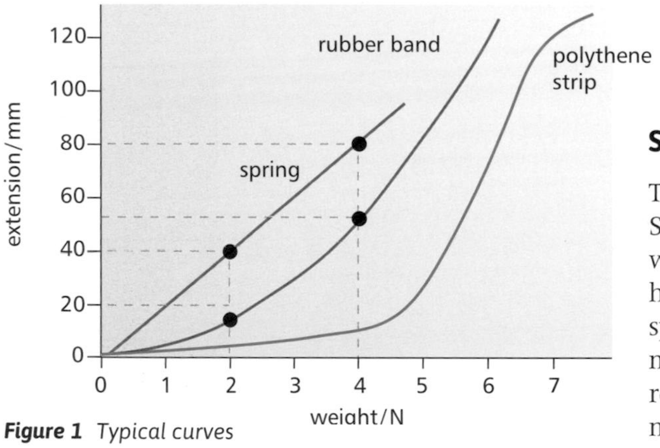

Figure 1 Typical curves

Stress and strain

The extension of a wire under tension may be measured using Searle's apparatus (Figure 2). A micrometer attached to the control wire is adjusted so the spirit level between the control and test wire is horizontal. When the test wire is loaded, it extends slightly causing the spirit level to drop on one side. The micrometer is then readjusted to make the spirit level horizontal again. The change of the micrometer reading is therefore equal to the extension. The extension may be measured for different values of tension by increasing the load (i.e. the test weights) in steps. At each step, the tension is equal to the total weight of the load.

For a wire of length L and area of cross-section A under tension:

- The **tensile stress** in the wire, $\sigma = \frac{F}{A}$, where F is the tension.

  The unit of stress is the **pascal (Pa)**, equal to $1\,\text{N}\,\text{m}^{-2}$.
- The **tensile strain** in the wire, $\varepsilon = \frac{x}{L}$, where x is the extension of the wire (i.e. change of length, ΔL). Strain is a ratio and therefore has **no unit**.

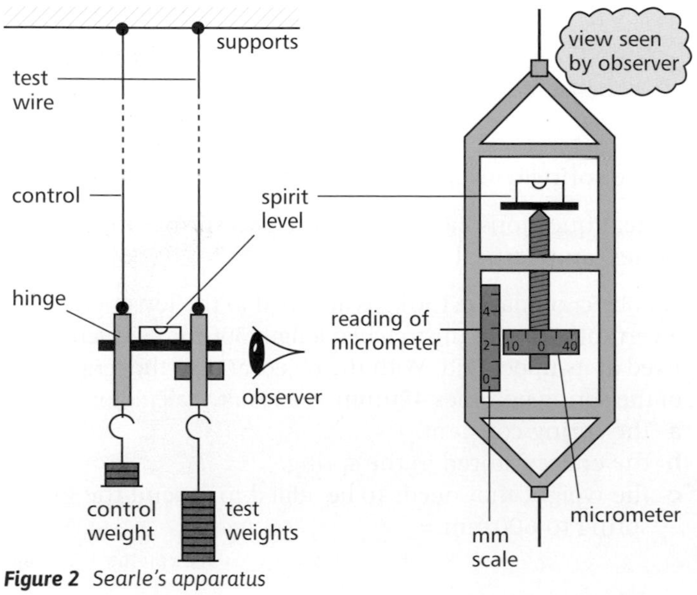

Figure 2 Searle's apparatus

Note

To calculate the stress in the wire, the diameter d of the wire should be measured then used in the formula $A = \frac{1}{4}\pi d^2$ to calculate A. The stress can then be calculated using $\sigma = F/A$.

Stress strain graphs

Figure 3 shows how the stress in a wire varies with strain.

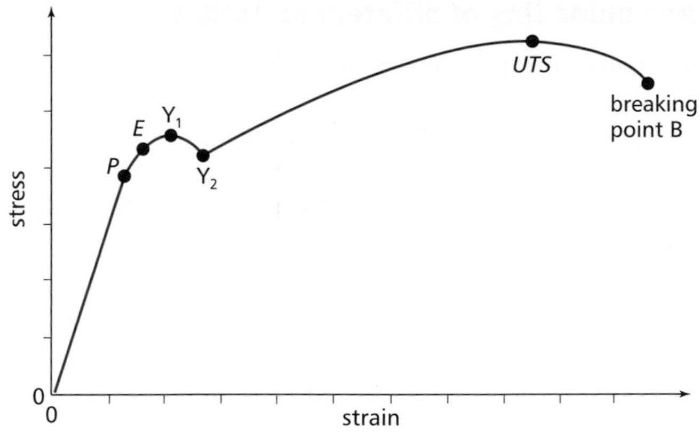

Figure 3 *Stress v. strain for a metal wire*

- **From 0 to the limit of proportionality P**, the stress is proportional to the strain.

The value of stress/strain is a constant, known as the **Young modulus** of the material:

$$\textbf{Young modulus, } E = \frac{\textbf{stress, } \sigma}{\textbf{strain, } \varepsilon} = \frac{\frac{F}{A}}{\frac{x}{L}} = \frac{FL}{Ax}$$

The unit of E is the pascal (Pa).

Because the line OP is straight and it passes through the origin, the gradient of this section is equal to stress ÷ strain. Hence the Young modulus is equal to the gradient of OP.

To determine the Young modulus of a metal in the form of a wire, the diameter and initial length of the wire are measured using a micrometer and metre ruler. The extension of the wire for different loads is then measured as described opposite. The corresponding stress and strain values can then be calculated and plotted as in Figure 3. The value of E is equal to the gradient of the initial straight section of the graph corresponding to section OP of Figure 3.

- **Beyond P**, the line curves and continues beyond the **elastic limit** E to the **yield point** $\mathbf{Y_1}$, which is where the wire weakens temporarily. The elastic limit is the point beyond which the wire is permanently stretched and suffers **plastic deformation**.
- **Beyond** $\mathbf{Y_2}$, a small increase in the stress causes a large increase in strain as the material of the wire undergoes **plastic flow**. Beyond maximum stress, or the **Ultimate Tensile Stress** (UTS), the wire loses its strength, extends and becomes narrower at its weakest point. Increase of stress occurs at this point due to the reduced area of cross-section, until the wire breaks at B. The stress at B is called the **breaking stress**.

Extension

Stress and strain curves for different materials

- The **stiffness** of different materials can be compared using the **gradient** of the stress–strain line which is equal to the Young modulus of the material. Thus steel is stiffer than copper (Figure 4).
- The **strength** of a material is its Ultimate Tensile Stress (UTS), which is its maximum stress. Steel is stronger than copper because its maximum stress is greater.
- A **brittle** material snaps without any noticeable yield. For example, glass breaks without any 'give'.
- A **ductile** material can be drawn into a wire. Copper is more ductile than steel.

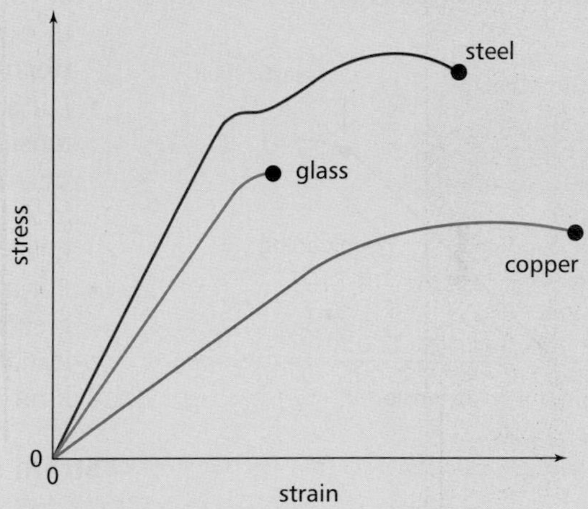

Figure 4 *Stress–strain curves*

Summary test 5.5

1 Calculate the stress in a wire of diameter 0.25 mm when the tension in the wire is 50 N.

2 A metal wire of diameter 0.23 mm and of unstretched length 1.405 m is suspended vertically from a fixed point. When a 40 N weight is suspended from the lower end of the wire, the wire stretches by an extension of 10.5 mm. Calculate the Young modulus of the wire material.

3 A vertical steel wire of length 2.5 m and diameter 0.35 mm supports a weight of 90 N. Calculate:

a the stress in the wire,

b the extension of the wire;

Young modulus of steel = 2.0×10^{11} Pa.

4 Use Figure 4 to decide:

a if copper is stronger than glass,

b if glass is stiffer than steel,

c if copper is more ductile than glass. Give a reason for each answer.

5.6 More about stress and strain

Learning outcomes

On these pages you will learn to:

- compare loading and unloading curves for a material under stress
- deduce the strain energy in a deformed material from the area under the force–extension graph
- calculate the elastic potential energy stored per unit volume in a material when its elastic limit is not exceeded

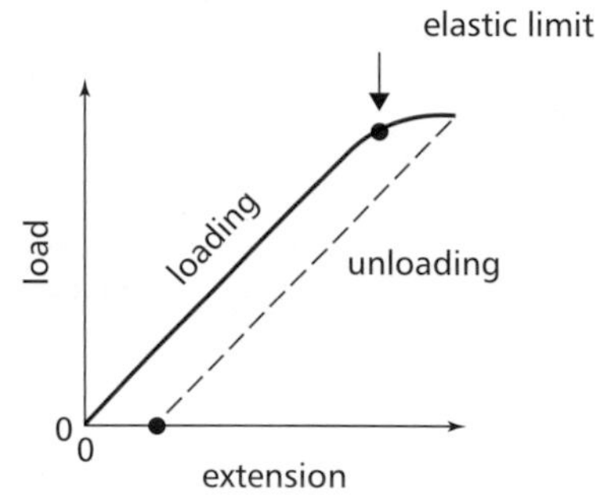

a *Metal wire*

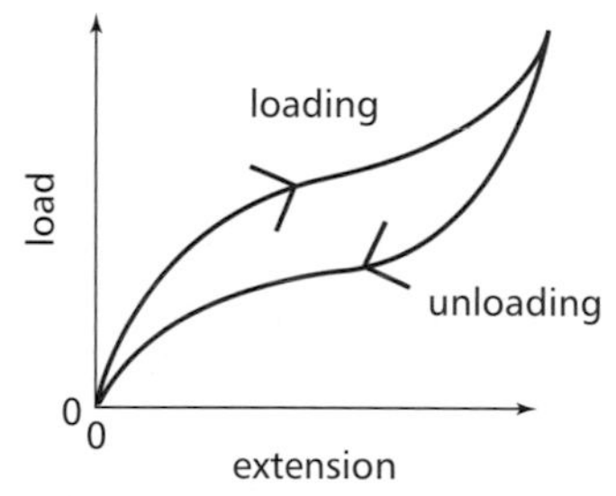

b *Rubber*

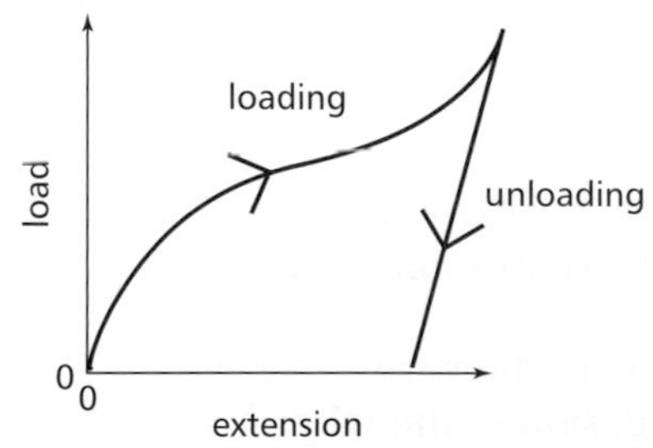

c *Polythene*

Figure 1 *Loading and unloading curves*

Investigating loading and unloading of different materials

How does the **stiffness** of a material change as a result of being stretched? To investigate this question, the tension in a strip of material is increased by increasing the weight it supports in steps. At each step, the extension of the material is measured. Typical results for different materials are shown in Figure 1. For each material, the loading curve and the subsequent unloading curve are shown.

- For a metal wire, its extension at any given tension when it is being unloaded is exactly the same as when it was being loaded (Figure 1a). The unloading curve and the loading curve are the same straight line. Provided its elastic limit is not exceeded, the wire returns to the same length when it has been completely unloaded as it had before it was loaded. If the wire is stretched beyond its elastic limit, the unloading line is parallel to the loading line. In this case, when it is completely unloaded, the wire is slightly longer so it has a **permanent extension**.
- For a rubber band (Figure 1b), the extension during unloading at any given tension is greater than during loading. The rubber band returns to the same unstretched length, but the unloading curve is below the loading curve except at zero extension and maximum extension. The rubber band remains elastic as it regains its initial length, but it has a **low limit of proportionality**.
- For a polythene strip (Figure 1c), the extension during unloading is also greater than during loading. However, the strip does not return to the same initial length when it is completely unloaded. The polythene strip has a low limit of proportionality and suffers **plastic deformation**.

Strain energy

The area under a graph of force v. extension is equal to the work done to stretch the wire. The work done to deform an object is referred to as **strain energy**. Consider the energy changes for each of the three materials in Figure 1 when each material is loaded then unloaded.

Metal wire (or spring)

Provided the limit of proportionality is not exceeded, the work done to stretch a wire to extension x is $\frac{1}{2}Fx$, where F is the tension in the wire at this extension. Because the elastic limit is not reached, the work done is stored as elastic potential energy in the wire. Therefore:

$$\textbf{Elastic potential energy stored in a stretched wire} = \tfrac{1}{2}Fx$$

Because the graph of tension against extension is the same for unloading as for loading, all the energy stored in the wire can be recovered when the wire is unloaded.

Since the volume of the wire = AL, where A is its cross-sectional area and L is its length:

$$\textbf{Elastic potential energy stored per unit volume} = \frac{1}{2}\frac{Fx}{AL} = \tfrac{1}{2} \times \textbf{stress} \times \textbf{strain}$$

Worked example

A steel wire of uniform diameter 0.35 mm and of length 810 mm is stretched to an extension of 2.5 mm. Calculate:

a the tension in the wire, **b** the elastic potential energy stored in the wire.

The Young modulus for steel = 2.1×10^{11} Pa

Solution

a Extension, $x = 2.5\,\text{mm} = 2.5 \times 10^{-3}\,\text{m}$

Area of cross-section of wire $= \dfrac{\pi(0.35 \times 10^{-3})^2}{4}$

$= 9.6 \times 10^{-8}\,\text{m}^2$

To find the tension, rearranging the Young modulus equation $E = \dfrac{FL}{Ax}$ gives

$$F = \frac{EAx}{L} = \frac{2.1 \times 10^{11} \times 9.6 \times 10^{-8} \times 2.5 \times 10^{-3}}{0.810} = 62\,\text{N}$$

b Elastic potential energy stored in the wire $= \frac{1}{2}Fx = 0.5 \times 62 \times 2.5 \times 10^{-3} = 7.8 \times 10^{-2}\,\text{J}$

Rubber band

The work done to stretch the rubber band is represented in Figure 1b by the area under the loading curve. The work done by the rubber band when it is unloaded is represented by the area under the unloading curve. The area between the loading curve and the unloading curve therefore represents the difference between energy stored in the rubber band when it is stretched and useful energy recovered from it when it unstretches. The difference is because some of the energy stored in the rubber band becomes internal energy of the molecules when the rubber band unstretches. Figure 2 shows how the area between the loading and unloading curve can be used to determine the internal energy retained when the rubber band is stretched then released.

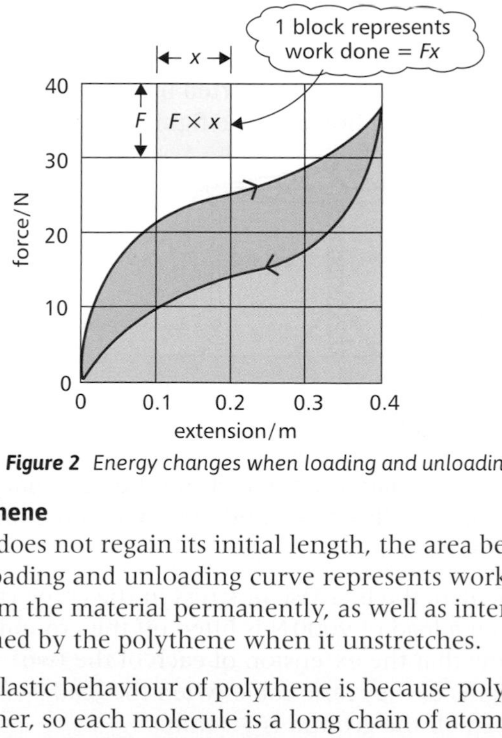

Figure 2 *Energy changes when loading and unloading rubber*

Polythene

As it does not regain its initial length, the area between the loading and unloading curve represents work done to deform the material permanently, as well as internal energy retained by the polythene when it unstretches.

The plastic behaviour of polythene is because polythene is a polymer, so each molecule is a long chain of atoms. Before being stretched, the molecules are tangled together or folded against each other. Weak bonds referred to as **cross-links** form between atoms where molecules are in contact with each other. When placed under tension, a thin sample of polythene easily stretches as the original weak cross-links break and the molecules align parallel to each other. New weak cross-links form in the stretched state and, when the tension is removed, the polythene strip remains stretched.

Extension

Notes Rubber is also a polymer, but its molecules are curled up and tangled together when it is unstretched. When placed under tension, its molecules are straightened out as its length increases more and more. When the tension is removed, its molecules curl up again and it regains its initial length.

A metal in the solid state is **crystalline** because it consists of tiny crystals or **grains** packed together. The atoms in each grain are arranged in layers in a regular pattern as in any crystal. When a metal is stretched beyond its elastic limit the layers of atoms in some of the grains slide past each other. When the distorting forces are removed, the displaced atoms do not return to their original places.

Solids such as glass are **amorphous**, which is neither crystalline nor polymeric. The atoms in an amorphous solid are randomly arranged without any regular pattern. Glass is brittle because stress causes any tiny cracks at its surface to propagate through the material.

Summary test 5.6

Young modulus for steel $= 2.1 \times 10^{11}\,\text{Pa}$; Young modulus for copper $= 1.3 \times 10^{11}\,\text{Pa}$

1 A vertical steel cable of diameter 24 mm and of length 18 m supports a weight of 1500 N attached to its lower end. Calculate:
  - **a** the tensile stress in the cable,
  - **b** the extension of the cable,
  - **c** the elastic potential energy stored in the cable, assuming its elastic limit has not been reached.

2 A vertical steel wire of diameter 0.28 mm and of length 2.0 m is fixed at its upper end and has a weight of 15 N suspended from its lower end. Calculate:
  - **a** the extension of the wire,
  - **b** the elastic potential energy stored in the wire.

3 A steel bar of length 40 mm and cross-sectional area $4.5 \times 10^{-4}\,\text{m}^2$ is placed in a vice, and compressed by 0.20 mm when the vice is tightened. Calculate:
  - **a** the compressive force exerted on the bar,
  - **b** the work done to compress it.

4 Figure 2 shows a force v. extension curve for a strip of rubber. Use the graph to determine:
  - **a** the work done to stretch the rubber to an extension of 0.40 m,
  - **b** the internal energy retained by the rubber when it unstretches.

5 Exam-style and Practice Questions

Launch additional digital resources for the chapter

$g = 9.81\,\text{m s}^{-2}$

1 A uniform copper wire of length 2.50 m has a diameter of 0.32 mm. Calculate:

a the volume of the wire,

b the mass of the wire,

c the mass per unit length of the wire.
Density of copper = $8900\,\text{kg m}^{-3}$

2 A hydraulic lift is used in a garage to raise a vehicle to enable its underside to be repaired. Figure 2.1 below shows how such a lift works. The maximum pressure of the fluid in the hydraulic lift is 1.2 MPa.

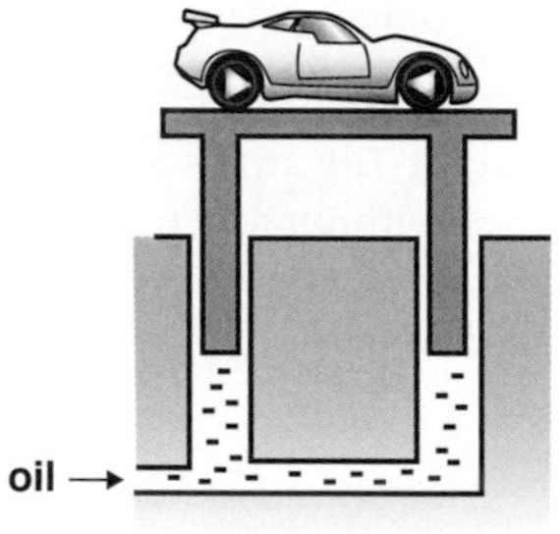

Figure 2.1

a The area of cross-section of each of the four 'legs' of the lift is $9.0 \times 10^{-3}\,\text{m}^2$. Calculate the maximum safe load the lift can raise if the lift platform has a weight of $1.5 \times 10^4\,\text{N}$.

b The lift is used to raise a vehicle of mass 2200 kg through a height of 2.4 m.
i Calculate the gain of potential energy of the vehicle.
ii Calculate the work done by the hydraulic system to lift the vehicle by 2.4 m.
iii Account for the difference between your answers to **i** and **ii**.

3 A steel spring of length 300 mm fixed at its upper end hangs vertically. When a 4.0 N weight is suspended from its lower end, the spring extends to an equilibrium length of 420 mm.

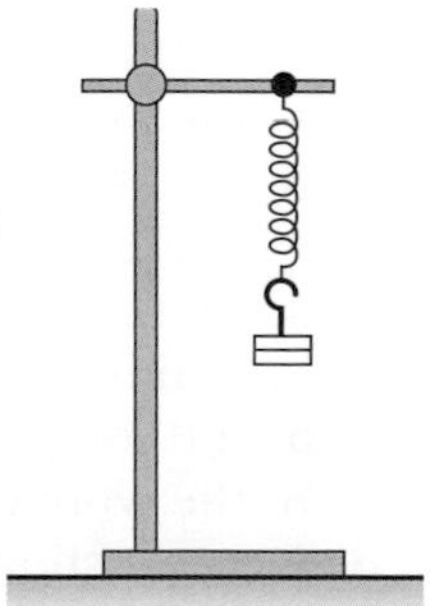

Figure 3.1

a Calculate:
i the spring constant of this spring,
ii the length of the spring when it supports a weight of 6.0 N.

b A second identical spring is suspended from the lower end of the first spring. Calculate the weight that would need to be suspended on the end of the lower spring to give a total extension for both springs equal to 90 mm.

4 a With the aid of a suitable example, explain what is meant by:
i a brittle material,
ii a ductile material.

b A steel guitar wire of diameter 0.28 mm and of length 800 mm is tightened by turning its tension key by 2 turns, each turn increasing the length of the wire by 1.8 mm. Calculate the increase of tension in the wire as a result.
Young modulus of steel = $2.1 \times 10^{11}\,\text{N m}^{-2}$

5 a Explain what is meant by the following terms:
i the elastic limit of a strip of material,
ii plastic behaviour.

b i Sketch a graph to show how the extension of a rubber band varies with the force used to stretch it.
ii When a car is in motion, each part of its tyres that make contact with the road is squashed and stretched as it passes through the point of contact with the road. In energy terms, explain why this repeated squashing and stretching causes the tyre to become warm.

6 A crane is designed to lift a load with a maximum weight of 9000 N to a maximum height of 65 m. The crane cable has an area of cross-section of $4.5 \times 10^{-4}\,\text{m}^2$ and is attached to a pulley block and hook of weight 600 N, so that the load is supported by two lengths of the crane cable (as shown in the figure below).

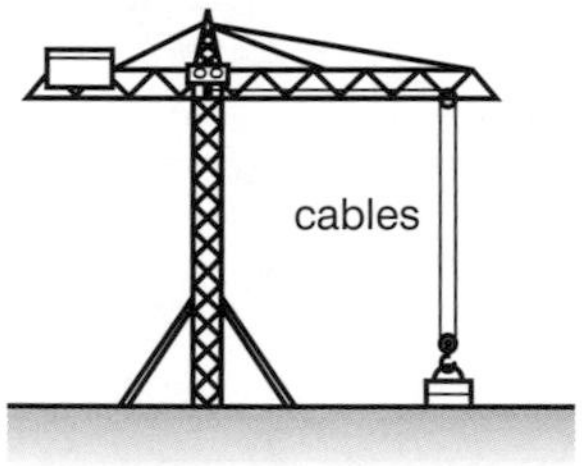

Figure 6.1

a Calculate the maximum weight of the suspended crane cable, pulley block and hook when the hook is just off the ground.

b i Calculate the increase of stress in the crane cables when a load of 9000 N is lifted off the ground.
ii Show that the extension of each of the two lengths of the crane cable when the load is raised off the ground is 3.3 mm.
Density of steel = $8000\,\text{kg m}^{-3}$
Young modulus of steel = $2.0 \times 10^{11}\,\text{N m}^{-2}$
iii Calculate the elastic energy stored in the two crane cables due to this extension.

7 The figure below shows a graph of the force used to stretch a metal wire against the extension of the wire for a wire of length 1620 mm and of cross-sectional area $1.40 \times 10^{-7}\,m^2$.

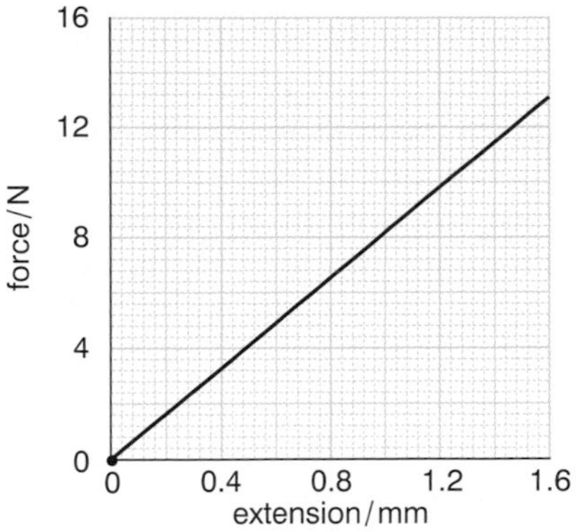

Figure 7.1

a Calculate the Young modulus of the wire.

b Calculate the energy stored in the wire when it extended from its unstretched length by 1.60 mm.

8 A spring which has a spring constant k hangs vertically from a fixed point. The spring supports a load of weight W attached to the lower end of the spring. When the load is at rest, the extension of the spring is within its elastic limit.

a **i** State what is meant by *elastic limit*.
ii State what is meant by *spring constant*.
iii Write down an expression in terms of k and W for the energy stored in the spring when the load is at rest.

b Figure 8.1 shows two arrangements P and Q of three identical springs used to support a load of weight W. The spring constant of each spring is k.

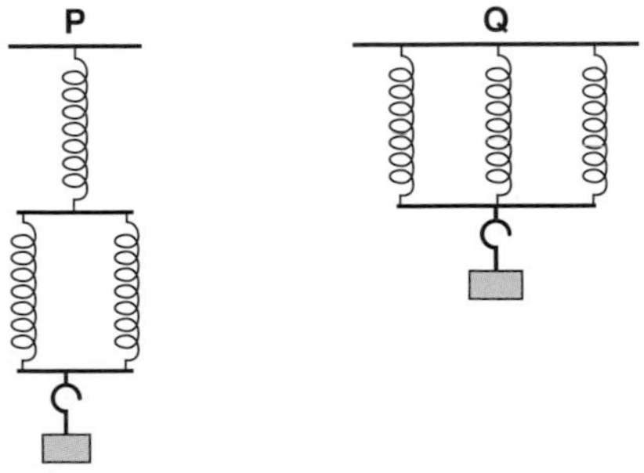

Figure 8.1

For each arrangement, determine:

i the extension in terms of e (the extension of a single spring for a load of weight W),
ii the effective spring constant in terms of k.

9 The variation of the tension T with extension e for a rubber band being stretched to an extension of 0.40 m is shown in Figure 9.1.

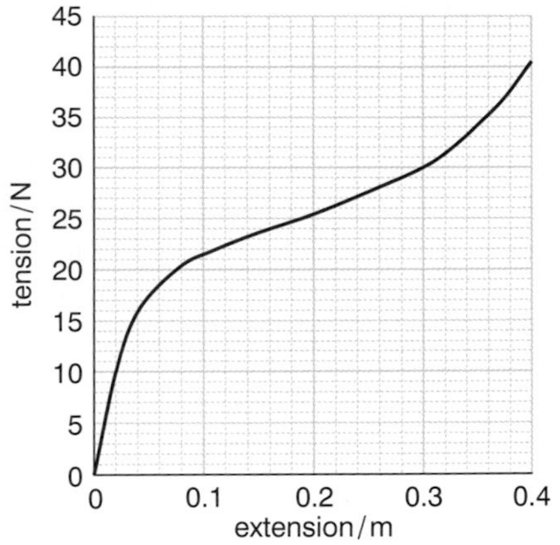

Figure 9.1

a Estimate the work done in extending the rubber band to an extension of 0.40 m.

b Describe how the stiffness of the rubber band changed as it was stretched.

10 In an investigation on upthrust, a metal cylinder of uniform cross-section was suspended vertically from a newton meter above a beaker of water, as shown in Figure 10.1.

Depth, d/mm	Newton meter reading/N	Upthrust, U/N
0	1.95	0
10	1.92	
21	1.88	
30	1.85	
39	1.82	
50	1.80	

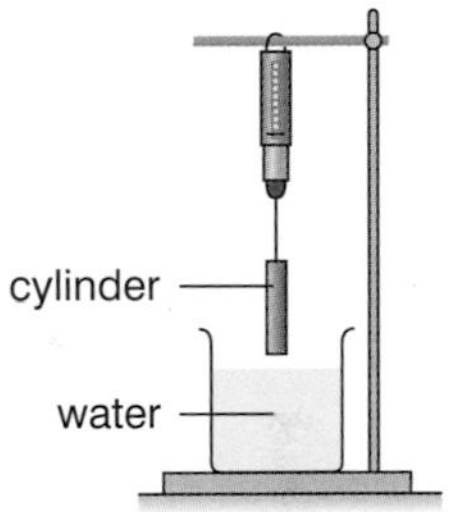

Figure 10.1

a The reading of the newton meter was 1.95 N. The cylinder was then partly lowered 50 mm into the water, causing the newton meter reading to decrease to 1.80 N.
i Explain why the newton meter reading decreased.
ii Calculate the upthrust on the cylinder.

b The cylinder was lowered to different measured depths in the water and the newton meter reading was recorded at each depth. See table above.
i Calculate the upthrust for the other depth measurements and copy and complete the table.
ii Plot a graph of the upthrust, U, against depth, d.
iii Determine the gradient of the graph.

c The length of the cylinder was 80 mm.
i Estimate the upthrust if the cylinder had been fully immersed.
ii Hence calculate the density of the cylinder. Density of water = $1000\,kg\,m^{-3}$

Key concepts: Forces

In studying chapters 1 to 5, you will have gained a wide appreciation of what forces do, as well as a deeper understanding of how physical principles relating to forces are combined with mathematical methods to predict and test their effects. The effects can then be compared with the predictions to find out if the mathematical methods give results consistent with observations. Such methods, developed by Galileo, Newton and other scientists, form the basis of many branches of physics and design engineering, from tunnels under the ground to robot vehicles on Mars. In addition, you should know now what we mean by 'conservation of energy', how we can use equations to calculate kinetic energy and potential energy changes and what we mean by 'efficiency' in relation to energy transfers. Momentum is another physical quantity you should now be familiar with – in particular, how you can use the relationship between force and momentum to calculate impact forces and what is meant by 'conservation of momentum' and how this important principle is used to work out what happens when objects collide or explode.

As you progressed through chapters 1–5, you will have developed your mathematical skills to a level where you can use equations and graphs to solve physical problems. In addition, you should now be able to carry out calculations accurately and express your answers in standard form in the correct units. You should also know that numerical answers should not be expressed to more significant figures than can be justified by the data provided.

Throughout the first five chapters, you will have had opportunities to develop your practical skills through measuring time intervals, distance, mass, weight and volume. Some practical experiments will have required you to use your measured data to calculate physical quantities such as density, enabling you to compare the accuracy of your results with accepted values. In addition, you should have become aware that each measurement you make has a degree of uncertainty and that uncertainties need to be estimated and used to give an overall uncertainty in a calculated quantity. Also, you will have plotted graphs, in particular straight-line graphs that can be linked to theoretical equations, from which the links can be tested. In addition, straight-line graphs can be used to determine physical quantities with less uncertainty than if a single set of measurements were used.

The following table provides an overview of the topics in chapters 1 to 5, showing where the key concepts have been developed.

Chapters 1–5: Key concepts

Topic			Key concepts	
1.1	introduction to vectors	use of vectors to add displacement and to add forces	use of maths	problem-solving
1.2–1.7	motion along a straight line	use of dynamics equations and graphs to solve mechanics problems involving motion along a straight line	use of maths	problem-solving, including the physical significance of gradients and areas under graphs
1.5	free fall	Galileo's inclined plane test	testing	concept of acceleration
		measurement of g	forces and fields	graph of $s = \frac{1}{2}gt^2$ to determine g
1.8–1.9	projectile motion	investigating projectile paths	testing	concept of vertical and horizontal motion
		calculations of vertical and horizontal motion	use of maths	problem-solving
2.1–2.3	force and motion	investigating force and acceleration	testing	verification of $F = ma$
		calculations using $F = ma$	use of maths	problem-solving using relevant equations
		resistive forces and terminal speed	forces and fields	concept of terminal speed
2.4–2.6	energy and power	calculations using the equations for kinetic energy, potential energy, power and work	energy and matter	conservation of energy
2.7–2.8	efficiency	causes of inefficiency and energy losses		concept of dissipative forces
3.1–3.6	forces in equilibrium	principle of moments	forces and fields	problem-solving using key principles regarding equilibrium
		conditions for equilibrium	use of maths	
4.1–4.5	momentum	Newton's second law of motion	forces and fields	problem-solving using $Ft = \Delta mv$
		momentum, impact forces and conservation of momentum	use of maths	problem-solving using conservation of momentum
5.1–5.4	properties of materials	density	energy and matter	practical skills, density measurements
		pressure, upthrust, Hooke's law	testing	Hooke's law
		calculations and graphs involving density, pressure, upthrust and Hooke's law	use of maths	problem-solving
5.5–5.6	deformation of solids	stress, strain	energy and matter	practical skills
		Young modulus	testing	measurement of Young modulus
		calculations and graphs involving Young modulus	use of maths	problem-solving

Question

*This question is about an investigation into the flight path of a ball projected horizontally from the edge of a table. It tests how we use models to make predictions about a physical system and how we then test these predictions with an experimental investigation. Notice also the use of mathematics to express the relationship between two perpendicular distances, **x** and **y**. In this case, the predicted relationship is of the form $y = kx^2$. So, by plotting a graph of **y** against x^2, we can tell if the relationship is valid by looking at whether or not the graph is linear (i.e. a straight-line graph) and passes through the origin. If the graph is linear, we can also determine the constant **k**, as it is equal to the gradient of the graph. You will notice as well that this investigation is about the behaviour of a particle in a field: in this case a uniform gravitational field. The path of such a particle is similar to that of a charged particle in a uniform electric field, because in both cases the acceleration of the particle is constant in magnitude and direction.*

Before answering the question, students should be given the opportunity to do this or a similar investigation.

1 A student wanted to measure the path of a small steel ball projected horizontally. Figure 1 shows the arrangement he used. The thin metal bar B was positioned at different horizontal distances from the edge of a flat laboratory bench. For each distance, the vertical position of the bar was adjusted until it was positioned so that the ball hit the bar after being released from rest inside the top of the cardboard tube T.

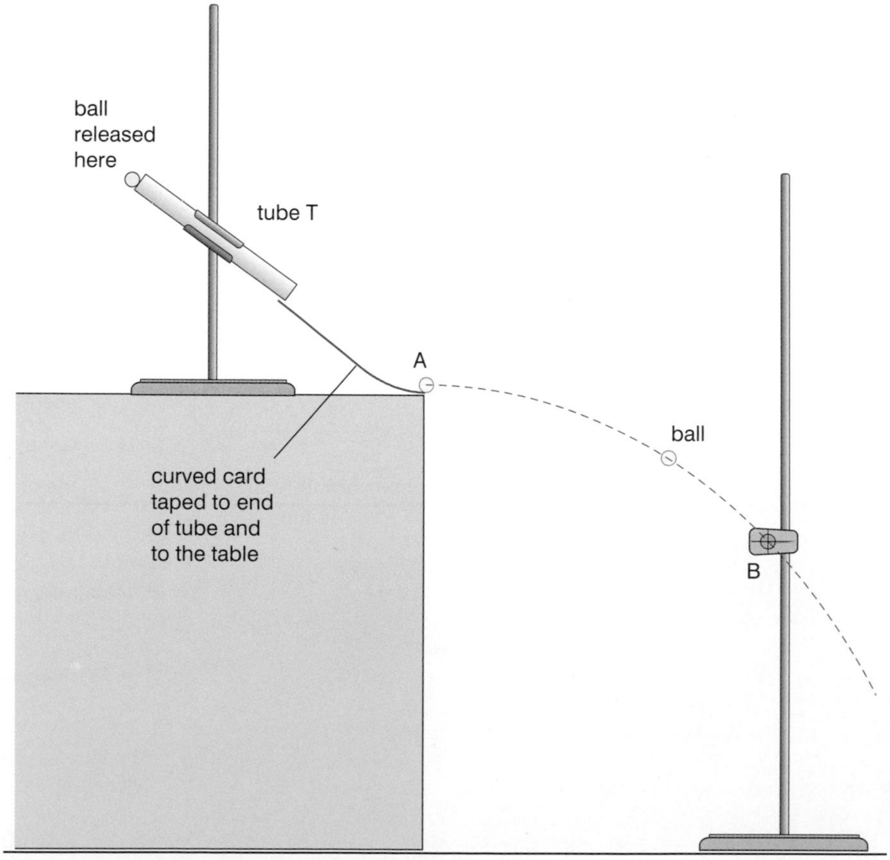

Figure 1

a i Why was it important to ensure that the ball was released at rest from the same position each time the test was carried out?

ii Give two reasons why a large steel ball would have been unsuitable. *(3 marks)*

b For each horizontal distance x from B to the point A where the ball left the bench, x was measured and the corresponding vertical distance y was measured in three separate tests. The results obtained are shown below.

x/m	0	0.205	0.402	0.596	0.803	0.995	1.205
y/mm	0	19	78	155	272	432	643
	0	20	65	164	277	438	631
		22	72	158	286	429	639
mean value of y/mm							
x^2/m^2							

i Copy and complete the table of results above.

The student reckoned that at time t after the ball left the bench, $y = \frac{1}{2}gt^2$ and $x = Ut$, where U is the horizontal speed of projection of the ball.

ii Use these equations to show that $y = kx^2$, where k is a constant, and derive an expression for k in terms of U and g.

iii Plot a graph of y against x^2 to find out if the student's hypothesis is correct.

iv Use the graph to determine U, the horizontal speed of projection of the ball. *(13 marks)*

c The student reckoned that the distances were measured to within 2 mm. Discuss this claim in terms of the above measurements. *(4 marks)*

6 Electric current

6.1 Electric charge

Learning outcomes

On these pages you will learn to:

- explain how to create and detect static charge
- explain what is meant by an electric field and a field line

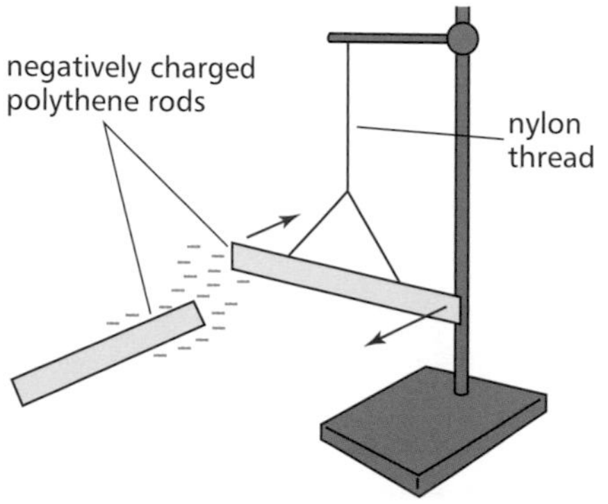

Figure 1 *Electrostatic forces*

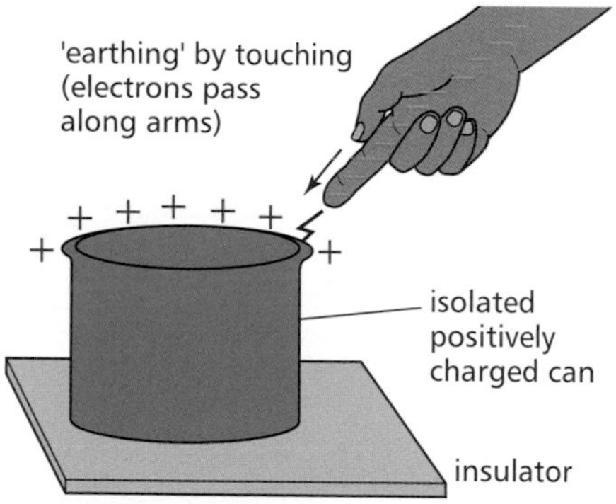

Figure 2 *Discharge to Earth*

Figure 3 *The electroscope*

Static electricity

Most plastic materials can be charged quite easily by rubbing with a dry cloth. When charged, they attract small bits of paper. Do charged pieces of plastic material attract one another? Figure 1 shows an arrangement to test for attraction. A charged perspex ruler will attract a charged polythene comb, but two charged rods of the same material always repel one another.

There are just two types of charge which are referred to as positive and negative charge. A charged object will always:

- repel another charged object that has the same type of charge (i.e. like charge)
- attract another charged object that has the other type of charge (i.e. unlike charge).

Like charges repel; unlike charges attract

Electrons are responsible for charging in most situations. An uncharged atom contains equal numbers of **protons** and electrons. An electron carries a tiny negative charge. A proton carries a tiny positive charge equal in magnitude to that of the electron.

- Add one or more electrons to an uncharged atom and it becomes negatively charged.
- Remove one or more electrons from an uncharged atom and it becomes positively charged.

An uncharged solid contains equal numbers of electrons and protons. When an uncharged perspex rod is rubbed with an uncharged dry cloth, electrons transfer from the rod to the cloth so the rod becomes positively charged and the cloth becomes negatively charged.

- Electrical conductors such as metals contain lots of free electrons. These are electrons which move about inside the metal and are not attached to any one atom. To charge a metal up, it must first be isolated from the Earth. Otherwise, any charge given to it is neutralised by electrons transferring between the conductor and the Earth. Then the isolated conductor can be charged by direct contact with any charged object. If an isolated conductor is charged positively then 'earthed', electrons transfer from the Earth to the conductor to neutralise or discharge it. See Figure 2.
- Insulating materials do not contain free electrons. All the electrons in an insulator are attached to individual atoms. Some insulators, such as perspex or polythene, are easy to charge because their surface atoms easily gain or lose electrons.

The gold leaf electroscope is used to detect charge. If a charged object is in contact with the metal cap of the electroscope, some of the charge on the object transfers to the electroscope. As a result, the gold leaf and the metal stem which is attached to the cap gain the same type of charge and the leaf rises because it is repelled by the stem.

If another object with the same type of charge is brought near the electroscope, the leaf rises further because the object forces some charge on the cap to transfer to the leaf and stem.

Field lines and patterns

Any two charged objects exert equal and opposite forces on each other without being directly in contact. An **electric field** is said to surround each charge. Suppose a small positive charge is placed as a test charge near a body with a much bigger charge which is also positive. If the test charge is free to move, it will follow a path away from the body with the bigger charge. The path a free test charge follows is called a **line of force** or a **field line**.

The direction of an electric field line is the direction in which a positive test charge would move. Figure 4 shows the patterns of fields around different charged objects. Each pattern is produced by semolina grains sprinkled on oil. An electric field is set up across the surface of the oil by connecting two metal conductors in the oil to the output terminals of a high voltage supply unit. The grains line up along the field lines, like plotting compasses in a magnetic field.

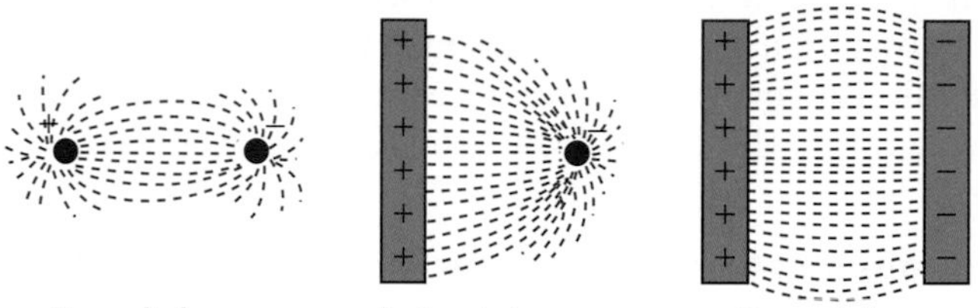

Figure 4 Electric field patterns

- Oppositely charged point objects create a field as shown in Figure 4a. The field lines become concentrated at the points. A positive test charge released from an off-centre position would follow a curved path to the negative point charge.
- A point object near an oppositely charged flat plate produces a field as shown in Figure 4b. The field lines are concentrated at the point object but they meet the plate at right angles to it.
- Two oppositely charged plates create a field as shown in Figure 4c. The field lines run parallel from one plate to the other, meeting the plates at right angles. The field is **uniform** between the plates as the field lines are parallel to each other.

Devices that use electric fields directly include:

- electrostatic precipitators which extract ash, dust and other fine particles from gases released in power stations and factories. The gases pass through a grid of wires at high voltage between earthed metal plates fixed vertically, as shown in Figure 5. The particles become charged as they pass through the grid and are then attracted onto the plates by the electric field between the grid wires and the plates.
- electrostatic sprays such as airborne crop sprays which produce charged droplets that spread out as they repel each other and are attracted onto the ground. Industrial paint sprays are used to coat metal panels with an even film of paint.

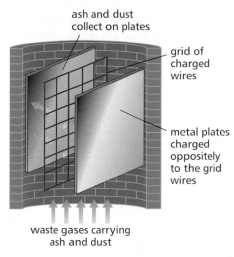

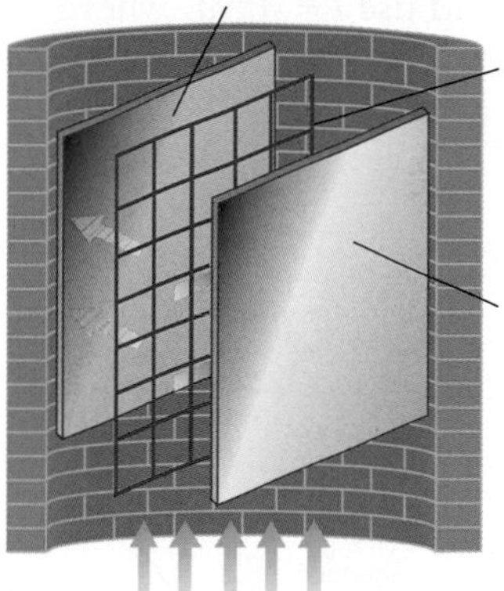

Figure 5 An electrostatic precipitator

Summary test 6.1

1 Explain each of the following observations in terms of transfer of electrons:

 a An insulated metal can is given a positive charge by touching it with a positively charged rod.

 b A negatively charged metal sphere suspended on a thread is discharged by connecting it to the ground using a wire.

2 An insulated metal conductor is earthed before a negatively charged object is brought near to it.

 a Explain why the free electrons in the conductor move as far away from the charged object as they can.

 b The conductor is then briefly earthed. The charged object is then removed from the vicinity of the conductor. Explain why the conductor is left with an overall positive charge.

3 A positively charged point object is placed near an earthed metal plate, as shown in Figure 6.

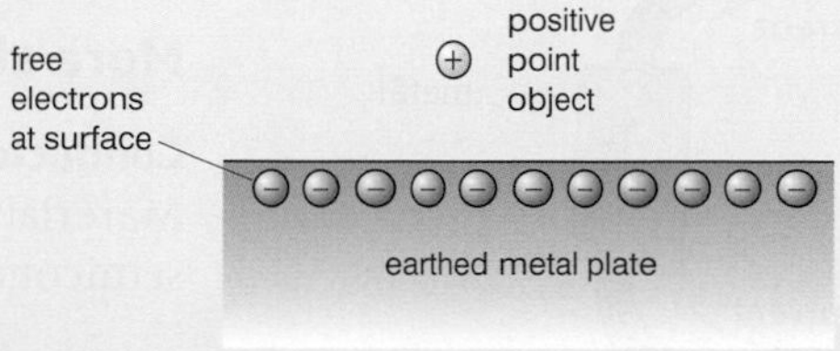

Figure 6

 a Explain why electrons gather at the surface of the metal plate near the object.

 b Explain why there is a force of attraction between the object and the metal plate.

4 Sketch the pattern of the field lines of the electric field:

 a between two oppositely charged parallel plates,

 b between a positively charged point object and an earthed metal plate.

6.2 Current and charge

Learning outcomes

On these pages you will learn to:

- recognise that an electric current is a flow of charge due to the passage of charged particles
- know that the charge on a charge carrier is quantised and that the charge carriers in a metal are electrons
- distinguish between a conductor, an insulator and a semiconductor
- state and use the relationship between the current and the charge flow in a certain time
- derive and use $I = Anvq$, where n is the number density of charge carriers

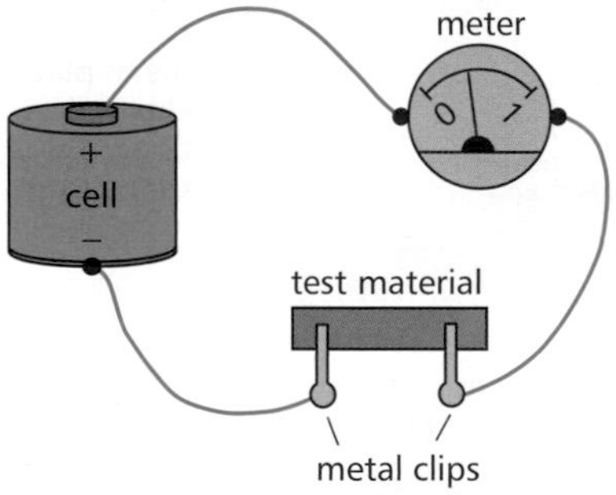

Figure 1 Testing for conduction

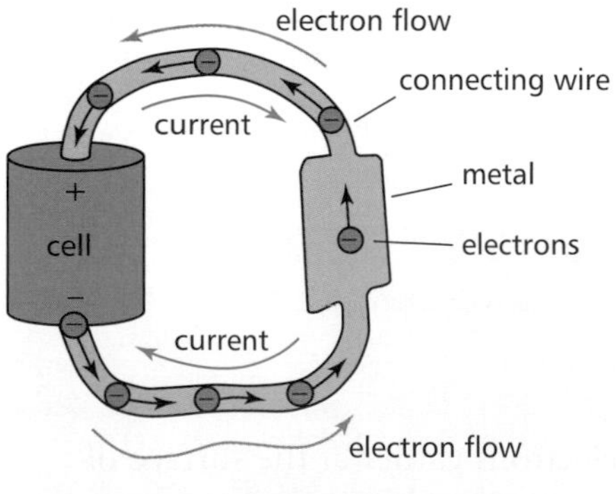

Figure 2 Convention for current

Electrical conduction

An electric current is a flow of charge due to the passage of charged particles. These charged particles are referred to as **charge carriers**.

- In metals, the charge carriers are **conduction electrons**. They move about inside the metal, repeatedly colliding with each other and the fixed positive ions in the metal.
- In comparison, when an electric current is passed through a salt solution, the charge is carried by ions which are charged atoms or molecules.
- Charge is quantised, which means it is only found in whole number multiples of 1.6×10^{-19} C. This basic amount is referred to as the 'quantum' of charge, denoted by the symbol e. For example, the charge on a calcium Ca^{2+} ion is equal to $+2e$.

A simple test for conduction of electricity is shown in Figure 1. If the test material is a metal, electrons pass round the circuit: they leave the battery at its negative terminal, pass though the metal, then re-enter the battery at its positive terminal.

The convention for the direction of current in a circuit is from **+ to –**, as shown in Figure 2. The convention was agreed long before the discovery of electrons. When it was set up, it was known that an electric current is the rate of flow of charge one way round a circuit. However, it was not known if the current was due to positive charge flowing round the circuit from + to –, or if it was due to negative charge flowing from – to +.

- The unit of current is the **ampere (A)**, which is defined in terms of the force between two parallel wires when they pass the same current. See Topic 16.3.
- The unit of charge is the **coulomb (C)**, equal to the charge flow in one second when the current is one ampere. The magnitude of the charge of the electron, e, is 1.6×10^{-19} C. This is sometimes referred to as the **elementary charge**.
- For a current I, the **charge flow Q** in time t is given by the equation

$$Q = It$$

- For charge flow Q in a time interval t, the current I is given by:

$$I = \frac{Q}{t}$$

The equation shows that a current of 1 A is due to a flow of charge of 1 coulomb per second. As the charge of the electron is 1.6×10^{-19} C, a current of 1 A along a wire must be due to 6.25×10^{18} electrons passing along the wire each second.

More about charge carriers

Conductors, insulators and semiconductors

Materials can be classified in electrical terms as conductors, insulators or semiconductors.

- In an **insulator**, each electron is attached to an atom and cannot move away from the atom. When a voltage is applied across an insulator, no current passes through the insulator because no electrons can move through the insulator.
- In a **metallic conductor**, most electrons are attached to atoms; but some are not and these are the **charge carriers** in the metal. When a voltage is applied across the metal, these **conduction electrons** move towards the positive terminal of the metal.
- In a **semiconductor**, the number of charge carriers increases with increase of temperature. The resistance of a semiconductor therefore decreases as its temperature is raised. A pure semiconducting material is referred to as an **intrinsic semiconductor**, because conduction is due to electrons that break free from the atoms of the semiconductor.

- In an **electrolyte**, the charge carriers are **positive and negative ions**. When a voltage is applied to the electrodes, the positive ions are attracted to the **cathode** (i.e. the negative electrode) and the negative ions are attracted to the **anode** (i.e. the positive electrode). The ions are discharged on reaching the relevant electrode. The process is known as **electrolysis**.

Drift velocity

The charge carriers in a conductor or a semiconductor move about at random inside the material. When a p.d. is applied across the material, in addition to their random motion, the charge carriers 'drift' to the positive end. Their mean **displacement** per unit time is their **drift velocity**. The current through a material depends on the drift velocity v of the charge carriers.

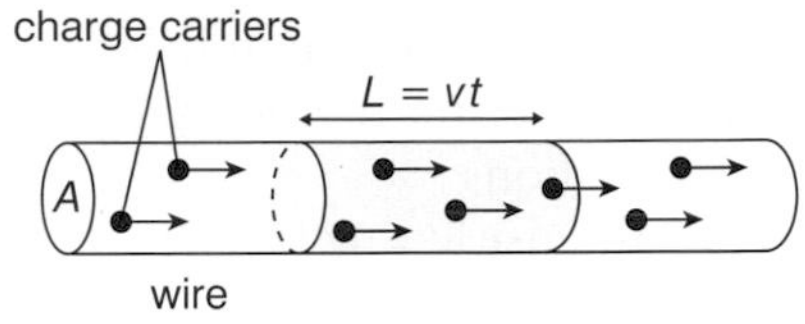

Figure 3 *Drift velocity*

Consider Figure 3, which shows charge carriers moving through a section of wire of cross-sectional area A and length L. The average time taken, t, for each charge carrier to move through it $= L \div v$.

If there are n charge carriers per unit volume in the wire,

- the number of charge carriers in this section $= n(A \times L)$ since the section volume $= A \times L$
- their total charge $Q =$ number of charge carriers × charge of each carrier (q) $= n(A \times L)q$

Therefore, the current through the section, $I = \dfrac{Q}{t} = \dfrac{nALq}{L/v}$

$$= Anvq$$

For electrons, $q = e$ gives $\boldsymbol{I = Anve}$

Physics and the Human Genome Project

To map the human genome, fragments of DNA are tagged with a C, G, A, or T base containing a dye. Each tagged fragment carries a negative charge. A voltage is applied across a strip of gel containing a spot of liquid containing tagged fragments. The fragments are attracted to the positive electrode. The smaller the fragment, the faster it moves; so the fragments separate out according to size, as they move to the positive electrode. The fragments pass through a spot of laser light, which causes the dye attached to each tag to **fluoresce** as it passes through the laser spot. Light sensors linked to a computer detect the glow from each tag. The computer is programmed to work out and display the sequence of bases in the DNA fragments.

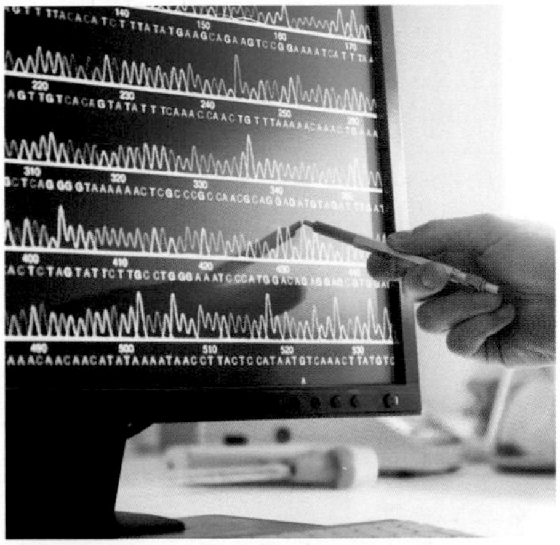

Figure 4 *Mapping the human genome*

Summary test 6.2

$e = 1.6 \times 10^{-19}$ C

1 **a** The current in a certain wire is 0.35 A. Calculate the charge passing a point in the wire:
i in 10 s, **ii** in 10 min.

b Calculate the average current in a wire through which a charge of 15 C passes in:
i 5 s, **ii** 100 s.

2 Calculate the number of electrons passing a point in the wire in 1 min when the current is:

a 1 μA **b** 5.0 A

3 A certain type of rechargeable battery is capable of delivering a current of 0.2 A for 4000 s, before its voltage drops and it needs to be recharged. Calculate:

a the total charge the battery can deliver before it needs to be recharged,

b the maximum time it could be used for without being recharged if the current through it were:
i 0.5 A **ii** 0.1 A

4 Estimate the drift velocity of the conduction electrons in a copper wire of diameter 0.4 mm when a current of 2.0 A passes through it. Assume copper contains 10^{29} electrons per m^3.

6.3 Potential difference and power

Learning outcomes

On these pages you will learn to:

- define potential difference (p.d.) and the volt
- define the electromotive force (e.m.f.) of an electrical source
- distinguish between e.m.f. and p.d. in terms of energy considerations
- recall and use $V = W / Q$
- recall and use $P = VI$

Energy and potential difference

When a torch bulb is connected to a battery, electrons deliver energy from the battery to the torch bulb. Each electron which passes through the bulb takes a fixed amount of energy from the battery and delivers it to the bulb. After delivering energy to the bulb, each electron re-enters the battery via the positive terminal to be resupplied with more energy to deliver to the bulb.

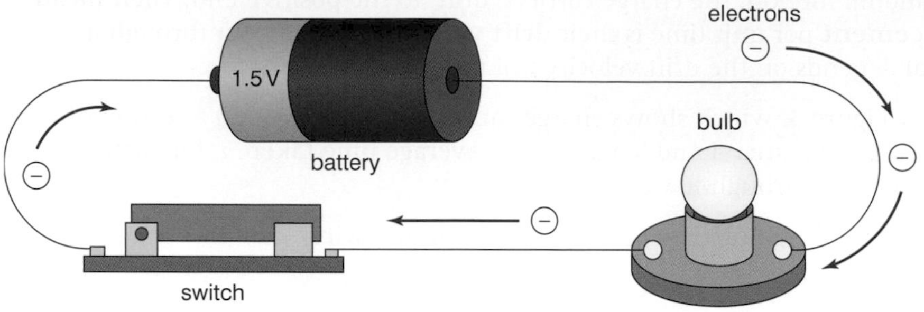

Figure 1 *Energy transfer by electrons*

Each electron in the battery has the **potential** to deliver energy, even if the battery is not part of a complete circuit. In other words, the battery supplies each electron with **electrical potential energy**. When the battery is in a circuit, each electron passing through the circuit component does work to pass through the component and therefore transfers some or all of its electrical potential energy to the component. The **work done** by an electron is equal to its loss of potential energy. The energy transfer (i.e. work done) per unit charge is defined as the **potential difference** (abbreviated as p.d.) or **voltage**, ***V***, across the component.

The potential difference across a component is defined as the energy transferred per unit charge.

The unit of p.d. is the **volt (V),** which is equal to 1 joule per coulomb.

If work W is done when charge Q flows through the component, the p.d. across the component, V, is given by:

$$V = \frac{W}{Q}$$

Rearranging this equation gives $W = QV$ for the work done or energy transfer when charge Q passes through a component which has a p.d. V across its terminals.

Examples

- If 30 J of work is done when 5 C of charge passes through a component, the p.d. across the component must be 6 V $\left(= \frac{30\,\text{J}}{5\,\text{C}}\right)$.
- If the p.d. across a component in a circuit is 12 V, then 3 C of charge passing through the component would transfer 36 J of energy from the battery to the component.

Energy transfer in different devices

An electric current has a **heating effect** in a component with **resistance**. It also has a **magnetic effect**, which is made use of in electric motors and loudspeakers.

- In a device that has resistance, such as an electrical heater, the work done on the device is transferred as **thermal energy**. This happens because the charge

Figure 2 *Electrical devices*

carriers repeatedly collide with atoms in the device and transfer energy to them, so the atoms **vibrate more** and the resistor becomes hotter.

- In an electric motor, the work done on the motor is transferred as **kinetic energy** of the motor. The charge carriers are electrons that need to be forced through the wires of the **spinning motor coil** against the opposing force on the electrons due to the motor's magnetic field.
- For a loudspeaker, the work done on the loudspeaker is transferred as **sound energy**. Electrons need to be forced through the wires of the **vibrating loudspeaker coil** against the force on them due to the loudspeaker magnet.

Electromotive force (e.m.f)

The e.m.f., *E*, of a source of electricity is defined as the electrical energy per unit charge produced inside the source.

The unit of e.m.f. is the **volt (V)**, the same as the unit of p.d., which is equal to 1 joule per coulomb.

For a source of e.m.f., E, in a circuit, the electrical energy produced when charge Q passes through the source is QE. This energy is transferred to other parts of the circuit, and some may be dissipated in the source itself due to the source's **internal resistance** (see Topic 7.3 for internal resistance).

Figure 3 *Power supplies*

Electrical power and current

Consider a component, or device, which has a potential difference V across its terminals and a current I passing through it. In time t:

- The charge flowing through it, $Q = It$
- The work done by the charge carriers, $W = QV = (It)V = IVt$

Work done, $W = IVt$

The energy transfer ΔE in the component or device is equal to the work done W. Therefore, because power $= \frac{\text{energy}}{\text{time}}$, the electrical power P supplied to the device is given by:

$$P = \frac{IVt}{t} = IV$$

Electrical power, $P = IV$

Notes

- This equation can be rearranged to give: $I = \frac{P}{V}$ or $V = \frac{P}{I}$
- The unit of power is the **watt (W)**. Therefore one volt is equal to one watt per ampere. For example, if the p.d. across a component is 4 V, then the power delivered to the component is 4 W per ampere of current.
- Energy supplied by mains electricity is measured in **kilowatt hours (kW h)**, usually referred to as 'units'. One kilowatt hour is the energy transfer when 1 kW of power is supplied for exactly 1 h.
  Therefore 1 kW h = 3.6 MJ (= 1000 W × 3600 s).
- The correct value of a **fuse** for an electrical appliance can be worked out using the equation $I = \frac{P}{V}$ if the power and voltage of the appliance are known.
  For example, a 230 V, 2.5 kW electric kettle would take a current of 11 A $\left(= \frac{2500\,\text{W}}{230\,\text{V}}\right)$. Therefore, given the choice of a 5 A or a 13 A fuse, a 13 A fuse should be used.

Worked example

A 12 V, 48 W electric heater is connected to a 12 V battery. Calculate: **a** the heater current, **b** the energy transfer in 300 s.

Solution

a Rearrange $P = IV$ to give $I = \frac{P}{V} = \frac{48\,\text{W}}{12\,\text{V}} = 4\,\text{A}$

b $\Delta E = IVt = 4\,\text{A} \times 12\,\text{V} \times 300\,\text{s} = 14\,400\,\text{J}$

Summary test 6.3

1 Calculate the energy transfer in 1200 s in a component when the p.d. across it is 12 V and the current is:
   a 2 A, **b** 0.05 A.

2 A 6 V, 12 W light bulb is connected to a 6 V battery. Calculate:
   a the current in the light bulb,
   b the energy transfer to the light bulb in 1800 s.

3 A 230 V electrical appliance has a power rating of 800 W.
   a Calculate:
      i the energy transfer in the appliance in 1 min,
      ii the current in the appliance.
   b Which of the following fuse values would be suitable for this appliance: 3 A, 5 A, or 13 A?

4 A battery has an e.m.f. of 9 V and a negligible internal resistance. It is capable of delivering a total charge of 1350 C. Calculate:
   a the maximum energy the battery could deliver,
   b the power it would deliver to the components of a circuit if the current in it was 0.5 A,
   c how long the battery would last if it supplies power at the rate calculated in part **b**.

6.4 Resistance

Learning outcomes

On these pages you will learn to:

- define resistance and the ohm
- recall and use $V = IR$
- sketch and discuss the I–V characteristics of a metallic conductor at constant temperature, a semiconductor diode and a filament lamp
- state Ohm's law
- recall and use $R = \rho L / A$ in resistivity calculations

Table 1 *Reminder about prefixes*

Prefix	Symbol	Value
nano	n	10^{-9}
micro	µ	10^{-6}
milli	m	10^{-3}
kilo	k	10^{+3}
mega	M	10^{+6}
giga	G	10^{+9}

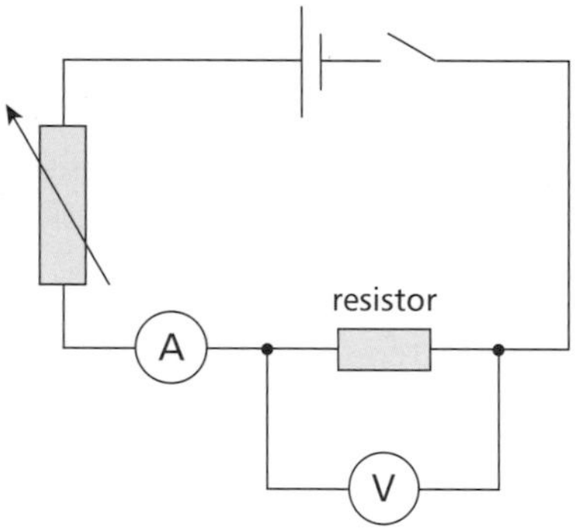

Figure 1 *Measuring resistance*

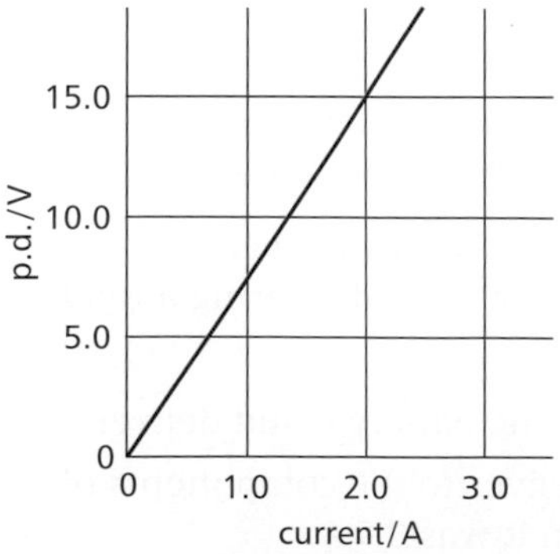

Figure 2 *Potential difference v. current for a resistor*

Definitions and laws

The **resistance** of a component in a circuit is a measure of the difficulty of making current pass through the component. Resistance is caused by the repeated collisions between the charge carriers in the material with each other and with the fixed positive ions of the material.

$$\textbf{Resistance of any component} = \frac{\textbf{p.d. across the component}}{\textbf{current through it}}$$

For a component in which the current is I when the p.d. across it is V, its resistance R is given by the equation

$$R = \frac{V}{I}$$

The unit of resistance is the **ohm (Ω)** which is equal to 1 volt per ampere.

Rearranging the above equation gives $V = IR$ or $I = \frac{V}{R}$.

Worked example

The current through a component is 2.0 mA when the p.d. across it is 12 V. Calculate:

a its resistance at this current,

b the p.d. across the component when the current is 50 µA, assuming that its resistance is unchanged.

Solution

a $R = \frac{V}{I} = \frac{12}{2.0 \times 10^{-3}} = 6000\,\Omega$

b $V = IR = 50 \times 10^{-6} \times 6000 = 0.30\,V$

Measurement of resistance

A **resistor** is a component designed to have a certain resistance which is the same regardless of the current through it. The resistance of a resistor can be measured using the circuit shown in Figure 1.

- The **ammeter**, A, is used to measure the current in the resistor. The ammeter must be in **series** with the resistor so that the same current passes through both the resistor and the ammeter.
- The **voltmeter**, V, is used to measure the p.d. across the resistor. The voltmeter must be in **parallel** with the resistor so that they have the same p.d. Also, **no current** should pass through the voltmeter, otherwise the ammeter will not record the exact current in the resistor. In theory, the voltmeter should have **infinite resistance**. In practice, a voltmeter with a sufficiently high resistance would be satisfactory.
- The **variable resistor** is used to adjust the current and p.d. as necessary. To investigate the variation of current with p.d., the variable resistor is adjusted in steps. At each step, the current and p.d. are recorded from the ammeter and voltmeter respectively. The measurements can then be plotted on a graph of p.d. against current, as shown in Figure 2.

Ohm's law

The graph for a resistor is a straight line through the origin. The resistance is the same, regardless of the current. The resistance is equal to the gradient of the graph, because the gradient is constant and at any point is equal to p.d./current.

The discovery that the potential difference across a metal wire is proportional to the current through it was made by Georg Ohm in 1826 and is known as Ohm's law.

Ohm's law states that the potential difference across a metallic conductor is proportional to the current through it, provided the physical conditions do not change.

Notes

Ohm's law is equivalent to the statement that the resistance of a metallic conductor under **constant physical conditions** (e.g. temperature) is constant.

For an **ohmic conductor**, $V = IR$ where R is constant.

Resistivity

For a conductor of length L and uniform cross-sectional area A, as shown in Figure 3, its resistance R is:

- proportional to L
- inversely proportional to A.

Hence $\boldsymbol{R = \frac{\rho L}{A}}$, where ρ is a constant for that material known as its **resistivity**.

Rearranging this equation gives the following equation, which can be used to calculate the resisitivity of a sample of material of length L and uniform cross-sectional area A:

$$\textbf{Resistivity, } \rho = \frac{RA}{L}$$

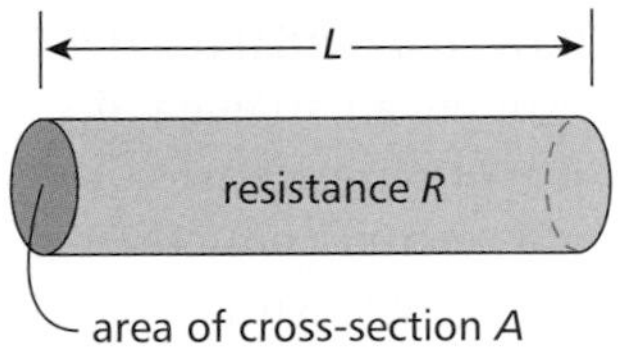

Figure 3 Resistivity

Table 2 Resistivity values of different materials at room temperature

Material	Resistivity / Ω m
Copper	1.7×10^{-8}
Constantan	5.0×10^{-7}
Carbon	3×10^{-5}
Silicon	2300.00
PVC	about 10^{14}

Notes

- The unit of resistivity is the **ohm metre (Ω m)**.
- For a conductor with a circular cross-section of diameter d:

$$A = \frac{\pi d^2}{4} \left(= \pi r^2 \text{ where radius } r = \frac{d}{2}\right)$$

Summary test 6.4

1 **a** Complete the table by calculating the missing value for each resistor:

	1	2	3	4	5
Current	2.0 A	0.45 A		5.0 mA	
Potential difference	12.0 V		5.0 V	0.80 V	50 kV
Resistance		22 Ω	40 kΩ		20 MΩ

b Use Figure 2 on the previous page to find the resistance of the resistor that gave the results shown.

2 Calculate the resistance of a uniform wire of diameter 0.32 mm and length 5.0 m. The resistivity of the material = 5.0×10^{-7} Ω m.

3 Calculate the resistance of a rectangular strip of copper of length 0.08 m, thickness 15 mm and width 0.80 mm. The resistivity of copper = 1.7×10^{-8} Ω m.

4 A wire of uniform diameter 0.28 mm and length 1.50 m has a resistance of 45 Ω. Calculate:

a its resistivity,

b the length of this wire that has a resistance of 1.0 Ω.

6.5 Components and their characteristics

Learning outcomes

On these pages you will learn to:

- recall the function and circuit symbols for commonly used components, including a diode, an LED, a thermistor and an LDR
- draw and interpret circuit diagrams with the above components in them
- describe how a potential divider is used to supply a variable p.d.
- describe how the resistance of a metal and a semiconductor thermistor changes with temperature

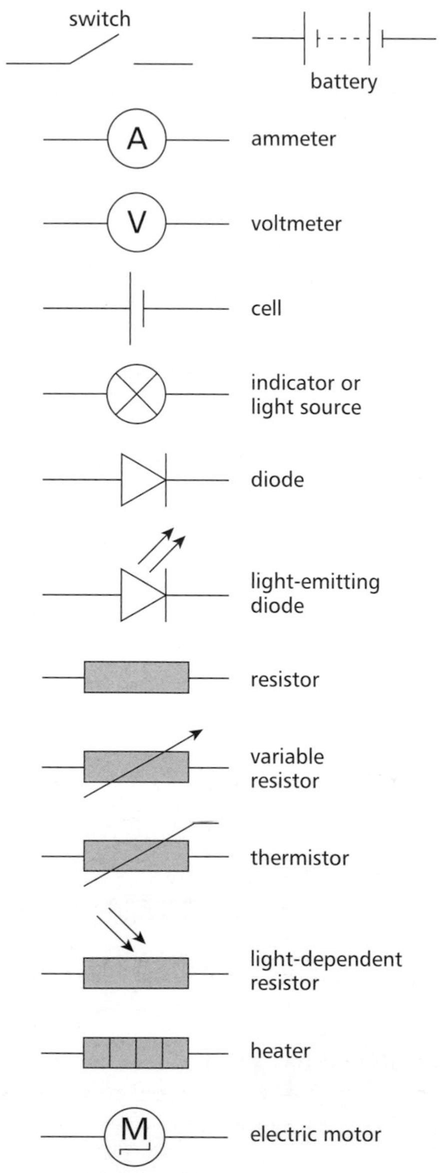

Figure 1 *Circuit components*

Circuit diagrams

Each type of component has its own **symbol**, which is used to represent the component in a **circuit diagram**. You need to recognise the symbols for different types of components to make progress – just like a motorist needs to know what different road signs mean. Note that, on a circuit diagram, the direction of the current is always shown from **+ to –** round the circuit.

The function of each of the components shown in Figure 1 is given in the list below:

- An **ammeter** measures the current in a circuit or a branch of a circuit.
- A **voltmeter** measures the p.d. between two points in a circuit.
- A **cell** is a source of electrical energy. Note that **a battery** is a combination of cells.
- The symbol for an **indicator** or any **light source** (including a filament lamp), except a light-emitting diode, is the same.
- A **diode** allows current to flow in one direction only. A **light-emitting diode** (or LED) emits light when it conducts. The direction in which the diode conducts is referred to as its 'forward' direction. The opposite direction is referred to as its 'reverse' direction.
- A **resistor** is a component designed to have a certain resistance.
- A **variable resistor** is used to control the current in a circuit.
- The resistance of a **thermistor** decreases with increase of temperature, if the thermistor is an intrinsic semiconductor such as silicon. Such a thermistor is described as a **negative temperature coefficient** (n.t.c.) thermistor.
- The resistance of a **light-dependent resistor** decreases with increase of light intensity.
- A **heater** is designed to transform electrical energy to heat.
- An **electric motor** is designed to transform electrical energy into kinetic energy.

Investigating the characteristics of different components

To measure the variation of current with p.d. for a component, use either:

- a **potential divider** to vary the p.d. from zero (Figure 2a), or
- a variable resistor to vary the current to a minimum (Figure 2b).

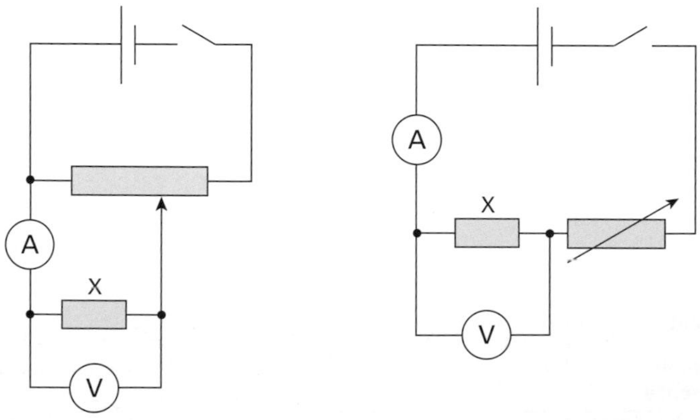

a Using a potential divider **b** Using a variable resistor

Figure 2 *Investigating component characteristics*

The advantage of using a potential divider rather than a variable resistor is that the current in the component and the p.d. across it can be reduced to zero. However, in the variable resistor circuit, when the resistance of the variable resistor is at a maximum, the current cannot be reduced any further unless a further resistor (not shown) is connected in series with the variable resistor.

The measurements for each type of component may be plotted on a graph of current (on the y-axis) against p.d. (on the x-axis). Typical graphs for a wire, a filament lamp, and a thermistor are shown in Figure 3. Note that the measurements are the same, regardless of which way the current passes through each of these components.

- A **wire** gives a straight line, with a constant gradient equal to 1/resistance R of the wire. Any resistor at constant temperature would give a straight line.
- A **filament bulb** gives a curve with a decreasing gradient, because its resistance increases as it becomes hotter.
- A **thermistor** at constant temperature gives a straight line. The higher the temperature, the greater the gradient of the line as the resistance falls with increase of temperature. The same result is obtained for a **light-dependent resistor** (LDR) in respect of light intensity. The greater the light intensity, the smaller is the resistance of the LDR.

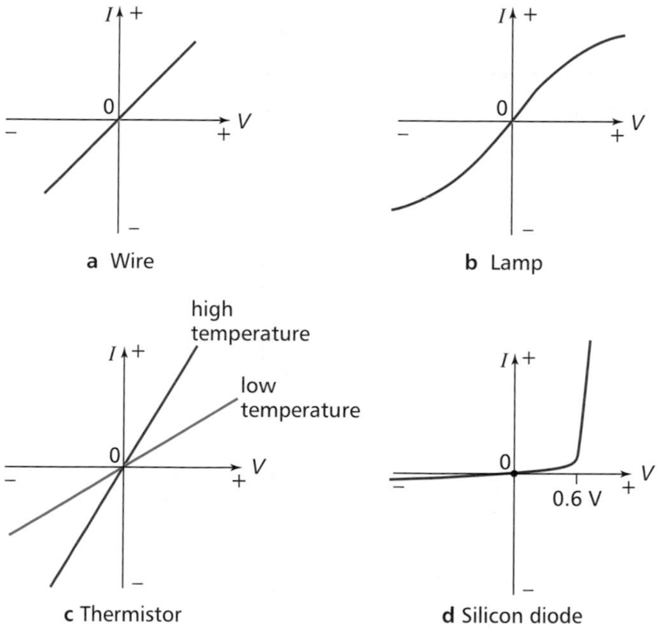

Figure 3 *Current v. potential difference for various components*

The diode

To investigate the characteristics of the **diode**, one set of measurements needs to be made with the diode in its 'forward direction' (i.e. forward biased) and another set with it in its 'reverse direction' (i.e. **reverse biased**). The current is very small when the diode is reverse biased and can only be measured using a milliammeter.

For typical results of a silicon diode, see Figure 3d. A silicon diode conducts easily in its 'forward' direction above a p.d. of about 0.6 V, and hardly at all below 0.6 V or in the opposite direction.

Resistance and temperature

- **The resistance of a metal** increases with increase of temperature. This is because the positive ions in the conductor vibrate more when its temperature is increased. The charge carriers (i.e. conduction electrons) therefore cannot pass through the metal as easily when a p.d. is applied across the conductor. A metal is said to have a **positive temperature coefficient** because its resistance increases with increase of temperature.
- **The resistance of a thermistor made from an intrinsic semiconductor** decreases with increase of temperature, as shown in Figure 4. This is because the number of charge carriers (i.e. conduction electrons) increases when the temperature is increased. A thermistor made from an intrinsic semiconductor therefore has a **negative temperature coefficient**.

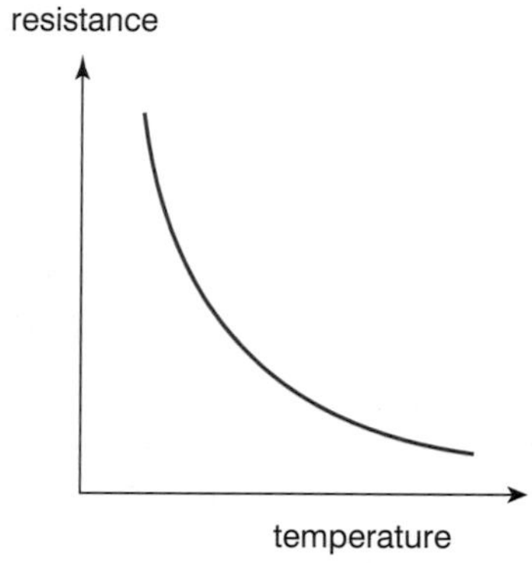

Figure 4 *Resistance v. termperature for a thermistor*

Summary test 6.5

1 A filament bulb is labelled '3.0 V, 0.75 W'.

a Calculate its current and its resistance at 3.0 V.

b State and explain what would happen to the filament bulb if the current was increased from the value in part **a**.

2 A certain thermistor has a resistance of 50 000 Ω at 20 °C and a resistance of 4000 Ω at 60 °C. It is connected in series with an ammeter and a 1.5 V cell. Calculate the ammeter reading when the thermistor is:

a at 20 °C, **b** at 60 °C.

3 A silicon diode is connected in series with a cell and a torch bulb.

a Sketch the circuit diagram showing the diode in its 'forward' direction.

b Explain why the torch bulb would not light if the polarity of the cell was reversed in the circuit.

4 The resistance of a certain metal wire increased from 25.3 Ω at 0 °C to 35.5 Ω at 100 °C. Assuming that the resistance over this range varies linearly with temperature, calculate:

a the resistance at 50 °C,

b the temperature when the resistance is 30.0 Ω.

Launch additional digital resources for the chapter

1 A rechargeable battery is capable of delivering a current of 0.25 A for 1800 s at a potential difference of 1.5 V. Calculate:

a the total charge the battery is capable of delivering without being recharged,

b the energy transferred from the battery as a result of delivering the charge calculated in part **a**.

2 A torch bulb is marked with the label '0.5 W, 1.5 V'. Calculate:

a the current in the torch bulb when it operates at normal brightness,

b the energy transferred to the torch bulb in 5 min when it operates at normal brightness.

3 a With the aid of a circuit diagram, describe how you would investigate the variation of p.d. with current for a 12 V filament light bulb.

b i Sketch a graph to show how the p.d. varies with current for a filament light bulb.
ii Describe how the resistance of a filament light bulb changes with brightness.
iii Explain why the resistance of a filament light bulb changes in the way described in part **ii**.

4 A 3.0 V battery, a diode, a switch, an ammeter and a 200 Ω resistor are connected in series.

a Sketch the circuit diagram for this arrangement with the diode in the forward direction.

b When the switch is closed, the ammeter records a current of 0.012 A. Calculate:
i the p.d. across the resistor,
ii the p.d. across the diode.

c The 200 Ω resistor was replaced by a 50 Ω resistor. Calculate the ammeter reading when the switch is closed.

5 a With the aid of a suitable circuit diagram, describe how you would measure the resistivity of a uniform wire of known cross-sectional area.

b A wire of length 1.20 m and area of cross-section $6.0 \times 10^{-7}\,m^2$ has a resistance of 1.1 Ω.

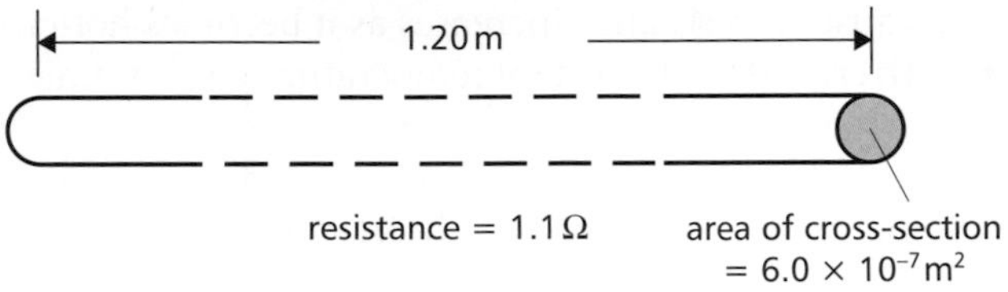

Figure 5.1

Calculate:
i the resistivity of the material of this wire,
ii the resistance of a wire consisting of this material of length 2.40 m and area of cross-section $3.0 \times 10^{-7}\,m^2$.

6 A thermistor, a 100 Ω resistor, a 9.0 V battery and a 0–100 mA ammeter are connected in series.

a Sketch the circuit diagram for this arrangement.

b i The ammeter reads 50 mA. State and explain how the ammeter reading would change if the temperature of the thermistor, which has a negative temperature coefficient, was raised.
ii What change would need to be made to the circuit to reduce the current to 20 mA without changing the temperature of the thermistor?

7 A copper wire in a domestic electric circuit has a diameter of 1.1 mm and a length of 12 m. The resistivity of copper = $1.7 \times 10^{-8}\,\Omega\,m$.

a Calculate:
i the resistance of this wire,
ii the p.d. across the ends of the wire when the current through it is 13 A,
iii the power dissipated by the wire due to its resistance when a current of 13 A is in it.

b Electrical safety regulations require that the p.d. across the ends of the wire should not exceed 6.0 V. Calculate the maximum safe current through the wire.

8 **a** Calculate the resistance of a 230 V, 100 W filament light bulb.

b The filament of the light bulb in part **a** consists of a coiled tungsten wire of diameter 0.05 mm. Calculate the total length of the wire of this filament. The resistivity of tungsten = $5.6 \times 10^{-8}\,\Omega\,m$

9 A light-dependent resistor has a resistance of 650 Ω in darkness. It is connected in series with a 4.5 V battery, a resistor, a milliammeter and a switch, as shown below.

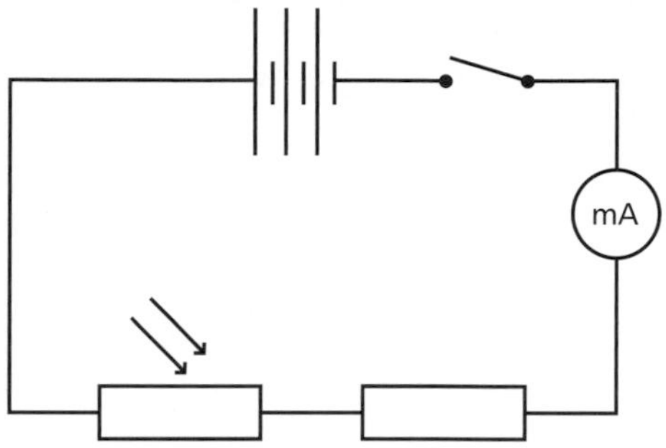

Figure 9.1

a The milliammeter reads 5.0 mA when the switch is closed and the LDR is in darkness. Calculate:
i the p.d. across the LDR,
ii the p.d. across the resistor,
iii the resistance of the resistor.

b Describe how, and explain why, the milliammeter reading changes when the LDR is exposed to light.

10 The following measurements were made in an investigation to measure the resisitivity of the material of a certain wire.

P.d. across the wire/V	0.00	2.0	4.0	6.0	8.0	10.0
Current through the wire/A	0.00	0.15	0.31	0.44	0.62	0.74

Length of wire = 1.60 m; Diameter of wire = 0.28 mm

a Plot a graph of the p.d. against the current.

b Use the graph to calculate the resistivity of the material of the wire.

11 A lamp X is connected in series with a variable resistor and an 18 V battery of negligible internal resistance. The variable resistor is adjusted until the lamp operates at its normal rating, which is 12 V and 24 W.

a Sketch the circuit diagram.

b Calculate the current in X and its resistance.

c Calculate the power supplied by the battery and the power dissipated in the variable resistor.

12 An overhead electric cable has 24 strands of aluminium wire of diameter 4.0 mm surrounding a core of 7 steel strands, each of diameter 3.0 mm (Figure 12.1).

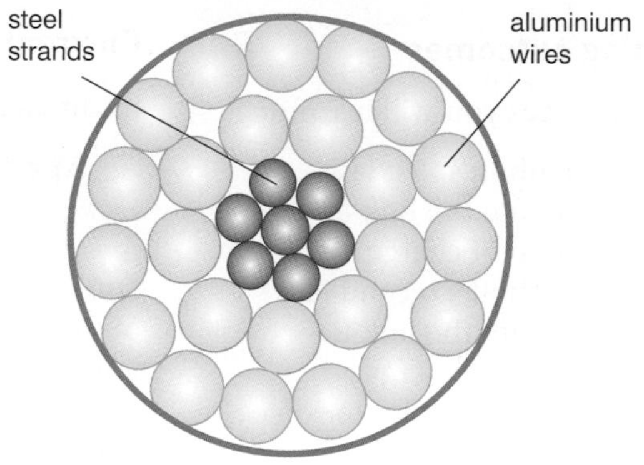

Figure 12.1

a **i** Calculate the resistance per metre of the aluminium strands if the resistivity of aluminium is $2.5 \times 10^{-8}\,\Omega\,m$.
ii Calculate the resistance per metre of the steel strands if the resistivity of steel is $1.6 \times 10^{-7}\,\Omega\,m$.

b When a potential difference is applied across a 1000 m length of the cable, a current of 100 A passes through it.
i Calculate the current in each of the two components of the cable.
ii Calculate the potential difference across the cable and the power dissipated in each of the two components of the cable.

13 In an experiment, the current and potential difference for a 12 V filament lamp are measured at 2 V steps. The measurements are shown in the table below.

Potential difference/V	0	2.01	4.00	5.98	8.02	10.1	11.8
Current/A	0	1.36	2.07	2.64	3.13	3.58	4.05

a Plot a graph of the potential difference against the current.

b Calculate the resistance of the filament lamp at:
i 1.00 A, **ii** 2.00 A, **iii** 3.00 A.

c The uncertainty in the current readings is ±0.02 A and the uncertainty in the potential difference readings is ±0.02 V. Estimate the uncertainty in your calculated value of resistance at 2.00 A.

d **i** Describe how the resistivity of the filament wire in the lamp changes as the wire becomes hotter.
ii Discuss the effect on the *drift velocity* of the charge carriers in the wire when the current in the wire is increased.

7 D.C. circuits

7.1 Circuit rules

Learning outcomes

On these pages you will learn to:

- recall Kirchhoff's first law and appreciate the link to conservation of charge
- recall Kirchhoff's second law and appreciate the link to conservation of energy
- apply Kirchhoff's laws to solve simple circuit problems

Currrent rules

Kirchhoff's first law

At any junction in a circuit, the total current leaving the junction is equal to the total current entering the junction.

For example, Figure 1 shows a junction of three wires where the current in two of the wires (Wire 1 and Wire 2) is given. The current in Wire 3 must be 1.0 A into the junction, because the total current into the junction (1.0 A along Wire 3 + 0.5 A along Wire 1) is the same as the total current out of the junction (1.5 A along Wire 2).

Kirchhoff's first law follows from the **conservation of charge**, as charge flow in and charge flow out of a junction are always equal. The current along a wire is the charge flow per second. In Figure 1, the charge entering the junction each second is 0.5 C along Wire 1 and 1.0 C along Wire 3. The charge leaving the junction each second must therefore be 1.5 C, as the junction does not retain charge.

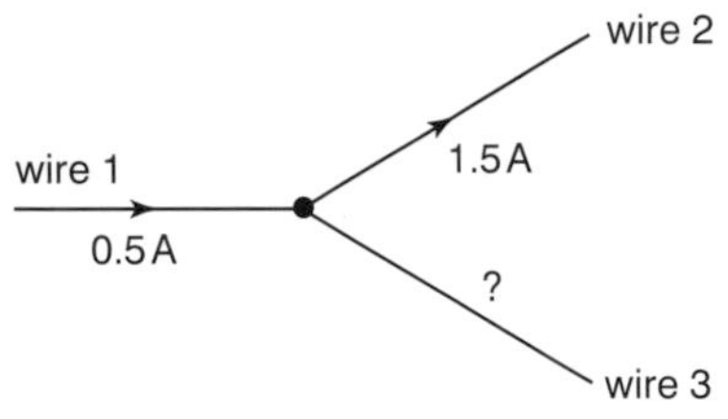

Figure 1 *Kirchhoff's first law*

Components in series

- **The current entering a component is the same as the current leaving the component.** In other words, components do not use up current. At any instant, the charge entering a component each second is equal to the charge leaving it because the same number of charge carriers enter and leave the component each second. In Figure 2, A_1 and A_2 show the same reading because they are measuring the same current.
- **The current passing through two or more components in series is the same through each component.** This is because each charge carrier passes through every component and the same number of charge carriers pass through each component each second. At any instant, charge flows at the same rate through each component.

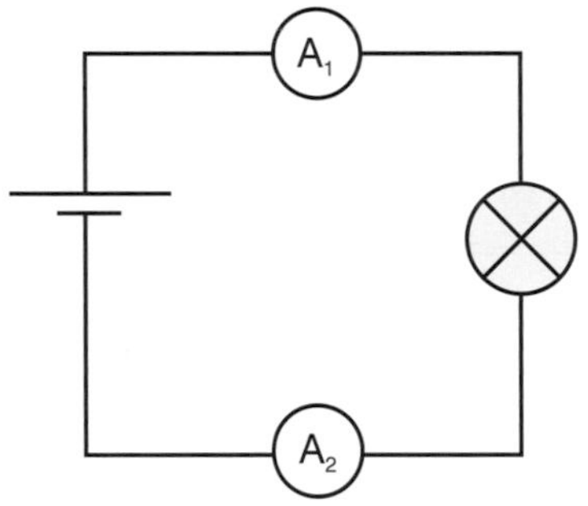

Figure 2 *Components in series*

Potential difference rules

Energy and potential difference

The potential difference, or voltage, between any two points in a circuit is defined as the energy transfer per coulomb of charge that flows from one point to the other.

If the charge carriers lose energy, the potential difference is a potential drop. If the charge carriers gain energy, which happens when they pass through a battery or cell, the potential difference is a potential rise equal to the e.m.f. of the battery or cell. The rules for potential differences are listed below with an explanation of each rule in energy terms.

- **For two or more components in series, the total p.d. across all the components is equal to the sum of the potential differences across each component.**

  Figure 3 shows a circuit consisting of a battery and three resistors in series. The p.d. across the battery terminals is equal to the sum of the potential differences across the three resistors. This is because each coulomb of charge from the battery delivers energy to each resistor as it flows round the circuit. The p.d. across each resistor is the energy delivered per coulomb of charge to

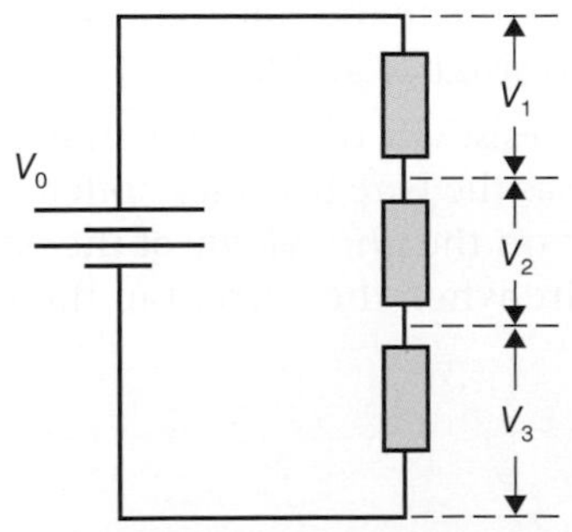

Figure 3 *Adding potential differences*

that resistor. So the sum of the potential differences across the three resistors is the **total energy** delivered to the resistors per coulomb of charge that passes through them, which is the p.d. across the battery terminals.

- **The p.d. across components in parallel is the same.**
  In Figure 4, charge carriers can pass through either of the two resistors in parallel. The same amount of energy is delivered by a charge carrier, regardless of which of the two resistors it passes through.

  If the variable resistor is adjusted so that the p.d. across it is 4 V, and if the battery p.d. is 12 V, the p.d. across the two resistors in parallel is 8 V (12 V − 4 V). This is because each coulomb of charge leaves the battery with 12 J of electrical energy, and uses 4 J on passing through the variable resistor. Therefore, each coulomb of charge has 8 J of electrical energy to deliver to either of the two parallel resistors.

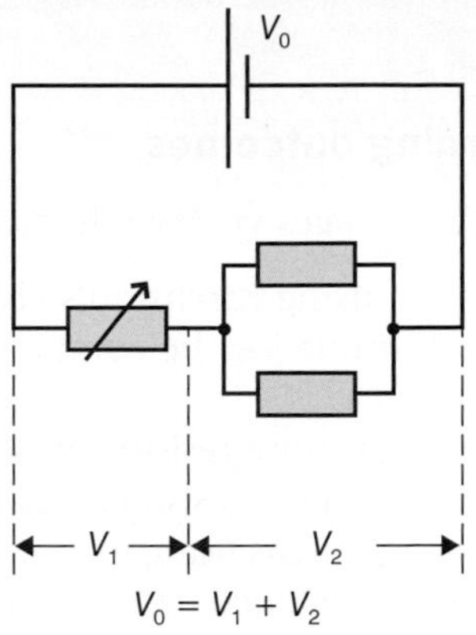

Figure 4 *Components in parallel*

- **For any complete loop of a circuit, the sum of the e.m.f.s round the loop is equal to the sum of the potential drops round the loop.**
  This statement is known as **Kirchhoff's second law**. It follows from the fact that the total e.m.f. in a loop is the total electrical energy per coulomb produced in the loop, and the sum of the potential drops is the electrical energy per coulomb delivered round the loop. The above statement follows, therefore, from the **conservation of energy**.

  For example, Figure 5 shows a 9 V battery connected to a 6 V light bulb in series with a variable resistor. If the variable resistor is adjusted so that the p.d. across the light bulb is 6 V, the p.d. across the variable resistor must be 3 V (9 V − 6 V). The only source of electrical energy in the circuit is the battery, so the sum of the e.m.f.s in the circuit is 9 V. This is equal to the sum of the p.ds round the circuit (3 V across the variable resistor + 6 V across the light bulb). In other words, the battery forces charge round the circuit. Every coulomb of charge leaves the battery with 9 J of electrical energy and supplies 3 J to the variable resistor and 6 J to the light bulb.

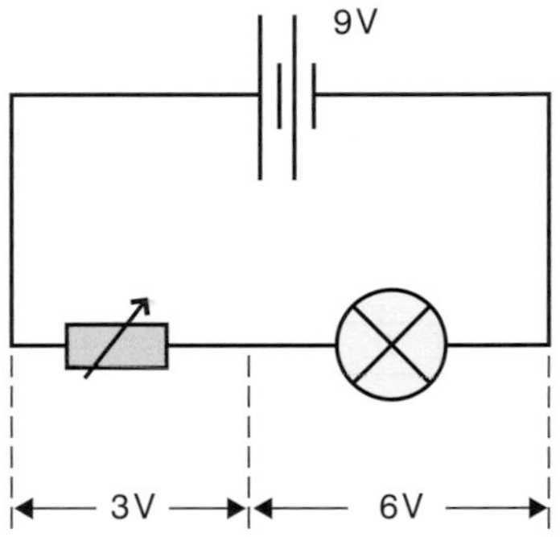

Figure 5 *Applying Kirchhoff's second law*

Summary test 7.1

1 A battery, which has an e.m.f. of 6 V and negligible internal resistance, is connected to a 6 V, 6 W light bulb in parallel with a 6 V, 24 W light bulb, as shown in Figure 6.

Calculate:

a the current in each light bulb,

b the current in the battery,

c the power supplied by the battery.

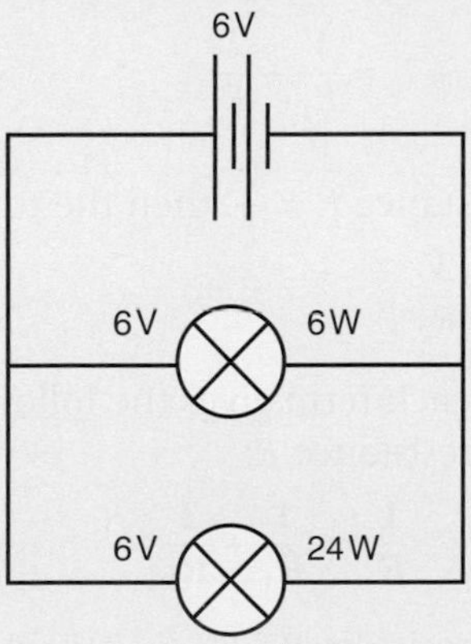

Figure 6

2 A 4.5 V battery is connected in series with a variable resistor and a 2.5 V, 0.5 W torch bulb.

a Sketch the circuit diagram for this circuit.

b The variable resistor is adjusted so that the p.d. across the torch bulb is 2.5 V. Calculate:
i the p.d. across the variable resistor,
ii the current in the torch bulb.

3 A 6.0 V battery is connected in series with an ammeter, a 20 Ω resistor and an unknown resistor R.

a Sketch the circuit diagram.

b The ammeter reads 0.20 A. Calculate:
i the p.d. across the 20 Ω resistor,
ii the p.d. across R,
iii the resistance of R.

4 In question **3**, when the unknown resistor is replaced with a torch bulb, the ammeter reads 0.12 A. Calculate:

a the p.d. across the torch bulb,

b the resistance of the torch bulb.

7.2 More about resistance

Learning outcomes

On these pages you will learn to:

- derive, using Kirchhoff's laws, the formula for the combined resistance of:
  - two or more resistors in series
  - two or more resistors in parallel
- solve problems using the formula for the combined resistance of:
  - two or more resistors in series
  - two or more resistors in parallel
- apply Kirchhoff's laws to solve simple circuit problems
- recall that rate of heat transfer in a resistor = I^2R

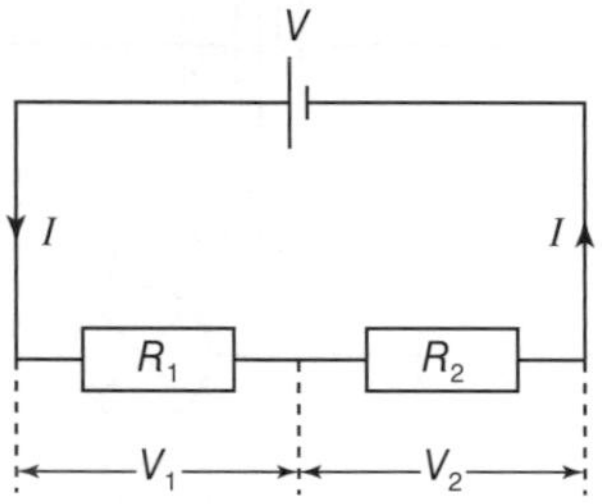

Figure 1 *Resistors in series*

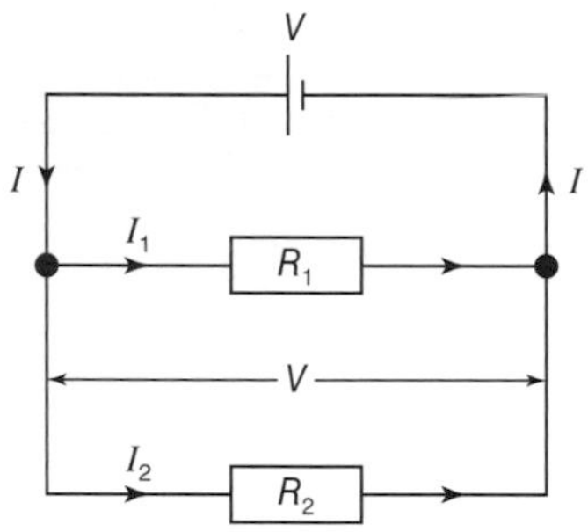

Figure 2 *Resistors in parallel*

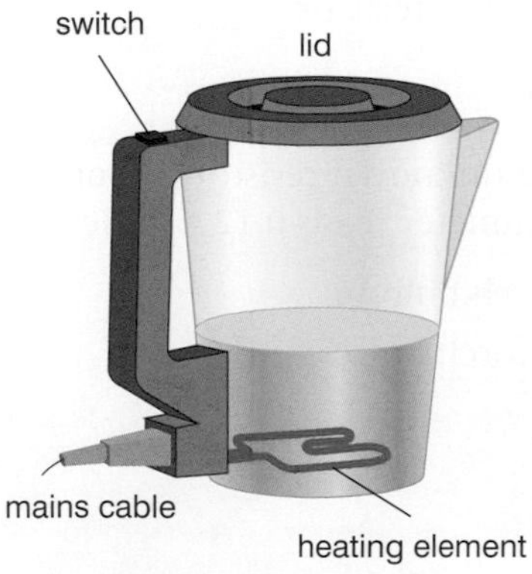

Figure 3 *Heating element in a kettle*

Resistor combination rules

Resistors in series

Resistors in series pass the same current, in accordance with Kirchhoff's first law. The total p.d. is equal to the sum of the individual potential differences.

- For two resistors R_1 and R_2 in series, as in Figure 1, when current I passes through the resistors:

  p.d. across R_1, $V_1 = IR_1$

  p.d. across R_2, $V_2 = IR_2$

  In accordance with Kirchhoff's second law, the cell p.d., V, is equal to the sum of the potential differences across the two resistors, $V = V_1 + V_2 = IR_1 + IR_2$

  therefore, the total resistance $R = \frac{V}{I} = \frac{IR_1 + IR_2}{I} = R_1 + R_2$

- For two or more resistors R_1, R_2, R_3, etc. in series, the theory can easily be extended to show that the **total resistance is equal to the sum of the individual resistances**:

$$\boldsymbol{R = R_1 + R_2 + R_3 + \ldots}$$

Resistors in parallel

Resistors in parallel have the same p.d. The current through a parallel combination of resistors is equal to the sum of the individual currents.

- For two resistors R_1 and R_2 in parallel, as in Figure 2, the p.d. across the combination is V. Applying Kirchhoff's second law to the circuit in Figure 2 gives $V = I_1R_1$ for the loop containing the cell and resistor R_1 and $V = I_2R_2$ for the loop containing the cell and resistor R_2.

  the current through resistor R_1, $I_1 = \frac{V}{R_1}$

  the current through resistor R_2, $I_2 = \frac{V}{R_2}$

  In accordance with Kirchhoff's first law, the total current through the combination, I = the sum of the currents through the two resistors

$$= I_1 + I_2 = \frac{V}{R_1} + \frac{V}{R_2}$$

  Since the total resistance $R = \frac{V}{I}$, then the total current $I = \frac{V}{R}$

  therefore $\frac{V}{R} = \frac{V}{R_1} + \frac{V}{R_2}$

  Cancelling V from each term gives the following equation, which is used to calculate the total resistance R:

$$\boldsymbol{\frac{1}{R} = \frac{1}{R_1} + \frac{1}{R_2}}$$

- For three or more resistors R_1, R_2, R_3, etc. in parallel, the theory can easily be extended to show that the total resistance R is given by:

$$\boldsymbol{\frac{1}{R} = \frac{1}{R_1} + \frac{1}{R_2} + \frac{1}{R_3} + \ldots}$$

Resistance heating

The heating effect of an electric current in any component is due to the resistance of the component. As explained in Topic 6.3, the charge carriers repeatedly collide with the positive ions of the conducting material. There is a **net transfer of energy** from the charge carriers to the positive ions as a result of these collisions.

After a charge carrier loses kinetic energy in such a collision, the force due to the p.d. across the material accelerates it until it collides with another positive ion. For a component of resistance R, when current I passes through it:

$$\text{p.d. across the component, } V = IR$$

Therefore the power supplied to the component, $P = IV = I^2R = \frac{V^2}{R}$

Hence the **energy per second** transferred to the component as thermal energy $= I^2R$

- If the component is at constant temperature, heat transfer to the surroundings takes place at the same rate. Therefore:

$$\textbf{Rate of heat transfer} = I^2R = \frac{V^2}{R}$$

- If the component heats up, its temperature rise depends on the power supplied to it (i.e. I^2R), the rate of heat transfer to the surroundings and the heat capacity of the component.
- The energy transfer per second to the component (i.e. the power supplied to it) does not depend on the direction of the current. For example, the heating effect of an alternating current at a given instant depends **only on the magnitude of the current** not on the direction of the current.

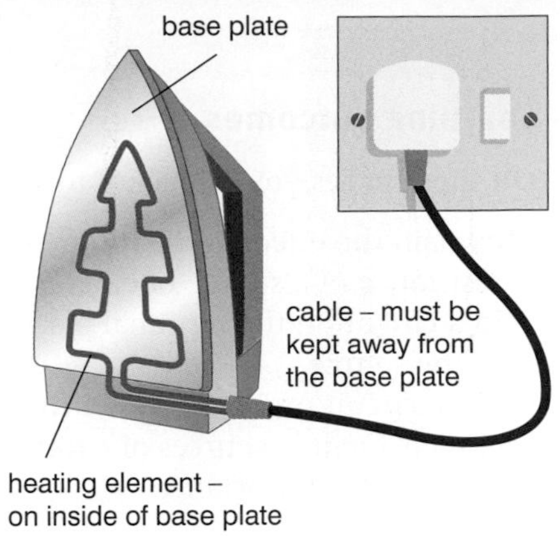

Figure 4 *Heating element in an iron*

Summary test 7.2

1 Calculate the total resistance of each of the resistor combinations in Figure 5.

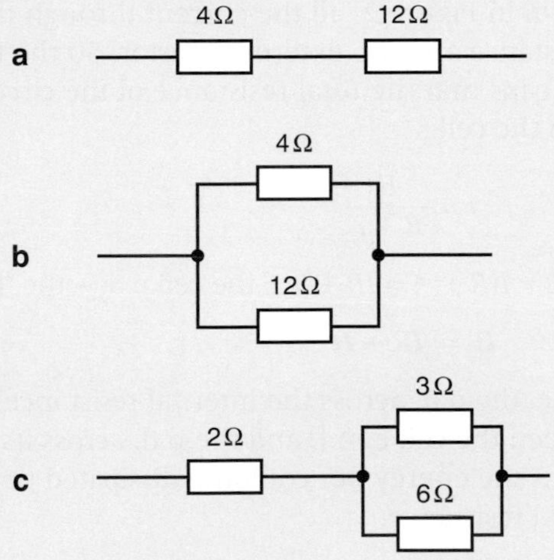

Figure 5

2 A 3 Ω resistor and a 6 Ω resistor are connected in parallel. The parallel combination is connected in series with a 6 V battery of negligible internal resistance and a 4 Ω resistor, as shown in Figure 6.

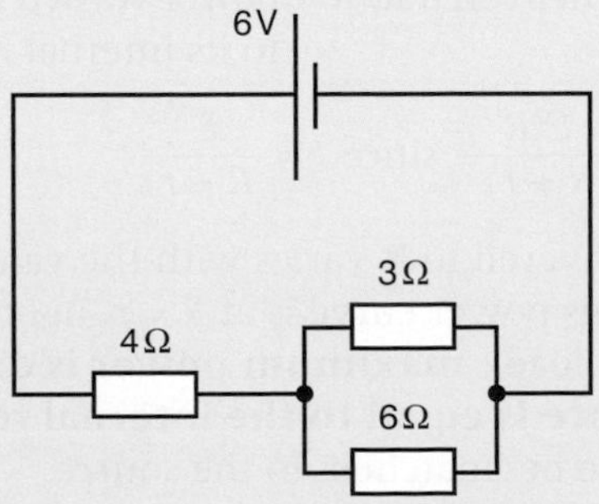

Figure 6

Calculate:

a the combined resistance of the 3 Ω resistor and the 6 Ω resistor in parallel,

b the total resistance of the circuit,

c the battery current,

d the power supplied to the 4 Ω resistor.

3 A 2 Ω resistor and a 4 Ω resistor are connected in series. The series combination is connected in parallel with a 9 Ω resistor and a 3 V battery of negligible internal resistance, as shown in Figure 7.

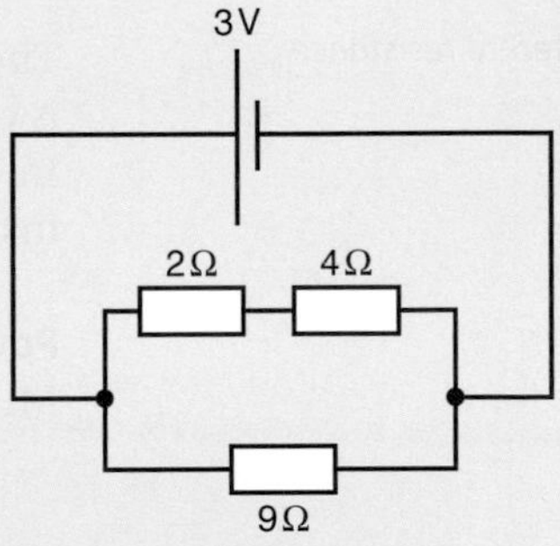

Figure 7

Calculate:

a the total resistance of the circuit,

b the battery current,

c the power supplied to each resistor,

d the power supplied by the battery.

4 Calculate:

a the power supplied to a 10 Ω resistor when the p.d. across it is 12 V,

b the resistance of a heating element designed to operate at 60 W and 12 V.

7.3 E.m.f. and internal resistance

Learning outcomes

On these pages you will learn to:

- explain the effect of the internal resistance of a source of e.m.f. in a circuit on the terminal p.d. of the source
- solve circuit problems in which circuits include sources of e.m.f. with internal resistance

Internal resistance

The **internal resistance** of a source of electricity is due to opposition to the flow of charge through the source. This causes electrical energy produced by the source to be dissipated inside the source when charge flows through it.

The **electromotive force (e.m.f.)** of the source is the electrical energy per unit charge produced by the source.

The **p.d. across the terminals** of the source is the electrical energy per unit charge that can be delivered by the source when it is in a circuit.

The **terminal p.d.** is less than the e.m.f. whenever current passes through the source. The difference is due to the internal resistance of the source.

The internal resistance of a source is the loss of potential difference per unit current in the source, when current passes through the source.

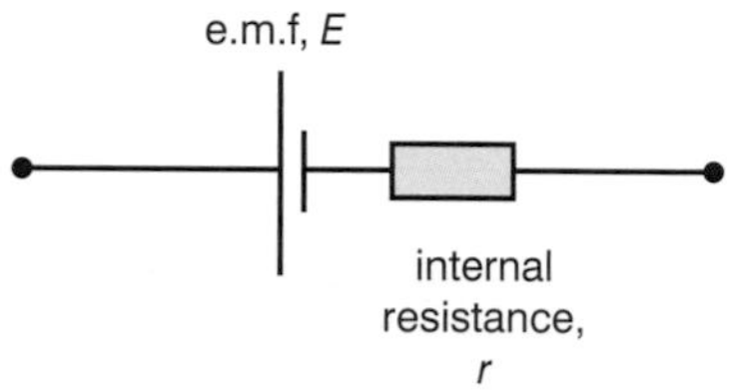

Figure 1 *Internal resistance*

In circuit diagrams, the internal resistance of a source may be shown as a resistor (labelled 'internal resistance') in series with the usual symbol for a cell or battery, as in Figure 1. If no internal resistance is shown, the symbol for the cell or battery should be labelled with symbols for its e.m.f. and its internal resistance.

When a cell of e.m.f. E and internal resistance r is connected to an external resistor of resistance R, as shown in Figure 2, all the current through the cell passes through its internal resistance and the external resistor. So the two resistors are in series, which means that the total resistance of the circuit is $r + R$. Therefore, the current through the cell:

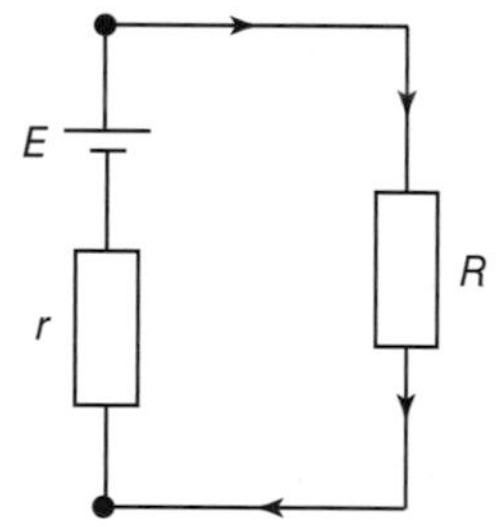

Figure 2 *E.m.f. and internal resistance*

$$I = \frac{E}{R + r}$$

In other words, the cell e.m.f., $E = I(R + r) = IR + Ir$ = the cell p.d. + the 'lost' p.d.

$$E = IR + Ir$$

The 'lost' p.d. inside the cell (i.e. the p.d. across the internal resistance of the cell) is equal to the difference between the cell e.m.f. and the p.d. across its terminals. In energy terms, the 'lost' p.d. is the energy per coulomb dissipated or wasted inside the cell due to its internal resistance.

Power

Multiplying each term of the above equation by the cell current I gives:

Power supplied by the cell, $IE = I^2R + I^2r$

In other words,

Power supplied by the cell = power delivered to R + power wasted in the cell due to its internal resistance

The power delivered to R $= I^2R = \dfrac{E^2R}{(R + r)^2}$ since $I = \dfrac{E}{R + r}$

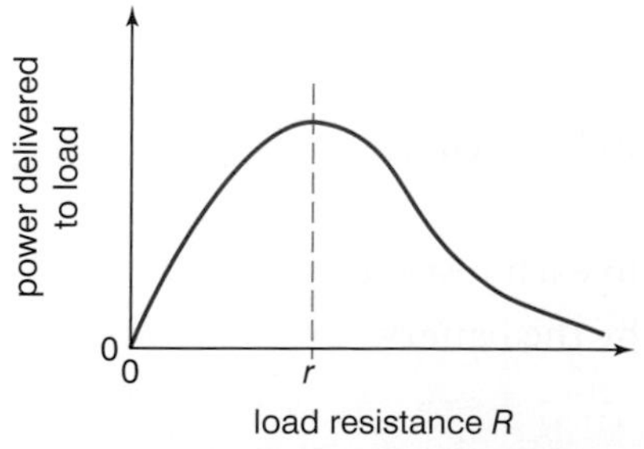

Figure 3 *Power delivered to a load v. load resistance*

Figure 3 shows how the power delivered to R varies with the value of R. It can be shown that the peak of this power curve is at $R = r$. In other words, when a source delivers power to a 'load', **maximum power is delivered to the load when the load resistance is equal to the internal resistance of the source.** The load is then said to be 'matched' to the source.

Measurement of internal resistance

The potential difference across the terminals of a cell when the cell is in a circuit can be measured by connecting a high-resistance voltmeter directly across the terminals of the cell. Figure 4 shows how the cell p.d. can be measured for different values of current. The current is changed by adjusting the variable resistor. The lamp limits the maximum current that can pass through the cell. The ammeter is used to measure the cell current.

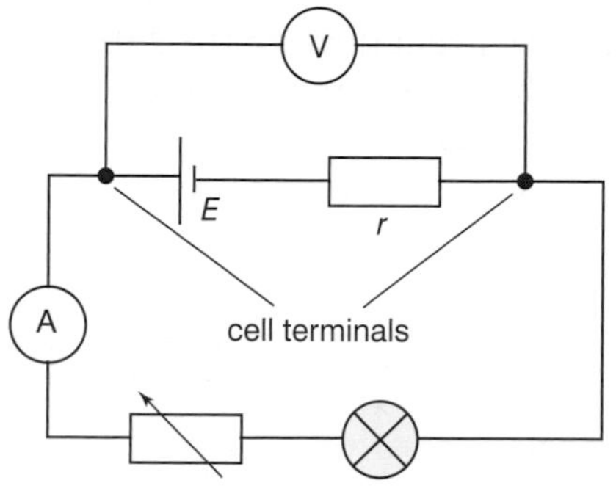

***Figure 4** Measuring internal resistance*

Graph of p.d. v. current

The measurements of cell p.d. and current for a given cell may be plotted on a graph, as shown in Figure 5.

The cell p.d. decreases as the current increases. This is because the 'lost' p.d. increases as the current increases.

- **The cell p.d. is equal to the cell e.m.f. at zero current.** This is because the 'lost' p.d. is zero at zero current.
- **The graph is a straight line with a negative gradient.** This can be seen by rearranging the equation

$$E = IR + Ir \text{ to become } IR = E - Ir$$

Because IR represents the cell p.d. V, then $V = E - Ir$. By comparison with the standard equation for a straight line $y = mx + c$, a graph of V on the y-axis against I on the x-axis gives a straight line with a gradient $-r$ and a y-intercept E. See Chapter 24, Mathematical skills.

Figure 5 shows the gradient triangle ABC, in which AB represents the lost p.d. and BC represents the current.

So the gradient $\frac{AB}{BC} = \frac{\text{lost voltage}}{\text{current}}$ = internal resistance r.

Note The internal resistance and the e.m.f. of a cell can be calculated, if the cell p.d. is measured for two different values of current. A pair of simultaneous equations can therefore be written, as follows:

- For current I_1, the cell p.d. $V_1 = E - I_1r$
- For current I_2, the cell p.d. $V_2 = E - I_2r$

Subtracting the first equation from the second gives:

$$V_1 - V_2 = (E - I_1r) - (E - I_2r) = I_2r - I_1r = (I_2 - I_1)r$$

Therefore, $r = \frac{(V_1 - V_2)}{(I_1 - I_2)}$

So r can be calculated from the above equation and then substituted into either equation for the cell p.d. to enable E to be calculated.

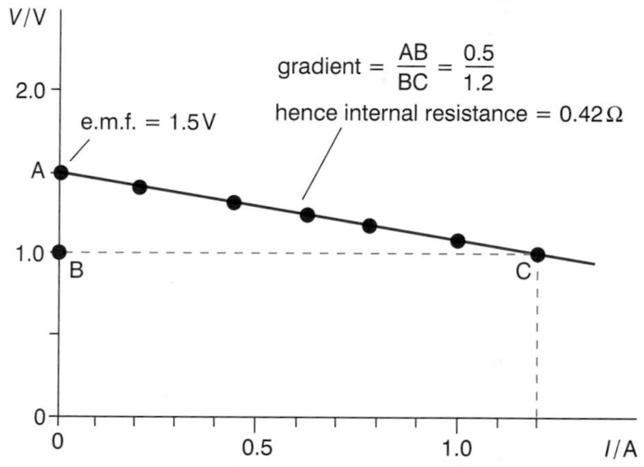

***Figure 5** A graph of cell p.d. v. current*

Summary test 7.3

1 A battery of e.m.f. 12 V and internal resistance 1.5 Ω is connected to a 4.5 Ω resistor. Calculate:

a the total resistance of the circuit,

b the current through the battery,

c the lost p.d.,

d the p.d. across the cell terminals.

2 A cell of e.m.f. 1.5 V and internal resistance 0.5 Ω is connected to a 2.5 Ω resistor. Calculate:

a the current,

b the terminal p.d.,

c the power delivered to the 2.5 Ω resistor,

d the power wasted in the cell.

3 The p.d. across the terminals of a cell is 1.1 V when the current from the cell is 0.20 A and 1.3 V when the current is 0.10 A. Calculate:

a the internal resistance of the cell,

b the cell's e.m.f.

4 A battery of unknown e.m.f., E, and internal resistance, r, is connected in series with an ammeter and a resistance box, R. The current was 2.0 A when $R = 4.0\,\Omega$, and 1.5 A when $R = 6.0\,\Omega$. Calculate E and r.

7.4 More circuit calculations

Learning outcomes

On these pages you will learn to:

- solve circuit problems in which circuits include two or more sources of e.m.f. in series or in parallel
- solve circuit problems in which circuits include diodes

Circuits with a single cell and one or more resistors

Here are some rules:

- Sketch the **circuit diagram** if it is not drawn.
- To calculate the **current** passing through the cell, calculate the total circuit resistance using the resistor combination rules. Don't forget to add on the internal resistance of the cell, if that is given:

$$\text{Cell current} = \frac{\text{cell e.m.f.}}{\text{total circuit resistance}}$$

- To work out the **current and p.d.** for each resistor, start with the **resistors in series** with the cell which pass the same current as the cell current:

$$\text{p.d. across each resistor in series with cell} = \text{current} \times \text{the resistance of each resistor}$$

- To work out the p.d. across the **parallel resistors**, find the difference between the cell e.m.f. and the sum of the p.d.s across the series resistors and the internal resistance.

  The current through any parallel resistor is then given by dividing the p.d. across the parallel resistors by the resistance of that resistor.

Circuits with two or more cells in series

The same rules as above apply, except the current through the cells is calculated by dividing the **overall (i.e. net) e.m.f.** by the total resistance.

- If the cells are connected in the **same** direction in the circuit (Figure 1a), the net e.m.f. is the **sum** of the individual e.m.fs. For example, in Figure 1a the net e.m.f. is 3.5 V.
- If the cells are connected in **opposite** directions to each other in the circuit (Figure 1b), the net e.m.f. is the **difference** between the e.m.fs in each direction. For example, in Figure 1b, the net e.m.f. is 0.5 V in the direction of the 2.0 V cell.

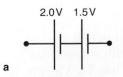

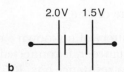

Figure 1 Cells in series

- The **total internal resistance** is the **sum** of the individual internal resistances. This is because the cells, and therefore the internal resistances, are in series.

Circuits with cells in parallel

Kirchhoff's laws are used to analyse circuits where cells and batteries are in parallel. For example, consider the circuit in Figure 2.

To determine the current through the 2.0 Ω resistor, let x and y represent the current through the 2.0 V and 1.5 V cells, which have internal resistances of 6.0 Ω and 4.0 Ω, respectively. The current through the 2.0 Ω resistor is therefore $x + y$.

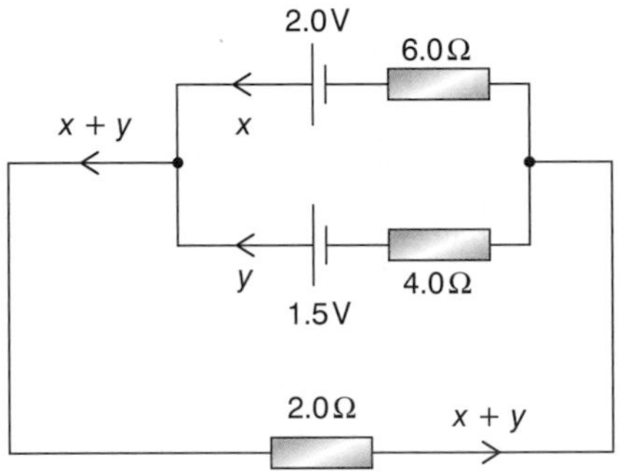

Figure 2

Applying Kirchhoff's second law

- to the outer loop (consisting of the 2.0 V cell with its internal resistance of 6.0 Ω and the 2.0 Ω resistor) gives
  $2.0 = 6.0x + 2.0\,(x + y)$ or $2.0 = 8.0\boldsymbol{x} + 2.0\boldsymbol{y}$
- to the middle loop (consisting of the 1.5 V cell with its internal resistance of 4.0 Ω and the 2.0 Ω resistor) gives
  $1.5 = 4.0y + 2.0\,(x + y)$ or $1.5 = 2.0\boldsymbol{x} + 6.0\boldsymbol{y}$

Solving these two simultaneous equations (see Chapter 24, Mathematical skills) gives $x = 0.205$ A and $y = 0.182$ A. Hence the current through the 2.0 Ω resistor $= x + y = 0.387$ A.

Note

In such calculations, a negative value for a current would mean the current is in the opposite direction to that assumed initially.

Diodes in circuits

Assume that a semiconductor diode has:

- **zero resistance in the forward direction** when the p.d. across it is 0.6 V or greater,
- **infinite resistance in the reverse direction** or at p.ds less than 0.6 V in the forward direction.

Therefore, in a circuit with one or more diodes as above:

- a p.d. of 0.6 V exists across a diode that is forward-biased and passing a current,
- a diode that is reverse-biased has infinite resistance.

For example, suppose a diode is connected in its forward direction in series with a 1.5 V cell and a 1.5 kΩ resistor, as in Figure 3.

The p.d. across the diode is 0.6 V, because it is forward-biased. Therefore, the p.d. across the resistor is 0.9 V (= 1.5 V − 0.6 V). The current through the resistor is therefore 6.0×10^{-4} A $\left(= \frac{0.9\,\text{V}}{1500\,\Omega}\right)$.

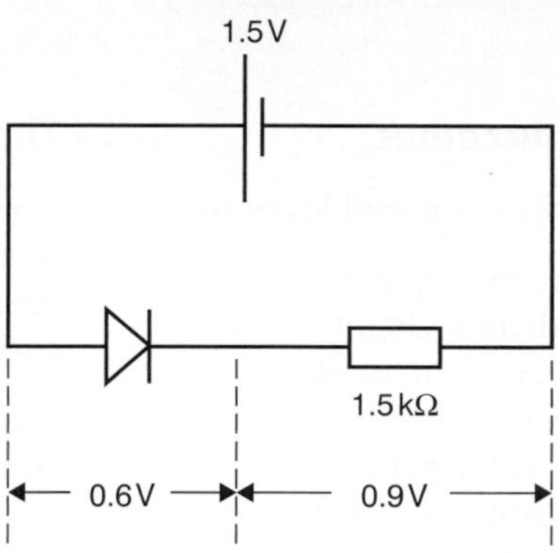

Figure 3 *Using a diode*

Summary test 7.4

1 A cell of e.m.f. 3.0 V, and negligible internal resistance, is connected to a 4.0 Ω resistor in series with a parallel combination of a 24.0 Ω resistor and a 12.0 Ω resistor (Figure 4).

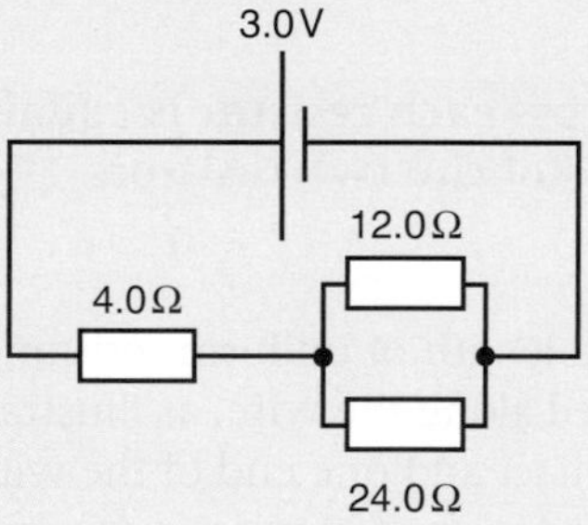

Figure 4

Calculate:

a the total resistance of the circuit,

b the cell current,

c the current and p.d. for each resistor.

2 A battery of e.m.f. 12.0 V, with an internal resistance of 3.0 Ω, is connected in series with a 15.0 Ω resistor and a battery of e.m.f. 9.0 V, which has an internal resistance of 2.0 Ω (Figure 5).

a Calculate:
  i the total resistance of the circuit,
  ii the battery current,
  iii the current and p.d. across the 15 Ω resistor.

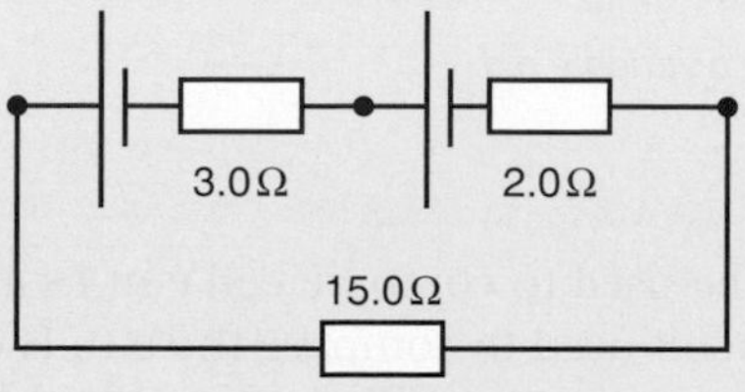

Figure 5

b A 1.0 Ω resistor is connected in parallel with the 12.0 V battery and the 3.0 Ω resistor. Calculate the current through: **i** the 1.0 Ω resistor, **ii** the 12.0 V battery.

3 a Two 8 Ω resistors and a battery of e.m.f. 12.0 V and internal resistance 8 Ω, are connected in series. Sketch the circuit diagram and calculate:
  i the power delivered to each external resistor,
  ii the power wasted due to internal resistance.

b The two 8 Ω resistors in part **a** are reconnected in parallel and are then connected to the same battery. Sketch the circuit diagram and calculate:
  i the power delivered to each external resistor,
  ii the power wasted due to internal resistance.

4 a For the circuit shown in Figure 6, calculate the p.d. and current for each resistor and diode.

b In the circuit (Figure 6), both diodes are reversed. Sketch the new circuit and calculate the p.d. and current for each resistor and diode for this new arrangement.

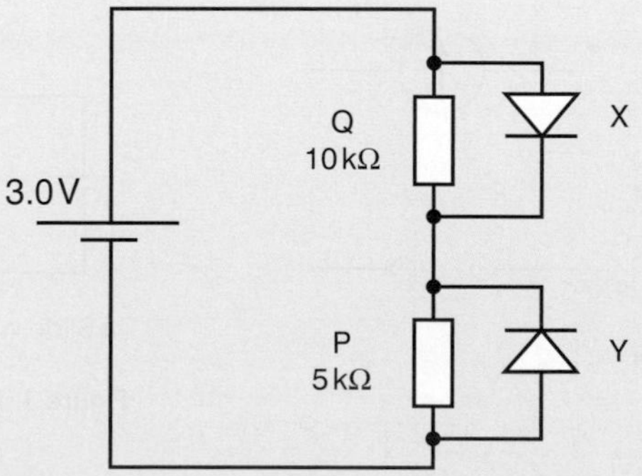

Figure 6

7.5 The potential divider

Learning outcomes

On these pages you will learn to:

- explain the principle of a potential divider circuit as a source of fixed p.d. or of variable p.d.
- recall and solve problems using the principle of the potentiometer as a means of comparing potential differences
- explain the operation of an electronic sensor consisting of a sensing device and a circuit that provides a voltage output
- explain the use of thermistors and light-dependent resistors in potential dividers

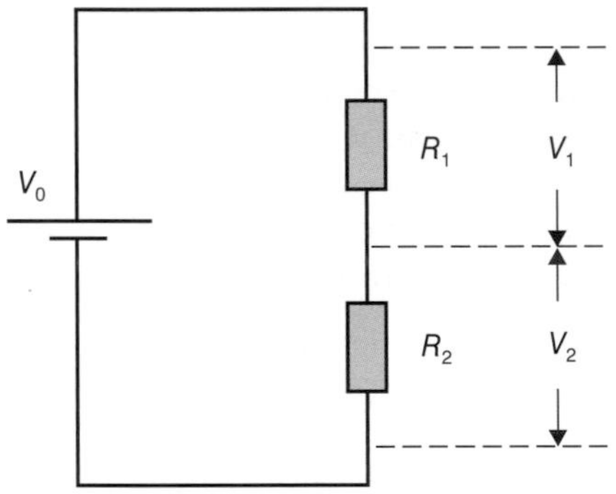

Figure 1 *A potential divider*

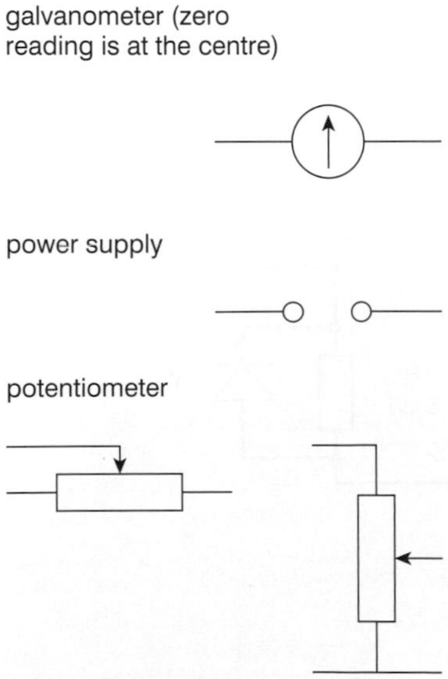

Figure 2 *More circuit components*

The theory of the potential divider

A **potential divider** consists of two or more resistances in series connected to a source of fixed potential difference. The source p.d. is divided beween the resistors, as they are in series with each other. A potential divider can be used to supply a p.d. of any value between zero and the source p.d.

Figure 1 shows a potential divider consisting of two resistors R_1 and R_2, in series connected to a source of fixed p.d. V_0.

Total resistance of the combination = $R_1 + R_2$

Therefore, current through the resistors, $I = \dfrac{\text{p.d. across the resistors}}{\text{total resistance}} = \dfrac{V_0}{R_1 + R_2}$

so the p.d. across resistor R_1, $V_1 = IR_1 = \dfrac{V_0 R_1}{R_1 + R_2}$

and the p.d. across resistor R_2, $V_2 = IR_2 = \dfrac{V_0 R_2}{R_1 + R_2}$

Dividing the equation for V_1 by the equation for V_2 gives:

$$\frac{V_1}{V_2} = \frac{R_1}{R_2}$$

This equation shows that:

The ratio of the p.ds across each resistor is equal to the resistance ratio of the two resistors.

To supply a variable p.d.

The source p.d. is connected to a fixed length of uniform resistance wire. A sliding contact on the wire can then be moved along the wire, as illustrated in Figure 3, giving a variable p.d. between the contact and one end of the wire. A uniform track of a suitable material may be used instead of resistance wire. The track may be linear or circular (Figure 3a,b). The circuit symbol for a variable potential divider which is also known as a **potentiometer** is shown in Figure 3c.

The variable potential divider in Figure 3c can be used to vary the brightness of a bulb by connecting the bulb between C and B. In contrast with using a variable resistor in series with the light bulb and the source p.d., the use of a potential divider enables the current through the light bulb to be reduced to zero. With a variable resistor at maximum resistance, there is a current through the light bulb.

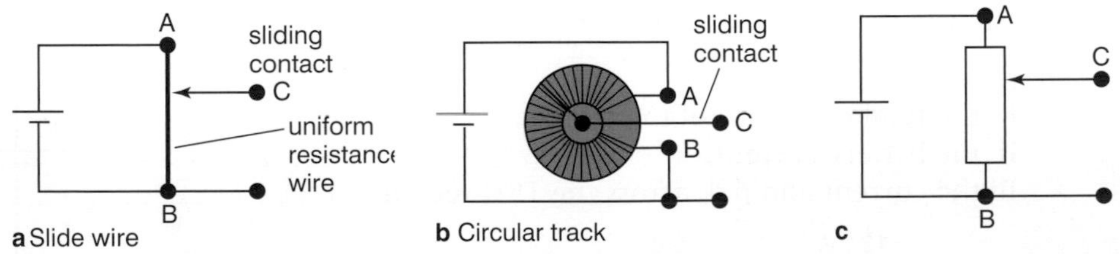

Figure 3 *Potential dividers used to supply a variable p.d.*

To compare potential differences

The slide-wire potentiometer can be used to compare cell e.m.f.s and potential differences. Figure 4 shows the circuit used to compare the e.m.fs of two cells, X and Y. One of the cells (X) is shown connected in series with **galvanometer** G (ie a centre-reading meter) between end B of the wire and the sliding contact C on the wire. The driver cell provides a constant p.d. across the wire.

The position of the sliding contact on the wire is adjusted until the galvanometer G reads zero (ie there is a null reading). The cell e.m.f. E_x is then opposed equally (or 'balanced') by the p.d. between B and C. The length L_X of the wire from B to C is then measured.

The procedure is repeated using the other cell (Y). The ratio of the cell e.m.fs, E_X / E_Y is equal to the ratio of their balance lengths, L_X / L_Y.

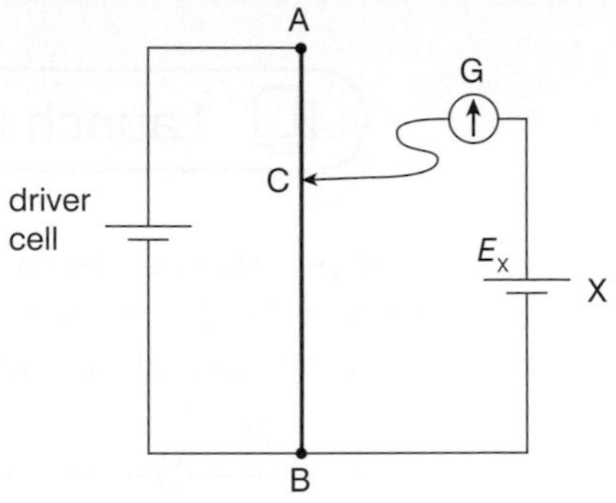

Figure 4 *Comparing cell e.m.f.s*

Sensor circuits

A **sensor circuit** produces an output p.d. which changes as a result of a change of a physical variable, such as temperature or pressure.

A temperature sensor

This consists of a potential divider made using a **thermistor** and a variable resistor (Figure 5).

With the temperature of the thermistor constant, the source p.d. is divided between the thermistor and the variable resistor. By adjusting the variable resistor, the p.d. across the thermistor can then be set at any desired value. When the temperature of the thermistor changes, its resistance changes so the p.d. across it changes. For example, suppose the variable resistor is adjusted so that the p.d. across the thermistor at 20 °C is exactly half the source p.d.; if the temperature of the thermistor is then raised, its resistance falls. As a result, the current increases, which means that the other resistor gets a bigger share of the source p.d.. So the thermistor p.d. therefore falls.

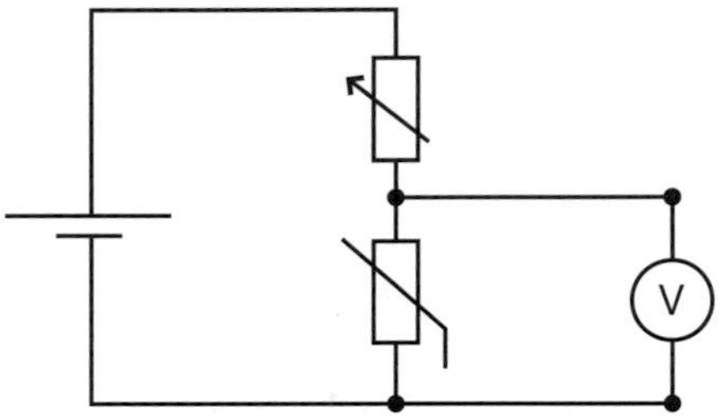

Figure 5 *A temperature sensor*

A light sensor

The circuit is similar to Figure 5 except a light-dependent resistor (LDR) is used instead of a thermistor. The p.d. across the LDR changes when the incident light intensity on the LDR changes. If the light intensity increases, the resistance of the LDR falls and the p.d. across the LDR falls.

Summary test 7.5

1 A potential divider consists of a 1.0 kΩ resistor in series with a 5.0 kΩ resistor and a battery of e.m.f. 4.5 V and negligible internal resistance.

a Sketch the circuit and calculate the p.d. across each resistor.

b A second 5.0 kΩ resistor is connected in the above circuit in parallel with the first 5.0 kΩ resistor. Calculate the p.d. across each resistor in this new circuit.

2 A 12 V battery, of negligible internal resistance, is connected to the fixed terminals of a variable potential divider (which has a maximum resistance of 50 Ω). A 12 V light bulb is connected between the sliding contact and the negative terminal of the potential divider. Sketch the circuit diagram and describe how the brightness of the light bulb changes when the sliding contact is moved from the negative to the positive terminal of the potential divider.

3 a A potential divider consists of an 8.0 Ω resistor in series with a 4.0 Ω resistor and a 6.0 V battery. Calculate:

i the current,

ii the p.d. across each resistor.

b In the circuit in part a, the 4 Ω resistor is replaced by a thermistor with a resistance of 8 Ω at 20 °C and a resistance of 4 Ω at 100 °C. Calculate the p.d. across the fixed resistor at: **i** 20 °C, **ii** 100 °C.

4 A light sensor consists of a 5.0 V cell, an LDR and a 5.0 kΩ resistor in series with each other. A voltmeter is connected in parallel with the resistor. When the LDR is in darkness, the voltmeter reads 2.2 V.

a Calculate:

i the p.d. across the LDR,

ii the resistance of the LDR when the voltmeter reads 2.2 V.

b Describe and explain how the voltmeter reading would change if the LDR was exposed to daylight.

7 Exam-style and Practice Questions

Launch additional digital resources for the chapter

1 Two resistors of resistances 6.0 Ω and 12.0 Ω, in parallel, are connected to a 2.0 Ω resistor and a 6.0 V battery. Figure 1.1 shows the circuit diagram.

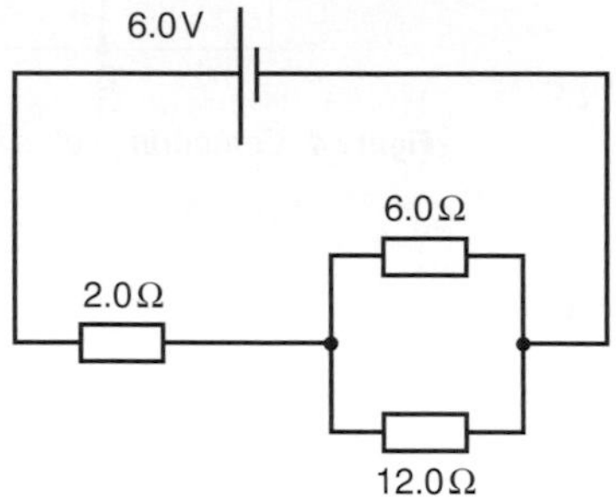

Figure 1.1

a Calculate the combined resistance of the three resistors.

b For each resistor, calculate:
i the current,
ii the p.d.,
iii the power dissipated.

2 A light bulb is connected in series with a resistor and a battery, of negligible internal resistance, as shown below. A variable resistor is connected in parallel with the light bulb.

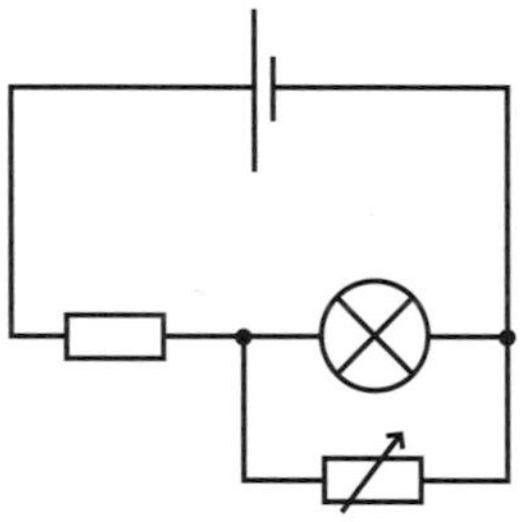

Figure 2.1

a Describe how the brightness of the light bulb changes as the resistance of the variable resistance is increased from zero.

b Explain why it is possible to reduce the bulb current to zero by adjusting the variable resistor.

3 A 15.0 Ω resistor and a 3.0 Ω resistor are connected in series with a 3.0 V battery, which has an internal resistance of 2.0 Ω, as shown in Figure 3.1.

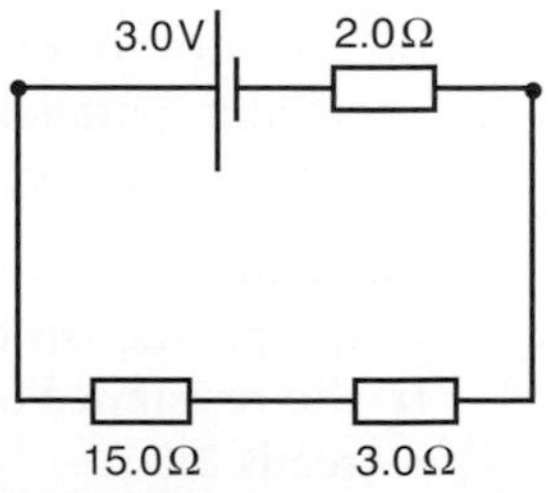

Figure 3.1

a Calculate:
i the total resistance of the circuit,
ii the p.d. across each of the external resistors,
iii the p.d. across the battery terminals.

b Calculate the power dissipated:
i in each resistor,
ii in the battery due to its internal resistance.

4 The circuit diagram below is for the rear windscreen heater of a car. It consists of 4 heating elements, each of resistance 6.0 Ω, connected to a 12.0 V battery of internal resistance 2.0 Ω.

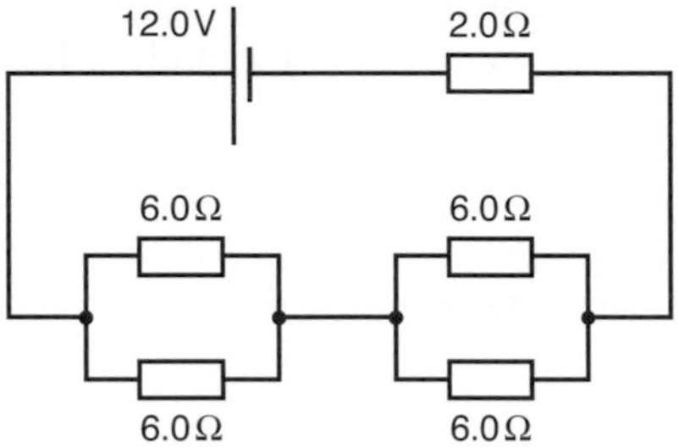

Figure 4.1

a Calculate:
i the current,
ii the p.d. across each heating element.

b Calculate the power dissipated:
i in each heating element,
ii in the battery due to its internal resistance.

c How would the operation of the heater differ if the heating elements were in parallel with each other, with the battery connected across the parallel combination?

5 A 6.0 V battery of unknown internal resistance is connected in series with a switch, a resistor of resistance 4.0 Ω and an ammeter. When the switch is closed, the ammeter reads 1.0 A.

a Calculate the internal resistance of the battery.

b A second 4.0 Ω resistor was connected in parallel with the first resistor. Calculate the ammeter reading with this second resistor in the circuit.

6 A hand-held hair dryer has two 570 Ω identical heating elements, each in series with a switch. The heating elements are connected in parallel with each other and a 230 V mains electricity supply. A 12 W, 230 V electric fan in the hair dryer is used to blow air over the heating elements.

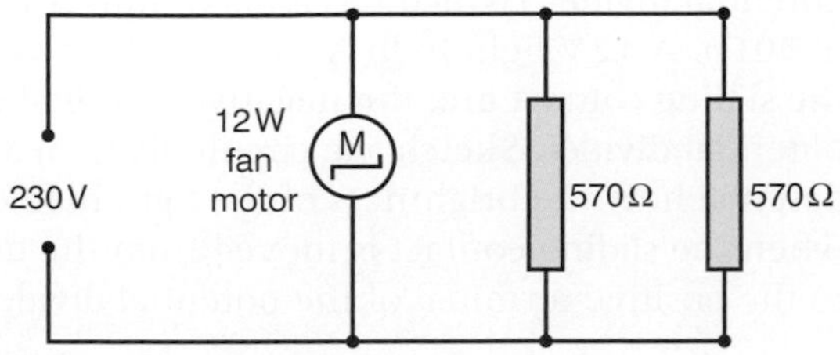

Figure 6.1

a When both heaters are being used, and the electric fan is on, calculate:
i the current passing through each heating element,
ii the power supplied by the electricity supply,
iii the total current passing through the hair dryer.

b When one heater only is on, calculate:
i the total current passing through the heater,
ii the power supplied by the electricity supply.

7 The diagram shows a potential divider circuit consisting of a 10 kΩ resistor R connected in series with an n.t.c. thermistor and a 6.0 V battery. A high resistance voltmeter is connected across the thermistor.

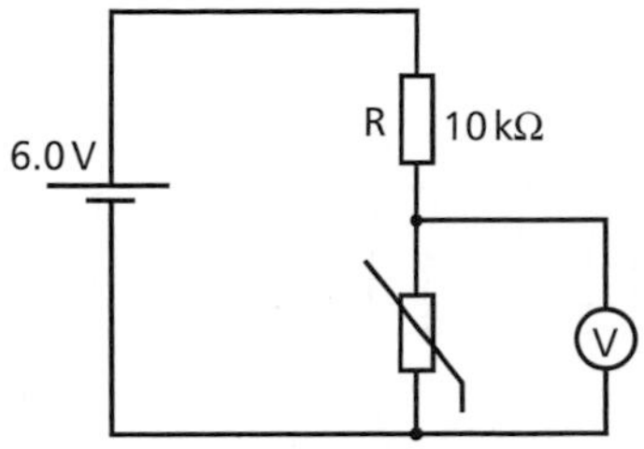

Figure 7.1

a The voltmeter reads 2.0 V when the thermistor's temperature is 20 °C. Calculate the resistance of the thermistor at this temperature.

b A second 10 kΩ resistor was connected in parallel with R. Calculate the voltmeter reading with this second resistor in the circuit when the thermistor is at the same temperature.

c Describe and explain how the voltmeter reading would change if the the temperature of the thermistor were increased.

8 The diagram shows a circuit in which two 5.0 kΩ resistors are connected in series with each other and a 5.0 V battery, of negligible internal resistance. A diode, is connected in parallel with each resistor, as shown.

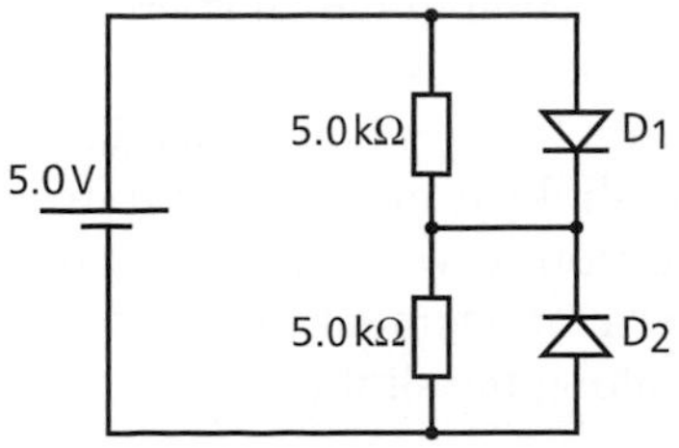

Figure 8.1

a Calculate:
i the p.d.,
ii the current through each resistor.

b Calculate the current through each resistor if the 5.0 V battery was replaced with a 9.0 V battery, also of negligible internal resistance.

9 In a potentiometer experiment using a slide-wire potentiometer as shown in Topic 7.5 Figure 3, a cell X is compared with a standard cell S of e.m.f. 1.50 ± 0.01 V. The balance length for S was 735 ± 3 mm and for X was 530 ± 3 mm.

a Calculate the e.m.f. of cell X.

b Calculate the percentage uncertainty in:
i the e.m.f. of S, **ii** each balance length.

c Hence calculate the uncertainty in the e.m.f. of X.

10 A slide wire potentiometer, as in Topic 7.5, Figure 3a, is used to compare the e.m.f. of a cell Y with the p.d. across the terminals of the same cell when a resistor S is connected across its terminals. The following measurements are made.
Without resistor S across Y, balance length $l_1 = 741$ mm
With resistor S across Y, balance length $l_2 = 625$ mm

a **i** Show that the ratio $\frac{l_1}{l_2} = \frac{R + r}{R}$
where r is the internal resistance of cell Y and R is the resistance of S.
ii The resistance of S is 5.00 Ω. Calculate the internal resistance r of cell Y.

b Immediately after the second measurement, the first measurement is repeated in case the e.m.f. of the driver cell changes during the measurements. The repeat measurement is the same as the first measurement. If the repeat measurement had been: **i** less than, **ii** greater than the first measurement, state one conclusion you would have drawn in each case.

11 a A cell of e.m.f., E_X, and negligible internal resistance, X, is connected to a network of identical resistors A, B, C, D, E and F as shown in Figure 11.1. Each resistor has resistance R.

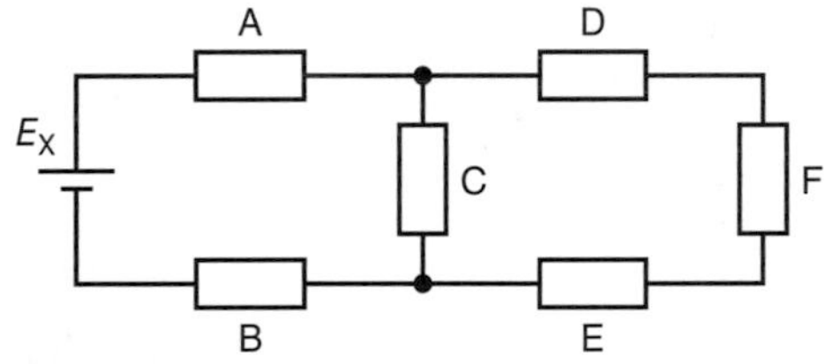

Figure 11.1

i Calculate the total resistance of the network in terms of R.
ii Determine the current in resistor F in terms of E_X and R.

b Resistor F is replaced by cell Y which is identical to X, as shown in Figure 11.2.

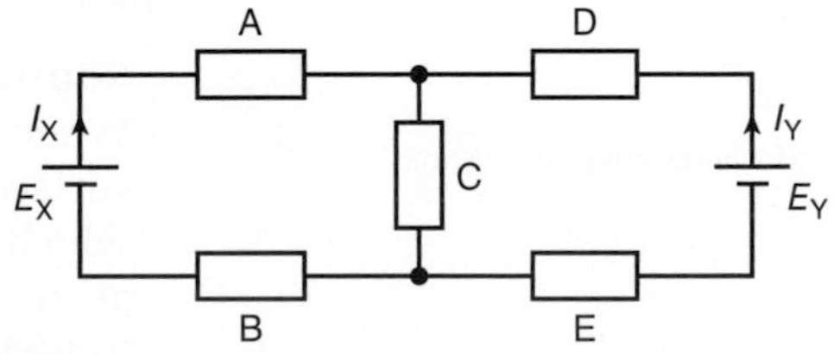

Figure 11.2

i Use Kirchhoff's laws to show that $E_X = (3I_X + I_Y)R$, where I_X and I_Y are the currents in cells X and Y respectively.
ii Derive a similar expression for E_Y, the e.m.f. of cell Y.
iii If each cell has an e.m.f. of 6.0 V and $R = 2.0\,\Omega$, calculate the potential difference across resistor C.

8 Waves

8.1 Waves and vibrations

Learning outcomes

On these pages you will learn to:

- recall that energy is transferred by a progressive wave
- describe wave motion in ropes and springs
- compare transverse and longitudinal waves
- analyse graphical representations of transverse and longitudinal waves

Types of wave

Waves transfer energy from the source that creates them, to the objects that absorb them. Waves that pass through a substance are **vibrations** of the particles of the substance. Sound waves, seismic waves and waves on strings are examples of waves that pass through a substance. These types of wave are often referred to as **mechanical waves**. When waves progress through a substance, the particles of the substance vibrate in a certain way which makes other particles vibrate in the same way, and so on.

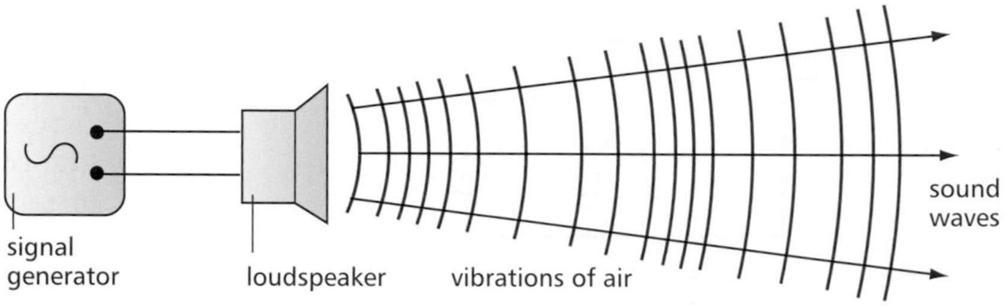

Figure 1 *Creating sound waves in air*

- **Electromagnetic waves** are vibrating **electric** and **magnetic fields** that progress through space, without the need for a substance. The vibrating electric field generates a vibrating magnetic field, which generates a vibrating electric field further away, and so on. Electromagnetic waves include radio waves, microwaves, infrared radiation, light, ultraviolet radiation, X-rays and gamma radiation. All electromagnetic waves travel with the same speed through free space. The full spectrum of electromagnetic waves is listed in Table 1. All types of waves transfer energy when they travel through a substance but only electromagnetic waves can transfer energy through a vacuum.

Table 1 *The electromagnetic spectrum*

Type	Wavelength range
radio	>0.1 m
microwave	0.1 m to 1 mm
infrared	1 mm to about 650 nm
visible	about 650 nm to 350 nm
ultraviolet	about 350 nm to 1 nm
X-rays	<1 nm
gamma rays	<1 nm

Longitudinal and transverse waves

Longitudinal waves

Longitudinal waves are waves in which the direction of vibration of the particles is **parallel** to (ie along) the direction in which energy is transferred by the waves. Sound waves, primary seismic waves and compression waves on a slinky are all longitudinal waves. Figure 2 shows how to send longitudinal waves along a 'slinky'. When one end of the slinky is moved to and fro repeatedly, each 'forward' movement causes a compression wave to pass along the slinky as the coils push into each other. Each 'reverse' movement causes the coils to move apart, so an 'expansion' wave passes along the slinky.

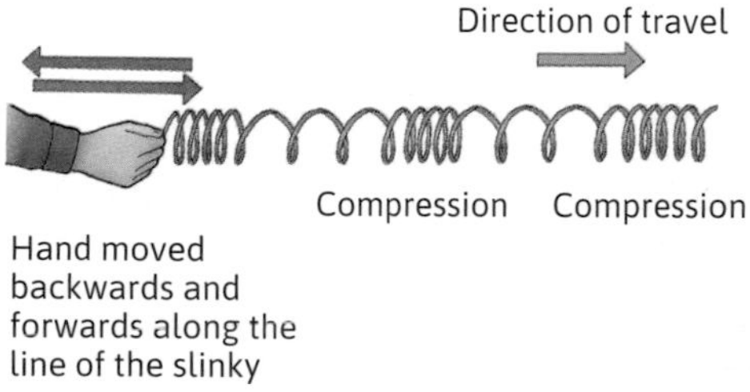

Figure 2 *Longitudinal waves on a slinky*

- **Sound waves** in air are created when a surface in contact with air vibrates. When the vibrating surface pushes on the air, the air molecules near the surface are pushed away from the surface, pushing on adjacent molecules which also push on adjacent molecules. This creates a wave of 'high density' air (i.e. a compression) that passes through the air. When the vibrating surface 'retreats', the air molecules near the surface move back into the space vacated by the surface, allowing adjacent air molecules to fall back and so on. This creates a wave of 'low density' air (referred to as a 'rarefaction') that passes through the air behind the compression wave. The vibrating surface therefore creates a series of **compressions** and **rarefactions** that pass through the air. The air molecules are therefore repeatedly pushed to and fro along the direction in which the sound travels.

Transverse waves

Transverse waves are waves in which the direction of vibration is **perpendicular** to the direction in which energy is transferred by the waves. Electromagnetic waves, secondary seismic waves and waves on a string or a wire are all transverse waves. Figure 3 shows transverse waves travelling along a rope. When one end of the rope is moved from side to side repeatedly, these sideways movements travel along the rope: each unaffected part of the rope is pulled sideways when the part next to it moves sideways.

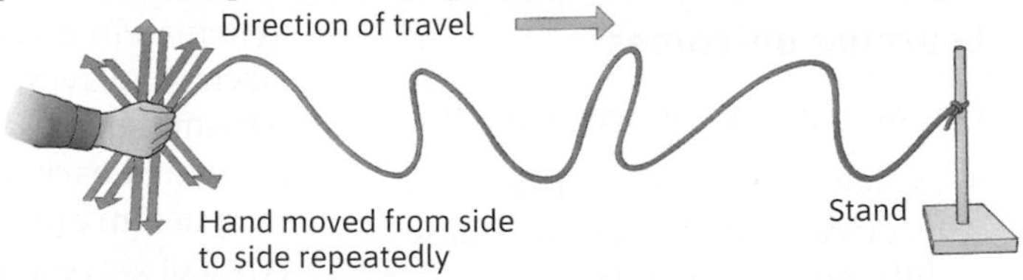

***Figure 3** Making rope waves*

Representations of transverse and longitudinal waves

Imagine a rope marked along its length into very short 'elements' of equal length. When a wave travels along the rope, each element is 'disturbed' by the wave and vibrates about its equilibrium (ie undisturbed) position along a line perpendicular to the direction of travel of the wave. Figure 4a shows a snapshot of a transverse wave on a rope. Figure 4b shows how the displacement of the rope at a certain time varies with distance from P along the rope. Figure 4c shows how the displacement of P varies with time.

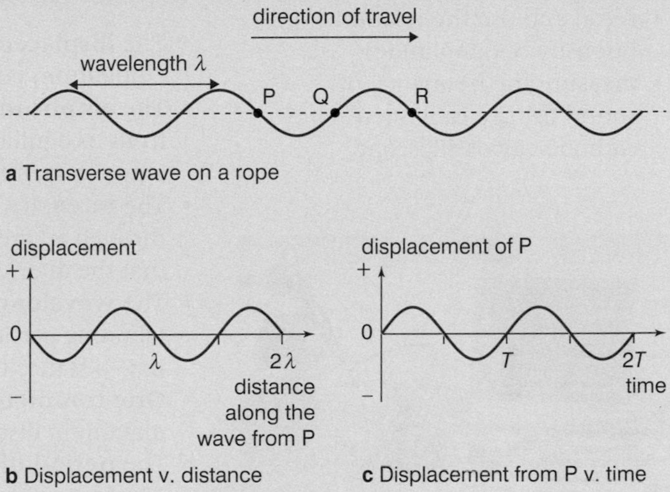

***Figure 4** Representing a transverse wave*

The graph is easy to interpret because the waves on the rope are transverse waves and the graph has the same shape as a 'snapshot' of the wave. For example, two elements either side of an element at zero displacement are displaced in opposite directions. Each point on the rope oscillates about its undisturbed position along a line perpendicular to the direction of travel of the waves. For example, point P in Figure 4a is at zero displacement in the snapshot and it then moves up to maximum positive displacement then back to zero displacement in the next half cycle. Figure 4c shows how the displacement of point P changes with time in the next two cycles. Note that in Figure 4c the symbol T represents the time for one cycle.

Compare the waves on a rope with longitudinal waves on a 'slinky'. Figure 5 shows a representation of a longitudinal wave travelling along the slinky. The compressions are where the coils are closest and the rarefactions are where they are furthest apart. A displacement against distance graph for Figure 5a would be similar to Figure 4b.

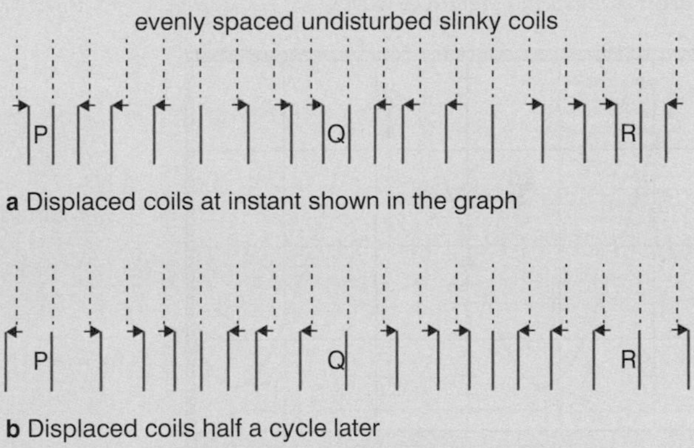

***Figure 5** Longitudinal waves on a slinky*

- The coil at the centre of each 'compression' has zero displacement (for example at P in Figure 5a) and the coils on either side near the centre are displaced towards the centre of the compression. On a displacement v. distance graph, this would be where the displacement changes from + to – along the wave.
- The coil at the centre of each 'rarefaction' has zero displacement (for example midway between P and Q) and the coils on either side near the centre are displaced away from the centre of the rarefaction. On a displacement v. distance graph, this would be where the displacement changes from – to + along the wave.

Summary test 8.1

1 Classify the following types of wave as either longitudinal or transverse:

radio waves	sound waves
secondary seismic waves	microwaves

2 Sketch a snapshot of a longitudinal wave travelling on a slinky coil, indicating the direction in which the waves are travelling and areas of compression and expansion.

3 Sketch a snapshot of a transverse wave travelling along a rope, indicating the direction in which the waves are travelling and the direction of motion of the particles at the peaks and troughs.

4 **a** Describe how the position of the coil marked Q in Figure 5a varies with time over the next cycle.

b Sketch a graph to illustrate your description in **a**.

8.2 Measuring waves

Learning outcomes

On these pages you will learn to:

- explain and use the terms displacement, amplitude, phase difference, period, frequency, wavelength and speed
- deduce, recall and use the wave equation $v = f\lambda$
- recall and use the relationship intensity $\propto$ (amplitude)^2
- measure the frequency of sound using a calibrated cathode-ray oscilloscope

Figure 1 Measuring electrical waves

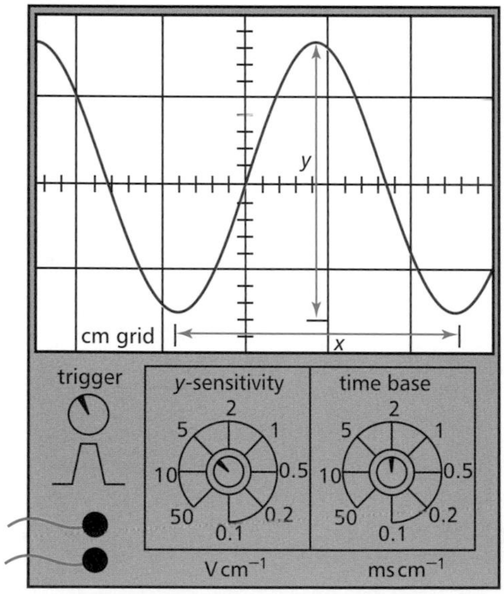

a Trace height y = 32 mm
∴ trace amplitude = 16 mm = 1.6 cm
given y-sensitivity = 5 V cm^{-1}
voltage amplitude = 5 × 1.6 = 8.0 volts

b x-distance from peak to peak = 32 mm = 3.2 cm
given time base 2 ms cm^{-1}
time period = 2 × 3.2 = 6.4 ms

$$\text{frequency } f = \frac{1}{\text{time period}} = \frac{1}{6.4 \times 10^{-3}\,\text{s}}$$

$f = 1.56 \times 10^2$ Hz = 156 Hz

Figure 3 Measuring a waveform

When an intercontinental phone call is made, sound waves are converted to electrical waves. These waves are carried by electromagnetic waves from ground transmitters to satellites in space and back to receivers on Earth, where they are converted back to electrical waves, then back to sound waves. The engineers who design and maintain communications systems need to measure the different types of wave at different stages to make sure the waves are not distorted.

Key terms

The following terms, illustrated in Figure 2, are used to describe waves:

- The **displacement** of a vibrating particle is its distance and direction from its equilibrium position.
- The **amplitude** of a wave is the maximum displacement of a vibrating particle from its equilibrium position. For a transverse wave, this is the height of a wave crest or the depth of a wave trough from the middle.
- The **intensity of a wave** is the **power per unit area** the waves would transfer through an area perpendicular to the direction of the waves. It can be shown that the intensity of a wave is proportional to the square of its amplitude.
- The **wavelength** of a wave is the least distance between two adjacent vibrating particles with the same displacement and velocity at the same time (e.g. distance between adjacent crests).
- **One complete cycle of a wave** is from a maximum displacement to the next maximum displacement (e.g. from one wave peak to the next).
- The **period of a wave** is the time for one complete wave to pass a fixed point.
- The **frequency** of a wave is the number of cycles of vibration of a particle per unit time. The unit of frequency is the **hertz (Hz)**. For waves of frequency f:

$$\textbf{Period of the wave} = \frac{1}{f}$$

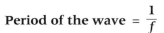

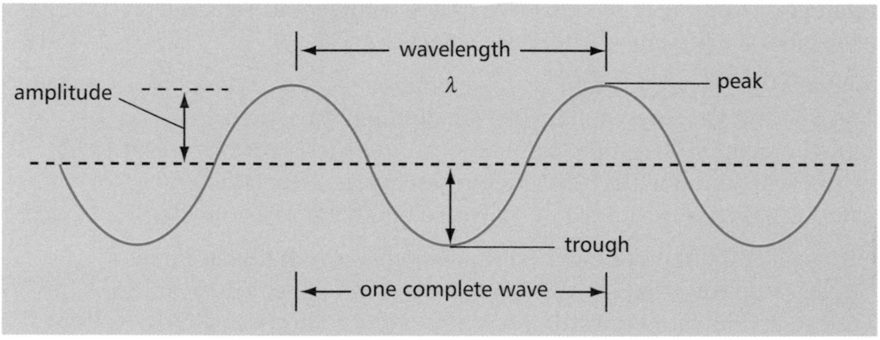

Figure 2 Parts of a wave

Using an oscilloscope to measure sound waves

The oscilloscope shown in Figure 3 is connected to a **microphone**, which detects sound waves emitted by a loudspeaker connected to a signal generator. The trace on the oscilloscope screen is the **waveform** of the sound waves produced by the loudspeaker. Figure 3 shows how to measure the amplitude and the frequency of this waveform using the oscilloscope controls. Note that the waveform **amplitude** is the distance **from the middle to the top**.

Wave speed

The higher the frequency of a wave, the shorter its wavelength. For example, if waves are sent along a rope, the higher the frequency at which they are produced, the closer together their wave peaks. The same effect can be seen in a ripple tank, when straight waves are produced at a constant frequency. If the frequency is raised to a higher value, the waves are closer together. Consider

Figure 4, which represents the crests of straight waves in a ripple tank travelling at constant speed.

- Each wave crest travels a distance equal to one wavelength (λ) in the time for one cycle.
- The time taken for one cycle $= \frac{1}{f}$, where f is the frequency of the waves.

Therefore, the speed of the waves:

$$v = \frac{\text{distance travelled in one cycle}}{\text{time taken for one cycle}} = \frac{\lambda}{1/f} = f\lambda$$

For waves of frequency f and wavelength λ:

$$\textbf{Wave speed, } v = f\lambda$$

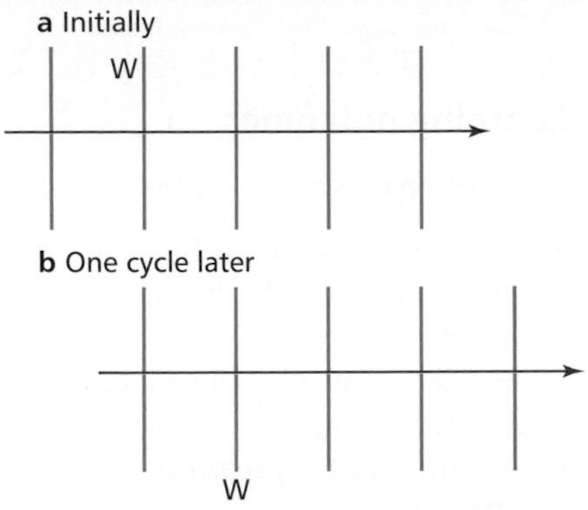

***Figure 4** Wave speed*

Note

The symbol c is used for the speed of electromagnetic waves.

Phase difference

The **phase difference** between two vibrating particles is the fraction of a cycle between the vibrations of the two particles, measured either in **degrees or radians** where:

$$1 \text{ cycle} = 360° = 2\pi \text{ radians}$$

For two points at distance d apart along a wave of wavelength λ,

$$\textbf{Phase difference, in radians} = \frac{2\pi d}{\lambda}$$

Figure 5 shows three successive snapshots of the particles of a transverse wave that progresses from left to right across the diagram. Particles O, P, Q, R and S are spaced approximately $\frac{1}{4}$ of a wavelength apart. Table 1 shows the phase difference between O and each of the other particles.

***Table 1** Phase differences*

	P	Q	R	S
Distance from O	$\frac{1}{4}\lambda$	$\frac{1}{2}\lambda$	$\frac{3}{4}\lambda$	λ
Phase difference relative to O/radians	$\frac{1}{2}\pi$	π	$\frac{3}{2}\pi$	2π

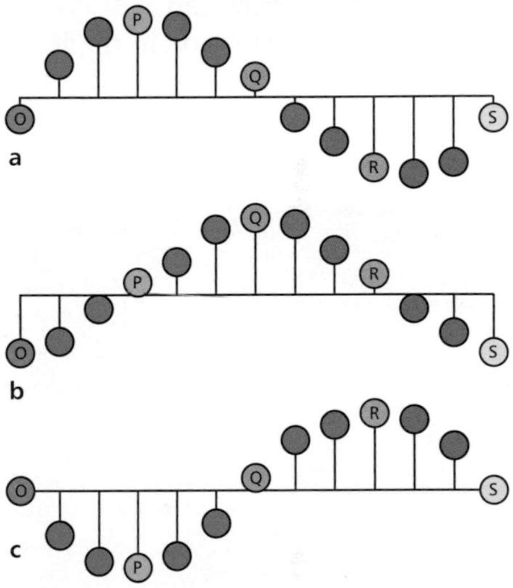

***Figure 5** Progressive waves*

Summary test 8.2

1 Sound waves, in air, travel at a speed of $340\,\text{m s}^{-1}$ at $20\,°\text{C}$. Calculate the wavelength of sound waves, in air, which have a frequency of:
  a 3400 Hz **b** 18 000 Hz.

2 Electromagnetic waves, in air, travel at a speed of $3.0 \times 10^8\,\text{m s}^{-1}$. Calculate the frequency of electromagnetic waves of wavelength:
  a 0.030 m **b** 600 nm.

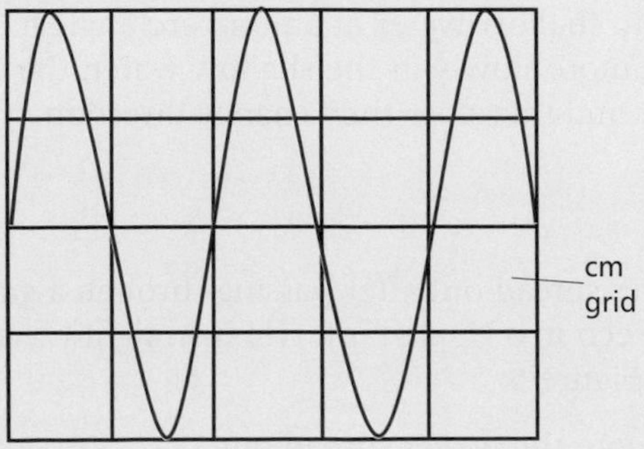

Figure 6

3 Figure 6 shows a waveform on an oscilloscope screen, when the y-sensitivity of the oscilloscope was $0.50\,\text{V cm}^{-1}$ and the time base was set at $0.5\,\text{ms cm}^{-1}$. Determine the amplitude and the frequency of this waveform.

4 For the waves in Figure 5:
  a determine:
    i the amplitude if the wavelength is 55 mm,
    ii the phase difference between P and R,
    iii the phase difference between P and S.
  b What would be the displacement and direction of motion of Q three-quarters of a cycle after the last snapshot?

8.3 Wave properties

Learning outcomes

On these pages you will learn to:

- know what is meant by reflection and refraction of waves
- describe examples of reflection and refraction
- explain what is meant by diffraction
- show an understanding of experiments that demonstrate diffraction, including the diffraction of water waves in a ripple tank with both a wide gap and a narrow gap
- describe wave motion in ripple tanks

Wave properties, such as reflection, refraction and diffraction, occur with many different types of wave. A **ripple tank** may be used to study these wave properties. The tank is a shallow transparent tray of water with sloping sides. The slopes prevent waves reflecting off the sides of tank. If they did reflect, it would be difficult to see the waves to be observed.

- The waves observed in a ripple tank are referred to as '**wave fronts**', which are lines of constant phase (e.g. crests).
- The direction in which a wave travels is at right angles to the wave front.

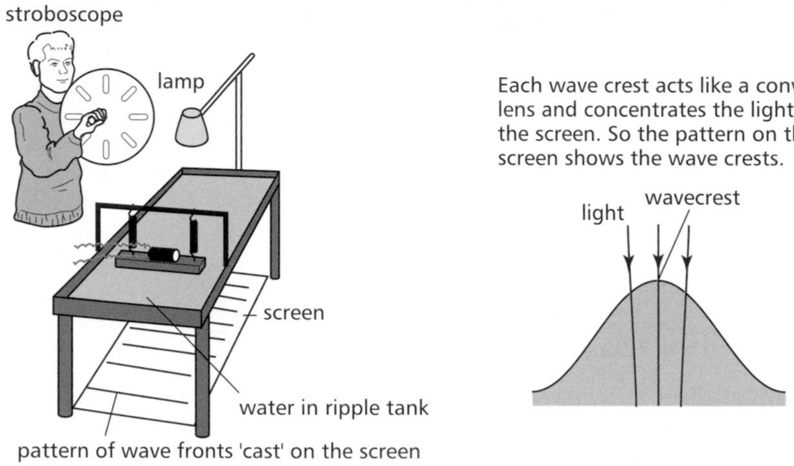

Figure 1 *The ripple tank*

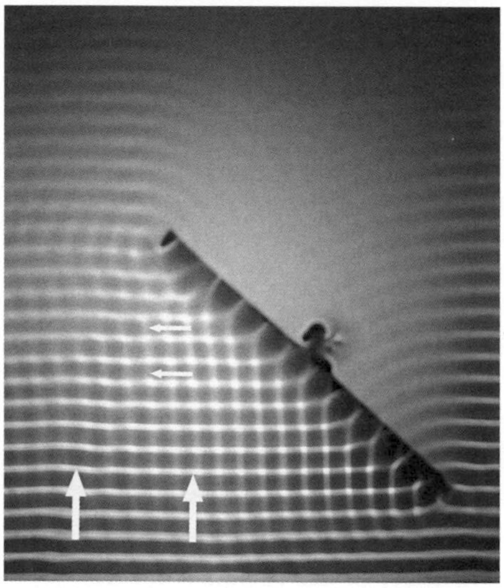

Figure 2 *Reflection of plane waves*

Reflection

- **Straight waves** directed at a certain angle to a hard flat surface (the '**reflector**') reflect off at the same angle, as shown in Figure 2. The angle between the reflected wave front and the surface is the **same** as the angle between the incident wave front and the surface. Therefore the direction of the reflected wave is at the same angle to the reflector as the direction of the incident wave. This same effect is observed when a light ray is directed at a plane mirror. The angle between the incident ray and the mirror is the **same** as the angle between the reflected ray and the mirror.

Refraction

When waves pass across a boundary at which the wave speed changes, the wavelength also changes. If the wave fronts are at a non-zero angle to the boundary, they change direction as well as changing speed. This effect is known as **refraction**.

Figure 4 shows the refraction of water waves in a ripple tank when they pass across a boundary, from deep to shallow water at a non-zero angle to the boundary. Because they move more slowly in the shallow water, the wavelength is smaller in the shallow water and therefore they change direction.

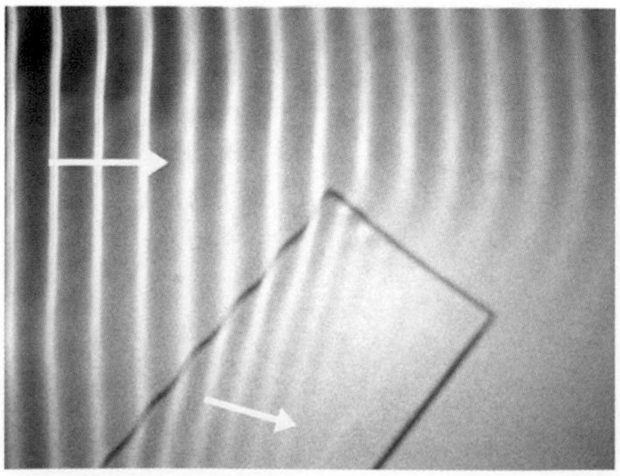

Figure 3 *Refraction*

Diffraction

Diffraction occurs when waves spread out after passing through a gap, or round an obstacle. The effect can be seen in a ripple tank when straight waves are directed at a gap, as shown in Figure 5.

- The narrower the gap, the more the waves spread out.
- The longer the wavelength, the more the waves spread out.

To explain why the waves are diffracted on passing through the gap, consider each point on a wave front as a secondary emitter of 'wavelets'. The wavelets from the points along a wave front travel only in the direction in which the wave is travelling, not in the reverse direction, and they combine to form a new wave front spreading beyond the gap.

Investigating diffraction

Use a ripple tank to direct plane waves continuously at a gap between two metal barriers.

1 Change the gap spacing and observe the effect on the diffraction of the waves that pass through the gap. You should find that the diffraction of the waves increases as the gap is made narrower, as shown in Figure 4.

2 Keep the gap spacing constant and change the wavelength of the waves by altering the frequency of the vibrating beam. Observe the effect on the diffraction of the waves. You should find that the smaller the wavelength of the waves, the less they are diffracted.

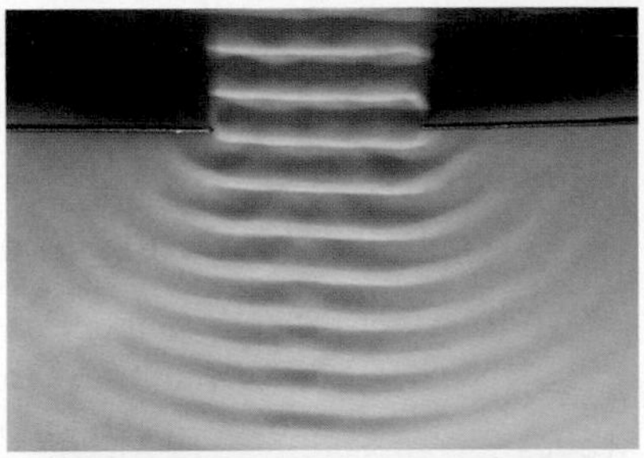

a) at a wide gap

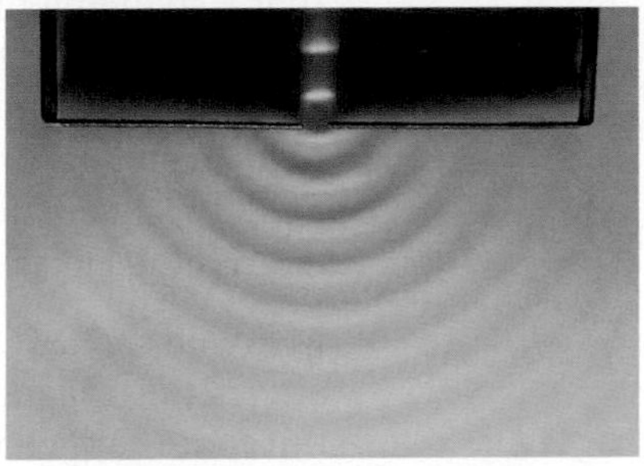

b) at a narrow gap

Figure 4 *The effect of the gap width*

Summary test 8.3

1 Copy and complete Figure 5, by showing the wave front after it has been reflected from the straight reflector. Also, show the direction of the reflected wave front.

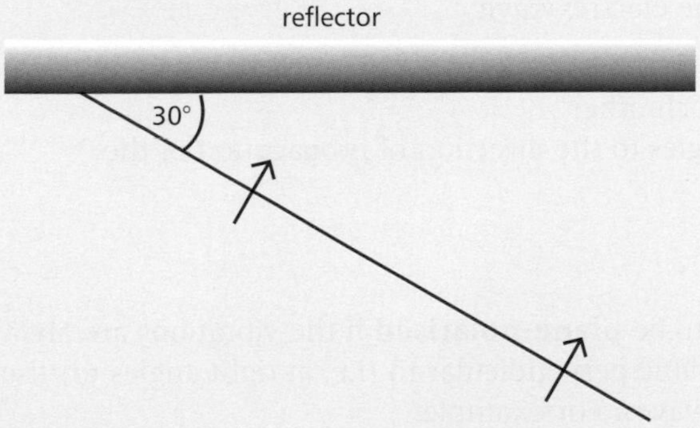

Figure 5

2 Copy and complete Figure 6, by showing the wave fronts after they have passed across the boundary and have been refracted. Also, show the direction of the refracted waves.

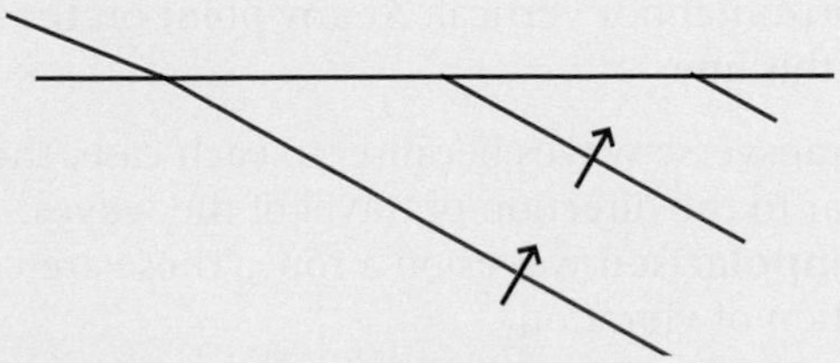

Figure 6

3 Water waves are diffracted on passing through a gap. How is the amount of diffraction changed as a result of:

a widening the gap without changing the wavelength,

b increasing the wavelength of the water waves without changing the gap width,

c increasing the wavelength of the water waves and reducing the gap width,

d widening the gap and increasing the wavelength of the waves?

4 Microwaves are reflected by a metal plate. In Figure 7, microwaves from a transmitter are directed at a gap between two metal plates. The detector placed on the other side of the plates receives a signal from the transmitter.

a Explain why the detector at the position shown receives a signal even though it is not directly in line with the transmitter.

b With the detector in the same position, describe how the signal would change if the gap was made wider.

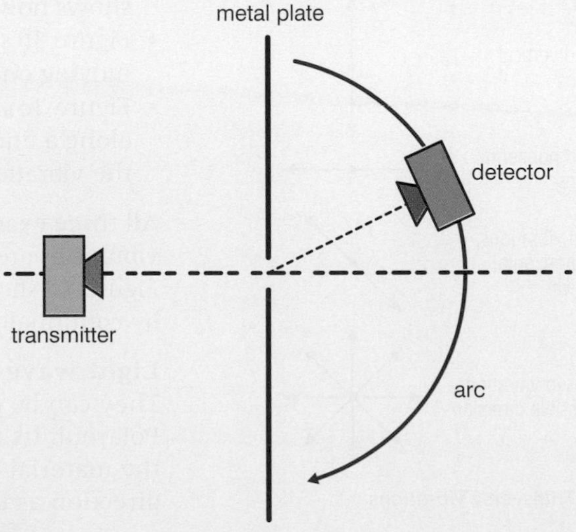

Figure 7

8.4 Electromagnetic waves and polarisation

Learning outcomes

On these pages you will learn to:

- recall that electromagnetic waves are transverse waves
- describe what an electromagnetic wave is
- explain what is meant by polarisation
- recall and use Malus's law

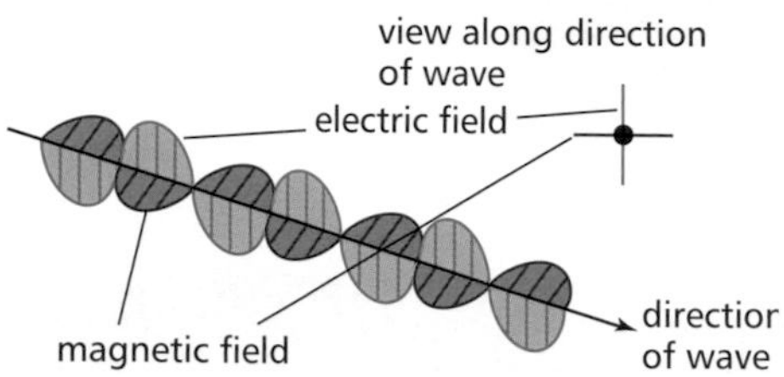

Figure 1 *Electromagnetic waves*

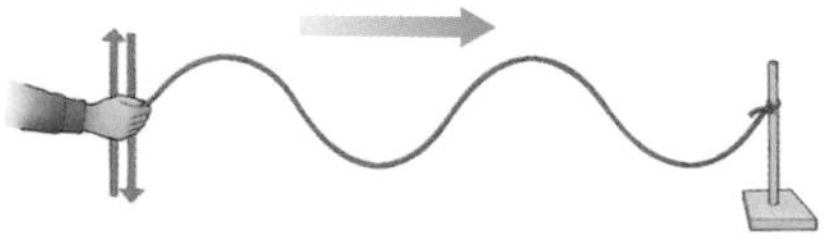

Figure 2 *Transverse waves*

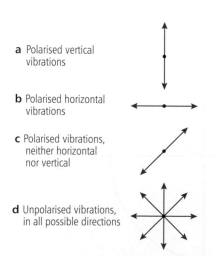

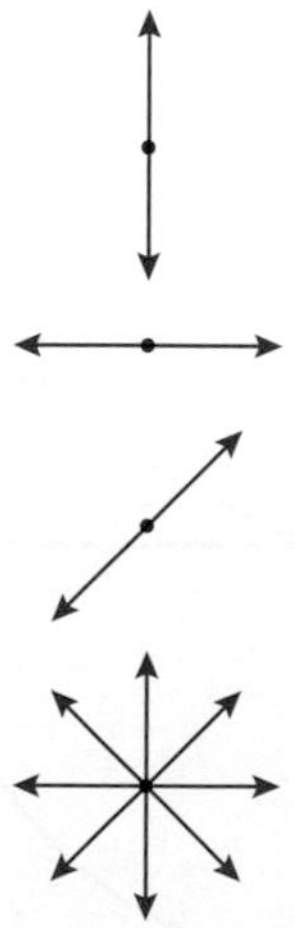

Figure 3 *Transverse vibrations*

The nature of electromagnetic waves

Electromagnetic waves were predicted by James Clerk Maxwell in 1862. Maxwell knew that a magnetic field is created round a wire when an electric current passes along the wire. He knew about Michael Faraday's discovery that a changing magnetic field in a coil of wire induces a voltage in the coil. He wondered if the two effects could be linked, and he used his mathematical skills to discover the link. In effect, Maxwell showed that the changing magnetic field created by an alternating current in a wire creates an alternating electric field, which creates an alternating magnetic field further away, which creates an alternating electric field, and so on.

Maxwell showed that the result is an **electromagnetic wave** consisting of an alternating magnetic field in phase with an alternating electric field. He worked out that the speed of an electromagnetic wave should be 300 000 km s^{-1}, the same as the speed of light in a vacuum! Maxwell realised, from his theory, that light must be an electromagnetic wave and that electromagnetic waves must exist beyond the visible spectrum. He knew that infrared and ultraviolet radiation are outside the visible spectrum, and he correctly predicted electromagnetic waves beyond these two invisible forms of radiation.

An electromagnetic wave consists of an **alternating magnetic field** (the magnetic wave) and an **alternating electric field** (the electric wave), in which the magnetic wave and the electric wave:

- are at right angles to each other
- vibrate in phase with each other
- both vibrate at right angles to the direction of propagation of the electromagnetic wave.

Polarisation

Tranverse waves are said to be **plane-polarised** if the vibrations are always along the same line in a plane perpendicular to (i.e. at right angles to) the direction of travel of the waves. For example:

- In Figure 2, each point on the rope vibrates in a vertical line only. Figure 3a shows how the rope would look if it was viewed end-on.
- Figure 3b shows an end-view of a rope made to vibrate horizontally by moving one end from side to side along a horizontal line.
- Figure 3c shows an end-view of the rope if one end is moved from side to side along a line which is neither horizontal nor vertical. At any point on the rope, the vibrations are always along this line.

All three examples are polarised transverse waves because, in each case, the vibrations are always perpendicular to the direction of travel of the waves. Figure 3d shows an end-view of **unpolarised** waves on a rope. These are created by continually changing the direction of vibration.

Light waves from a lamp bulb are unpolarised transverse waves. They can be polarised by passing them through special material called Polaroid. Its molecules are all lined up in the same direction. As a result, the material only allows light waves through that vibrate in the same direction as the molecules (see Figure 4).

By convention, the plane of polarisation of an electromagnetic wave is defined as the plane in which the electric field oscillates.

Investigating polarisation

If unpolarised light is passed through two Polaroid filters as in Figure 5, the transmitted light intensity changes if one polaroid is rotated relative to the other one. The filters are said to be 'crossed' when the transmitted intensity is a minimum. At this position, the polarised light from the first filter (the 'polariser') cannot pass through the second filter (the analyser), as the alignment of molecules in the second filter is at 90° to the alignment in the first filter. This is like passing rope waves through two 'letter boxes' at right angles to each other, as shown in Figure 5.

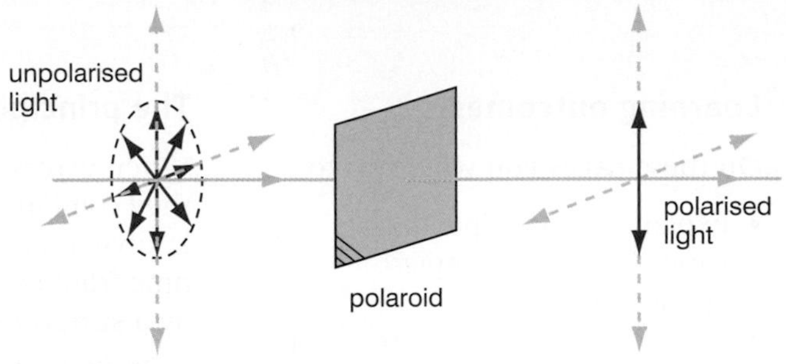

Figure 4 *Polarising a light beam using a Polaroid filter*

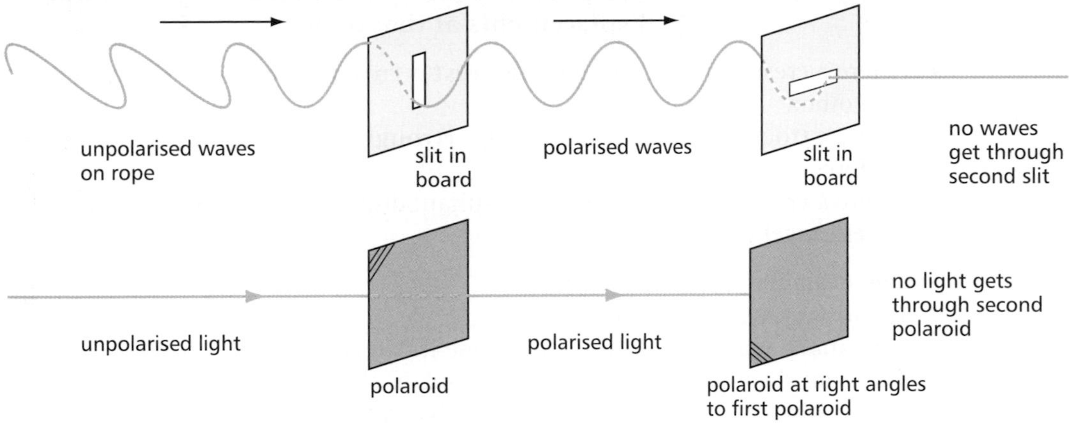

Figure 5 *Investigating polarisation*

The intensity of the polarised light transmitted through the analyser changes as the analyser is rotated about the transmitted beam. If the analyser is rotated from the position of maximum transmitted intensity I_0, the intensity I for an angle of rotation θ is given by the equation

$$I = I_0 \cos^2\theta$$

This equation is known as **Malus's law**. The equation follows from the fact that the amplitude A of the wave at an angle of rotation θ is equal to $A_0 \cos\theta$. Since the intensity is proportional to the amplitude A, it follows that the intensity is proportional to $\cos^2\theta$. Figure 6 shows how the intensity of the transmitted beam changes as the angle of rotation increases from 0 to 360°.

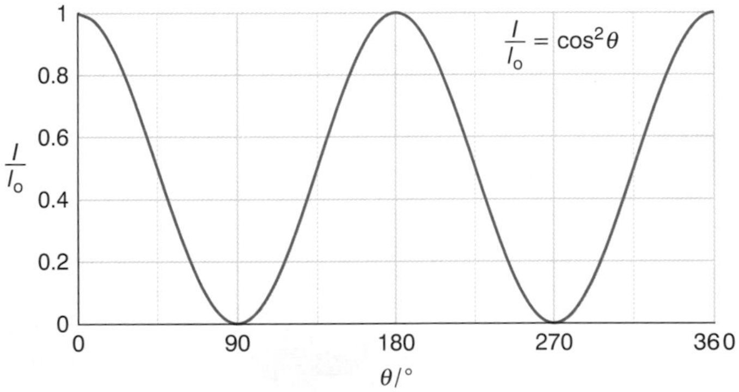

Figure 6 *Graph of* $\frac{\text{intensity, } I}{\text{maximum intensity, } I_0}$ *against angle of rotation,* θ

Summary test 8.4

1 a Describe an electromagnetic wave.

b Explain what is meant by electromagnetic waves that are:

i polarised,

ii unpolarised.

2 A light source is observed through two Polaroid filters. Describe and explain what you would expect to observe when one of the filters is rotated through 180° about the line of sight.

3 When the aerial of a portable radio is turned away from a vertical position, the radio signal becomes weaker. Explain this effect.

4 a State Malus's law.

b A polarised beam of light is directed through a Polaroid filter. Calculate the percentage reduction in the intensity of the transmitted light when the filter is rotated about the beam by an angle of 10° from the position of maximum intensity.

8.5 Superposition

Learning outcomes

On these pages you will learn to:

- explain the principle of superposition in simple applications
- use the principle of superposition in simple applications, including interference and stationary waves
- describe and explain experiments that demonstrate two-source interference using water ripples, sound waves and microwaves

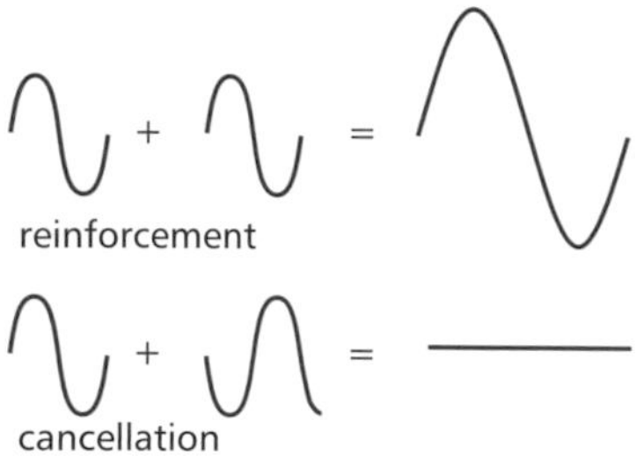

***Figure 1** Superposition*

***Figure 2** Making stationary waves*

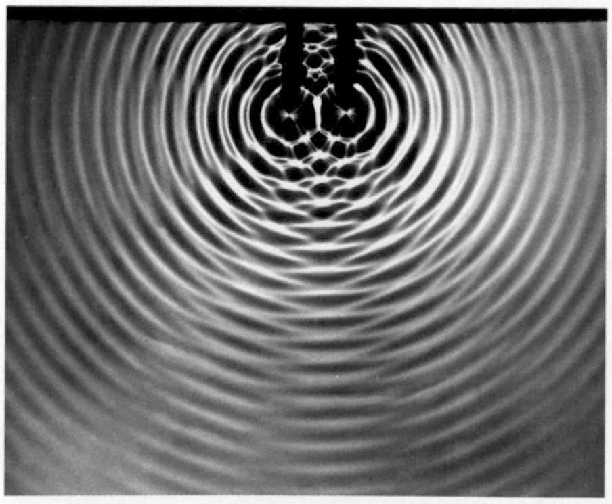

***Figure 3** Interference of water waves*

The principle of superposition

When waves meet, they pass through each other. At the point where they meet, they combine for an instant before they move apart. This combining effect is known as **superposition**. Imagine a boat hit by two wave crests at the same time from different directions. Anyone on the boat would know it had been hit by a **supercrest**, the combined effect of two wave crests.

The principle of superposition states that when two waves meet, the total displacement at a point is equal to the sum of the individual displacements at that point.

- Where a crest meets a crest, a **supercrest** is created; the two waves reinforce each other.
- Where a trough meets a trough, a **supertrough** is created; the two waves reinforce each other.
- Where a crest meets a trough, the resultant displacement is **zero**; the two waves cancel each other out.

Further examples of superposition

1 Stationary waves on a rope

Stationary waves are formed on a rope if two people send waves continuously along a rope from either end, as shown in Figure 2. The two sets of waves are referred to as **progressive waves**, to distinguish them from stationary waves. They combine at fixed points along the rope to form points of no displacement, or **nodes**, along the rope. At each node, the two sets of waves are always 180° out of phase so they cancel each other out. Stationary waves are described in more detail in Topic 8.6.

2 Water waves in a ripple tank

A vibrating dipper on a water surface sends out circular waves. Figure 3 shows a snapshot of two sets of circular waves produced in this way in a ripple tank. The waves pass through each other continuously:

- Points of **cancellation** are created where a crest from one dipper meets a trough from the other dipper. These points of cancellation are seen as gaps in the wave fronts.
- Points of **reinforcement** are created where a crest from one dipper meets a crest from the other dipper, or where a trough from one dipper meets a trough from the other dipper.

As the waves are continuously passing through each other at constant frequency and at a constant phase difference, cancellation and reinforcement occur at fixed positions. This effect is known as **interference**. The two dippers are said to be **coherent** emitters of waves, because they vibrate with a constant phase difference. If the phase difference changed at random, the points of cancellation and reinforcement would move about at random and no interference pattern would be seen. Interference of light is described in more detail in Topic 9.1.

Note

The points of cancellation and reinforcement would be further apart if the wavelength were increased or the dippers were closer together.

Tests using microwaves

A microwave transmitter and receiver can be used to demonstrate reflection, refraction, diffraction, interference and polarisation of microwaves. The transmitter produces microwaves of wavelength 3.0 cm. The receiver can be connected to a suitable meter, which gives a measure of the intensity of the microwaves at the receiver.

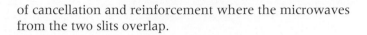

- Place the receiver in the path of the microwave beam from the transmitter. Move the receiver gradually away from the transmitter and note that the receiver signal decreases with distance from the transmitter. This shows that the microwaves become weaker as they travel away from the receiver.
- Place a metal plate between the transmitter and the receiver to show that microwaves cannot pass through metal.
- Use two metal plates to make a narrow slit and show that the receiver detects microwaves that have been diffracted as they pass through the slit. Show that if the slit is made wider, less diffraction occurs.
- Use a narrow metal plate with the two plates from above to make a pair of slits, as in Figure 4. Direct the transmitter at the slits and use the receiver to find points of cancellation and reinforcement where the microwaves from the two slits overlap.

Interference of sound waves

To demonstrate interference of sound, a signal generator connected to two loudspeakers in series may be used. The two loudspeakers in series are coherent emitters of sound waves because their phase difference is constant. With the speakers operating at about 2 kHz about a metre apart, a careful listener walking slowly in front of the speakers ought to be able to locate points of cancellation and reinforcement. If the speakers are moved further apart, the points of cancellation and reinforcement will be closer together.

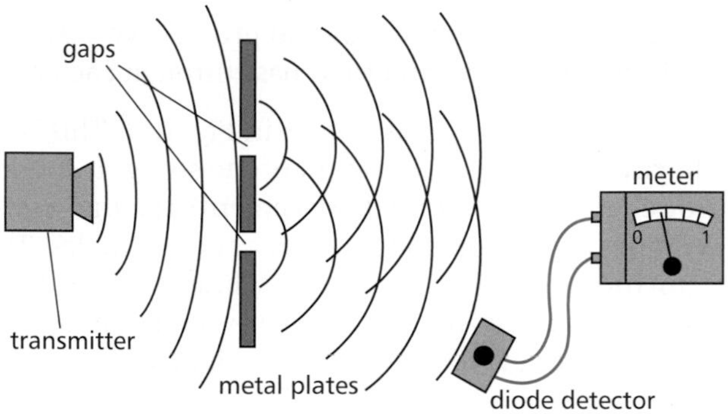

Figure 4 *Interference of microwaves*

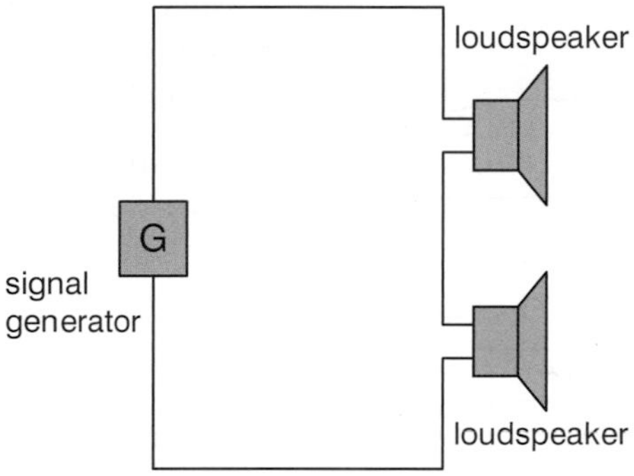

Figure 5

Summary test 8.5

1 Figure 6 shows two wave pulses on a rope travelling towards each other. Sketch a snapshot of the rope:

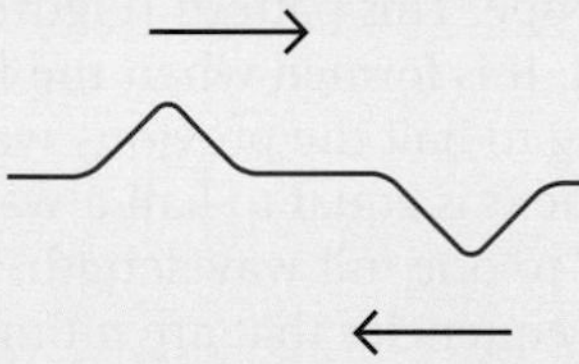

Figure 6

a when the two waves are passing through each other,

b when the two waves have passed through each other.

2 How would you expect the interference pattern in Figure 3 to change if:

a the two dippers are moved further apart,

b the frequency of the waves produced by the dippers is reduced?

3 Microwaves from a transmitter are directed at a narrow slit between two metal plates. A receiver is placed in the path of the diffracted microwaves.

How would you expect the receiver signal to change if:

a the receiver is moved directly away from the slit,

b the slit is then made narrower?

4 Microwaves, from a transmitter, are directed at two parallel slits in a metal plate. A receiver is placed on the other side of the metal plate. When the receiver is moved a short distance along a line AB parallel to the plate, the receiver signal decreases then increases again.

a Explain why the signal decreased and then increased when it was moved from A along the line AB.

b Explain why the signal increased as it moved towards B.

Stationary and progressive waves

Learning outcomes

On these pages you will learn to:

- compare progressive and stationary waves in terms of particle vibrations
- explain the formation of a stationary wave on a string and identify nodes and antinodes
- describe and explain experiments that demonstrate stationary waves using microwaves and stretched strings

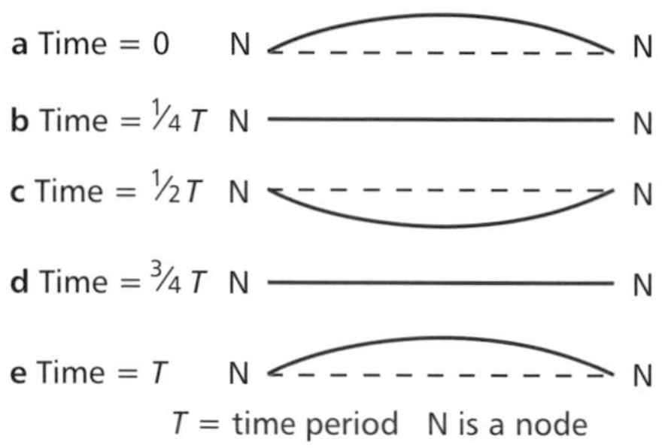

Figure 1 *Fundamental or first harmonic vibrations*

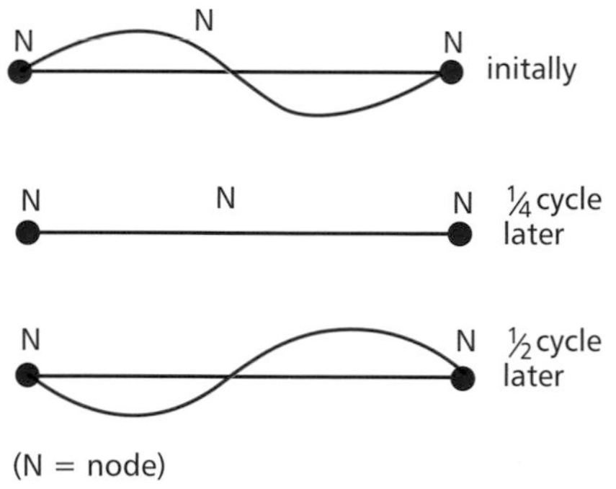

Figure 2 *A stationary wave of two loops*

Note

Stationary waves do not transfer energy. The amplitude of vibration is zero at the nodes, so there is no energy at the nodes. The amplitude of vibration is a maximum at the antinodes, so there is maximum energy at the antinodes. Because the nodes and antinodes are at fixed positions, no energy is transferred in a stationary wave pattern.

Formation of stationary waves

When a guitar string is plucked, the sound produced depends on the way in which the string vibrates. If the string is plucked gently at its centre, a stationary wave of constant frequency is set up on the string. The sound produced therefore has a constant frequency. If the guitar string is plucked harshly, the string vibrates in a more complicated way and the note produced contains other frequencies as well as the frequency produced when it is plucked gently.

As explained in Topic 8.5, a stationary wave is formed when two progressive waves pass through each other. This can be achieved on a string in tension by:

- Sending progressive waves along the string from either end.
- Fixing one end of the string and sending progressive waves along it from the other end. The waves reflect at the fixed end and pass through progressive waves moving towards the fixed end.
- Fixing both ends and making the middle part vibrate, so progressive waves travel towards each end, reflect at the ends, and then pass through each other.

The simplest stationary wave pattern on a string is shown in Figure 1. This is the **fundamental** or **first harmonic** mode of vibration of the string. It consists of a single loop that has a **node** (i.e. a point of no displacement) at either end. The string vibrates with maximum amplitude mid-way between the nodes. This position is referred to as an **antinode**. In effect, the string vibrates from side-to-side repeatedly. For this pattern to occur, the distance between the nodes at either end (i.e. the length of the string) must be equal to one half-wavelength of the waves on the string:

$$\textbf{Distance between adjacent nodes} = \tfrac{1}{2}\lambda$$

If the frequency of the waves sent along the rope from either end is raised steadily, the pattern in Figure 1 disappears: a new pattern is observed with two equal loops along the rope. This pattern (Figure 2) has a node at the centre as well as at either end. It is formed when the frequency is twice as high as in Figure 1, corresponding to half the previous wavelength. Because the distance from one node to the next is equal to half a wavelength, the length of the rope is therefore equal to one full wavelength. As explained in the next topic, the patterns occur at frequencies that are n times the fundamental frequency where n is a whole number, each pattern corresponding to n loops.

Explanation of stationary waves

Consider a snapshot of two progressive waves passing through each other:

- When they are in phase, they reinforce each other to produce a large wave (Figure 3a).
- A quarter of a cycle later, the two waves have each moved one quarter of a wavelength in opposite directions. They are now in **antiphase**, so they cancel each other (Figure 3b).
- After a further quarter cycle, the two waves are back in phase. The resultant is again a large wave as in Figure 3a, except reversed.

The points where there is no displacement (i.e. the nodes) are fixed in position throughout. Between these points, the stationary wave oscillates between the nodes.

In general, in any stationary wave pattern:

The amplitude of a vibrating particle in a stationary wave pattern varies with position from:

- zero at a node,
- to maximum amplitude at an antinode.

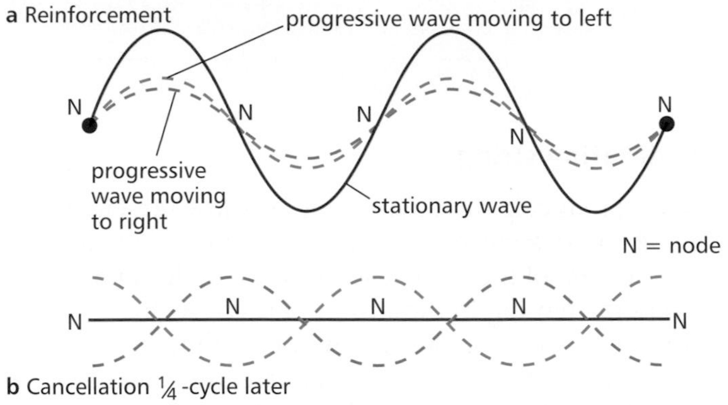

Figure 3 *Explaining stationary waves*

The phase difference between two vibrating particles:

- is zero if the two particles are between adjacent nodes or separated by an even number of nodes,
- is 180° if the two particles are separated by an odd number of nodes.

Table 1 *Comparison between stationary waves and progressive waves in terms of particle vibrations*

	Stationary waves	Progressive waves
Frequency	All particles, except at the nodes, vibrate at the the same frequency.	All particles vibrate at same frequency.
Amplitude	The amplitude varies from zero at the nodes to a maximum at the antinodes.	The amplitude is the same for all particles.
Phase difference between two particles	$m\pi$, where m is the number of nodes between the two particles.	$2\pi x/\lambda$, where x = distance apart and λ is the wavelength.

More examples of stationary waves

Sound in a pipe

Sound resonates at certain frequencies in an air-filled tube or pipe. In a pipe closed at one end, these resonant frequencies occur when there is an antinode at the open end and a node at the other end. (See Topic 8.8.)

Using microwaves

Microwaves from a transmitter are directed normally at a metal plate, which reflects the microwaves back towards the transmitter. When a detector is moved along the line between the transmitter and the metal plate, the detector signal is found to be zero at equally spaced positions along the line. The reflected waves and the waves from the transmitter form a stationary wave pattern. The positions where no signal is detected are where nodes occur. They are spaced at intervals of one half of a wavelength.

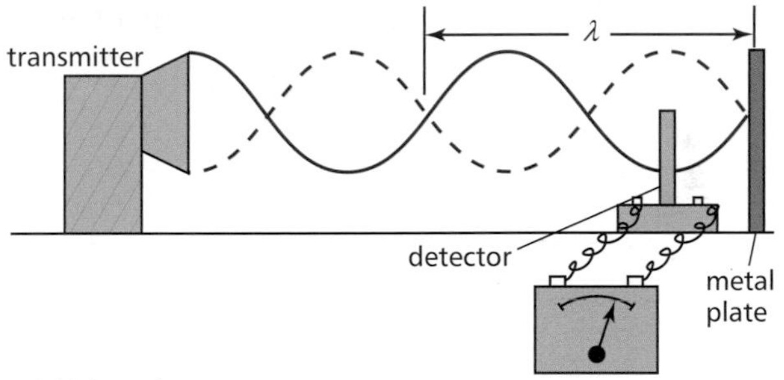

Figure 4 *Using microwaves*

Summary test 8.6

1 **a** Sketch the stationary wave pattern seen on a rope, when there is a node at either end and an antinode at its centre.

b If the rope in part **a** is 4.0 m in length, calculate the wavelength of the waves on the rope for the pattern you have drawn.

2 The stationary wave pattern shown in Figure 5 is set up on a rope of length 3.0 m.

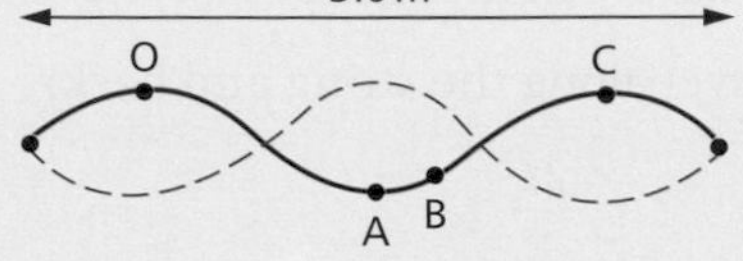

Figure 5

a Calculate the wavelength of these waves.

b State the phase difference between the particle vibrating at O and the particle vibrating at:

i A

ii B

iii C

3 State two differences between a stationary wave and a progressive wave in terms of the vibrations of the particles.

4 Microwaves from a transmitter are reflected back to the transmitter using a flat metal plate. A detector is moved along the line between the transmitter and the metal plate. The detector signal is zero at positions 15 mm apart.

a Explain why the signal is zero at certain positions.

b Calculate the wavelength of the microwaves.

8.7 More about stationary waves on strings

Learning outcomes

On these pages you will learn to:

- explain the stationary wave patterns formed on a string
- know how the frequencies of the stationary wave patterns on a string relate to each other

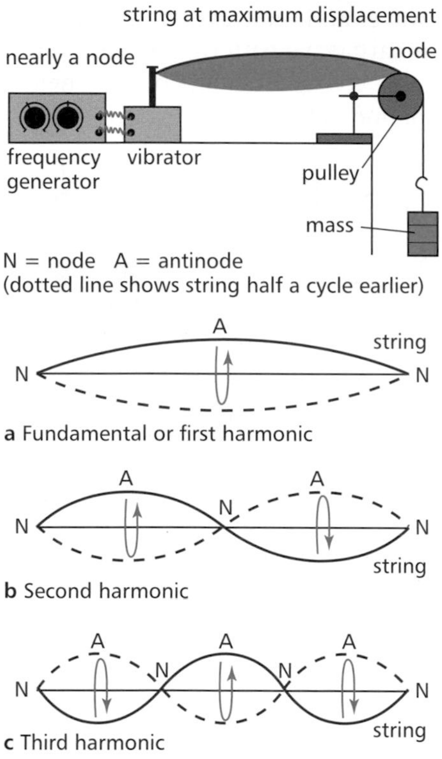

Figure 1 *Stationary waves on a string*

Stationary waves on a vibrating string

A controlled arrangement for producing stationary waves is shown in Figure 1. A string or wire is tied at one end to a mechanical vibrator, connected to a frequency generator. The other end of the string passes over a pulley and supports a mass, which keeps the tension in the string constant. As the frequency of the generator is increased from a very low value, different stationary wave patterns are seen on the string. In every case, the length of string between the pulley and the vibrator has a node at either end.

- The **fundamental** or **first harmonic pattern** of vibration is seen at the lowest possible frequency. This has an antinode at the middle, as well as a node at either end. Because the length L of the vibrating section of the string is between adjacent nodes, the wavelength of the waves that form this pattern, the fundamental wavelength λ_1, is equal to $2L$. Therefore, the fundamental or first harmonic frequency:

$$f_1 = \frac{v}{2L}$$

where v is the speed of the progressive waves on the wire.

- The next stationary wave pattern, the second harmonic, is where there is a node at the middle so the string is in two loops. The wavelength of the waves that form this pattern λ_2 is L, because each loop has a length of half a wavelength. Therefore, the frequency of the second harmonic vibrations:

$$f_2 = \frac{v}{\lambda_2} = \frac{v}{L} = 2f_1$$

- The next stationary wave pattern, the **third harmonic**, is where there are nodes at a distance of $\frac{1}{3}L$ from either end and an antinode at the middle. The wavelength of the waves that form this pattern λ_3 is $\frac{2}{3}L$, because each loop has a length of half a wavelength. Therefore, the frequency of the third harmonic vibrations:

$$f_3 = \frac{v}{\lambda_3} = \frac{3v}{2L} = 3f_1$$

In general, stationary wave patterns occur at frequencies f_1, $2f_1$, $3f_1$, $4f_1$, etc., where f_1 is the frequency of the first harmonic vibrations. This is the case in any vibrating linear system that has a node at either end.

Explanation of the stationary wave patterns on a vibrating string

In the arrangement shown in Figure 1, consider what happens to a wave sent out by the vibrator. The crest reverses its phase when it reflects at the fixed end, and travels back along the string as a trough. When it reaches the vibrator, it reflects and reverses phase again, travelling away from the vibrator once more as a crest. If this crest is reinforced by a crest created by the vibrator, the amplitude of the wave is increased. This is how a stationary wave is formed. The key condition is that the time taken for a wave to travel along the string and back should be equal to the time taken for a whole number of cycles of the vibrator:

- The time taken for a wave to travel along the string and back:

$$t = \frac{2L}{v}$$

where v is the speed of the waves on the string.

- The time taken for the vibrator to pass through a whole number of cycles:

$$t = \frac{m}{f}$$

where f is the frequency of the vibrator and m is a whole number (i.e. $m = 1$ or 2 or 3, etc.).

Therefore the key condition may be expressed as $\frac{2L}{v} = \frac{m}{f}$

Rearranging this equation gives:

$$f = \frac{mv}{2L} = mf_1 \text{ and } \lambda = \frac{v}{f} = \frac{2L}{m}$$

In other words:

- Stationary waves are formed at frequencies f_1, $2f_1$, $3f_1$, etc.
- The length of the vibrating section of the string $L = \frac{m\lambda}{2}$ = whole number of half wavelengths.

Making music

A guitar produces sound when its strings vibrate as a result of being plucked. In an electronic guitar, the vibrations of the string are detected by a microphone which produces electrical waves. These are amplified and converted back to sound waves in a loudspeaker. In an acoustic guitar, the string vibrations make the guitar surfaces vibrate and send out sound waves.

When a stretched string or wire vibrates, its **pattern of vibration** is a mix of its fundamental (i.e. first harmonic) mode of vibration and the harmonics of higher frequencies. The sound produced is the same mix of frequencies which change with time as the pattern of vibration changes. A **spectrum analyser** can be used to show how the intensity of a sound varies with frequency and with time. Combined with a **sound synthesiser**, the original sound can be altered by amplifying or suppressing different frequency ranges.

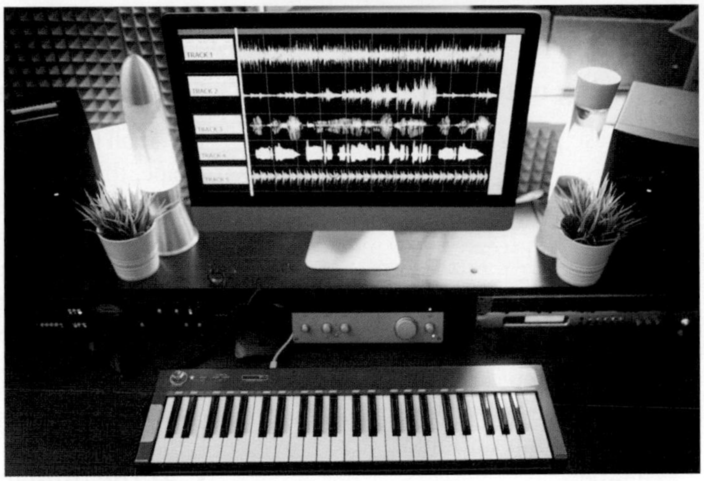

Figure 2 *A sound synthesiser*

The **pitch** of a note produced by a stretched string can be altered by changing the **tension** of the string, or by altering its **length**. The pitch is raised by raising the tension or shortening the length. Lowering the tension or increasing the length lowers the pitch. By changing the length or altering the tension, a vibrating string or wire can be tuned to the same pitch as a vibrating tuning fork. However, the sound from a vibrating string includes the frequencies of higher harmonics as well as the first harmonic (i.e. fundamental) frequency, whereas a tuning fork vibrates only at a single frequency. The wire is tuned when its first harmonic frequency is the same as the tuning fork frequency. A simple visual check is to balance a small piece of paper on the wire at its centre, and place the base of the vibrating tuning fork on one end of the wire. If the wire is tuned to the tuning fork, it will vibrate in its first harmonic mode and the piece of paper will fall off.

Summary test 8.7

1 A stretched wire of length 0.80 m vibrates in its first harmonic mode at a frequency of 256 Hz. Calculate:

a the wavelength of the progressive waves on the wire,

b the speed of the progressive waves on the wire.

2 The first harmonic frequency of vibration of a stretched wire is inversely proportional to the length of the wire. For the wire in question **1**, calculate the length of the wire needed to produce a frequency of:

a 512 Hz

b 384 Hz

Assume that the tension of the wire is not changed.

3 Describe how you would expect the note from a vibrating guitar string to change if the string is:

a shortened,

b tightened.

4 The speed, v, of the progressive waves on a stretched wire varies with the tension T in the wire, in accordance with the equation $v = \left(\frac{T}{\mu}\right)^{1/2}$, where μ (Greek *mu*) is the mass per unit length of the wire. Use this formula to explain why a nylon wire and a steel wire, of the same length, diameter and tension, produce notes of different pitch. State, with a reason, which wire would produce the higher pitch.

8.8 Stationary waves in pipes

Learning outcomes

On these pages you will learn to:

- explain the formation of a stationary wave in a pipe closed at one end
- explain the formation of a stationary wave in a pipe open at both ends
- identify nodes and antinodes
- determine the wavelength of sound using stationary waves in a pipe

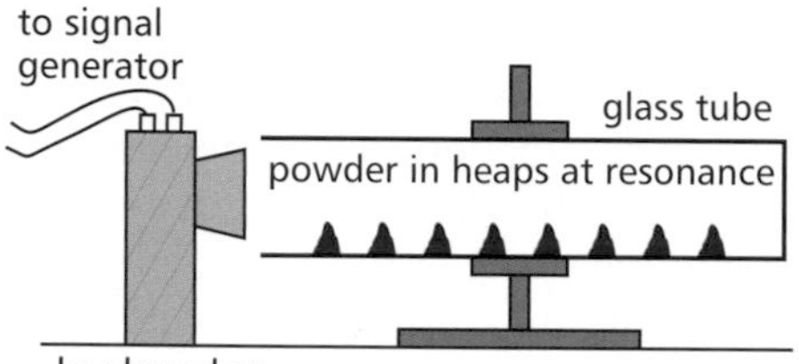

***Figure 1** Acoustic resonance*

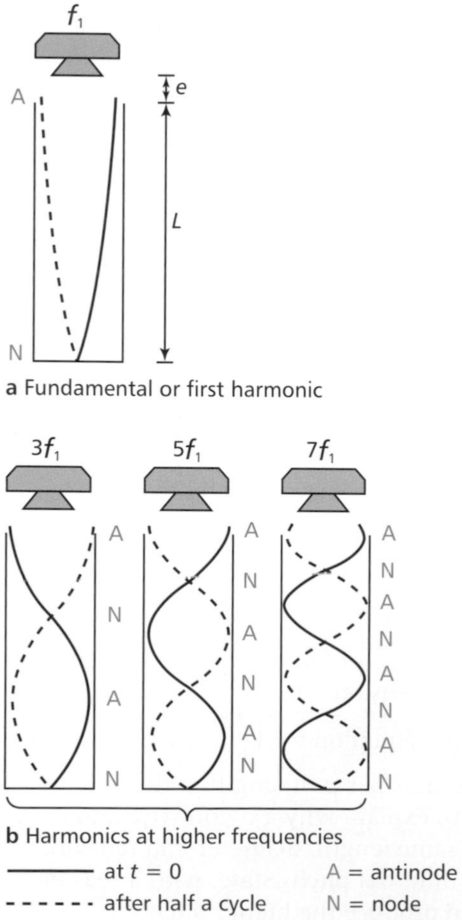

***Figure 2** Resonances in a pipe closed at one end*

Stationary waves in a pipe closed at one end

If a small loudspeaker connected to a signal generator is placed near the open end of a pipe, the pipe resonates with sound at certain frequencies as the signal frequency is changed. Each **resonance** is due to stationary sound waves in the pipe. Sound waves from the transmitter at the open end are reflected at the other end of the pipe. The reflected waves, and the waves directly from the transmitter, pass along the pipe in opposite directions and therefore form a stationary wave pattern at certain frequencies. Figure 1 shows how this effect can be seen, as well as heard. The powder forms small heaps at the nodes, where there are no vibrations. Antinodes occur mid-way between the nodes and just beyond the open end. For greatest effect, the position of the loudspeaker needs to be at the antinode just beyond the open end.

- **At the fundamental frequency** or **first harmonic**, the lowest frequency at which the pipe resonates with sound, there is a node at the closed end and an antinode at the open end with no nodes or antinodes between. As shown in Figure 2, the pipe length $L = \frac{1}{4}\lambda_1$ (ignoring the small end-correction, e, at the open end). Therefore, the first harmonic wavelength $\lambda_1 = 4L$, the first harmonic frequency, $f_1 = \dfrac{v}{4L}$, where v is the speed of sound in the pipe.
- **Harmonics at higher $3f_1$, $5f_1$, $7f_1$, etc.** In each case, there is a node at the closed end and an antinode at the open end, with one or more equally spaced nodes or antinodes between. As shown in Figure 2:

1 At the **third harmonic**, the pipe length $L = \frac{3}{4}\lambda_1$ (ignoring the small end-correction at the open end).

Therefore, the third harmonic wavelength $\lambda_2 = \dfrac{4L}{3}$, where L is the pipe length, the third harmonic frequency, $f_2 = \dfrac{3v}{4L} = 3f_1$

2 At the **fifth harmonic**, the pipe length $L = \frac{5}{4}\lambda_3$ (ignoring the small end-correction).

Therefore, the fifth harmonic wavelength $\lambda_3 = \dfrac{4L}{5}$, where L is the pipe length,

and the fifth harmonic frequency, $f_3 = \dfrac{5v}{4L} = 5f_1$

Further resonances occur at $7f_1$, $9f_1$, etc., corresponding to an odd number of quarter wavelengths equal to the pipe length. Note there are no even harmonics for a pipe closed at one end.

To determine the wavelength of sound in a pipe closed at one end, a vertical glass tube is used as the pipe, as shown in Figure 3. The length of the air column in the pipe is varied by using the valves to alter the level of water in the pipe. A small loudspeaker (or a tuning fork) is used to direct sound waves of constant frequency into the open end of the pipe.

- The water level in the pipe is gradually lowered from near the top of the pipe until the pipe resonates at its shortest length. The length, L_1, of the air column at this resonance is then measured. This length plus the small end-correction, e, is equal to one-quarter of a wavelength.
- The water level is then lowered further until the pipe resonates with sound at the third harmonic. The length, L_2, of the air column at this resonance is then measured. This length plus the small end-correction is equal to three-quarters of a wavelength.

The difference between the two length measurements, $L_2 - L_1$, is equal to half of the wavelength of the sound waves in the pipe. Hence the wavelength λ $(= 2(L_2 - L_1))$ can be calculated.

Stationary waves in a pipe open at both ends

Sound waves travelling along the pipe partially reflect at the open end because the speed changes at the exit. An antinode is formed at either end. Therefore the pipe resonates with sound at any frequency corresponding to the pipe length L equal to a whole number of half wavelengths (the distance between two adjacent antinodes). Figure 4 shows the stationary wave patterns in this situation.

- **At the fundamental frequency** or **first harmonic**, there is an antinode at either end of the pipe with a node mid-way. As shown in Figure 4, the pipe length $L = \frac{1}{2}\lambda_1$ (ignoring the small end-corrections at each end).
  Therefore, the first harmonic wavelength $\lambda_1 = 2L$, where L is the pipe length,

  $$\text{the first harmonic frequency, } f_1 = \frac{v}{2L},$$

  where v is the speed of sound in the pipe.
- **Harmonics at higher frequencies** occur at frequencies $2f_1$, $3f_1$, $4f_1$, etc. In each case, there is an antinode at either end, with one or more equally spaced nodes and antinodes between, as shown in Figure 4,

1 At the **second harmonic**, the wavelength λ_2 = the pipe length L (ignoring the small end-corrections).

Therefore, the first second harmonic frequency, $f_2 = \frac{v}{L} = 2f_1$

2 At the **third harmonic**, the wavelength λ_3 is such that the pipe length $L = \frac{3}{2}\lambda_3$(ignoring the small end-corrections).

Therefore, the third harmonic wavelength $\lambda_3 = \frac{2L}{3}$, where L is the pipe length,

the third harmonic frequency, $f_3 = \frac{3v}{2L} = 3f_1$

Further resonances occur at $4f_1$, $5f_1$, etc., corresponding to a whole number of half-wavelengths equal to the pipe length.

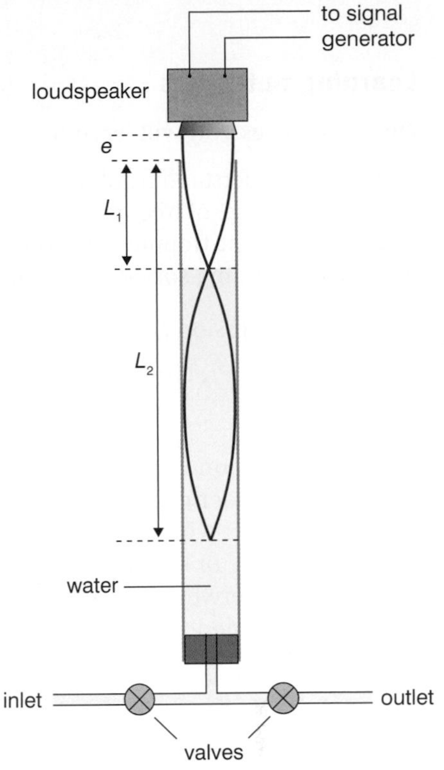

Figure 3 *Using stationary waves to measure the wavelength of sound*

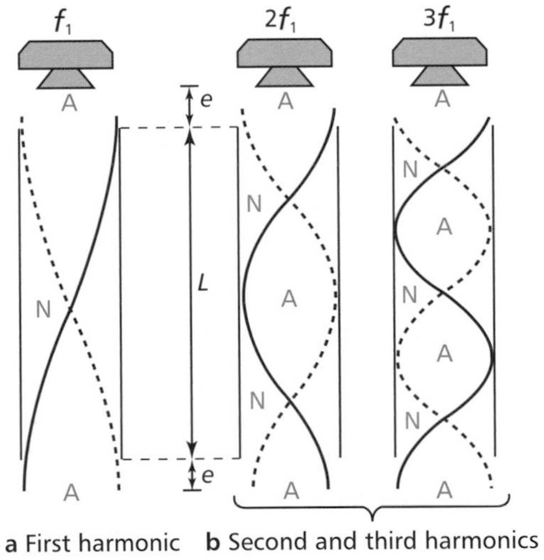

Figure 4 *Stationary waves in an open pipe*

Summary test 8.8

1 A pipe, of length 0.60 m, is closed at one end and open at the other end. The speed of sound in the pipe is 340 m s^{-1}. Estimate:

a its first harmonic wavelength and frequency,

b the frequency of its third harmonic.

2 A pipe of length 1.50 m, closed at one end and open at the other end, resonates at a frequency of 170 Hz. The speed of sound in the pipe is 340 m s^{-1}. Calculate:

a the wavelength of the sound waves in the pipe,

b the first harmonic frequency of this pipe.

3 A pipe, of length 2.40 m, is open at both ends. The speed of sound in the pipe is 340 m s^{-1}. Calculate:

a its first harmonic frequency,

b its second harmonic frequency.

4 A wind organ has pipes open at both ends, which are of lengths from 0.25 m to 2.50 m. Calculate the range of first harmonic frequencies from these pipes. The speed of sound in the pipes is 340 m s^{-1}.

8.9 The Doppler effect

Learning outcomes

On these pages you will learn to:

- show an understanding of the Doppler effect and recall that the change of frequency is called a Doppler shift
- appreciate that Doppler shift is observed with all waves, including sound and light
- use the expression $f_0 = \dfrac{f_s v}{v \pm v_s}$ for a source of sound waves moving relative to a stationary observer

When a vehicle sounding its horn speeds past, the note heard by a bystander changes pitch. When approaching, the pitch is higher; when moving away, the pitch is lower. This is an example of the Doppler effect: when there is relative motion between a source of sound and an observer, the observed frequency differs from the emitted frequency.

The Doppler effect occurs with any form of wave motion, provided there is relative motion between the observer and the source of the waves along the line between them. To understand the cause of the Doppler effect, consider a source of waves moving at constant speed along a straight line, as in Figure 1, emitting waves of a constant frequency f_s.

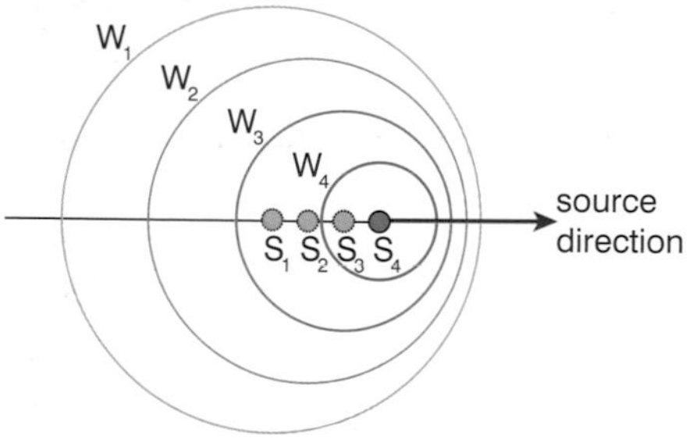

Figure 1 *The Doppler effect*

In Figure 1, the source has moved from position S_1 to S_4, where it emits wavefront W_4. It emitted wavefront W_1 when it was at S_1, wavefront W_2 when it was at S_2 and wavefront W_3 when it was at S_3. Because the source is moving in one direction, the wavefronts ahead of the source are bunched together and the wavefronts behind it are further apart.

More generally, if the wave speed is v and the speed of the source is v_s, then in one second the source emits f_s wavefronts and moves through a distance v_s, the leading wavefront having moved through a distance v.

Ahead of the source, the waves are bunched up because the source is moving in the same direction as the waves. So the distance from the leading wavefront to the source is equal to $v - v_s$ after the source has emitted f_s wavefronts. As this distance is equal to f_s wavelengths, the wavelength λ of the waves moving in the same direction as the source is given by:

$$\lambda = \frac{v - v_s}{f_s}$$

A stationary observer ahead of the source would therefore detect waves of frequency f_0, where:

$$f_0 = \frac{\text{wave speed}}{\text{wavelength}} = \frac{v}{\left(\dfrac{v - v_s}{f_s}\right)} = \frac{v f_s}{(v - v_s)}$$

Note that the observed frequency f_0 is greater than the source frequency f_s because:

$$\frac{v}{v - v_s} > 1$$

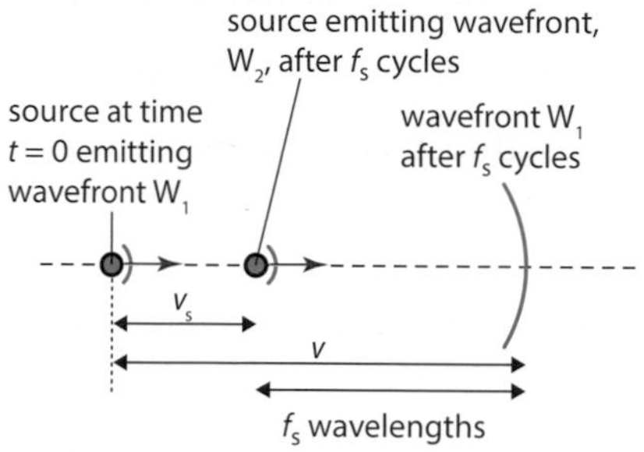

Figure 2 *Ahead of the source*

Behind the source, the waves are stretched out because the waves are moving in the opposite direction to the source direction. So the distance from the leading wavefront to the source is equal to $v + v_s$ after the source has emitted f_s wavefronts. As this distance is equal to f_s wavelengths, the wavelength λ of the waves moving in the opposite direction to the source is given by:

$$\lambda = \frac{v + v_s}{f_s}$$

A stationary observer behind the source would therefore detect waves of frequency f, where:

$$f_0 = \frac{\text{wave speed}}{\text{wavelength}} = \frac{v}{\left(\dfrac{v + v_s}{f_s}\right)} = \frac{v f_s}{(v + v_s)}$$

The equation shows that the observed frequency f_o is less than the source frequency f_s because:

$$\frac{v}{(v + v_s)} < 1$$

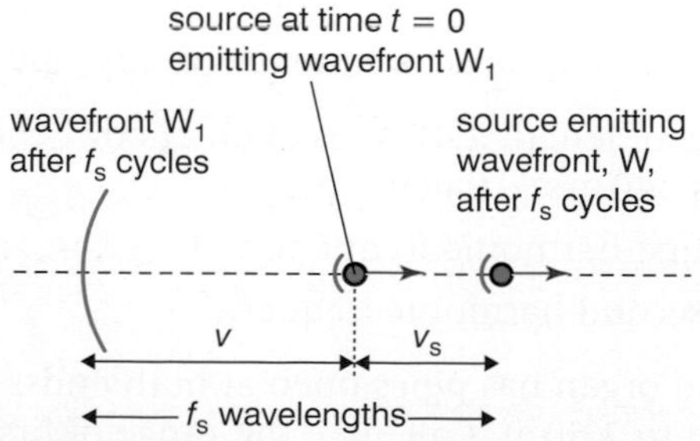

Figure 3 *Behind the source*

Notes

1 The above equations may be summarised as a single equation:

$$f_0 = \frac{vf_s}{(v \pm v_s)}$$

where the – sign applies ahead of the source and the + sign applies behind it.

2 The change of frequency:

$$\Delta f = f_0 - f_s = \frac{vf_s}{(v \pm v_s)} - f_s$$

$$= \frac{vf_s}{(v \pm v_s)} - \frac{(v \pm v_s)f_s}{(v \pm v_s)} = \frac{(v_s f_s)}{(v \pm v_s)}$$

where the – sign applies when the observer is ahead of the source (i.e. the source is moving towards the observer) and the + sign applies when the observer is behind the source (i.e. the source is moving away from the observer). Notice the difference in the top line of this equation and the equation in Note 1.

3 For electromagnetic waves, where the speed of the source is much less than the speed of the waves, it can be shown that the change of frequency (or **Doppler shift**)

$$\Delta f = \frac{v_s f_s}{c},$$

where c is the speed of electromagnetic waves. You will meet this in Chapter 23 Astrophysics and cosmology'.

Worked example

A train sounds its horn as it travels at a speed of $30\,\text{m}\,\text{s}^{-1}$ towards a level crossing. The sound is emitted from the horn at a frequency of 840 Hz. Calculate:

a the wavelength of the sound waves travelling towards the level crossing

b the observed frequency of the sound waves at the level crossing.

The speed of sound in air = $340\,\text{m}\,\text{s}^{-1}$.

a $\lambda = \frac{v - v_s}{f_s} = \frac{340 - 30}{840} = 0.37\,\text{m}$

b $f_0 = \frac{v}{\lambda} = \frac{340}{0.37} = 920\,\text{Hz}$

Speed measurement using the Doppler effect

A radar speed camera emits pulses of microwave radiation and detects any pulses reflected from a vehicle back to the camera. If the vehicle is moving towards or away from the camera, the frequency of such a reflected pulse differs from the emitted pulse. The difference in frequency is proportional to the speed of the vehicle and so can be used to measure the vehicle speed.

The same principle is used in:

- the Doppler anemometer, which uses pulses of laser light to measure wind speed by detecting pulses reflected by small particles in the wind
- Doppler echocardiography, which uses ultrasound pulses to create an image of the heart showing the speed and direction of blood flow inside it.

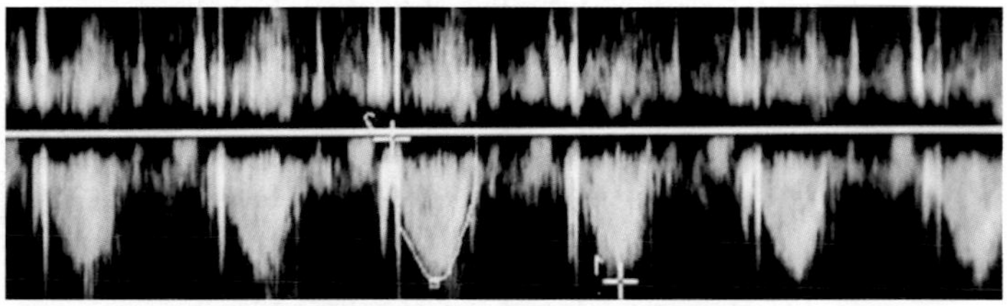

Figure 4 *An echocardiogram*

Summary test 8.9

The speed of sound in air = $340\,\text{m}\,\text{s}^{-1}$.

The speed of electromagnetic waves in a vacuum = $3.0 \times 10^8\,\text{m}\,\text{s}^{-1}$.

1 A car on a motorway sounds its horn as it approaches an overhead bridge at a speed of $28\,\text{m}\,\text{s}^{-1}$. The horn emits sound waves at a frequency of 1100 Hz.

a Calculate the frequency of the sound heard by an observer on the bridge as the car approaches.

b State and explain what the frequency of the sound heard by the observer would be when the car was directly below the bridge.

2 A train is travelling at a speed of $40\,\text{m}\,\text{s}^{-1}$ on a straight section of track between two overhead bridges, X and Y, when it sounds its horn. An observer on bridge X hears the sound from the horn at a lower frequency than does an observer on bridge Y.

a State whether the train is moving towards X or towards Y and explain why.

b The observer at X hears sound of frequency 950 Hz from the horn. Calculate the frequency of the sound emitted by the horn.

3 The Sun has an equatorial diameter of $1.4 \times 10^9\,\text{m}$ and spins at its equator at a steady rate of one revolution every 25 days.

a Calculate the speed of rotation of an atom at the Sun's equator.

b Calculate the change in frequency observed in light of frequency $5.0 \times 10^{14}\,\text{Hz}$ emitted by atoms on the Sun's equator when they are moving away from the Earth at the speed calculated in **a**.

4 The Doppler anemometer uses pulses of laser light to measure wind speed by detecting pulses reflected by small particles in the wind. Explain why: **a** ultrasound, and **b** microwaves would be unsuitable for use instead of laser pulses.

Launch additional digital resources for the chapter

1 a i Explain the difference between longitudinal waves and transverse waves.

ii State one example of a longitudinal wave.

iii State one example of a transverse wave.

b With the aid of a diagram, explain what is meant by:

i diffraction of a wave,

ii refraction of a wave.

2 a Explain what is meant by a 'polarised' transverse wave.

b Explain why sound waves cannot be polarised, whereas light waves can.

3 An oscilloscope is used to display the signal from a microphone when sound waves of constant frequency are directed at the microphone (see graph below).

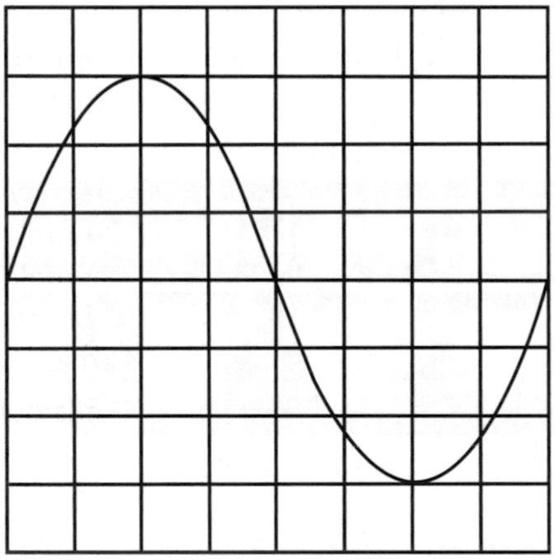

Figure 3.1

a The oscilloscope time base is set at $0.5\,\mathrm{ms\,cm^{-1}}$. Calculate the frequency of the sound waves.

b Sketch the display you would observe on the oscilloscope if the frequency of the sound waves was doubled with no change of loudness.

4 a Explain what is meant by the 'superposition' of waves.

b Two small loudspeakers, 1.2 m apart, are connected to a signal generator, as shown opposite. The signal generator is adjusted to produce an alternating voltage of constant amplitude, at a frequency of 1.0 kHz.

i An observer moves along the line XY at a perpendicular distance of 2.0 m from the loudspeakers, as shown. The intensity of the sound rises and falls as she moves along the line. Explain this effect.

ii In part **i**, the observer notes that successive minima are further apart along the line XY when the test is repeated at a lower frequency. Explain this observation.

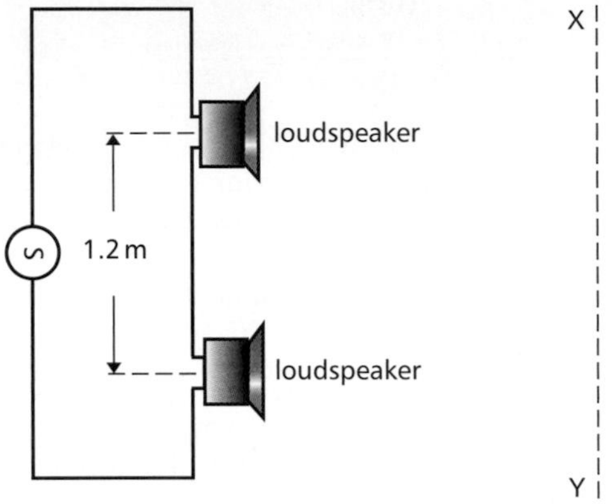

Figure 4.1

5 Microwaves from a transmitter were directed at two narrow gaps between three metal plates. A detector was placed on the other side of the plates at P, as shown below.

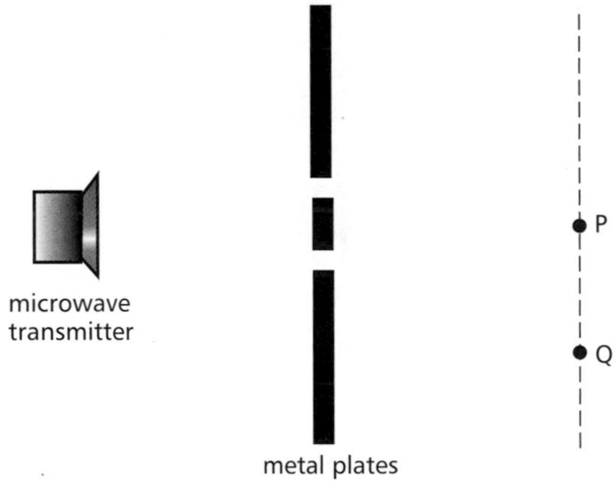

Figure 5.1

a The detector was moved along a line parallel to the plates to Q. The signal from the detector decreased from a maximum at P to a minimum mid-way between P and Q, then to a maximum at Q.

i Explain why there was a maximum at P and at Q.

ii Explain why there was a minimum mid-way between P and Q.

b Explain how the detector signal would change when the detector is mid-way between P and Q and a metal plate is placed over one of the gaps.

6 A stationary wave pattern of a wire, of length 0.60 m, vibrating at a frequency of 300 Hz is shown below.

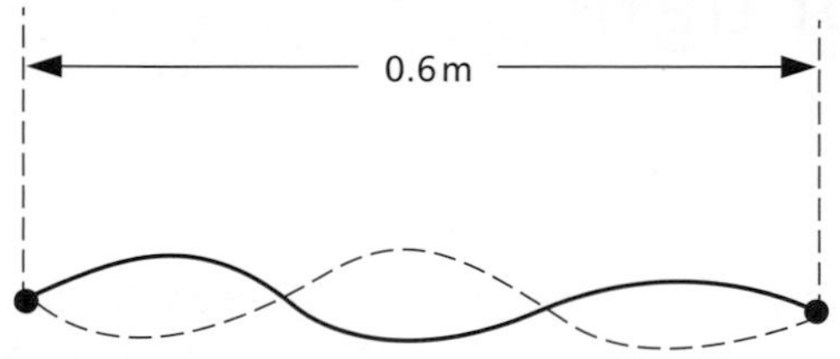

Figure 6.1

a In terms of amplitude and phase difference, compare the motion of the wire 0.10 m from the left-hand end with:

i its motion at the midpoint,

ii 0.10 m from the other end.

b i Calculate the wavelength of the waves on the wire.

ii The frequency of vibrations is increased to 400 Hz, and a different stationary wave pattern is produced. Sketch the pattern you would expect to observe at 400 Hz.

7 Sound from a small loudspeaker connected to a signal generator was directed into a vertical glass tube containing some water, as shown below. The length of the air column in the tube could be varied by altering the level of water in the tube.

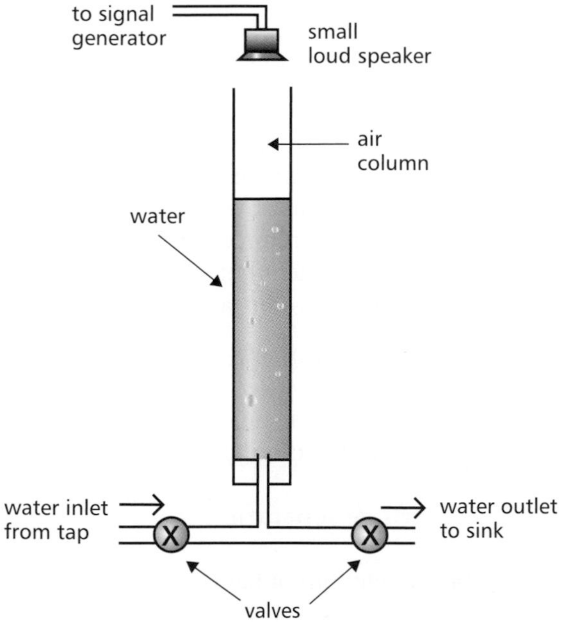

Figure 7.1

a The length of the air column was gradually increased from a few centimetres, by lowering the level of water in the tube slowly. The tube resonated with sound at certain positions of the water level. Explain why this effect happened only at certain lengths of the air column.

b The tube resonated with sound of frequency 300 Hz when the length of the air column was 270 mm and 820 mm.

i Diagram **a** shows how the amplitude of the sound waves in the tube varied with position along the tube when the air column length was 270 mm. Using a copy of diagram **b**, sketch the corresponding pattern when the air column length was 820 mm.

ii Show that the wavelength of the sound in the tube was 1.10 m, and hence calculate the speed of sound in the tube.

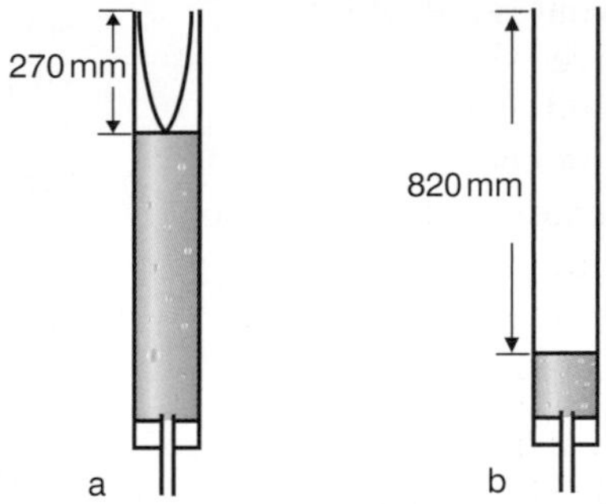

Figure 7.2

8 An open-ended pipe, of length 0.80 m, resonates with sound at a frequency of 200 Hz. The speed of sound in the pipe is $330\,\text{m}\,\text{s}^{-1}$.

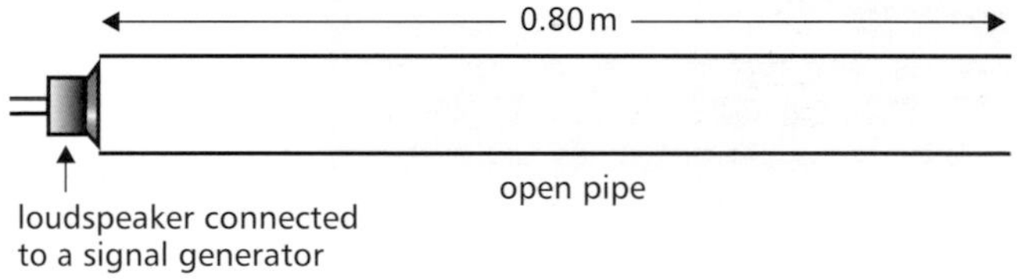

Figure 8.1

a i Calculate the wavelength of sound waves of frequency 200 Hz in the pipe.

ii Sketch the stationary wave pattern in the pipe when it resonates at 200 Hz.

b i Calculate the next highest frequency at which the pipe would resonate if the frequency was gradually increased.

ii Explain why the pipe resonates at certain frequencies only.

9 a State two differences between the vibrations of a stationary wave and those of a progressive wave.

b i Figure 9.1 shows a progressive wave travelling from left to right. Describe how the vibrations of the particle at point P compare with the vibrations of the particle at Q, and state the phase difference between their vibrations.

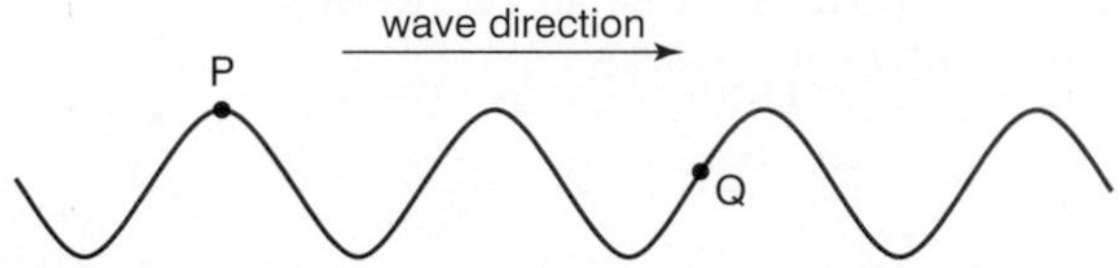

Figure 9.1

ii Describe two essential conditions for the formation of a stationary wave pattern from two progressive waves.

9 Interference

9.1 Interference of light

Learning outcomes

On these pages you will learn to:

- explain what is meant by interference
- describe and explain the double-slit experiment using light to demonstrate two-source interference
- recall and solve problems using the equation $\lambda = ax/D$ for double-slit interference using light

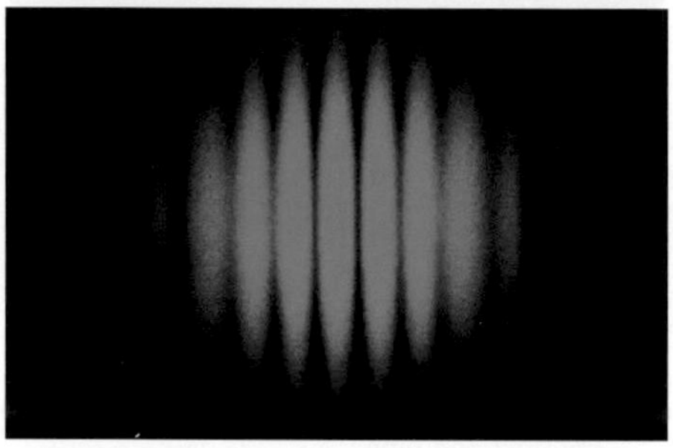

Figure 1 *Double-slit interference pattern as seen on a screen*

The wave nature of light

Young's double-slit experiment

The wave nature of light was first suggested by Christiaan Huygens in the seventeenth century, but it was rejected at the time in favour of Sir Isaac Newton's corpuscular theory of light. Newton considered that light was composed of tiny particles he referred to as corpuscles and he was able to explain reflection and refraction using his theory. Huygens was also able to explain reflection and refraction using his wave theory. However, the two theories differed about whether or not light in a transparent substance travels faster (as predicted by Newton's theory) or slower (as predicted by Huygens' theory) than in air. Because of Newton's much stronger scientific reputation, Newton's theory of light remained unchallenged for over a century, until 1803, when Thomas Young at the Royal Institution first demonstrated interference of light. Even so, Newton's theory of light was not rejected in favour of Huygens' wave theory until several decades later when the speed of light in water was measured and found to be slower than in air.

An arrangement like the one used by Thomas Young to observe interference is shown in Figure 2. Young would have used a candle instead of a light bulb to illuminate a narrow single slit. A pair of narrow slits, referred to as the 'double slits', is illuminated by light from the single slit. Alternate bright and dark fringes can be seen using a microscope, or on a white screen placed where the diffracted light from the double slits overlaps. The fringes are evenly spaced and parallel to the double slits.

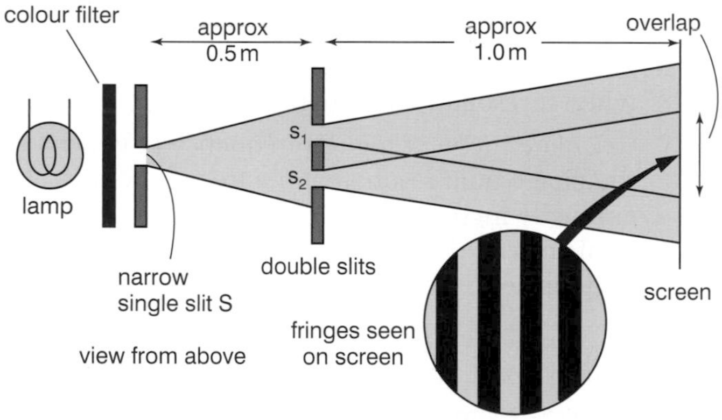

Figure 2 *Young's double-slit experiment*

The fringes are formed due to **interference of light** from the two slits:

- Where a **bright fringe** is formed, the light from one slit reinforces the light from the other slit. In other words, the light waves from each slit arrive **in phase** with each other.
- Where a **dark fringe** is formed, the light from one slit cancels the light from the other slit. In other words, the light waves from the two slits arrive **180° out of phase**.

The distance from the centre of a bright fringe to the centre of the next bright fringe is called the **fringe separation,** x. This depends on the slit spacing a and the distance D from the slits to the screen, in accordance with the equation

$$\textbf{fringe separation, } x = \frac{\lambda D}{a}$$

where λ is the wavelength of light.

Note

The slit spacing is the distance between the **centres** of the slits.

The equation shows that the fringes become more widely spaced if:

- the distance D from the slits to the screen is increased,
- the wavelength λ of the light used is increased,
- the slit spacing, a, is reduced.

The theory of the double-slit equation

Consider the two slits S_1 and S_2 shown in Figure 3. At a point P on the screen where the fringes are observed, light emitted from S_1 arrives later than light from S_2 emitted at the same time. This is because the distance S_1P is greater than the distance S_2P. The difference between distances S_1P and S_2P is referred to as the **path difference**.

- **For reinforcement at P**, the path difference $S_1P - S_2P = m\lambda$, where $m = 0, 1, 2$, etc.

  Therefore, light emitted simultaneously from S_1 and S_2 arrives in phase at P, if reinforcement occurs at P.
- **For cancellation at P**, the path difference $S_1P - S_2P = (m + \frac{1}{2})\lambda$, where $m = 0, 1, 2$, etc.

  Therefore, light emitted simultaneously from S_1 and S_2 arrives at P out of phase by 180°, if cancellation occurs at P.

In Figure 3, a point Q along line S_1P has been marked such that $QP = S_2P$. Therefore, the path difference $S_1P - S_2P$ is represented by the distance S_1Q.

Because triangles S_1S_2Q and MOP are very nearly similar in shape, where M is the midpoint between the two slits and O is the midpoint of the central bright fringe of the pattern, then:

$$\frac{S_1Q}{S_1S_2} = \frac{OP}{OM}$$

If P is the mth bright fringe from the centre (where $m = 0$, 1, 2, etc.), then $S_1Q = m\lambda$ and $OP = mx$, where x is the distance between centres of adjacent bright fringes.

Also, OM = distance D and S_1S_2 = slit spacing a.

Therefore, $$\frac{m\lambda}{a} = \frac{mx}{D}$$

Rearranging this equation gives: $$\boldsymbol{\lambda = \frac{ax}{D}}$$

By measuring the slit spacing, a, the fringe separation, x, and the slit–screen distance D, the wavelength λ of the light used can be calculated. The formula is valid only if the fringe separation, a, is much less than the distance D from the slits to the screen. This condition is to ensure the triangles S_1S_2Q and MOP are very nearly similar in shape.

Note

Light wavelengths are usually expressed in **nanometres (nm)**, where $1\,\text{nm} = 10^{-9}\,\text{m}$.

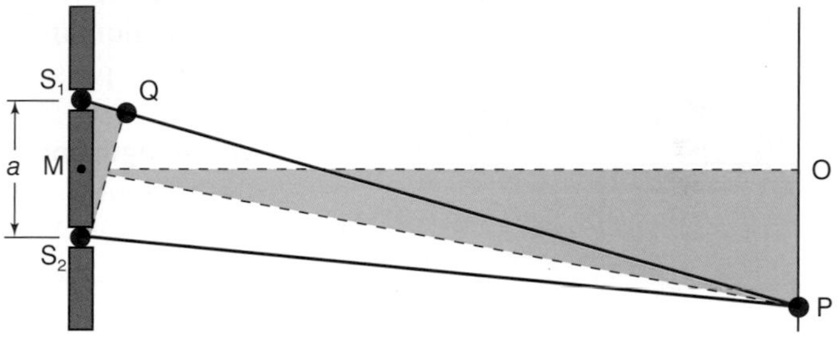

Figure 3 *The theory of the double-slit experiment*

Summary test 9.1

1 In a double-slit experiment using red light, a fringe pattern is observed on a screen at a fixed distance from the double slits. How would the fringe pattern change if:

   a the screen was moved closer to the slits,

   b one of the double slits was blocked completely?

2 The following measurements were made in a double-slit experiment:

   Slit spacing, $a = 0.4$ mm; fringe separation, $x = 1.1$ mm; slit–screen distance, $D = 0.80$ m.

   Calculate the wavelength of light used.

3 In question **2**, the double slits were replaced by a pair of slits with a slit spacing of 0.5 mm. Calculate the fringe separation for the same slit–screen distance and wavelength.

4 The following measurements were made in a double-slit experiment:

   Slit spacing, $a = 0.4$ mm; fringe separation, $x = 1.1$ mm; wavelength of light used, $\lambda = 590$ nm.

   Calculate the distance from the slits to the screen.

9.2 More about interference

Learning outcomes

On these pages you will learn to:

- explain the terms monochromatic light and coherence
- show an understanding of the conditions required to observe two-source interference
- describe two-source interference fringes using monochromatic light and white light

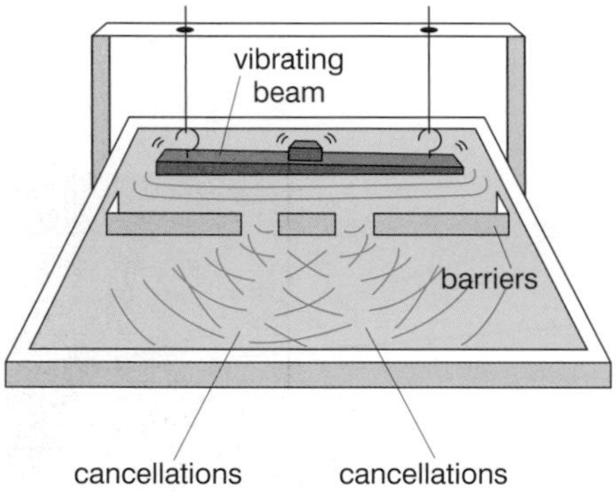

Figure 1 *Interference of water waves*

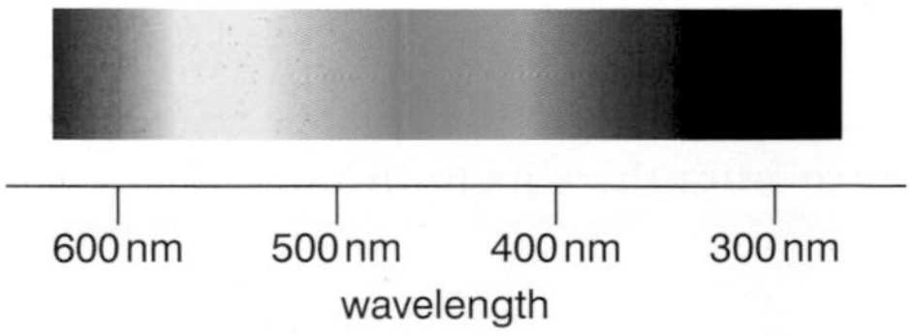

Figure 2 *Wavelength and colour*

Coherence

The double slits are described as **coherent sources,** because they emit light waves with a constant phase difference. This is because each wave crest, or wave trough, from the single slit always passes through one of the double slits a fixed time after it passes through the other slit. The double slits therefore emit wave fronts with a constant phase difference.

The arrangement is like the ripple tank demonstration in Figure 1. Straight waves from the beam vibrating on the water surface diffract after passing through the two gaps in the barrier, and produce an interference pattern where the diffracted waves overlap. If one gap is closer to the beam than the other, each wave front from the beam passes through the nearer gap first. However, the time interval between the same wave front passing through the two gaps is always the same, so the waves emerge from the gaps with a constant phase difference.

Light from two separate light bulbs could not form an interference pattern, because the two light sources emit light waves at random. The points of cancellation and reinforcement would change at random, so no interference pattern is possible.

Wavelength and colour

In the double-slit experiment, the fringe separation depends on the colour of light used. White light is composed of a continuous spectrum of colours, corresponding to a continuous range of wavelengths from about 350 nm for violet light to about 650 nm for red light. Each colour of light has its own wavelength, as shown in Figure 2.

The fringe patterns shown in Figure 3 show that the fringe separation is greater for red light than for blue light. This is because red light has a longer wavelength than blue light. The fringe spacing, x, depends on the wavelength, λ, of the light according to the formula $\frac{x}{D} = \frac{\lambda}{a}$, as explained in Topic 9.1.

Rearranging this formula gives $x = \frac{\lambda D}{a}$. Thus the longer the wavelength of the light used, the greater the fringe separation is.

Light sources

- **Vapour lamps and discharge tubes** produce light with a dominant colour. For example, the sodium vapour lamp produces a yellow/orange glow which is due to light of wavelength 590 nm. Other wavelengths of light are also emitted from a sodium vapour lamp, but the colour due to light of wavelength 590 nm is much more intense than any other colour. A sodium vapour lamp is, in effect, a **monochromatic** light source because its spectrum is dominated by light of a certain colour.
- **Light from a filament lamp or from the Sun** is composed of the colours of the spectrum, and therefore covers a continuous range of wavelengths from about 350 nm to about 650 nm.
- **Light from a laser** is of a specific wavelength, and therefore laser light is highly monochromatic. For example, a helium–neon laser produces red light

Figure 3 *The double slits fringe pattern*

of wavelength 635 nm only. Because a laser beam is almost perfectly parallel and monochromatic, a convex lens can focus it to a very fine spot. The beam power is then concentrated in a very small area. This is why a laser beam is very dangerous if it enters the eye. The **eye lens** would focus the beam on a tiny spot on the **retina** and the intense concentration of light at that spot would destroy the retina.

Always wear safety goggles in the presence of a laser beam. Never look along a laser beam, even after reflection.

Observing Young's fringes

Provided the light source is **not** a laser, a microscope may be used to observe the fringe pattern and to measure the fringe spacing. If so, the plane of viewing of the fringe pattern must be located to measure the distance D from the slits to the fringes accurately. Figure 4 shows how this can be done.

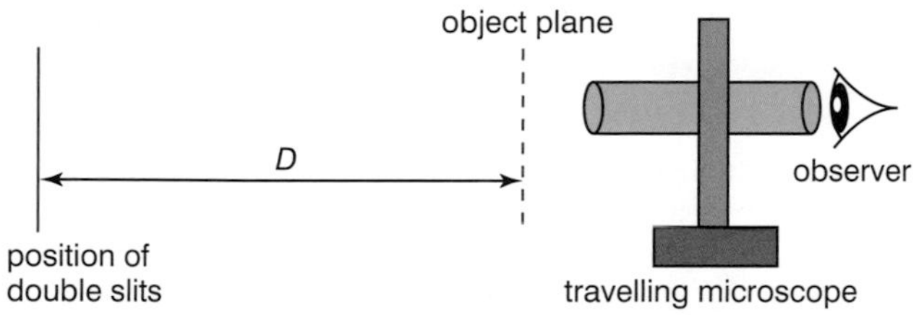

Figure 4 *Measuring the fringes using a microscope*

The contrast between the bright and dark fringes can be improved by narrowing the single slit (see Topic 9.1 Figure 2). If this slit is too wide, each part of it produces a fringe pattern which is displaced slightly from the pattern due to adjacent parts of the single slit. As a result, the dark fringes of the double slit pattern become narrower than the bright fringes, and contrast is lost between the dark and the bright fringes.

If a laser is used as the light source, a screen **must** be used on which to observe the fringe pattern produced by a laser. Never look along a laser beam, even after reflection. Safety goggles must always be worn in the presence of a laser beam.

White light fringes

Figure 3 shows the fringe patterns observed with blue light and with red light. As explained above, the blue light fringes are closer together than the red light fringes. The fringe pattern produced by white light is shown in Figure 5. Each component colour of white light produces its own fringe pattern, each pattern centred on the screen at the same position. As a result:

- The central fringe is **white**, because every colour contributes at the centre of the pattern.
- The inner fringes are tinged with **blue on the inner side and red on the outer side**. This is because the red fringes are more spaced out than the blue fringes, and the two fringe patterns do not overlap exactly.
- The outer fringes merge into an indistinct **background of white light**. This is because, where the fringes merge, different colours reinforce and therefore overlap.

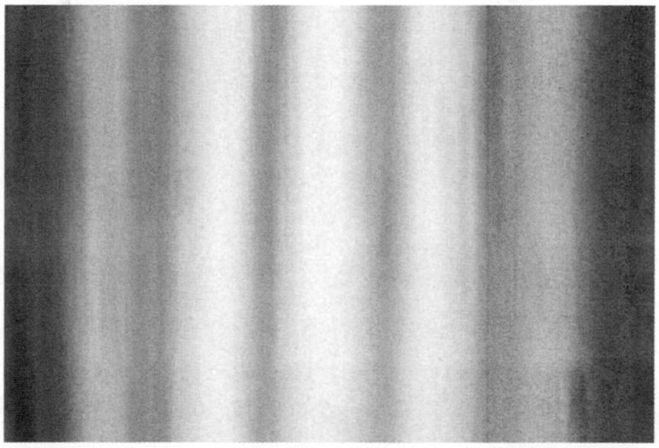

Figure 5 *White light fringes*

Summary test 9.2

1 a Sketch an arrangement that may be used to observe the fringe pattern produced when light from a narrow slit, illuminated by a sodium vapour lamp, is passed through a pair of double slits.

b Describe the fringe pattern you would expect to observe in part **a**.

2 In question **1**, describe how the fringe pattern would change if:

a one of the double slits is blocked,

b the narrow single slit is replaced by a wider slit.

3 Double slit interference fringes are observed using light of wavelength 590 nm and a pair of double slits of slit spacing 0.50 mm. The fringes are observed on a screen at a distance of 0.90 m from the double slits. Calculate the fringe separation of these fringes.

4 Describe and explain the fringe pattern that would be observed in question **3**, if the light source were replaced by a white light source.

The diffraction grating

Learning outcomes

On these pages you will learn to:

- explain how a diffraction grating works
- recall and solve problems using the formula $d \sin\theta = n\lambda$
- describe the use of a diffraction grating to determine the wavelength of light

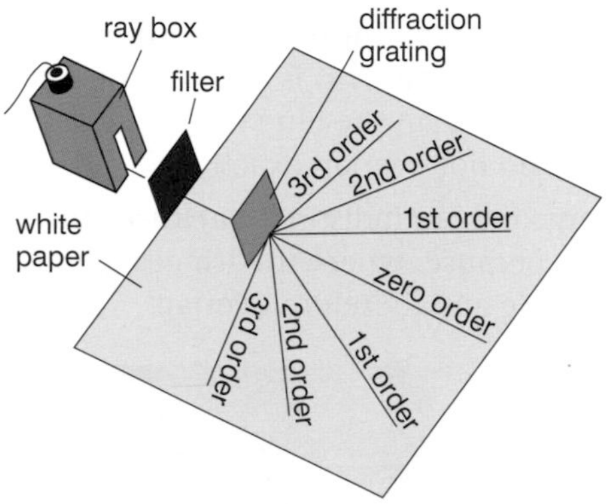

Figure 1 *The diffraction grating*

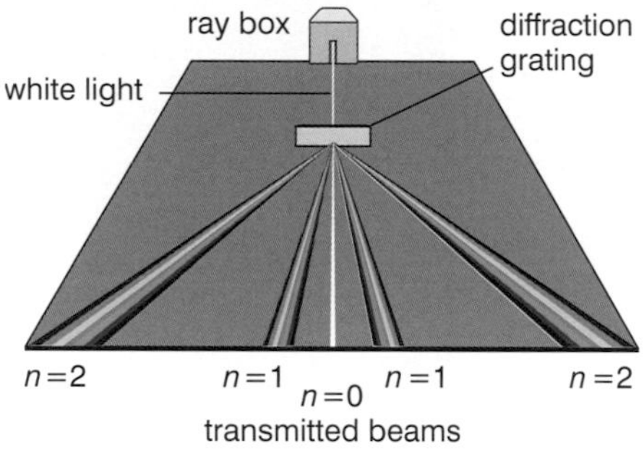

Figure 2 *Using white light*

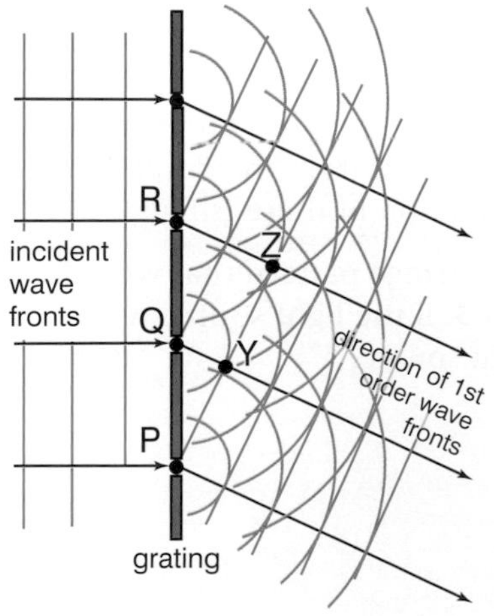

Figure 3 *Formation of the first order wavefronts*

Investigating the diffraction grating

A **diffraction grating** consists of a transparent plate with many closely spaced parallel slits ruled on it. When a parallel beam of monochromatic light is incident normally on a diffraction grating, as shown in Figure 1, light is transmitted by the grating in certain directions only. This is because:

- the light passing through each slit is diffracted
- the diffracted light waves from adjacent slits reinforce each other in certain directions only, including the incident light direction, and cancel out in all other directions.

The central beam, referred to as the 'zero order beam', is in the same direction as the incident beam. The other transmitted beams are numbered outwards from the zero order beam and are referred to as 'first order', 'second order', etc.

The **angle of diffraction**, θ, between each transmitted beam and the central beam increases if:

- light of a longer wavelength is used (e.g. by replacing a blue filter with a red filter)
- a grating with closer slits is used.

Note

Figure 2 shows what happens if white light instead of monochromatic light is directed at the grating. The central transmitted beam is white and each non-central transmitted beam is spread out as a spectrum of colour from violet to red. This is because white light consists of a continuous range of wavelengths from about 350 nm (violet) to about 650 nm (red). Therefore, in each order, red light is diffracted more than violet light as it has a longer wavelength than violet light.

The diffraction grating equation

The wavelength of the light in each transmitted beam can be calculated by measuring the angle of diffraction, θ, of the beam and using the **diffraction grating equation**

$$d \sin\theta = n\lambda$$

where n is the order number of the beam and d is the slit spacing of the grating.

To prove the diffraction grating equation, consider a magnified view of part of a diffraction grating, as shown in Figure 3. Each slit diffracts the light waves that pass through it. As each diffracted wavefront emerges from a slit, it reinforces a wavefront from each of the adjacent slits. For example, in Figure 3, the wavefront emerging at P reinforces the wavefront emitted from Q one cycle earlier which reinforces the wavefront emitted from R one cycle earlier, etc. The effect is to form a new wavefront PYZ which travels in a certain direction and contributes to the first order diffracted beam.

Figure 4 shows the formation of a wavefront of the nth order beam. The wavefront emerging from slit P reinforces a wavefront emitted n cycles earlier by the adjacent slit Q. This earlier wavefront therefore must have travelled a distance of n wavelengths from the slit. Thus the perpendicular distance QY from the slit to the wavefront is equal to $n\lambda$, where λ is the wavelength of the light waves.

Since the angle of diffraction of the beam, θ, is equal to the angle between the wavefront and the plane of the slits, it follows that $\sin\theta = \frac{QY}{QP}$, where QP is the grating spacing (i.e. the centre-to-centre distance d between adjacent slits).

Substituting d for QP and $n\lambda$ for QY therefore gives $\sin\theta = n\lambda/d$. Rearranging this equation gives the diffraction grating equation above.

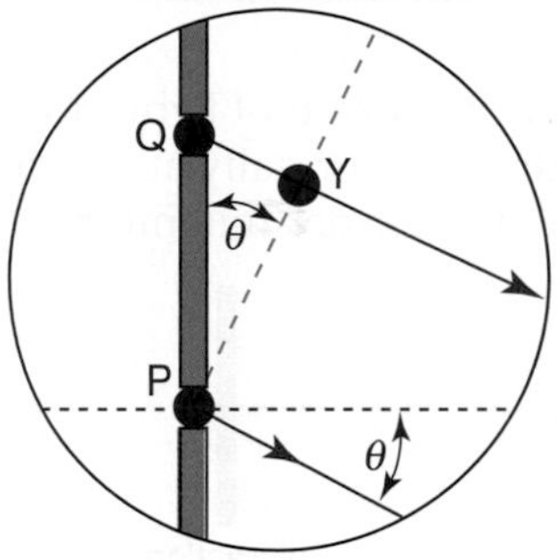

Figure 4 *The nth order wavefront*

Notes

1. The number of slits per metre on the grating, $N = 1/d$ where d is the grating spacing.
2. For a given order and wavelength, the smaller the value of d, the greater the angle of diffraction is. In other words, the larger the number of slits per metre, the bigger the angle of diffraction.
3. Fractions of a degree are usually expressed either as a decimal or in minutes (abbreviated ′), where $1° = 60'$.
4. To find the maximum number of orders produced, substitute $\theta = 90°$ ($\sin 90° = 1$) into the diffraction grating equation and calculate n using $n = d/\lambda$.

The maximum number of orders is given by the value of d/λ rounded down to the nearest whole number.

Worked example

For a diffraction grating with 600 slits per millimetre, calculate the maximum order number for light of wavelength 580 nm.

Solution

$d = 1/N = 1/600\,\text{mm} = 1.67 \times 10^{-3}\,\text{mm}$

$$\frac{d}{\lambda} = \frac{1.67 \times 10^{-3}\,\text{mm}}{580\,\text{nm}} = \frac{1.67 \times 10^{-6}\,\text{m}}{580 \times 10^{-9}\,\text{m}} = 2.88$$

Therefore the maximum order number = 2 (as 2.88 rounded down to the nearest whole number = 2).

Note There would be a first order beam and a second order beam either side of the zero order beam.

Diffraction gratings in action

We can use a diffraction grating in a **spectrometer** to study the spectrum of light from any light source and to measure light wavelengths very accurately.

- A filament lamp produces a continuous spectrum of colour from deep violet at wavelengths of about 350 nm to deep red at about 650 nm, as shown in Topic 9.2 Figure 2.
- A glowing gas in a vapour lamp or a discharge tube produces a spectrum consisting of narrow vertical lines of different colours as shown in Figure 5. The wavelengths of the lines are characteristic of the chemical elements that produce the light.

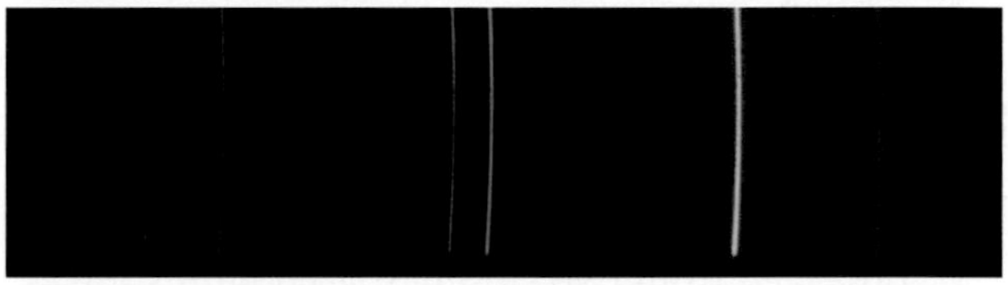

Figure 5 *A line spectrum*

A spectrometer is designed to measure angles to within 1 arc minute which is a sixtieth of a degree. The angle of diffraction of a diffracted beam can be measured very accurately. Using light of a known wavelength, the grating spacing of a diffraction grating can therefore be measured very accurately with a spectrometer. The spectrometer and the grating can then be used to measure light of any wavelength.

Summary test 9.3

1. A laser beam of wavelength 630 nm is directed normally at a diffraction grating which has 300 lines per millimetre. Calculate:
   - **a** the angle of diffraction of each of the first two orders,
   - **b** the number of diffracted orders produced.
2. Light directed normally at a diffraction grating contains wavelengths of 580 nm and 586 nm only. The grating has 600 lines per mm.
   - **a** How many diffracted orders are observed in the transmitted light?
   - **b** For the highest order, calculate the angle between the two diffracted beams.
3. Light of wavelength 430 nm is directed normally at a diffraction grating. The first order transmitted beams are at 28° to the zero order beam. Calculate:
   - **a** the number of slits per millimetre on the grating,
   - **b** the angle of diffraction for each of the other diffracted orders of the transmitted light.
4. A diffraction grating is used in a spectrometer to view light of wavelength 430 nm. The first order diffracted beam is observed at an angle of 14° 55′ to the zero order beam.
   - **a** Calculate the number of lines per mm on the grating.
   - **b** **i** Determine the order number of the highest order beam.
     ii Calculate the angle of diffraction of this order.

   (Reminder: 1 degree = 60 minutes of arc)

9 Exam-style and Practice Questions

Launch additional digital resources for the chapter

1 An interference pattern of bright and dark fringes is observed when two closely spaced slits are illuminated by a parallel beam of monochromatic light, as shown below:

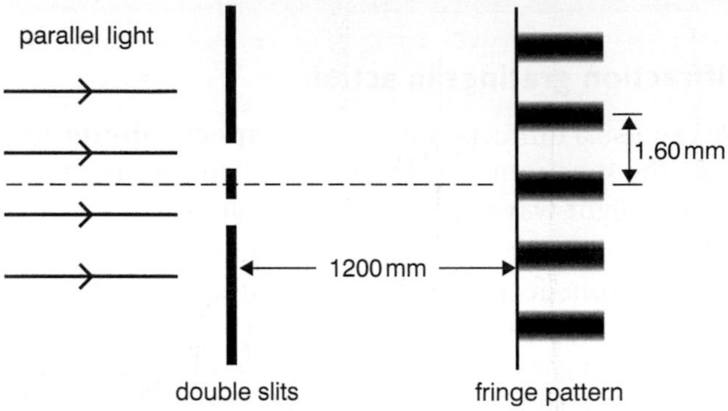

Figure 1.1

a Explain the formation of:
i a bright fringe,
ii a dark fringe.

b The two slits act as coherent emitters of light.
i Explain what is meant by this statement.
ii Explain why the fringe pattern could not be observed from two closely spaced light sources, such as two nearby filaments.

c The two slits were at a distance of 0.4 mm apart. The spacing between two adjacent bright fringes was 1.60 mm, when the fringes were observed at a distance of 1200 mm from the slits. Calculate the wavelength of the monochromatic light.

2 **a** Plane waves are incident on a slit as shown in Figure 3.1.

i Complete Figure 3.1 to show the two preceding wavefronts that have passed through the gap.

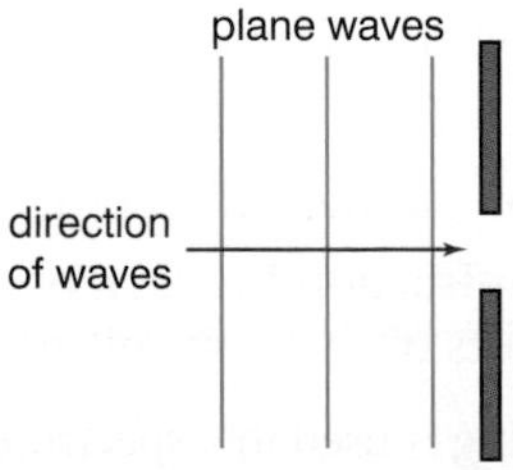

Figure 3.1

ii Describe how the two preceding wavefronts would differ if the slit was significantly wider.

b A narrow parallel beam of light of wavelength 630 nm is directed normally at a diffraction grating which has 600 lines per millimetre.

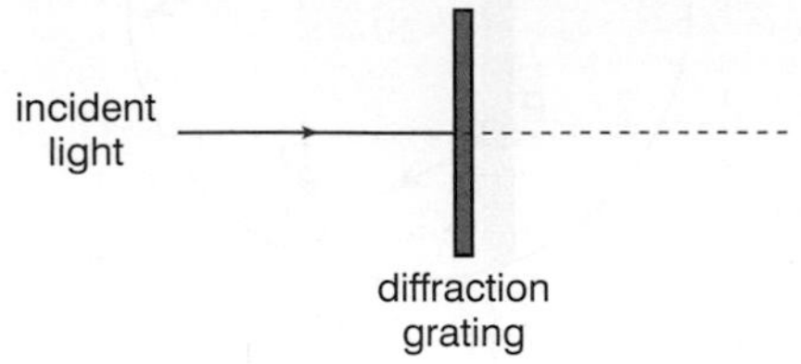

Figure 3.2

i Determine the number of orders of diffracted light that can be observed each side of the zero order.
ii Calculate the angle of diffraction of the highest order.

3 In a double-slit experiment, when light of a certain wavelength was directed normally at the slits, an interference pattern was observed on a screen placed as shown in Figure 4.1.

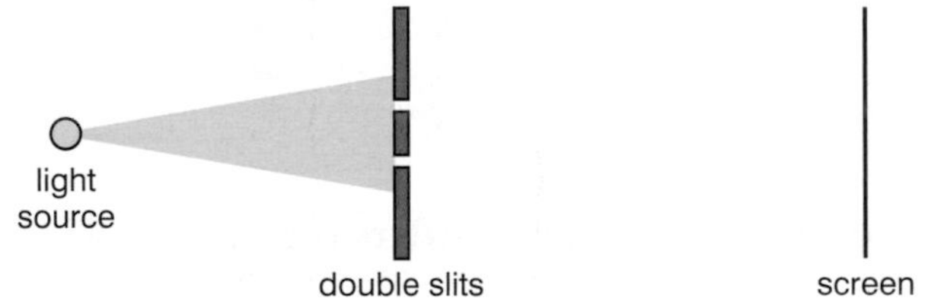

Figure 4.1

a **i** The following measurements were made:
Distance across five fringes = 4.8 mm
Distance between the centres of each slit (i.e. the slit spacing) = 0.40 mm
Distance from the slits to the screen = 810 mm
Calculate the wavelength of the light.

ii The slit–screen distance was estimated to have an uncertainty of ± 5 mm. The uncertainty in the spacing across the five fringes was estimated to have an uncertainty of ± 0.5 mm. The uncertainty in the slit spacing was estimated at ± 0.02 mm. Calculate the percentage uncertainty in each measurement and state and explain which of the three measurements is the least precise.

b Describe and explain how the fringe pattern in **a** would have differed if a beam of white light had been used instead.

4 A narrow beam of white light was directed normally, as shown in Figure 5.1, at a diffraction grating having 600 lines per millimetre. Two diffracted orders of the white light spectrum were observed on each side of a zero order beam.

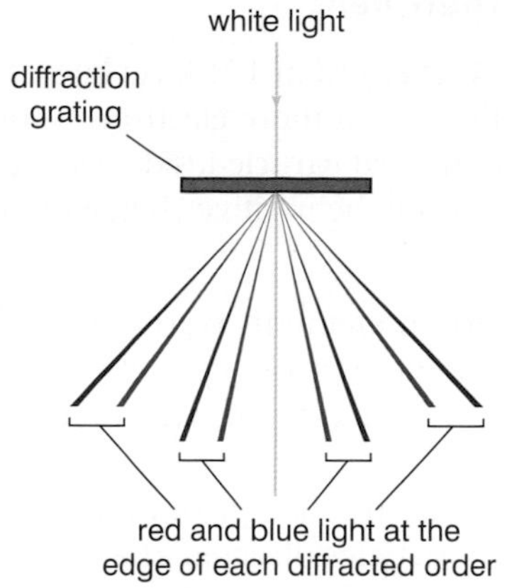

Figure 5.1

a Describe the nature of white light.

b The second diffracted order on each side produced a white light spectrum with angles of diffraction of 48° for the red and 31° for the blue light. Use this data to calculate the wavelength of:

i the red light,

ii the blue light.

c Part of a third order spectrum was observed on each side of the zero order beam. Explain why only part of the third order spectrum was observed.

d A semi-circular glass block was placed against the diffraction grating, as shown in Figure 5.2.

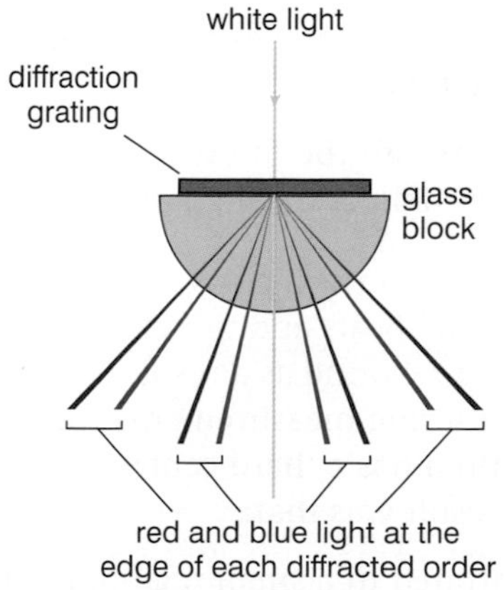

Figure 5.2

The angle of diffraction of the second order red light was 30°, and the angle of diffraction of the second order blue light was 20°. The reduction in the angle of diffraction due to the glass block is because light in glass has a smaller wavelength than in air.

i Use the information above to calculate the wavelength of each colour in glass.

ii For each light colour, calculate the ratio:

$$\frac{\text{the wavelength in air}}{\text{the wavelength in glass}}$$

iii The uncertainty in each measured angle was $\pm 1°$. Discuss whether or not it is reasonable to conclude that the above ratio for red light differs from the ratio for blue light.

5 Light of wavelength 560 nm is directed normally at four parallel narrow slits spaced 0.12 mm apart. The two outer slits are blocked off and a screen is placed 1800 mm from the slits perpendicular to the direction of the incident beam.

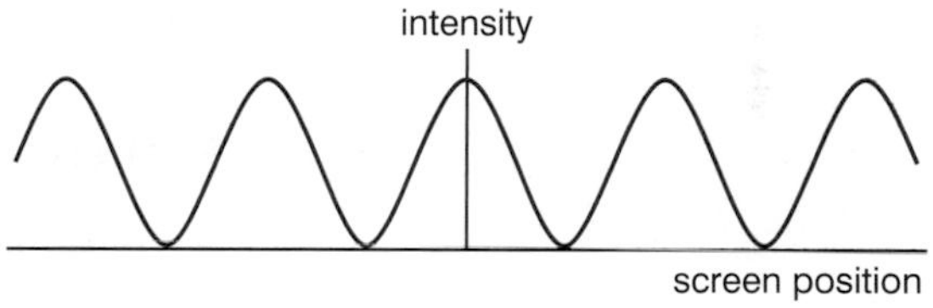

Figure 6.1 *The variation of intensity with position across the screen.*

a Calculate the fringe spacing.

b Copy the diagram and sketch on your copy the variation of intensity across the screen if the two outer slits had been used instead of the two inner slits. Give an explanation for this second fringe pattern.

c Discuss one aspect of the fringe pattern that would differ from either of the previous patterns if all four slits had been used.

6 A diffraction grating with 500 lines per millimetre was used to measure the wavelengths of a certain line emission spectrum.

a A fourth order blue line was measured at an angle of diffraction of 59°25′ and a prominent orange line was measured nearby at an angle of 62°15′.

i Calculate the wavelength of the line at 59°25′.

ii State the order number of the other line and calculate its wavelength.

b Describe the difference between a line emission spectrum and the spectrum of white light.

10 Particle physics

10.1 The discovery of the nucleus

Learning outcomes

On these pages you will learn to:

- describe the basic principles of Rutherford's α-particle scattering experiment
- describe what Rutherford discovered about the scattering of the alpha particles
- state and explain the key conclusions that Rutherford came to about the structure of the atom

Rutherford's alpha scattering experiment

The nucleus was discovered by Ernest Rutherford in 1914. He knew from the work of J.J. Thomson that every atom contains one or more electrons. Thomson had shown that the electron is a negatively charged particle inside every atom but no one knew, until Rutherford's discovery, how the positive charge in the atom was distributed.

Rutherford knew that the atoms of certain elements were unstable and emitted radiation. It had been shown that there were three types of such radiation, referred to as **alpha radiation** (symbol α), **beta radiation** (symbol β) and gamma radiation (symbol γ). Rutherford knew that α-radiation consisted of fast-moving positively charged particles. He used this type of radiation to probe the atom. He reckoned that a beam of the particles directed at a thin metal foil might be scattered slightly by the atoms of the foil if the positive charge was spread out throughout each atom. He was astonished when he discovered that some of the particles bounced back from the foil – in his own words 'as incredible as if you fired a 15-inch naval shell at tissue paper and it came back'.

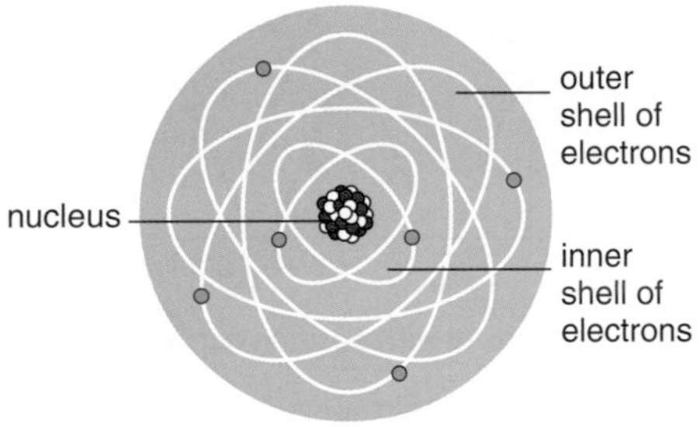

Figure 1 *The structure of an atom*

Rutherford's alpha scattering experiment in more detail

Rutherford used a narrow beam of alpha particles, all of the same kinetic energy, in an evacuated container to probe the structure of the atom. The diagram shows an outline of the arrangement he used. A thin metal foil was placed in the path of the beam. Alpha particles scattered by the metal foil were detected by a detector which could be moved round at a constant distance from the point of impact of the beam on the metal foil. See Figure 2.

Rutherford used a microscope to observe the pinpoints of light emitted by alpha particles hitting a fluorescent screen. He measured the number of alpha particles reaching the detector per minute for different angles of deflection from zero to almost 180°. His measurements showed that:

1 most alpha particles pass straight through the foil with little or no deflection
2 a small percentage of alpha particles deflect through angles of more than 90°.

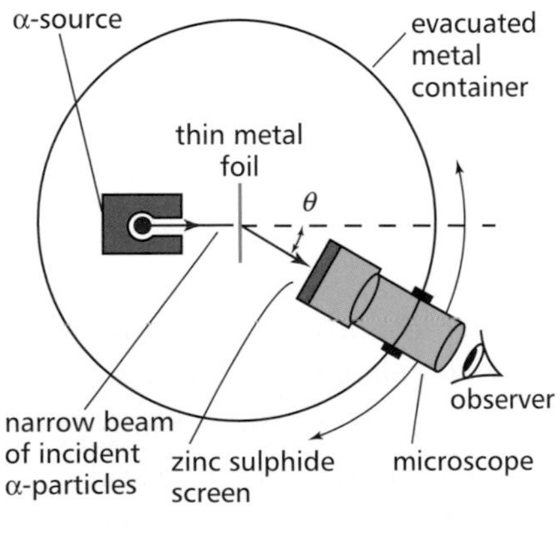

Figure 2 *Rutherford's α-scattering apparatus*

Imagine throwing tennis balls at a row of vertical posts separated by wide gaps. Most of the balls would pass between the posts and therefore would not be deflected much. However, some would rebound as a result of hitting a post. Rutherford realised that the alpha scattering measurements could be explained in a similar way by assuming every atom has a 'hard centre' much smaller than the atom. His interpretation of each result was that:

1 most of the atom's mass is concentrated in a small region, the **nucleus**, at the centre of the atom,
2 the nucleus is positively charged because it repels alpha particles (which carry positive charge) that approach it too closely.

Figure 3 shows the paths of some alpha particles which pass near a fixed nucleus. The closer an alpha track deflection passes to a nucleus, the greater its deflection is because the alpha particle and the nucleus repel each other as they carry the same type of charge. Using **Coulomb's law of force** (i.e. the law of force between charged objects) and Newton's laws of motion, Rutherford used his nuclear model to explain the exact pattern of the results. By testing foils of different metal elements, he also showed that the magnitude of the charge of a nucleus is $+Ze$, where e is the charge of the electron and Z is the **atomic number** of the element.

Notes

1 The alpha particles must have the same speed otherwise slow alpha particles would be deflected more than faster alpha particles on the same initial path.

2 The tube must be evacuated or the alpha particles would be stopped by air molecules.

3 The activity of the source (i.e. the number of alpha particles emitted each second) decreases because more and more nuclei become stable and the number of radioactive nuclei in the source decreases. The source of the alpha particles must therefore have a long half-life, otherwise later readings would be lower than earlier readings due to radioactive decay of the source nuclei.

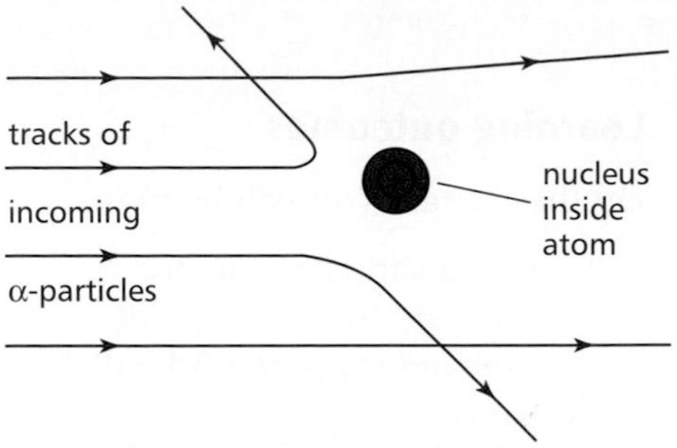

Figure 3 *α-scattering paths*

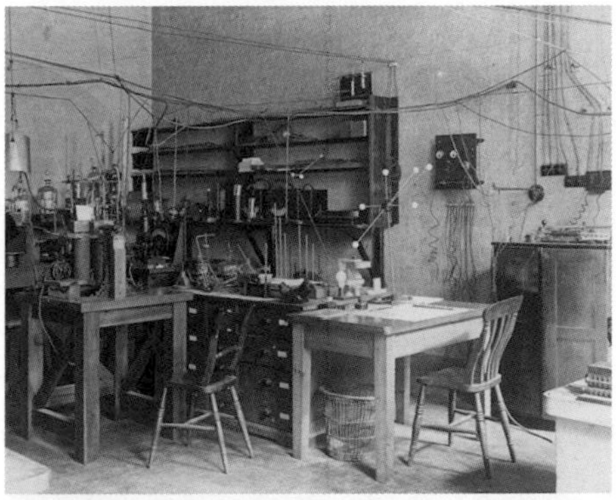

Figure 4 *Rutherford's Cavendish laboratory, 1920s*

Ernest Rutherford 1871–1937

Ernest Rutherford arrived in Britain from New Zealand in 1895. By the age of 28, he was a professor. He made important discoveries about radioactivity and was awarded the Nobel Prize for chemistry in 1908 for his investigations into the disintegration of radioactive substances. He worked in the universities of Montreal, Manchester and Cambridge. He put forward the nuclear model of the atom and proved it experimentally using α-scattering experiments. He was knighted in 1914 and made Lord Rutherford of Nelson in 1931. His co-worker Otto Hahn described him as a 'very jolly man'. In 1915, he expressed the hope that 'no one discovers how to release the intrinsic energy of radium until man has learned to live at peace with his neighbour'. After his death in 1937, his ashes were placed close to Newton's tomb in Westminster Abbey.

Summary test 10.1

1 a In the Rutherford α-particle scattering experiment, most of the alpha particles passed straight through the metal foil. What did Rutherford deduce about the atom from this discovery?

b A small fraction of the alpha particles were deflected through large angles. What did Rutherford deduce about the atom from this discovery?

2 In Rutherford's α-particle scattering experiment, why was it essential that:

a the apparatus was in an evacuated chamber,

b the foil was very thin,

c the α-particles in the beam all had the same speed,

d the beam was narrow?

3 An alpha particle collided with a nucleus and was deflected by it, as shown in Figure 5.

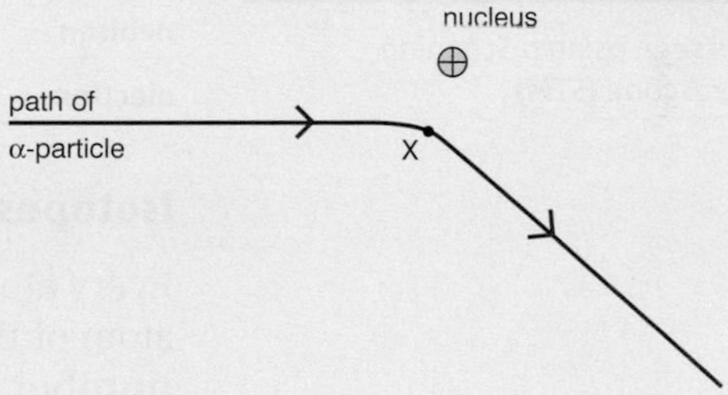

Figure 5

a Copy the diagram and show on it the direction of the force on the alpha particle when it was at the position marked X.

b Describe how:

i the kinetic energy of the alpha particle, and

ii the potential energy of the alpha particle changed during this interaction.

4 Explain why the radioactive source in Rutherford's alpha scattering experiment needed to have:

a a very long half-life,

b to emit alpha particles with the same kinetic energy.

10.2 Inside the atom

Learning outcomes

On these pages you will learn to:

- describe a simple model for the nuclear atom in terms of protons, neutrons and orbital electrons
- distinguish between nucleon number and proton number
- represent different nuclides in terms of mass number and proton number using the accepted notation
- state what is meant by the term isotope and explain why an element can exist in various isotopic forms

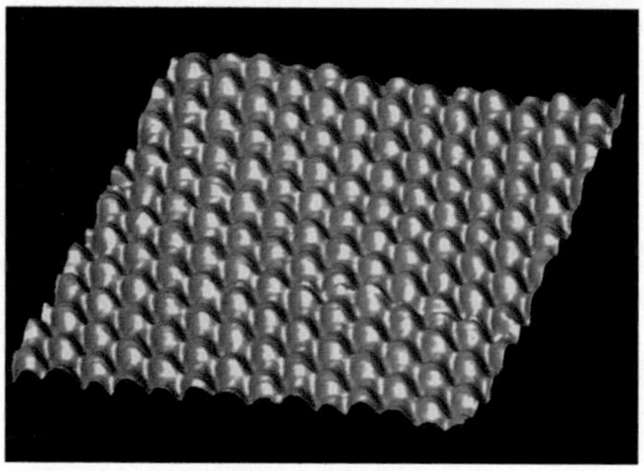

Figure 1 *Atoms seen using a Scanning Tunnelling Microscope (STM)*

The structure of the atom

Atoms are so small, less than a millionth of a millimetre in diameter, that we can only see images of them using the latest electron microscopes. We can't see inside them yet we know, from Rutherford's alpha scattering investigations, that every atom contains a positively charged nucleus composed of:

- **protons** and **neutrons**,
- **electrons** that surround the nucleus.

Each electron has a negative charge and is held in the atom by the electrostatic force of attraction between it and the nucleus because the nucleus is positively charged. Rutherford's investigations showed that the nucleus contains most of the mass of the atom and its diameter is of the order of 0.000 01 times the diameter of a typical atom.

Table 1 shows the charge and the mass of the proton, the neutron and the electron in SI units (i.e. coulombs for charge and kilograms for mass) and relative to the charge and mass of the proton. Notice that:

1. the electron has a much smaller mass than the proton and the neutron,
2. the proton and the neutron have almost equal mass,
3. the electron has equal and opposite charge to the proton. The neutron is uncharged.

Table 1 *Inside the atom*

	Charge		Mass	
	/C	/charge of the proton	/kg	/mass of the proton
proton	$+1.60 \times 10^{-19}$	1	1.67×10^{-27}	1
neutron	0	0	1.67×10^{-27}	1
electron	-1.60×10^{-19}	−1	9.11×10^{-31}	0.0005

Isotopes

Every atom of a given element has the same number of protons as any other atom of the same element. The proton number is usually called the **atomic number** (symbol Z) of the element. For example:

- $Z = 6$ for carbon because every carbon atom has 6 protons in its nucleus,
- $Z = 92$ for uranium because every uranium atom has 92 protons in its nucleus.

The atoms of an element can have different numbers of neutrons. Atoms of the same element with different numbers of neutrons are called **isotopes**. For example, the most abundant isotope of natural uranium contains 146 neutrons and the next most abundant contains 143 neutrons.

Isotopes are forms of the same element with different numbers of neutrons and the same number of protons in their nuclei.

A proton or a neutron in the nucleus is referred to as a **nucleon**. The total number of protons and neutrons in an atom is called the **nucleon number** (symbol A) or sometimes the **mass number** of the atom. This is because it is almost numerically equal to the mass of the atom in relative units (where the mass of a proton or neutron is approximately 1).

The isotopes of an element are labelled according to their atomic number *Z*, their mass number *A* and the chemical symbol of the element. Figure 2 shows how we do this. Notice that:

- *Z* is at the bottom left of the element symbol and gives the number of protons in the nucleus,
- *A* is at the top left of the element symbol and gives the number of protons and neutrons in the nucleus,
- The number of neutrons in the nucleus = $A - Z$.

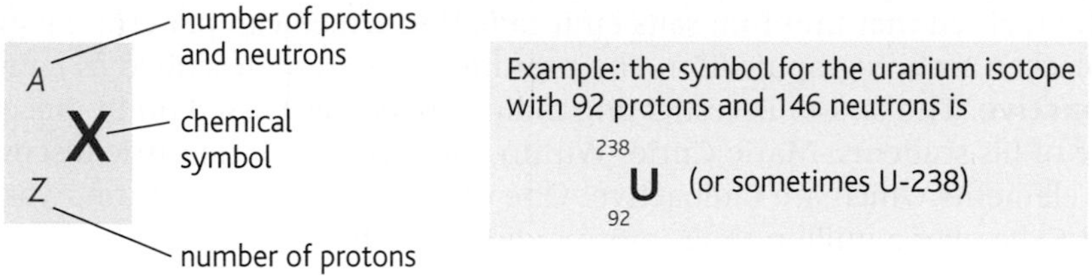

Figure 2 *Isotope notation*

Each type of nucleus is called a **nuclide** and is labelled using the isotope notation. For example, a nuclide of the carbon isotope $^{12}_{6}C$ has two fewer neutrons and two fewer protons than a nuclide of the oxygen isotope $^{16}_{8}O$.

About neutrons

From his α-scattering results, Rutherford deduced that the charge of a nucleus is $+Ze$, where *Z* is the atomic number of the element and the mass of the nucleus is Am_u, where *A* is the mass number of the nucleus. The $^{1}_{1}H$ nucleus is the smallest known nucleus, and physicists concluded that it is a single particle, which became known as the proton. They knew that *Z* and *A* are integers and that for all nuclei larger than the $^{1}_{1}H$ nucleus, *A* is greater than *Z*. So they put forward the hypothesis that the difference between *A* and *Z* could be due to the existence of an uncharged particle (which therefore did not contribute to *Z*) with about the same mass as the proton (as the difference is always an integer). They called this neutral particle the neutron, even though there was no direct evidence for it at the time. Such evidence was eventually found by Sir James Chadwick, who was one of Rutherford's former students.

Summary test 10.2

You will need to use data from Table 1 to answer some of the questions below.

1 State the number of protons and the number of neutrons in a nucleus of:

a $^{12}_{6}C$ **b** $^{16}_{8}O$ **c** $^{235}_{92}U$ **d** $^{24}_{11}Na$ **e** $^{63}_{29}Cu$

2 Name the particle in an atom which:

a has zero charge,

b has the largest charge per unit mass,

c when removed leaves a different isotope of the element.

3 a A $^{63}_{29}Cu$ atom loses two electrons. For the ion formed:

i calculate its charge in C,

ii state the number of nucleons it contains.

An ion has a mass of 2.67×10^{-26} kg and a negative charge of 3.2×10^{-19} C.

b The ion has eight protons in its nucleus. How many neutrons and how many electrons does it have?

4 A block of wax contains hydrogen and carbon atoms. When a beam of neutrons is directed at a block of wax, the neutrons knock protons out of the block. Explain why protons are knocked out of the block rather than carbon nuclei.

10.3 The properties of α, β and γ radiation

Learning outcomes

On these pages you will learn to:

- show an understanding of the nature and properties of α, β and γ radiations

Figure 1 *Marie Curie (1867–1934)*

Marie Curie established the nature of radioactive materials. She showed how radioactive compounds could be separated and identified. She and her husband Pierre won the 1903 Nobel Prize for their discovery of two new elements, polonium and radium. After Pierre's death in 1906 she continued her painstaking research and was awarded a second Nobel Prize in 1911 – an unprecedented honour.

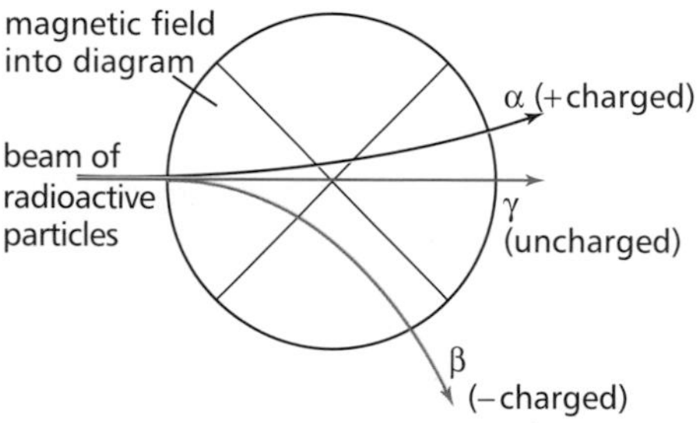

Figure 2 *Deflection by a magnetic field*

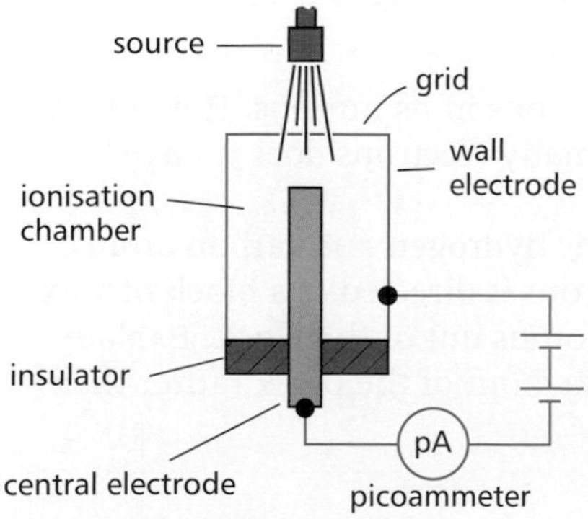

Figure 3 *Investigating ionisation*

The discovery of radioactivity

In 1896, Henri Becquerel was investigating materials that glow when placed in an X-ray beam. He wanted to find out if strong sunlight could make uranium salts glow. He prepared a sample and placed it in a drawer on a wrapped photographic plate, ready to test the salts on the next sunny day. When he developed the plate, he was amazed to see the image of a key. He had put the key on the plate in the drawer and then put the uranium salts on top of the key. He realised that uranium salts emit radiation which can penetrate paper and blacken a photographic film. The uranium salts were described as being **radioactive**. The task of investigating radioactivity was passed on by Becquerel to one of his students, Marie Curie. Within a few years, Marie Curie discovered other elements which are radioactive. One of these elements, radium, was found to be over a million times more radioactive than uranium.

Rutherford's investigations into radioactivity

Rutherford wanted to find out what the radiation emitted by radioactive substances was and what caused it. He found that the radiation:

- Ionised air, making it conduct electricity. He made a detector which could measure the radiation from its ionising effect.
- Was of two types. One type which he called **alpha** (α) radiation was easily absorbed. The other type which he called **beta** (β) radiation was more penetrating. A third type of radiation, called **gamma** (γ) radiation, even more penetrating than β radiation, was discovered a year later.

Further tests showed that a magnetic field deflects α and β radiation in opposite directions and has no effect on γ radiation. From the deflection direction, it was concluded that α radiation consists of positively charged particles and β radiation consists of negatively charged particles. γ radiation was later shown to consist of high-energy **photons**.

1 Ionisation

The ionising effect of each type of radiation can be investigated using an ionisation chamber and a picoammeter, as shown in Figure 3. The chamber contains air at atmospheric pressure. Ions created in the chamber are attracted to the oppositely charged electrode where they are discharged. Electrons pass through the picoammeter as a result of ionisation in the chamber. The current is proportional to the number of ions per second created in the chamber.

- α radiation causes strong ionisation. However, if the source is moved away from the top of the chamber, ionisation ceases beyond a certain distance. This is because α radiation has a range in air of no more than a few centimetres.
- β radiation has a much weaker ionising effect than air. Its range in air varies up to a metre or more. A β-particle, therefore, produces fewer ions per millimetre along its path than an α-particle does.
- γ radiation has a much weaker ionising effect than either α or β radiation. This is because photons carry no charge so they have less effect than α- or β-particles do.

2 Cloud chamber observations

A cloud chamber contains air saturated with a vapour at a very low temperature. Due to ionisation of the air, an α- or a β-particle passing through the cloud chamber leaves a visible track of minute condensed vapour droplets. This is because the air space is supersaturated. When an ionising particle passes through the supersaturated vapour, the ions produced trigger the formation of droplets.

- α-particles produce straight tracks that radiate from the source and are easily visible. The tracks from a given isotope are all approximately the same length, indicating that the α-particles have the same range.
- β-particles produce wispy tracks that are easily deflected as a result of collisions with air molecules. The tracks are not as easy to see as α-particle tracks because β-particles are less ionising than α-particles.

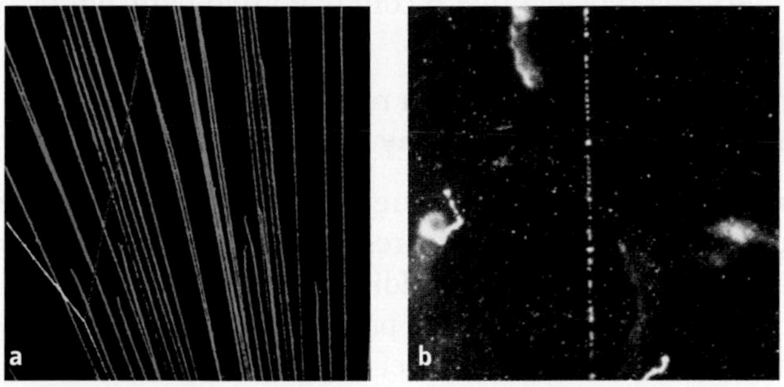

Figure 4 *Cloud chamber photographs* ***a*** *α-particle tracks* ***b*** *β-particle tracks*

3 Absorption tests

Figure 5 shows how a Geiger tube and a counter may be used to investigate absorption by different materials. Each particle of radiation that enters the tube is registered by the counter as a single count or 'click'. The clicks occur randomly which indicates that radioactive emission is a **random** process. The number of counts in a given time is measured and used to work out the **count rate** which is the number of counts divided by the time taken. Before the source is tested, the count rate due to **background radioactivity** must be measured. This is the count rate without the source present.

- The count rate is then measured with the source at a fixed distance from the tube without any absorber present. The background count rate is then subtracted from the count rate with the source present to give the **corrected (i.e. true) count rate** from the source.
- The count rate is then measured with the absorber in a fixed position between the source and the tube. The corrected count rates with and without the absorber present can then be compared.

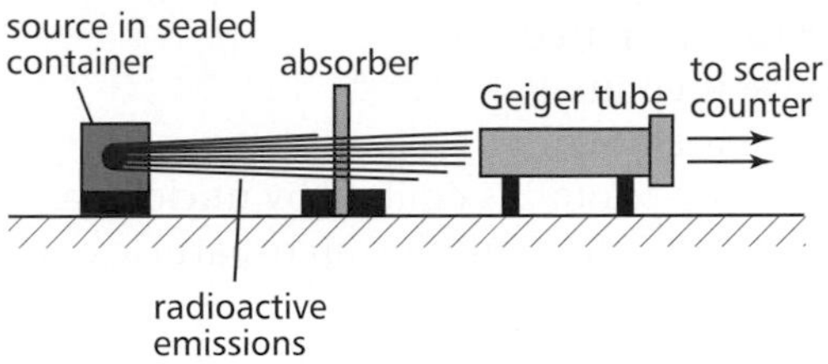

Figure 5 *Investigating absorption*

By using absorbers of different thicknesses of the same material, the effect of the absorber thickness can be investigated. Figure 6 shows a typical set of measurements for the absorption of β radiation by aluminium.

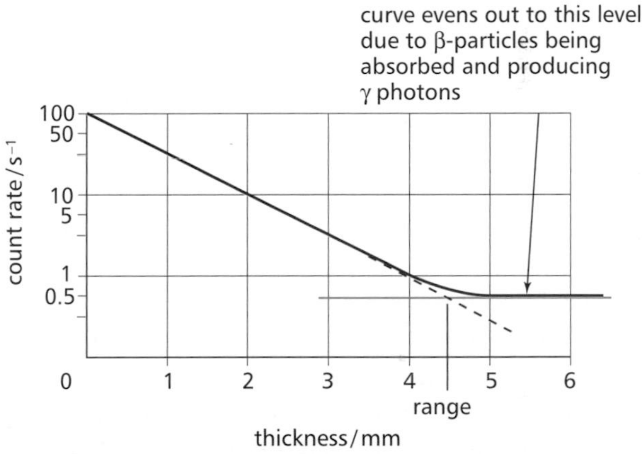

Figure 6 *Count rate v. absorber thickness*

Summary test 10.3

1 A beam of radiation from a radioactive substance passes through paper and is then stopped by an aluminium plate of thickness 5 mm.

   a What type of particles are in this beam?

   b Describe a further test you could do to check your answer in **a**.

2 a What type of radioactivity was responsible for the image of the key seen by Becquerel in the effect described on p.150?

   b Explain why an image of the key was produced on the photographic plate.

3 a Which type of radiation from a radioactive source is: **i** least ionising, **ii** most ionising?

   b When an α-emitting source above an ionisation chamber grid was moved gradually away from the grid, the ionisation current suddenly dropped to zero. Explain why the current suddenly dropped to zero.

4 In an absorption test, as shown in Figure 5 using a β-emitting source and a Geiger counter, a count rate of 8.2 counts per second was obtained without the absorber present and a count rate of 3.7 counts per second was obtained with the absorber present. The background count rate was 0.4 counts per second. What percentage of the β-particles hitting the absorber:

   a pass through it,

   b are stopped by the absorber?

10.4 More about α, β and γ radiation

Learning outcomes

On these pages you will learn to:

- state that (electron) antineutrinos and (electron) neutrinos are produced during β^- and β^+ decay
- use equations to represent the nuclear change when:
  - an α-particle is emitted
  - a β^- particle is emitted
  - a β^+ particle is emitted
- recognise that a γ-photon is emitted by nuclei that have excess energy after emitting an α-particle or a β^- particle or a β^+ particle

The range of α, β and γ radiation in air

The arrangement in Topic 10.3 Figure 5 without the absorbers may be used to investigate the range of each type of radiation in air. The corrected count rate is measured for different distances between the source and the tube, starting with the source close to the tube.

- α radiation has a range of several centimetres in air. The count rate decreases sharply once the tube is beyond the range of the α-particles. This can be seen in Topic 10.3 Figure 4 as the α-particle tracks are the same length indicating that the particles from the source have the same range and, therefore, the same initial kinetic energy. The range differs from one source to another indicating that the initial kinetic energy differs from one source to another.
- β radiation has a range in air of up to a metre or so. The count rate gradually decreases with increasing distance until it is the same as the background count rate at a distance of about 1 metre. The reason for the gradual decrease of count rate as the distance increases is that the β-particles from any given source have a range of initial kinetic energies up to a maximum. Faster β-particles travel further in air than slower β-particles as they have greater initial kinetic energy.
- γ radiation has an unlimited range in air. The count rate gradually decreases with increasing distance because the radiation spreads out in all directions so the proportion of the γ-photons from the source entering the tube decreases.

The nature of α, β and γ radiation

Alpha radiation consists of positively charged particles. Each α-particle is composed of two protons and two neutrons, the same as the nucleus of a helium atom.

Some years before his discovery that every atom contains a nucleus, Rutherford discovered that neutralised α-particles are the same as helium atoms. After he established the nuclear model of the atom, it was realised that the nucleus of the hydrogen atom, the lightest known atom, was a single positively charged particle which became known as the **proton**. Rutherford realised that other nuclei contain protons and he predicted the existence of neutral particles of similar mass, **neutrons**, in the nucleus. For example, the helium nucleus carries twice the charge of the hydrogen nucleus and therefore contains two protons. However, its mass is four times the mass of the hydrogen nucleus so Rutherford predicted that it contained two neutrons as well as two protons.

β-radiation is emitted by a nucleus with either too many neutrons or too many protons.

A nucleus with too many neutrons emits a negative β-particle (symbol β^-) as a result of one of its neutrons changing into a proton. In addition, an antineutrino (symbol $\bar{\nu}$) is emitted. The β^- particles were shown to be fast-moving electrons by deflecting them in electric and magnetic fields in order to measure their specific charge (their charge/mass value). This was found to be the same as the specific charge of the electron.

A nucleus with too many protons emits a positive β-particle (symbol β^+), also called a positron, as a result of one of its protons changing into a neutron. In addition, a neutrino (symbol ν)is emitted. See Figure 1 for more about the discovery of the positron.

Particles and antiparticles

For every known type of particle, there is a corresponding antiparticle with equal and opposite charge. The theory of antiparticles was predicted in 1928 by English physicist Paul Dirac. In his theory, he also predicted:

- **pair production**, whereby a photon of sufficient energy could produce a particle and its corresponding antiparticle
- **annihilation**, when a particle and its corresponding antiparticle collide, annihilate each other and produce photons.

Antiparticles were discovered in 1932 by American physicist Carl Anderson, who used a cloud chamber and a camera to photograph the trails produced by cosmic rays. He found trails that were curved by a magnetic field in the opposite direction to the trails produced by β^- particles, so they must have been produced by positively charge particles. He measured the specific charge of these particles and found it was the same as that of the β^- particles. So he concluded that each such positively charged particle was a **positron**, the antiparticle of the electron.

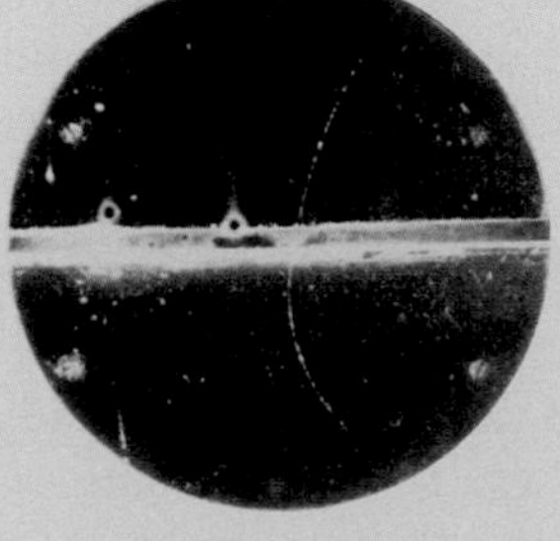

Figure 1 The discovery of the positron

An elusive particle

When the energy spectrum of beta particles was first measured, scientists were puzzled when they discovered the kinetic energy of β-particles varied up to a maximum even though the nucleus always lost a certain amount of energy in the process of emitting a β-particle. Either energy was not conserved in the change or some of it was carried away by unknown uncharged particles and antiparticles, which they called **neutrinos** and **antineutrinos**. This latter hypothesis was eventually shown to be correct when antineutrinos from a nuclear reactor were detected as a result of their interaction with cadmium nuclei in water. Now we know that billions of these elusive particles from the Sun sweep though our bodies every second without interacting.

γ radiation consists of photons with a wavelength of the order of a fraction of a nanometre or less. This discovery was made by using a crystal to diffract a beam of γ radiation in a similar way to the diffraction of light by a diffraction grating.

The equations for radioactive change

A nuclide ${}^{A}_{Z}X$ contains Z protons and $A - Z$ neutrons.

- its charge = $+Ze$, where e is the magnitude of the charge of an electron
- its mass in **atomic mass units, u,** = A approximately.

Note

The mass of a proton is 1.00728 u and the mass of a neutron is 1.00866 u, where $1\,u = \frac{1}{12}$ of the mass of a ${}^{12}_{6}C$ atom.

1 α-emission

An α-particle is represented by the symbol ${}^{4}_{2}\alpha$. Its charge = $+2e$ so $Z = 2$ and it consists of 2 neutrons and 2 protons so $A = 4$.

When a nucleus ${}^{A}_{Z}X$ emits an α-particle, it loses two protons and two neutrons. Therefore, its proton number (Z) is reduced by 2 and its mass number (A) is reduced by 4.

$${}^{A}_{Z}X \rightarrow {}^{4}_{2}\alpha + {}^{A-4}_{Z-2}Y$$

2 β-emission

β^- and β^+ particles are represented in equations by the symbols ${}^{0}_{-1}\beta$ and ${}^{0}_{+1}\beta$, respectively, as their mass is much less than 1 u and their charge is $-e$ and $+e$, respectively. Antineutrinos and neutrinos are uncharged and have very little mass compared with β-particles.

When a nucleus ${}^{A}_{Z}X$ emits a β^- particle (i.e. an electron), a neutron in the nucleus changes into a proton and an antineutrino, $\bar{\nu}$, is also emitted. Therefore, the proton number of the nucleus increases by 1 and the mass number is unchanged.

$${}^{A}_{Z}X \longrightarrow {}^{A}_{Z+1}Y + {}^{0}_{-1}\beta + \bar{\nu}$$

When a nucleus ${}^{A}_{Z}X$ emits a β^+ particle (i.e. a positron), a proton in the nucleus changes into a neutron and a neutrino (ν) is also emitted. Therefore, the proton number of the nucleus decreases by 1 and the mass number is unchanged.

$${}^{A}_{Z}X \longrightarrow {}^{A}_{Z-1}Y + {}^{0}_{+1}\beta + \nu$$

3 γ-emission

A γ-photon is emitted if a nucleus has excess energy after it has emitted an α- or a β-particle.

No change occurs in the number of protons or neutrons of a nucleus when it emits a γ-photon.

Note

That in all the above changes, the nucleon number A and the charge are both conserved. In other words, the total number of protons and neutrons after the change is the same as before the change and the total charge after the change is the same as before the change.

Summary test 10.4

1 Copy and complete each of the following equations representing α-emission.

a ${}^{238}_{92}U \rightarrow {}_{90}Th +$ b ${}_{90}Th \rightarrow {}^{224}_{88}Ra +$

2 Copy and complete each of the following equations representing β-emission.

a ${}^{64}_{29}Cu \rightarrow {}_{30}Zn + {}^{0}\beta$ b ${}_{15}P \rightarrow {}^{32}S + {}^{0}_{-1}\beta$

3 The bismuth isotope ${}^{213}_{83}Bi$ decays by emitting a β-particle to form an unstable isotope of polonium (Po) which then decays by emitting an α-particle to form an unstable isotope of lead (Pb). This isotope then decays by emitting a β-particle to form a stable isotope of bismuth.

a Write down the symbol for each of the three product nuclides in this sequence.

b Write down the number of protons and the number of neutrons in a nucleus of:
i the bismuth isotope ${}^{213}_{83}Bi$,
ii the stable bismuth isotope.

4 The polonium isotope, ${}^{205}_{84}Po$, emits α radiation and decays into nuclei of an isotope of a lead (Pb) isotope which is unstable and decays into thallium, emitting β radiation in the form of positrons in the process.

a Write down equations for the change when:
i a nucleus of the polonium isotope emits an α particle, **ii** the lead nucleus formed in **i** emits a positron when it changes into a thallium nucleus.

b State two differences in the physical properties of a positron and a β^- particle.

EXTENSION

10.5 The dangers of radioactivity

Learning outcomes

On these pages you will learn to:

- explain what is meant by background radioactivity
- recall different types of ionising radiation
- recognise the hazards of ionising radiation and how it may be monitored
- describe procedures used to ensure safe use of radioactive materials

Figure 1 *A radioactive warning sign*

Table 1 *Sources of background radioactivity*

Source	Typical annual dose*
Natural radioactivity in the air	800
Ground and buildings	380
Food and drink	370
Cosmic rays	310
Nuclear weapons testing	10
Air travel	8
Nuclear power	3

* in microsieverts, a measure of the effect of radioactivity on cells.

The hazards of ionising radiation

Ionising radiation is hazardous because it damages living cells. **Ionising radiation** is any form of radiation that creates ions in substances it passes through. Such radiation includes **X-rays**, protons and neutrons as well as α, β and γ radiation. Ionising radiation affects living cells because:

- it can destroy cell membranes which causes cells to die, or
- it can damage vital molecules such as DNA directly or indirectly by creating 'free radical' ions which react with vital molecules. Normal cell division is affected and nuclei become damaged. Damaged DNA can cause cells to divide and grow uncontrollably, causing a tumour which may be cancerous. Damaged DNA in a sex cell (i.e. an egg or a sperm) can cause a mutation which might be passed on to future generations.

As a result of exposure to ionising radiation, living cells die or grow uncontrollably or mutate, affecting the health of the affected person (somatic effects) and possibly affecting future generations (genetic effects). High doses of ionising radiation kill living cells. Cell mutation and cancerous growth occur at low doses as well as at high doses. There is no evidence of the existence of a threshold level of ionising radiation below which living cells would not be damaged.

Background radioactivity occurs naturally due to cosmic radiation and from radioactive materials in rocks, soil and in the air. Everyone is exposed to background radioactivity which varies with location due to local geological features. In addition, radon gas, which is radioactive, can accumulate in poorly ventilated areas of buildings in such locations. Table 1 shows the many sources of background radioactivity that occur in most populated locations.

Radiation monitoring

The Geiger tube

The Geiger tube is a sealed metal tube that contains argon gas at low pressure. The thin mica window at the end of the tube allows α- and β-particles to enter the tube. γ-photons can enter the tube through the tube wall as well. A metal rod down the middle of the tube is at a positive potential as shown in Figure 2. The tube wall is connected to the negative terminal of the power supply and is earthed.

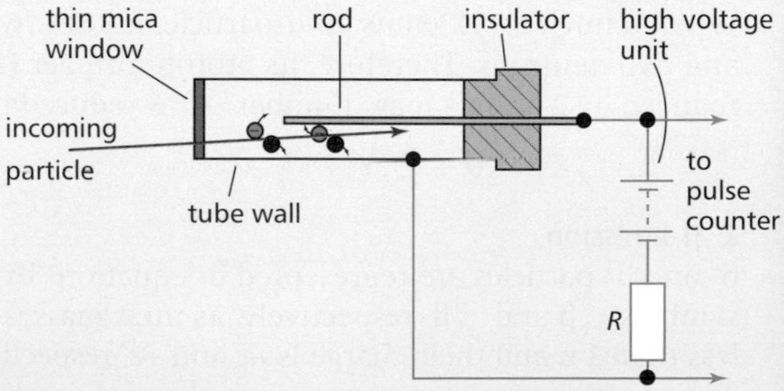

Figure 2 *A Geiger tube*

When a particle of ionising radiation enters the tube, the particle ionises the gas atoms along its track. The negative ions are attracted to the rod and the positive ions to the wall. The ions accelerate and collide with other gas atoms, producing more ions which produce further ions in the same way. Within a very short time, many ions are created and discharged at the electrodes. A pulse of charge passes round the circuit through resistor R, causing a voltage pulse across R which is recorded as a single count by the pulse counter.

The **dead time** of the tube, the time taken to regain its non-conducting state after an ionising particle enters it, is typically of the order of 0.2 ms. Another particle that enters the tube in this time will not cause a voltage pulse. Therefore, the count rate should be no greater than about $5000\,s^{-1}$ ($= \frac{1}{0.2\,ms}$).

The film badge

Anyone using equipment that produces ionising radiation must wear a film badge to monitor his or her exposure to ionising radiation. The badge contains a strip of photographic film in a light-proof wrapper. Different areas of the wrapper are covered by absorbers of different materials and different thicknesses. When the film is developed, the amount of exposure to each form of ionising radiation can be estimated from the blackening of the film. If the badge is overexposed, the wearer is not allowed to continue working with the equipment.

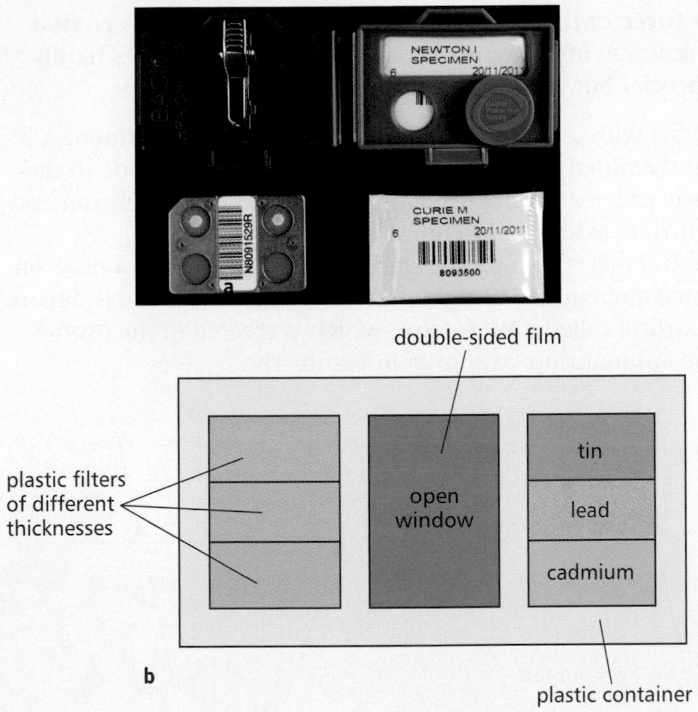

Figure 3 ***a*** *A film badge* ***b*** *Inside a film badge*

Safe use of radioactive materials

Because radioactive materials produce ionising radiation, they must be stored and used with care. In addition, disposal of a radioactive substance must be done in accordance with specific regulations. Only approved institutions are allowed to use radioactive materials. Approval is subject to regular checks and approved institutions are categorised according to purpose.

1 **Storage of radioactive materials** should be in lead-lined containers. Most radioactive sources produce γ radiation as well as α or β radiation so the lead lining of a container must be thick enough to absorb all the γ radiation from the sources in the container. In addition, regulations require that the containers are under 'lock and key' and a record of the sources is kept.

2 **When using radioactive materials**, established rules and regulations must be followed. No source should be allowed to come into contact with the skin.
   - Solid sources should be transferred using handling tools such as tongs or a glove-box or using robots. The handling tools ensure the material is as far from the user as practicable so the intensity of the γ radiation from the source at the user is as low as possible and the user is beyond the range of α or β radiation from the source.
   - Liquid and gas sources and solids in powder form should be in sealed containers. This is to ensure that radioactive gas cannot be breathed in and radioactive liquid cannot be splashed on the skin or drunk.

3 **Disposal** requires long-term storage until the radioactive material is no longer radioactive. The key aim is to ensure that people and the environment are not put at risk. As explained in the next section, radioactive half-lives differ according to the isotope. The half-life of a radioactive isotope is the time it takes for the activity of the isotope to decrease to half. For example, an isotope with a half-life of 5 years would decay to 6.25% of its initial activity after 20 years.

Summary test 10.5

1 a What is meant by ionisation?

b Explain why a source of α radiation is not as dangerous as a source of β radiation provided the sources are outside the body.

2 a Discuss the reasons why ionising radiation is hazardous to a person exposed to the radiation.

b i What is the purpose of a film badge worn by a radiation worker?

ii With the aid of a diagram, describe what is in a film badge and how the film badge is tested.

3 Explain why a radioactive source should be:

i kept in a lead-lined storage box when not in use,

ii transferred using a pair of tongs with long handles.

4 Discuss the precautions you would take when carrying out an experiment using a source of γ radiation.

10.6 Fundamental particles

Learning outcomes

On these pages you will learn to:

- appreciate that protons and neutrons are not fundamental particles since they contain quarks
- describe a simple quark model of hadrons in terms of up, down and strange quarks and their respective antiquarks
- describe protons and neutrons in terms of a simple quark model
- appreciate that there is a weak interaction between quarks, giving rise to β decay
- describe β^- and β^+ decay in terms of a simple quark model
- appreciate that electrons and neutrinos are leptons

The fundamental forces of nature

In your physics course, you will meet four fundamental forces:

- the **electromagnetic force**, which acts between charged particles. This force is due to the charged particles creating and exchanging 'virtual photons', and its range is unlimited.
- the **strong nuclear force**, which holds the nucleus together and acts between neutrons and protons. Its range is of the order of 2–3 femtometres.
- the **weak nuclear force**, which causes a neutron to change into a proton or a proton to change into a neutron. Its range is of the order of a fraction of a femtometre.
- the **gravitational force**, which acts between any two matter particles. Its range is unlimited.

Physicists now know that at very high energies, the electromagnetic force and the weak nuclear force are unified as the 'electroweak' force. At even higher energies, the electroweak force and the strong nuclear force may possibly be unified. But the nature of gravity, the commonest force because we experience it all the time, may be beyond unification!

Forces and interactions

A stable isotope has nuclei that do not disintegrate, so there must be a force holding them together. We call this force the **strong nuclear force** or the 'strong interaction', because it overcomes the electrostatic force of repulsion between the protons in the nucleus and keeps the protons and neutrons together, except in unstable nuclei. Its range is no more than about 3–4 femtometres (fm), where $1\,\text{fm} = 10^{-15}\,\text{m}$. In comparison, the electrostatic force between two charged particles has an infinite range.

The strong nuclear force holds the neutrons and protons in a nucleus together. However, it doesn't cause a neutron to change into a proton in β^- decay or a proton to change into a neutron in β^+ decay. These changes cannot be due to the electromagnetic force, as the neutron is uncharged. There must be a different force at work in the nucleus causing these changes. It must be weaker than the strong nuclear force, otherwise it would affect stable nuclei; hence we refer to it as the **weak nuclear force** or the 'weak interaction'. When a nucleus emits a beta particle, it also emits a neutrino or an antineutrino. These particles hardly interact with other particles but they sometimes do. For example:

- a neutrino can interact with a neutron and make it change into a proton. A β^- particle is created and emitted at the same time. The interaction is due to the exchange of a particle called a **W^- boson**, which is created in the neutron and absorbed by the neutrino, as shown in Figure 1a.
- an antineutrino can interact with a proton and make it change into a neutron. A β^+ particle is created and emitted at the same time. The interaction is due to the exchange of a particle called a **W^+ boson**, which is created in the proton and absorbed by the antineutrino, as shown in Figure 1b.

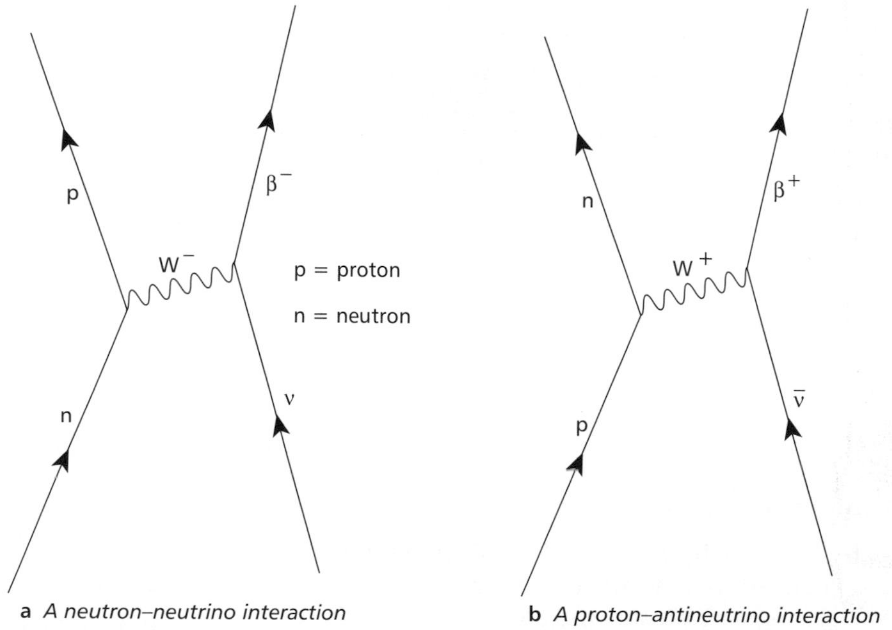

a *A neutron–neutrino interaction* **b** *A proton–antineutrino interaction*

Figure 1 *Weak interactions*

Collisions and colliders

Cosmic rays are high-energy particles that travel through space from the stars, including the Sun. When a cosmic particle enters the Earth's atmosphere, it interacts with the nuclei of atoms in the atmosphere and creates new short-lived particles and antiparticles. When they were first discovered, most physicists thought cosmic rays were from terrestial radioactive substances. This theory was disproved when physicist and amateur balloonist Victor Hess found that the ionising effect of the rays was significantly greater at 5000 m than at the ground.

Further investigations showed that most cosmic rays are fast-moving protons or small nuclei. They collide with the nuclei of gas atoms in the atmosphere, making them unstable and creating showers of particles and antiparticles that can be detected at ground level. By using cloud chambers and other detectors, new types of short-lived particles and antiparticles were discovered, including:

- the **muon** or 'heavy electron' (symbol μ), a negatively charged particle with a rest mass over 200 times the rest mass of the electron. Muons and antimuons decay into electrons and antineutrinos or positrons and neutrinos, respectively.
- the **pion** or 'π meson', a particle that can be positively charged (π^+), negatively charged (π^-) or neutral (π^0) and has a rest mass greater than a muon but less than a proton. Charged π mesons decay into muons and antineutrinos or antimuons and neutrinos.
- the **kaon** or 'K meson', which also can be positively charged (K^+), negatively charged (K^-) or neutral (K^0) and has a rest mass greater than a pion but still less than a proton. Like π mesons, kaons are produced through the strong interaction when protons moving at high speed crash into nuclei. Because they are produced in pairs through the strong interaction and decay through the weak interaction, they are called **strange** particles.

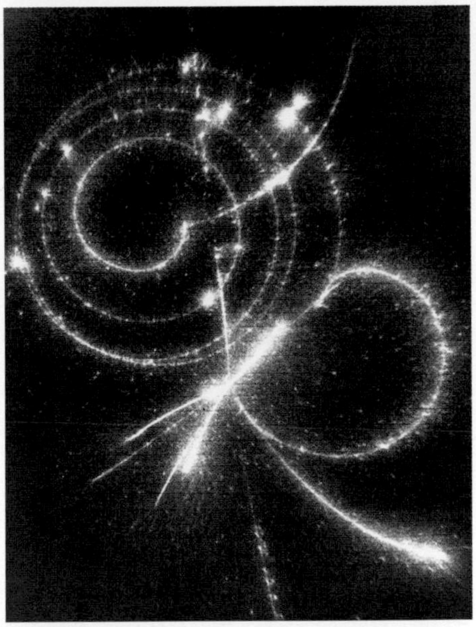

Figure 2 *Creation and decay of a π meson*

More strange and non-strange particles were discovered using high-energy accelerators such as:

- the Stanford Linear Accelerator in California. It accelerates electrons over a distance of 3 km through a potential difference (p.d.) of 50 000 million volts. The energy of an electron accelerated through this p.d. is 50 000 MeV (= 50 GeV). When the electrons collide with a target, they can create lots of particle–antiparticle pairs.
- the Large Hadron Collider at CERN near Geneva. It is designed to accelerate charged particles to energies of more than 7000 GeV. Unlike a linear accelerator, this accelerator is a 28 km diameter ring constructed in a circular tunnel below the ground. It is used by physicists from many countries to find out more about the fundamental nature of matter and radiation.

Figure 3 *The Large Hadron Collider*

Hadrons and leptons

Protons, neutrons, π mesons and K mesons and their antiparticles are examples of particles and antiparticles that we call **hadrons**, because they experience the strong interaction. Electrons, muons and neutrinos and their antiparticles are examples of particles and antiparticles that we call **leptons**, because they experience the weak interaction.

All leptons are fundamental particles in that they are elementary. Except for the proton, which is stable, hadrons decay in one or more stages into leptons and/or protons. Hadrons that include a proton in their decay products are referred to as **baryons**. Thus, a hadron is either a baryon or a **meson**, according to whether or not its decay includes a proton.

- Leptons do not interact through the strong interaction. They interact through the weak interaction and, if charged, through the electromagnetic interaction.

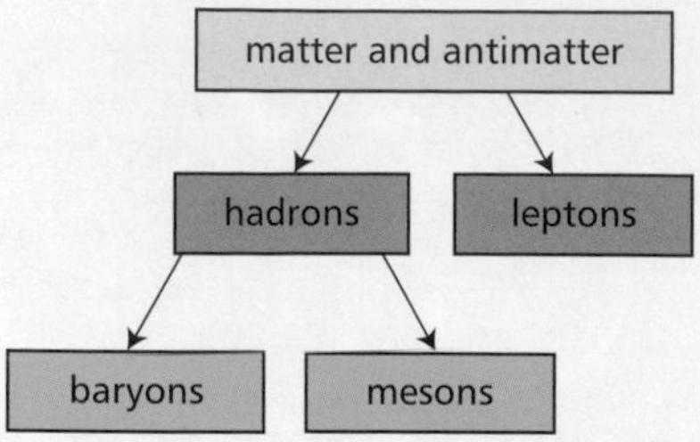

Figure 4 *Hadrons and leptons*

- Hadrons interact through the strong interaction and, if charged, through the electromagnetic interaction. Apart from the proton, which is stable, hadrons decay through the weak interaction.

Quarks

The properties of hadrons – can be explained by assuming they are composed of smaller particles known as **quarks** and **antiquarks**. Six different types of quarks and their corresponding antiquarks are necessary in the standard **quark model**. These six types are referred to as up, down, strange, charm, top and bottom quarks and are denoted by the symbols u, d, s, c, t, and b respectively. The six types are collectively called quark **flavours**. The equivalent antiquarks are denoted by relevant quark symbol with a bar over the top, for example $\bar{u}$ (pronounced 'u bar' denotes the antiparticle of the up quark. The properties of these six quarks are shown in Table 1.

Table 1 *Quark properties*

	Quarks						Antiquarks					
Type	u	d	c	s	t	b	$\bar{u}$	$\bar{d}$	$\bar{c}$	$\bar{s}$	$\bar{t}$	$\bar{b}$
Charge	$+\frac{2}{3}$	$-\frac{1}{3}$	$+\frac{2}{3}$	$-\frac{1}{3}$	$+\frac{2}{3}$	$-\frac{1}{3}$	$-\frac{2}{3}$	$+\frac{1}{3}$	$-\frac{2}{3}$	$+\frac{1}{3}$	$-\frac{2}{3}$	$+\frac{1}{3}$

The rules for combining quarks and antiquarks to form baryons and mesons are astonishingly simple.

Mesons: each meson is composed of a quark and an antiquark (e.g. a π^- meson is the $\bar{u}d$ combination).

Baryons: each baryon is composed of three quarks (e.g. the proton is the uud combination).

Antibaryons: each antibaryon is composed of three antiquarks (e.g. the antiproton is the $\bar{u}\bar{u}\bar{d}$ combination).

Quark combinations

Mesons are hadrons that each consist of a quark and an antiquark. Figure 6 shows all nine different quark–antiquark combinations and the meson each combination forms in each case. Notice that:

- A π^0 meson can be any quark-corresponding antiquark combination.
- Each pair of charged mesons is a particle–different antiparticle pair.
- There are two uncharged K mesons, the K^0 meson and the $\overline{K^0}$ meson.
- A strange meson contains a strange quark or antiquark. A strange baryon contains one or more strange quarks. A strange antibaryon contains one or more strange antiquarks.

The antiparticle of any meson is a quark–antiquark pair and is therefore a meson.

Baryons and antibaryons are hadrons that consist of three quarks or three antiquarks:

- a proton is the uud combination
- a neutron is the udd combination.

The proton is the only stable baryon. A free neutron decays into a proton.

Evidence for quarks

Experimental evidence for the three-quark model was first obtained using the Stanford Linear Accelerator to make electrons collide at high speed with protons. The results showed that the electrons were scattered by **three** scattering centres in each proton.

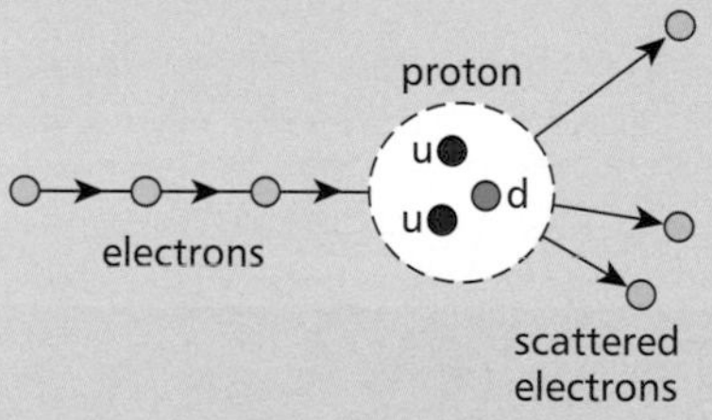

Figure 5 *Evidence for quarks*

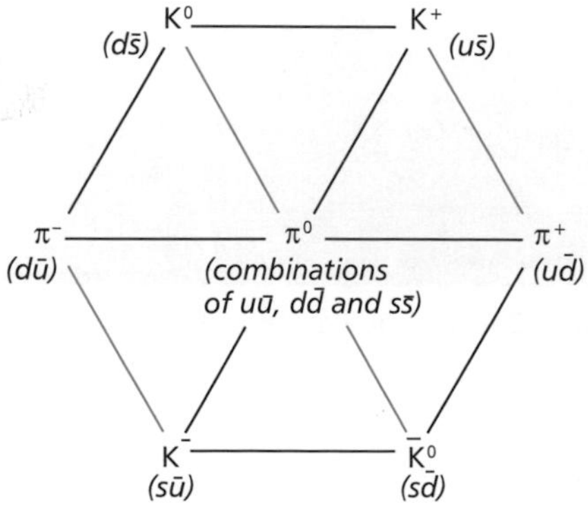

Figure 6 *Quark combinations for the mesons*

Quarks and beta decay

In β^- decay, a neutron in a neutron-rich nucleus changes into a proton and an electron and an antineutrino are released. In quark terms, a down quark changes to an up quark. In the process, a W^- boson is created and emitted from the down quark. The W^- boson then decays into a β^- particle and an antineutrino. A simple diagram for this change is shown in Figure 7(a).

In β^+ decay, a proton in a proton-rich nucleus changes into a neutron and a positron and an neutrino are released. In quark terms, an up quark changes to a down quark. In the process, a W^+ boson is created and emitted from the up quark. The W^+ boson then decays into a β^+ particle and a neutrino. A simple diagram for this change is shown in Figure 7(b).

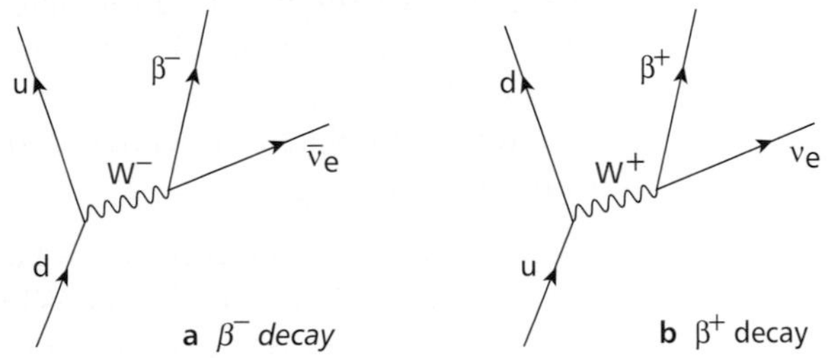

Figure 7 *Quark changes in beta decay*

Summary test 10.6

1 a State two differences between a W boson and a virtual photon.

b State the approximate range of a W boson and estimate its lifetime, given it cannot travel faster than the speed of light, which is $3.0 \times 10^8\,\text{m}\,\text{s}^{-1}$.

2 a State the difference between a lepton and a hadron in terms of their interactions.

b Give an example of: **i** a baryon, **ii** a non-strange meson, **iii** a strange meson.

3 a State the quark composition of: **i** a proton, **ii** a neutron.

b With the aid of a diagram, explain in terms of quarks the changes that take place when a proton changes into a neutron in a nucleus.

4 The omega minus (Ω^-) consists of three strange quarks.

a Determine its charge relative to the proton.

b It decays into a π^- meson and a baryon X, which contains two strange quarks and another quark.

i State the quark composition of the π^- meson.

ii Determine the quark composition of X. Give a reason for your answer.

10 Exam-style and Practice Questions

Launch additional digital resources for the chapter

$c = 3.00 \times 10^8\,\mathrm{m\,s^{-1}}$, $e = 1.60 \times 10^{-19}\,\mathrm{C}$,
$h = 6.63 \times 10^{-34}\,\mathrm{J\,s}$

1 A beam of α-particles was directed at normal incidence towards a thin metal foil. A detector was used to measure the number of α-particles per second scattered by different angles.

a Explain why:

i most α-particles passed through the metal foil with little or no deflection,

ii some α-particles were scattered through very large angles.

b The figure below shows the path of two α-particles moving at the same speed in the same direction as they approached a nucleus of an atom. Copy the diagram and complete the paths of these particles.

Figure 1.1

2 An α-particle collided head-on with a nucleus, as shown below.

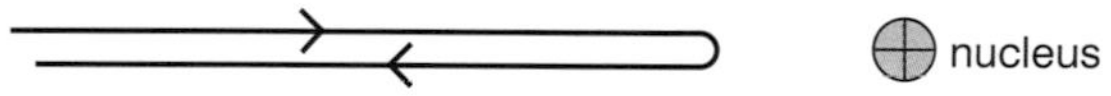

Figure 2.1

a Describe and explain how the kinetic energy of the α-particle changed as it approached then moved away from the nucleus.

b **i** Discuss the factors that determine how closely the α-particle can approach the nucleus.

ii Explain why an α-particle with more kinetic energy on such a track might not rebound from the nucleus.

3 In a nuclear reaction, a neutron collided with a nucleus of the lithium isotope $^{6}_{3}\mathrm{Li}$. A tritium nucleus $^{3}_{1}\mathrm{H}$ and another nucleus X was formed as a result.

Write down an equation that represents this reaction and identify the nucleus X.

4 A narrow beam of ionising radiation from a radioactive substance was directed at a detector at a distance of 30 cm from the source. The detector reading was unchanged when an aluminium plate 5 mm thick was placed between the source and the detector.

a What type of radiation was emitted by this source?

b With the source at a constant distance from the detector, discuss the effect on the detector reading of:

i placing a second identical aluminium plate in the path of the beam,

ii placing a lead plate of thickness 10 mm in the path of the beam.

5 **a** A Geiger tube in a room recorded a background count in 5 minutes of 130 counts. When a point source of γ radiation was placed at a distance of 0.25 m from the Geiger tube, the Geiger counter recorded 2450 counts in exactly 5 minutes. Calculate:

i the count rate in counts per second due to the source at this distance,

ii the number of counts in 5 minutes with the source at 0.50 m from the tube.

b When an experiment is carried out using a radioactive source in a school laboratory, explain why:

i the source should be kept in a lead-lined container when it is not in use,

ii the user should keep as far from the source as possible and keep the time of use as short as possible.

6 **a** A $^{238}_{92}\mathrm{U}$ nucleus emits an α-particle. The nucleus then emits a β^--particle. Calculate the number of protons and neutrons in the nucleus formed immediately after the emission of:

i the α-particle,

ii the β^--particle.

b A copper disc is placed in the core of a nuclear reactor where the copper nuclei are bombarded by neutrons. As a result, the sample becomes radioactive and emits β^--particles.

Copy and complete the following equations representing these changes:

$$^{63}_{29}\mathrm{Cu} + {}^{1}_{0}\mathrm{n} \rightarrow \mathrm{X}$$

$$\mathrm{X} \rightarrow \mathrm{Zn} + {}^{0}_{-1}\beta$$

7 A radioactive nucleus decays with the emission of a beta particle and a gamma-ray photon.

a Describe the changes that occur in the proton number and the nucleon number of the nucleus.

b Comment on the relative penetrating powers of the two types of ionising radiation.

c Gamma rays from a point source are travelling towards a detector. The distance from the source to the detector is changed from 1.0 m to 4.0 m. Calculate:

$$\frac{\text{intensity of radiation at } 4.0\,\text{m}}{\text{intensity of radiation at } 1.0\,\text{m}}$$

8 In α-scattering experiments, α-particles with the same kinetic energy in a narrow beam are directed normally at a very thin metal foil.

Figure 8.1 shows the path of two α-particles, X and Y, that are deflected by the nucleus of an atom.

a Explain why the deflection of X is greater than the deflection of Y.

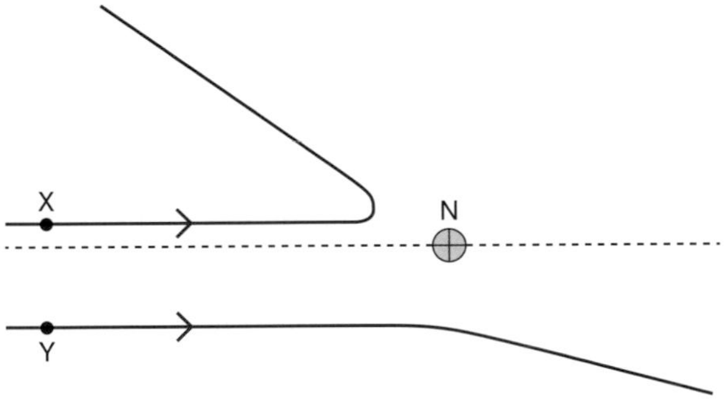

Figure 8.1

b A small proportion of the incident α-particles undergo very large deflections. What conclusions about the nature of atoms can be drawn from this observation?

c i State one reason why the metal foil needs to be very thin.

ii Explain why the α-particles in the beam need to have the same initial kinetic energy.

9 The bismuth isotope $^{209}_{83}\text{Bi}$ has an atomic mass of 209 u and a radius of 7.1 fm. (See Topic 10.6 for explanation of fm.)

a State the number of: **i** protons, **ii** neutrons in this nucleus.

b Calculate the density of this nucleus.
$1\,\text{u} = 1.661 \times 10^{-27}\,\text{kg}$

c Bismuth has a density of $9800\,\text{kg}\,\text{m}^{-3}$. Explain why this is much smaller than the density of a bismuth nucleus.

10 The thorium isotope $^{232}_{90}\text{Th}$ is radioactive and undergoes a series of α– and β–decays which terminate in the formation of the stable lead isotope $^{208}_{82}\text{Pb}$.

a The first two decays are an α–emission followed by a β-emission. The following equations represent these two changes. Determine the missing values *a* to *e* in these equations.

i $^{232}_{90}\text{Th} \rightarrow {}^{4}_{a}\alpha + {}^{b}_{c}\text{Ra}$ **ii** ${}^{b}_{c}\text{Ra} \rightarrow {}^{0}_{d}\beta + {}^{228}_{e}\text{Ac}$

b After the first two decays, the nucleus in **a** may undergo eight further α- and β-emissions of which five are α-emissions. Calculate the number of β-emissions in the series.

Key concepts: Electricity, waves and radioactivity

In studying electricity in Chapters 6 and 7, you will have gained a clear understanding of how the 'conduction electron' model is used to explain why metals conduct and insulators do not and why the resistivity of an intrinsic semiconductor decreases as its temperature is increased. Electrons are negatively charged particles, and their movement round an electric circuit causes energy to transfer from the battery or power supply to the components in the circuit. Conservation of charge and of energy are important concepts in electric circuit theory because these two principles underpin Kirchoff's laws and other circuit rules. In order to apply these rules to electric circuits, the characteristics of the components in a circuit need to be known before the currents and potential differences within the circuit can be determined. Such component characteristics include knowledge of its resistance and its variation, if any, with temperature and whether or not the component is an ohmic conductor. Internal resistance is an important characteristic of a battery or power source, as internal resistance transfers energy by heating from the power source to the surroundings and is therefore a source of inefficiency in the transfer of energy in an electric circuit.

You should also have developed your mathematical skills in studying electricity as a result of using circuit rules to calculate potential differences, currents and power supplied to the components in a circuit. Conservation of energy applied to a circuit means that the power supplied by a power source less the power dissipated to its internal resistance should be equal to the sum of the powers supplied to all the components.

Electricity is a practical subject, and you should now know how to measure currents and potential differences accurately and to use variable resistors and potential dividers to control currents and potential differences. The correct use of a potential divider in a sensor circuit requires a very clear grasp of the concepts of potential difference and resistance in order to predict how a sensor works.

Waves transfer energy and information. After studying chapters 8 and 9, you should understand the main differences between longitudinal waves and transverse waves and the difference between mechanical waves and electromagnetic waves. You should also know what is meant by wave characteristics such as wavelength, frequency and phase difference. You should also be able to describe wave properties such as reflection, refraction, diffraction and interference and be able to explain two-source interference. You should know how a diffraction grating works and how it is used to measure light wavelengths. Also, you should be able to describe stationary wave patterns and be able to explain their formation. A clear grasp of the principle of superposition is important in gaining a good understanding of interference and also of the formation of stationary waves.

The final chapter on particle physics develops previous knowledge from earlier chapters about the structure of the atom and about conservation of charge. In studying the nuclear model of the atom, the role of experimentation should be recognised as being crucial to our understanding of matter. Rutherford used his experimental skills to obtain measurements of the deflection of alpha particles and then successfully applied his theoretical knowledge of mechanics and electric fields to his nuclear model of the atom. Using the nuclear model of the atom, you should have a clear understanding of what is meant by an isotope and how equations based on conservation principles are used to represent nuclear changes. The nuclear model of the atom enables us to understand the nature and properties of alpha, beta and gamma radiation as well as gaining knowledge and awareness of the practical methods and devices used in determining such properties. In such practical work you should have become aware of the hazards of ionising radiation and of the importance of adhering to vital safety rules when radioactive materials are used.

The topic on fundamental particles provides awareness of how experimental discoveries have led to our present knowledge of the structure of matter, including matter and antimatter, the quark model and the weak nuclear force. Conservation rules on nuclear equations developed earlier provide the basic principles that underpin such knowledge.

The following table provides an overview of the topics in chapters 6 to 10, showing where key concepts have been developed.

Chapters 6–10: Key concepts

Topic			Key concepts	
6.1–6.2	conductors and insulators	electrons as charge carriers	models	drift velocity
		calculations	use of maths	problem-solving
	electric fields	nature of an electric field	forces and fields	action at a distance
6.3–6.5	potential difference and resistance	potential difference	energy and matter	energy transfer per unit charge
		I–V characteristics		
		resistance and resistivity		practical skills using an ammeter and a voltmeter and graph work
		I–V graphs		
		calculations involving resistance and resistivity		problem-solving
7.1–7.2	Kirchhoff's laws	currents at a junction	models	conservation of charge and of energy applied to electric circuits
		p.d. round a circuit	energy and matter	
		resistor rules		
		circuit calculations	use of maths	problem-solving
7.3–7.4	e.m.f. and internal resistance	measurement of e.m.f. and internal resistance	testing	practical skills involving electrical measurements and graph work
		conservation of energy	energy and matter	
		maximum power		
		circuit calculations	use of maths	application of Kirchoff's laws
7.5	the potential divider	theory and use of the potential divider	testing	practical skills involving measurements of p.d.
		sensor circuits	models	predicting output changes in sensor circuits
8.1–8.4	wave properties	longitudinal and transverse waves	energy and waves	amplitude, phase difference
		polarisation of transverse waves		
		reflection, refraction, diffraction, interference	testing	practical skills involved in measuring amplitudes and frequencies
8.5–8.8	stationary waves	stationary wave formation, nodes	energy and waves	principle of superposition
8.9	Doppler effect	Doppler shift	use of maths	problem-solving
		calculations using Doppler effect equations		

continued on p164

Chapters 6–10: Key concepts *(continued)*

Topic			Key concepts	
9.1–9.2	interference of light	theory of two-slit interference	models	principle of superposition
			energy and waves	path difference
				coherence
		calculations involving use of the two-slit interference equation	use of maths	problem-solving
9.3	diffraction grating	theory of the diffraction grating	energy and waves	superposition of diffracted wavefronts
		observation and measurement of spectra	testing	practical skills
		calculation of wavelength	use of maths	problem-solving

10.1–10.2	structure of the atom	Rutherford's alpha scattering experiment	models	nuclear model of the atom
		isotopes		
		calculation of nuclear size and nuclear density	use of maths	problem-solving
10.3–10.5	α, β and γ radiation	properties	testing	investigation of range, absorption, ionisation
		ionisation		
		background radioactivity		
		nuclear equations		conservation rules
10.6	fundamental particles	matter and antimatter	energy and matter	
		strong and weak nuclear forces	forces and fields	
		classification of particles		
		quarks	models	the quark model

Question

This question is about an investigation into the heating effect of an electric current and the use of physics equations to predict and test the results. It tests how we use models to make predictions about a physical system and how we then test these predictions with an experimental investigation. It also tests practical skills, when electrical and temperature measurements are used to plot a graph. Accuracy in making temperature measurements is often more difficult to achieve than in electrical or mechanical measurements such as distance or time intervals and simple practical techniques such as stirring the water being heated are important.

Note the use of physics theory and mathematics to predict the relationship between the current I and the temperature change, ΔT, of the water being heated. In this case, the predicted relationship is of the form $\Delta T = kI^2$. So by plotting a graph of ΔT against I^2, we can tell if the relationship is valid from whether or not the graph is linear (i.e. a straight-line graph). If the graph is linear, we can also determine the constant **k**, *as it is equal to the gradient of the graph.*

This investigation involves a predicted relationship of the same form as that in the Key concepts 'projectile motion' question on p.92. In both situations, although the measurements give a straight-line graph, the spacing between the plotted points increases along the line. This uneven spacing causes unnecessary difficulty in judging the best-fit line. In practice, the values of **I** *(or* **x** *in the previous question) should be selected to give approximately even intervals between successive values of I^2 rather than between successive values of* **I** *(or x^2 in the previous question).*

Before answering the question, students should be given the opportunity to do this or a similar investigation.

1 A student carried out an investigation to find out how the heating effect of an electric current changes when the current is increased. A 12 V heater in series with an ammeter, a variable resistor, a switch and a 12 V battery was used to heat some water in an insulated beaker. A thermometer was used to measure the temperature change of the water. The test was repeated several times with a different current each time. In each test, the same amount of water was heated for 600 seconds.

a **i** Sketch the circuit diagram.

ii In each test, the current was kept constant. Describe how this was achieved using the circuit you sketched. *(2 marks)*

b **i** Outline a test you would carry out to ensure that the beaker insulation was effective.

ii Why was it important to heat the same volume of water for the same length of time in each test? *(4 marks)*

c The table below shows the results of these tests.

current I / A	1.2	2.0	3.1	3.7	4.5
temperature increase ΔT / °C	3.0	6.5	16.5	23.5	35.0
I^2 / A^2					

The student predicted that the temperature rise ΔT should be directly proportional to the square of the current.

i Give a justification for this prediction.

ii Copy the table and complete the third row.

iii Plot a suitable graph to find out if the prediction is valid. *(9 marks)*

d The student noticed that in each test the resistance of the variable resistor had to be decreased to keep the current constant during the time the water was heated. Give an explanation of this observation. *(2 marks)*

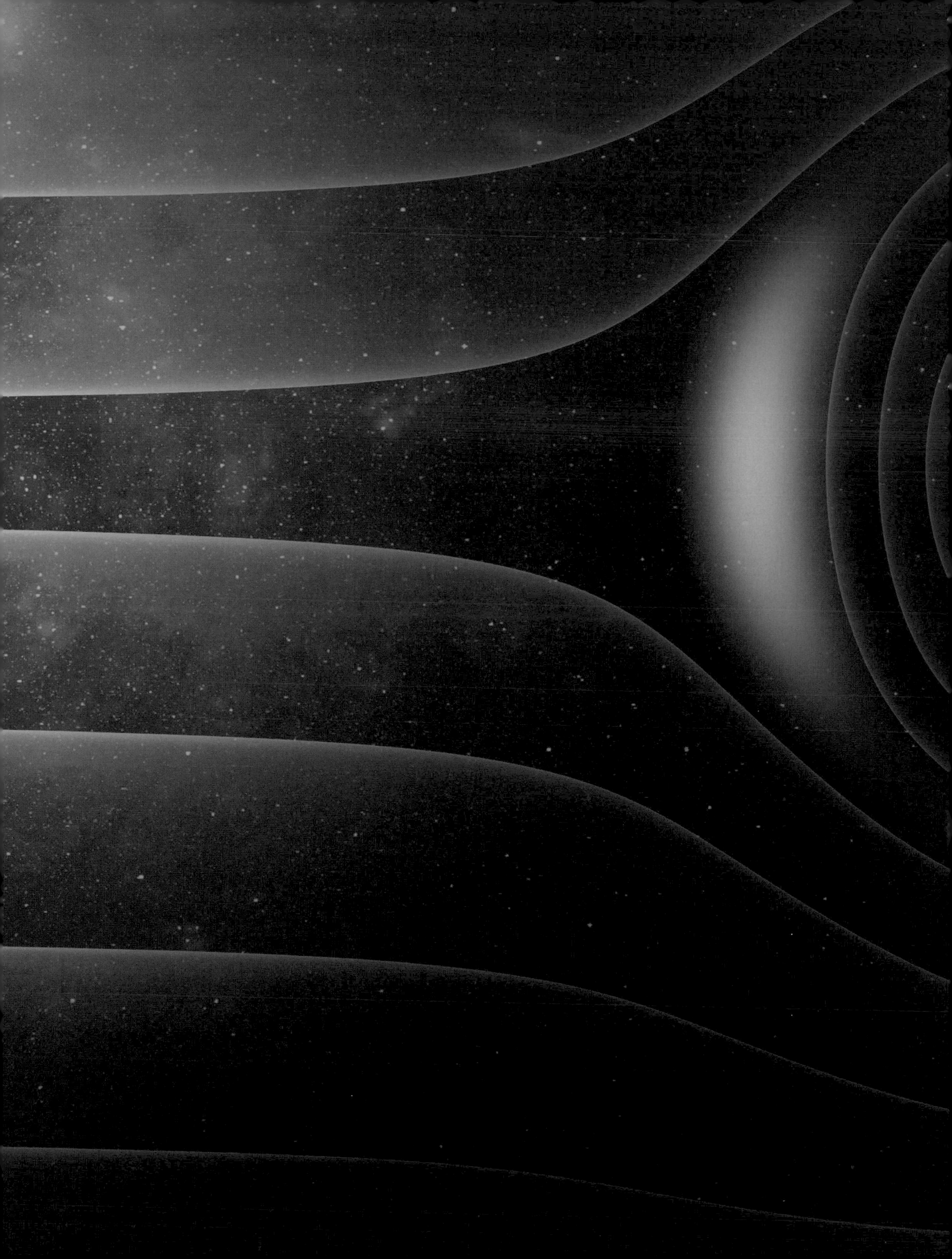

A Level

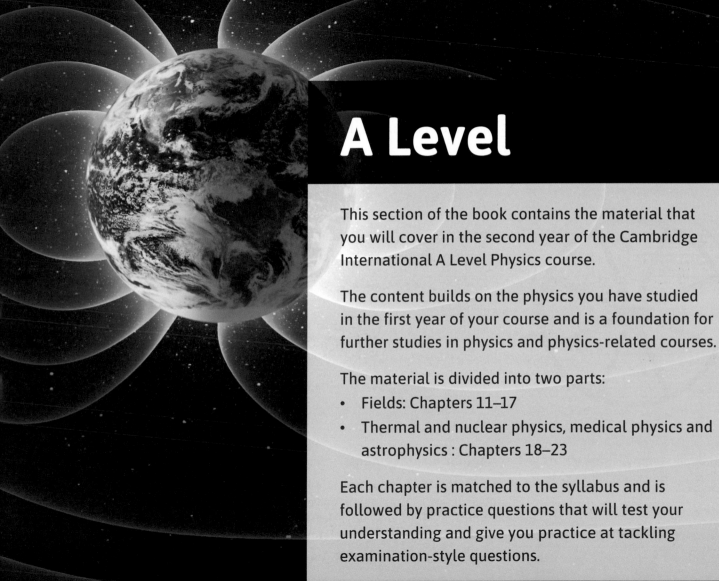

This section of the book contains the material that you will cover in the second year of the Cambridge International A Level Physics course.

The content builds on the physics you have studied in the first year of your course and is a foundation for further studies in physics and physics-related courses.

The material is divided into two parts:

- Fields: Chapters 11–17
- Thermal and nuclear physics, medical physics and astrophysics : Chapters 18–23

Each chapter is matched to the syllabus and is followed by practice questions that will test your understanding and give you practice at tackling examination-style questions.

11 Motion in a circle

11.1 Uniform circular motion

Learning outcomes

On these pages you will learn to:

- explain what is meant by angular displacement and express angular displacement in radians
- explain what is meant by angular speed and angular velocity
- use the concept of angular speed to solve problems
- recall and use $v = r\omega$ to solve problems

In a cycle race, the cyclists pedal furiously at top speed. The speed of the perimeter of each wheel is the same as the cyclist's speed, provided the wheels do not slip on the ground. If the cyclist's speed is constant, the wheels must turn at a steady rate. An object rotating at a steady rate is said to be in **uniform circular motion**.

Consider a point on the perimeter of a wheel of radius r rotating at a steady speed.

- The circumference of the wheel = $2\pi r$
- The frequency of rotation $f = \frac{1}{T}$, where T is the time for one rotation.

The speed of a point on the perimeter $= \frac{\text{circumference}}{\text{time for 1 rotation}} = \frac{2\pi r}{T} = 2\pi r f$

Worked example

A cyclist is travelling at a speed of $13\,\text{m}\,\text{s}^{-1}$ on a bicycle which has wheels of radius 390 mm. Calculate:

a the time for one rotation of the wheel,

b i the frequency of rotation of the wheel.

ii the number of rotations of the wheel in 1 minute.

Solution

a Rearranging speed $v = \frac{2\pi r}{T}$ gives the time for 1 rotation, $T = \frac{2\pi r}{v}$

Therefore, $T = \frac{2\pi \times 0.39\,\text{m}}{13\,\text{m}\,\text{s}^{-1}} = 0.19\,\text{s}$

b i Frequency $f = \frac{1}{T} = \frac{1}{0.19\,\text{s}} = 5.3\,\text{Hz}$

ii Number of rotations in 1 minute = 60 × 5.3 = 318

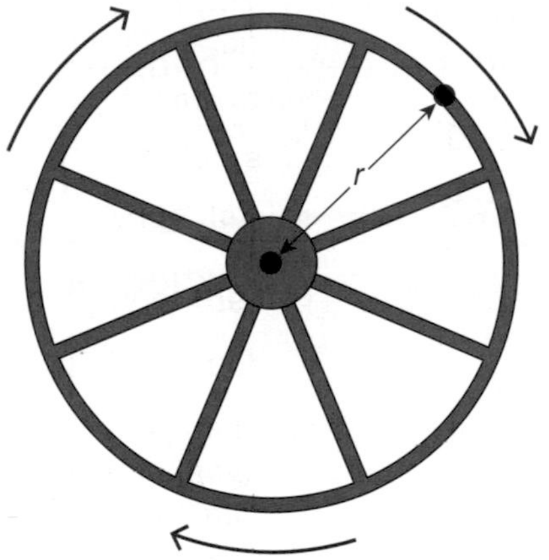

Figure 1 In uniform circular motion

Angular displacement

In the UK, the London Eye is a very popular tourist attraction. The wheel has a diameter of 130 m and takes passengers high above the surrounding buildings, giving a glorious view on a clear day. Each full rotation of the wheel takes 30 minutes. Each capsule therefore takes its passengers through an angle of 0.2° each second or $= \frac{0.2\pi}{180}$ radians each second as 360° = 2π radians.

For any object in uniform circular motion, the object turns through an angle of $\frac{2\pi}{T}$ radians each second, where T is the time taken for 1 complete rotation.

In other words, the angular displacement of the object each second is $\frac{2\pi}{T}$.

Therefore, the **angular displacement** of the object in time t is given by:

$$\theta \text{ (in radians)} = \frac{2\pi t}{T} = 2\pi f t$$

Figure 2 The London Eye

Angular velocity

The angular velocity of an object moving in uniform circular motion is defined as its angular displacement per second.

The unit of angular velocity is the radian per second ($rad\,s^{-1}$).

Using the previous equation $\theta = \frac{2\pi t}{T}$ therefore gives:

$$\text{angular velocity, } \omega = \frac{\text{angular displacement, } \theta}{\text{time taken, } t} = \frac{2\pi}{T}$$

The object travels once round the circle in its **time period**. For a circle of radius r, its circumference is equal to $2\pi r$. Hence its speed v is given by

$$v = \frac{\text{circumference of the circle}}{\text{time period}} = \frac{2\pi r}{T}$$

Substituting $\omega = \frac{2\pi}{T}$ gives:

$$v = \omega r$$

Figure 3 shows the angular displacement of the object in time t which is less than the time period T.

1 The angular displacement of the object, $\theta = \omega t$.
2 The object travels along an arc of the circle of length $s = vt = \frac{2\pi rt}{T} = r\theta$

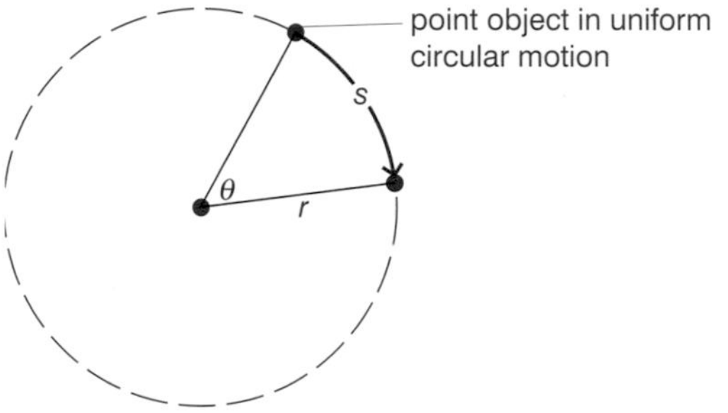

Figure 3 *Arcs and angles*

Worked example

A cyclist travels at a speed of $12\,m\,s^{-1}$ on a bicycle which has wheels of radius 0.40 m. Calculate:

a the frequency of rotation of each wheel,

b the angular velocity of each wheel,

c the angle the wheel turns through in 0.10 s:

i in radians, **ii** in degrees.

Solution

a Circumference of wheel $= 2\pi r = 2\pi \times 0.4 = 2.5\,m$

Time for 1 wheel rotation, $T = \frac{\text{circumference}}{\text{speed}} = \frac{2.5}{12} = 0.21\,s$

Frequency $= \frac{1}{T} = \frac{1}{0.21} = 4.8\,Hz$

b Angular velocity $= \frac{2\pi}{T} = 30\,rad\,s^{-1}$

c **i** Angle the wheel turns through in 0.10 s, $\theta = \frac{2\pi t}{T} = \frac{2\pi \times 0.10}{0.21}$ = 3.0 radians

ii $\theta = 3.0 \times \frac{360}{2\pi} = 172°$

Note

One radian is defined as the angle subtended by a circular arc of length equal to the radius of the circle. It is based on a scale of 2π radians being equal to 360°. This is because the formula for the circumference of a circle $2\pi r$ (where r is the circle radius) then leads to the following useful formula for the length, s, of an arc of a circle:

$$s = r\theta$$

where θ is the angle in radians subtended by the arc at the centre of the circle. See Figure 3.

Summary test 11.1

1 Calculate the angular displacement in radians of the tip of the minute hand of a clock in:

a 1 second,

b 1 minute,

c 1 hour.

2 An electric motor turns at a frequency of 50 Hz. Calculate:

a its time period,

b the angle it turns through in radians in:
i 1 ms,
ii 1 second.

3 The Earth takes exactly 24 hours for 1 full rotation. Calculate:

a the speed of rotation of a point on the Equator,

b the angle the Earth turns through in 1 second in:
i degrees, **ii** radians.

The radius of the Earth = 6 400 km.

4 A satellite in a circular orbit of radius 8 000 km takes 120 minutes per orbit. Calculate:

a its speed,

b its angular displacement in 1.0 s in:
i degrees, **ii** radians.

11.2 Centripetal acceleration

Learning outcomes

On these pages you will learn to:

- describe uniform circular motion
- understand centripetal acceleration in the case of uniform circular motion
- recall and use centripetal acceleration equations $a = r\omega^2$ and $a = v^2/r$
- recall and use centripetal force equations $F = mr\omega^2$ and $F = mv^2/r$

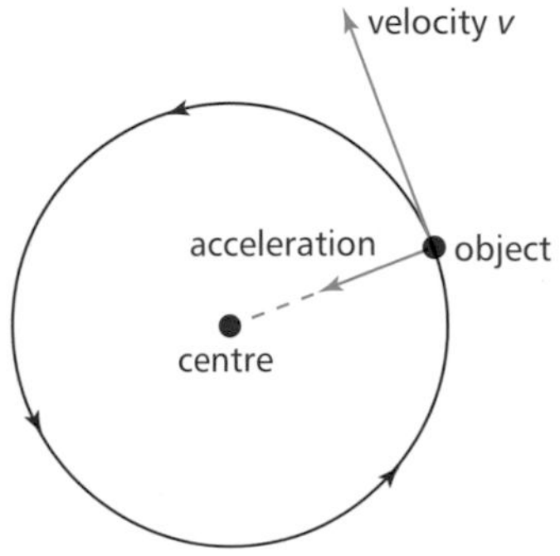

Figure 1 *Centripetal acceleration*

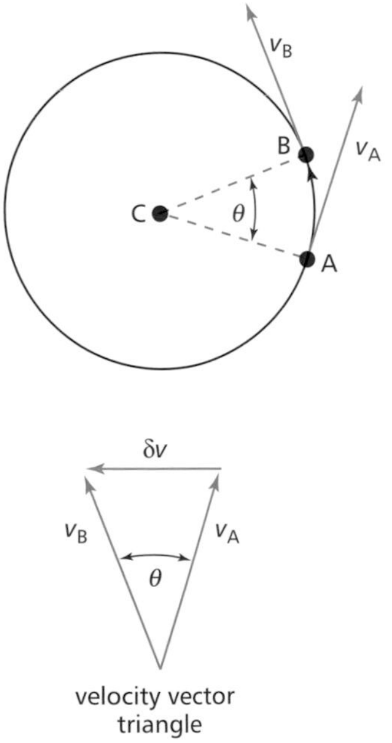

Figure 2 *Proving* $a = \frac{v^2}{r}$

The **velocity** of an object moving round a circle at constant speed continually changes direction. Because its velocity changes, the object therefore accelerates. If this seems odd because the speed is constant, remember that acceleration is change of velocity per second. Passengers on the London Eye might not notice they are being accelerated but if the wheel was made to rotate much faster, they undoubtedly would notice.

The velocity of an object in uniform circular motion at any point is along the tangent to the circle at that point. The direction of the velocity changes continuously as the object moves round on its circular path. The change in the direction of the velocity is towards the centre of the circle. So its acceleration is towards the centre of the circle and is referred to as **centripetal acceleration**. Centripetal means 'towards the centre of the circle'.

For an object moving at constant speed v in a circle of radius r, it can be shown that:

$$\textbf{centripetal acceleration, } a = \frac{v^2}{r}$$

Proof of this equation is not required for the specification. However, a proof is given to provide a better understanding of the idea of centripetal acceleration.

Proof of $a = \frac{v^2}{r}$

- Consider an object in uniform circular motion at speed v moving in a short time interval δt from position A to position B along its path. Therefore the distance AB along the circle, $\delta s = v\delta t$. Figure 2 shows the idea.
- The line from the object to the centre of the circle at C turns through angle θ when the object moves from A to B. The velocity direction of the object turns through the same angle θ, as shown in Figure 2.
- The change of velocity, δv = velocity at B − velocity at A, is shown in the velocity vector triangle in Figure 2.
- The triangles ABC and the velocity vector triangle have the same shape because they both have two sides of equal length with the same angle θ between the two sides.

  Provided θ is small, then $\frac{\delta v}{v} = \frac{\delta s}{r}$

  Because $\delta s = v\,\delta t$, then $\frac{\delta v}{v} = \frac{v\,\delta t}{r}$

  Therefore, acceleration, $a = \frac{\text{change of velocity } \delta v}{\text{time taken } \delta t}$

  $= \frac{v^2}{r}$ towards the centre.

Note

Since $v = \omega r$, then $a = \frac{v^2}{r} = \frac{(\omega r)^2}{r} = \omega^2 r$

Centripetal force

To make an object move round on a circular path, it must be acted on by a resultant force which changes its direction of motion. Figure 3 shows a 'hammer' being whirled round in a circle. The tension in the cable pulls on the ball and changes its direction continuously. When the cable is released, the ball flies off at a tangent.

The resultant force on an object moving round a circle at constant speed is referred to as the **centripetal force** because it acts towards the centre of the circle.

- For an object whirling round on the end of a string, the tension in the string is the centripetal force.
- For a satellite moving round the Earth, the force of gravity between the satellite and the Earth is the centripetal force. See Figure 4.
- For a planet moving round the Sun, the force of gravity between the planet and the Sun is the centripetal force.
- For a capsule on the London Eye, the centripetal force is the resultant of the support force on the capsule and the force of gravity on it.
- In Chapter 16, you will meet the use of a magnetic field to bend a beam of charged particles (e.g. electrons) in a circular path. The magnetic force on the moving charged particles is the centripetal force.

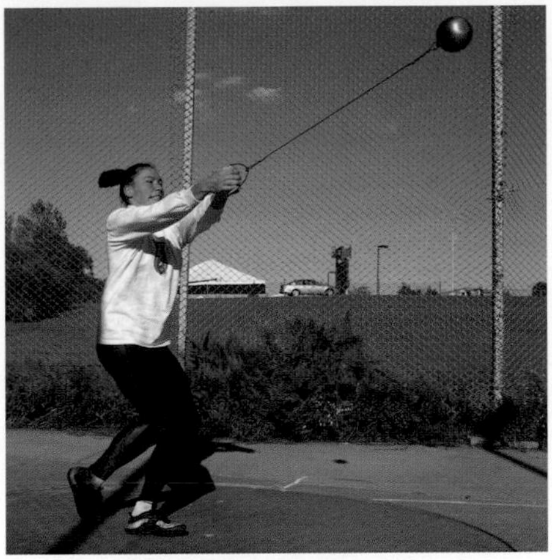

Figure 3 *Centripetal force in action*

Any object that moves in circular motion is acted on by a resultant force which always acts towards the centre of the circle. The resultant force is the centripetal force and therefore causes a centripetal acceleration.

Notes

1 If the object is acted on by a single force only (e.g. a satellite in orbit round the Earth), that force is the centripetal force and causes the centripetal acceleration.

2 The centripetal force is at right angles to the direction of the object's velocity. Therefore, no work is done by the centripetal force on the object because there is no displacement in the direction of the force. The kinetic energy of the object is therefore constant so its speed is unchanged.

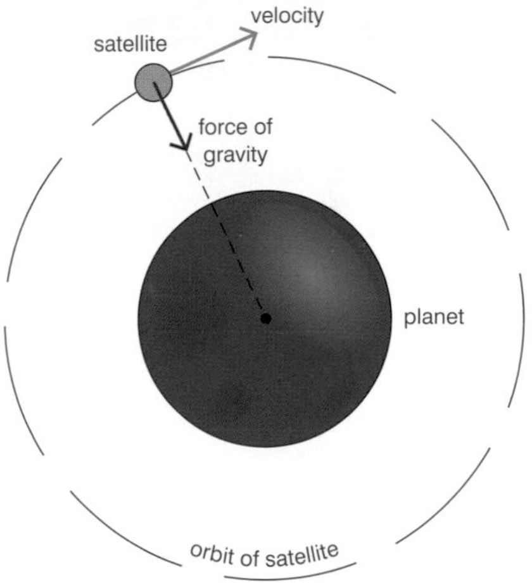

Figure 4 *A satellite in uniform circular motion*

Equation for centripetal force

For an object moving at constant speed v along a circular path of radius r, its centripetal acceleration $a = \frac{v^2}{r}$ $(= \omega^2 r)$

Therefore, applying Newton's second law for constant mass in the form '$F = ma$' gives

$$\textbf{centripetal force } F = \frac{mv^2}{r} = m\omega^2 r$$

Summary test 11.2

1 The wheel of the London Eye has a diameter of 130 m and takes 30 minutes for a full rotation. Calculate:
  a the speed of a capsule,
  b i the centripetal acceleration of a capsule,
  ii the centripetal force on a person of mass 65 kg in a capsule.

2 An object of mass 0.15 kg moves round a circular path of radius 0.42 m at a steady rate once every 5.0 seconds. Calculate:
  a the speed and acceleration of the object,
  b the centripetal force on the object.

3 a The Earth moves round the Sun on a circular orbit of radius 1.5×10^{11} m, taking $365\frac{1}{4}$ days for each complete orbit. Calculate:
  i the speed, ii the centripetal acceleration of the Earth on its orbit round the Sun.
  b A satellite is in orbit just above the surface of a spherical planet which has the same radius as the Earth and the same acceleration of free fall at its surface. Calculate:
  i the speed,
  ii the time for 1 complete orbit of this satellite.

The radius of the Earth = 6400 km
Acceleration of free fall = 9.81 m s^{-2}

4 A hammer thrower whirls a 2.0 kg hammer on the end of a rope in a circle of radius 0.8 m. The hammer took 0.6 s to make one full rotation just before it was released. Calculate:
  a the speed of the hammer just before it was released,
  b its centripetal acceleration,
  c the centripetal force on the hammer just before it was released.

11.3 On the road

Learning outcomes

On these pages you will learn to:

- recall the equations for angular speed, centripetal acceleration and centripetal force
- solve problems on uniform circular motion in the context of road vehicles and aircraft

Even on a very short journey, the effects of circular motion can be important. For example, a vehicle that turns a corner too fast could skid or topple over. A vehicle that goes over a curved bridge too fast might even lose contact briefly with the road surface. To make any object move on a circular path, the object must be acted on by a resultant force which is always towards the centre of curvature of its path.

Examples

1 Over the top of a hill

Consider a vehicle of mass m moving at speed v along a road that passes over the top of a hill or over the top of a curved bridge.

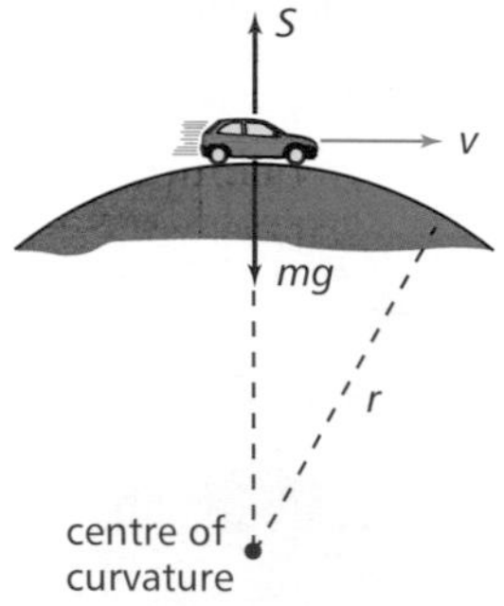

Figure 1 Over the top

At the top of the hill, the support force S from the road on the vehicle is directly upwards in the opposite direction to its weight, mg. The resultant force on the vehicle is the difference between the weight and the support force. This difference acts towards the centre of curvature of the hill as the centripetal force. In other words,

$$mg - S = \frac{mv^2}{r}, \text{ where } r \text{ is the radius of curvature of the hill}$$

The vehicle would lose contact with the road if its speed is equal to or greater than a certain speed, v_0. If this happens, then $S = 0$ so $mg = \frac{mv_0^2}{r}$

Therefore, the vehicle speed should not exceed v_0, where $v_0^2 = gr$, otherwise the vehicle will lose contact with the road surface at the top of the hill. Prove for yourself that a vehicle that travels over a curved bridge of radius of curvature 5 m would lose contact with the road surface if its speed exceeded 7 m s^{-1}.

2 On a roundabout

Consider a vehicle of mass m moving at speed v in a circle of radius r as it moves round a roundabout on a level road. The centripetal force is provided by the force of friction between the vehicle's tyres and the road surface. In other words,

$$\text{force of friction } F = \frac{mv^2}{r}$$

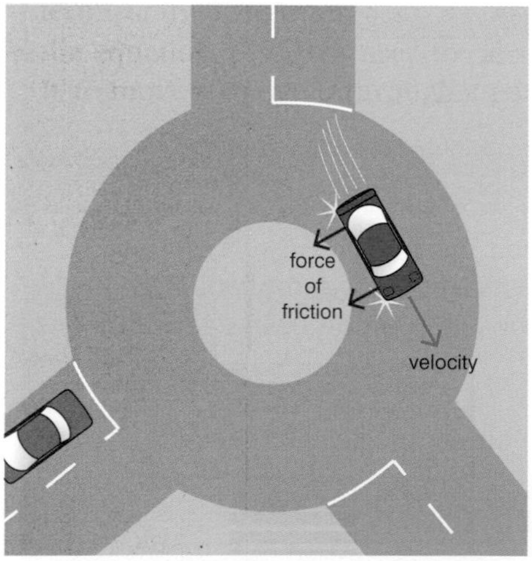

Figure 2 On a roundabout

For no skidding or slipping, the force of friction between the tyres and the road surface must be less than a limiting value F_0 which is proportional to the vehicle's weight.

Therefore, for no slipping, the speed of the vehicle must be less than a certain value v_0 which is given by the equation,

$$\text{limiting force of friction } F_0 = \frac{mv_0^2}{r}$$

3 On a banked track

A race track is often banked where it curves. Motorway slip roads in cities often bend in a tight curve. Such a road is usually banked to enable vehicles to drive round without any sideways friction on the tyres. Rail tracks on curves are usually banked to enable trains to move round the curve without slowing down too much.

- Without any banking, the centripetal force is provided only by sideways friction between the vehicle wheels and the road surface. As explained in the previous example, the vehicle on a bend slips outwards if its speed is too high.

- On a banked track, the speed can be higher. To understand why, consider Figure 3 which represents the front-view of a racing car of mass m on a banked track, where θ = the angle of the track to the horizontal. For there to be no sideways friction on the tyres due to the road, the horizontal component of the support forces N_1 and N_2 must act as the centripetal force.

Resolving these forces into horizontal components ($= (N_1 + N_2)\sin\theta$) and vertical components ($= (N_1 + N_2)\cos\theta$),

- because $(N_1 + N_2)\sin\theta$ acts as the centripetal force, then $(N_1 + N_2)\sin\theta = \frac{mv^2}{r}$
- because $(N_1 + N_2)\cos\theta$ balances the weight (mg), then $(N_1 + N_2)\cos\theta = mg$.

Therefore $$\tan\theta = \frac{(N_1 + N_2)\sin\theta}{(N_1 + N_2)\cos\theta} = \frac{mv^2}{mgr}$$

Simplifying this equation gives the condition for no sideways friction; $\tan\theta = \frac{v^2}{gr}$

In other words, there is no sideways friction if the speed v is such that $\boldsymbol{v^2 = gr\tan\theta}$.

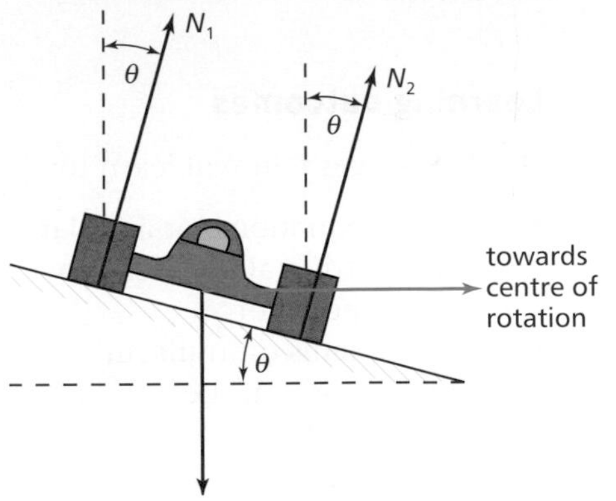

Figure 3 *A racing car taking a bend*

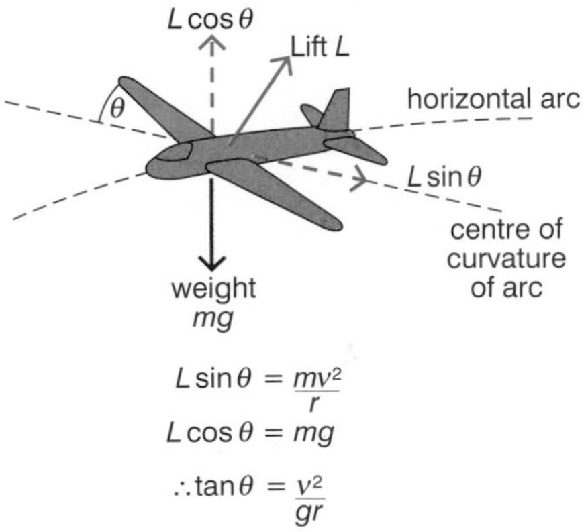

Figure 4 *An aircraft circling*

Notes

1 An aircraft banks when it turns in a circle. The horizontal component of the lift force acts as the centripetal force. Figure 4 shows the idea. The radius of the circle, r, depends on the speed and the angle of banking in accordance with the equation $\boldsymbol{v^2 = gr\tan\theta}$.

2 The **conical pendulum** in Figure 5 shows an object in uniform circular motion due to the horizontal component of a force, in this case the tension in the supporting thread. Resolving the thread tension T into horizontal and vertical components gives:

- vertical component $T\cos\theta = mg$
- horizontal component $T\sin\theta = mv^2/r$, where the circle radius $r = L\sin\theta$

Hence $\tan\theta = \frac{\sin\theta}{\cos\theta} = \frac{mv^2/r}{mg} = \frac{v^2}{gr}$

Summary test 11.3

$g = 9.81\,\text{m}\,\text{s}^{-2}$

1 A vehicle of mass 1200 kg passes over a bridge of radius of curvature 15 m at a speed of $10\,\text{m}\,\text{s}^{-1}$. Calculate:

  a the centripetal acceleration of the vehicle on the bridge,

  b the support force on the vehicle when it is at the top.

2 The maximum speed for no skidding of a vehicle of mass 750 kg on a roundabout of radius 20 m is $9.0\,\text{m}\,\text{s}^{-1}$. Calculate, for this speed:

  a the centripetal acceleration,

  b the centripetal force on the vehicle.

3 Explain why a circular athletics track that is banked is suitable for sprinters but not for marathon runners.

4 At a racing car circuit, the track is banked at an angle of 25° to the horizontal on a bend which has a radius of curvature of 350 m.

  a Calculate the speed of a vehicle on the bend if there is to be no sideways friction on its tyres.

  b Discuss and explain what could happen to a vehicle that took the bend too fast.

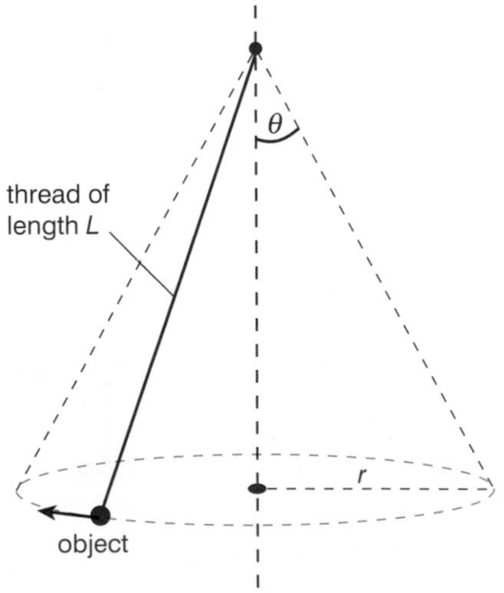

Figure 5 *The conical pendulum*

11.4 At the fairground

Learning outcomes

On these pages you will learn to:

- recall the equations for angular speed, centripetal acceleration and centripetal force
- solve problems on uniform circular motion in the fairground rides

Many of the rides at a fairground take people round in circles. Some examples are analysed below. It is worth remembering that centripetal acceleration values of more than 2–3g can be dangerous to the average person.

Examples

1 The Big Dipper

A ride that takes you at high speed through a big dip pushes you into your seat as you pass through the dip. The difference between the support force on you (acting upwards) and your weight acts as the centripetal force.

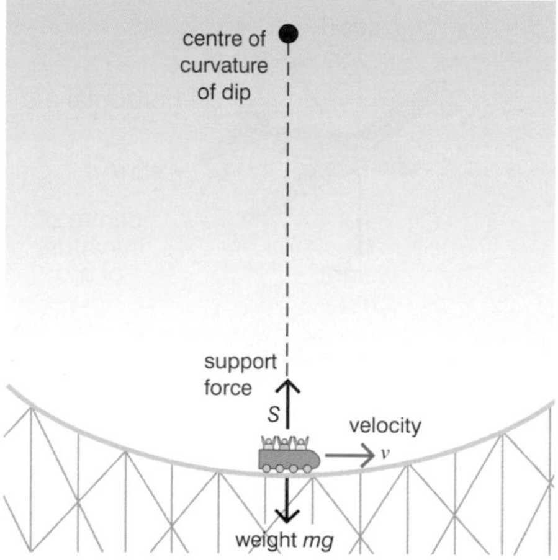

Figure 1 In a dip

At the bottom of the dip, the support force S on you is vertically upwards, as shown in Figure 1.

Therefore, for a speed v at the bottom of a dip of radius of curvature r,

$$S - mg = \frac{mv^2}{r}$$

So the support force $S = mg + \frac{mv^2}{r}$

The extra force you experience due to your motion is therefore $\frac{mv^2}{r}$.

2 The very long swing

Consider a person of mass m on a very long swing of length L, released from height h above the equilibrium position. The maximum speed is when the swing passes through the lowest point. This can be worked out by equating the loss of potential energy to the gain of kinetic energy.

$$\tfrac{1}{2}mv^2 = mgh$$

where v is its speed as it passes through the lowest point.

$$\text{Therefore } v^2 = 2gh$$

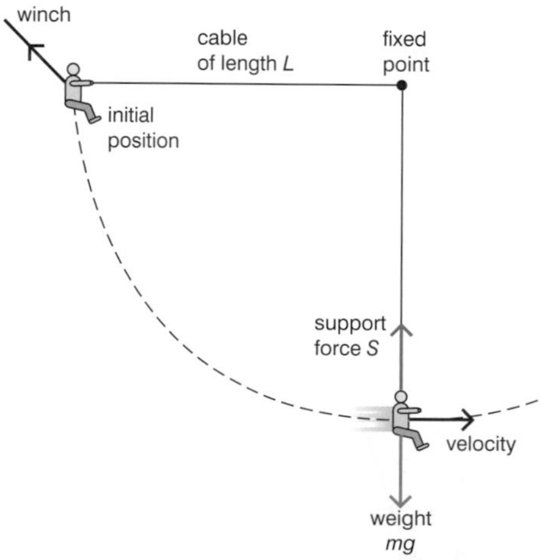

Figure 2 The very long swing

The person on the swing is on a circular path of radius L. At the lowest point, the support force S on the person due to the rope is in the opposite direction to the person's weight, mg. The difference, $S - mg$, acts towards the centre of the circular path and provides the centripetal force. Therefore

$$S - mg = \frac{mv^2}{L}$$

Because $v^2 = 2gh$, $S - mg = \frac{2mgh}{L}$

In other words, $\frac{2mgh}{L}$ represents the extra support force the person experiences due to circular motion. Prove for yourself that for $h = L$ (i.e. a 90° swing), the extra support force is equal to twice the person's weight.

3 The Big Wheel

This ride takes its passengers round in a vertical circle on the inside of the circumference of a very large wheel. The wheel turns fast enough to stop the passengers falling out as they pass through the highest position.

At maximum height, the reaction R from the wheel on each person acts downwards. Therefore, the resultant force at this position = $mg + R$. This reaction force and the weight provide the centripetal force. Therefore, at the highest position when the wheel speed is v,

$$mg + R = \frac{mv^2}{r}, \qquad \text{where } r \text{ is the radius of the wheel}$$

$$\therefore \quad R = \frac{mv^2}{r} - mg$$

At a certain speed v_0 such that $v_0{}^2 = gr$, then $R = 0$ so there would be no force on the person due to the wheel.

A person in a Big Wheel with capsules (e.g. the London Eye) would be unsupported at speed v_0. Such a wheel must turn more slowly otherwise passengers would lose contact with the capsule floor.

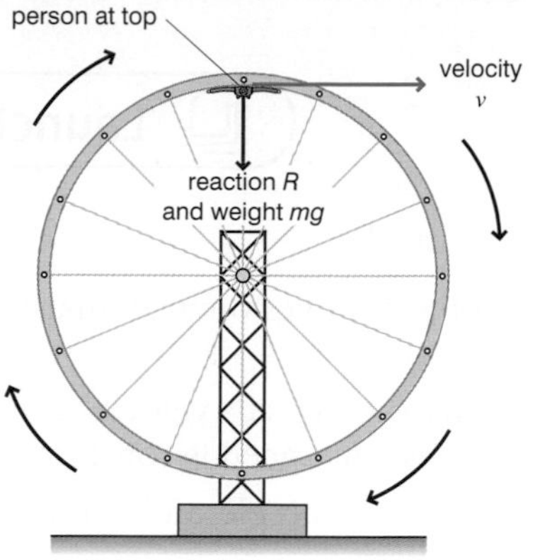

***Figure 3** The Big Wheel*

Summary test 11.4

$g = 9.81\,\text{m}\,\text{s}^{-2}$

1 A train on a fairground ride is initially stationary before it descends through a height of 45 m into a dip which has a radius of curvature of 78 m, as shown in Figure 4.

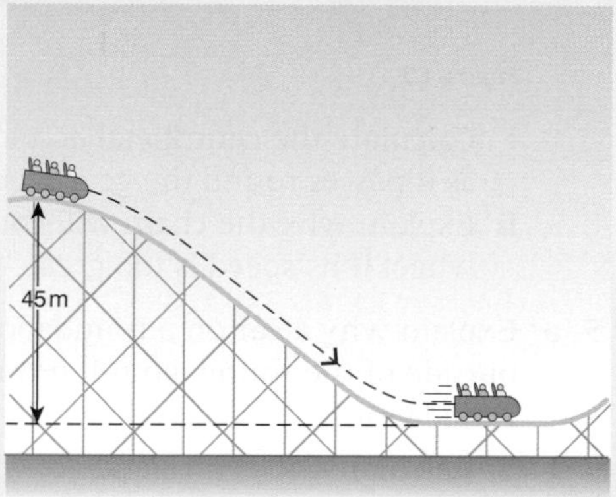

Figure 4

a Calculate the speed of the train at the bottom of the dip, assuming air resistance and friction are negligible.

b Calculate:

i the centripetal acceleration of the train at the bottom of the dip,

ii the extra support force on a person of weight 600 N in the train.

2 A very long swing at a fairground is 32 m in length. A person of mass 69 kg on the swing descends from a position when the swing is horizontal. Calculate:

a the speed of the person at the lowest point,

b the centripetal acceleration at the lowest point,

c the support force on the person at the lowest point.

3 The Big Wheel at a fairground has a radius of 12.0 m and rotates in a vertical plane once every 6.0 seconds. Calculate:

a the speed of rotation of the perimeter of the wheel,

b the centripetal acceleration of a person on the perimeter,

c the support force on a person of mass 72 kg at the highest point.

4 The wheel of the London Eye has a diameter of 130 m and takes 30 minutes to complete one revolution. Calculate the change, due to rotation of the wheel, of the support force on a person of weight 500 N in a capsule at the top of the wheel.

Chapter Summary

1 For an object of mass m in uniform circular motion on a circle of radius r:

- its **speed** $v = \frac{2\pi r}{T}$, where T is the time for one rotation
- its **frequency of rotation** $f = \frac{1}{T}$
- its **angular displacement**, in radians, in time $t = 2\pi f t = \frac{2\pi t}{T}$
- its **centripetal acceleration**, $a = \frac{v^2}{r}$ towards the centre of the circle
- the **centripetal force** on the object $= \frac{mv^2}{r}$

2 Examples:

- Car of mass m going over the top of a hill: support force $S = mg - \frac{mv^2}{r}$
- Object of mass m going through the bottom of a dip: support force $S = mg + \frac{mv^2}{r}$
- Person on the inside of a vertical fairground wheel at the top: reaction force $R = \frac{mv^2}{r} - mg$
- Banked track at angle θ to horizontal: $v^2 = gr\tan\theta$ for no sideways friction.

11 Exam-style and Practice Questions

Launch additional digital resources for the chapter

$g = 9.81\,m\,s^{-2}$

1 An object moves on a circular path at constant speed. Explain:

a why the object's velocity continually changes even though its speed is constant,

b the object accelerates even though its speed is constant.

2 The Moon moves round the Earth once every $27\frac{1}{3}$ days on a circular orbit of radius 380 000 km. Calculate:

a the speed of the Moon,

b its centripetal acceleration.

3 A pulley wheel of diameter 24 mm fitted to an electric motor in a machine rotates at a frequency of 30 Hz when the machine is in normal operation. A belt fitted to the wheel is used to drive a drum in the machine, as shown in Figure 3.1.

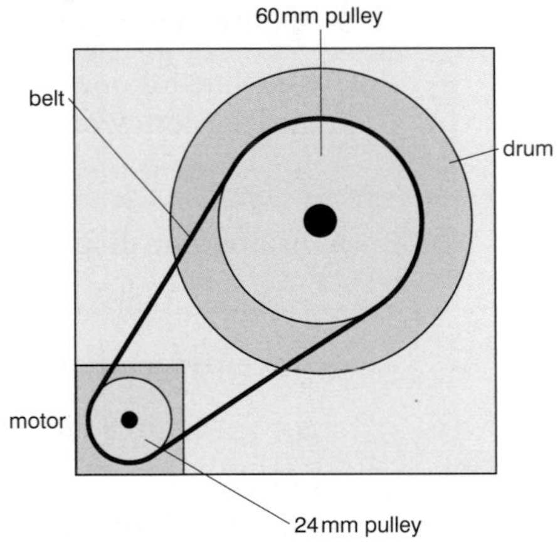

Figure 3.1

a Calculate:

i the speed of the belt on the wheel,

ii the centripetal acceleration of the belt attached to the wheel as it moves round the pulley wheel.

b The belt drives the drum via a second pulley wheel of diameter 60 mm attached to the drum axle. Calculate:

i the frequency of rotation of the second pulley wheel,

ii the centripetal acceleration of the belt as it passes round the second pulley wheel.

4 A cyclist travels at a speed of $15\,m\,s^{-1}$ on a bicycle fitted with wheels of diameter 850 mm.

a Calculate:

i the frequency of rotation of each wheel,

ii the centripetal acceleration of the tyre on each wheel.

b The rear wheel of the bicycle is driven by a chain which passes round a gear wheel of diameter 55 mm attached to the axle of the rear wheel.

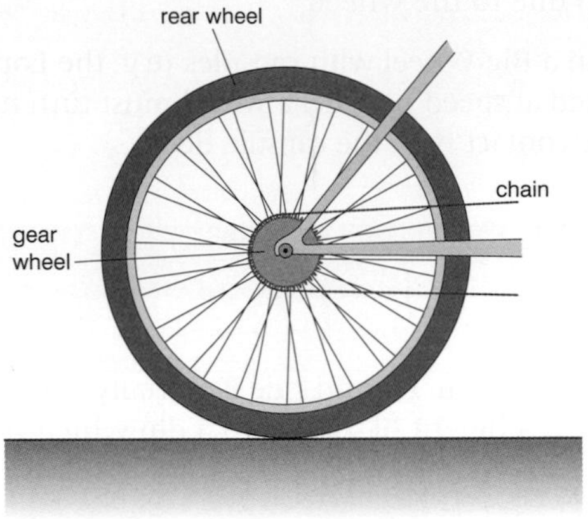

Figure 4.1

i Calculate the centripetal acceleration of the chain as it passes round the gear wheel.

ii Explain why the chain will come off the gear wheel if its speed is too great.

5 a Explain why a car on a roundabout will slide to the outside of the roundabout if it travels too fast on the roundabout.

b i Calculate the centripetal acceleration of a vehicle travelling round a roundabout of diameter 40 m when the speed of the vehicle is $8\,m\,s^{-1}$.

ii For no slippage on the roundabout, the frictional force on the vehicle tyres must be less than $0.6 \times$ the vehicle weight. Calculate the maximum speed of the vehicle on the roundabout for no slippage.

6 On a fairground ride, a train of mass 1500 kg moving at a speed of $1.5\,m\,s^{-1}$ descends from the highest point of the track through a height of 42 m to the bottom of a dip and then passes over a 'hill' which is 15 m higher than the bottom of the dip.

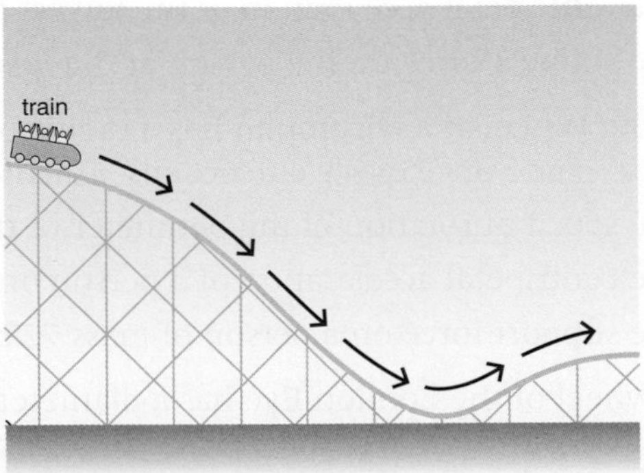

Figure 6.1

a i Calculate the speed of the train at the bottom of the dip.

ii Show that the speed of the train as it passes over the top of the hill is $23\,\mathrm{m\,s^{-1}}$. (Assume air resistance and friction are negligible.)

b i The dip has a radius of curvature of 65 m. Show that a person of mass 80 kg in the train experiences an extra support force of 1020 N on passing through the bottom of the dip.

ii The passengers in the train momentarily leave their seats when the train passes over the hill. Show that the radius of curvature of the hill is 54 m.

7 a Explain why a train can travel round a horizontal curve at a higher speed if the track is banked rather than flat.

b Discuss why the train would leave the track if it travelled round the curve too fast.

8 Figure 8.1 below shows a cross-section of an automatic brake fitted to a rotating shaft. The brake pads are held on the shaft by springs.

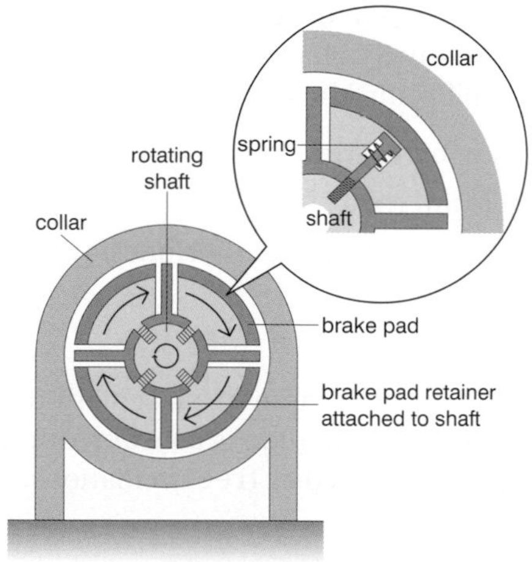

Figure 8.1

a Explain why the brake pads press against the inner surface of the stationary collar if the shaft rotates too fast.

b Each brake pad and its retainer has a mass of 0.30 kg and its centre of mass is 60 mm from the centre of the shaft. The tension in the spring attached to each pad is 250 N. Calculate the maximum frequency of rotation of the shaft for no braking.

9 a Define *angular velocity* for an object moving in uniform circular motion.

b A ball B of mass 0.16 kg and diameter 24 mm is attached to a string of length 670 mm. The ball is whirled round about a fixed point P in a vertical circle as shown in Figure 9.1.

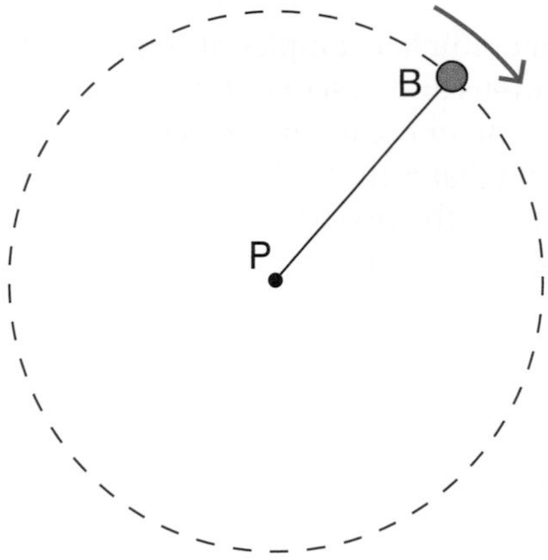

Figure 9.1

The ball rotates 24 times in 10.0 s at a constant frequency.

i Calculate the angular frequency of rotation of the ball.

ii Calculate the speed of the ball.

iii Explain why the tension in the string is greatest when the ball is at its lowest position and calculate the tension at this position.

iv The string would break if the tension in it exceeded 37 N. If the frequency of rotation was gradually increased, calculate the maximum frequency of rotation of the ball.

12 Oscillations

12.1 Measuring oscillations

Learning outcomes

On these pages you will learn to:

- describe simple examples of free oscillations and describe the motion of an oscillating object
- explain what is meant by the amplitude, the period, the frequency and the angular frequency for an oscillating object
- explain what is meant by phase difference between two objects oscillating at the same frequency
- relate the frequency and angular frequency of an oscillating object to its time period

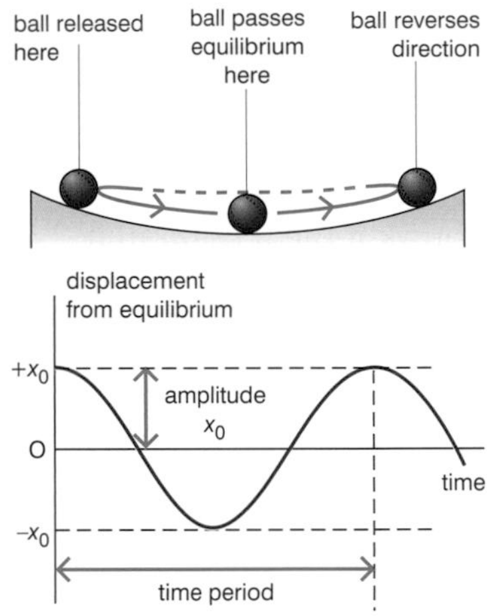

***Figure 1** Oscillating motion*

There are many examples of oscillations in everyday life. A car that travels over a bump bounces up and down for a short time afterwards. Every microcomputer has an electronic oscillator to drive its internal clock. A child on a swing moves forwards then backwards repeatedly. In this simple example, one full cycle of motion is from maximum height at one side to maximum height on the other side then back again. The lowest point is referred to as the **equilibrium** position as it is where the child eventually comes to a standstill. The child in motion is said to **oscillate** about equilibrium.

Further examples of oscillating motion include:

- an object on a spring moving up and down repeatedly,
- a pendulum moving to-and-fro repeatedly,
- a ball bearing rolling from side-to-side,
- a small boat rocking from side to side.

Displacement v. time for an oscillating object

An oscillating object moves repeatedly one way then in the opposite direction through its equilibrium position. The **displacement** of the object (i.e. magnitude and direction) from equilibrium continually changes during the motion. In one full cycle after passing through equilibrium, the displacement of the object:

- increases as it moves away from equilibrium, then
- decreases as it returns to equilibrium, then
- reverses and increases as it moves away from equilibrium in the opposite direction, then
- decreases as it returns to equilibrium.

The **amplitude** of the oscillations is the maximum displacement of the oscillating object from equilibrium. If the amplitude is constant and no frictional forces are present, the oscillations are described as **free** oscillations. See Topic 12.5.

The **time period** of the oscillating motion is the time for one complete cycle of oscillations. One full cycle after passing through any position, the object passes through that same position in the same direction.

The **frequency** of oscillations is the number of cycles per second made by an oscillating object.

The unit of frequency is the hertz (Hz) which is 1 cycle per second.

For oscillations of frequency f, the time period $\mathbf{T_P = \frac{1}{f}}$

The **angular frequency** ω of the oscillating motion is defined as $\frac{2\pi}{T_P}$ (= $2\pi f$).

The unit of ω is the radian per second (rad s^{-1}).

Phase difference

Imagine two children on adjacent identical swings. The time period, T_P of the oscillating motion is the same, as the swings are identical. If one child reaches maximum displacement on one side a certain time, Δt, later than the other child, they oscillate out of phase. Their **phase difference** stays the same as they oscillate, always corresponding to a fraction of a cycle equal to $\frac{\Delta t}{T_P}$. For example, if the time period is 2.0 s and one child reaches maximum displacement on one side 0.5 seconds later than the other child, the later child will always be a quarter of a cycle $\left(= \frac{0.5\,s}{2.0\,s}\right)$ behind the other child. Their phase difference, in radians, is therefore $0.5\pi \left(= 2\pi\frac{\Delta t}{T_P}\right)$.

In general, for two objects oscillating at the same frequency,

$$\textbf{their phase difference, in radians, } = 2\pi\frac{\Delta t}{T_P}$$

where Δt is the time between successive instants when the two objects are at maximum displacement in the same direction.

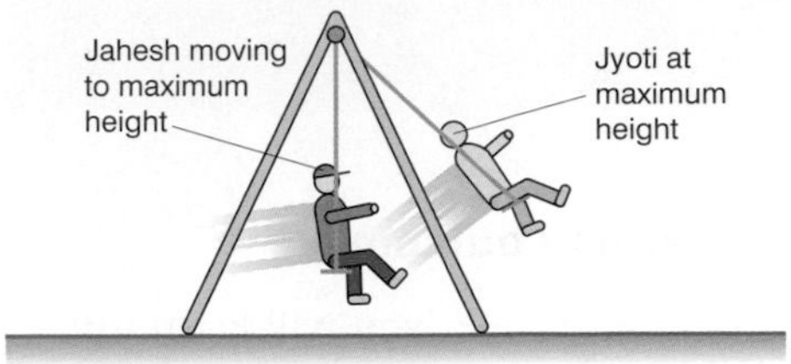

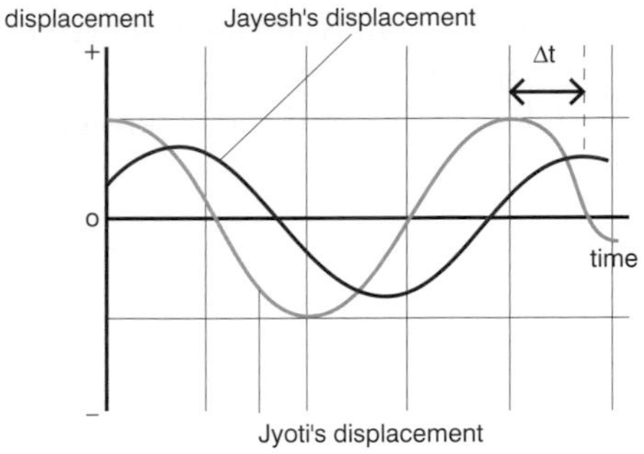

Figure 2 *Phase difference*

Notes

1 2π radians = 360° so the phase difference in degrees is $360\,\frac{\Delta t}{T_P}$.

2 The two objects oscillate in phase if $\Delta t = T_P$. The phase difference of 2π is therefore equivalent to zero.

3 Table of phase differences

Δt	0	$0.25\,T_P$	$0.50\,T_P$	$0.75\,T_P$	T_P
Phase difference in radians	0	$\frac{\pi}{2}$	π	$\frac{3\pi}{2}$	2π
Phase difference in degrees	0	90	180	270	360

Summary test 12.1

1 Describe how the velocity of a bungee jumper changes from the moment he jumps off the starting platform to the moment he next returns to the platform.

2 **a** What is meant by *free oscillations*?

b A metre rule is clamped to a table so that part of its length projects at right angles from the edge of the table, as shown in Figure 3. A 100 g mass is attached to the free end of the rule. When the free end of the rule is depressed downwards then released, the mass oscillates. Describe how you would find out if the oscillations of the mass are free oscillations.

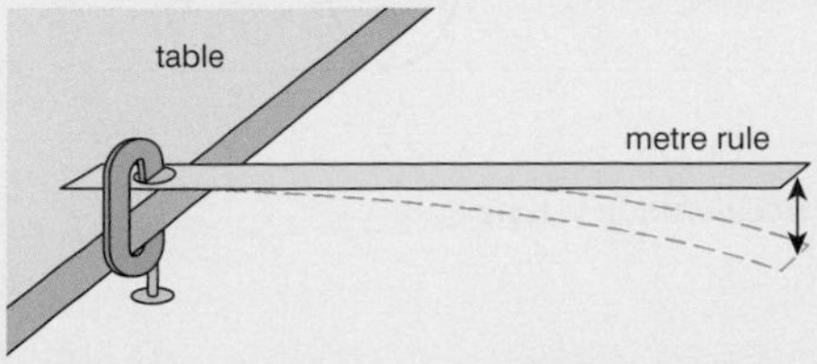

Figure 3

3 An object suspended from the lower end of a vertical spring is displaced downwards from equilibrium by a distance of 20 mm then released. It takes 9.6 s to undergo 20 complete cycles of oscillation. Calculate:

a its time period,

b its frequency of oscillation.

4 Two identical pendulums X and Y each consist of a small metal sphere attached to a thread of a certain length. Each pendulum makes 20 complete cycles of oscillation in 16 s. State the phase difference, in radians, between the motion of X and that of Y if:

a X passes through equilibrium 0.2 s after Y passes through equilibrium in the same direction,

b X reaches maximum displacement at the same time as Y reaches maximum displacement in the opposite direction.

12.2 The principles of simple harmonic motion

Learning outcomes

On these pages you will learn to:

- draw and describe graphs to show the changes in displacement, velocity and acceleration during simple harmonic motion
- define simple harmonic motion

An oscillating object speeds up as it returns to equilibrium and it slows down as it moves away from equilibrium. Figure 1 shows one way to record the displacement of an oscillating pendulum.

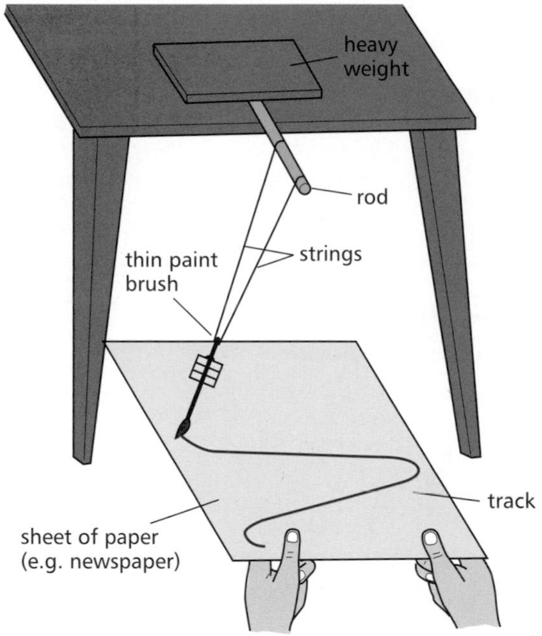

Figure 1 *Investigating oscillations*

The variation of displacement with time is shown in Figure 2 (i). Provided friction is negligible, the amplitude of the oscillations is constant.

The variation of velocity with time is given by the gradient of the displacement v. time graph, as shown by Figure 2 (ii).

- The velocity is greatest where the gradient of the displacement v. time graph is greatest (i.e. at zero displacement when the object passes through equilibrium).
- The velocity is zero where the gradient of the displacement v. time graph is zero (i.e. at maximum displacement).

The variation of acceleration with time is given by the gradient of the velocity v. time graph, as shown by Figure 2 (iii).

- The acceleration is greatest where the gradient of the velocity v. time graph is greatest. This is when the velocity is zero and occurs at maximum displacement.
- The acceleration is zero where the gradient of the velocity v. time graph is zero. This is when the displacement is zero.

By comparing Figure 2 (i) and (iii) directly, it can be seen that:

the acceleration is always in the opposite direction to the displacement.

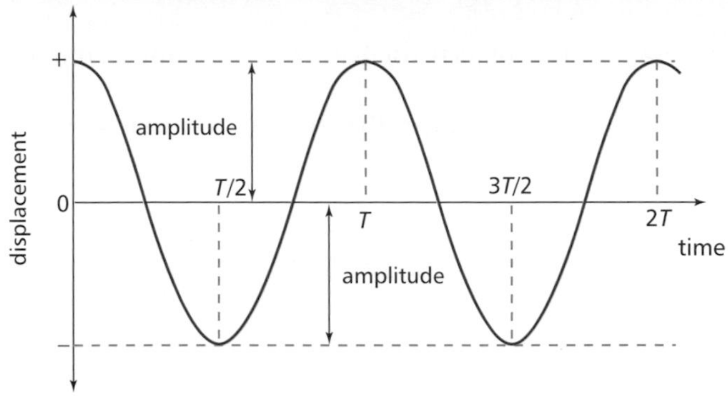

Figure 2 (i) *Displacement v. time*

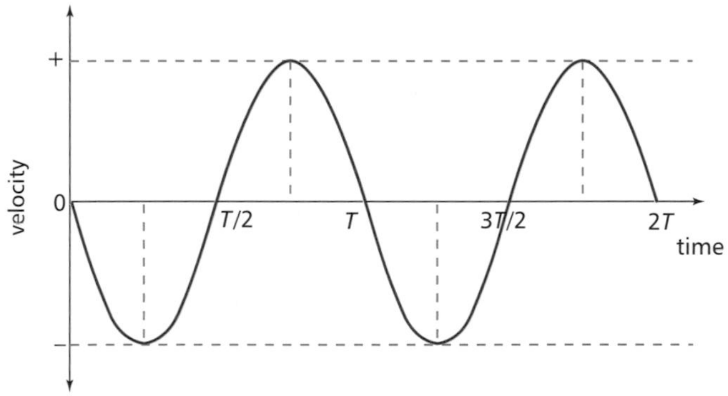

Figure 2 (ii) *Velocity v. time*

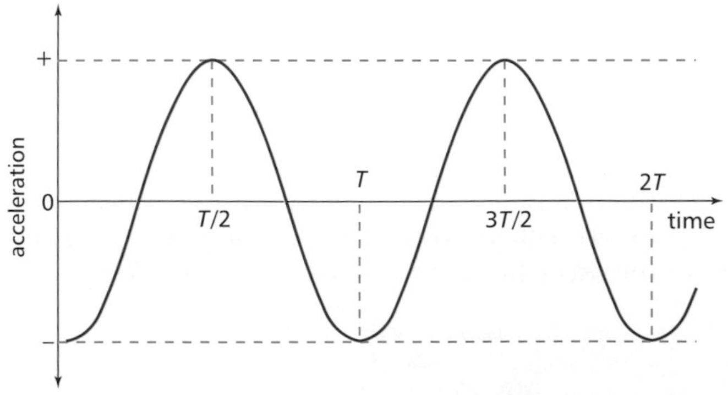

Figure 2 (iii) *Acceleration v. time*

In other words, if one direction is referred to as the positive direction and the other as the negative direction, the acceleration direction is always the opposite sign to the displacement direction.

Simple harmonic motion is defined as oscillating motion in which the acceleration is:

1 proportional to the displacement,
2 always in the opposite direction to the displacement.

Acceleration, a = – constant × displacement x

The constant depends on the time period T_P of the oscillations. The shorter the time period, the faster the oscillations which means the larger the acceleration at any given displacement. So the constant is greater the shorter the time period. As shown in Topic 12.3, the constant in this equation is ω^2, where the angular frequency, $\omega = \frac{2\pi}{T_P}$ $(= 2\pi f)$.

Therefore the defining equation for simple harmonic motion is

acceleration, $\boldsymbol{a = -\omega^2 x}$, where x = displacement

ω = angular frequency $= \frac{2\pi}{T_P}$

This equation may also be written as

$\boldsymbol{a = -(2\pi f)^2 x}$, where f = frequency

Notes

1 The time period is independent of the amplitude of the oscillations.
2 Maximum displacement $x_{max} = \pm x_0$, where x_0 is the amplitude of the oscillations. Therefore,
- when $x_{max} = +x_0$, the acceleration $a = -(2\pi f)^2 x_0$, and
- when $x_{max} = -x_0$, the acceleration $a = +(2\pi f)^2 x_0$.

Summary test 12.2

1 A small object attached to the end of a vertical spring (Figure 3) oscillates with an amplitude of 25 mm and a time period of 2.0 s. The object passes through equilibrium moving upwards at time $t = 0$. What is the displacement and direction of motion of the object:

a $\frac{1}{4}$ cycle later, **b** $\frac{1}{2}$ cycle later,
c $\frac{3}{4}$ cycle later, **d** 1 cycle later?

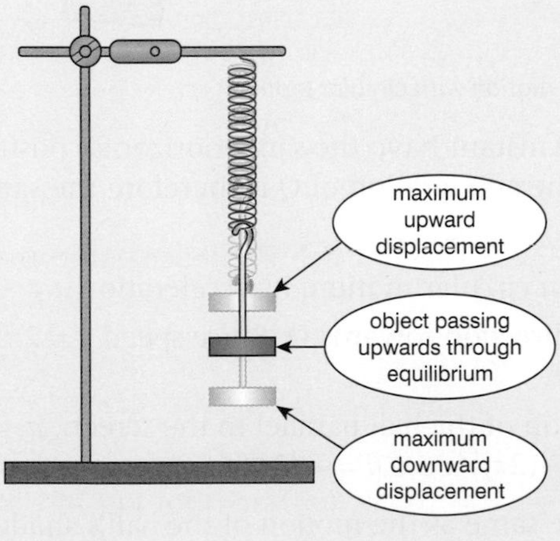

Figure 3

2 For the oscillations in **1**, calculate:

a the frequency,
b the acceleration of the object when its displacement is:
i +25 mm, **ii** 0, **iii** –25 mm.

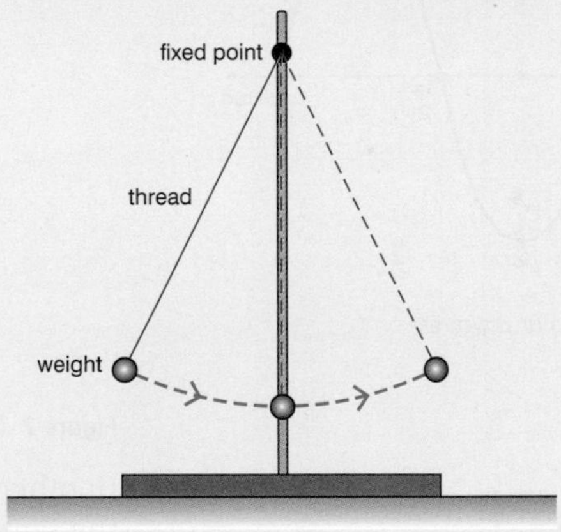

Figure 4

3 A simple pendulum consists of a small weight on the end of a thread. The weight is displaced from equilibrium and released. It oscillates with an amplitude of 32 mm, taking 20 s to execute 10 oscillations. Calculate:

a its frequency,
b its initial acceleration.

4 For the oscillations in **3**, the object is released at time $t = 0$. State the displacement and calculate the acceleration when:

a $t = 1.0$ s,
b $t = 1.5$ s.

12.3 More about sine waves

Learning outcomes

On these pages you will learn to:

- recall and use the equation $a = -\omega^2 x$ to solve simple harmonic motion problems
- recall and use $x = x_0 \sin \omega t$ as a solution to the equation $a = -\omega^2 x$

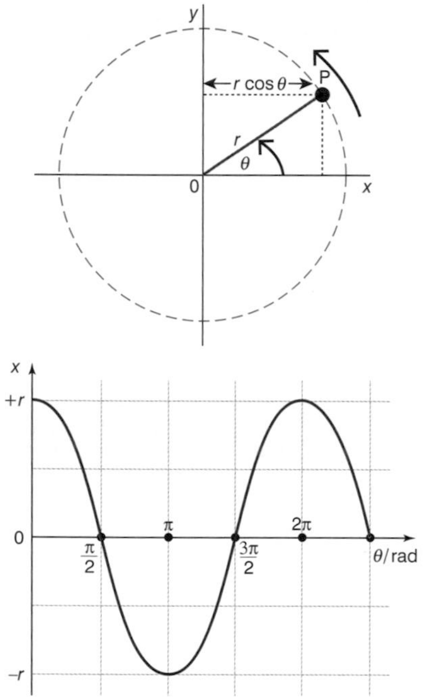

Figure 1 *Circles and waves*

Circles and waves

Consider a small object P in uniform circular motion, as shown in Figure 1. Measured from the centre of the circle at O, the coordinates of P are therefore $x = r \cos \theta$ and $y = r \sin \theta$, where θ is the angle between the x-axis and the radial line OP. Figure 1 shows how the x-coordinate changes as angle θ changes. The shape of the curve is called a **sinusoidal wave**. It has the same shape as the simple harmonic motion curves in Topic 12.2 Figure 2.

Oscillating shadows

To see the link between simple harmonic motion and sine curves, consider the motion of the ball and the pendulum bob in Figure 2. A projector is used to cast a shadow of the ball (P) in uniform circular motion on to a screen alongside the shadow of the bob (Q) of a simple pendulum oscillating above the circle. The two shadows keep up with each other exactly when their time periods are matched.

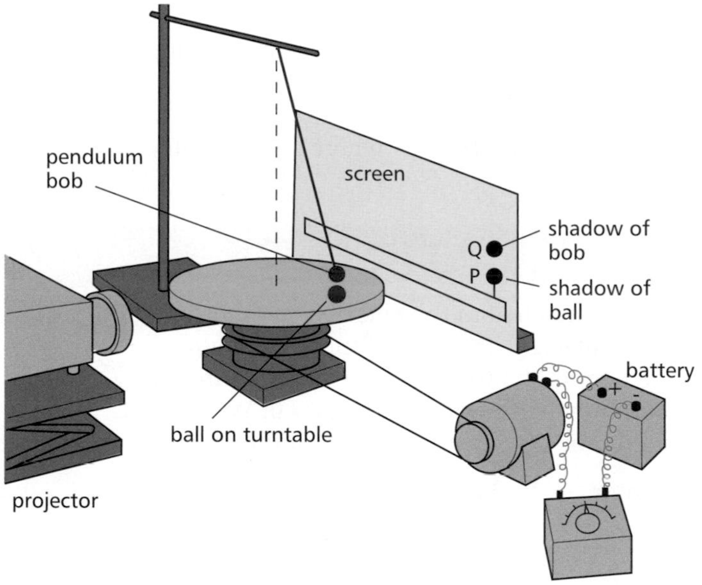

Figure 2 *Comparing simple harmonic motion with circular motion*

In other words, P and Q at any instant have the same horizontal position and the same horizontal motion. The acceleration of Q is therefore the same as the acceleration of P.

- Because the ball is in uniform circular motion, its acceleration $a = -\frac{v^2}{r}$, where the minus sign indicates its direction towards O. Since speed $v = 2\pi r f$ (see Topic 11.1), $a = -(2\pi f)^2 r$.
- The component of acceleration of the ball parallel to the screen, $a_x = a \cos \theta$, $\therefore$ the acceleration of P, $a_x = -(2\pi f)^2 r \cos \theta = -(2\pi f)^2 x$

Because the bob's motion is the same as the motion of the ball's shadow,

$$\text{the acceleration of the bob, } a_x = -(2\pi f)^2 x$$

This is the defining equation for simple harmonic motion and it shows why the constant of proportionality is $(2\pi f)^2$ or ω^2 where $\omega = 2\pi f$.

Note

The bob oscillates along the x-axis between $x = -r$ and $x = +r$. Its amplitude of oscillation, $x_0 = r$.

Sine wave solutions

For any object oscillating at frequency f in simple harmonic motion, its acceleration a at displacement x is given by

$$a = -(2\pi f)^2 x = -\omega^2 x$$

The variation of displacement with time depends on the initial displacement and the initial velocity (i.e. the displacement and velocity at time $t = 0$). For example:

1 If $x = 0$ when $t = 0$ and the object is moving to maximum displacement $+x_0$, then:

- Its displacement at time t is given by $x = x_0 \sin \omega t$. This is how the displacement of the bob's shadow Q varies with time if $t = 0$ is when the ball is nearest the screen.
- Its velocity at time t is given by $v = v_0 \cos \omega t$ where its maximum velocity $v_0 = \omega x_0$. This is how the velocity of the bob's shadow varies with time if time $t = 0$ when the ball is nearest the screen.

2 If $x = +x_0$ when $t = 0$ and the object has zero velocity at that instant, then its displacement at time t later is given by $x = x_0 \cos \omega t$. This is how the displacement of Q varies with time if $t = 0$ had been at the instant when Q was at maximum positive displacement.

Notes

1 In the CIE exams, you are expected to be able to use the equations $x = x_0 \sin \omega t$ and $v = v_0 \cos \omega t$, which give the displacement and velocity at time t after zero initial displacement.

2 The general solution is $x = x_0 \sin(\omega t + \phi)$, where ϕ is the phase difference between the instants when $t = 0$ and when $x = 0$. The general solution is not required in this physics specification.

3 The time period T_P does not depend on the amplitude of the oscillating motion. For example, the time period of an object oscillating on a spring is the same, regardless of whether the amplitude is large or small.

4 The symbol A is sometimes used for amplitude instead of x_0.

Worked example

A small object on a spring oscillates with a time period of 0.48 s and an amplitude of 15 mm. Its displacement, x, from equilibrium at time t is given by $x = x_0 \cos \omega t$. Calculate:

a its frequency,

b its displacement and acceleration at $t = 0.20$ s.

Solution

a $f = \dfrac{1}{T_P} = \dfrac{1}{0.48} = 2.08\,\text{Hz}$

b $x_0 = 15\,\text{mm} = 0.015\,\text{m}$,
$\omega = 2\pi f = 2\pi \times 2.08 = 13.1\,\text{rad}\,\text{s}^{-1}$

$x = 0.015 \cos \omega t = 0.015 \cos(13.1 \times 0.20) = -1.30 \times 10^{-2}\,\text{m}$

$a = -\omega^2 x = -(13.1^2 \times -1.30 \times 10^{-2}) = 2.23\,\text{m}\,\text{s}^{-2}$

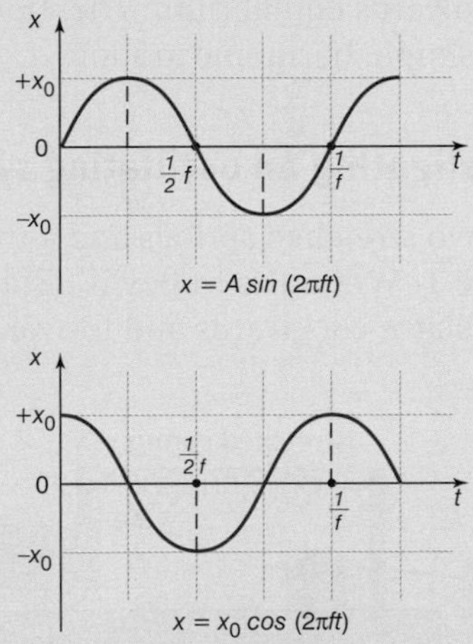

Figure 3 *Graphical solutions*

Summary test 12.3

1 An object oscillates in simple harmonic motion with a time period of 3.0 s and an amplitude of 58 mm. Calculate:

a its frequency,

b its maximum acceleration.

2 The displacement of an object oscillating in simple harmonic motion varies with time in accordance with the equation $x/\text{mm} = 12 \sin 10t$, where t is the time in seconds after the object's displacement was at its maximum positive value.

a Determine:
i the amplitude, **ii** the time period.

b Calculate the displacement of the object at $t = 0.1$ s.

3 An object on a spring oscillates with a time period of 0.48 s and a maximum acceleration of $9.81\,\text{m}\,\text{s}^{-2}$. Calculate:

a its frequency,

b its amplitude.

4 An object oscillates in simple harmonic motion with an amplitude of 12 mm and a time period of 0.27 s. Calculate:

a its frequency,

b its displacement and its velocity:
i at 0.10 s,
ii at 0.20 s after its displacement was zero and it was moving towards the maximum positive displacement.

12.4 Applications of simple harmonic motion

Learning outcomes

On these pages you will learn to:

- describe how to measure the period of the oscillations of a loaded spring and a simple pendulum
- use the equations for the period of the oscillations of a loaded spring and a simple pendulum to solve problems

For any oscillating object, the resultant force acting on the object acts towards the equilibrium position. The resultant force is described as a restoring force as it always acts towards equilibrium. Provided the restoring force is proportional to the displacement from equilibrium, the acceleration is proportional to the displacement and always acts towards equilibrium. Therefore, the object oscillates with simple harmonic motion.

Investigating an oscillating system

Use two stretched springs and a trolley, as shown in Figure 1. When the trolley is displaced and then released, it oscillates backwards and forwards.

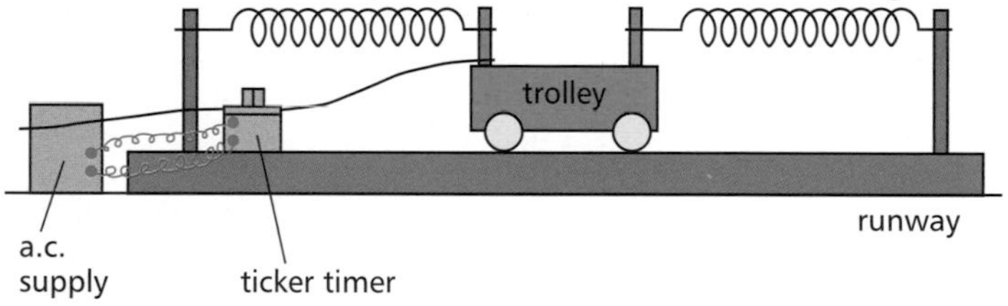

Figure 1 *Investigating oscillations*

- The first half-cycle of the trolley's motion can be recorded using a length of ticker tape attached at one end to the trolley. When the trolley is released, the tape is pulled through a ticker timer that prints dots on the tape at a rate of 50 dots per second.
  A graph of displacement v. time for the first half-cycle can be drawn using the tape, as shown in Figure 2. The graph can be used to measure the time period which can be checked as explained on the next page if the trolley mass m and the combined spring constant k are known.
- A motion sensor linked to a computer can be used to record the oscillating motion of the trolley. (See p.27.)

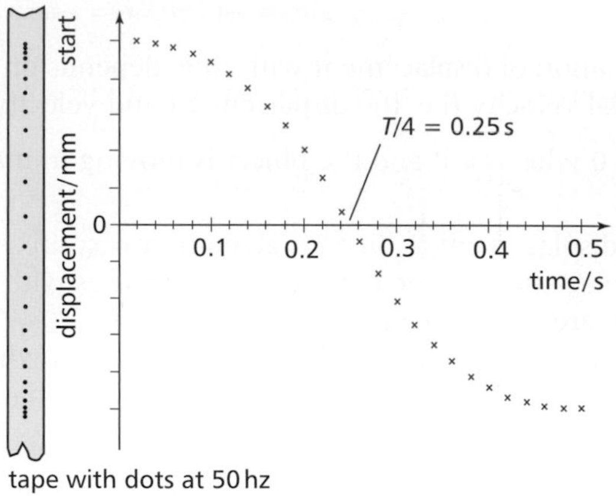

Figure 2 *Displacement–time curve from a tickertape*

What determines the frequency of oscillation of a loaded spring?

In the above investigation, the frequency of oscillation of the trolley can be changed by loading the trolley with extra mass or by replacing the springs with springs of different stiffness. The frequency is reduced by:

1. **Adding extra mass:** This is because the extra mass increases the inertia of the system. At a given displacement, the trolley would therefore be slower than if the extra mass had not been added. Each cycle of oscillation would therefore take longer.
2. **Using weaker springs:** The restoring force on the trolley at any given displacement would be less, as a weaker spring has a smaller spring constant. So the trolley's acceleration and speed at any given displacement would be less than it would have been if the weaker springs had not been added. Each cycle of oscillation would therefore take longer.

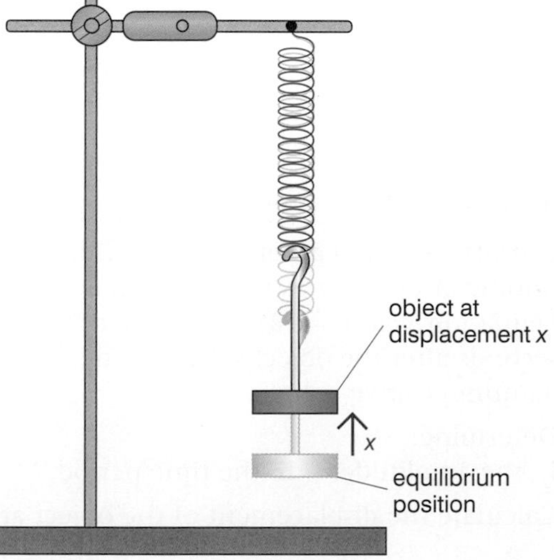

Figure 3 *The oscillations of a loaded spring*

To see exactly how the mass and the spring constant affect the frequency, consider a small object of mass m attached to a spring (see Figure 3).

- Assuming the spring obeys **Hooke's law**, the tension T in the spring is in proportion to its extension e from its unstretched length. This relationship can be expressed by means of the equation $T = ke$, where k is the spring constant.
- When the object is oscillating and is at displacement x from its equilibrium position, the change of tension in the spring provides the restoring force on the object. Using the equation $T = ke$, the change of tension ΔT from equilibrium is therefore given by $\Delta T = -kx$, where the minus sign represents the fact that the change of tension always tries to restore the object to its equilibrium position. Hence, the restoring force on the object = $-kx$.
- Therefore, the acceleration,

$$a = \frac{\text{restoring force}}{\text{mass}} = -\frac{kx}{m} = -\omega^2 x$$

where $\omega^2 = \dfrac{k}{m}$

The object therefore oscillates in simple harmonic motion with a time period

$$T_P = \frac{2\pi}{\omega} = 2\pi\sqrt{\frac{m}{k}}$$

Notes

1 The formula for the time period shows that T_P is increased by increasing m (i.e. using a larger mass) or by reducing k (i.e. using a weaker spring).

2 The time period does not depend on g. A mass–spring system on the Moon would have the same time period as it would on Earth.

3 The tension T in the spring varies from $mg + kx_0$ to $mg - kx_0$, where x_0 = amplitude.

Maximum tension is when the spring is stretched as much as possible (i.e. $x = -x_0$)

Minimum tension is when the spring is stretched as little as possible (i.e. $x = +x_0$).

Worked example

$g = 9.81\,\text{m}\,\text{s}^{-2}$

A spring of natural length 300 mm hangs vertically with its upper end attached to a fixed point. When a small object of mass 0.20 kg is suspended from the lower end of the spring in equilibrium, the spring is stretched to a length of 379 mm. Calculate:

a **i** the extension of the spring at equilibrium, **ii** the spring constant,

b the time period of oscillations that the mass on the spring would have if the mass was to be displaced downwards slightly then released.

Solution

a **i** Extension of spring at equilibrium,

$e_0 = 79\,\text{mm} = 0.0790\,\text{m}$

ii Spring constant $k = mg / e_0 = 0.20 \times 9.81 / 0.079 = 24.8\,\text{N}\,\text{m}^{-1}$

b $T_P = 2\pi\sqrt{\dfrac{0.20}{24.8}} = 0.56\,\text{s}$

Investigating the oscillations of a loaded spring

To verify the formula $T_P = 2\pi\sqrt{\dfrac{m}{k}}$ for a loaded spring:

- measure the time period for at least six different known masses,
- time 20 complete cycles of oscillation three times for each mass to give an average value of the time period for each mass.

A graph of T_P^2 on the vertical axis against m on the horizontal axis should give a straight line through the origin. This is because $T_P^2 = \dfrac{4\pi^2}{k}m$, which means that a graph of T_P^2 against m should be a straight line through the origin with a gradient equal to $\dfrac{4\pi^2}{k}$.

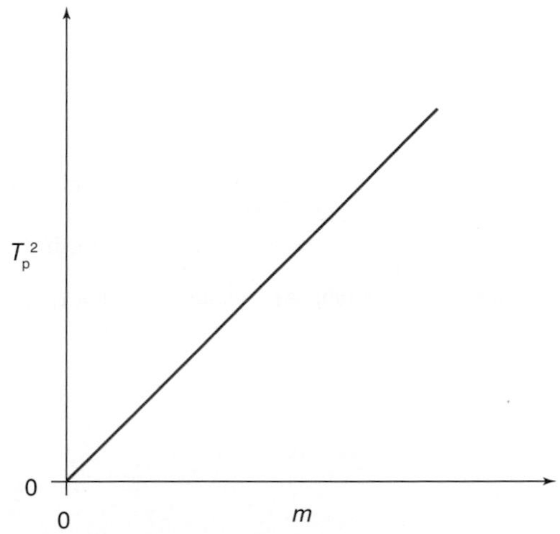

Figure 4 *T_P^2 against m for a loaded spring*

Field patterns

1 **A radial field** is where the field lines are like the spokes of a wheel, which for a gravitational field are always directed to the centre. Figure 2 shows an example of a radial field. The force of gravity on a small mass near a much larger spherical mass is always directed to the centre of the larger mass. For example, the force on a small object near a spherical planet always acts towards the centre of the planet, regardless of the position of the object. The magnitude of g in a radial field decreases with increased distance from the massive body.

2 **A uniform field** is where the gravitational field strength is the same in magnitude and direction throughout the field. The field lines are therefore parallel to one another.

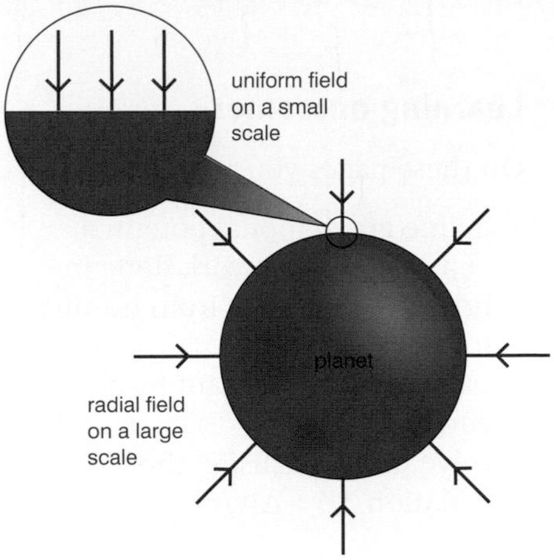

Figure 2 *Field patterns*

Is the Earth's gravitational field uniform or radial? The force of gravity due to the Earth on a small mass decreases with distance from the Earth so the gravitational field strength of the Earth falls with increasing distance from the Earth. The field is therefore radial. However, over small distances which are much less than the Earth's radius, the change of gravitational field strength is insignificant. In other words, over such small distances, the acceleration of free fall, g, (i.e. gravitational field strength) is constant. For example, the measured value of g has the same magnitude ($= 9.81\,\text{N}\,\text{kg}^{-1}$) and direction (downwards) 100 m above the Earth as it has on the surface. In theory, g is smaller higher up, but the difference is too small to be noticeable – provided we don't go too high! Only over distances which are small compared with the Earth's radius can the Earth's field be considered uniform.

Summary test 13.1

1 **a** What is meant by a *field line* or a *line of force* of a gravitational field?

b With the aid of a diagram in each case, explain what is meant by:
i a radial field, **ii** a uniform field.

2 **a** Calculate the gravitational force on:
i an object of mass 3.5 kg in a gravitational field at a position where $g = 9.5\,\text{N}\,\text{kg}^{-1}$,
ii an object of mass 100 kg in a gravitational field at a position where $g = 1.6\,\text{N}\,\text{kg}^{-1}$.

b Calculate the gravitational field strength at a position in a gravitational field where:
i an object of mass 2.5 kg experiences a force of 40 N,
ii an object of mass 18 kg experiences a force of 72 N.

3 Show that the acceleration of an object falling freely in a gravitational field is equal to g, where g is the gravitational field strength at that position.

4 Figure 3 represents a small part of the Earth's surface. Sketch the lines of force near this part of the Earth's surface:

a if the density of the Earth in this part is uniform,

b if there is a large mass of dense matter under this part of the surface.

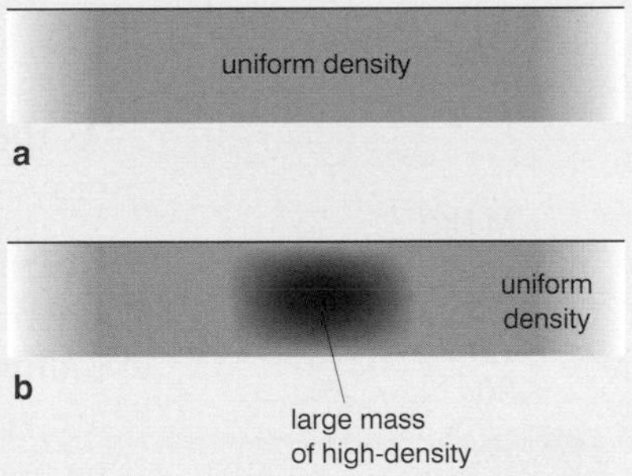

Figure 3

13.2 Gravitational potential

Learning outcomes

On these pages you will learn to:

- define gravitational potential at a point as the work done in bringing unit mass from infinity to the point
- explain what is meant by an equipotential
- solve problems using the equation $\Delta\phi = \Delta W/m$

Figure 1 Into space

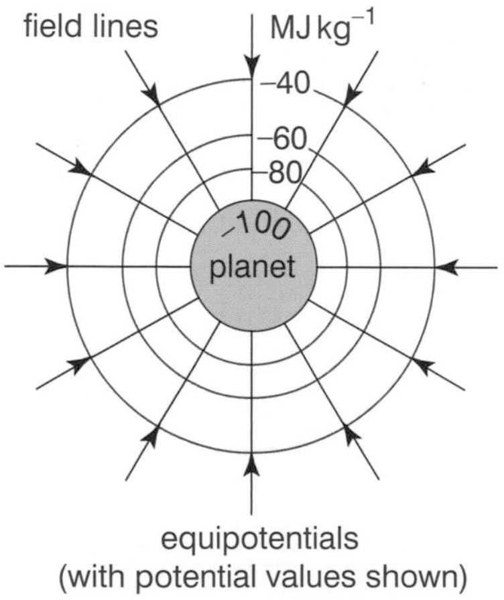

Figure 2 Equipotentials near a planet

Working in a gravitational field

Gravitational potential energy is the energy of an object due to its position in a gravitational field. The position for zero gravitational potential energy is at infinity – in other words, the object would be so far away that the gravitational force on it is negligible. A rocket climbing out of a planet's gravitational field needs to increase its gravitational potential energy to zero to escape completely. At the surface, its gravitational potential energy was negative so it needs to do work to escape from the field completely.

The gravitational potential at a point is defined as the work done per unit mass to move a small object from infinity to that point. In this book, the symbol for gravitational potential is ϕ (pronounced 'phi'). Note that ϕ is a scalar quantity and its unit is the joule per kilogram (J kg^{-1}).

For a small object of mass m at a position where the gravitational potential is ϕ, the work W that must be done on the object to enable it to escape completely is given by $W = m\phi$. Rearranging this gives

$$\phi = \frac{W}{m}$$

Suppose a rocket has a 'payload' mass of 1000 kg and the gravitational potential at the surface of the planet is −100 MJ kg^{-1}. Assume the fuel is used quickly to boost the rocket to high speed. For the rocket to escape completely, the gravitational potential energy of the 1000 kg payload must increase from −100 × 1000 MJ to zero. So the work done on the payload must be at least 100 000 MJ to escape. If the rocket payload is only given 40 000 MJ of kinetic energy from the fuel, then it can only increase its gravitational potential energy by 40 000 MJ. So it can only reach a position in the field where the gravitational potential is −60 MJ kg^{-1}.

Note

In general, if a small object of mass m is moved from gravitational potential ϕ_1 to gravitational potential ϕ_2, its change of gravitational potential energy $\Delta E_P = m(\phi_2 - \phi_1) = m\Delta\phi$ where $\Delta\phi = (\phi_2 - \phi_1)$.

As the work done ΔW to move it from ϕ_1 to ϕ_2 is equal to the change of its gravitational potential energy, then $\boldsymbol{\Delta W = m\Delta\phi}$.

Near the Earth's surface, we can use $\Delta E_P = mg\Delta h$ because g is effectively unchanged from its surface value provided Δh is much smaller than the Earth's radius. Remember that $\Delta E_P = mg\Delta h$ can only be applied for values of Δh which are very small compared with the Earth's radius, whereas $\Delta E_P = m\Delta\phi$ can always be applied.

Equipotentials

Equipotentials are lines of constant potential. Hillwalkers ought to know all about equipotentials, since a map contour line is a line of constant potential. A contour line joins points of equal height above sea level. So a hillwalker following a contour line has constant potential energy. Sensible hillwalkers take great care where the contour lines are very close to one another. One slip and their gravitational potential energy might fall dramatically!

The equipotentials near a spherical planet are circles as shown in Figure 2. At increasing distance from the surface, the gravitational field becomes weaker, so the gain of gravitational potential energy per metre of height gain becomes less. In other words, away from the surface, the equipotentials for equal increases of potential are spaced further apart.

However, near the surface over a small region, the equipotentials are horizontal (i.e. parallel to the ground) as shown in Figure 3. This is because the gravitational field over a small region is uniform. A 1 kg mass raised from the surface of the Earth by 1 m gains 9.81 J of gravitational potential energy; if it is raised another 1 m, it gains another 9.81 J. So its gravitational potential energy rises by 9.81 J for every metre of height it gains above the surface.

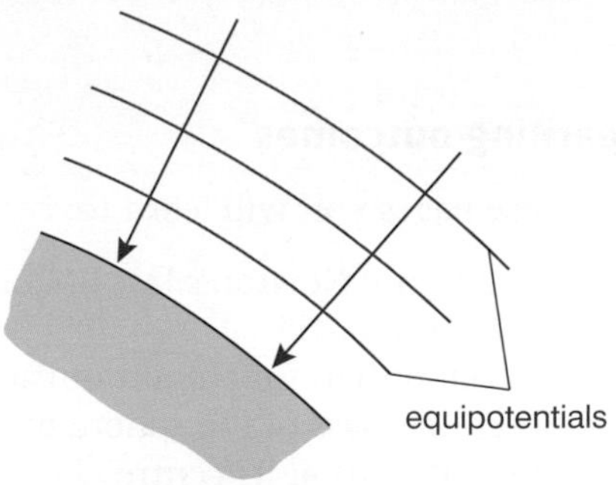

***Figure 3** Equipotentials near a surface*

Summary test 13.2

$g = 9.81\,\mathrm{N\,kg^{-1}}$

1 a Calculate the gain of gravitational potential energy of an object of mass 12 kg when its centre of mass is raised through a height of 2.0 m.

b Show that the gravitational potential difference between the Earth's surface and a point 2.0 m above the surface is $19.6\,\mathrm{J\,kg^{-1}}$.

2 A rocket of mass 35 kg launched from the Earth's surface gains 70 MJ of gravitational potential energy when it reaches its maximum height.

a Calculate the gravitational potential difference between the Earth's surface and the highest point reached by the rocket.

b The gravitational potential of the Earth's gravitational field at the surface of the Earth is $-63\,\mathrm{MJ\,kg^{-1}}$. Calculate: **i** the gravitational potential at the highest point reached by the rocket, **ii** the work that would need to have been done by the rocket to escape from the Earth's gravitational field.

3 Figure 4 shows the equipotentials near a non-spherical object.

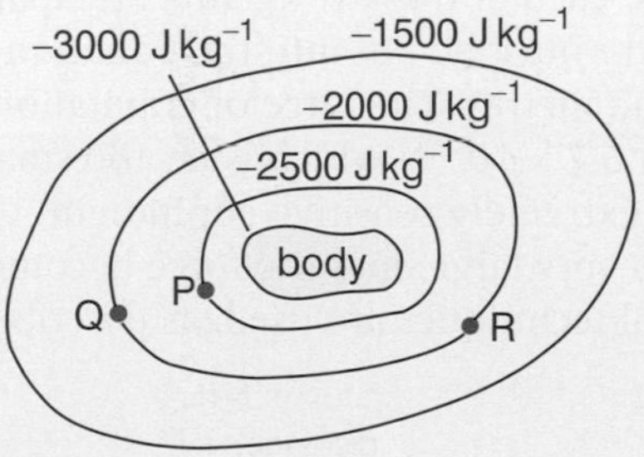

Figure 4

a Calculate the gravitational potential energy of a 0.1 kg object at: **i** P, **ii** Q, **iii** R.

b How much work must be done on the object to move it from: **i** P to Q, **ii** Q to R?

4 Figure 5 shows equipotentials at a spacing of 1.0 km near a planet. The point labelled X is on the $-500\,\mathrm{kJ\,kg^{-1}}$ equipotential.

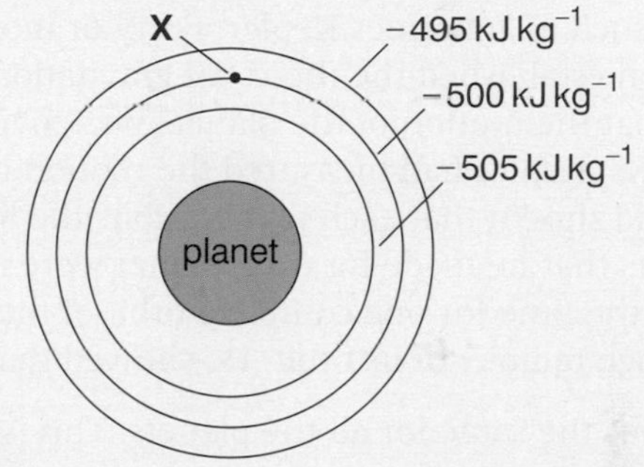

Figure 5

a Show that the work done to move a 1 kg object from X to a position 10 m higher is 50 J.

b Hence calculate the gravitational field strength at X.

c Calculate the work that would need to be done to remove an object of mass 50 kg from X to infinity.

13.3 Newton's law of gravitation

Learning outcomes

On these pages you will learn to:

- recall and use Newton's law of gravitation in the form $F = Gm_1m_2/r^2$
- understand that, for a point outside a uniform sphere, the mass of the sphere may be considered to be a point mass at its centre

We owe our understanding of gravitation to Isaac Newton. 'The notion of gravity was occasioned by the fall of an apple!', said Newton when asked what made him develop the idea of gravity. Newton's theory of gravitation was an enormous leap forward because it explains events from the 'down-to-earth' falling apple to the motion of the planets. Like any good theory, it can be used to make predictions; for example, the return of a comet and its exact path can be calculated using the law of gravitation.

Newton realised that gravity is universal. Any two masses exert a force of attraction on each other. He knew about the careful measurements of planetary motion made by astronomers such as Johannes Kepler. Forty or more years before Newton established the theory of gravitation, Kepler had shown that the motion of the planets was governed by a set of laws. Kepler had measured the motion of each planet and had shown that each planet orbits the Sun. The measurements that he made for each planet were its time period T (i.e. the time for one complete orbit of the Sun) and the average radius r of its orbit. He showed that the value of $\frac{r^3}{T^2}$ was the same for all the planets. This is known as **Kepler's third law**.

	Mercury	Venus	Earth	Mars	Jupiter	Saturn
Average radius r of orbit/10^{10} m	6	11	15	23	78	143
Time T for one orbit/10^7 s	0.8	1.95	3.2	5.9	37.4	93.0
r^3/T^2/ 10^{16} m^3 s^{-2}	337	350	330	349	340	338

Figure 1 *Kepler's third law*

To explain Kepler's third law, Newton started by assuming that the planets and the Sun were point masses. A scale model of the Solar System with the Sun represented by a marble would put the Earth about a metre away, represented by a grain of sand! Newton assumed that the force of gravitation between a planet and the Sun varied inversely with the square of their distance apart.

In other words, if the force is F at distance d apart, then:

- at distance $2d$ apart, the force is $\frac{F}{4}$,
- at distance $3d$ apart, the force is $\frac{F}{9}$,
- at distance $4d$ apart, the force is $\frac{F}{16}$.

Using this **inverse-square law of force**, Newton was able to prove that $\frac{r^3}{T^2}$ was the same for all the planets. The actual proof is outlined in Topic 13.5. Newton then went on to use the inverse-square law of force to explain and make predictions for many other events involving gravity.

Newton's law of gravitation assumes that the gravitational force between any two point objects is:

- always an attractive force,
- proportional to the mass of each object,
- proportional to $\frac{1}{r^2}$, where r is their distance apart.

These three requirements can be summarised as

$$\textbf{gravitational force } F = -\frac{Gm_1m_2}{r^2}$$

where m_1 and m_2 = masses of the two objects. The minus sign in the equation is because the force is always an attractive force.

The constant of proportionality, G, in the above equation, is called the **universal constant of gravitation**. The unit of G can be worked out from $F = -\frac{Gm_1m_2}{r^2}$; rearranged, the equation gives $G = -\frac{Fr^2}{m_1m_2}$. So G can be given units of $\mathrm{N\,m^2\,kg^{-2}}$.

The value of G is $6.67 \times 10^{-11}\,\mathrm{N\,m^2\,kg^{-2}}$.

Work out for yourself the gravitational force between two point masses, each of mass 10 kg at 0.1 m apart. The values of m_1, m_2 and r must be put into the equation in units of kilograms and metres. The force of gravitational attraction works out at 6.7×10^{-7} N which is far too small to notice except with extremely sensitive equipment. Only if one of the masses is very large does the force become noticeable, unless special techniques are used, as described later.

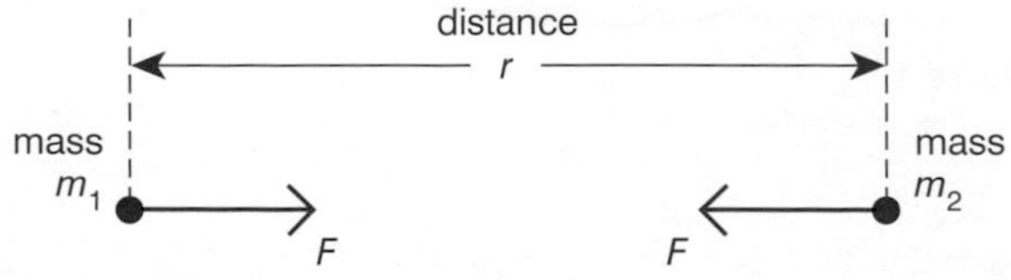

Figure 2 *Newton's law of gravitation*

Worked example

The distance from the centre of the Sun to the centre of the Earth is 1.5×10^{11} m. The mass of the Sun is 2.0×10^{30} kg and the mass of the Earth is 6.0×10^{24} kg.

a The Earth has a diameter of 1.3×10^{7} m. The Sun has a diameter of about 1.4×10^{9} m. Why is it reasonable to consider the Sun and the Earth at a distance of 1.5×10^{11} m apart as point masses on this distance scale?

b Calculate the force of gravitational attraction between the Sun and the Earth.

$G = 6.67 \times 10^{-11}\,\mathrm{N\,m^2\,kg^{-2}}$

Solution

a On a scale model where the centre of the Sun was 1 m away from the centre of the Earth, the Sun would be a sphere of diameter about 1 cm and the Earth would be a sphere of diameter about 0.1 mm, no larger than a dot. The distance from the Earth to any part of the Sun is, therefore, the same to within 1%.

b $F = -\dfrac{6.67 \times 10^{-11} \times 2.0 \times 10^{30} \times 6.0 \times 10^{24}}{(1.5 \times 10^{11})^2}$

$= -3.6 \times 10^{22}\,\mathrm{N}$

Cavendish's measurement of *G*

The first accurate measurement of *G* was made by Henry Cavendish in 1798. He devised a torsion balance made of two small lead balls at either end of a rod. The rod was suspended horizontally by a torsion wire, as in Figure 3. The wire was calibrated by measuring the couple required to twist it per degree. Then, with the rod at rest in equilibrium, two massive lead balls were brought near the torsion balance to make the wire twist. By measuring the angle through which it twisted, the force of attraction between each massive lead ball and the small ball nearest to it was calculated. The distance between the centres of the small and large masses was also measured. Then *G* was calculated using the equation for the law of gravitation.

***Figure 3** Cavendish's measurement of G*

Summary test 13.3

$G = 6.67 \times 10^{-11}\,\mathrm{N\,m^2\,kg^{-2}}$

1 a Calculate the force of gravitational attraction between two 'point objects' of masses 60 kg and 80 kg at a distance of 0.5 m apart.

b Calculate the distance between two identical point objects, each of mass 0.20 kg, that exert a force of 9.0×10^{-8} N on each other.

2 a Calculate the force of gravitational attraction between the Earth and an object of mass 80 kg on the surface of the Earth, where $g = 9.81\,\mathrm{N\,kg^{-1}}$.

b Use the result of your calculation in **a** to estimate the mass of the Earth. Assume that the mass of the Earth is concentrated at its centre.

The radius of the Earth = 6.4×10^{6} m

3 The Sun exerted a force of 6.0 N on a 1000 kg comet when it was at a distance of 1.5×10^{11} m from the Sun. Calculate the force due to the Sun on the comet when it was at a distance of:

a 0.5×10^{11} m from the Sun,

b 7.5×10^{11} m from the Sun.

4 A space rocket of mass 1500 kg travelled from the Earth to the Moon, a distance of 3.8×10^{8} m.

a When the space rocket was mid-way between the Earth and the Moon, calculate the force of gravitational attraction on it:

i due to the Earth,

ii due to the Moon.

b Calculate the magnitude and direction of the force of gravity of the Earth and the Moon on the space rocket when it was mid-way between the Earth and the Moon.

The mass of the Earth = 6.0×10^{24} kg

The mass of the Moon = 7.4×10^{22} kg

13.4 Planetary fields

Learning outcomes

On these pages you will learn to:

- derive the equation $g = \frac{GM}{r^2}$ for the gravitational field strength of a point mass
- recall and solve problems using the above equation
- solve problems using the equation $\phi = -\frac{GM}{r}$ for the potential in the field of a point mass

Note

The gravitational field strength at the surface of a sphere of radius R and mass M is given by, $g_S = \frac{GM}{R^2}$ as the distance from the centre $r = R$ at the surface.

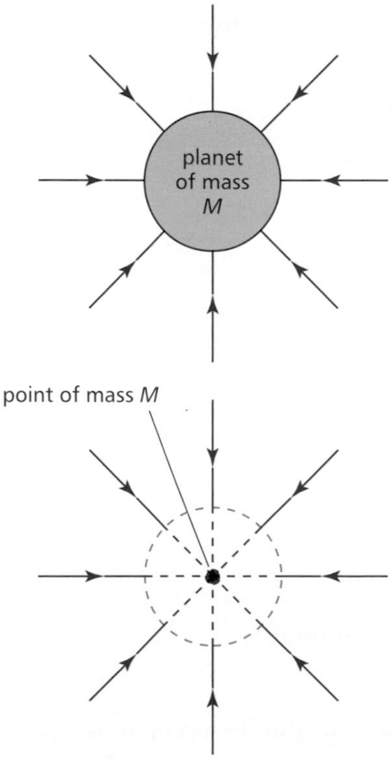

Figure 1 *Comparing fields*

Gravitational field strength near a spherical planet

The law of gravitation can be used to determine the gravitational field strength at any point in the field of a planet or any other spherical mass. Newton showed that the field of a spherical mass is the same as if it were a 'point mass' with all its mass M concentrated at its centre. The field lines of a spherical mass are always directed towards the centre, so the field pattern is just the same as for a point mass, as shown in Figure 1.

- For a point mass M, the force of attraction on a 'test mass' m (where $m \ll M$) at distance r from M is given by Newton's law of gravitation $F = \frac{GMm}{r^2}$.
  Therefore, the gravitational field strength at distance r is given by $g = \frac{F}{m} = \frac{GM}{r^2}$.
- For a spherical mass M of radius R, the force of attraction on a 'test mass' m at distance r **from the centre of M** is the same as if mass M was concentrated at its centre. Therefore,

  the force of attraction between m and M, $F = \frac{GMm}{r^2}$.

  Therefore, the gravitational field strength at distance r is given by $g = \frac{F}{m} = \frac{GM}{r^2}$,

  provided distance r is greater than or equal to the radius R of the sphere.

  Gravitational field strength, $g = \frac{GM}{r^2}$

  at distance r from a point object or the centre of a sphere of mass M.

Worked example

$G = 6.67 \times 10^{-11}\,\text{N m}^2\,\text{kg}^{-2}$

The gravitational field strength at the surface of the Earth is $9.81\,\text{N kg}^{-1}$. Calculate:

a the mass of the Earth,

b the gravitational field strength of the Earth at a height of 1000 km above the surface.

The radius of the Earth = 6400 km

Solution

a Rearranging $g_S = \frac{GM}{R^2}$ gives $M = \frac{g_S R^2}{G} = \frac{9.81 \times (6400 \times 10^3)^2}{6.67 \times 10^{-11}} = 6.0 \times 10^{24}\,\text{kg}$

b At height $h = 1000\,\text{km}$, $r = R + h = 7400\,\text{km}$

$$\therefore g = \frac{GM}{r^2} = \frac{6.67 \times 10^{-11} \times 6.0 \times 10^{24}}{(7400 \times 10^3)^2} = 7.3\,\text{N kg}^{-1}$$

The variation of *g* with distance from the centre of a spherical planet (or star)

1 **At and beyond the surface of a spherical planet** of mass M and radius R,

$$g = \frac{GM}{r^2},$$

where r is the distance from the centre of the sphere.

Because the surface gravitational field strength, $g_S = \frac{GM}{R^2}$, then $GM = g_S R^2$.

Therefore, $\boldsymbol{g = g_S \frac{R^2}{r^2}}$.

The equation and Figure 2 shows how g changes with increase of distance r. The shape of the curve beyond $r = R$ is an **inverse-square law** curve.

2 Inside the planet

From the equation $g = \frac{GM}{r^2}$, you might think that g inside the Earth becomes ever larger and larger as r becomes smaller and smaller. However, inside the planet, only the mass in the sphere of radius r contributes to g. The rest of the mass outside r up to the surface gives no resultant force. So, as r becomes smaller, the mass M which contributes to g becomes smaller too. At the centre, the mass that contributes to g is zero. So g is zero at the centre. Figure 2 also shows how g varies with distance from the centre inside the planet, assuming its density is uniform.

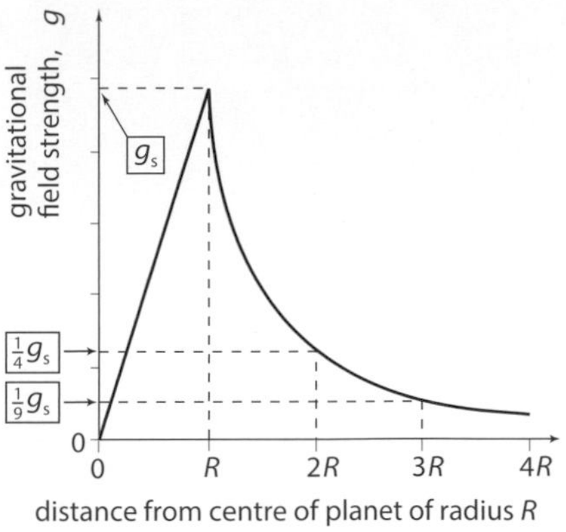

Figure 2 Gravitational field strength

Gravitational potential near a spherical planet

At or beyond the surface of a spherical planet, the gravitational potential ϕ at distance r from the centre of the planet of mass M is given by:

$$\phi = -\frac{GM}{r}$$

Applying this formula to the surface of the Earth with $M = 6.0 \times 10^{24}$ kg and $r = 6.4 \times 10^7$ m gives a value of $-63\,\text{MJ kg}^{-1}$. This means that 63 MJ of work needs to be done to remove a 1 kg mass from the surface of the Earth to infinity.

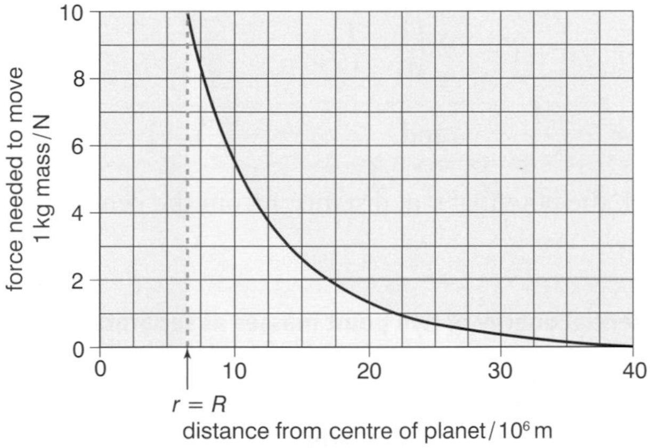

Figure 3 Work done

Figure 3 shows how the force of gravity on a 1 kg mass varies with distance r from the centre of the Earth. As explained previously, the mathematical equation for this curve is $g = g_S R^2/r^2$. The area under the curve represents the work done to move the 1 kg mass from infinity to the surface.

- Each grid square in Figure 3 represents a 1 N force acting for a distance of 2.5×10^6 m, and therefore represents 2.5 MJ (= 1 N × 2.5×10^6 m) of work done.
- The work done to move the 1 kg mass from infinity to the surface can therefore be estimated by counting the number of grid squares under the curve and multiplying this number by 2.5 MJ.

By counting part-filled squares that are half-filled or more as wholly-filled squares and neglecting part-filled squares that are less than half-filled, show for yourself that this method gives an estimate for the work done which is very close to the value of 63 MJ determined above.

In effect, the area method used above is an application of 'work done = force × distance moved' with a variable force $F = GMm/r^2$ and $m = 1$ kg.

Note

The equation $g = \frac{g_S R^2}{r^2}$ shows how g changes with increase of distance r.

- At distance $r = 2R$, $g = \frac{g_S R^2}{(2R)^2} = \frac{g_S}{4}$
- At distance $r = 3R$, $g = \frac{g_S R^2}{(3R)^2} = \frac{g_S}{9}$
- At distance $r = 4R$, $g = \frac{g_S R^2}{(4R)^2} = \frac{g_S}{16}$

Consider one small step in moving the 1 kg mass from infinity to the surface. Suppose the distance from the centre of the Earth changes from r_1 to r_2 in making this small step of distance Δr.

The work done ΔW to make this small step is given by

$$\Delta W = F\Delta r = \frac{GMm\Delta r}{r^2}, \quad \text{where } \Delta r = r_2 - r_1.$$

In Figure 3, the work done ΔW in making each small step $\Delta r = 2.5 \times 10^6$ m is represented by each column of grid squares under the curve. Thus the total work done to move from infinity to the surface is given by the total area under the curve.

Proof of the formula for gravitational potential $\phi = -GM/r$

Although the 'proof' below is not required in the specification, it is provided to give some further insight into the use of maths in physics.

The change of potential $\Delta\phi$ for the small step (from $r_1 = r$ to $r_2 = r - \Delta r$) $= \phi_2 - \phi_1$, where ϕ_2 is the potential at r_2 and ϕ_1 is the potential at r_1.

$$\Delta\phi = \frac{\Delta W}{m} = \frac{F\Delta r}{m} = -\frac{GM\Delta r}{r^2},$$

where the minus sign indicates a decrease in r.

As $$\frac{1}{r_1} - \frac{1}{r_2} = \frac{r_2 - r_1}{r_1 r_2} = \frac{(r - \Delta r) - r}{r(r - \Delta r)} = -\frac{\Delta r}{r^2} \quad \text{provided } \Delta r \ll r$$

then $$\Delta\phi = \phi_2 - \phi_1 = GM\left(\frac{1}{r_1} - \frac{1}{r_2}\right)$$

Hence $$\phi_2 = -\frac{GM}{r_2} \quad \text{and} \quad \phi_1 = -\frac{GM}{r_1}$$

Therefore, in general, the potential ϕ at distance r from the centre of a planet is given by $\phi = -\dfrac{GM}{r}$

The gravitational potential energy of two point masses at separation r

When two point masses M and m are at distance r apart, their gravitational potential energy E_P is given by the following equation:

$$E_P = -\frac{GMm}{r}$$

To prove this equation, the gravitational potential ϕ of M at distance r from M is $-GMm/r$ and is zero at infinity. So if m is moved from infinity, the work done by their mutual force of gravitational attraction changes their potential energy from zero to $m\phi$ or $m \times (-GM/r)$. So at separation r their gravitational potential energy $= -GMm/r$. The same equation is obtained if M is moved from infinity to a distance r from m.

The variation of gravitational potential with distance from the centre of a spherical planet

The gravitational potential ϕ near a spherical planet is inversely proportional to the distance r from the centre of the planet, as given by the equation $\phi = -\dfrac{GM}{r}$, where M is the mass of the planet. Figure 4 shows how the gravitational potential of the Earth varies with distance. Note that the potential curve is a $1/r$ curve not an inverse square (i.e. $1/r^2$) curve like the field strength curve in Figure 3.
So the potential:

- at distance $2R$ from the centre is $0.50 \times$ the potential at distance R from the centre,
- at distance $3R$ from the centre is $0.33 \times$ the potential at distance R from the centre,

- at distance $4R$ from the centre is $0.25 \times$ the potential at distance R from the centre, etc.

The gradient of the potential curve at any point is equal to $-g$, where g is the gravitational field strength at the point.

Consider moving a small mass m a distance Δr away from the planet.

The work done on $m = m\Delta\phi$ where $\Delta\phi$ is the change of potential.

The force applied to m = work done ÷ distance moved in the direction of the force

$$= m\frac{\Delta\phi}{\Delta r}$$

As the gravitational force F is equal and opposite to the applied force, $F = -m\frac{\Delta\phi}{\Delta r}$.

Hence $$g = \frac{F}{m} = -\frac{\Delta\phi}{\Delta r} = -\text{gradient of the potential curve.}$$

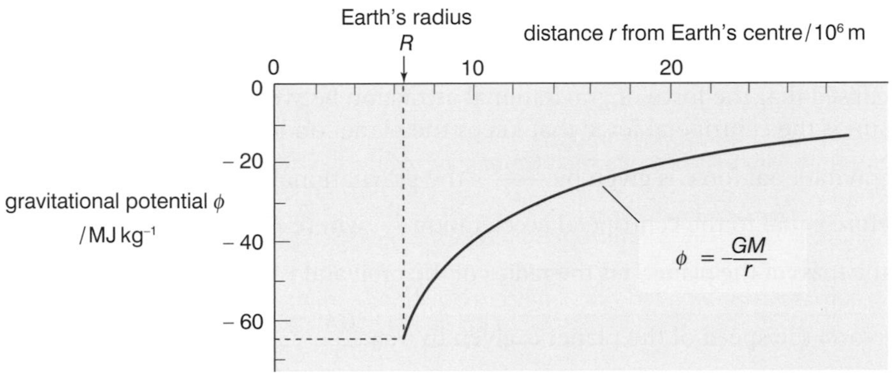

Figure 4 *Gravitational potential near the Earth*

Summary test 13.4

$G = 6.67 \times 10^{-11}\,\text{N m}^2\,\text{kg}^{-2}$

1 The Moon has a radius of 1740 km and its surface gravitational field strength is $1.62\,\text{N kg}^{-1}$ to three significant figures.

a Calculate the mass of the Moon.

b The Moon's gravitational pull on the Earth causes the ocean tides. Show that the gravitational pull of the Moon on the Earth's oceans is approximately three-millionths of the gravitational pull of the Earth on its oceans. The distance from the Earth to the Moon is 3.8×10^8 m.

2 The Sun has a mass of 2.0×10^{30} kg and a mean radius of 1.4×10^9 m. Calculate:

a its gravitational field strength at:

i its surface,

ii the Earth's orbit which is at a distance of 1.5×10^{11} m from the Sun.

b The Earth has a mass of 6.0×10^{24} kg. Show that the gravitational field strength of the Earth is equal and opposite to the gravitational field strength of the Sun at a distance of 260 000 km from the centre of the Earth.

3 The tip of the tallest mountain on the Earth, Mount Everest, is 9 km above sea level. The mean radius of the Earth to the nearest kilometre is 6378 km.

a Calculate the difference between the gravitational field strength of the Earth at sea level and the top of Mount Everest.

b Discuss if it is reasonable to assume that the Earth's gravitational field is uniform between the surface and a height of 10 km above the surface.

c Calculate the gain of potential energy of a mountaineer of mass 80 kg who travels to the top of the mountain from sea level.

4 Use the data in **1** to calculate the gravitational potential at the surface of the Moon and hence calculate the work done to launch a 500 kg rocket from the surface so that it escapes from the Moon's gravitational field.

13.5 Satellite motion

Learning outcomes

On these pages you will learn to:

- analyse circular orbits in inverse square law fields by relating the gravitational force to the centripetal acceleration it causes
- use the equation $r^3/T^2 = GM/4\pi^2$ for circular orbits
- explain what is meant by a geostationary orbit

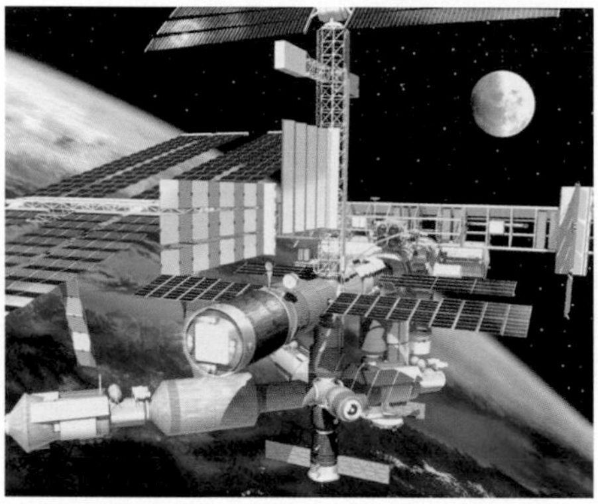

Figure 1 *Space station in orbit*

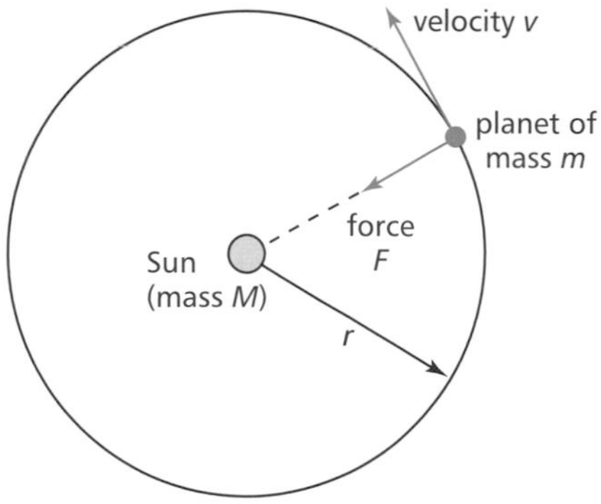

Figure 2 *Explanation of planetary motion*

On any clear night you ought to be able to see satellites passing overhead in the night sky. Although they are pinpoints of light, they are noticeable because they move steadily through the constellations. Websites supply information to enable you to identify some of them from their directions. However, **satellite motion** is not confined to artificial satellites orbiting the Earth. Any small mass which orbits a larger mass is a satellite. The Moon is the Earth's only natural satellite. Mars has two moons, *Phobos* and *Deimos*. Jupiter has more than 60 satellites including the four innermost satellites, *Io*, *Callisto*, *Ganymede* and *Europa*, first observed by Galileo four centuries ago.

From Kepler to Newton

Newton knew that the time period, T, of a planet orbiting the Sun depends on the mean radius, r, of the orbit in accordance with Kepler's third law:

$$\frac{r^3}{T^2} = \text{constant}$$

He realised that the force of gravitational attraction between each planet and the Sun is the centripetal force that keeps the planet on its orbit. By assuming the gravitational force is given by $\frac{GMm}{r^2}$, the gravitational field strength, $\frac{GM}{r^2}$, is therefore equal to the centripetal acceleration $\frac{v^2}{r}$, where M is the mass of the Sun, m is the mass of the planet, r is the radius of the orbit and v is the speed of the planet.

Therefore, the speed of the planet is given by $v^2 = \frac{GM}{r}$

Because speed, $v = \frac{\text{circumference of the orbit}}{\text{time period}} = \frac{2\pi r}{T}$

then $\frac{(2\pi r)^2}{T^2} = \frac{GM}{r}$

Rearranging this equation gives $\boldsymbol{\frac{r^3}{T^2} = \frac{GM}{4\pi^2}}$

As $\frac{GM}{4\pi^2}$ is the same for all the planets, $\frac{r^3}{T^2}$ is the same for all the planets.

So, by assuming the force of attraction F varies with distance according to the inverse-square law (i.e. $F \propto \frac{1}{r^2}$), Newton was able to prove Kepler's third law.

Newton's theory not only explains Kepler's laws, but it also allows the mass M to be calculated if the value of G is known. The Earth orbits the Sun once per year on a circular orbit of radius 1.5×10^{11} m, so you can prove for yourself that the value of r^3/T^2 for any planet is $3.4 \times 10^{18}\,\text{m}^3\,\text{s}^{-2}$. Given $G = 6.67 \times 10^{-11}\,\text{N}\,\text{m}^2\,\text{kg}^{-2}$, show that the mass of the Sun is 2.0×10^{30} kg.

Geostationary satellites

A geostationary satellite orbits the Earth from west to east directly above the Equator and has a time period of exactly 24 hours. It therefore remains in a fixed position above the Equator because it has exactly the same time period as the Earth's rotation.

The radius of orbit of a geostationary satellite can be calculated as follows using the equation $\frac{r^3}{T^2} = \frac{GM}{4\pi^2}$

$$T = 24 \text{ hours} = 24 \times 3600\,\text{s} = 86\,400\,\text{s}$$

$$r^3 = \frac{GM}{4\pi^2}T^2 = \frac{6.67 \times 10^{-11} \times 6.0 \times 10^{24} \times (86\,400)^2}{4\pi^2} = 7.6 \times 10^{22}\,\text{m}^3$$

$$\therefore \quad r = 4.23 \times 10^7\,\text{m} = 42\,300\,\text{km}$$

The radius of the Earth is 6400 km. Therefore, the height of a geostationary satellite above the Earth is 36 000 km (= 42 300 − 6400 km to two significant figures).

Chapter Summary

Definitions

Gravitational field strength, g, is the force per unit mass on a small test mass placed in the field.

Gravitational potential, ϕ, at a point, is the work done per unit mass to move a small object from infinity to that point.

A line of force or **a field line** is the line followed by a small mass acted on by no other forces than the force of gravity.

A uniform field exists in a region where g is the same in magnitude and direction everywhere in the region.

Equations

1. $g = \frac{F}{m}$, where F is the gravitational force on a small mass m.

   $\phi = \frac{W}{m}$ where W is the work done to move a small mass m from infinity.
2. Newton's law of gravitation; the gravitational force F between two point masses m_1 and m_2 at distance r apart is given by $F = \frac{Gm_1m_2}{r^2}$
3. At distance r from a point mass M,

   $g = \frac{GM}{r^2}$ $\quad \phi = -\frac{GM}{r}$
4. At or beyond the surface of a sphere of mass M,

   $g = \frac{GM}{r^2}$

   $\phi = -\frac{GM}{r}$

   where r is the distance to the centre.
5. At the surface of a sphere of mass M and radius R,

   $g_S = \frac{GM}{R^2}$
6. For a satellite in a circular orbit, its centripetal acceleration $\frac{v^2}{r} = g$
7. For point masses m and M at separation r,

   $E_P = -\frac{GMm}{r}$

Summary test 13.5

$G = 6.67 \times 10^{-11}\,\text{N}\,\text{m}^2\,\text{kg}^{-2}$

The radius of the Earth = 6400 km,
$g = 9.81\,\text{N}\,\text{kg}^{-1}$ at the Earth's surface.

1 a Two satellites X and Y are seen from the ground crossing the night sky at the same time. Satellite X crosses the sky faster than Y. State with a reason which satellite is higher.

b Explain why satellite TV dishes must be aligned carefully so they always point to the same position above the equator.

2 A space probe moving at a speed of $3.2\,\text{km}\,\text{s}^{-1}$ is in a circular orbit about a planet of mass M. The time period of the satellite is 110 minutes. Calculate:

a the radius of the orbit,

b the centripetal acceleration of the satellite,

c the mass of the planet.

3 a A satellite moves at speed v in a circular orbit of radius r.

i Write down an expression for the centripetal acceleration of the satellite.

ii Show that the speed of the satellite is given by the equation $v^2 = gr$, where g is the gravitational field strength at the orbit.

b A satellite orbits the Earth in a circular orbit at a height of 100 km. Calculate:

i the gravitational field strength of the Earth at this distance,

ii the speed of the satellite,

iii the time period of the satellite.

4 a Show that the speed, v, of a satellite in a circular orbit of radius r about a planet of mass M is given by the equation $v^2 = \frac{GM}{r}$.

b A weather satellite is in a polar orbit at a height of 1600 km.

i Show that its speed is $7.1\,\text{km}\,\text{s}^{-1}$.

ii Calculate its time period.

iii Explain why such a satellite can survey global weather patterns every day.

13 Exam-style and Practice Questions

Launch additional digital resources for the chapter

$G = 6.67 \times 10^{-11}\ \text{N m}^2\text{kg}^{-2}$

The Earth: radius = 6400 km, mass = 6.0×10^{24} kg, $g = 9.81\ \text{N kg}^{-1}$ at the surface.

1 a Sketch the pattern of lines of force of the gravitational field surrounding a uniform sphere.

b On your diagram:

i mark two points X and Y where the gravitational field strength has the same magnitude but is in opposite directions,

ii mark a point Z where the gravitational field strength is 0.25 times the gravitational field strength at X and in the same direction.

2 Jupiter's mass is 318 times the mass of the Earth.

a Calculate the gravitational force between the Earth and Jupiter when Jupiter is 6.3×10^{11} m from the Earth.

b Calculate the gravitational field strength of Jupiter at a distance of 6.3×10^{11} m from its centre.

3 a Sketch a graph to show how the gravitational field strength of the Earth varies with height from the surface to a height of 13 000 km above its surface.

b Calculate the gravitational field strength of the Earth at a distance of 5000 km above the surface.

4 a The Moon is 380 000 km from the Earth. Calculate the gravitational field strength of the Earth at the Moon.

b Use your answer to **a** to calculate:

i the centripetal acceleration of the Moon,

ii the speed of the Moon,

iii the time period of the Moon.

5 The mean radius of Jupiter's orbit round the Sun is 7.8×10^{11} m. Jupiter has a mass of 1.9×10^{27} kg. The Sun has a mass of 2.0×10^{30} kg.

a Calculate the distance from Jupiter to the point along the line between the Sun and Jupiter at which the gravitational field strength of Jupiter is equal and opposite to the gravitational field strength of the Sun.

b The asteroid belt consists of minor planets and smaller bodies in orbits round the Sun between the Sun and Jupiter. Discuss the view that the asteroids could be the remains of a larger body which was pulled apart by the gravitational force of Jupiter and of the Sun.

6 a Calculate the gravitational force of the Earth on an astronaut who weighs 690 N on the Earth's surface when the astronaut is in a spacecraft orbiting the Earth at a height of 100 km.

b Discuss whether or not an astronaut in an orbiting spacecraft is weightless.

7 A spy satellite orbits the Earth in a polar orbit at a height of 300 km.

a Calculate the gravitational field strength of the Earth at this height.

b i Show that the satellite has a speed of $7.7\ \text{km s}^{-1}$.

ii Calculate the time period of the satellite at this height.

8 a i What is meant by a geostationary orbit?

ii A communications satellite is in a geostationary orbit directly above the equator. Explain why it is advantageous for such a satellite to be in a geostationary orbit.

b A communications network company proposes to place 12 satellites in the same orbit, equally spaced along the orbit. The orbit is to be 500 km above the Earth.

i Calculate the gravitational field strength of the Earth at this height.

ii Show that the time period of a satellite in this orbit is 1 hour and 35 minutes.

iii Explain why a mobile phone at ground level using this system would need to switch to a different satellite every 8 minutes.

9 a Define *gravitational field strength*.

b Io is one of the four moons of the planet Jupiter that were first observed by Galileo. Io orbits Jupiter once every 1.77 days at an average orbital radius r of 4.22×10^5 km.

i Calculate the average speed, v, of Io about Jupiter.

ii Show that $v = \sqrt{\frac{GM}{r}}$, where M is the mass of Jupiter, and hence calculate the mass of Jupiter.

iii The radius of Jupiter is 6.99×10^4 km. Calculate the mean density of Jupiter.

c In 1979, Io was discovered to be volcanic. Scientists think it is continually being distorted by the varying gravitational forces due to Jupiter and its other moons. Europa, the nearest moon to Io, has a mass of 4.92×10^{22} kg and orbits Jupiter once every 3.8 days at an average orbital radius of 6.71×10^5 km.

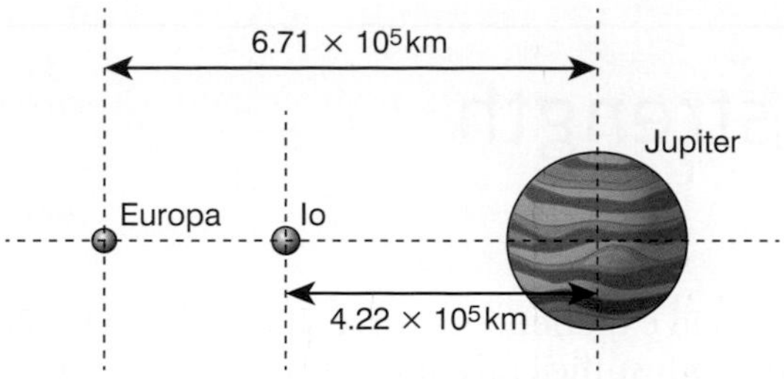

Figure 9.1

Use this data to discuss whether or not the gravitational force on Io due to Europa has a significant effect on Io compared with the gravitational force on Io due to Jupiter. Assume that the two moons orbit Jupiter in the same plane.

10a i State Newton's law of gravitation.

ii Use Newton's law of gravitation to show that the gravitational field strength g at height h above the surface of a spherical planet is equal to $\frac{g_S}{(1+\frac{h}{R})^2}$, where g_S is the gravitational field strength at the surface of the planet and R is the radius of the planet.

iii Draw a graph on a copy of the axes below to show how the Earth's gravitational field strength, g, varies with height, h, above the surface.

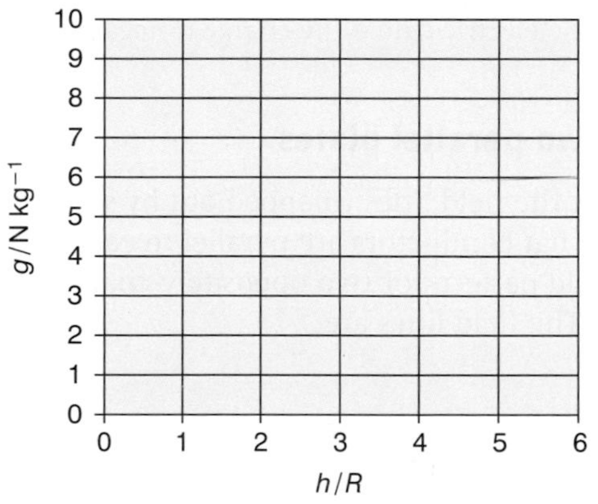

Figure 10.1

b The Earth's mass is 5.974×10^{24} kg. Its shape is not quite spherical. Its polar diameter is 12 714 km and its equatorial diameter is 12 756 km. Its gravitational field strength g at its poles is 9.83 N kg^{-1} and 9.78 N kg^{-1} at the Equator.

i The difference between the polar and equatorial values of g is partly due to the rotation of the Earth causing any object at the Equator to experience a centripetal acceleration. Calculate the centripetal acceleration of an object at the Equator and discuss whether or not the rotation of the Earth is a significant factor in explaining the difference between the polar and equatorial values of g.

ii Another factor that may contribute to the difference is that the Earth's equatorial diameter is 42 km greater than its polar diameter. Discuss whether or not this difference is a significant factor.

11a Define *gravitational potential* at a point.

b A spherical planet has a radius R and a mass M.

i Write down expressions for the gravitational potential, ϕ, above the surface at distance r from the centre of the planet and for the surface gravitational potential, ϕ_S.

ii Draw a graph on a copy of the axes below to show how the ratio ϕ/ϕ_S at and above the surface varies with distance r from the centre of the planet.

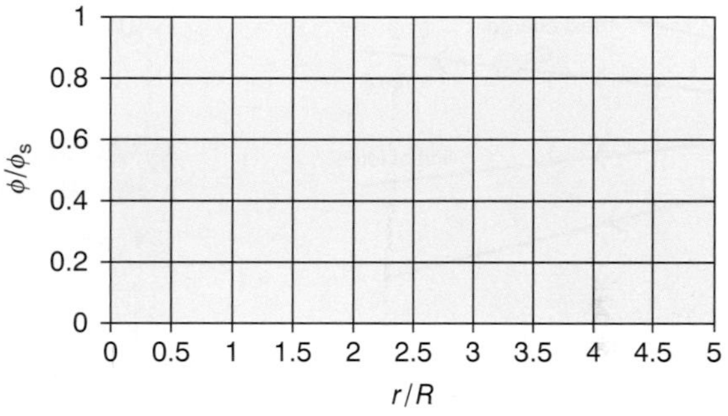

Figure 11.1

iii Show that the gravitational potential ϕ at distance r from the centre of the planet is given by the equation $\phi = -gr$, where g is the magnitude of the gravitational field strength at distance r from the centre of the planet.

c i The surface gravitational field strength of the Earth's moon is 1.62 N kg^{-1} and its diameter is 1740 km. Calculate the surface gravitational potential due to the Moon at the surface of the Moon.

ii The escape speed from the Moon's surface is 2.38 km s^{-1}. Use this value to calculate the gravitational potential needed to escape from the Moon.

iii State one reason why the answers to **i** and **ii** differ.

14 Electric fields

14.1 Electric field strength

Learning outcomes

On these pages you will learn to:

- define electric field strength and use $F = QE$
- recall and use $E = \Delta V/\Delta d$ to calculate the field strength of the uniform field between charged parallel plates

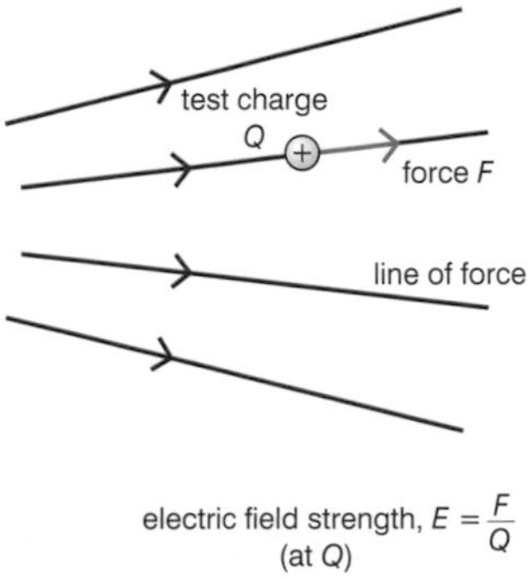

Figure 1 *Electric field strength*

Note

The unit of *E* may be written as the newton per coulomb ($N C^{-1}$) or the volt per metre ($V m^{-1}$).

The link between the two can be seen because $F = QE = \frac{Q\Delta V}{\Delta d}$.

Rearranging this equation gives $\frac{F}{Q} = \frac{\Delta V}{\Delta d}$ (= E). Therefore, the newton per coulomb and the volt per metre are both acceptable as the unit of E.

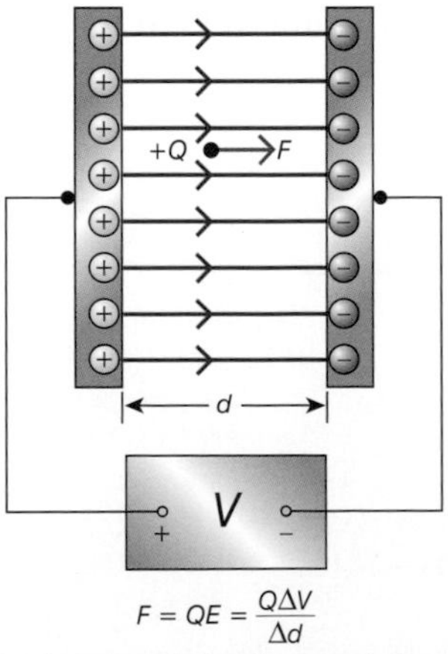

Figure 2 *The electric field strength between two parallel plates*

Inside an electric field

A charged object in an electric field experiences a force due to the field. Provided the object's size and charge are both sufficiently small, the object may be used as a 'test' charge to measure the strength of the field at any position in the field.

The electric field strength, *E*, at a point in the field is defined as the force per unit charge on a positive test charge placed at that point.

The unit of E is the newton per coulomb ($N C^{-1}$).

In Figure 1, a positive test charge Q at a certain point in an electric field is acted on by force F due to the electric field. The electric field strength, E, at that point is given by the equation

$$E = \frac{F}{Q}$$

Notes

1. Rearranging this equation gives $\boldsymbol{F = QE}$ for the force F on a test charge Q at a point in the electric field where the electric field strength is E.
2. Electric field strength is a vector which is in the same direction as the force on a positive test charge. In other words, the direction of a field line at any point is the direction of the electric field strength at that point. The force on a small charge in an electric field is:
   - in the same direction as the electric field if the charge is positive,
   - in the opposite direction to the electric field if the charge is negative.

The electric field between two parallel plates

Figure 4c in Topic 6.1 shows that the field lines (mapped out by semolina grains) between two oppositely charged flat conductors are parallel to each other and at right angles to the plates. The field pattern for two oppositely charged flat plates is similar, as shown in Figure 2. The field lines are:

- parallel to each other,
- at right angles to the plates,
- from the positive plate to the negative plate.

The field between the plates is **uniform**. This is because the electric field strength has the same magnitude and direction everywhere between the plates. The electric field strength E can be calculated from the potential difference ΔV between the plates and their separation Δd using the equation

$$E = \frac{\Delta V}{\Delta d}$$

To prove this equation, consider a small charge Q between the plates, as in Figure 2.

1. The force F on a small charge Q in the field is given by $F = QE$, where E is the electric field strength between the plates.
2. If the charge is moved from the positive to the negative plate, the work done W by the field on Q is given by W = force $F \times$ distance moved $= QE\Delta d$.
3. By definition, the potential difference between the plates, ΔV is the work done per unit charge when a small charge is moved through potential difference ΔV.

Therefore, $\Delta V = \frac{W}{Q} = \frac{QE\Delta d}{Q} = E d$, so rearranging $\Delta V = E\Delta d$ gives $E = \frac{\Delta V}{\Delta d}$

Deflection of a beam of charged particles in a uniform electric field

A charged particle moving across a uniform electric field experiences a constant acceleration because the force on it is constant. Its motion is similar to that of a small object in a small region of the Earth's gravitational field, where g is constant.

Figure 3 shows an electron beam deflected by an electric field produced by applying a p.d. between the metal deflecting plates P and Q. The beam is produced by attracting electrons from a heated filament towards the positively charged anode; it is deflected towards the positive plate Q. By adjusting the strength of the field, the extent of the deflection can be controlled.

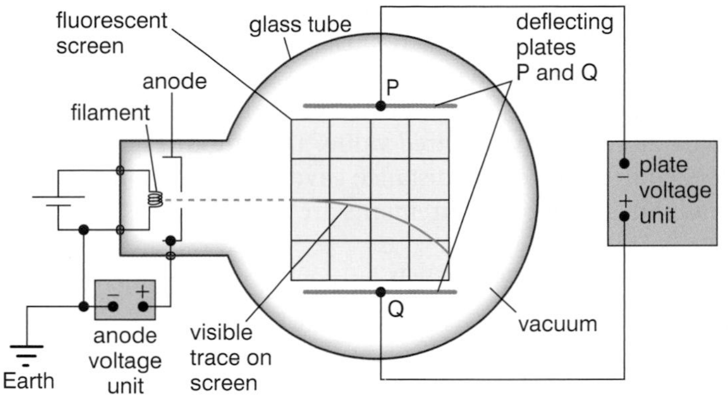

Figure 3 Deflection of an electron beam; the hole in the anode lets some electrons through and the electron beam can be seen as it passes along the screen

The beam curves in a parabolic path just as a projectile projected horizontally does. The projectile equations from Topic 1.8 can be used to determine its deflection, provided g is replaced by acceleration $a = \frac{eE}{m}$ where m is the mass of the electron and e is the charge of the electron.

Notes

1 If the p.d. between plates P and Q is ΔV_P, each electron experiences a force $F = eE = e\frac{\Delta V_P}{\Delta d}$, where Δd is the perpendicular distance between plates P and Q.

2 The acceleration of each electron towards the positive plate, $a = \frac{F}{m} = \frac{e}{m} \times \frac{\Delta V_P}{\Delta d}$, where m is the mass of the electron.

3 The time taken, t, by each electron to cross the field $= \frac{L}{v}$, where L is the length of each plate and v is the initial speed of the electron on entry to the field.

4 The deflection, y, of the electron on leaving the field is given by the equation $y = \frac{1}{2}at^2$.

Using the above equations, it can be shown that y is directly proportional to the plate p.d. ΔV_P.

Summary test 14.1

$e = 1.6 \times 10^{-19}$ C

1 A +40 nC point charge Q_1 is placed in an electric field.

a Calculate the magnitude of the force on Q_1 if the electric field strength where Q_1 is placed is 3.5×10^4 V m^{-1}.

b Q_1 is moved to a different position in the electric field. The force on Q_1 at this position is 1.6×10^{-3} N. Calculate the magnitude of the electric field strength at this position.

2 Figure 4 shows the path of a charged dust particle in an electric field.

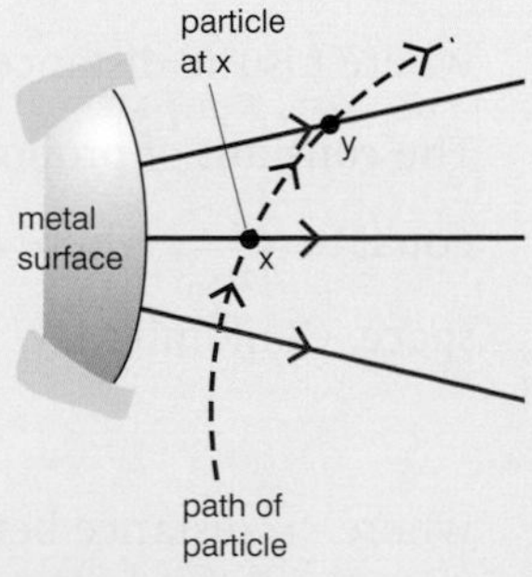

Figure 4

a The electric field strength at X is 65 kV m^{-1}. The force due to the field on the particle when it is at X is 8.2×10^{-3} N towards the metal surface.

i What type of charge does X carry?

ii Calculate the charge carried by the particle.

b i Calculate the magnitude of the force on the particle when it is at Y where the electric field strength is 58 kV m^{-1}.

ii State the direction of the force on the particle when it is at Y.

3 A high voltage supply unit is connected across a pair of parallel plates which are at a separation of 50 mm.

a The voltage is adjusted to 4.5 kV. Calculate:

i the electric field strength between the plates,

ii the electric force on a droplet in the field that carries a charge of 8.0×10^{-19} C.

b The separation between the plates is altered without changing the p.d. between the plates. The droplet in **a** is now acted on by a force of 4.5×10^{-14} N. Calculate the new separation between the plates.

4 A beam of electrons moving horizontally at speed v enters a vertical uniform electric field of width x and strength E. In terms of the charge e and the mass m of the electrons, show that the deflection y of the electrons on leaving the field is given by an equation of the form $y = kx^2$ and determine an expression for k.

14.2 Coulomb's law

Learning outcomes

On these pages you will learn to:

- recall and use Coulomb's law in the form $F = Q_1Q_2/4\pi\varepsilon_0 r^2$ for the force between two point charges in free space or air
- understand that, for any point outside a spherical conductor, the charge on the sphere may be considered to act as a point charge at its centre

Like charges repel and unlike charges attract. The force between two charged objects depends on how close they are to each other. The exact link was first established by Charles Coulomb in France in 1784. He devised a very sensitive torsion balance to measure the force between charged pith balls. Figure 1 shows the arrangement. A needle with a ball made of pith (a substance obtained from plants) at one end and a counterweight at the other end was suspended horizontally by a vertical wire. Another pith ball on the end of a thin vertical rod could be placed in contact with the first ball.

Figure 1 Coulomb's torsion balance

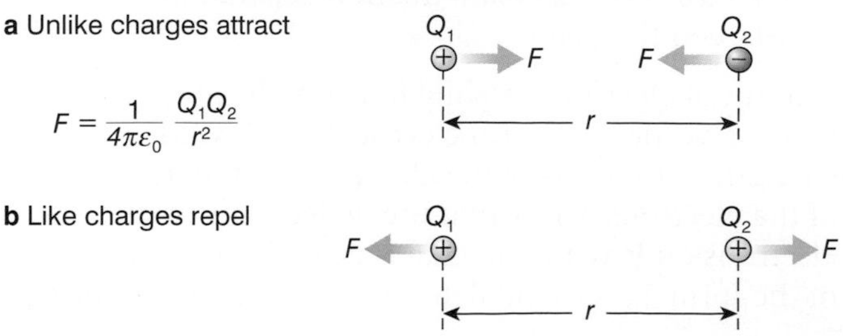

Figure 2 Coulomb's law

The pith balls were small enough to be considered as point objects. The ball on the rod was charged and then placed in contact with the other ball on the needle. The contact between them charged the second ball which was then repelled by the ball fixed on the rod. This caused the wire to twist until the electrical repulsion was balanced by the twist built up in the wire. By turning the torsion head at the top of the wire, the distance between the two balls could be set at any required value. The amount of turning needed to achieve that distance gave the force. Some of Coulomb's many measurements are below.

Table 1 *Some of Coulomb's results*

Distance, r	36	18	8.5
Force, F	36	144	567

Measurements for both variables were actually made in degrees, so the above values are in relative units. Can you make out a pattern for these measurements? Halving the distance from 36 to 18 makes the force increase by a factor of 4. Halving the distance from 18 to 8.5 (near enough 9) increases the force again by a factor of about 4. The measurements fit the link that the force, F, is proportional to $\frac{1}{r^2}$. All the other measurements made by Coulomb fitted the same link.

Because the force is also proportional to the charge on each ball, Coulomb deduced the following equation, known as Coulomb's law, for the force, F, between two 'point charges', Q_1 and Q_2:

$$F = k\frac{Q_1Q_2}{r^2},$$

where r is the distance between the charges.

The constant of proportionality, k, can be shown to be equal to $\frac{1}{4\pi\varepsilon_0}$, where ε_0 is the absolute permittivity of free space. Coulomb's law is therefore written as

$$\boldsymbol{F = \frac{1}{4\pi\varepsilon_0}\frac{Q_1Q_2}{r^2}},$$

where r = distance between two point charges Q_1 and Q_2
A method of measuring ε_0 is outlined in Topic 15.2. The accepted value of ε_0 is $8.85 \times 10^{-12}\,\mathrm{F\,m^{-1}}$ so

$$\frac{1}{4\pi\varepsilon_0} = 9.0 \times 10^{9}\,\mathrm{m\,F^{-1}}$$

Worked example

$e = 1.6 \times 10^{-19}\,\text{C}$, $\frac{1}{4\pi\varepsilon_0} = 9.0 \times 10^9\,\text{m F}^{-1}$

Calculate the force between a proton and an electron at a separation of $3.0 \times 10^{-10}\,\text{m}$.

Solution

$$F = \frac{1}{4\pi\varepsilon_0}\frac{Q_1Q_2}{r^2} = \frac{9.0 \times 10^9 \times 1.6 \times 10^{-19} \times 1.6 \times 10^{-19}}{(3.0 \times 10^{-10})^2}$$
$$= 2.6 \times 10^{-9}\,\text{N}$$

Note on $k = \frac{1}{4\pi\varepsilon_0}$

If Coulomb's law in the form $F = k\frac{Q_1Q_2}{r^2}$ is applied to the force on a 'test' charge q at distance r from a point charge Q, the force on the test charge $F = \frac{kQq}{r^2}$, so the electric field strength at distance r, $E = \frac{F}{q} = \frac{kQ}{r^2}$.

By introducing $\frac{1}{4\pi\varepsilon_0}$ as k, the equation $E = \frac{kQ}{r^2}$ may be written as $\frac{Q}{4\pi r^2} = \varepsilon_0 E$ or $\frac{Q}{A} = \varepsilon_0 E$. In this form, E is the electric field strength at the surface of a sphere of radius r and surface area $A(= 4\pi r^2)$ which has a charge Q evenly distributed on its surface. The equation $\frac{Q}{A} = \varepsilon_0 E$ gives the surface charge density $\frac{Q}{A}$ needed to produce an electric field of strength E at the surface. Thus ε_0 represents the charge per unit area on a surface in a vacuum that produces an electric field of strength 1 volt per metre above the surface. The equation applies to any arrangement, for example, as in Figure 3. The explanation here is not part of this specification and is provided to give a better understanding of Coulomb's law.

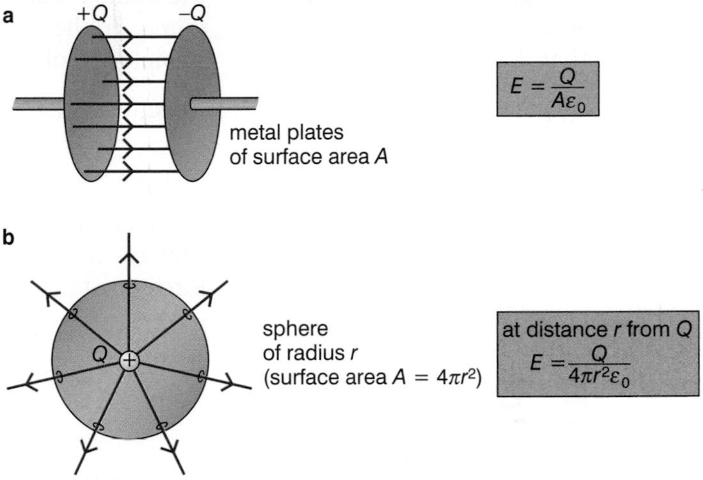

Figure 3 *Comparison of surface electric field strengths*

Summary test 14.2

$\varepsilon_0 = 8.85 \times 10^{-12}\,\text{F m}^{-1}$, $\frac{1}{4\pi\varepsilon_0} = 9.0 \times 10^9\,\text{m F}^{-1}$,
$e = 1.6 \times 10^{-19}\,\text{C}$

1 Calculate the force between an electron and:

a a proton at a distance of $2.5 \times 10^{-9}\,\text{m}$,

b a nucleus of a nitrogen atom (charge $+7e$) at a distance of $2.5 \times 10^{-9}\,\text{m}$.

2 a Two point charges $Q_1 = +6.3\,\text{nC}$ and $Q_2 = -2.7\,\text{nC}$ exert a force of $3.2 \times 10^{-5}\,\text{N}$ on each other when they are at a certain distance, d, apart. Calculate:

i the distance, d, between the two charges,

ii the force between the two charges if they are moved to distance $3d$ apart.

b A charge of $+4.0\,\text{nC}$ is added to each charge in **a**. Calculate the force between Q_1 and Q_2 when they are at separation d.

3 A $+30\,\text{nC}$ point charge is at a fixed distance of 6.2 mm from a point charge Q. The charges attract each other with a force of $4.3 \times 10^{-2}\,\text{N}$.

a Calculate the magnitude of charge Q and state whether Q is a positive or a negative charge.

b The two charges are moved 2.5 mm further apart. Calculate the force between them in this new position.

4 Two point objects, X and Y, carry equal and opposite amounts of charge at a fixed separation of $3.6 \times 10^{-2}\,\text{m}$. The two objects exert a force on each other of $5.1 \times 10^{-5}\,\text{N}$.

a Calculate the magnitude, Q, of each charge and state whether the charges attract or repel each other.

b The charge of each object is increased by adding a positive charge of $+2Q$ to each object. Calculate the separation at which the two objects would exert a force of $5.1 \times 10^{-5}\,\text{N}$ on each other and state whether the objects attract or repel each other.

14.3 Electric potential

Learning outcomes

On these pages you will learn to:

- define electric potential at a point in terms of the work done in bringing unit positive charge from infinity to the point
- recall the definition of electric field strength and the use of the equation $F = QE$
- state that the electric field strength of the field at a point is equal to the negative of potential gradient at that point

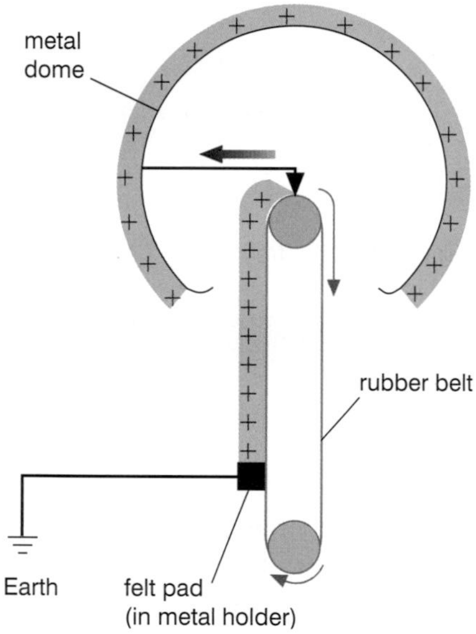

Figure 1 *The Van de Graaff generator*

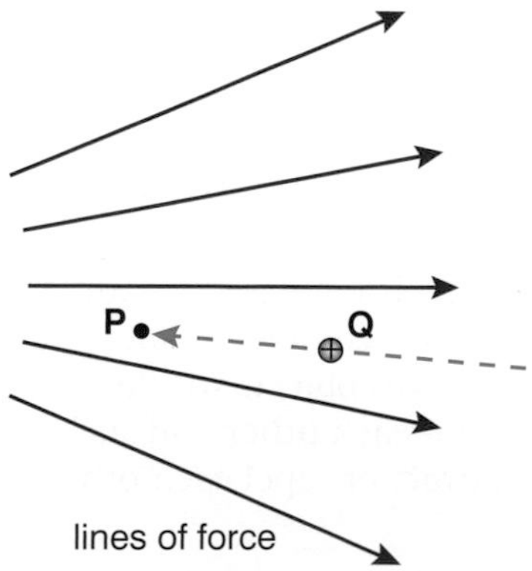

Figure 2 *Moving charge Q to position P*

The Van de Graaff generator

A Van de Graaff generator can easily produce sparks in air several centimetres in length. Figure 1 shows how a Van de Graaff generator works. Charge created when the rubber belt rubs against a pad is carried by the belt up to the metal dome of the generator. As charge gathers on the dome, the potential difference between the dome and Earth increases until sparking occurs.

A spark suddenly transfers energy from the dome. Work must be done to charge the dome because a force is needed to move the charge on the belt up to the dome. So the electric potential energy of the dome increases as it charges up. Some or all of this energy is transferred from the dome when a spark is created.

In general, work must be done to move a charged object A towards another object B that has the same type of charge. Their electric potential energy increases as A moves towards B.

The electric potential energy of A increases from zero if it is moved from infinity towards B. The electric field of B causes a force of repulsion to act on A and this force must be overcome to move A closer to B.

The electric potential at a certain position in any electric field is defined as the work done per unit positive charge on a 'positive test charge' (i.e. a small positively charged object) when it is moved from infinity to that position.

By definition, the position of zero potential energy is infinity. Thus the electric potential at a certain position is the potential energy per unit positive charge of a 'positive test charge' at that position P.

The unit of electric potential is the volt (V), equal to $1\,\mathrm{J\,C^{-1}}$.

Consider a positive test charge Q placed at a position in an electric field where its electric potential energy is E_P. The electric potential V at this position is given by

$$V = \frac{E_P}{Q}$$

Note that rearranging this equation gives $E_P = QV$

Note

If a test charge $+Q$ is moved in an electric field from one position where the electric potential is V_1 to another where the electric potential is V_2, the work done ΔW on it is given by

$$\Delta W = Q(V_2 - V_1)$$

Potential gradients

Equipotentials are lines of constant potential. A test charge moving along an equipotential has constant potential energy. No work is done by the electric field on a test charge moving along an equipotential because the force due to the field is at right angles to the equipotential. In other words, the lines of force of the electric field cross the equipotential lines at right angles.

The equipotentials for an electric field are like equipotentials for a gravitational field; both are lines of constant potential energy for the appropriate test object, in one case a test charge and in the other case a test mass.

Figure 3 shows the equipotentials of the electric field due to two positively charged objects.

Suppose a +2.0 μC test charge is moved from X to Y.

The potential at X, V_X, is +1000 V so the test charge at X has potential energy equal to $+2.0 \times 10^{-3}$ J

$$(= QV_X = +2.0 \times 10^{-6}\,\text{C} \times +1000\,\text{V})$$

The potential at Y, V_Y, is +400 V so the test charge at Y has potential energy equal to $+8.0 \times 10^{-4}$ J

$$(= QV_Y = +2.0 \times 10^{-6}\,\text{C} \times +400\,\text{V})$$

Therefore, moving the test charge from X to Y lowers its potential energy by 1.2×10^{-3} J.

Note that if the test charge is moved from Y to Z along the +400 V equipotential, its potential energy remains constant at $+8.0 \times 10^{-4}$ J.

The **potential gradient** at any position in an electric field is the change of potential per unit change of distance in a given direction.

1 If the field is non-uniform as in Figure 3, the potential gradient varies according to position and direction. The closer the equipotentials are, the greater the potential gradient. The potential gradient is at right angles to the equipotentials.
2 If the field is uniform, such as the field between the two oppositely charged parallel plates shown in Figure 4, the equipotentials **between the plates** are equally spaced lines parallel to the plates. Figure 4 also shows how the potential relative to the negative plate changes with perpendicular distance x from the negative plate.

The graph shows that the potential relative to the negative plate is proportional to distance x. In other words, the potential gradient is constant (such that the potential increases in the opposite direction to the electric field) and equal to $\frac{\Delta V}{\Delta d}$.

The electric field strength E between the plates is equal to $\frac{\Delta V}{\Delta d}$ and is directed from the + to the – plate. In other words:

The electric field strength is equal to the negative of the potential gradient.

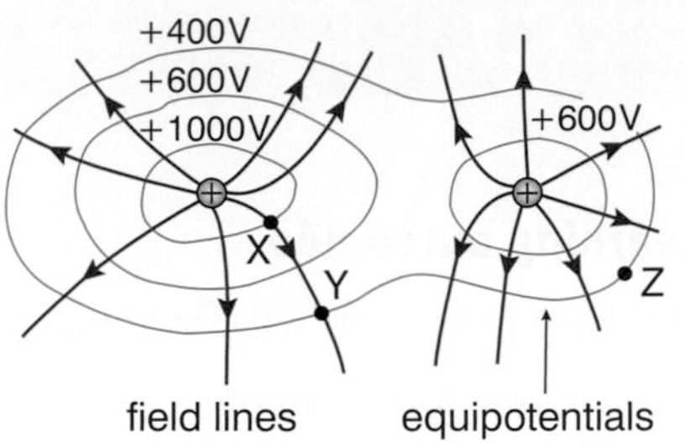

Figure 3 *Equipotentials*

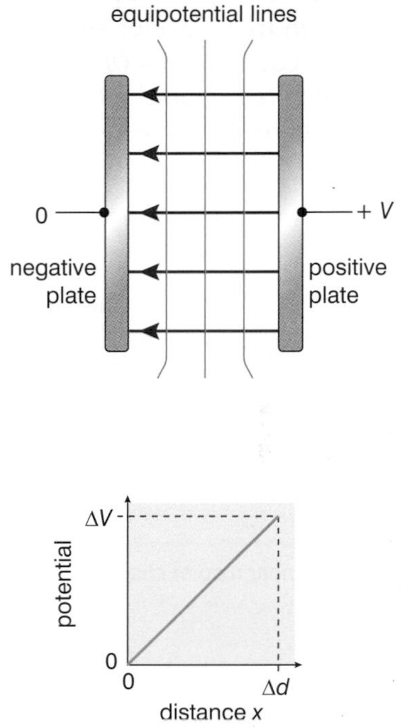

Figure 4 *A uniform potential gradient*

Summary test 14.3

$e = 1.6 \times 10^{-19}$ C

1 An electron in a beam is accelerated from a potential of −50 V to a potential of +450 V. Calculate:
   - **a** the potential energy of the electron at:
     - **i** −50 V,
     - **ii** +450 V,
   - **b** the change of potential energy of the electron.

2 In Figure 3 above, a test charge q is moved from X to Z. Calculate the change of potential energy of the test charge:
   - **a** if q = +3.0 μC,
   - **b** if q = −2.0 μC.

3 An oil droplet carrying a charge of +2e is in air between two parallel metal plates separated by a distance of 20 mm. The p.d. between the plates is 5.0 V.
   - **a** Calculate:
     - **i** the potential gradient between the two plates,
     - **ii** the force on the droplet.
   - **b** Calculate the change of electrical potential energy of the oil droplet if it moves from the midpoint of the plates to the negative plate.

4 **a** Define electric potential and state its unit.
   - **b** Two parallel horizontal metal plates are placed one above the other at a separation of 20 mm. A potential difference of +60 V is applied between the plates with the top plate positive.
     - **i** Calculate the electric field strength between the plates.
     - **ii** Sketch a graph to show how the electric potential V between the plates varies with height h above the lower plate.

14.4 Electric field strength and potential of point charges

Learning outcomes

On these pages you will learn to:

- recall the definition of electric field strength and use $F = QE$
- recall and use $E = Q/4\pi\varepsilon_0 r^2$ for the field strength of a point charge in free space or air
- use the equation $V = Q/4\pi\varepsilon_0 r$ for the potential in the field of a point charge

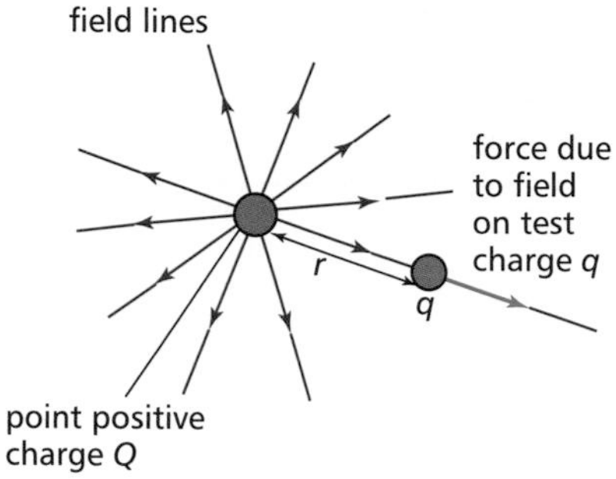

Figure 1 *Force near a point charge Q*

A point charge is a convenient expression for a charged object in a situation where distances under consideration are much greater than the size of the object. The same idea applies to a distant star which is considered as a point object because its diameter is much smaller than the distance to it from the Earth. A 'test' charge in an electric field is a point charge that does not alter the electric field in which it is placed. This would happen if an object with a sufficiently large charge is placed in an electric field and it causes a change in the distribution of charge that creates the field.

Consider the electric field due to a point charge $+Q$, as shown in Figure 1. The field lines radiate from the point charge because a test charge $+q$ in the field would experience a force directly away from Q wherever it was placed. Coulomb's law gives the force F on the test charge q as

$$F = \frac{1}{4\pi\varepsilon_0}\frac{Qq}{r^2}$$

Therefore, as electric field strength $E = \frac{F}{q}$ by definition, the electric field strength at distance r from Q is given by

$$\boldsymbol{E = \frac{Q}{4\pi\varepsilon_0 r^2}}$$

Note that, if Q is negative, the above formula gives a negative value of E corresponding to the field lines pointing inwards towards Q.

Worked example

$\frac{1}{4\pi\varepsilon_0} = 9.0 \times 10^9\,\text{m F}^{-1}$, $e = 1.6 \times 10^{-19}\,\text{C}$

Calculate the electric field strength due to a nucleus of charge $+82e$ at a distance of 0.35 nm.

Solution

$$E = \frac{Q}{4\pi\varepsilon_0 r^2} = \frac{9.0 \times 10^9 \times (+82 \times 1.6 \times 10^{-19})}{(0.35 \times 10^{-9})^2} = 9.6 \times 10^{11}\,\text{V m}^{-1}$$

Electric field strength as a vector

If a test charge is in an electric field due to several point charges, each charge exerts a force on the test charge. The resultant force F on the test charge gives the resultant electric field strength at the position of the test charge. Consider the following situations:

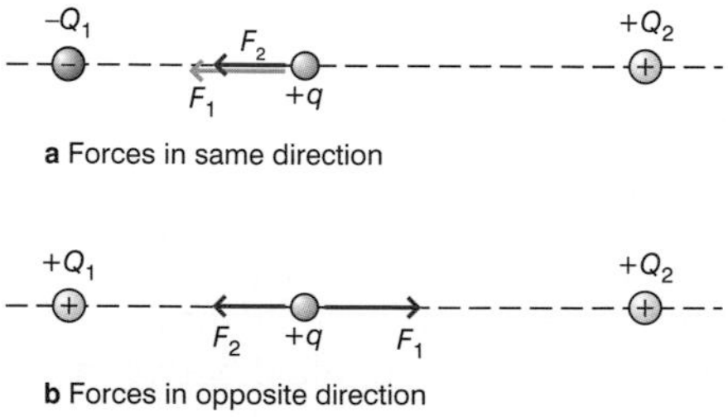

Figure 2 *Forces along the same line*

- **Forces in the same direction**: Figure **2a** shows a test charge $+q$ on the line between a negative point charge Q_1 and a positive point charge Q_2. The test charge experiences a force $F_1 = qE_1$ where E_1 is the electric field strength due to Q_1 and a force $F_2 = qE_2$ where E_2 is the electric field strength due to Q_2. The two forces act in the same direction because Q_1 attracts q and Q_2 repels q. So the resultant force $F = F_1 + F_2 = qE_1 + qE_2$. Therefore, the resultant electric field strength $E = \frac{F}{q} = \frac{qE_1 + qE_2}{q} = E_1 + E_2$
- **Forces in opposite directions**: Figure **2b** shows a test charge $+q$ on the line between two positive point charges Q_1 and Q_2. The test charge experiences a force $F_1 = qE_1$ where E_1 is the electric field strength due to Q_1 and a force $F_2 = qE_2$ where E_2 is the electric field strength due to Q_2. The two forces act in opposite directions because Q_1 repels q and Q_2 repels q. Assuming F_1 is greater than F_2, the resultant force $F = F_1 - F_2 = qE_1 - qE_2$.

  Therefore, the resultant electric field strength $E = \frac{F}{q} = \frac{qE_1 - qE_2}{q} = E_1 - E_2$

- **Forces at angle θ to each other**: Figure 3 shows a test charge $+q$ on perpendicular lines from two positive point charges Q_1 and Q_2. The test charge experiences a force $F_1 = qE_1$ where E_1 is the electric field strength due to Q_1 and a force $F_2 = qE_2$ where E_2 is the electric field strength due to Q_2. The two forces are at angle θ to each other, F_1 along the line between q and Q_1 and F_2 along the line between q and Q_2. The magnitude of the resultant force F is given by the formula $F^2 = F_1^2 + F_2^2 + 2F_1F_2\cos\theta$ as explained in Topic 1.1. Therefore, the resultant electric field strength $E = \frac{F}{q}$.

In general, the resultant electric field strength is the vector sum of the individual electric field strengths.

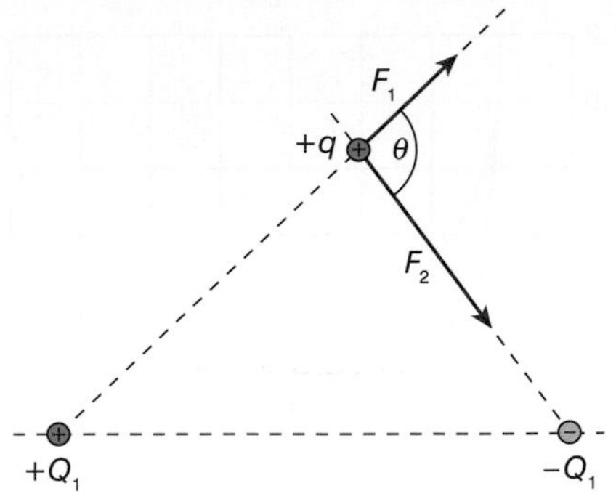

Figure 3 *Forces at angle θ*

Worked example

$$\frac{1}{4\pi\varepsilon_0} = 9.0 \times 10^9\,\text{m F}^{-1}$$

A +65 μC point charge Q_1 is at a distance of 50 mm from a +38 μC charge Q_2.

A +12 pC charge q is placed at M, midway between Q_1 and Q_2.

Calculate:

a the resultant electric field strength at M,

b the magnitude and direction of the force on q.

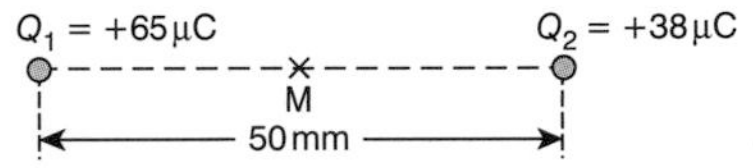

Figure 4

Solution

a The electric field strength due to Q_1 at M, $E_1 = \frac{Q_1}{4\pi\varepsilon_0 r_1^2}$, where r_1 = 25 mm.

Therefore, $E_1 = 9.0 \times 10^9 \times \frac{65 \times 10^{-6}}{(25 \times 10^{-3})^2} = 9.4 \times 10^8\,\text{V m}^{-1}$ away from Q_1.

The electric field strength due to Q_2 at M, $E_2 = \frac{Q_2}{4\pi\varepsilon_0 r_2^2}$, where r_2 = 25 mm.

Therefore, $E_2 = 9.0 \times 10^9 \times \frac{38 \times 10^{-6}}{(25 \times 10^{-3})^2} = 5.5 \times 10^8\,\text{V m}^{-1}$ away from Q_2.

As E_1 and E_2 are in opposite directions, the resultant electric field strength $E = E_1 - E_2 = 9.4 \times 10^8 - 5.5 \times 10^8 = 3.9 \times 10^8\,\text{V m}^{-1}$ away from Q_1 towards Q_2.

b The resultant force on $q = qE = +12 \times 10^{-12} \times 3.9 \times 10^8 = 4.7 \times 10^{-3}$ N.

Electric potential near a point charge

Because Coulomb's law $F = \frac{1}{4\pi\varepsilon_0}\frac{Q_1Q_2}{r^2}$ and Newton's law $F = G\frac{m_1m_2}{r^2}$ are both inverse-square relationships, the forces vary with distance in the same way. Therefore the formula for the electrical potential near a point charge Q is of the same form as the formula for the gravitational potential near a point mass (or spherical mass), i.e. $\phi = \frac{-GM}{r}$, which was derived in Topic 13.4.

Using this analogy and replacing M by Q and G by $\frac{1}{4\pi\varepsilon_0}$ in the gravitational potential formula above therefore gives the following formula for the electric potential V at distance r from a point charge Q:

$$V = \frac{Q}{4\pi\varepsilon_0 r}$$

The equation shows that the electric potential V is inversely proportional to the distance r. The curve is not an 'inverse-square law' curve as V is proportional to $1/r$. See Figure 6.

More about radial fields

The electric field surrounding a charged metal sphere is radial, like the field due to a point charge. The charge on the metal surface is evenly distributed on its surface. The field has the same strength as if all the charge was concentrated at the centre of the sphere.

Note

A test charge inside the sphere would experience zero force due to the charge on the sphere, because the electric field due to the charge on the sphere is in all directions from any point inside the sphere.

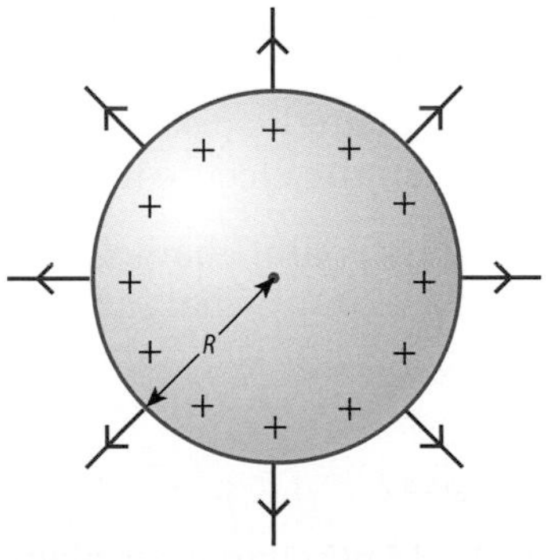

Figure 5 *The electric field near a charged hollow metal sphere*

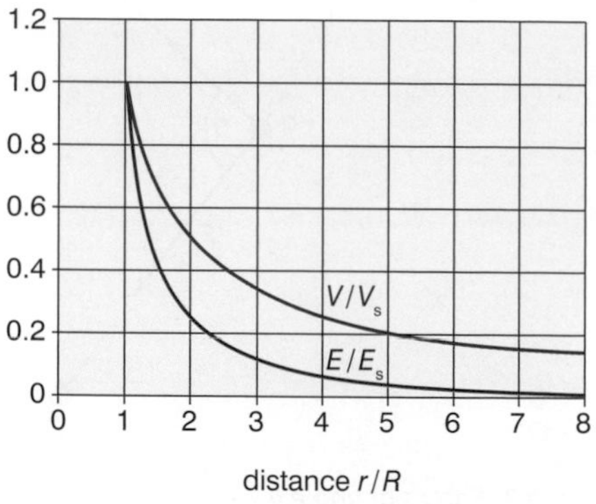

***Figure 6** The electric field strength and potential near a charged sphere*

Notes

1. Gravitational potential is always negative as the force is always attractive, whereas electric potential in the electric field near a point charge Q can be positive or negative depending on whether Q is a positive or a negative charge.
2. The potential energy E_P of two point charges Q_1 and Q_2 at distance d apart is given by

$$E_P = QV = \frac{Q_1 Q_2}{4\pi\varepsilon_0 d}$$

For a sphere of radius R with a charge Q,

- Outside the sphere at distance r from the centre of the sphere, the electric field strength $E = \frac{Q}{4\pi\varepsilon_0 r^2}$ and the electric potential $V = \frac{Q}{4\pi\varepsilon_0 r}$
- At the surface of the sphere, the electric field strength $E_s = \frac{Q}{4\pi\varepsilon_0 R^2}$ and the electric potential $V_s = \frac{Q}{4\pi\varepsilon_0 R}$
- If the sphere is hollow, there is no electric field inside it. Therefore the electric field strength is zero and the electric potential is V_s.

Figure 6 shows how the electric field strength and potential vary with distance from the centre of the sphere.

Outside the sphere:

- the field strength curve is an 'inverse-square law' curve as E is proportional to $\frac{1}{r^2}$.
- the potential curve is an 'inverse' curve as V is proportional to $\frac{1}{r}$.

Mathematical note

The electric field strength is equal to the negative of the potential gradient.

- The potential V at distance r from a point charge Q is $\frac{Q}{4\pi\varepsilon_0 r}$.
- The potential at distance $r + \Delta r$ is $\frac{Q}{4\pi\varepsilon_0 (r + \Delta r)}$.

Therefore from r to $r + \Delta r$,
the change of potential $\Delta V = \frac{Q}{4\pi\varepsilon_0 (r + \Delta r)} - \frac{Q}{4\pi\varepsilon_0 r} = \frac{-Q\Delta r}{4\pi\varepsilon_0 r(r + \Delta r)}$.

For $\Delta r << r$, Δr can be neglected in the denominator so $\Delta V = \frac{-Q\Delta r}{4\pi\varepsilon_0 r^2} = -E\Delta r$.

Hence the electric field strength $E = -\frac{\Delta V}{\Delta r}$, which is the negative of the potential gradient.

The equation applies along a field line to small changes of distance in a non-uniform field and to any change of distance in a uniform field.

Summary test 14.4

$\frac{1}{4\pi\varepsilon_0} = 9.0 \times 10^9\,\text{m F}^{-1}$

1 **a** Calculate the electric field strength at a distance of 3.2 mm from a +6.0 nC point charge.

b Calculate the distance from the point charge in **a** at which the electric field strength is $5.4 \times 10^5\,\text{V m}^{-1}$.

2 A + 25 μC point charge Q_1 is at a distance of 60 mm from a +100 μC charge Q_2.

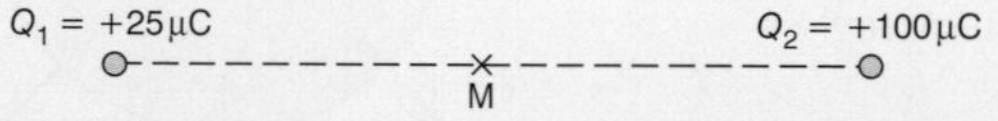

Figure 7

a A +15 pC charge q is placed at M, 25 mm from Q_1 and 35 mm from Q_2. Calculate:

i the resultant electric field strength at M,

ii the magnitude and direction of the force on q.

b Show that the electric field strength due to Q_1 and Q_2 is zero at the point which is 20 mm from Q_1 and 40 mm from Q_2.

3 A +15 μC point charge Q_1 is at a distance of 20 mm from a +10 μC charge Q_2.

a Calculate the resultant electric field strength:

i at M, the midpoint between the two charges,

ii at the point P along the line between Q_1 and Q_2 which is 25 mm from Q_1 and 45 mm from Q_2,

iii at a point S which is 20 mm from Q_1 and 20 mm from Q_2.

b **i** Explain why there is a point along the line between the two charges at which the electric field strength is zero.

ii Calculate the distance from this point to Q_1 and to Q_2.

4 A +15 μC point charge Q_1 is at a distance of 30 mm from a −30 μC charge Q_2.

a Calculate the electric potential at the midpoint of the two charges.

b **i** Show that the electric potential is zero at a point between the two charges which is 10 mm from Q_1 and 20 mm from Q_2.

ii Calculate the electric field strength at this position and state its direction.

Comparison between electric and gravitational fields

14.5

The similarities and differences between the two types of fields are listed in the table below. In the mid-nineteenth century, James Maxwell showed that electric and magnetic forces are different manifestations of the electromagnetic force. Towards the end of the twentieth century, physicists proved that the electromagnetic force and the nuclear force responsible for radioactive decay are different manifestations of a more fundamental force, the electroweak force. At the present time, the force of gravity remains outside this theoretical framework, despite repeated attempts to establish a unified framework. The fundamental nature of the force of gravity remains mysterious even though we use it in everyday situations more than any other force.

Table 1

	Gravitational force	Electric force
Similarities		
Line of force or a field line	Path of a free 'test' mass in the field	Path of a free 'test' charge in the field
Inverse-square law of force	Newton's law of gravitation $F = \frac{Gm_1m_2}{r^2}$	Coulomb's law of force $F = \frac{1}{4\pi\varepsilon_0}\frac{Q_1Q_2}{r^2}$
Field strength	Force per unit mass, $g = F/m$	Force per unit + charge, $E = F/q$
Unit of field strength	$N\,kg^{-1}$ or $m\,s^{-2}$	$N\,C^{-1}$ or $V\,m^{-1}$
Potential	Gravitational potential energy per unit mass	Electric potential energy per unit + charge
Unit of potential	$J\,kg^{-1}$	V (= $J\,C^{-1}$)
Potential energy between two point masses or charges	$E_P = -\frac{Gm_1m_2}{r}$	$E_P = -\frac{Q_1Q_2}{4\pi\varepsilon_0 r}$
Uniform fields	g is the same everywhere, field lines parallel	E is the same everywhere, field lines are parallel
Radial fields	Due to a point mass or a uniform spherical mass M, $g = \frac{GM}{r^2}$, $\varphi = -\frac{Gm}{r}$	Due to a point charge or a charged metal sphere of charge Q, $E = \frac{Q}{4\pi\varepsilon_0 r^2}$, $V = \frac{Q}{4\pi\varepsilon_0 r}$
Differences		
Action at a distance	Between any two masses	Between any two charged objects
Force	Attracts only	Unlike charges attract, like charges repel
Constant of proportionality in force law	G	$\frac{1}{4\pi\varepsilon_0}$

Learning outcomes

On these pages you will learn to:

- recognise the analogy between certain qualitative and quantitative aspects of electric fields and gravitational fields

Chapter Summary

A **line of force** or a **field line** of an electric field is a line which a free positive 'test' charge would follow in the field.

The electric field strength, E, at a point in the field is defined as the force per unit charge on a positive test charge placed at that point.

Electric potential, V, at a point is defined as the work done per unit charge to move a positive test charge from infinity to that point.

A uniform electric field is one where the electric field strength has the same magnitude and direction everywhere between the plates.

Electric field strength $E = \frac{F}{Q}$

Coulomb's law of force

$$F = \frac{1}{4\pi\varepsilon_0}\frac{Q_1Q_2}{r^2}$$

Uniform electric field $E = \frac{\Delta V}{\Delta d}$

where V is the p.d. between two oppositely charged parallel plates at separation d.

Radial electric field

$$E = \frac{Q}{4\pi\varepsilon_0 r^2} \qquad V = \frac{Q}{4\pi\varepsilon_0 r}$$

where r is the distance from a point charge Q or from the centre of a metal sphere carrying a charge Q.

Electric potential

$$V = \frac{E_p}{Q}$$

Potential energy between two point charges

$$E_P = -\frac{Q_1\,Q_2}{4\pi\varepsilon_0 r}$$

Launch additional digital resources for the chapter

$\frac{1}{4\pi\varepsilon_0} = 9.0 \times 10^9\,\mathrm{m\,F^{-1}}$, $e = 1.6 \times 10^{-19}\,\mathrm{C}$,
$g = 9.81\,\mathrm{N\,kg^{-1}}$, $G = 6.67 \times 10^{-11}\,\mathrm{N\,m^2\,kg^{-2}}$

1 a Explain what is meant by a *uniform electric field*.

b A beam of electrons is directed horizontally into a uniform electric field which acts vertically downwards, as shown in Figure 1.1 below.

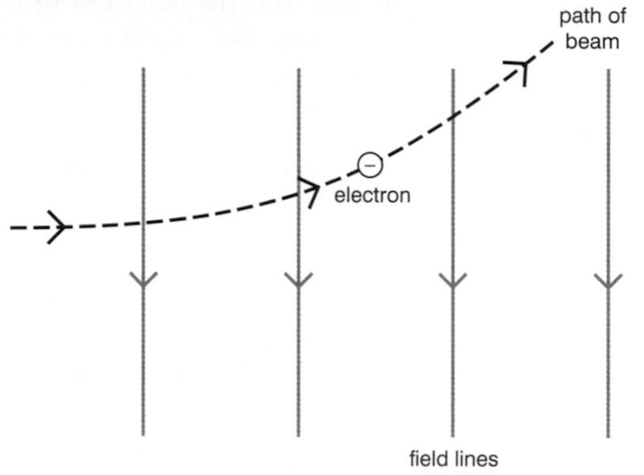

Figure 1.1

i Explain why the beam follows a path which curves upwards.

ii Each electron in the beam experiences a constant force of 1.8×10^{-13} N. Calculate the strength of the electric field.

2 A small metal sphere on an insulating rod is placed on a top pan balance, as shown in Figure 2.1 below. An identical metal sphere carrying a charge of $2Q$ on an insulating rod is brought into contact with the sphere on the balance so that each sphere acquires the same charge.

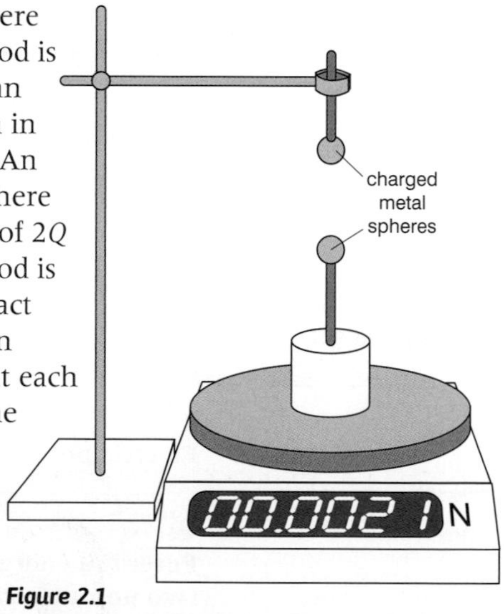

Figure 2.1

a When the two spheres are at a separation of 68 mm, the top pan balance reading increases by 2.1×10^{-3} N. Calculate the charge, Q, on each sphere.

b Explain why the electric field strength is zero at the midpoint between the two charges.

3 A +36 nC point charge X is at a distance of 75 mm from a −9 nC point charge Y.

a i Sketch the pattern of the electric field lines between the point charges.

ii Calculate the electric field strength at the midpoint between X and Y.

b At a certain position along the line XY, the resultant electric field strength is zero.

i Explain why this position is more than 75 mm further from X than it is from Y.

ii Calculate the distance from this position to X and to Y.

4 a Explain why an air molecule can be ionised by an electric field that is sufficiently strong.

b The spherical dome of a Van de Graaff machine has a diameter of 0.30 m. Sparks are produced from the dome if the electric field strength at the surface reaches $40\,\mathrm{kV\,m^{-1}}$. Calculate:

i the maximum charge that can be stored on the dome,

ii the electric field strength at a distance of 1.0 m from the surface of the dome when the surface electric field strength is $40\,\mathrm{kV\,m^{-1}}$.

5 Two identical small conducting spheres are supported at either end of an insulating fibre which is attached at its midpoint to a fixed support. When the two spheres are in contact with each other, they are charged by contact with a charged rod. The two spheres repel each other as shown in Figure 5.1 opposite.

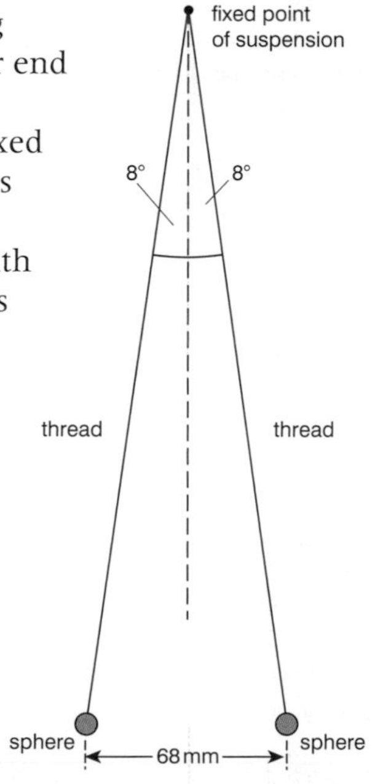

Figure 5.1

a In equilibrium, each section of the thread makes an angle of 8° to the vertical. Show that the electrostatic force of repulsion on each sphere $F = mg\tan 8°$, where m is the mass of each sphere.

b The mass of each sphere is 3.4×10^{-3} kg. Their separation at equilibrium is 68 mm. Calculate the charge on each sphere.

6 a State two similarities and two differences between an electric field and a gravitational field.

b Show that the electrostatic force of repulsion between two protons is 1.2×10^{36} × the gravitational force between them at the same distance.

Mass of a proton = 1.67×10^{-27} kg

7 a Define *electric field strength.*

b Figure 7.1 shows how the electric potential near a positively charged metal sphere of radius 0.100 m varies with distance from the surface of the sphere when the sphere is at a potential of 32.0 kV.

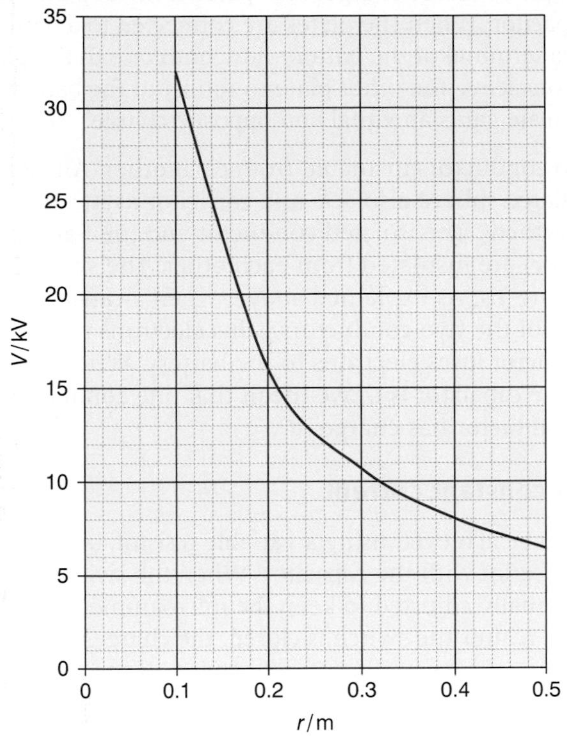

Figure 7.1

i Calculate the charge on the sphere and hence determine the charge per unit area on the sphere.

ii Calculate the electric field strength at a distance of 0.100 m from the surface of the sphere.

c An insulated uncharged metal rod XY of length 0.100 m is placed as shown in Figure 7.2, with X at a distance of 0.100 m from the surface of the sphere.

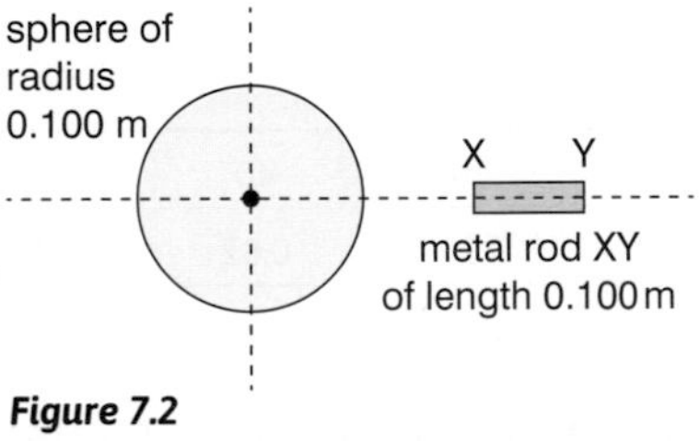

Figure 7.2

i Explain why there is a force of attraction between the rod and the sphere even though the rod is uncharged.

ii Sketch a graph on a copy of Figure 7.1 to show how the electric potential varies with position along the radial line from the sphere through XY to beyond Y.

iii Describe how the magnitude of the electric field strength along a radial line from the sphere through XY to beyond Y compares with the field strength when the rod was absent.

8 a Define the *electric field strength* of an electric field.

b Figure 8.1 shows electric field lines between a negatively charged metal plate A at a fixed distance from a larger positively charged metal plate B.

i Mark the direction of the field lines.

ii Mark a region where the electric field is uniform. Label this region U.

iii Describe how the electric field strength changes between A and B along the top curved field line.

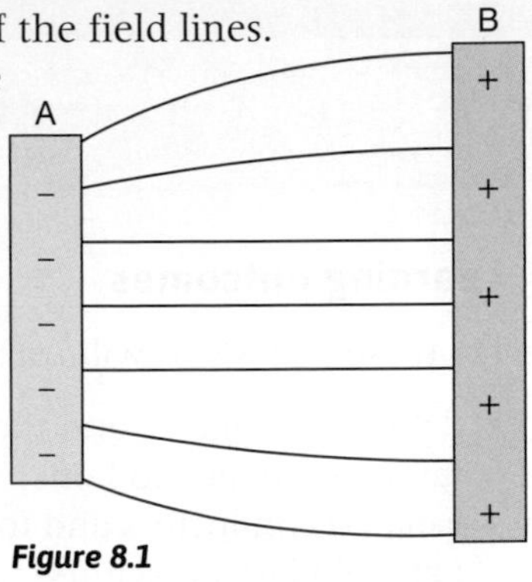

Figure 8.1

c A polar molecule is a molecule in which the centre of its negative charge (N) is not at the same position as the centre of its positive charge (P) and P and N are at a fixed distance apart, as in Figure 8.2.

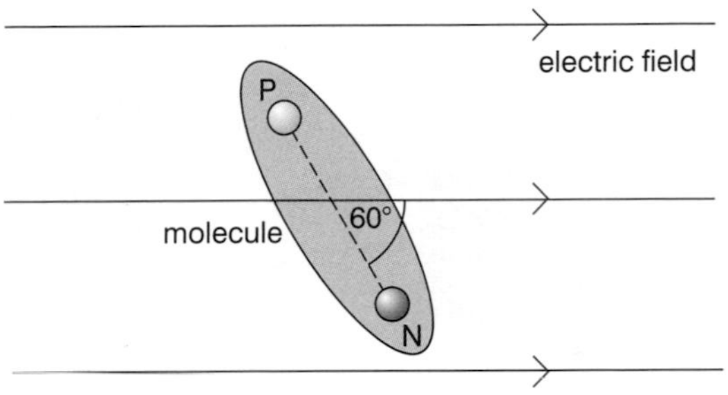

Figure 8.2

In a particular polar molecule, the magnitude of the charge at P and N is 1.6×10^{-19} C and their distance apart, PN, is 3.5×10^{-9} m. The molecule is subjected to a uniform electric field of strength 50 kV m^{-1} such that the line PN is at an angle of 60° to the electric field, as shown in Figure 8.2.

i Draw an arrow at P and an arrow at N to show the direction of the electric force on P and on N.

ii Calculate the torque on the molecule due to these electric forces.

9 a Define *electric potential.*

b A hydrogen atom consists of a single proton and a single electron. In a simple model of the atom, the electron moves in a circular orbit around the proton.

When the atom is in its lowest energy state, the electron is at a mean distance of 0.053 nm from the proton.

i Calculate the force of attraction between the proton and the electron in this state.

ii Calculate the potential energy of the atom in this state.

iii Assuming the proton is at rest and the electron moves at constant speed v on a circular orbit of radius r, write down an equation relating the electrostatic force on the electron to its centripetal acceleration. Hence show that the kinetic energy of the electron is equal to 0.5 times the magnitude of the potential energy of the atom.

iv Calculate the total energy E_0 of the atom in electron volts in its lowest energy state.

15 Capacitors

15.1 Capacitance

Learning outcomes

On these pages you will learn to:

- define capacitance and the farad, as applied to both isolated conductors and to parallel plate capacitors
- state and explain how the p.d. across a capacitor changes when the charging current is constant
- recall and use $C = Q/V$

A capacitor is a device designed to store charge. Two parallel metal plates placed near each other form a capacitor. When the plates are connected to a battery, electrons from the negative terminal of the battery flow onto one of the plates. An equal number of electrons leave the other plate to return to the battery via its positive terminal. So each plate gains an equal and opposite charge.

A capacitor consists of two conductors insulated from each other. An isolated conductor such as an insulated sphere also acts as a capacitor as it can store charge because it is not 'earthed'. The isolated conductor and the Earth (which acts as the second conductor) are insulated from each other. The symbol for a capacitor is shown in Figure 1b. As explained above, when a capacitor is connected to a battery, one of the two conductors gains electrons from the battery and the other conductor loses electrons to the battery. When we say that the charge stored by the capacitor is Q, we mean that one conductor has charge $+Q$ and the other conductor has charge $-Q$.

a

b

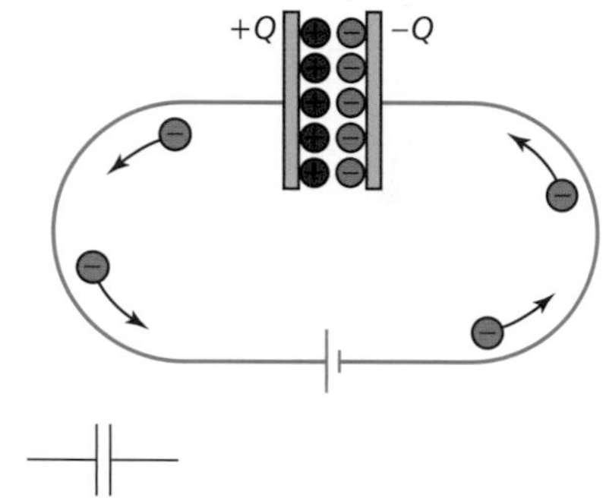

Figure 1 ***a*** *Storing charge* ***b*** *Capacitor symbol*

Charging a capacitor at constant current

Figure 2 shows how this can be achieved using a variable resistor, a switch, a microammeter and a cell in series with the capacitor. When the switch is closed, the variable resistor is continually adjusted to keep the microammeter reading constant. At any given time, t, after the switch is closed, the charge Q on the capacitor can be calculated using the equation $Q = It$, where I is the current.

A high-resistance voltmeter connected in parallel with the capacitor enables the capacitor p.d. to be measured. To investigate how the capacitor p.d. changes with time for constant current, use the variable resistor to keep the current constant and either:

- use a stopwatch and measure the voltmeter reading at measured times, or
- use a datalogger as shown in Figure 2b.

Typical readings for a current of 15 μA are shown in the following table. The charge Q has been calculated using $Q = It$.

Table 1 *Current = 15 mA*

Time, t/s	0	20	40	60	80	100
p.d., V/volts	0	0.29	0.62	0.90	1.22	1.50
charge, Q/μC	0	300	600	900	1200	1500

The graph of charge stored, Q, against p.d., V, for these measurements is shown in Figure 3. The measurements define a straight line passing through the origin. Therefore, the charge stored, Q, is proportional to the p.d., V. In other words, the charge stored per volt is constant.

The capacitance C of a capacitor is defined as the charge stored per unit p.d.

The unit of capacitance is the farad (F), equal to 1 coulomb per volt. Note that $1\,\mu F = 10^{-6}\,F$.

For a capacitor which stores charge Q at p.d. V, its capacitance can be calculated using the equation

$$C = \frac{Q}{V}$$

a

b

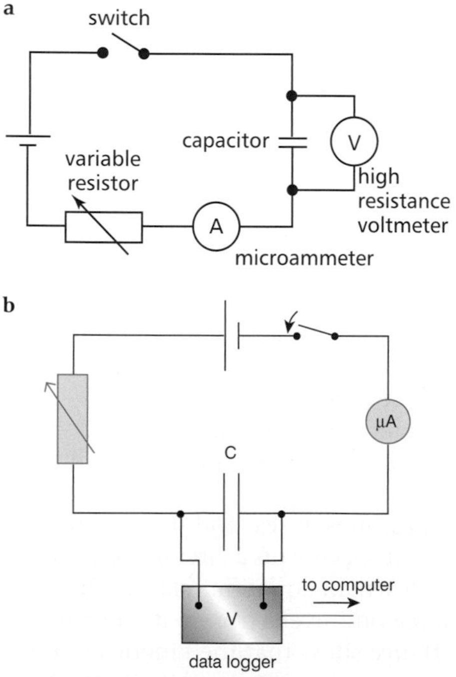

Figure 2 *Investigating capacitors*
a *Circuit diagram* ***b*** *Using a data logger*

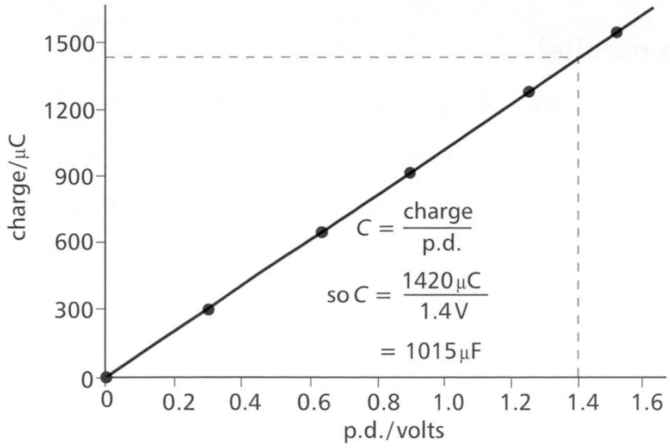

Figure 3 *Graph of results*

Worked example

A capacitor which is initially uncharged is charged at a constant current of 16 μA for 40 s. At the end of this time, the capacitor p.d. increases to 5.8 V. Calculate:

a the charge stored by the capacitor,
b the capacitance of the capacitor.

Solution

a $Q = It = 16 \times 10^{-6} \times 40 = 6.4 \times 10^{-4}\,C$

b $C = \frac{Q}{V} = \frac{6.4 \times 10^{-4}}{5.8\,V} = 1.1 \times 10^{-4}\,F\ (= 110\,\mu F)$

Summary test 15.1

1 Complete the following table

	a	b	c	d	e	f
charge/μC	60	330		6.30	52	
p.d./V	12		9.0	4.5		50
capacitance/μF		150	1100		4.7	68

2 A 22 μF capacitor is charged by means of a constant current of 2.5 μA to a p.d. of 12.0 V. Calculate:

a the charge stored on the capacitor at 12.0 V,

b the time taken.

3 A capacitor is charged by means of a constant current of 0.5 μA to a p.d. of 5.0 V in 55 s. Calculate:

a the charge stored, **b** the capacitance of the capacitor.

4 A capacitor is charged by means of a constant current of 24 μA to a p.d. of 4.2 V in 38 s. The capacitor is then charged from 4.2 V by means of a current of 14 μA in 50 s. Calculate:

a the charge stored at a p.d. of 4.2 V,

b the capacitance of the capacitor,

c the extra charge stored at a current of 14 μA,

d the p.d. after the extra charge was stored.

Note

Rearranging $C = \frac{Q}{V}$ gives $Q = CV$ or $V = \frac{Q}{C}$.

Extension

Practical capacitors

Most practical capacitors are designed using two strips of aluminium foil as plates separated by a thin layer of **dielectric** which is an insulating material that increases the charge that can be stored for a given p.d. Capacitances of less than about 10 μF contain dielectrics such as mica or polyester.

Figure 4 shows the construction of such a capacitor in which the foil and the dielectric are rolled up into a convenient size. See Topic 15.2 for more about dielectrics.

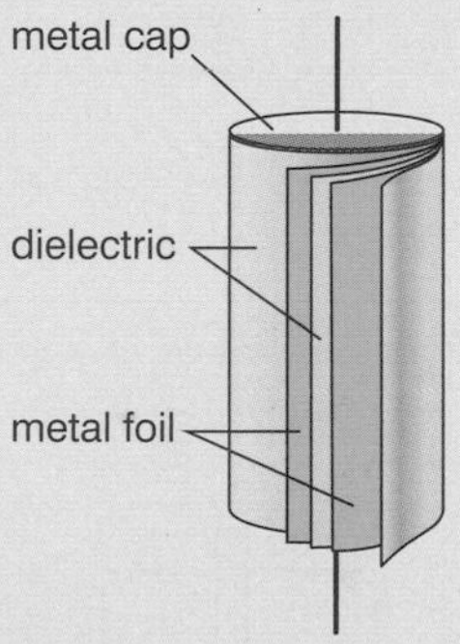

Figure 4 *A practical capacitor*

Electrolytic capacitors contain paper soaked with an electrolyte such as aluminium borate between the metal strips. When used in a d.c. circuit, a thin oxide layer which is a dielectric forms on the positive plate. As will be explained in Topic 15.3, because the dielectric layer is very thin, the capacitance can be much greater than that of non-electrolytic capacitors. Electrolytic capacitors must be connected with the correct polarity otherwise the dielectric is damaged.

15.2 Capacitors in series and parallel

Learning outcomes

On these pages you will learn to:

- derive formulae for the total capacitance of capacitors in series and in parallel
- solve problems using the capacitance formulae for capacitors in series and in parallel

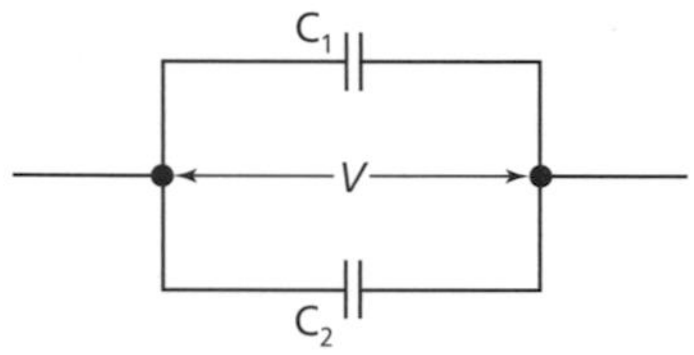

Figure 1 *Capacitors in parallel*

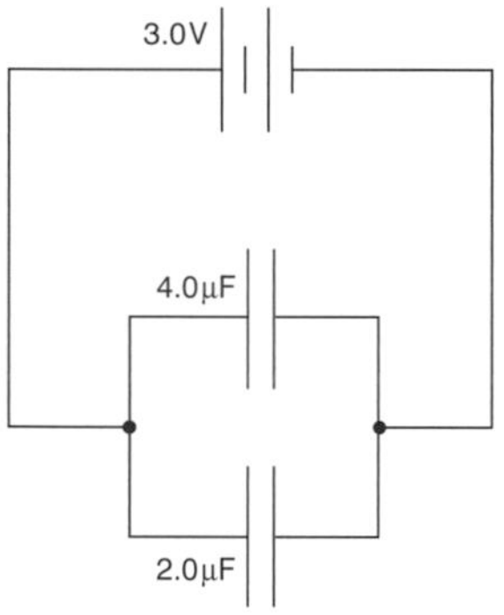

Figure 2

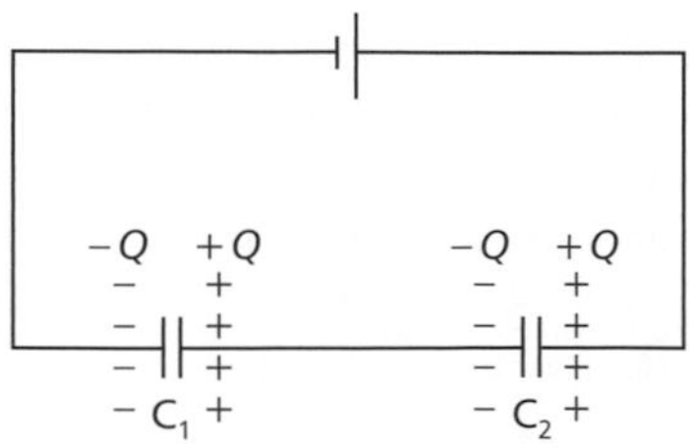

Figure 3 *Capacitors in series*

Capacitors in parallel

For two or more capacitors of capacitances C_1, C_2, C_3 ... in parallel,

the combined capacitance $C = C_1 + C_2 + C_3 + \ldots$

Consider two capacitors C_1 and C_2 in parallel, as shown in Figure 1. Components in parallel have the same p.d. so the charge stored by each capacitor, for p.d. V, is

- $Q_1 = C_1 V$ for capacitor C_1, and
- $Q_2 = C_2 V$ for capacitor C_2.

Therefore, the total charge stored $Q = Q_1 + Q_2 = C_1V + C_2V$

So the combined capacitance $C = \dfrac{Q}{V} = \dfrac{C_1V + C_2V}{V} = C_1 + C_2$

Note

For two capacitors in parallel, the ratio of the charge stored by each capacitor,

$$\frac{Q_1}{Q_2} = \frac{C_1V}{C_2V} = \frac{C_1}{C_2}$$

So the ratio of the charge stored by each capacitor is equal to the ratio of the capacitances, $\dfrac{C_1}{C_2}$.

Worked example

A 3.0 V battery is connected to a 4.0 μF capacitor and a 2.0 μF capacitor in parallel with each other. Calculate:

a the combined capacitance of the two capacitors,
b i the charge stored on each capacitor,
ii the total charge stored.

Solution

a $C = C_1 + C_2 = 4.0 + 2.0 = 6.0\,\mu F$
b i For the 4.0 μF capacitor, $Q_1 = C_1V = 4.0\,\mu F \times 3.0\,V = 12.0\,\mu C$.
For the 2.0 μF capacitor, $Q_2 = C_2V = 2.0\,\mu F \times 3.0\,V = 6.0\,\mu C$.
ii The total charge stored $Q = CV = 6.0\,\mu F \times 3.0\,V = 18\,\mu C$.

Note The total charge stored, Q, could be calculated using $Q = Q_1 + Q_2$

Capacitors in series

Components in series store the same charge. Figure 3 shows why this is so for two capacitors. The battery is connected to the left-hand plate of C_1 and to the right-hand plate of C_2. Electrons transfer from the negative terminal of the battery to the left-hand plate of C_1 and from the right-hand plate of C_2 to the battery. Transfer of electrons takes place to the left-hand plate of C_2 from the right-hand plate of C_1 but there is no transfer of electrons to either of these plates from the battery. Each plate stores the same magnitude of charge Q so the total charge stored is Q.

For two or more capacitors of capacitances C_1, C_2, C_3 ... in series,

the combined capacitance $\dfrac{1}{C} = \dfrac{1}{C_1} + \dfrac{1}{C_2} + \dfrac{1}{C_3} + \ldots$

Consider two capacitors C_1 and C_2 in series, as shown in Figure 3. Because they store the same amount of charge Q, the p.d. across each capacitor is:

- $V_1 = \frac{Q}{C_1}$ for capacitor C_1, and
- $V_2 = \frac{Q}{C_2}$ for capacitor C_2.

The capacitors are in series, so the total p.d. $V = V_1 + V_2$

$$\therefore \quad V = \frac{Q}{C_1} + \frac{Q}{C_2}$$

Because the combined capacitance $C = \frac{Q}{V}$,

$$\frac{1}{C} = \frac{V}{Q} = \frac{\left(\frac{Q}{C_1} + \frac{Q}{C_2}\right)}{Q} = \frac{1}{C_1} + \frac{1}{C_2}$$

Worked example

A 3.0 V battery, a 4.0 μF capacitor C_1 and a 2.0 μF capacitor C_2 are connected in series with each other. Calculate:

a the combined capacitance of the two capacitors,
b **i** the charge stored on each capacitor,
ii the p.d. across each capacitor.

Solution

a $\frac{1}{C} = \frac{1}{4.0} + \frac{1}{2.0} = 0.25 + 0.50 = 0.75\,\mu F^{-1}$

Therefore $C = \frac{1}{0.75} = 1.33\,\mu F$.

b **i** $Q = CV = 3.0\,V \times 1.33\,\mu F = 4.0\,\mu C$

ii For C_1, $V_1 = \frac{Q}{C_1} = \frac{4.0\,\mu C}{4.0\,\mu F} = 1.0\,V$. For C_2, $V_2 = \frac{Q}{C_2} = \frac{4.0\,\mu C}{2.0\,\mu F} = 2.0\,V$.

Note The p.d. across C_2, V_2, could have been calculated using $V = V_1 + V_2$.

Combinations with capacitors in series and parallel

Figure 5 shows two arrangements where three capacitors are connected to a battery.

- In **a**, two capacitors C_1 and C_2 in parallel are connected in series to a third capacitor C_3 and the battery.
- In **b**, the two capacitors C_1 and C_2 in series are connected in parallel with the third capacitor C_3 and the battery.

The steps below show how to work out the charge and p.d. of each capacitor, given the capacitance values and the battery p.d., V_0.

1 Calculate the total capacitance C by using the appropriate combination rule to find the combined capacitance C' of C_1 and C_2 then use the other combination rule to work out the combined capacitance of C_3 and C'.
2 Work out the total charge stored Q using $Q = CV$.
3 For **a**, charge Q is stored by C_3 and the same amount of charge is shared between C_1 and C_2 in proportion to their capacitances. After working out the charge stored by each capacitor, use $V = \frac{Q}{C}$ to work out the p.d. across each capacitor.
For **b**, the battery p.d., V_0, is across C_3 and is shared between C_1 and C_2 in inverse proportion to their capacitances. After working out the p.d. across each capacitor, use $Q = CV$ to work out the charge stored by each capacitor.

Notes

For two capacitors in series,
1 the ratio of the p.ds,
$$\frac{V_1}{V_2} = \frac{Q}{C_1} \div \frac{Q}{C_2} = \frac{C_2}{C_1}$$
So the ratio of the p.ds across the two capacitors is equal to the inverse ratio of the capacitances.
2 $\frac{1}{C} = \frac{1}{C_1} + \frac{1}{C_2} = \frac{C_1 + C_2}{C_1 C_2}$
Making C the subject of this expression gives
$$C = \frac{C_1 C_2}{C_1 + C_2} = \frac{\text{product of the capacitances}}{\text{sum of the capacitances}}$$

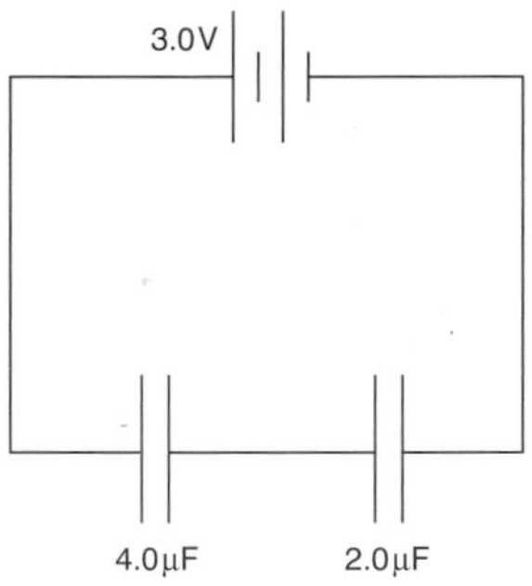

Figure 4

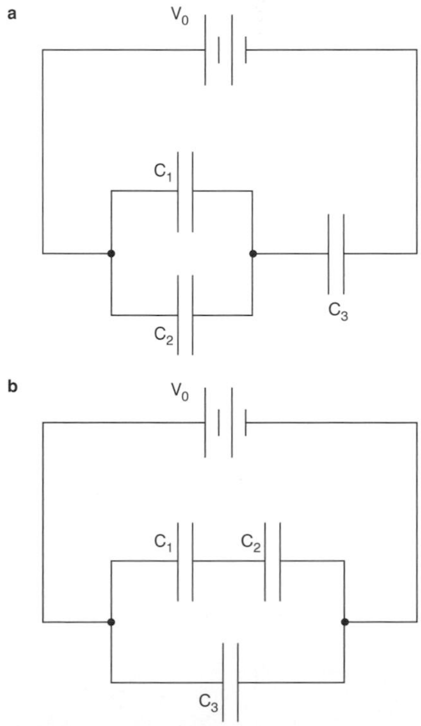

Figure 5

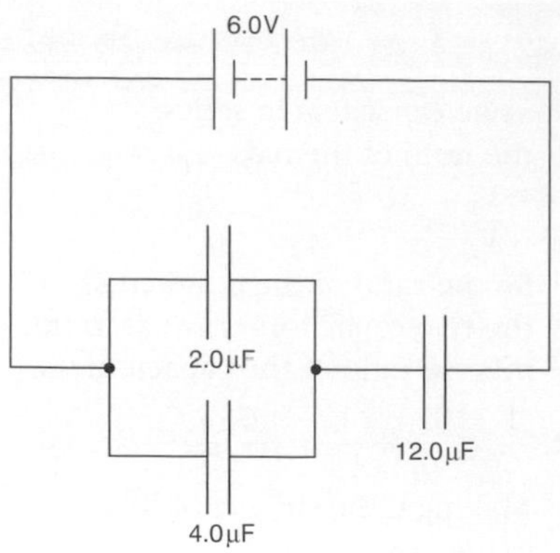

Figure 6

Worked example

A 2.0 μF capacitor and a 4.0 μF capacitor are in parallel with each other. The parallel combination is in series with a 12.0 μF capacitor and a 6.0 V battery, as shown in Figure 6. Calculate:

a the total capacitance of the combination,
b the charge stored by the combination,
c the charge and p.d. for each capacitor.

Solution

a $C' = 2.0 + 4.0 = 6.0\,\mu F$

$\therefore$ the total capacitance C is given by $\frac{1}{C} = \frac{1}{6.0} + \frac{1}{12.0} = 0.25\,\mu F^{-1}$

$\therefore C = \frac{1}{0.25} = 4.0\,\mu F$

b The charge stored $Q = CV = 4.0 \times 6.0 = 24\,\mu C$

c C_3: charge stored $= Q = 24\,\mu C$. $\therefore$ p.d. across $C_3 = \frac{Q}{C_3} = \frac{24\,\mu C}{12\,\mu F} = 2.0\,V$

C_1 and C_2: $\frac{C_1}{C_2} = \frac{2\,\mu F}{4\,\mu F} = 0.5$, then $\frac{Q_1}{Q_2} = \frac{C_1}{C_2} = 0.5$

As $Q = Q_1 + Q_2 = 24\,\mu C$, then $Q_1 = 8\,\mu C$ and $Q_2 = 16\,\mu C$

$\therefore \quad V_1 = \frac{Q_1}{C_1} = \frac{8\,\mu C}{2\,\mu F} = 4\,V$

$V_2 = \frac{Q_2}{C_2} = \frac{16\,\mu C}{4\,\mu F} = 4\,V$

Note $V_1 = V_2 = 6.0\,V - V_3$

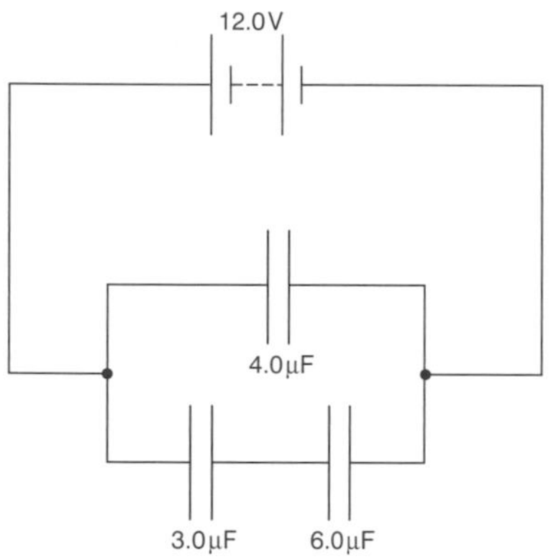

Figure 7

Worked example

A 3.0 μF capacitor and a 6.0 μF capacitor are in series with each other. The series combination is in parallel with a 4.0 μF capacitor and a 12.0 V battery, as shown in Figure 7. Calculate:

a **i** the total capacitance of the combination,
ii the charge stored by the combination,
b the charge and p.d. for each capacitor.

Solution

a **i** $\frac{1}{C'} = \frac{1}{3.0} + \frac{1}{6.0} = 0.5\,\mu F^{-1}$, $\therefore C' = \frac{1}{0.5} = 2.0\,\mu F$

$\therefore$ the total capacitance $C = C' + C_3 = 2.0 + 4.0 = 6.0\,\mu F$

ii The charge stored $Q = CV = 6.0 \times 12.0 = 72\,\mu C$

b C_3: p.d. V_3 = battery p.d. = 12 V

charge stored $Q_3 = C_3 V_3 = 4.0\,\mu F \times 12.0\,V = 48\,\mu C$.

C_1 and C_2: charge stored by the combination $= Q - Q_3 = 72 - 48 = 24\,\mu C$

They store the same amount of charge as they are in series with each other.

$\therefore Q_1 = Q_2 = 24\,\mu C$

$\therefore V_1 = \frac{Q_1}{C_1} = \frac{24\,\mu C}{3.0\,\mu F} = 8.0\,V$

$V_2 = \frac{Q_2}{C_2} = \frac{24\,\mu C}{6.0\,\mu F} = 4.0\,V$

Note $V_1 + V_2 = V_3 = 12\,V$

Extension

More about dielectrics

When a dielectric is placed between oppositely charged metal plates, the electrons in each dielectric molecule are pulled a little towards the positive plate. Each molecule of the dielectric becomes **polarised**, as shown in Figure 8. The surface of the dielectric facing the positive plate gains a layer of negative charge due to polarisation and the surface facing the negative plate gains positive charge. The p.d. between the plates is equal to the battery p.d. and is constant. As a result:

- the positive charge on the dielectric surface facing the negative plate attracts electrons from the battery onto the negative plate,
- the negative charge on the dielectric surface facing the positive plate pushes electrons back to the battery from the negative plate.

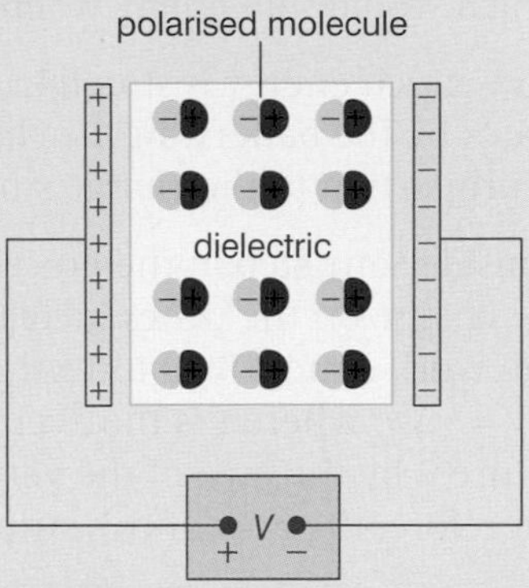

Figure 8 Dielectrics

So the dielectric increases the charge stored on the plates without changing the plate p.d. In other words, the dielectric increases the capacitance of the parallel plates.

The relative permittivity ε_r or dielectric constant, may be defined as

$$\varepsilon_r = \frac{C}{C_0}$$

where C is the capacitance with a dielectric completely filling the space between the plates and C_0 is the capacitance with empty space between the plates.

Typical values for ε_r are 7 for mica, 2.3 for polythene, 2.7 for paper and 81 for water.

The capacitance C of a pair of parallel metal plates depends on:

- the perpendicular distance Δd between the plates,
- the surface area A of the plates,
- the presence of a dielectric between the plates.

For a constant p.d. ΔV between the plates, it can be shown using Coulomb's law, as outlined in Topic 14.2, that in the absence of a dielectric between the plates, the electric field strength $E = \frac{Q}{A\varepsilon_0}$, where Q is the magnitude of the charge on each plate Q.

Since $E = \frac{\Delta V}{\Delta d}$, rearranging $\frac{\Delta V}{\Delta d} = \frac{Q}{A\varepsilon_0}$ gives the following formula for the capacitance C_0 of the 'empty' parallel plates:

$$C_0 = \frac{A\varepsilon_0}{\Delta d}$$

If the space between the plates is completed filled with a dielectric of relative permittivity ε_r, the capacitance $C = \varepsilon_r C_0 = \frac{A\varepsilon_r\varepsilon_0}{\Delta d} = \frac{A\varepsilon}{\Delta d}$, where the permittivity of the dielectric $\varepsilon = \varepsilon_r\varepsilon_0$.

Note ε_0 may be determined by measuring the capacitance C_0 of 'empty' parallel plates of known area A and known spacing Δd. then using the equation for C_0 above to calculate ε_0.

Summary test 15.2

1 A 3.0 V battery is connected to a 2.0 μF capacitor in parallel with a 3.0 μF capacitor. Sketch the circuit diagram and calculate:

a the combined capacitance of the two capacitors,

b the charge stored and the p.d. across each capacitor.

2 **a** A 4.5 V battery is connected to a 6.0 μF capacitor in series with a 4.0 μF capacitor. Calculate:

i the combined capacitance of the two capacitors,

ii the charge stored and the p.d. across each capacitor.

b A 10 μF capacitor is connected in parallel with the 6.0 μF capacitor.

i Sketch the new circuit diagram and calculate the total capacitance of the three capacitors.

ii Calculate the charge and p.d. across each capacitor.

3 A student is given three capacitors of values 10 μF, 22 μF and 47 μF. Sketch the eight different possible combinations using all three capacitors and calculate the capacitance of each combination.

4 A 4.0 μF capacitor in series with a 10.0 μF capacitor are connected to a 6.0 V battery. A 2.0 μF capacitor is then connected to the battery in parallel with the two capacitors in series.

a Sketch the circuit diagram for this arrangement and calculate its total capacitance.

b Calculate the charge and p.d. for each capacitor in the arrangement.

15.3 Energy stored in a charged capacitor

Learning outcomes

On these pages you will learn to:

- derive from the area under a potential–charge graph the equations $E = \frac{1}{2}QV = \frac{1}{2}CV^2$ for the energy stored in a capacitor
- solve problems using the above equations

When a capacitor is charged, energy is stored in it. A charged capacitor discharged across a torch bulb will release its energy in a brief flash of light from the bulb, as long as the capacitor has been charged initially to the operating p.d. of the bulb. Charge flow is rapid enough to give a large enough current to light the bulb, but only for a brief time. The bulb could be replaced by a miniature electric motor which would spin briefly when the capacitor is discharged through it.

How much energy is stored in a charged capacitor? The charge is forced onto the plates by the battery. In the charging process, the p.d. across the plates increases in proportion to the charge stored, as shown in Figure 2.

Consider one step in the process of charging a capacitor of capacitance C when the charge on the plates increases by a small amount Δq from q to $q + \Delta q$. The work done ΔW to force the extra charge Δq on to the plates is given by $\Delta W = v\Delta q$, where v is the average p.d. during this step. $v\Delta q$ is represented in Figure 2 by the area of the vertical strip of width Δq and height v under the line. Therefore, the area of this strip represents the work done ΔW in this small step.

Now consider all the small steps from zero p.d. to the final p.d. V. The total work done W is obtained by adding up the work done for each small step. In other words, W is represented by the total area under the line from zero p.d. to p.d. V. As this area is a triangle of height V and base length Q $(= CV)$, the total work done W = triangle area = $\frac{1}{2}$ × height × base = $\frac{1}{2}VQ$

Because the energy stored in the capacitor, or **capacitor energy**, is equal to the work done on it to charge it, the energy stored = $\frac{1}{2}VQ$.

Energy stored by the capacitor, $W = \frac{1}{2}QV$

Note

Using $Q = CV$, the above equation may be written as $W = \frac{1}{2}CV^2 = \frac{1}{2}\frac{Q^2}{C}$

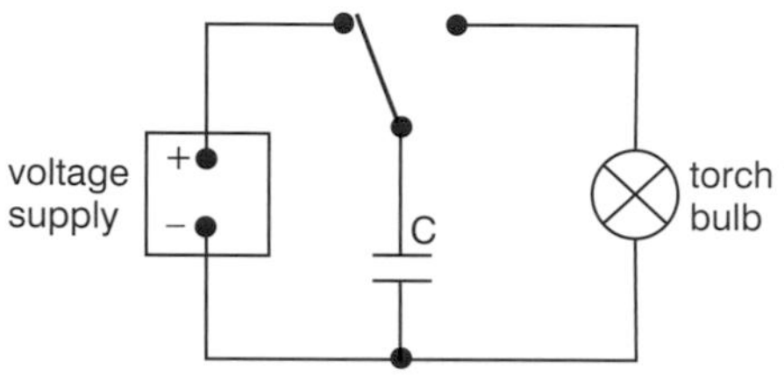

Figure 1 Releasing stored energy

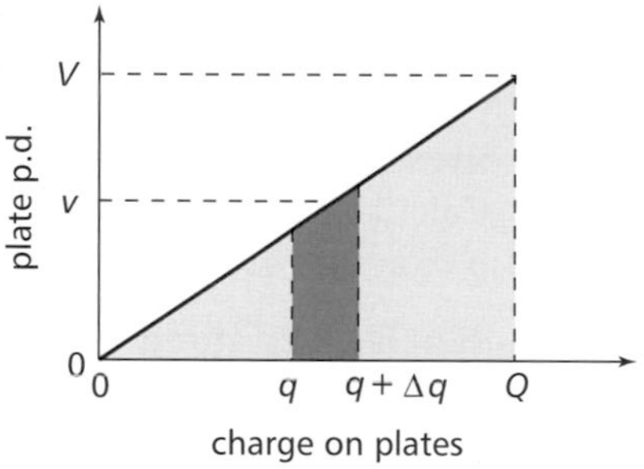

Figure 2 Energy stored in a capacitor

Measuring the energy stored in a charged capacitor

A joulemeter is used to measure the energy transfer from a charged capacitor to a light bulb when the capacitor discharges. The capacitor p.d., V, is measured and the joulemeter reading recorded before the discharge starts. When the capacitor has discharged, the joulemeter reading is recorded again. The difference of the two joulemeter readings is the energy transferred from the capacitor during the discharge process. This is the total energy stored in the capacitor before it discharged. This can be compared with the calculation of the energy stored using $W = \frac{1}{2}CV^2$.

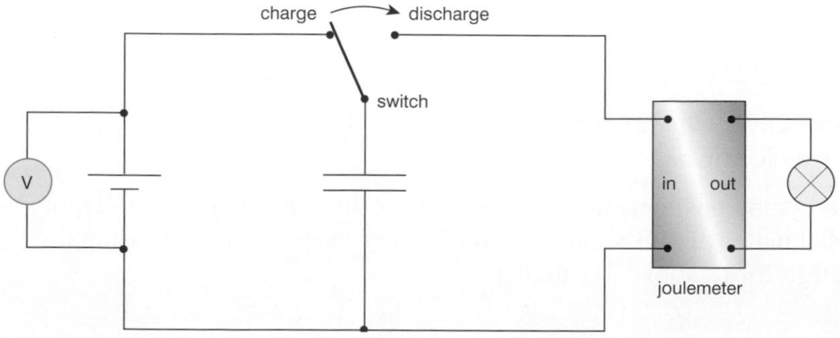

Figure 3 Measuring energy stored

The energy stored in a thundercloud

Imagine a thundercloud and the Earth below like a pair of charged parallel plates. Because the thundercloud is charged, a strong electric field exists between the thundercloud and the ground. The potential difference between the thundercloud and the ground, $V = Ed$, where E is the electric field strength and d is the height of the thundercloud above the ground.

- For a thundercloud carrying a constant charge Q, the energy stored $W = \frac{1}{2}QV = \frac{1}{2}QEd$.
- If the thundercloud is forced by winds to rise to a new height d', the energy stored $W' = \frac{1}{2}QEd'$.
- As the electric field strength is unchanged (since it depends on the charge per unit area), the increase in the energy stored $= W' - W = \frac{1}{2}QEd' - \frac{1}{2}QEd = \frac{1}{2}QE\,\Delta d$, where $\Delta d = d' - d$.

This increase in the energy stored is because work is done by the force of the wind to overcome the electrical attraction between the thundercloud and the ground and make the charged thundercloud move away from the ground.

The insulating property of air breaks down if it is subjected to an electric field strength more than about $300\,\mathrm{kV\,m^{-1}}$. Prove for yourself that, for every metre rise of the thundercloud carrying a maximum charge of 20 C, the energy stored would increase by 3 MJ. At a height of 500 m, the energy stored would be 1500 MJ.

Summary test 15.3

1 Calculate the charge and energy stored in a $10\,\mu\mathrm{F}$ capacitor charged to a p.d. of:

a 3.0 V,

b 6.0 V.

2 A $50\,000\,\mu\mathrm{F}$ capacitor is charged from a 9 V battery then discharged through a light bulb in a flash of light lasting 0.2 s. Calculate:

a the charge and energy stored in the capacitor before discharge,

b the average power supplied to the light bulb.

3 A $2.2\,\mu\mathrm{F}$ capacitor is connected in series with a $10\,\mu\mathrm{F}$ capacitor and a 3.0 V battery. Calculate the charge and energy stored in each capacitor.

4 In Figure 4, a $4.7\,\mu\mathrm{F}$ capacitor is charged from a 12.0 V battery by connecting the switch to X. The switch is then reconnected to Y to charge a $2.2\,\mu\mathrm{F}$ capacitor from the first capacitor.

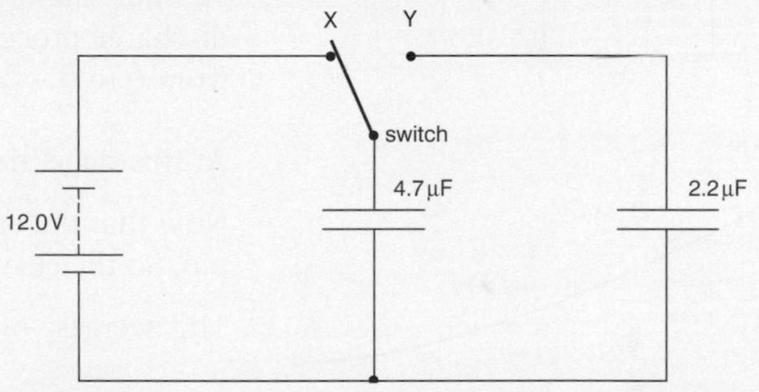

Figure 4

Calculate:

a the initial charge and energy stored in the $4.7\,\mu\mathrm{F}$ capacitor,

b the combined capacitance of the two capacitors,

c the final p.d. across the two capacitors,

d the final energy stored in each capacitor. Account for the loss of energy stored.

15.4 Capacitor discharge

Learning outcomes

On these pages you will learn to:

- draw a graph to show how the p.d. across a capacitor decreases as the capacitor discharges through a fixed resistor
- recall and use the equation $V = V_0 e^{-t/RC}$
- explain what is meant by the time constant of a circuit that has a capacitor in series with a resistor
- recall and use the time constant formula $\tau = RC$
- show an understanding of the function of capacitors in simple circuits

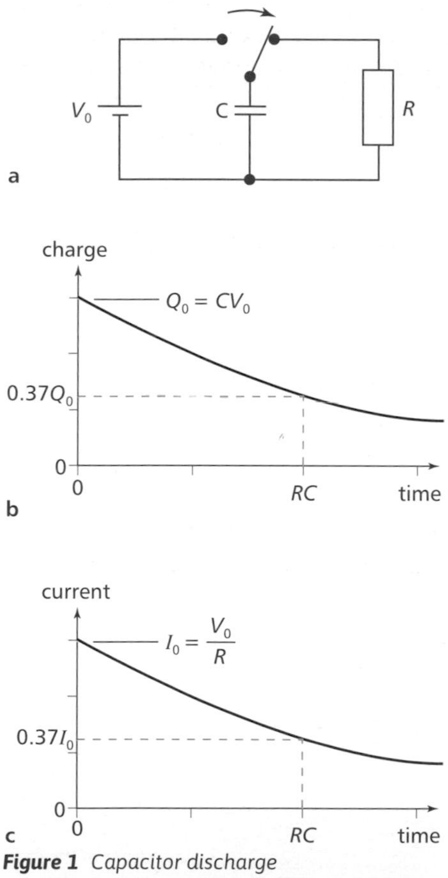

***Figure 1** Capacitor discharge*

Note

A graph of capacitor p.d. V against time t has the same shape as the graphs above. This is because V is proportional to the charge Q (because $Q = CV$) and it is also proportional to the current I (because $V = IR$).

When a capacitor discharges through a fixed resistor, the discharge current decreases gradually to zero. Figure 1 shows a circuit in which a capacitor is discharged through a resistor when the switch is changed over. The reason why the current decreases gradually is that the p.d. across the capacitor decreases as it loses charge. Because the resistor is connected directly to the capacitor, the resistor current (= p.d./resistance) decreases as the p.d. decreases.

The situation is not unlike water emptying through a pipe at the bottom of the container. When the container is full, the flow rate out of the pipe is high because the water pressure at the pipe is high. As the container empties, the water level falls so the water pressure at the pipe falls and the flow rate drops.

The graphs in Figure 1 show how the current and the charge decrease with time. Both curves have the same shape because both the current and charge (and p.d.) decrease **exponentially**. This means that any of these quantities decreases by the same factor in equal intervals of time. For example, for initial charge Q_0, if the charge is $0.9Q_0$ after a certain time t_1, the charge will be:

- $0.9 \times 0.9\,Q_0$ after time $2t_1$,
- $0.9 \times 0.9 \times 0.9\,Q_0$ after time $3t_1$ …
- $0.9^n Q_0$ after time $n\,t_1$

About exponential decrease

To understand why the decrease is exponential, consider one small step in the discharge process of a capacitor C through a resistor R when the charge decreases from Q to $Q - \Delta Q$ in time Δt.

At this stage, the current $I = \dfrac{\text{p.d. across the plates, } V}{\text{resistance, } R} = \dfrac{Q}{CR}$ as $V = \dfrac{Q}{C}$

Note that the current is proportional to the charge which is proportional to the p.d. So the curves all have the same shape.

The decrease of charge $\Delta Q = -I\,\Delta t$ (– as Q decreases).

Therefore, $\Delta Q = -\dfrac{Q}{CR}\Delta t$, which gives

$$\frac{\Delta Q}{Q} = -\frac{\Delta t}{CR}$$

The equation tells us that the fractional drop of charge $\dfrac{\Delta Q}{Q}$ is the same in any short interval of time Δt during the discharge process. For example, suppose $\Delta t = 10\,\text{s}$ and $CR = 100$. Then $\dfrac{\Delta Q}{Q} = -0.1 \left(= -\dfrac{\Delta t}{CR}\right)$. So the charge decreases to 0.9 of its value at the start of the 10 s interval. So, if the initial charge is Q_0, the charge still on the plates will be:

- $0.9\,Q_0$ after 10 s,
- $0.9 \times 0.9\,Q_0$ after a further 10 s,
- $0.9 \times 0.9 \times 0.9\,Q_0$ after a further 10 s, etc.

In theory, the charge on the plates never becomes zero.

Exponential changes occur whenever the rate of change of a quantity is proportional to the quantity itself. Rearranging the equation $\dfrac{\Delta Q}{Q} = -\dfrac{\Delta t}{CR}$ gives

$$\frac{\Delta Q}{\Delta t} = -\frac{Q}{CR}$$

For very short intervals of time (i.e. $\Delta t \to 0$), $\dfrac{\Delta Q}{\Delta t}$ represents the rate of change of charge and is written $\dfrac{dQ}{dt}$.

Therefore, $\frac{dQ}{dt} = -\frac{Q}{CR}$

The graphical solution to this equation is shown in Figure 1b.
The mathematical solution is

$Q = Q_0 e^{-t/RC}$, where Q_0 = initial charge, and
e = the exponential function (sometimes written 'exp')

The quantity RC is called the **time constant**, τ, for the circuit. At time τ after the start of the discharge, the charge falls to 0.37 (= e^{-1}) of its initial value.

Time constant = RC, where R = circuit resistance, C = capacitance

The unit of RC is the second. This is because 1 ohm = 1 $\frac{\text{volt}}{\text{ampere}}$ and 1 farad = 1 $\frac{\text{coulomb}}{\text{volt}}$ so the unit of $RC = \frac{\text{volt}}{\text{ampere}} \times \frac{\text{coulomb}}{\text{volt}} = \frac{\text{coulomb}}{\text{ampere}}$ = the second.

Notes

1 The current, the p.d. and the charge are all proportional to one another. All three quantities decrease exponentially in capacitor discharge in accordance with the equation $x = x_0 e^{\frac{-t}{CR}}$, where x represents either the current or the charge or the p.d.

2 The inverse function of e^x is $\ln x$, where ln is the natural logarithm. To calculate t, given x, x_0, R and C, use of the inverse function of e^x gives $\ln x = \ln x_0 - \left(\frac{t}{RC}\right)$.

3 The function $e^z = 1 + z + \frac{z^2}{2 \times 1} + \frac{z^3}{3 \times 2 \times 1} + \ldots$, etc. It can be shown mathematically that the rate of change of this function with respect to z is the same function. This is why the function appears whenever the rate of change of a quantity is proportional to the quantity itself. Note that $z = 1$ gives $e = 1 + 1 + \frac{1}{2} + \frac{1}{6} + \frac{1}{24}$ etc. = 2.718. You can check this on your calculator by keying in 'e^x' then pressing 1 to give 2.718. Keying in −1 instead of 1 gives 0.37 for e^{-1}.

Worked example

A 47 μF capacitor charged to 12.0 V is then discharged through a 2.2 MΩ resistor.

a Calculate the time taken for the capacitor p.d. to decrease to below 3.0 V.

b i Calculate the percentage of the energy stored in the capacitor that remains when its p.d. is 3.0 V.

ii Discuss where the energy transferred from the capacitor is transferred to.

Solution

a $3.0 = 12.0\, e^{-t/RC}$. Rearranging this gives $e^{-t/RC} = \frac{3.0}{12.0} = 0.25$. Taking natural logarithms of both sides gives $\ln 0.25 = -\frac{t}{RC}$ so $\frac{t}{RC} = -\ln 0.25 = 1.386$.
Therefore $t = 1.386\, RC = 1.386 \times 2.2\,\text{M}\Omega \times 47\,\mu\text{F} = 143\,\text{s}$.

b i Energy stored initially $= \frac{1}{2}CV^2 = 0.5 \times 2.2\,\mu\text{F} \times (12.0\,\text{V})^2 = 158.4\,\mu\text{J}$
Energy stored at 3.0 V $= 0.5 \times 2.2\,\mu\text{F} \times (3.0\,\text{V})^2 = 9.9\,\mu\text{J}$
% of initial energy stored $= \left(\frac{9.9\,\mu\text{J}}{158.4\,\mu\text{J}}\right) \times 100\% = 6.25\%$

ii Energy is transferred to the wires by the resistance heating effect of the current in the wires. The energy is then dissipated to the surroundings.

Worked example

A 2200 μF capacitor is charged to a p.d. of 9.0 V then discharged through a 100 kΩ resistor using a circuit as shown in Figure 1a.

a Calculate:
i the initial charge stored by the capacitor,
ii the time constant of the circuit.

b Calculate the p.d. after:
i a time equal to the time constant, **ii** 300 s.

Solution

a i $Q_0 = CV_0$
$= 2200 \times 10^{-6}\,\text{F} \times 9.0\,\text{V}$
$= 2.0 \times 10^{-2}\,\text{C}$,

ii Time constant $= RC$
$= 100 \times 10^3\,\Omega \times 2.2 \times 10^{-3}\,\text{F}$
$= 220\,\text{s}$

b i When $t = RC$, $V = V_0 e^{-1}$
$= 0.37 \times 9.0 = 3.33\,\text{V}$

ii When $t = 300\,\text{s}$,
$\frac{t}{RC} = \frac{300}{220} = 1.36$
$\therefore V = V_0 e^{-t/RC} = 9.0\, e^{-1.36}$
$= 2.3\,\text{V}$

See Chapter 24 Mathematical skills for more about exponential decrease.

Note

A quicker and easier method is to recognise that capacitor energy is proportional to the square of the capacitor p.d. As the capacitor p.d. decreases to $\frac{1}{4}$ of the initial p.d., the energy stored decreases to $\frac{1}{16}$ of the initial energy stored (i.e. 6.25%).

Investigating capacitor discharge

Figure 2 shows how to measure the p.d. across a capacitor as it discharges through a fixed resistor. An oscilloscope is used as it has a very high resistance so the discharge current from the capacitor passes only through the fixed resistor.

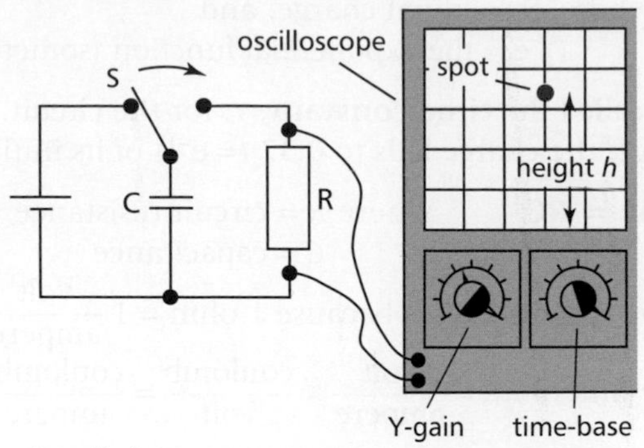

Figure 2 *Measuring capacitor discharge*

The oscilloscope is used to measure the capacitor p.d. at regular intervals. A data logger or a digital voltmeter could be used instead of the oscilloscope.

The measurements may be used to plot a graph of voltage against time. The time taken for the voltage to decrease to 37% (= 1/e) of the initial value can be measured from the graph and compared with the calculated value of RC.

The significance of the time constant

The time constant RC is the time taken, in seconds, for the capacitor to discharge to 37% of its initial charge. Given values of R and C, the time constant can be quickly calculated and used as an approximate measure of how quickly the capacitor discharges. In the worked example in the margin of the previous page, the time constant of 220 s gives a 'rule of thumb' estimate of the time taken to discharge significantly but not completely. Also, $5RC$ gives a 'rule of thumb' estimate for the time taken to discharge by over 99%. Prove for yourself that $t = 5RC$ gives a value which is less than 1% of the initial value.

Applications of capacitor discharge

1 **Any electronic timing circuit or time-delay circuit** makes use of capacitor discharge through a fixed resistor. Figure 3 shows an alarm circuit where the alarm rings if the input voltage to the electronic circuit drops below a certain value after the switch is reset. The time delay between resetting the switch and the alarm ringing can be increased by increasing the resistance R or the capacitance C. Such a change to the circuit would make the discharge of C through R slower so increasing the time for the capacitor voltage to decrease sufficiently to make the alarm ring.

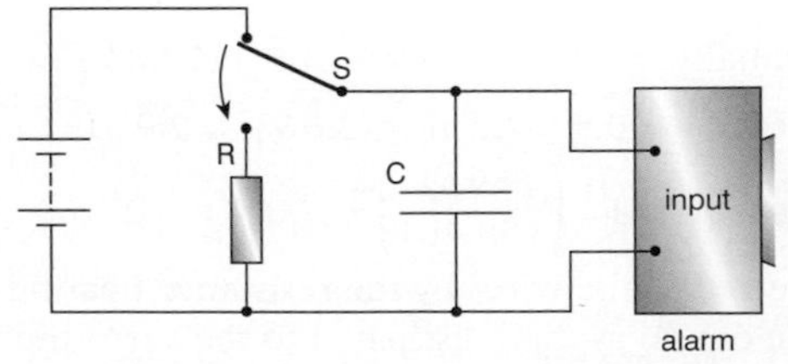

Figure 3 *A time-delayed alarm circuit*

2 **Capacitor smoothing** is used in applications where sudden voltage variations or 'glitches' can have undesirable effects. For example, mains appliances being switched on or off in a building could affect computers connected to the mains supply in the building. A large capacitor in a computer supplies current if the mains supply is interrupted so the computer circuits continue to function normally.

Summary test 15.4

1 A 50 μF capacitor is charged by connecting it to a 6.0 V battery then discharged through a 100 kΩ resistor.

a Calculate:

i the charge stored in the capacitor immediately after it has been charged,

ii the time constant of the circuit.

b i Estimate how long the capacitor would take to discharge to about 2 V.

ii Estimate the resistance of the resistor that you would use in place of the 100 kΩ resistor if the discharge is to be 99% completed within about 5 s.

2 A 68 μF capacitor is charged to a p.d. of 9.0 V then discharged through a 20 kΩ resistor.

a Calculate:

i the charge stored by the capacitor at a p.d. of 9.0 V,

ii the initial discharge current.

b Calculate the p.d. and the discharge current 5.0 s after the discharge started.

3 A 2.2 μF capacitor is charged to a p.d. of 6.0 V and then discharged through a 100 kΩ resistor. Calculate:

a the charge and energy stored in this capacitor at 6.0 V,

b the p.d. across the capacitor 0.5 s after the discharge started,

c the energy stored at this time.

4 A 4.7 μF capacitor is charged to a p.d. of 12.0 V and then discharged through a 220 kΩ resistor. Calculate:

a the energy stored in this capacitor at 12.0 V,

b the time taken for the p.d. to fall from 12.0 V to 3.0 V,

c the energy lost by the capacitor in this time.

Chapter Summary

The **capacitance** of a capacitor is defined as the charge stored per unit p.d.

The **unit of capacitance** is the farad (F), equal to 1 coulomb per volt. Note that 1 μF = 10^{-6} F.

Capacitor equation $C = \frac{Q}{V}$

Capacitor combination rules

1 Capacitors in parallel: combined capacitance $C = C_1 + C_2 + C_3 + \ldots$

2 Capacitors in series: combined capacitance $\frac{1}{C} = \frac{1}{C_1} + \frac{1}{C_2} + \frac{1}{C_3} + \ldots$

Energy stored by the capacitor, $W = \frac{1}{2}QV = \frac{1}{2}CV^2$

Capacitor discharge

1 **Time constant,** $\tau = RC$

2 **Exponential decrease equation** for current or charge or p.d.: $x = x_0\, e^{-t/RC}$

Launch additional digital resources for the chapter

1 **a** Define the capacitance of a capacitor.

b **i** With the aid of a circuit diagram, describe how a capacitor may be charged using a battery and a variable resistor to keep the charging current constant.

ii Figure 1.1 below shows how the p.d. changes with time when a capacitor is charged at a constant current of 16 μA. Calculate the charge stored after 20 s and calculate the capacitance of the capacitor.

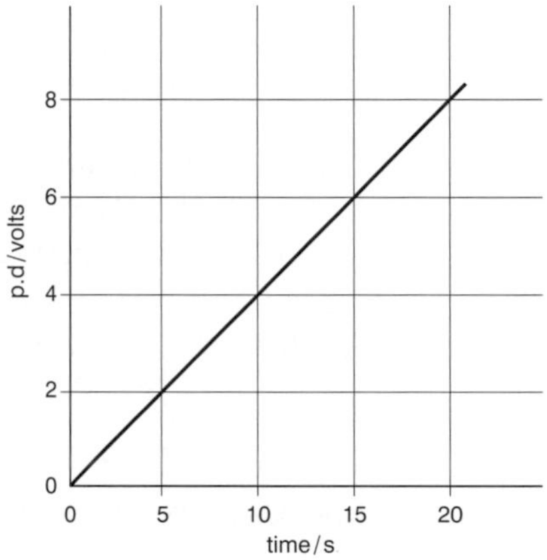

Figure 1.1

2 **a** A 2.0 μF capacitor is connected in series with a 4.0 μF capacitor and a 6.0 V battery.

i Sketch the circuit diagram and calculate the combined capacitance of the two capacitors.

ii Calculate the charge, p.d. and energy stored for the 2.0 μF capacitor.

b A capacitor *C* is connected in parallel with the 4.0 μF capacitor in the circuit in **a**. The p.d. across the 2.0 μF capacitor changes to 5.0 V. Sketch the new circuit diagram and calculate the capacitance of *C*.

3 A 4.0 μF capacitor is charged by connecting it to a 9.0 V battery, as shown below. The capacitor is then disconnected from the battery and connected to an uncharged 6.0 μF capacitor.

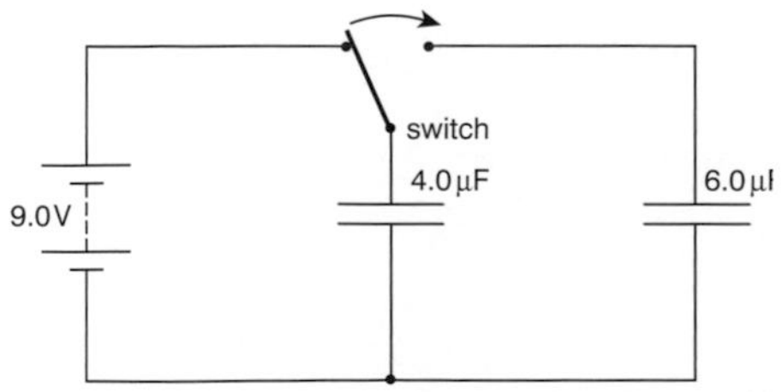

Figure 3.1

a Calculate the charge and energy stored in the 4.0 μF capacitor when the p.d. across its terminals is 9.0 V.

b The p.d. across the two capacitors becomes the same when they are connected together.

i Calculate the combined capacitance of the two capacitors.

ii Show that the p.d. across the two capacitors is 3.6 V after they have been connected together.

iii Calculate the total energy stored by the two capacitors after they have been connected together and explain why it differs from the initial energy stored.

4 The flashlight circuit of a camera includes two 47 000 μF capacitors in parallel with each other. The capacitors are charged using a 4.5 V battery. A flash of light is emitted from the flashbulb when the charged capacitors are connected across the flashbulb.

a **i** Calculate the energy stored in the capacitor when the p.d. across its terminals is 4.5 V.

ii The duration of the discharge is approximately 50 ms. Estimate the power of the flashlight, stating any assumptions made in your estimate.

b Explain why less energy would be stored in the capacitors if they were in series with each other.

5 A capacitor *C* of capacitance 470 μF is connected to a 12 V battery then discharged through a 100 kΩ resistor *R*.

a Calculate the charge and energy stored in *C* when the p.d. across its terminals is 12 V.

b **i** Calculate the time constant of the discharge circuit.

ii Show that the p.d. across the capacitor decreases to 3.3 V in 60 s.

iii Calculate the energy transferred from the capacitor to the resistor when the p.d. decreases from 12 V to 3.3 V.

6 A capacitor circuit consists of 4 identical 10 mF capacitors connected in parallel to a 12 V voltage supply unit.

a Calculate the total charge and energy stored.

b The circuit is used as a 'back-up' supply in a data capture device in case the p.d. from the voltage supply unit is interrupted. The device switches off if the voltage supplied to it falls below 10 V. In normal operation, the voltage supply unit supplies a current of 80 mA at 12 V.

i Show that the circuit resistance in normal operation is approximately 150 Ω.

ii If the voltage supply unit is switched off, estimate how much longer the device would continue to operate for.

7 An insulated metal sphere X of radius R has a positive charge Q at potential V.

a Show that the capacitance C of the sphere is equal to $4\pi\varepsilon_0 R$.

b An insulated uncharged metal sphere Y of radius $0.20R$ at a fixed distance from X is charged from X using an insulated conductor, as shown in Figure 7.1. Y is then moved far away from X.

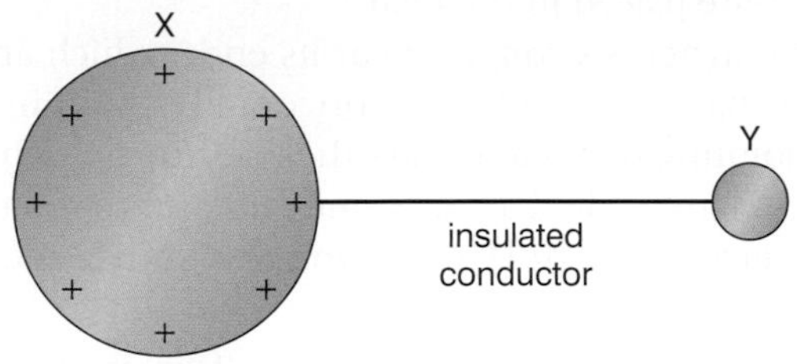

Figure 7.1

i In terms of electrons, describe how Y becomes charged when the conductor is connected between X and Y.

ii Explain why the potential of Y increases and that of X decreases until X and Y are at the same potential V_f.

iii Show that $V_f = 0.83V$.

c The energy stored by the charged sphere can be calculated using a capacitor energy equation.

i Use this equation to show that the energy dissipated in the conductor when Y is charged from X is one-sixth of the initial energy stored in X.

ii State the reason why energy is dissipated when Y becomes charged from X.

8 In an experiment to measure the capacitance C of a capacitor, the circuit in Figure 8.1 was used to charge the capacitor and then discharge it through a resistor of known resistance R.

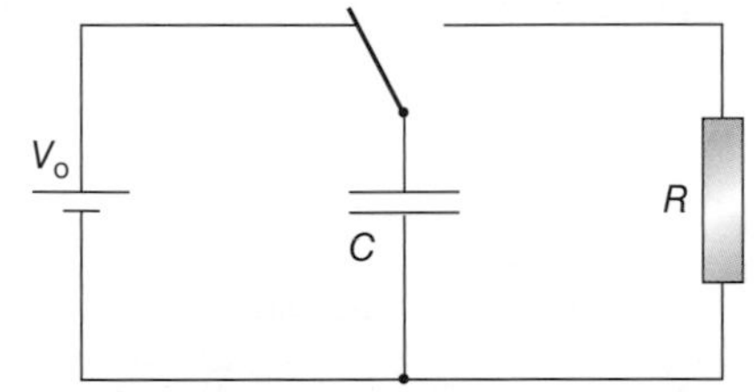

Figure 8.1

a The capacitor p.d., V, at time t after the discharge commenced is given by $V = V_0 e^{-t/CR}$.

Show that this equation can be rearranged into an equation of the form $\ln V = a - bt$, where a and b are constants, and determine expressions for a and b.

b As the capacitor discharged, its p.d. was measured every 30 seconds using a digital voltmeter. The measurements were repeated twice as shown in Table 1.

Table 1

t/s	0	30	60	90	120	150	180	210	240	270	300
V/V	4.50	3.82	3.26	2.78	2.33	2.00	1.70	1.43	1.23	1.04	0.89
	4.51	3.81	3.25	2.77	2.35	2.10	1.72	1.43	1.25	1.02	0.90
	4.50	3.83	3.25	2.76	2.34	1.98	1.69	1.42	1.22	1.04	0.87
mean V/V	4.503	3.820	3.253	2.760	2.340	2.027	1.703				
ln V	1.505	1.340	1.180	1.017	0.850	0.707	0.532				

i Write the missing entries for Table 1.

ii Use the measurements to plot a graph of $\ln V$ on the y-axis against t on the x-axis.

iii Use your graph to determine the time constant of the discharge circuit.

iv The resistance R of the resistor was 68 kΩ. Determine the capacitance C of the capacitor.

c i Discuss the reliability of the measurements.

ii Estimate the accuracy of your value of capacitance, given the resistor value is accurate to within 1%.

9 A 2.2 μF capacitor is discharged from a 2.0 V cell and the capacitor is then discharged through a 470 kΩ resistor.

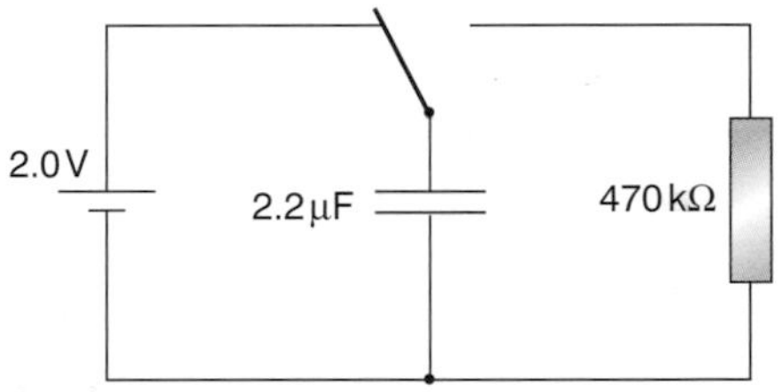

Figure 9.1

a i Calculate the initial current in the resistor when the capacitor starts to discharge through the resistor.

ii Explain why the current in the resistor decreases as the capacitor discharges.

b i Calculate the time constant of the discharge circuit.

ii Calculate the current in the resistor 5.0 s after the discharge started.

c i Sketch a graph to show how the current decreases with time in the first 5.0 s of the discharge.

ii Explain why the charge on the capacitor decreases at the same rate as the current decreases.

16 Magnetic fields

16.1 Magnetic field patterns

Learning outcomes

On these pages you will learn to:

- recognise that a magnetic field is an example of a field of force produced either by current-carrying conductors or by permanent magnets
- explain what is meant by a magnetic field line
- sketch magnetic field patterns due to a bar magnet, a long straight wire, a flat circular coil and a long solenoid

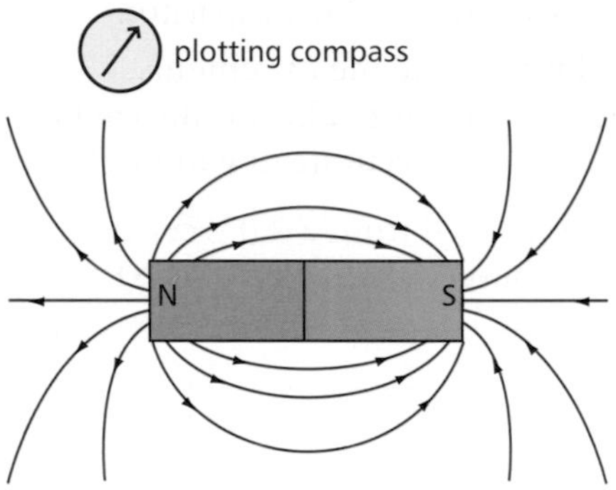

Figure 1 *The magnetic field near a bar magnet*

Lines of force

- A **magnetic field** is a force field produced either by a magnet or by moving charges such as when a current is in a wire. The force field acts on any other magnet or current-carrying wire placed in the field.
- The magnetic field of a bar magnet is strongest near its ends, which are referred to as '**poles**'. A bar magnet free to turn horizontally about its centre aligns itself with one end pointing north and the other pointing south. This occurs because the Earth's magnetic field attracts one end and repels the other end. The poles are referred to as **north-seeking** and **south-seeking**, according to the direction in which each pole points.
- A line of force of a magnetic field is a line along which a 'free' north pole would move in the field. Lines of force are often referred to as '**magnetic field lines**'. Note that the lines of force of a permanent magnet loop round from the north pole to the south pole of the magnet. A plotting compass points in the direction of a line of force.

The force between two magnets

Two bar magnets placed end-to-end attract or repel, depending on whether the nearest poles are:

- **like polarity**, in which case they repel, or
- **unlike polarity**, in which case they attract.

Electromagnetism

A magnetic field is created round a wire whenever **a current passes** along the wire. The pattern of the magnetic field lines for a long straight wire, a solenoid and a flat coil are shown opposite. Note that the lines of force are **complete loops**. Also, the **direction of the lines of force** depends on the direction of current. If the current is reversed, the direction of the lines of force is reversed.

- **For a long straight wire**, the lines of force are circles centred on the wire in a plane perpendicular to the wire. The direction of the lines of force depends on the current direction and can be worked out using the 'corkscrew rule' as shown in Figure 2.
- **For a solenoid**, the lines of force pass through the solenoid along its axis and loop round outside the solenoid. The direction of the lines of force depends on the current direction and can be worked out using the right hand grip rule or the solenoid rule, as shown in Figure 3.

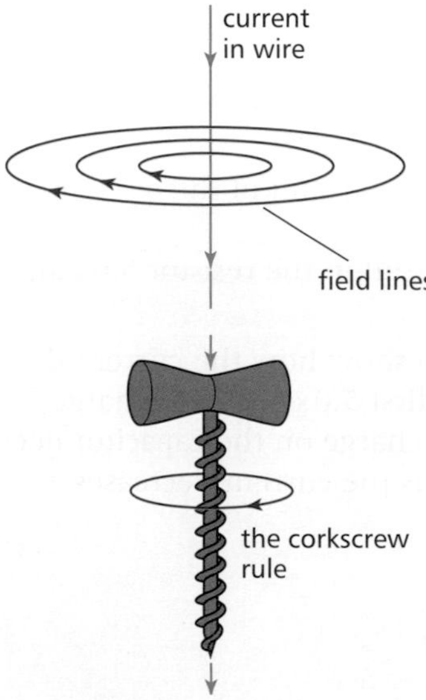

Figure 2 *The magnetic field near a wire*

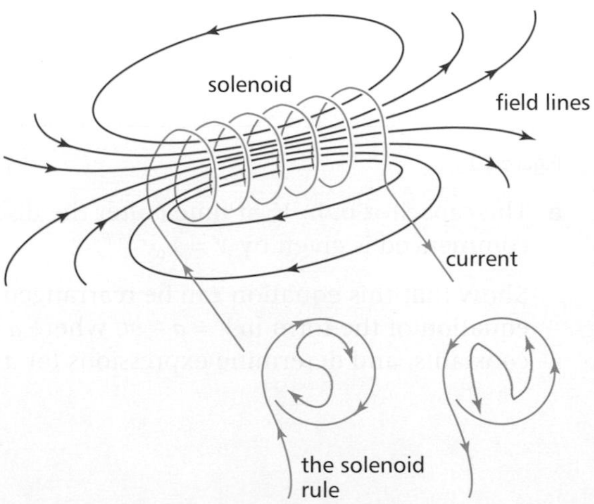

Figure 3 *The magnetic field of a solenoid*

- **For a flat coil**, the lines of force are lines that pass through the coil and loop round outside the coil. The direction of the lines of force can be worked out from the current direction using the solenoid rule, as explained on the previous page.

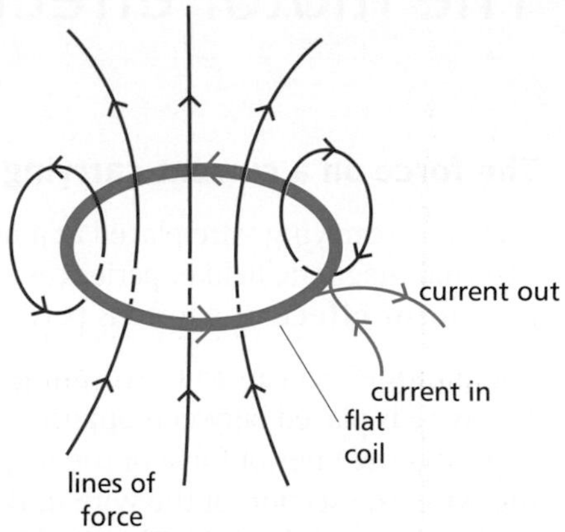

Figure 4 *The magnetic field of a flat coil*

Summary test 16.1

1 A plotting compass is placed at the intersection of two perpendicular lines drawn on a sheet of paper. The plotting compass points north along one of the lines. When a bar magnet is placed along the other line near the plotting compass, as shown in Figure 5, the plotting compass points north-east.

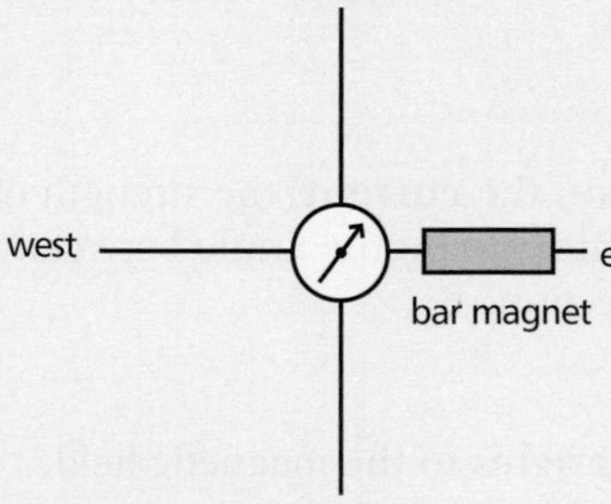

Figure 5

a What is the polarity of the pole of the magnet nearest the plotting compass?

b If the magnet is turned round, what direction will the plotting compass then point to?

2 An underground cable is aligned horizontally in an east-west direction. The cable carries a direct current from west to east, as shown in Figure 6.

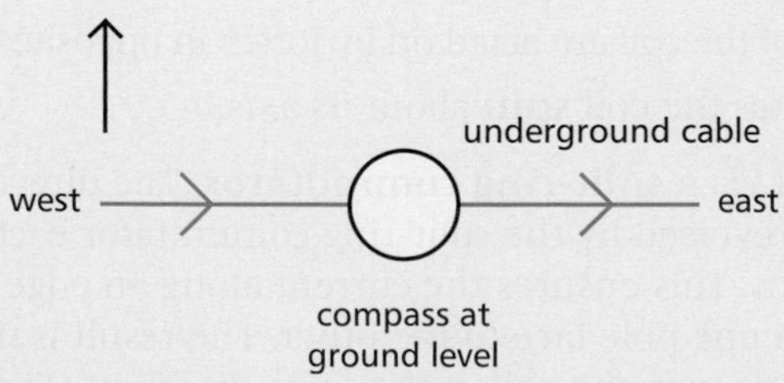

Figure 6

a What would be the direction of a magnetic compass directly above the wire, assuming the magnetic field due to the cable is much stronger than the Earth's magnetic field?

b How would the direction of the compass change if the current in the cable is gradually reduced to zero?

3 A plotting compass is placed at point P at the end of a solenoid. When a direct current is passed through the solenoid, the compass points into the solenoid, as shown in Figure 7.

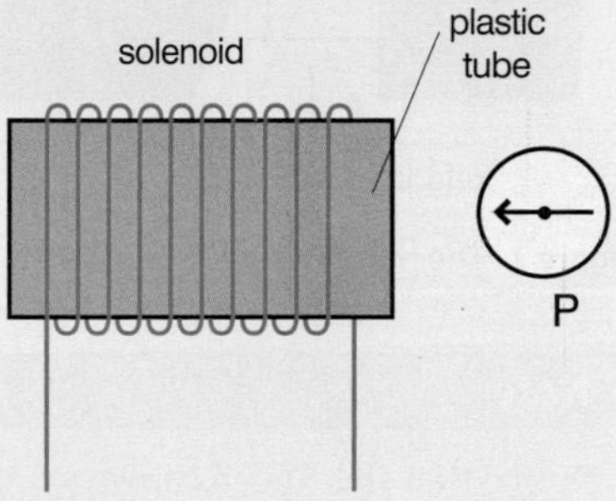

Figure 7

a State the direction of the current round the solenoid, as seen by an observer looking directly at the end of the solenoid at P.

b How would the direction of the plotting compass change if the current in the solenoid was reversed?

4 A student makes a model ammeter using a pair of flat coils in series with each other and a plotting compass, as shown in Figure 8.

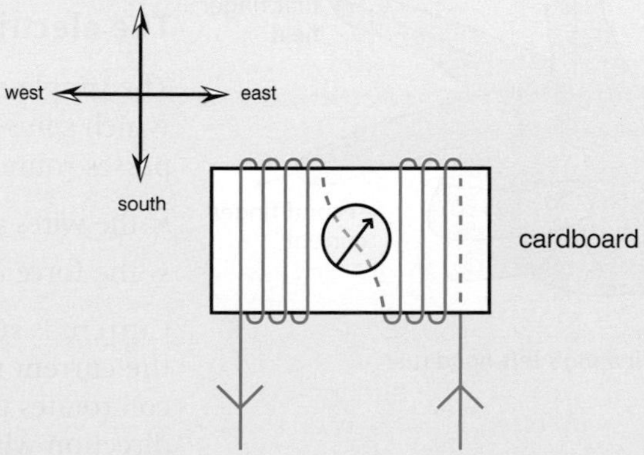

Figure 8

a Explain why the compass needle deflects when a direct current is passed through the coils.

b Explain why the compass needle cannot deflect more than 90° no matter how much current is passed through the coils.

16.2 The motor effect

Learning outcomes

On these pages you will learn to:

- recognise that a force acts on a current-carrying conductor placed in a magnetic field at a non-zero angle to the field lines
- describe the operation of a simple electric motor

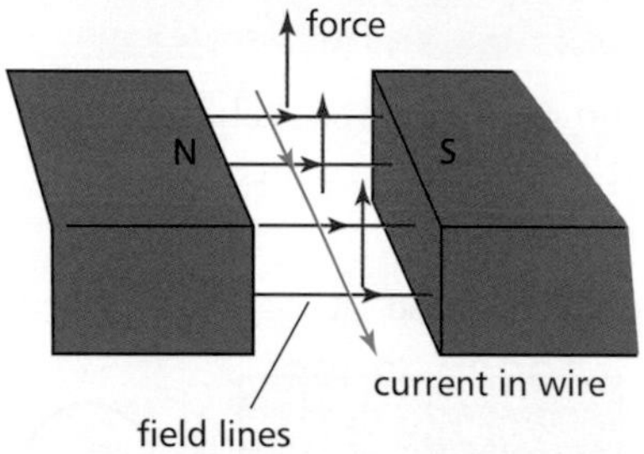

Figure 1 The motor effect

Note

Notice that the force is always perpendicular to the field lines, unlike electric forces on point charges (and gravitational forces on small objects) which are always parallel to the field lines.

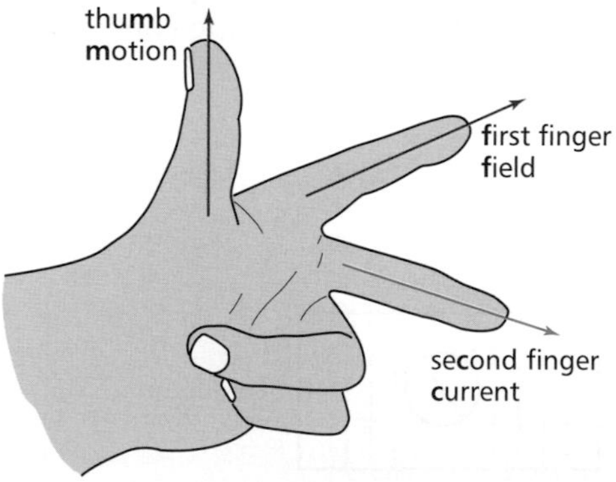

Figure 3 Fleming's left-hand rule

The force on a current-carrying wire in a magnetic field

A current-carrying wire placed at a non-zero angle to the lines of force of an external magnetic field experiences a force due to the field. This effect is known as the **motor effect.** The force is perpendicular to the wire and to the lines of force.

The motor effect can be tested using the simple arrangement shown in Figure 1. The wire is placed between opposite poles of a U-shaped magnet so it is at right angles to the lines of force of the magnetic field. When a current is passed through the wire, the section of the wire in the magnetic field experiences a force that pushes it out of the field. The combined magnetic field due to the wire and the magnet is stronger on one side of the wire than on the opposite side. The wire is pushed in the direction where the combined field is weakest, as shown in Figure 2.

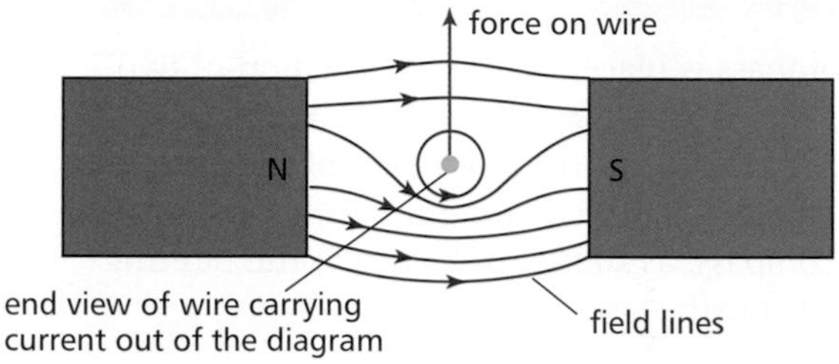

Figure 2 A field pattern

Force factors

The magnitude of the force depends on the **current**, the strength of the **magnetic field**, the **length** of the wire and on the **angle** between the lines of force of the field and the current direction.

The force is:

- greatest when the wire is at **right angles** to the magnetic field,
- zero when the wire is **parallel** to the magnetic field.

The direction of the force is **perpendicular** to the direction of the field and to the direction of the current, as indicated by Fleming's left-hand rule, shown in Figure 3. If the current is reversed, or if the magnetic field is reversed, the direction of the force is reversed.

The electric motor

The simple electric motor consists of a coil of insulated wire, the **armature**, which spins between the poles of a U-shaped magnet. When a direct current passes round the coil:

- the wires at opposite edges of the coil are acted on by forces in opposite directions,
- the force on each edge makes the coil **spin** about its axis.

Current is supplied to the coil via a **split-ring commutator**. The direction of the current round the coil is reversed by the split-ring commutator each time the coil rotates through half a turn. This ensures the current along an edge changes direction when it moves from one pole face to the other. The result is that the force on each edge continues to turn the coil in the same direction (Figure 4).

The **direction of rotation** of the motor is reversed by either reversing the current **or** the direction of the magnetic field. If both the current and the magnetic field are reversed, the direction of rotation is unchanged.

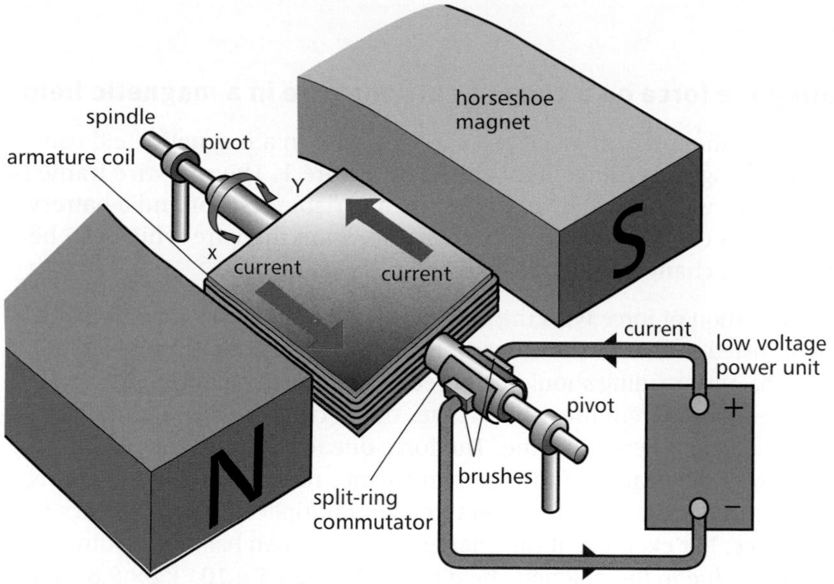

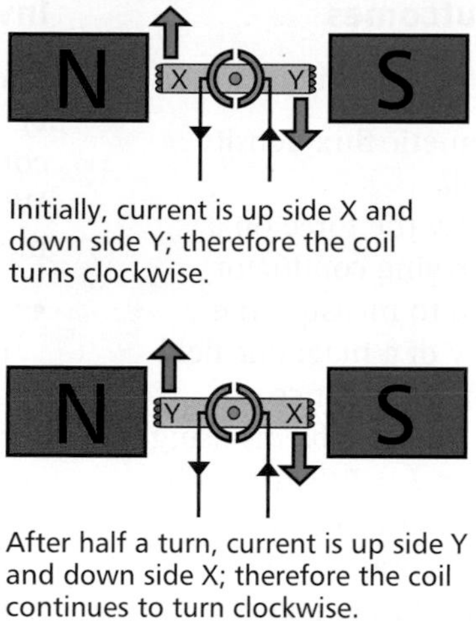
N
X
Y
S
Initially, current is up side X and down side Y; therefore the coil turns clockwise.
N
Y
X
S
After half a turn, current is up side Y and down side X; therefore the coil continues to turn clockwise.

Figure 4 *The simple electric motor*

The **a.c. motor** has an electromagnet instead of a permanent magnet connected to the same a.c. supply as the armature. The direction of rotation does not change when the current reverses because the magnetic field of the electromagnet reverses as well.

The **speed of rotation** depends on the **current**, the **strength** of the magnet and the **number of turns** of the coil. The strength of the magnetic field is increased if an armature with an iron core is used.

A practical electric motor

A practical electric motor has an armature with several **evenly spaced coils** wound on it. Each coil is connected to its own section of the commutator. The result is that each coil in sequence experiences a turning effect when it is connected to the voltage supply, so the motor runs smoothly.

Figure 5 *A practical electric motor*

Summary test 16.2

1 A fixed vertical wire is in a horizontal magnetic field. State the direction of the force on the wire if:

a the current is upwards and the magnetic field lines are from east to west,

b the current is upwards and the magnetic field lines are from south to north,

c the current is downwards and the magnetic field lines are from east to west.

2 A fixed horizontal wire lies along a line from east to west in a magnetic field. State the direction of the magnetic field lines if the current in the wire is from east to west and the force on it is:

a vertically up,

b horizontal and due north.

3 A rectangular coil carries a current in a magnetic field.

a Explain why the coil experiences a turning effect when the plane of the coil is parallel to the magnetic field lines, as in Figure 6a.

b Explain why the coil experiences no turning effect when the plane of the coil is perpendicular to the magnetic field lines, as in Figure 6b.

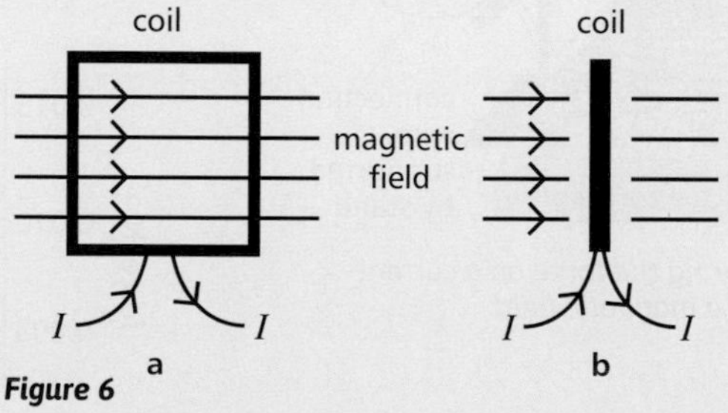

Figure 6

4 a What is the function of the split-ring commutator in a d.c. electric motor?

b A simple electric motor containing a permanent magnet is connected to a battery and a variable resistor. What would be the effect on the motor of:

i increasing the current,

ii reversing the current,

iii using an a.c. supply instead of the battery?

Learning outcomes

On these pages you will learn to:

- define magnetic flux density and the tesla
- describe how the force on a current-carrying conductor can be used to measure the flux density of a magnetic field using a current balance
- recall and solve problems using the equation $F = BIl\sin\theta$, with directions as interpreted by Fleming's left-hand rule
- recognise that the magnetic flux density in a solenoid is increased if a ferrous material is placed in the solenoid
- explain the forces between current-carrying conductors and predict the direction of the forces

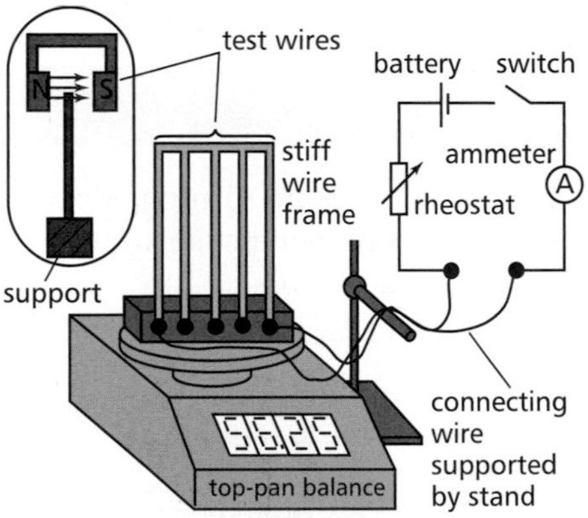

Figure 1 *Measuring the force on a current-carrying wire in a magnetic field*

Investigating the force on a current-carrying wire in a magnetic field

The magnitude of the force on a current-carrying wire in a magnetic field can be investigated using the arrangement shown in Figure 1. The stiff wire frame is connected in series with a switch, an ammeter, a variable resistor and a battery. When the switch is closed, the magnet exerts a force on the wire which can be measured from the change of the top-pan balance reading.

- To test the variation of force with the current through the wire, the variable resistor is adjusted to change the current. Before switching the current on, the top-pan balance reading should be noted. The top-pan balance reading is then measured for different measured values of the current. The length of the test wire in the field is kept the same. The force due to the magnetic field is worked out from the change of the top-pan balance reading. If this is in grams, the reading must be converted to kilograms then multiplied by g (= $9.81\,\text{m}\,\text{s}^{-2}$) to give the force. For example, if the change of the top-pan balance reading is 20.5 g, the force due to the magnetic field is 0.20 N (= $20.5 \times 10^{-3}\,\text{kg} \times 9.81\,\text{m}\,\text{s}^{-2}$).

  A graph of a typical set of results is shown in Figure 2.
- To test the variation of force with the length of the wire, the current is kept the same throughout by using the variable resistor as necessary. The length of the wire in the field is changed by reconnecting the wires to the frame. For each length, the reading of the top-pan balance is noted and the force due to the magnetic field is calculated. A graph of a typical set of results is shown in Figure 3.
- To test the variation of the force with the angle between the wire and the magnetic field lines, the magnet can be turned gradually. This test shows that the force is a maximum when the wire is perpendicular to the magnetic field lines.

The tests above show that the force F on the wire is proportional to:

- the current I,
- the length l of the wire.

Magnetic flux density

The magnetic flux density, B, is defined as the force per unit current per unit length on a current-carrying wire placed perpendicular to the field lines.

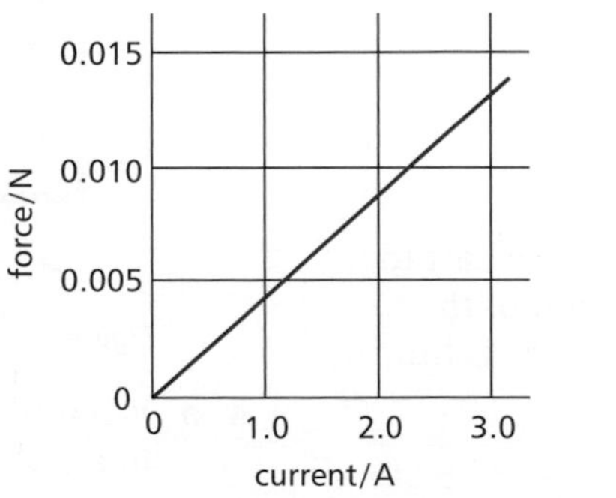

length of wire in the magnetic field = 30 mm

Figure 2 *Force v. current*

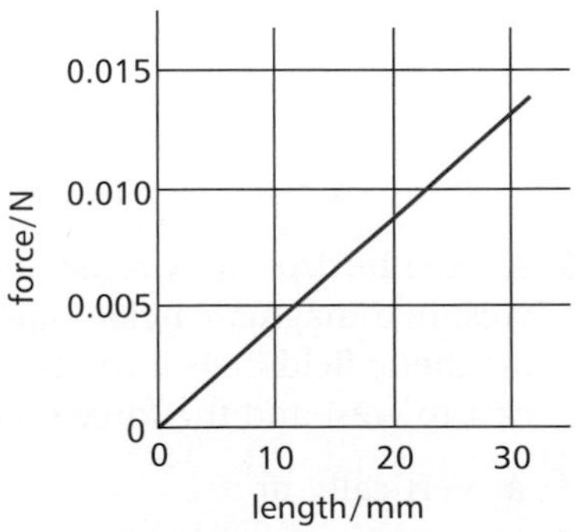

current = 3.0 A

Figure 3 *Force v. length*

The unit of B is the **tesla (T)**, equal to $1\,\mathrm{N\,A^{-1}\,m^{-1}}$. The magnetic flux density is sometimes also referred to as the **magnetic field strength**.

For a wire of length l at right angles to a uniform magnetic field of flux density B, the force on the wire when current I passes through it is given by:

$$F = BIl$$

Worked example

A horizontal wire of length 0.050 m is in a uniform magnetic field directed vertically upwards. The wire lies along a line from north to south. When a current of 4.0 A is passed along the wire, a force of 5.6×10^{-2} N is exerted on the wire, as shown in Figure 4.

a Calculate the magnetic flux density of the magnetic field.
b State the direction of the force on the wire if the current in the wire was from north to south.

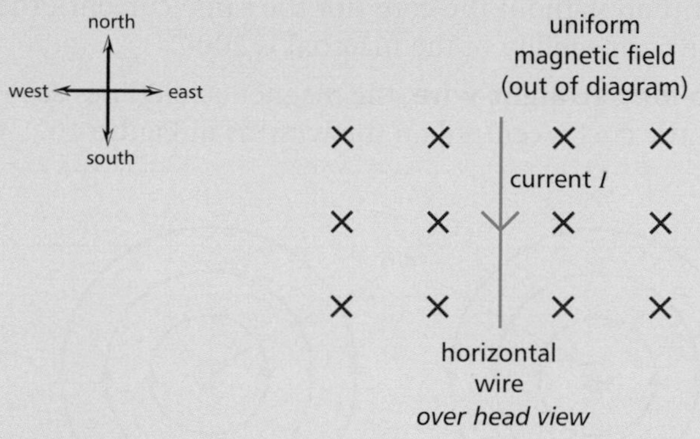

Figure 4

Solution

a $B = \dfrac{F}{Il} = \dfrac{5.6 \times 10^{-2}}{4.0 \times 0.050} = 0.28\,\mathrm{T}$

b Using Fleming's left-hand rule (Topic 16.2, Figure 3), gives due west for the direction of the force.

For a straight wire that is not at right angles to the magnetic field lines, the force on the wire due to the field is determined using the component of the magnetic field at right angles to the wire.

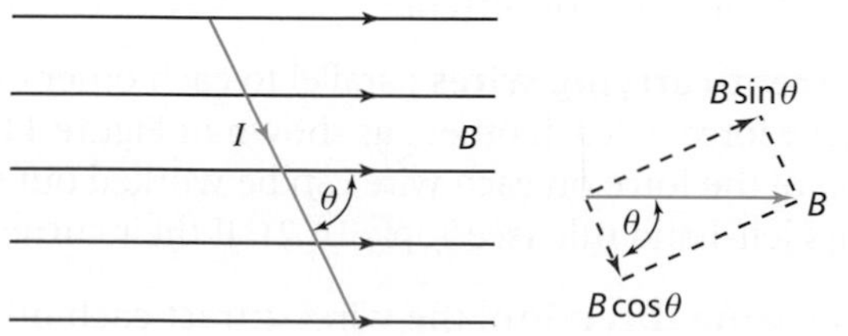

Figure 5 $F = BIl\sin\theta$

For angle θ between the wire and the field lines:

- the component of B perpendicular to the wire $= B\sin\theta$,
- the component of B parallel to the wire $= B\cos\theta$.

Because no force acts on the wire due to the parallel component,

1 the magnitude of the force on the wire, $F = (B\sin\theta)Il = BIl\sin\theta$

2 the direction of the force on the wire is given by Fleming's left hand rule, where the field direction is the direction of the perpendicular component of B at the wire.

For a wire of length l carrying a current I in a uniform magnetic field B at angle θ to the field lines,

the force on the wire, $\boldsymbol{F = BIl\sin\theta}$

Worked example

A straight horizontal wire XY of length 5.0 m is in a uniform horizontal magnetic field of magnetic flux density 120 mT. The wire is at an angle of 30° to the field lines, as shown in Figure 6. When the wire conducts a current of 14 A from X to Y, calculate the magnitude of the force on the wire and state its direction.

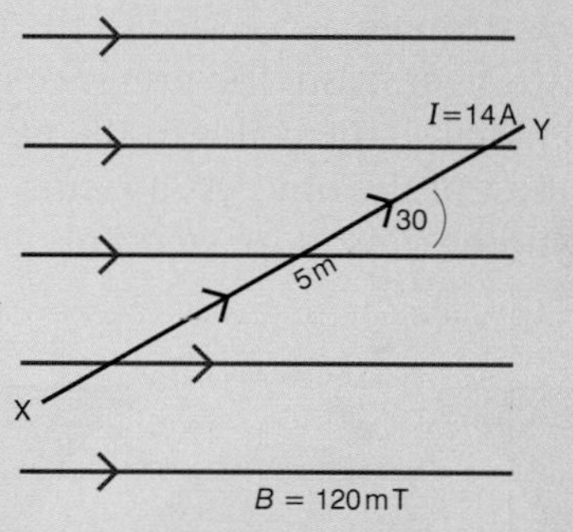

Figure 6

Solution

$\boldsymbol{F = BIl\sin\theta} = 0.12\,\mathrm{T} \times 14\,\mathrm{A} \times 5.0\,\mathrm{m} \times \sin 30 = 4.2\,\mathrm{N}$

The force on the wire is vertically downwards.

The couple on a coil in a magnetic field

Consider a rectangular current-carrying coil in a uniform horizontal magnetic field, as shown in Figure 7. The coil has n turns of wire and can rotate about a vertical axis.

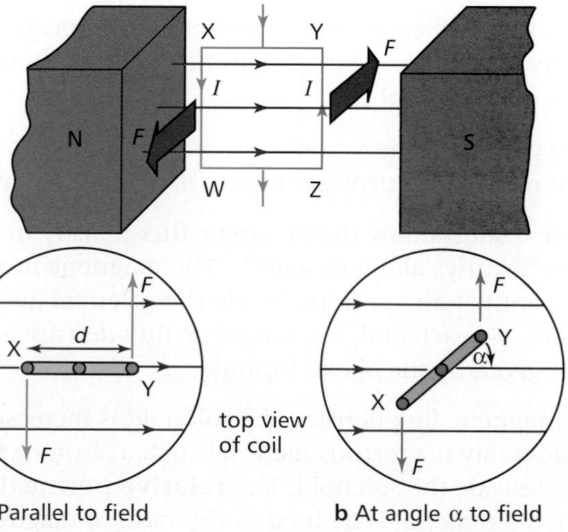

Figure 7 *Couple on a coil*

- The long sides of the coil are vertical and of length l. Each side therefore experiences a horizontal force $F = BIln$ in opposite directions at right angles to the field lines.
- The pair of forces acting on the long sides form a couple as the forces are not directed along the same line. The torque of the couple $= Fd$, where d is the perpendicular distance between the line of action of the forces on each side. If the plane of the coil is at angle α to the field lines, then $d = w\cos\alpha$ where w is the width of the coil.
- Therefore, the torque $= Fw\cos\alpha = BIlnw\cos\alpha$ $= BIAn\cos\alpha$, where $A = lw =$ the coil area.
  If $\alpha = 0$ (i.e. the coil is parallel to the field) the torque $= BIAn$ since $\cos 0 = 1$.
  If $\alpha = 90°$ (i.e. the coil is perpendicular to the field) the torque $= 0$ since $\cos 90° = 0$.

In an electric motor, the coil is wound on an iron core. The field is much stronger as a result so the torque on the motor is much stronger. Also, the presence of the core makes the field radial so the coil is in the plane of the field (i.e. $\alpha = 0$) for most of the time. As a result, the torque is steady and the motor runs more smoothly.

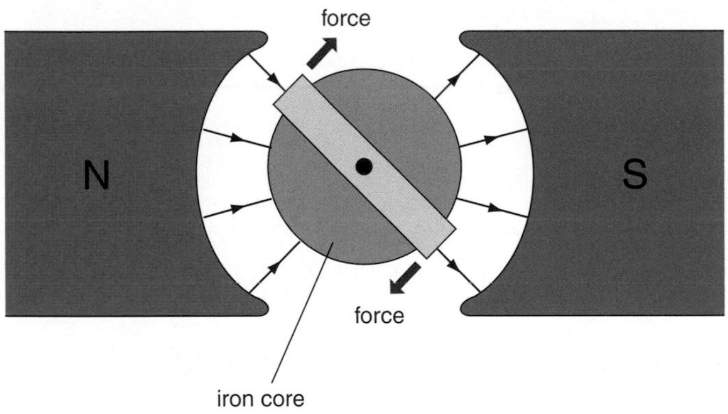

Figure 8 *In an electric motor*

Magnetic flux density due to current-carrying conductors

In a solenoid, experiments show that the magnetic flux density B inside a solenoid does not vary with position and is proportional to:

- the current I, and
- the number of turns per metre, n, on the solenoid.

Figure 9 shows how the magnetic flux density in a solenoid varies along its length. The magnetic flux density is constant at any position inside the solenoid away from its ends. At each end, the magnetic flux density is half the value given by the above formula.

The magnetic flux density of a solenoid is increased considerably if a ferrous material such as iron or steel is placed inside the solenoid. The **relative permeability** μ_r of such a material is defined as the ratio of magnetic flux

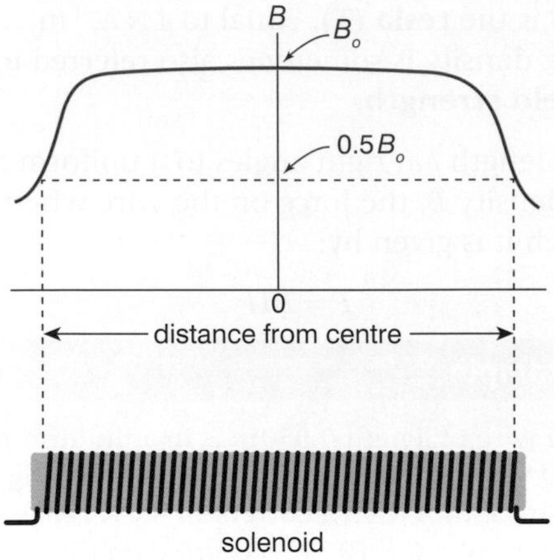

Figure 9 *B in a solenoid*

density with and without the solenoid core filled with the material. For example, if a solenoid with an iron core produces a magnetic flux density which is 2000 times greater than without the core (for the same current), the relative permeability of the material is 2000.

Near a long straight wire, the magnetic field lines are concentric circles centred on the wire, as in Figure 10.

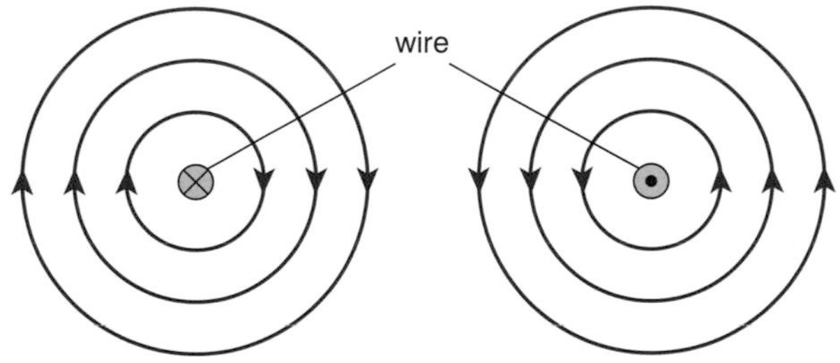

a Clockwise field lines for the current into the plane of diagram

b Anticlockwise field lines for the current into the plane of diagram

Figure 10 *Magnetic field lines near a current-carrying wire*

The magnetic field strength near the wire decreases with distance from the wire. The field lines are clockwise for a current into the plane of the diagram and anticlockwise for the opposite direction of current.

Two current-carrying wires parallel to each other exert a magnetic force on each other , as shown in Figure 11. The direction of the force on each wire can be worked out using Fleming's left-hand rule (see Topic 16.2). If their currents are:

- **in the same direction**, the wires attract each other. This is because each wire experiences a magnetic force towards the other one due to the magnetic field of the other wire. See Figure 12a.
- **in opposite directions**, the wires repel each other. This is because each wire experiences a magnetic force away from the other one due to the magnetic field of the other wire. See Figure 12b.

In Figure 12a, both currents are in the same direction and the wires attract each other. So why do the forces on both X and Y reverse in direction when the current in Y is reversed? This is because reversing the current reverses the magnetic field direction. As a result:

- The force on wire X reverses because the magnetic field acting on X has reversed and X's current is in the same direction.
- The force on wire Y reverses because its current has been reversed and the magnetic field acting on Y is unchanged in direction.

Note that the force *F* on each wire:

- decreases the greater the separation of the wires. This is because the magnetic flux density of each wire decreases with distance from the wire.
- is proportional to the product of their currents. In other words, *F* is proportional to $I_X \times I_Y$, where I_X is the current in X and I_Y is the current in Y.

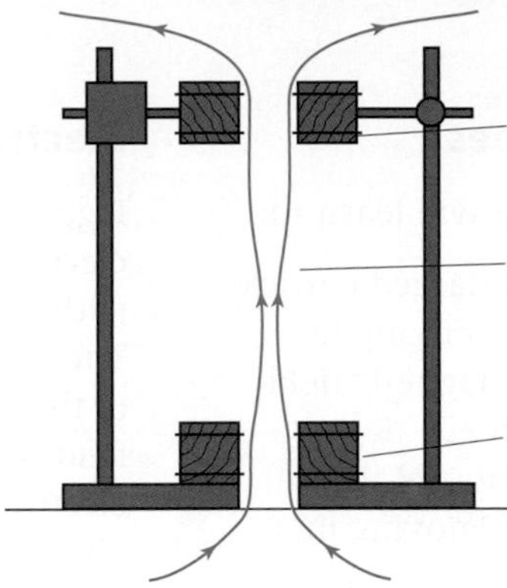

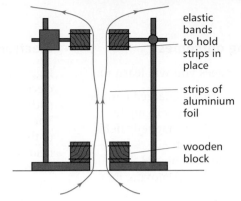

Figure 11 *The force between two current-carrying conductors*

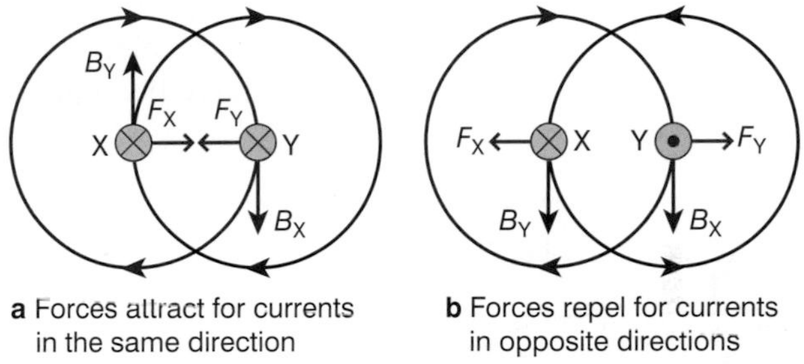

Figure 12 *The magnetic force between two parallel current-carrying wires*

Summary test 16.3

1 a Use Figure 2 to work out the magnetic flux density of the magnet that was used in the test.

b Calculate the magnitude of the force on a straight horizontal wire of length 0.10 m carrying a current of 4.0 A in a uniform horizontal magnetic field of flux density 55 mT when the angle between the wire and the field lines is:
i 0, **ii** 30°, **iii** 90°.

2 Table 1 relates the force on a current-carrying wire, at right angles to the lines of force of a magnetic field, to the magnetic flux density and the current. Complete the table by working out the missing data in each column.

Table 1

	a	b	c	d
B/T	0.20 T vertically down	0.20 T vertically down	?	0.1 T horizontal due?
I/A	3.0 A horizontal due north	?	3.0 A horizontal due north	2.0 A vertically up
l/m	0.040 m	0.040 m	0.040 m	0.040 m
F/N	?	0.036 N horizontal due south	0.024 N horizontal due west	? horizontal due east

3 At a certain location on the Earth's surface, the magnetic flux density *B* of the Earth is 70 μT in a direction due north at 70° to the horizontal. A horizontal cable of length 52 m aligned from east to west carries a current of 28 A.

Calculate the magnitude of the magnetic force on the cable and state its direction.

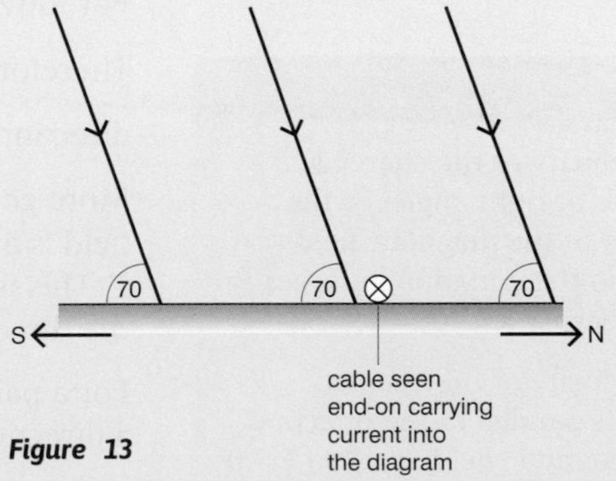

Figure 13

4 A rectangular coil of width 60 mm and of length 80 mm has 50 turns. The coil is placed horizontally in a uniform horizontal magnetic field of flux density 85 mT with its shorter side parallel to the field lines. A current of 8.0 A was passed through the coil. Sketch the arrangement and determine the force on each side of the coil.

16.4 Moving charges in a magnetic field

Learning outcomes

On these pages you will learn to:

- recognise that a charged particle moving at a non-zero angle to the lines of a magnetic field experiences a force
- predict the direction of the force on a charge moving in a magnetic field
- recall and solve problems using $F = BQv\sin\theta$
- derive and use the equation $V_H = \frac{BI}{ntq}$ for the Hall voltage

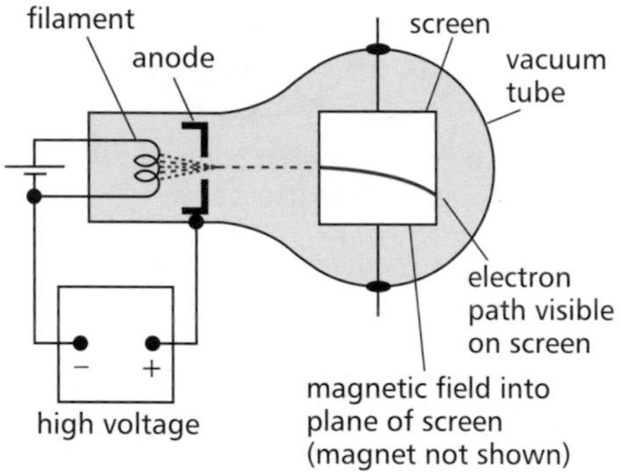

Figure 1 An electron deflection tube

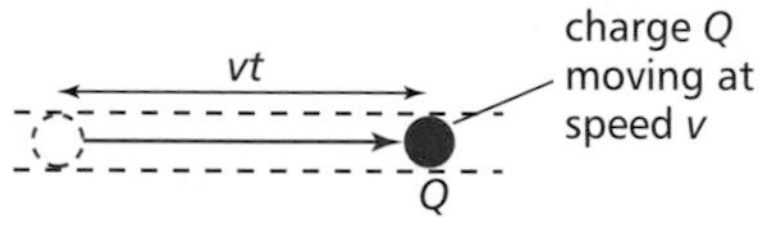

Figure 3 Force on a moving charge

Note

1. If the velocity of the charged particle is at right angles to the direction of the magnetic field, $\theta = 90°$ so the equation becomes $F = BQv$ because $\sin 90° = 1$.
2. If the velocity of the charged particle is parallel to the direction of the magnetic field, $\theta = 0$ so $F = 0$ because $\sin 0 = 0$.

Electron beams

Figure 1 shows a vacuum tube designed to show the effect of a magnetic field on an electron beam. The production of the electron beam is outlined in Topic 14.1. The path of the beam can be seen where it passes over the fluorescent screen in the tube. The beam is deflected downwards when a magnetic field is directed into the plane of the screen. Each electron in the beam experiences a force due to the magnetic field. The beam follows a circular path because the direction of the force on each electron is perpendicular to the direction of motion of the electron (and to the field direction). The direction of the force on an electron in the beam can also be worked out using Fleming's left hand rule, provided we remember the convention that the current direction is opposite to the direction in which the electrons move.

The reason why a current-carrying wire in a magnetic field experiences a force is that the electrons moving along the wire are pushed to one side by the field. If the electrons in Figure 2 had been confined to a wire, the whole wire would have been pushed downwards.

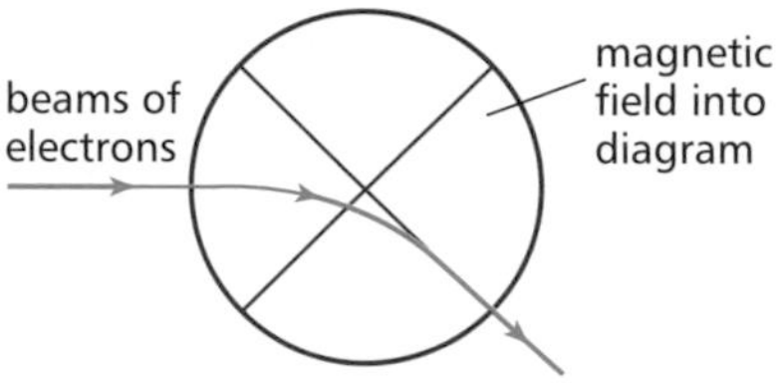

Figure 2 Electrons in a magnetic field

Force on a moving charge in a magnetic field

A beam of charged particles crossing a vacuum tube is an electric current across the tube. Suppose each charged particle has a charge Q and moves at speed v. In a time interval t, each particle travels a distance vt. Its passage is equivalent to current $I = \frac{Q}{t}$ along a wire of length $l = vt$. See Figure 3.

If the particles pass through a uniform magnetic field in a direction at right angles to the field lines, each particle experiences a force F due to the field. If the particles were confined to a wire, the force would be given by $F = BIl$.

For moving charges, the same equation applies where $I = \frac{Q}{t}$ and $l = vt$.

Therefore, for a charged particle moving across a uniform magnetic field in a direction at right angles to the field, $F = BIl = B \times \frac{Q}{t} \times vt = BQv$.

More generally, if the direction of motion of a charged particle in a magnetic field is at angle θ to the lines of the field, then $B\sin\theta$ is used in the equation for F. This is because $B\sin\theta$ is the component of the magnetic field perpendicular to the direction of motion of the charged particle.

For a particle of charge Q moving through a uniform magnetic field at speed v in a direction at angle θ to the field, the force on the particle is given by

$$\boldsymbol{F = BQv\sin\theta}$$

The Hall probe

Hall probes are used to measure magnetic field strength and also as magnetic field sensors.

A Hall probe contains a slice of semiconducting material. Figure 4 shows the slice in a magnetic field with the field lines perpendicular to the flat side of the slice. A constant current passes through the slice, as shown. The **charge carriers** (which are electrons in an n-type semiconductor) are initially deflected by the field. As a result, a potential difference is created between the top and bottom edges of the slice. This effect is known as the **Hall Effect** after its discoverer.

The p.d., referred to as the Hall voltage, is proportional to the magnetic flux density, provided the current is constant. This is because each charge carrier passing through the slice is subjected to a magnetic force $F_{mag} = BQv$, where v is the speed of the charge carrier. Once the Hall voltage has been created, the magnetic deflection of a charge carrier entering the slice is opposed by the force on it due to the electric field created by the Hall voltage so each charge carrier passes through undeflected. The electric field force $F_{elec} = \frac{QV_H}{d}$, where V_H represents the Hall voltage and d is the distance between the top and bottom edges of the slice. Therefore, $\frac{QV_H}{d} = BQv$ gives $V_H = Bvd$.

For constant current I passing through a slice of cross-sectional area A, $I = nAvq$, where n is the number of charge carriers per unit volume and v is their drift velocity. See Topic 6.2.

Combining this equation with the equation $V_H = Bvd$ gives:

$$V_H = Bvd = \frac{B(nAvq)d}{nAq} = \frac{BId}{nAq}$$

As $A = d \times t$, where t is the thickness of the slice:

$$V_H = \frac{BId}{nAq} = \frac{BId}{n(d \times t)q}$$

Therefore $$\boldsymbol{V_H = \frac{BI}{ntq}}$$

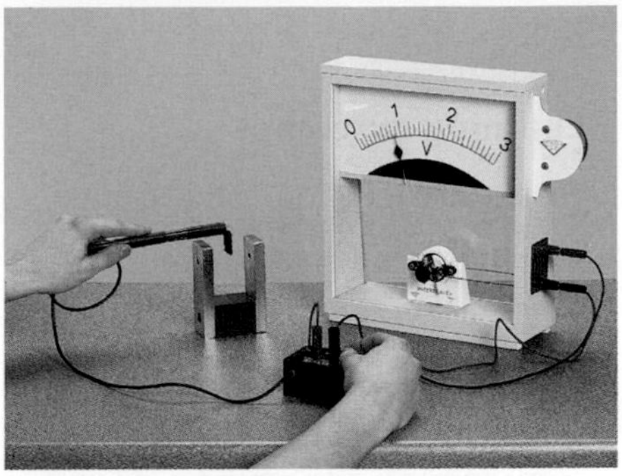

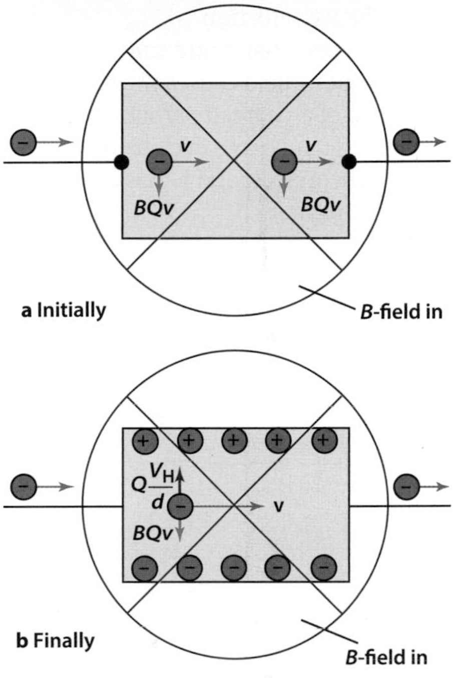

Figure 4 The Hall Effect

Summary test 16.4

$e = 1.6 \times 10^{-19}$ C

1 In Figure 2, how would the force on the electrons in the magnetic field differ if:
  i the magnetic field was reversed in direction,
  ii the magnetic field was reduced in strength,
  iii the speed of the electrons was increased?

2 Calculate the force on an electron that enters a uniform magnetic field of flux density 150 mT at a velocity of 8.0×10^6 m s^{-1} at an angle of:
  i 90°, **ii** 30° to the field.

3 A beam of protons moving at constant speed is directed into a uniform magnetic field in the same direction as the field.
  a Explain why the beam is not deflected by the field.
  b Describe and explain how the path of the beam in the field would have differed if the beam had been directed into the field at a slight angle to the field lines.

4 In a Hall probe, electrons passing through the semiconductor slice experience a force due to a magnetic field.
  a Explain why a potential difference is created across the slice as a result of the application of the magnetic field.
  b When the magnetic flux density is 90 mT, each electron moving through the slice experiences a force of 6.4×10^{-20} N due to the magnetic field. Calculate:
    i the mean speed of the electrons passing through the slice,
    ii the force on each electron if the magnetic flux density is increased to 120 mT.

16.5 Charged particles in circular orbits

Learning outcomes

On these pages you will learn to:

- describe and analyse the deflection of beams of charged particles by uniform electric and uniform magnetic fields
- explain how electric and magnetic fields can be used in velocity selection
- recognise that a uniform magnetic field causes the circular orbit of a charged particle in devices such as a cyclotron, a synchrotron and a mass spectrometer

Magnetic fields are used to control beams of charged particles in many devices, from television tubes to high energy accelerators. The force of the magnetic field on a moving charged particle is at right angles to the direction of motion of the particle.

- No work is done by the magnetic field on the particle as the force acts at right angles to the velocity of the particle. Its direction of motion is changed by the force but not its speed. The kinetic energy of the particle is unchanged by the magnetic field.
- In accordance with Fleming's left-hand rule, the magnetic force is perpendicular to the velocity at any point along the path. The force therefore acts towards the centre of curvature of the circular path.
- The particle moves on a circular path. The force causes a centripetal acceleration because it is perpendicular to the velocity. Figure 1 shows the deflection of a beam of electrons in a uniform magnetic field. The path is a complete circle because the magnetic field is uniform and the particle remains in the field.

The radius, r, of the circular orbit in Figure 1 depends on the speed v of the particles and the magnetic flux density B.

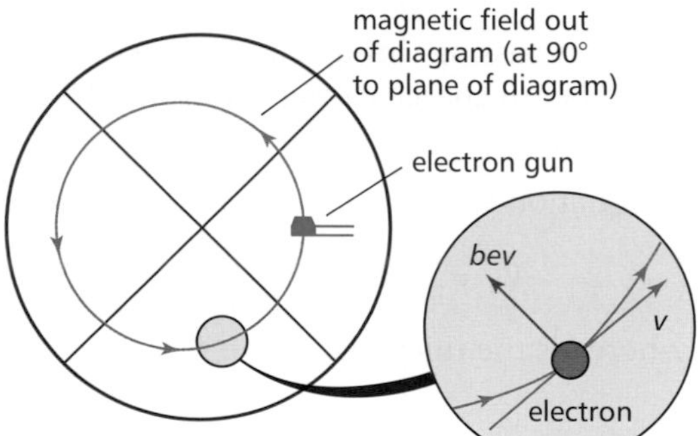

Figure 1 *A circular orbit in a magnetic field*

Figure 2 *Charged particles in a magnetic field. The large spiral is due to a charged particle created by a collision (not shown) between a fast-moving incoming particle and the nucleus of an atom. The charged particle is forced onto a curved path by the magnetic field. It spirals inwards because its kinetic energy and momentum decreases gradually as it transfers energy to the atoms it passes through. Therefore it is deflected more and more by the field as its speed decreases.*

At any point on the orbit, the particle is acted on by a magnetic force $F = BQv$ and it experiences a centripetal acceleration, $a = \frac{v^2}{r}$ towards the centre of the circle. Applying Newton's second law in the form $F = ma$ gives

$$BQv = \frac{mv^2}{r}$$

Rearranging this equation gives

$$r = \frac{mv}{BQ}$$

The equation for r shows that:

1 r decreases if B is increased,

2 r increases if v is increased,

3 particles in a beam with different values of specific charge, $\frac{Q}{m}$, are separated by a magnetic field.

Applications

The following applications make use of the essential principle that charged particles move on circular paths when in a magnetic field and moving at right angles to the field lines.

1 Electrons moving in a circle

The fine beam tube shown in Figure 3 contains hydrogen gas at low pressure. When a beam of electrons from the electron gun in the tube passes through the gas, the atoms along the beam path emit light due to collisions with electrons. So the path of the beam is seen as a fine trace of light in the tube. A pair of coils, placed either side of the tube, is supplied with a direct current to produce a uniform magnetic field through the tube.

Provided the initial direction of the beam is at right angles to the magnetic field lines, the beam curves round in a circle.

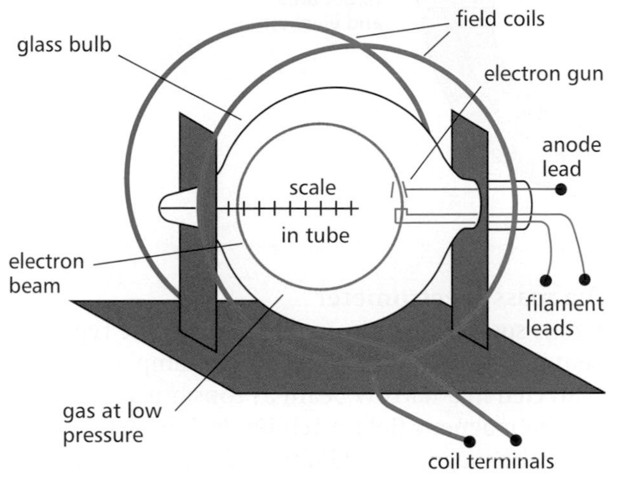

Figure 3 *Using a fine beam tube*

As explained opposite, the radius of the circle $r = \frac{mv}{Be}$, where B is the magnetic flux density and v is the speed of the electrons.

The equation shows that the radius of the circle can be reduced by:

- increasing the magnetic flux density, or
- reducing the anode voltage of the electron gun since this reduces the speed of the electrons.

The fine beam tube may be used to measure e/m, the specific charge of the electron. This was first determined by J. J. Thomson in 1895. Its value is $1.76 \times 10^{11}\,\mathrm{C\,kg^{-1}}$.

Before Thomson made this measurement, the hydrogen ion was known to have the largest specific charge of any charged particle. Thomson showed that the electron's specific charge is 1860 times larger than that of the hydrogen ion. However, Thomson could not conclude that the electron has a much smaller mass than the hydrogen ion as the charge of the electron was not known at that time. The charge of the electron was measured by R. Millikan in 1915.

2 The cyclotron

The cyclotron was invented in 1930 by E.O. Lawrence. It consists of two hollow D-shaped electrodes (referred to as 'dees') in a vacuum chamber. A high-frequency alternating voltage is applied between the dees. A beam of charged particles is directed into one of the dees near the centre of the cyclotron. A uniform magnetic field is applied perpendicular to the plane of the dees.

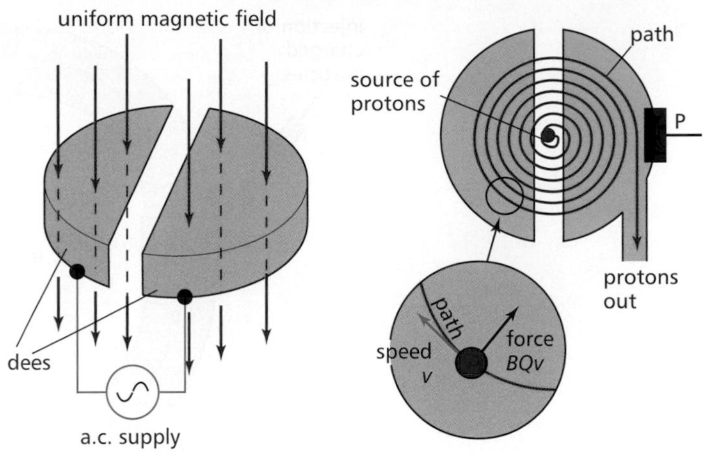

Figure 4 *The cyclotron*

The charged particles are forced on a circular path by the magnetic field, causing the particles to emerge from the dee they were directed into. The particles emerging from the dee when the alternating voltage reverses are accelerated into the other dee where they are forced on a circular path by the magnetic field. When they emerge from this dee, the alternating voltage reverses again and accelerates the particles into the first dee where the process is repeated. This occurs because the time taken, T, by a particle to move round its semi-circular path in the dee $= \frac{\pi r}{v}$, where r is the radius of the path and v is the particle speed.

As explained previously,

$$r = \frac{mv}{BQ} \text{ so } T = \frac{m\pi}{BQ}.$$

For an alternating voltage of frequency f, the time for one half cycle $= \frac{1}{2f}$.

Therefore,

$$\frac{1}{2f} = \frac{m\pi}{BQ} \text{ so } f = \frac{BQ}{2\pi m}.$$

The equation shows that the frequency is independent of the radius and the speed so the charged particles cross between the dees each time the voltage reverses.

The particles gain speed each time they are accelerated from one dee to the other. The radius of the circular path, therefore, is larger each time a particle travels into and out of a dee. The particles emerge from the cyclotron when the radius of orbit is equal to the dee radius R. As $v = \frac{BQr}{m}$ the speed of the particles on exit from the cyclotron $= \frac{BQR}{m}$.

The **synchrotron** accelerates charged particles to much higher energies than a cyclotron. The magnetic field is increased to keep the particles in an orbit of constant radius as they are boosted to higher and higher speeds.

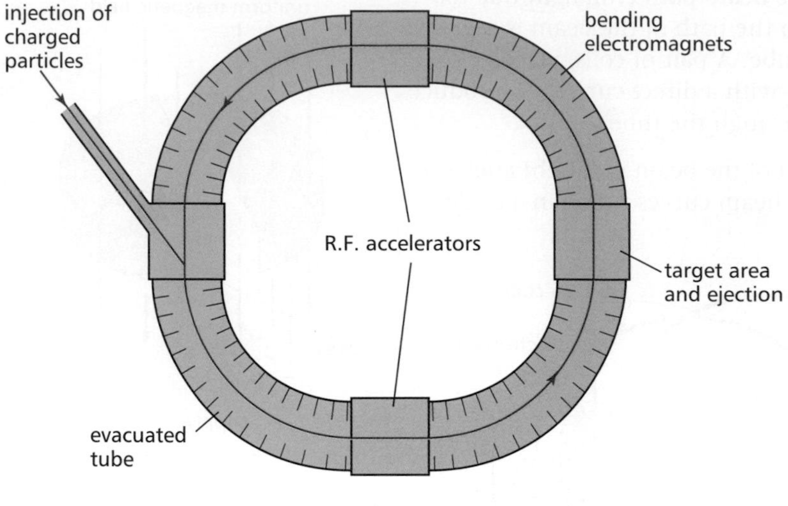

Figure 5 *The synchrotron*

As the radius of orbit is constant, semi-circular dees are not necessary and the particles move round a ring-shaped tube in a magnetic field created by a large number of electromagnets positioned round the ring. A high-frequency alternating voltage, applied between electrodes positioned in the ring, is used to accelerate the charged particles in the ring to high energies. In operation, the particles are injected into the ring and are boosted in 'bursts' to high energies each time the magnetic field is increased.

Extension

The Van Allen belts

Charged particles from space are trapped in belts above the Earth's atmosphere by the Earth's magnetic field. These belts were predicted by Van Allen and were discovered by Geiger counters aboard the US Explorer 1 satellite in 1958. The charged particles in the belts spiral around the field lines, bouncing back near the poles. The inner belt consists of protons and other charged particles with energies of 10 to 100 MeV that stream from the Sun in the 'solar wind' or as a result of cosmic rays (i.e. high energy particles from space) colliding with atoms in the upper atmosphere. The outer belt stretches out to about 20 000 km above the Earth and consists mostly of electrons with energies of 10 to 100 keV from the Sun.

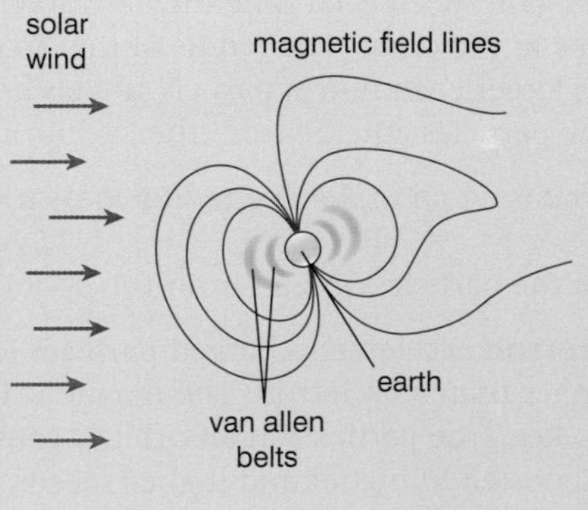

Figure 6 *The Van Allen belts*

3 The mass spectrometer

The mass spectrometer is used to analyse the type of atoms present in a sample. The atoms of the sample are ionised and directed in a narrow beam at constant velocity into a uniform magnetic field. Each ion is deflected in a semi-circle by the magnetic field onto a detector, as shown in Figure 7. The radius of curvature of the path of each ion depends on the specific charge $\frac{Q}{m}$ of the ion in accordance with the equation $r = \frac{mv}{BQ}$. Each type of ion is deflected by a different amount onto the detector. The detector is linked to a computer which is programmed to show the relative abundance of each type of ion in the sample.

The ions in the beam enter the magnetic field at the same speed. When they are produced from the sample, they have a continuous range of speeds. Before they enter the magnetic field, they are formed into a beam and directed through a **velocity selector,** as shown in Figure 7. The velocity selector consists of a magnet and a pair of parallel plates at spacing d and voltage V_p due to a high voltage supply. The magnet and the plates are aligned so each ion passing through the velocity selector is acted on by an electric field force, $F_{elec} = \frac{QV_P}{d}$, in the opposite direction to the magnetic field force $F_{mag} = B_SQv$, where B_S is the magnetic flux density of the magnet in the velocity selector. Ions moving at a certain speed such that $B_SQv = \frac{QV_P}{d}$ experience equal and opposite forces and so pass through undeflected. All other ions are deflected and do not pass through the collimator slit. So the beam emerging from the collimator consists of different types of ions, all with the same speed $v = \frac{V_P}{B_S d}$.

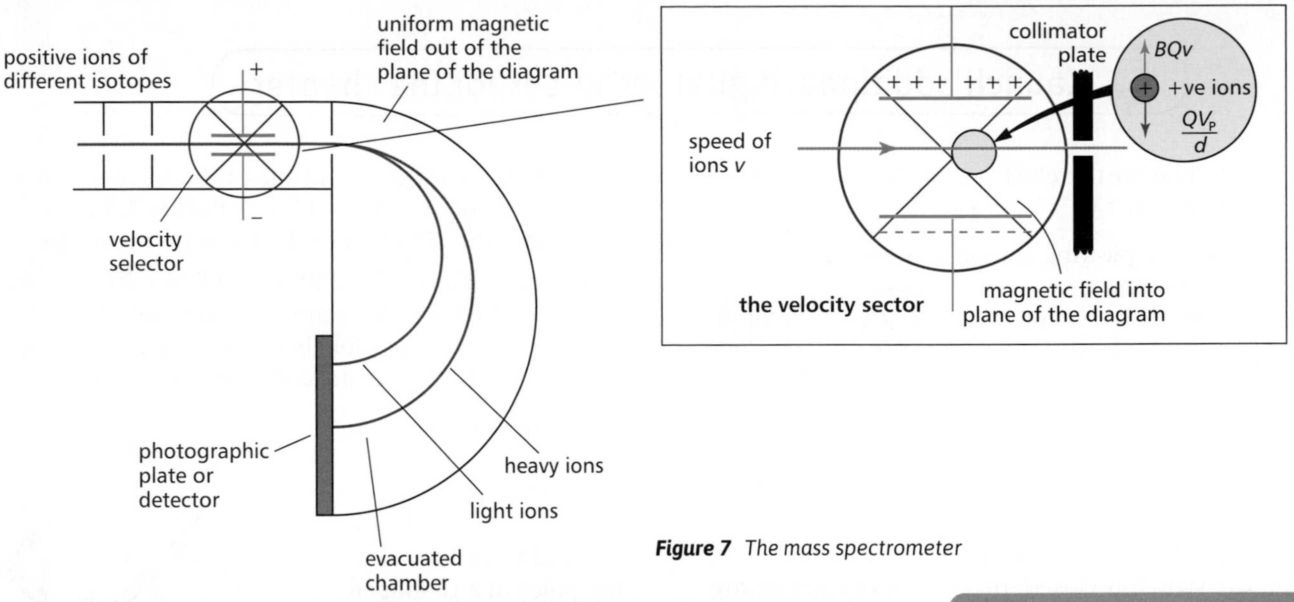

Figure 7 *The mass spectrometer*

Summary test 16.5

$e = 1.6 \times 10^{-19}$ C, e/m for the electron $= 1.76 \times 10^{11}$ C kg^{-1}

1 A beam of electrons at a speed of 3.2×10^7 m s^{-1} is directed into a uniform magnetic field of flux density 8.5 mT in a direction perpendicular to the field lines. The electrons move on a circular orbit in the field.

a i Explain why the electrons move on a circular orbit.

ii Calculate the radius of the orbit.

b The flux density is adjusted until the radius of orbit is 65 mm. Calculate the flux density for this new radius.

2 In a fine beam tube, electrons are accelerated from rest through a certain p.d. before being directed at a speed of 2.9×10^7 m s^{-1} in a narrow beam into a uniform magnetic field.

a The beam follows a circular path of radius 35 mm in the magnetic field. Calculate the flux density of the magnetic field.

b The speed of the electrons in the beam was halved as a result of reducing the anode voltage. Calculate the new radius of curvature of the beam in the field.

3 The first cyclotron, used to accelerate protons, was 0.28 m in diameter and was in a magnetic field of flux density 1.1 T.

a Show that protons emerged from this cyclotron at a maximum speed of 1.5×10^7 m s^{-1}.

b Calculate the maximum kinetic energy, in MeV, of a proton from this accelerator.

The mass of a proton $= 1.67 \times 10^{-27}$ kg.

1 MeV $= 1.6 \times 10^{-13}$ J

4 In a mass spectrometer, a beam of different ions moving at a speed of 7.6×10^4 m s^{-1} was directed into a uniform magnetic field of flux density 680 mT, as shown in Figure 7.

a An ion was deflected through 180° to a position on the detector which was 28 mm from where it entered the field. Calculate the specific charge of the ion.

b A different type of ion was deflected onto the same position on the detector when the magnetic flux density was changed to 400 mT. Calculate the specific charge of this ion.

Chapter Summary

1 a $\boldsymbol{F = BIl \sin \theta}$ gives the force F on a current-carrying wire of length l in a uniform magnetic field B at angle θ to the field lines, where I is the current.

b The direction of the force is given by Fleming's left-hand rule where the field direction is the direction of the field component perpendicular to the wire.

2 a $\boldsymbol{F = BQv \sin \theta}$ gives the force F on a particle of charge Q moving through a uniform magnetic field B at speed v in a direction at angle θ to the field.

b If the velocity of the charged particle is perpendicular to the field, $\boldsymbol{F = BQv}$.

c The direction of the force is given by Fleming's left hand rule, provided the current is in the direction in which positive charge would flow.

3 $\boldsymbol{BQv = \frac{mv^2}{r}}$ gives the radius of the orbit of a charge moving in a direction at right angles to the lines of a magnetic field.

16 Exam-style and Practice Questions

Launch additional digital resources for the chapter

$e = 1.6 \times 10^{-19}$ C, e/m for the electron $= 1.76 \times 10^{11}$ C kg^{-1}
The mass of a proton is 1.67×10^{-27} kg

1 Figure 1.1 shows a plotting compass mid-way between a bar magnet and one end of a solenoid. The solenoid is connected in series with a battery, a variable resistor and a switch.

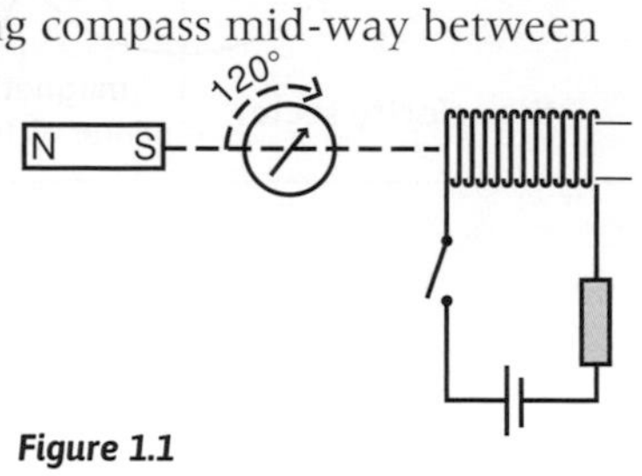

Figure 1.1

a When the switch is open, the plotting compass points directly towards the bar magnet. When the switch is closed, the needle of the plotting compass turns through 120°. Explain why the needle turns when the switch is closed.

b With the switch closed, the variable resistor is adjusted, making the needle turn back by 30°.

i What must have been the effect of the adjustment of the variable resistor on the solenoid current?

ii What would be the effect on the direction of the compass needle if the magnet is moved away from the plotting compass?

2 A U-shaped magnet is placed on the pan of a top-pan balance, as shown below. A straight wire is placed horizontally between the poles of the magnet, which are of length 32 mm. The wire is connected to a variable resistor, an ammeter, a switch and a battery.

a When the switch is closed, the reading of the top-pan balance changes by 0.028 N when the ammeter reads 3.8 A. Calculate the magnetic flux density between the poles of the magnet.

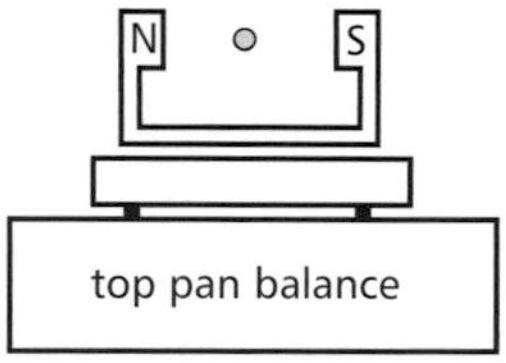

Figure 2.1

b Calculate the force on the wire if the current is increased to 7.0 A.

3 The Earth's magnetic field at a certain location has a downward vertical component of 58 μT, and a horizontal component of 18 μT in a direction due north. A horizontal cable of length 50 m lies along a line from north to south. The cable carries a direct current of 26 A from south to north.

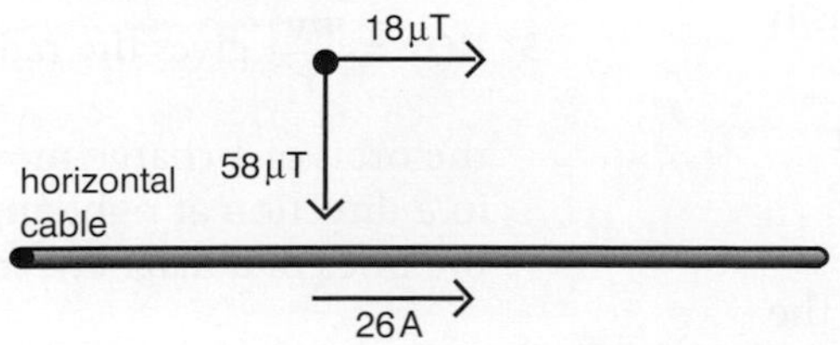

Figure 3.1

a i Show that the magnitude of the force on the cable due to the Earth's magnetic field is 7.5×10^{-2} N.

ii State the direction of the force on the cable.

b Without further calculation, explain why the force on the cable for the same current would have been different, had the cable been aligned along a line from east to west instead of from north to south.

4 The armature coil of an electric motor has 100 turns and is of length 0.12 m, as shown below. The coil spins between the poles of a U-shaped magnet, where the magnetic flux density is 0.18 T.

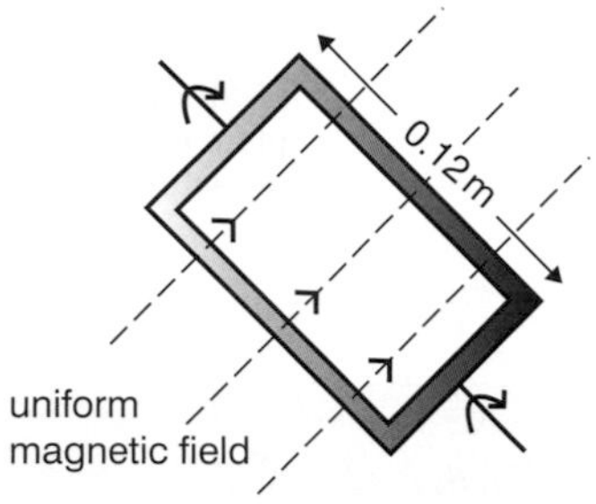

Figure 4.1

a i Calculate the force on each side of the coil when the current through it is 0.8 A.

ii Discuss how the force on each side of the coil changes during one complete rotation of the coil.

b Explain why the force acting on each side of the coil has its maximum turning effect when the plane of the coil is parallel to the lines of force of the magnetic field.

5 a The magnetic flux density between the poles of a U-shaped magnet was determined by measuring the force on a wire of length 32 mm positioned at right angles to the field lines. When a current of 4.5 A was passed along the wire, the force was 0.013 N. Calculate the magnetic flux density at the wire.

b Calculate the angle the wire would need to be moved through to reduce the force on the wire by 5% when the current is 4.5 A.

6 The Earth's magnetic field at a certain location has a flux density of 0.070 mT in a direction due north at an angle of 70° to the surface, as shown in Figure 6.1. A straight wire of length 35 mm carrying a current of 8.2 A upwards is placed vertically in the field.

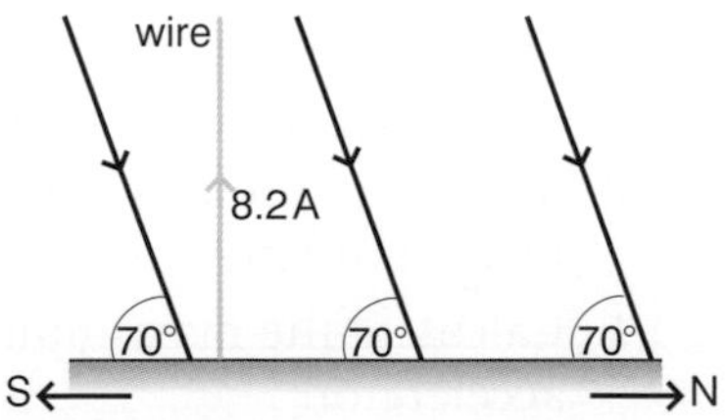

Figure 6.1

a Show that the wire experiences a force of 6.9×10^{-6} N.

b Determine the direction of the force.

7 A flat rectangular coil of length 65 mm and width 42 mm consists of 20 turns of insulated wire. The coil is placed in a horizontal uniform magnetic field of flux density 95 mT with its plane parallel to the field and with its shorter sides horizontal.

a Sketch this arrangement and explain why the coil experiences a torque when it carries a direct current.

b The current in the coil is adjusted to 4.2 A. Calculate the force on each of the longer sides due to the magnetic field and hence show that the torque on the coil is 0.022 N m.

8 In a d.c. electric motor, the coil consists of 60 turns of insulated wire wound on a rectangular frame of length 25 mm and width 20 mm. The coil spins in a magnetic field of magnetic flux density 110 mT.

a Calculate the maximum torque on the coil due to the magnetic field when a current of 0.80 A passes through the coil.

b A 1.0 N weight is attached to a thread wrapped round a pulley of diameter 5 mm fitted to the motor spindle. Discuss if this weight could be raised by the motor when the coil current is 0.80 A.

9 A beam of electrons in a uniform magnetic field of magnetic flux density 3.6 mT travel on a circular path of radius 55 mm.

a Explain why the electrons travel at constant speed in the field.

b Calculate the speed of the electrons.

10 A beam of electrons moving at a speed of $2.7 \times 10^7\,\mathrm{m\,s^{-1}}$ is directed horizontally into a uniform magnetic field of flux density 8.6 mT which is directed vertically upwards.

a i Explain why the beam moves on a circular orbit in the field.

ii Calculate the radius of curvature of the beam in the field.

b The magnetic flux density is reduced steadily to zero then reversed and increased steadily to 8.6 mT in the opposite direction. Describe and explain the effect on the beam of these changes.

11 A cyclotron has a diameter 0.80 m. Protons are accelerated by the cyclotron to a maximum kinetic energy of 4.2 MeV.

a Show that the maximum speed of the protons is $2.8 \times 10^7\,\mathrm{m\,s^{-1}}$.

b i Calculate the magnetic flux density of the magnet used in this cyclotron.

ii Show that the frequency of the alternating voltage applied to the cyclotron was 11 MHz.

12 In a mass spectrometer, a beam of different ions travelling at the same speed is directed into a uniform magnetic field at right angles to the field.

a Explain why the magnetic field separates the ions according to their specific charge $\frac{Q}{m}$.

b i In a test of a mass spectrometer, protons at a speed of $4.8 \times 10^6\,\mathrm{m\,s^{-1}}$ were directed into the field. The protons moved in the field in a semi-circular orbit of radius 60 mm. Calculate the magnetic flux density of the magnetic field.

ii Calculate the specific charge of an ion that moved on an orbit of radius 420 mm in the same flux density.

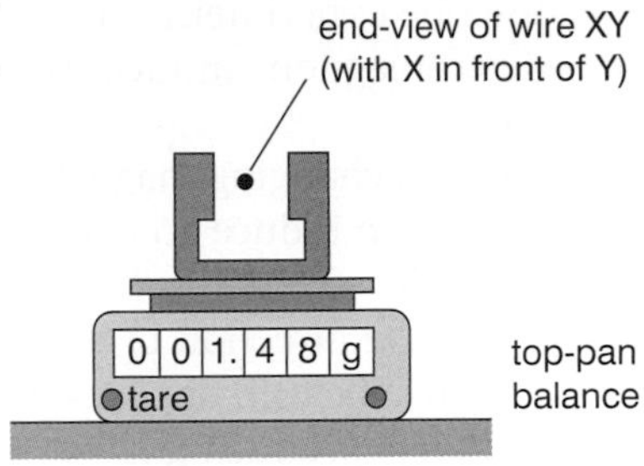

Figure 13.1

13 **a** Define the *magnetic flux density* of a magnetic field.

b Figure 13.1 shows an arrangement used to measure the magnetic flux density B between the poles of a U-shaped magnet.

The wire is horizontal and perpendicular to the field which is also horizontal. The top-pan balance is used to measure the magnetic force on the length of wire between the poles when there is a current in the wire. The balance reading was set to zero (by pressing 'tare') before the current was switched on.

i The following balance readings were recorded for different values of current in the wire.

Current/A	0	0.49	1.01	1.48	2.02	2.50
Balance reading/g	0	0.37	0.70	1.02	1.47	1.82
Force/10^{-3} N	0					

i Copy and complete the table above by calculating the force on the wire.

ii Plot a graph of the magnetic force on the wire against the current.

iii Use the graph to calculate the force per unit current on the wire.

iv The length of wire between the poles was 42 mm. Calculate the magnetic flux density between the poles.

c i Assuming the readings did not fluctuate, the uncertainty in the current is ±0.01 A and the uncertainty in the balance reading is ±0.01 g. Estimate the uncertainty in your answer to **biii**.

ii The uncertainty in the length measurement was ±2 mm. Estimate the uncertainty in the magnetic flux density.

17 Electromagnetic induction

17.1 Generating electricity

Learning outcomes

On these pages you will learn to:

- describe experiments on electromagnetic induction to show:
  - how a changing magnetic field can induce an e.m.f. in a circuit
  - that the direction of the induced e.m.f. opposes the change producing it
  - the factors affecting the magnitude of the induced e.m.f.
- explain in terms of electrons why an e.m.f. is induced in a conductor when it cuts across the lines of a magnetic field

To generate electricity, all you need is a magnet and some wire which is connected to a sensitive meter, as shown in Figure 1. When the magnet is moved near the wire, a small current passes through the meter. This happens because an e.m.f. is **induced** in the wire. This effect, known as **electromagnetic induction**, occurs whenever a wire cuts across the lines of a magnetic field. If the wire is part of a complete circuit, the induced e.m.f. causes an induced current in the circuit as it forces electrons round the circuit. The induced e.m.f. can be increased by:

- moving the wire faster,
- using a stronger magnet,
- making the wire into a coil and pushing the magnet in or out of the coil, as in Figure 3 below. The more turns there are in the coil, the greater the induced e.m.f. is.

No e.m.f. is induced in the wire if the wire moves parallel to the magnetic field lines as it moves through the field. The wire must cut across the lines of the magnetic field for an e.m.f. to be induced in the wire.

Other methods of generating induced e.m.fs include:

1 **Using an electric motor in reverse**, as in Figure 2. The falling weight makes the motor coil turn between the poles of the magnet in the motor. The e.m.f. induced in the coil forces a current round the circuit and so causes the lamp to light. The faster the coil turns, the brighter the lamp is.
2 **Using a dynamo**, as in Figure 3. When the magnet in the dynamo spins, an e.m.f. is induced in the coil. If the coil is connected to a lamp, the lamp lights because the e.m.f. forces a current round the circuit.

In both examples above, an e.m.f. is induced because there is relative motion between the coil and the magnet. In the electric motor in reverse, the coil spins and the magnet is fixed. In the dynamo, the magnet spins and the coil is fixed.

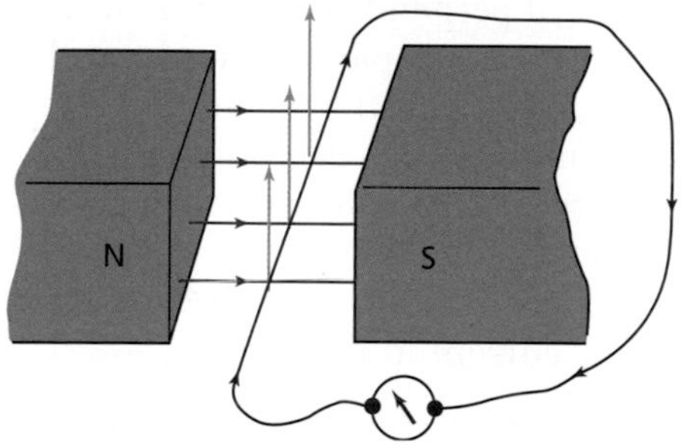

Figure 1 Generating an electric current

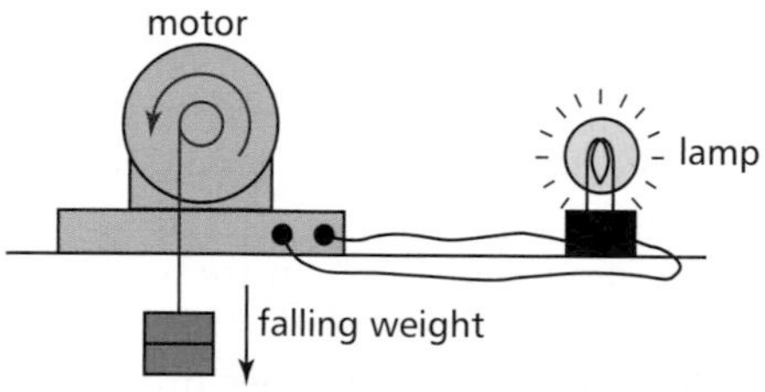

Figure 2 A motor as a generator

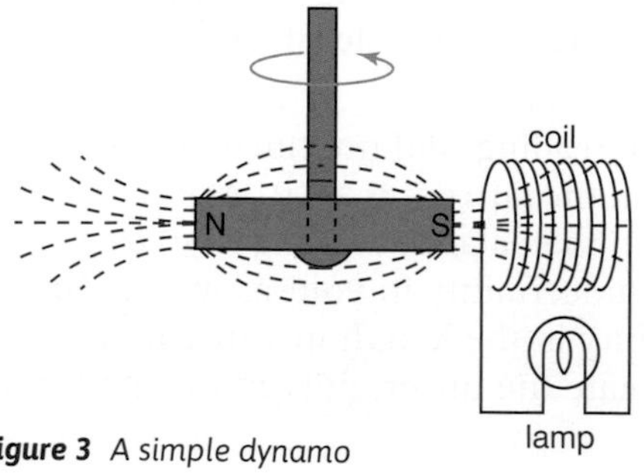

Figure 3 A simple dynamo

Energy changes

When a magnet is moved relative to a conductor (e.g. a wire or a coil), an e.m.f. is induced in the circuit. If the conductor is part of a complete circuit which has no other sources of e.m.f., a current passes round the circuit just as if the circuit included a battery. However, unlike the e.m.f. of a battery which is constant, the induced e.m.f. becomes zero when the relative motion between the magnet and the wires ceases.

An electric current transfers energy from the source of the e.m.f. in a circuit to the other components in the circuit. For example, when a dynamo is used to light a lamp, energy is transferred from the dynamo to the lamp. The current through the dynamo coil causes a reaction force on the coil due to the magnet. Work must therefore be done to keep the magnet spinning. The energy transferred from the coil to the lamp is equal to the work done on the coil to keep it spinning.

The rate of transfer of energy from the source of e.m.f. to the other components of the circuit is equal to the product of the induced e.m.f. and the current. This is because:

- the induced e.m.f. is the energy transferred from the source per unit charge that passes through the source,
- the current is the charge flow per second.

So the induced e.m.f. × the current = energy transferred per unit charge from the source × the charge flow per second = energy transferred per second from the source.

Michael Faraday 1791–1861

Electromagnetic induction was discovered by Michael Faraday in 1831 at the Royal Institution, London. Faraday knew that a current passing along a wire produced a magnetic field near the wire and he wanted to know if a magnet could be used to produce a current. Using a magnetic compass near a loop of wire as a detector of current, he showed that the compass deflected whenever the magnet was moved in or out of the wire. He used the term **'electromotive force' (e.m.f.)** to describe the voltage induced in a wire. When he demonstrated his discoveries to an invited audience at the Royal Institution, he was asked the question 'What use is electricity, Mr Faraday?'. He replied with another question 'What use is a new baby?'. No one can tell what can grow from a new discovery.

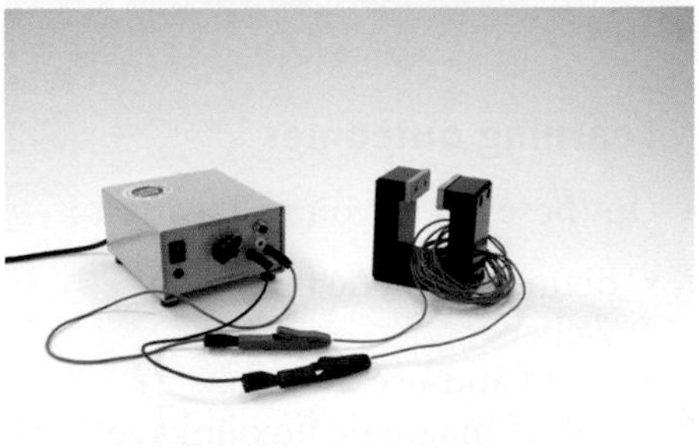

***Figure 4** Apparatus for electromagnetic induction*

Understanding electromagnetic induction

When a beam of electrons is directed across a magnetic field, each electron experiences a force at right angles to its direction of motion and to the field direction. A metal rod is like a tube containing lots of free electrons. If the rod is moved across a magnetic field, as shown in Figure 5, the magnetic field forces the free electrons in the rod to one end away from the other end. So, one end of the rod becomes negative and the other end positive. In this way, an e.m.f. is induced in the rod. The same effect happens if the magnetic field is moved and the rod is stationary. As long as there is relative motion between the rod and the magnetic field, an e.m.f. is induced in the rod. If the relative motion ceases, the induced e.m.f. becomes zero because the magnetic field no longer exerts a force on the electrons in the rod. Note that when the rod is part of a complete circuit, the electrons are forced round the circuit. In other words, the induced e.m.f. drives a current round the circuit.

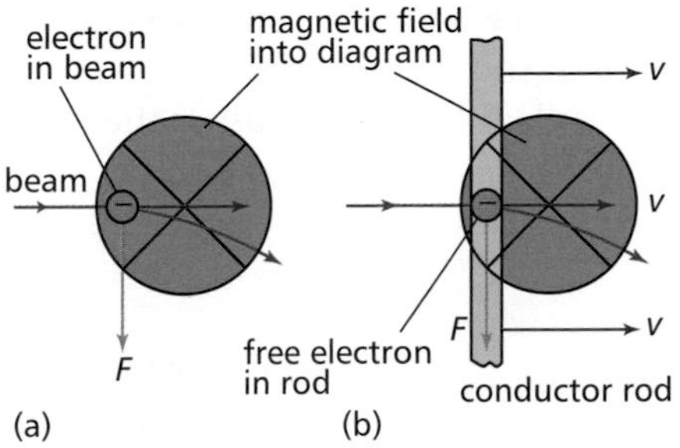

***Figure 5** Deflection of electrons in a magnetic field*

The dynamo rule

In Figure 5, the magnetic field is into the plane of the diagram and the motion of the conductor is towards the right. The electrons in the rod are forced downwards. The direction of the induced current can also be worked out using Fleming's right-hand rule, also referred to as the **dynamo rule**, as shown in Figure 6. The direction of the induced current is, in accordance with the current convention, opposite to the direction of the flow of electrons in the conductor.

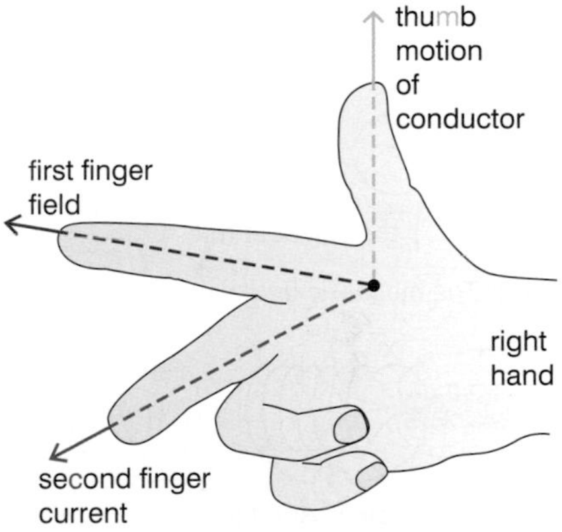

***Figure 6** Fleming's right-hand rule*

Summary test 17.1

1 A coil of wire is connected to a sensitive meter.

a Explain why the meter shows a brief reading when a magnet is pushed into the coil.

b State two ways in which the meter reading could be made larger.

2 An electric motor consists of a coil of wire between the poles of a magnet. The motor is connected to a lamp. A thread wrapped round the motor spindle is used to support a weight, as shown in Figure 2.

a Explain why the lamp lights when the weight descends.

b What difference would it have made if the magnet had been much stronger?

3 a Explain why a lamp connected to a dynamo lights when the dynamo turns.

b Why is the dynamo easier to turn when the lamp is disconnected?

4 A horizontal rod aligned along a line from east to west is dropped through a horizontal magnetic field which is directed from south to north.

a i What is the direction of the velocity of the rod?

ii Determine which end of the rod is positive. Explain your answer.

b Explain why no e.m.f. is induced in the rod if it is aligned from north to south then dropped in the field.

17.2 The laws of electromagnetic induction

Learning outcomes

On these pages you will learn to:

- define magnetic flux and the weber
- recall and use $\Phi = BA$
- define magnetic flux linkage
- recall Faraday's law of electromagnetic induction and Lenz's law
- use the above laws to explain and solve problems
- explain Lenz's law using the principle of conservation of energy

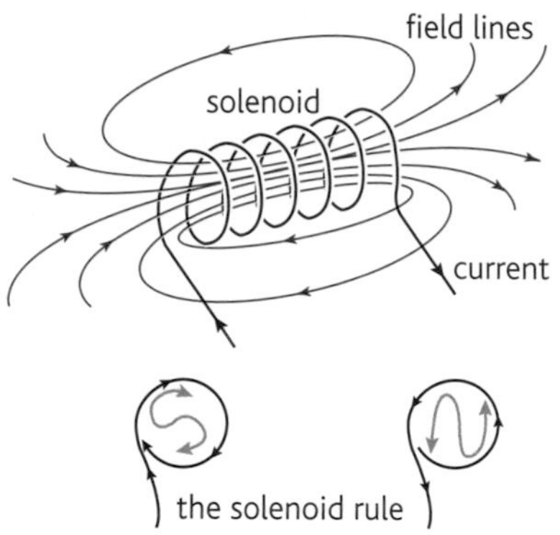

Figure 1 *The magnetic field near a solenoid*

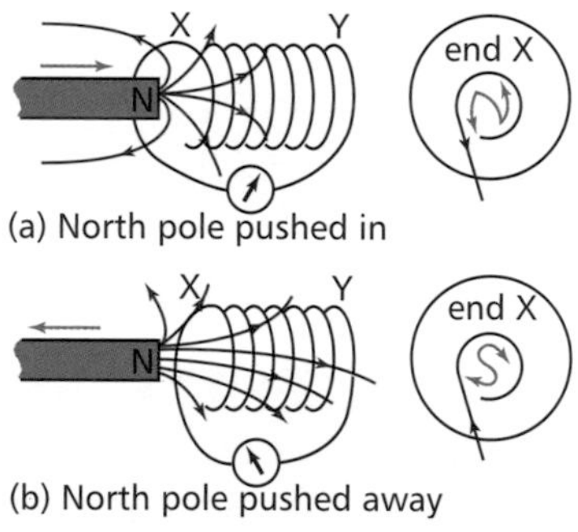

Figure 2 *Lenz's law*

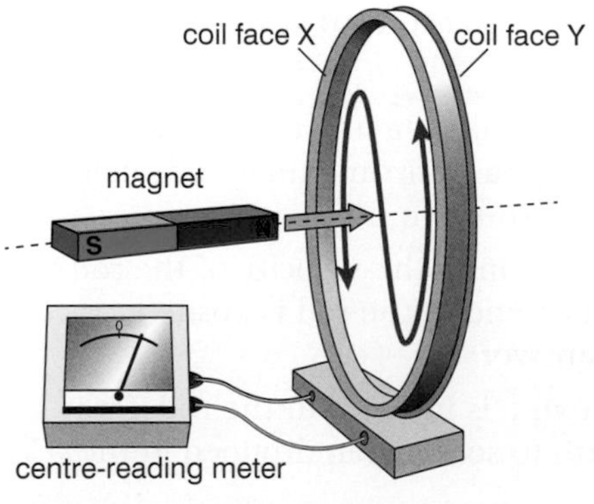

Figure 3 *Induced polarity of a coil*

Coils, currents and fields

A magnetic field is produced in and around a coil when an electromagnet is connected to a battery and a current is passed through it. A magnetic compass near the electromagnet is deflected when current passes through the coil. For a long coil or solenoid, the pattern of the magnetic field lines is like the pattern for a bar magnet – except the magnetic field lines near a bar magnet loop round from the north pole to the south pole of the magnet. Figure 1 shows the magnetic field pattern of a current-carrying solenoid. The field lines pass through the solenoid and loop round outside the solenoid from one end (the north pole) to the other end (the south pole). If each end in turn is viewed from outside the solenoid:

- current passes a**N**ticlockwise (or cou**N**terclockwise) round the '**N**orth pole' end,
- current passes clockwise round the 'south pole' end.

Lenz's law

When a bar magnet is pushed into a coil connected to a meter, the meter deflects. If the bar magnet is pulled out of the coil, the meter deflects in the opposite direction. What determines the direction of the induced current? Consider the north pole of a bar magnet approaching end X of a coil, as shown in Figure 2.

The induced current passing round the circuit creates a magnetic field due to the coil. The coil field must act against the incoming north pole, otherwise it would pull the north pole in faster, making the induced current bigger, pulling the north pole in even faster still, etc. Clearly, conservation of energy forbids this creation of kinetic and electrical energy from nowhere. So, the induced current creates a magnetic field in the coil which opposes the incoming north pole. The induced polarity of end X must therefore be a north pole so as to repel the incoming north pole. Therefore, the current must go round end X of the coil in an anticlockwise direction, as shown.

If the magnet is removed from inside the coil, the induced current passes round end X of the coil in a clockwise direction. This corresponds to an induced south pole at end X which, therefore, opposes the magnet moving away.

Lenz's law states that the direction of the induced current is always such as to oppose the change that causes the current.

The explanation of Lenz's law is that energy is never created or destroyed. The induced current could never be in a direction to help the change that causes it; that would mean producing electrical energy from nowhere, which is forbidden!

An induced current is generated if the coil is moved instead of the magnet and the magnet is at rest. In general, whenever the magnet and the coil move relative to each other, an induced current is generated and Lenz's law can be used to work out the direction of the current.

Some further examples of the applications of Lenz's law are given below.

1 **A bar magnet moves towards and through a flat coil**, as shown in Figure 3.
   - When the bar magnet approaches face X of the coil with its north pole leading, the induced polarity of face X of the coil must be a north pole in order to oppose the movement of the bar magnet. So the induced current in the coil is anticlockwise, as seen in face X.
   - After passing through the coil, the bar magnet moves away from face Y of the coil with its south pole trailing. The induced polarity of face Y of the coil must be a north pole in order to oppose the south pole of the bar magnet moving away.

So the induced current in the coil is anticlockwise, as seen in face Y and therefore clockwise, as seen in face X, opposite in direction to when the magnet was approaching the coil.

2 **A flat coil or a solenoid (P) carrying a direct current moves towards another flat coil (Q)**, as shown in Figure 4. The direction of the current in P determines its magnetic polarity.
When P and Q are approaching each other, the induced polarity in Q is the same as the 'leading' polarity of P (and therefore opposite to the 'trailing' polarity of P). The same ideas apply if Q is replaced by a large solenoid.

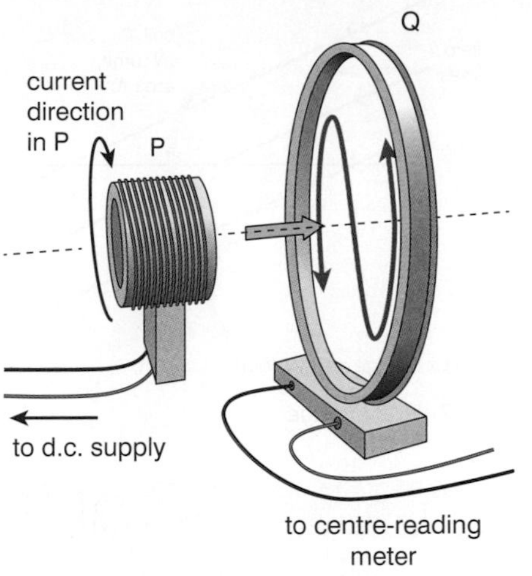

Figure 4 *Induced polarity due to another coil*

Faraday's law of electromagnetic induction

Consider a conductor of length L, which is part of a complete circuit cutting through the lines of a magnetic field of flux density B.

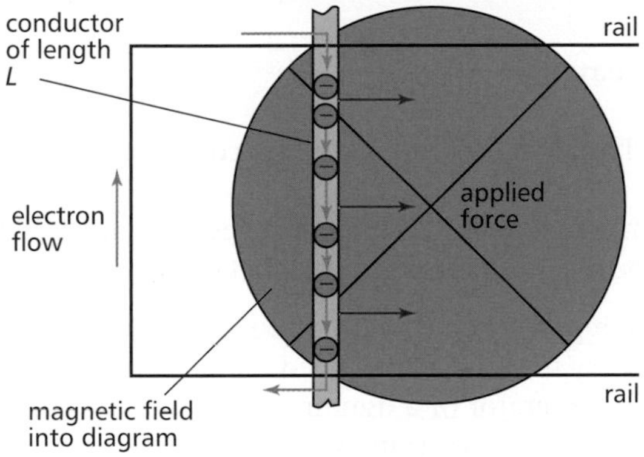

Figure 5 *Induced e.m.f. in a conductor*

Figure 6 *An electric car*

An e.m.f., E is induced in the conductor and an induced current I passes round the circuit.

The conductor experiences a force $F = BIL$ due to carrying a current in a magnetic field. The force opposes the motion of the conductor and so an equal and opposite force must be applied to the conductor to keep it moving in the field. If the conductor moves a distance Δs in time Δt:

- the work done W by the applied force is given by $W = F\,\Delta s = BIL\,\Delta s$
- the charge transfer along the conductor in this time $Q = I\Delta t$.

Therefore, the induced e.m.f. $$E = \frac{W}{Q} = \frac{BIL\,\Delta s}{I\Delta t} = \frac{BL\,\Delta s}{\Delta t}$$

As $L\Delta s = \Delta A$ 'swept out' in time Δt,

$$E = \frac{B\Delta A}{\Delta t} = \frac{\Delta BA}{\Delta t} = \frac{\Phi}{\Delta t}$$

where $\Phi = BA$ = magnetic flux.

The product of the magnetic flux density, B, and the area, A, is called the **magnetic flux**. The concept of magnetic flux is very useful for calculating induced e.m.fs. The example of the conductor cutting across the field lines shows that the induced e.m.f. is equal to the magnetic flux swept out by the conductor each second. Michael Faraday was the first person to show how induced e.m.fs could be calculated from magnetic flux changes.

- **Magnetic flux, $\Phi = BA$.**
- **Flux linkage through a coil of N turns = $N\Phi = NBA$** where B is the magnetic flux density perpendicular to area A.
- The unit of magnetic flux is the **weber** (Wb), equal to $1\,\mathrm{T\,m^2}$.

Regenerative braking

A battery-powered electric vehicle contains an **alternator** that can be used as an electric motor or as a generator. When the alternator is used as an electric motor, it is driven by the batteries. When the brakes are applied, the alternator is used to generate electricity which is used to recharge the battery. Some of the kinetic energy is transferred to electrical energy in the battery. The induced current through the alternator coil creates a magnetic field that acts against the magnetic field of the alternator. So, the alternator experiences a braking force which helps to slow the vehicle down.

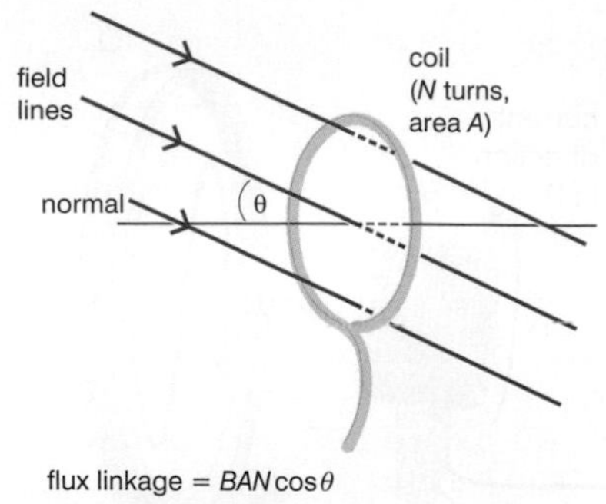

Figure 7 *Flux linkage*

Notes

1 The minus sign represents the fact that the induced e.m.f. acts in such a direction as to oppose the change that causes it.
2 The unit of flux change per second, the weber per second, is the same as the volt. Therefore, the weber is equal to 1 volt second.

Note that magnetic flux density B (in teslas) is the flux per unit area passing through an area at right angles to the area (i.e. normally). Therefore 1 tesla = 1 weber per square metre.

In general, when the magnetic field is at angle θ to the normal at the coil face, as shown in Figure 7, the flux linkage through the coil $N\Phi = BAN\cos\theta$

1 When the magnetic field is along the normal (i.e. perpendicular) to the coil face, the flux linkage = BAN.
2 When the coil is turned through 180°, the flux linkage = $-BAN$.
3 When the magnetic field is parallel to the coil area, the flux linkage = 0 as no field lines pass through the coil area.

Faraday's law of electromagnetic induction states that the induced e.m.f. in a circuit is proportional to the rate of change of flux linkage through the circuit.

$$\textbf{Induced e.m.f., } E = -N\frac{\Delta\Phi}{\Delta t}$$

where $N\frac{\Delta\Phi}{\Delta t}$ is the change of flux linkage per second.

Whenever the flux linkage through a circuit changes, an e.m.f. is induced in the circuit. The flux can be due to a permanent magnet or due to a current-carrying wire.

- If the flux is due to a permanent magnet, motion of the magnet relative to the circuit is necessary to cause an induced e.m.f. This is how an e.m.f. is generated in an a.c. generator or a dynamo.
- If the flux is due to a current-carrying wire, changing the current in the wire causes an induced e.m.f. in the circuit. This is how an e.m.f. is generated in a **transformer** or an induction coil.

Examples

1 A moving conductor in a magnetic field

An e.m.f. is induced in the conductor provided the conductor cuts across the lines of the magnetic field. The direction of motion of the conductor in Figure 5 is at right angles to the field lines.

As explained on the previous page, the magnitude of the induced e.m.f. $E = \frac{Bl\Delta s}{\Delta t}$,

where l is the length of the conductor and Δs is the distance it moves in time Δt. Note that the change of flux in this time, $\Delta\Phi = Bl\Delta s$ so the change of flux per second, $\frac{\Delta\Phi}{\Delta t}$, is equal to the magnitude of the induced e.m.f.

Because the speed of the conductor, $v = \frac{\Delta s}{\Delta t}$, the induced e.m.f. $E = \frac{Bl\Delta s}{\Delta t} = Blv$.

$$\textbf{Induced e.m.f., } E = Blv$$

2 A rectangular coil moving into a uniform magnetic field

Consider a rectangular coil of N turns, length l and width w moving into a uniform magnetic field of flux density B at constant speed v, as shown in Figure 8. Suppose the coil enters the field at time $t = 0$.

- The time taken by the coil to enter the field completely = $\frac{\text{coil width}}{\text{speed}} = \frac{w}{v}$.
  During this time, the flux linkage $N\Phi$ increases steadily from 0 to $BNlw$. Therefore, the induced e.m.f. = change of flux linkage per second, $\frac{N\Delta\Phi}{\Delta t} = \frac{BNlw}{w/v} = BlvN$
- When the coil is completely in the field, the flux linkage through it (= $BNlw$) does not change so the induced e.m.f. is zero.

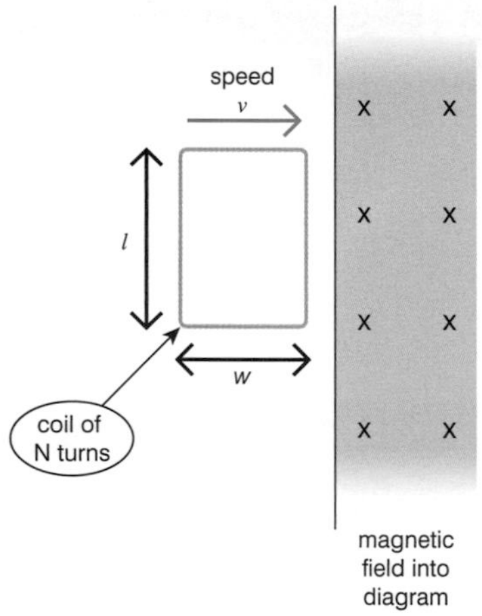

Figure 8 *Flux changes*

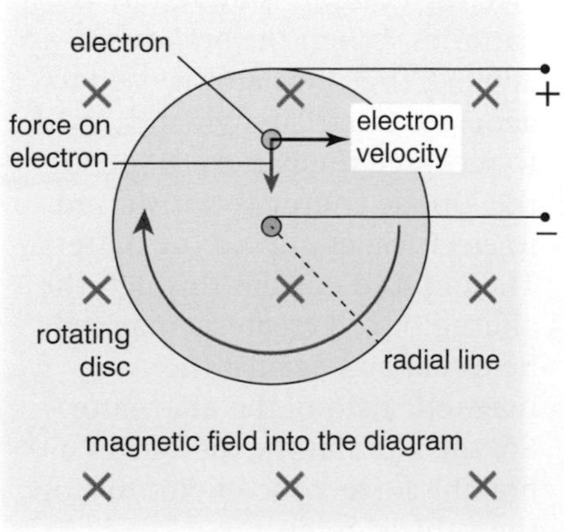

Figure 9 *The rotating disc*

3 A spinning disc in a uniform magnetic field

When the disc spins at constant frequency, a constant e.m.f. is induced between the centre of the disc and its edge. In Figure 9, the conduction electrons in the rotating disc are forced to the centre by the magnetic field so the induced e.m.f. is negative at the centre and positive at the edge. Reversing the field or rotation direction reverses the induced e.m.f.

In one revolution, each radial line on the disc sweeps out an area $A = \pi r^2$ where r is the disc radius. Therefore, the magnetic flux swept out:

- in one revolution is $B \times \pi r^2$ ($= B \times A$),
- in one second is $B \times \pi r^2 \times f$, where f is the frequency of rotation of the disc.

Therefore, the induced e.m.f., E = change of magnetic flux per second $= B \times \pi r^2 \times f$

4 A fixed coil in a changing magnetic field

Figure 10 shows a small coil on the axis of a current-carrying solenoid. The magnetic field of the solenoid passes through the small coil. If the current in the solenoid changes, an e.m.f. is induced in the small coil. This is because the magnetic field through the coil changes so the flux linkage through it changes, causing an induced e.m.f. Because the induced e.m.f. is proportional to the rate of change of flux linkage through the coil and the flux linkage is proportional to the current I in the solenoid, the magnitude of the induced e.m.f is, therefore, proportional to the rate of change of current in the solenoid.

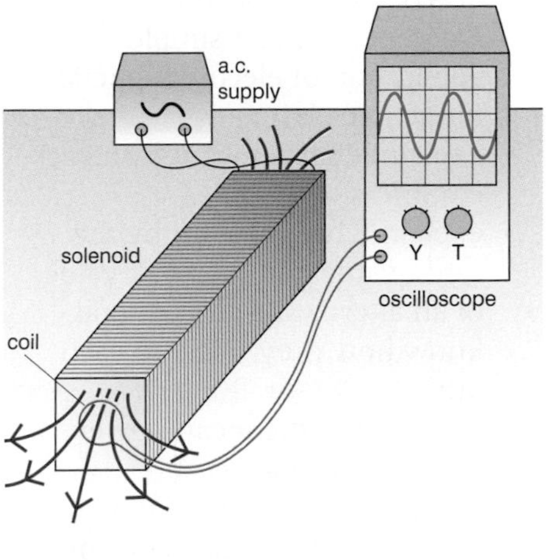

Figure 10 *A changing magnetic field*

Summary test 17.2

1 A uniform magnetic field of flux density 72 mT is confined to a region of width 60 mm, as shown in Figure 11. A rectangular coil of length 50 mm and width 20 mm has 15 turns. The coil is moved into the magnetic field at a speed of $10\,\text{mm}\,\text{s}^{-1}$ with its longer edge parallel to the edge of the magnetic field.

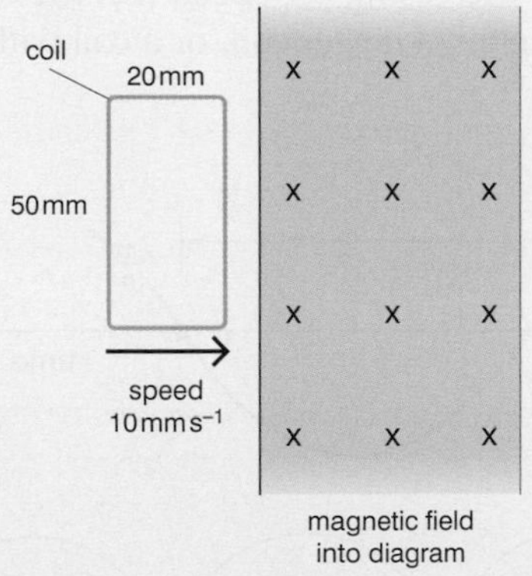

Figure 11

a Calculate:
  i the flux linkage through the coil when it is completely in the field,
  ii the time taken for the flux linkage to increase from zero to its maximum value,
  iii the induced e.m.f. in the coil as it enters the field.

b **i** Sketch a graph to show how the flux linkage through the coil changes with time from the instant the coil enters the field to when it leaves the field completely.
  ii Sketch a graph to show how the induced e.m.f. in the coil varies with time.

2 A rectangular coil of length 40 mm and width 25 mm has 20 turns. The coil is in a uniform magnetic field of flux density 68 mT.

a Calculate the flux linkage through the coil when the coil is at right angles to the field lines.

b The coil is removed from the field in 60 ms. Calculate the mean value of the induced e.m.f.

3 A circular coil of diameter 24 mm has 40 turns. The coil is placed in a uniform magnetic field of flux density 85 mT with its plane perpendicular to the field lines.

a Calculate:
  i the area of the coil in m^2,
  ii the flux linkage through the coil.

b The coil was reversed in a time of 95 ms. Calculate:
  i the change of flux linkage through the coil,
  ii the magnitude of the induced e.m.f.

4 A small circular coil of diameter 15 mm and 25 turns is placed in a fixed position on the axis of a solenoid, as shown in Figure 10. The magnetic flux density of the solenoid at this position varies with current according to the equation $B = kI$, where $k = 1.2 \times 10^{-3}\,\text{T}\,\text{A}^{-1}$.

a Calculate the flux linkage through the coil when the current in the solenoid is 1.5 A.

b The current in the solenoid was reduced from 1.5 A to zero in 0.20 s. Calculate the magnitude of the induced e.m.f. in the small coil.

17.3 Alternating current

Learning outcomes

On these pages you will learn to:

- describe and explain the a.c. generator as a simple application of electromagnetic induction
- explain what is meant by a back e.m.f.
- explain what is meant by the r.m.s. value and the peak value of an alternating current and know how they relate to each other
- deduce that the mean power in a resistive load is half the maximum power for a sinusoidal alternating current
- represent a sinusoidally alternating current or voltage by an equation of the form $x = x_0 \sin \omega t$
- describe rectifier circuits with and without a smoothing capacitor

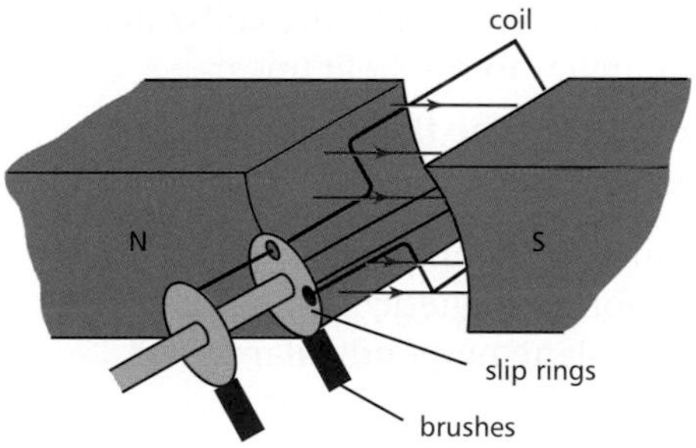

Figure 1 *The a.c. generator*

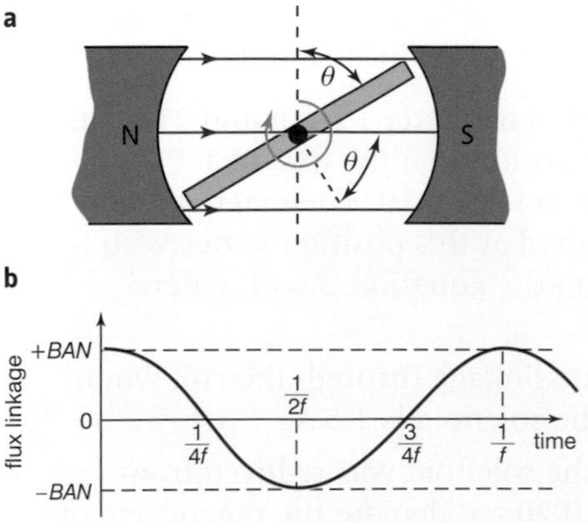

Figure 2 *Flux linkage in a spinning coil*

Alternating current generators are used in power stations and in mobile and emergency generators. In this section, the simple a.c. generator is considered as an application of electromagnetic induction.

The simple a.c. generator

The simple a.c. generator consists of a rectangular coil that spins in a uniform magnetic field, as shown in Figure 1. When the coil spins at a steady rate, the flux linkage changes continuously. At an instant when the normal to the plane of the coil is at angle θ to the field lines, the flux linkage through the coil, $N\Phi = BAN\cos\theta$, where B is the magnetic flux density, A is the coil area and N is the number of turns on the coil.

For a coil spinning at a steady frequency, f, $\theta = 2\pi ft$ at time t after $\theta = 0$. So the flux linkage $N\Phi$ ($= BAN\cos 2\pi ft$) changes with time as shown in Figure 2.

- The gradient of the graph is the change of flux linkage per second, $N\frac{\Delta\Phi}{\Delta t}$, so it represents the induced e.m.f. It can be shown mathematically that the induced e.m.f. alternates in accordance with the equation $E = E_0 \sin 2\pi ft$, where f is the frequency of rotation of the coil and E_0 is the peak e.m.f. Substituting the angular frequency ω for $2\pi f$ therefore gives $\mathbf{E = E_0 \sin \omega t}$. Figure 3 shows how the induced e.m.f., E, varies with time t.
- The induced e.m.f is zero when the sides of the coil move parallel to the field lines. At this position, the rate of change of flux is zero and the sides of the coil do not cut the field lines.
- The induced e.m.f. is a maximum when the sides of the coil cut at right angles across the field lines. At this position, the e.m.f. induced in each wire of each side $= Blv$, where v is the speed of each wire and l is its length. So, for N turns and two sides, the induced e.m.f. at this position $\mathbf{E_0 = 2NBlv}$. The equation shows that the peak e.m.f. can be increased by increasing the speed (i.e. the frequency of rotation) or by using a stronger magnet, a bigger coil, or a coil with more turns.

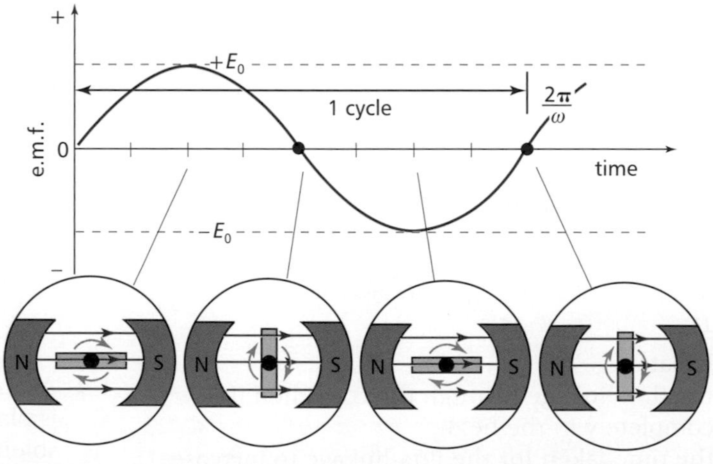

Figure 3 *E.m.f. v. time for an a.c. generator*

Note An a.c. generator in a power station has three coils at 120° to each other. Each coil produces an alternating voltage 120° out of phase with the voltage from the other coils.

Back e.m.f.

An e.m.f. is induced in the spinning coil of an electric motor because the flux linkage through the coil changes. The induced e.m.f. E is referred to as a **back e.m.f.** because it acts against the p.d. V applied to the motor in accordance with Lenz's law. At any instant, $V - E = IR$, where I is the current through the motor coil and R is the circuit resistance.

Because the induced e.m.f. is proportional to the speed of rotation of the motor, the current changes as the motor speed changes.

- At low speed, the current is high because the induced e.m.f. is small.
- At high speed, the current is low because the induced e.m.f. is high.

Multiplying the equation $V - E = IR$ by I throughout gives $IV - IE = I^2R$. Rearranging this equation gives:

Electrical power supplied by the source		Electrical power transferred to mechanical power		Electrical power wasted due to the circuit resistance
	=		+	
IV		IE		I^2R

The above power equation shows that electrical power supplied to the motor that is not used as mechanical power is wasted due to the resistance heating effect of the current.

- When the motor spins without driving a **load**, it spins at high speed so the current is very small. Its speed is limited by friction in the bearings and air resistance. It uses little or no power.
- When the motor is used to drive a load, its speed is less than when it is 'off-load' so the current is much larger. The power it uses from the voltage source that is not transferred as mechanical power to the load is wasted due to the resistance heating effect of the current.

Rectifier circuits

Diodes are used to 'rectify' or convert alternating current to direct current.

Half-wave rectification is produced using a single diode, as shown in Figure 4. The resistor is needed to limit the current otherwise the diode would be destroyed by overheating. The diode conducts every other half-cycle when the a.c. supply across it is in the forward direction. The current is negligible when the supply is in the reverse direction.

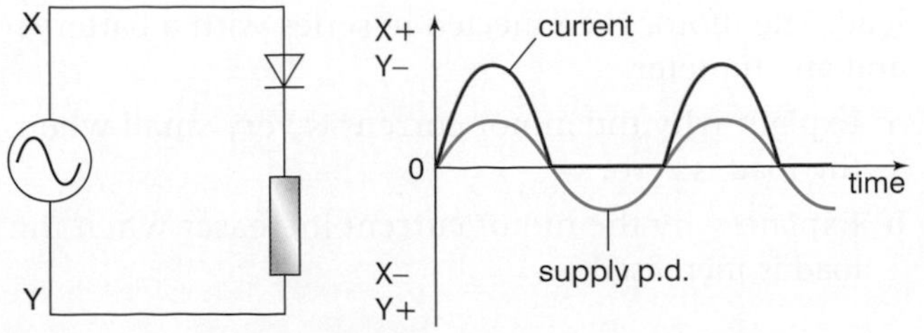

Figure 4 *Half-wave rectification*

Full-wave rectification is produced using four diodes in a **bridge rectifier** circuit as shown in Figure 5. Diodes D_1 and D_3 conduct when X is positive (and Y is negative); diodes D_4 and D_2 conduct when Y is positive (and X is negative). As a result, the direction of the current through R is the same in both half-cycles. The current waveform in Figure 5 shown in purple is the full-wave rectified wave form.

The waveform can be made smoother by connecting a capacitor across the output terminals of the bridge rectifier. The capacitor charges up from the bridge rectifier as the output p.d. increases then discharges through the resistor as the output p.d. decreases. In this way, the current through the resistor is smoothed out, as shown in red in Figure 5. Note that the capacitance C of the capacitor should be sufficiently large in relation to the load resistance R so that the time constant RC is much greater than the time period of the a.c. supply. The effect would then be to reduce the decrease in the current (from the peak value) so the ripple is much smaller.

The effect of increasing the time constant can be calculated using the capacitor discharge equation $I = I_0\, e^{-t/RC}$. For example, in each half-cycle at 50 Hz, the discharge time would be 0.01 s (= ½ × 0.02 s) so for $R = 100\,\Omega$ and

- C = **1.0 mF**, the time constant RC would be 0.1 s giving $I = I_0\, e^{-0.01\,s/0.1\,s} = I_0 e^{-0.1} = 0.905\, I_0$, which is a decrease of $0.095\, I_0$ or a 9.5% drop.
- C = **100 mF**, the time constant RC would be 10 s giving $I = I_0 e^{-0.01s/10\,s} = I_0 e^{-0.001} = 0.999\, I_0$, which is a decrease of $0.001\, I_0$ or a 0.1% drop.

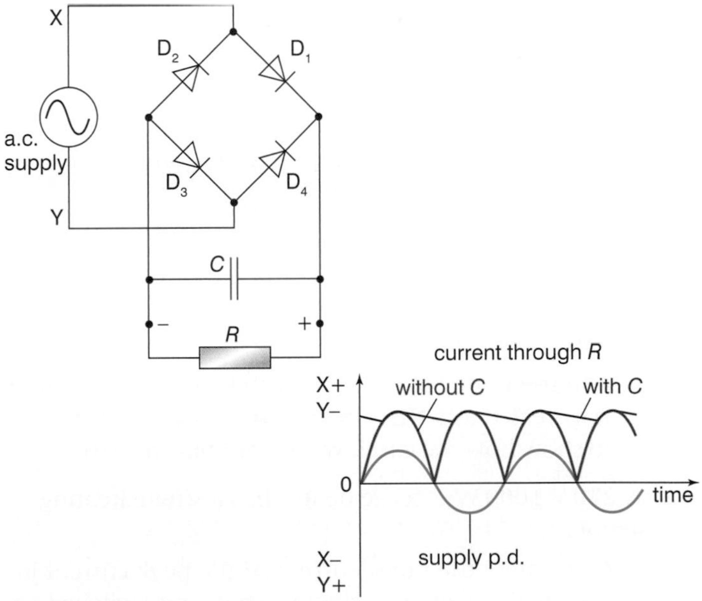

Figure 5 *Full-wave rectification*

The heating effect of an alternating current

Imagine an electric heater supplied with alternating current at a very low frequency. The heater would heat up and then cool down repeatedly as the current changed.

As explained in Topic 7.2, the heating effect of an electric current varies according to the square of the current. This is because the electrical power P supplied to the heater for a current I is given by

$P = IV = I^2R = I_0{}^2R \sin^2 \omega t$, where R is the resistance of the heater element and the current $I = I_0 \sin \omega t$.

Figure 6 shows how the power ($= I^2R$) varies with time.

- At the peak current I_0 in either direction, maximum power is supplied equal to $I_0{}^2R$.
- At zero current, zero power is supplied.

For a sinusoidal current, the **mean power** over a full cycle is half the peak (i.e. maximum) power. This can be seen from the symmetrical shape of the power curve in Figure 6 about the mean power. The mean power is therefore $\frac{1}{2}I_0{}^2R$.

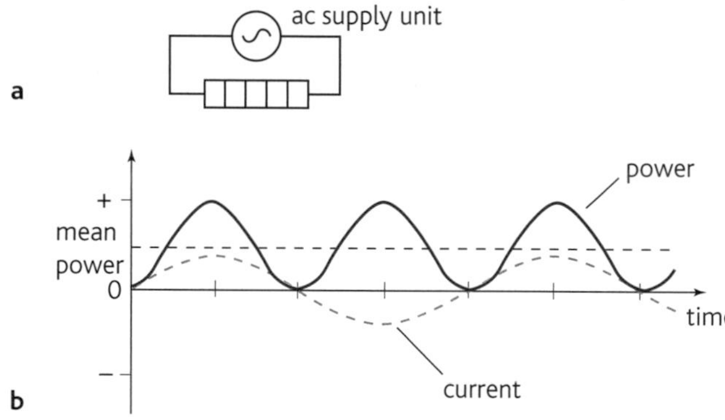

***Figure 6** Variation of power with time for an alternating current*

The direct current that would give the same power as the mean power is called the **root mean square value** of the alternating current, $I_{\text{r.m.s.}}$.

The root mean square value of an alternating current is the value of direct current that would givethe same heating effect as the alternating current in the same resistor.

Therefore, $I_{\text{r.m.s.}}^2 R = 0.5I_0{}^2R$

Cancelling R from this equation and rearranging the equation gives $I_{\text{r.m.s.}}^2 = 0.5I_0{}^2$

Therefore $I_{\text{r.m.s.}} = \frac{1}{\sqrt{2}}I_0$

Also, the root mean square value of an alternating p.d. is given by $V_{\text{r.m.s.}} = \frac{1}{\sqrt{2}}V_0$

The root mean square value of an alternating current or p.d. = $\frac{1}{\sqrt{2}}$ × the peak value

For example, if the peak voltage of an a.c. supply is 50 V, the r.m.s. value will be 35 V.

Note

For any resistor of known resistance in an alternating circuit, if we know the r.m.s. p.d. or current for the resistor, we can calculate the mean power supplied to it using the r.m.s. values.

$$P = I_{\text{r.m.s.}}^2 R = V_{\text{r.m.s.}}^2 / R = I_{\text{r.m.s.}} V_{\text{r.m.s.}}$$

For example, if an alternating current of r.m.s. value 4 A is passed through a 5 Ω resistor, the mean power supplied to the resistor = $4^2 \times 5$ = 80 W.

Summary test 17.3

1 a An a.c. generator produces an alternating e.m.f. with a peak value of 8.0 V and a frequency of 20 Hz. Sketch a graph to show how the e.m.f. varies with time.

b The frequency of rotation of the a.c. generator in **a** is increased to 30 Hz. On the same axes, sketch a graph to show how the e.m.f. varies with time at 30 Hz.

2 A 230 V 1000 W electric heater has a single heating element.

a Calculate: **i** the r.m.s. current, **ii** the peak current in the heating element when the heater is switched on.

b Calculate the peak power supplied to the heater when it is on.

The potential difference of the electricity supply decreases at times of high demand. Estimate the percentage drop in the power supplied to the heater when it is on and the p.d. drops by 5%.

3 The coil of an a.c. generator has 80 turns, a length of 65 mm and a width of 38 mm. It spins in a uniform magnetic field of flux density 130 mT at a constant frequency of 50 Hz.

a Calculate the maximum flux linkage through the coil.

b i Show that each side of the coil moves at a speed of 6.0 m s^{-1}.

ii Show that the peak voltage is 8.1 V.

4 An electric motor is to be used to move a variable load. The motor is connected in series with a battery and an ammeter.

a Explain why the motor current is very small when the load is zero.

b Explain why the motor current increases when the load is increased.

The transformer

Learning outcomes

On these pages you will learn to:

- describe and explain the transformer as a simple application of electromagnetic induction
- recall and use the transformer rule $V_S/V_P = N_S/N_P$
- calculate the efficiency of a transformer from the primary and secondary current and p.d.
- explain why transmission of electrical power is more efficient at high voltage than at low voltage

A transformer changes an alternating p.d. to a different peak value. Any transformer consists of two coils: the primary coil and the secondary coil. The two coils have the same iron core. When the primary coil is connected to a source of alternating p.d., an alternating magnetic field is produced in the core. The field passes through the secondary coil. So, an alternating e.m.f. is induced in the secondary coil by the changing magnetic field. The symbol for the transformer is shown in Figure 1.

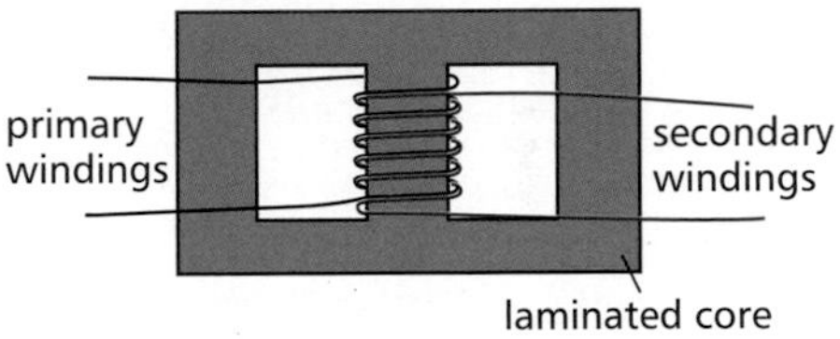

a Practical arrangement

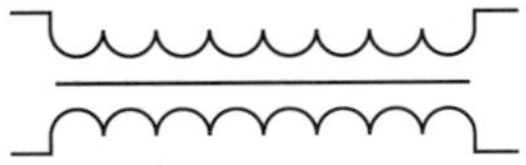

b Transformer symbol

Figure 1 *The transformer*

The transformer rule

A transformer is designed so that all the magnetic flux produced by the primary coil passes through the secondary coil.

Let Φ = the flux in the core passing through each turn at an instant when an alternating p.d. V_P is applied to the primary coil.

- The flux linkage in the secondary coil = $N_S\Phi$, where N_S is the number of turns on the secondary coil. From Faraday's law, the induced e.m.f. in the secondary coil, $V_S = N_S\dfrac{\Delta\Phi}{\Delta t}$
- The flux linkage in the primary coil = $N_P\Phi$, where N_P is the number of turns on the primary coil. From Faraday's law, the induced e.m.f. in the primary coil = $N_P\dfrac{\Delta\Phi}{\Delta t}$. The induced e.m.f. in the primary coil opposes the p.d. applied to the primary coil, V_P.

  Assuming the resistance of the primary coil is negligible, the applied p.d. $V_P = N_P\dfrac{\Delta\Phi}{\Delta t}$

  Dividing the equation for V_S by the equation for V_P gives $\dfrac{V_S}{V_P} = N_S\dfrac{\Delta\Phi}{\Delta t}\Big/ N_P\dfrac{\Delta\Phi}{\Delta t}$

  Cancelling $\dfrac{\Delta\Phi}{\Delta t}$ from this equation gives the **transformer rule**

$$\frac{V_S}{V_P} = \frac{N_S}{N_P}$$

- **A step-up transformer** has more turns on the secondary coil than on the primary coil. So the secondary voltage is stepped up compared with the primary voltage (i.e. $N_S > N_P$ so $V_S > V_P$)
- **A step-down transformer** has fewer turns on the secondary coil than on the primary coil. So the secondary voltage is stepped down compared with the primary voltage (i.e. $N_S < N_P$ so $V_S < V_P$).

Transformer efficiency

$$\text{The efficiency of a transformer} = \frac{\text{power delivered by the secondary coil}}{\text{power supplied to the primary coil}} = \frac{I_S V_S}{I_P V_P}\ (\times 100\%)$$

When a device (e.g. a lamp) is connected to the secondary coil, because the efficiency of a transformer is almost equal to 100%,

the electrical power supplied to the primary coil = the electrical power supplied by the secondary

Therefore, the current ratio $\dfrac{I_S}{I_P} = \dfrac{V_P}{V_S} = \dfrac{N_P}{N_S}$

- In a step-up transformer, the voltage is stepped up and the current is stepped down.
- In a step-down transformer, the voltage is stepped down and the current is stepped up.

The grid system

Electricity from power stations in most countries is fed into a national grid system which supplies electricity to most parts of the country. The grid is a network of cables, either underground or on pylons. Each power station generates alternating current in three phases (see Topic 17.3) at a precise frequency of 50 Hz (or 60 Hz in some countries) at about 25 kV.

Step-up transformers at the power station increase the alternating voltage to 400 kV or more for long-distance transmission via the grid system. Step-down transformers operate in stages, as shown in Figure 2. Factories are supplied with all three phases at either 33 kV or 11 kV. Homes are supplied via a local transformer sub-station with single-phase a.c. at 220–240 V in most countries except in America (110–130 V), Japan (100 V) and a few other countries.

Transmission of electrical power over long distances is much more efficient at high voltage than at low voltage. The reason is that the current needed to deliver a certain amount of power is reduced if the voltage is increased. So power wasted due to the heating effect of the current through the cables is reduced. To deliver power P at voltage V, the current required $I = \frac{P}{V}$. If the resistance of the cables is R, the power wasted through heating the cables is $I^2R = \frac{P^2R}{V^2}$. Therefore, the higher the voltage, the smaller the ratio of the wasted power to the power transmitted.

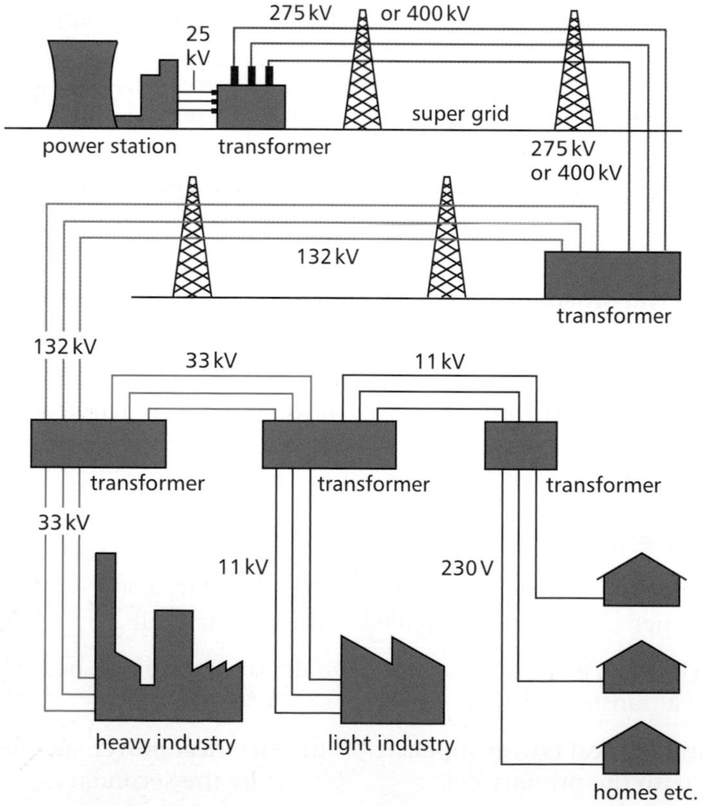

Figure 2 *The grid system*

For example, for transmission of 1 MW of power through cables of resistance 500 Ω at 25 kV, the current necessary would be 40 A (= 1 MW/25 kV) so the power wasted would be 0.8 MW (= $I^2R = 40^2 \times 500$ W). Prove for yourself that at 400 kV, the power wasted would be about 3 kW.

Voltage adaptors

Portable electronic devices such as laptop computers contain batteries that need to be recharged using a voltage adaptor. Figure 3 shows the circuit of such an adaptor. It supplies a constant low voltage when it is plugged into a mains socket.

- The transformer steps down the alternating p.d. from the mains to an alternating p.d. with a much smaller peak p.d. The turns ratio of the transformer is chosen to give a suitable p.d. from the transformer.
- The bridge rectifier converts the alternating p.d. from the transformer to a full-wave direct p.d.
- The capacitor across the bridge rectifier smooths out the full-wave p.d. to give a constant output p.d, as explained in Topic 17.3. The greater the capacitance of C, the smaller the variation in the output p.d. as current is drawn from the voltage adaptor.

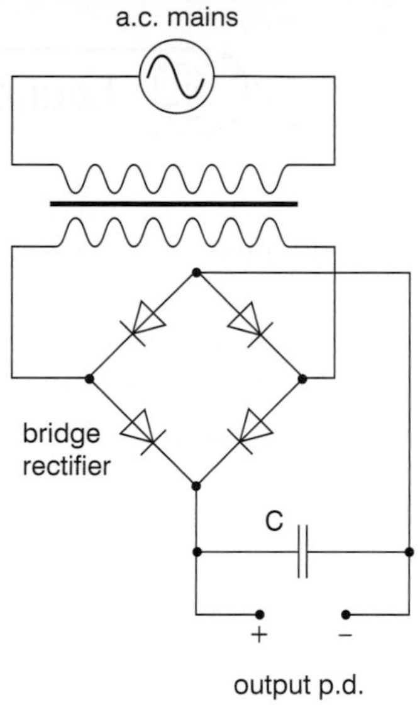

Figure 3 *A voltage adaptor*

Summary test 17.4

1 a Explain why an alternating e.m.f. is induced in the secondary coil of a transformer when the primary coil is connected to an alternating voltage supply.

b In terms of electrical power, explain why the current through the primary coil of a transformer increases when a device is connected to the secondary coil.

2 a Explain why a transformer is designed so that as much of the magnetic flux produced by the primary coil of a transformer as possible passes through the secondary coil.

b Explain why a transformer works using alternating current but not using direct current.

3 A transformer has a primary coil with 120 turns and a secondary coil with 2400 turns.

a Calculate the primary voltage needed for a secondary voltage of 230 V.

b A 230 V, 60 W lamp is connected to its secondary coil. Calculate the current through:

i the secondary coil,

ii the primary coil. State any assumptions made in this calculation.

4 a Explain why transmission of electrical power over a long distance is more efficient at high voltage than at low voltage.

b A power cable of resistance 200 Ω is to be used to deliver 2.0 MW of electrical power at 120 kV from a power station to an industrial estate. Calculate:

i the current through the cable,

ii the power wasted in the cable.

Chapter Summary

Magnetic flux, $\Phi = BA$

Flux linkage through a coil of N turns = $N\Phi = NBA$, where B is the magnetic flux density perpendicular to area A.

Lenz's law states that the direction of the induced current is always such as to oppose the change that causes the current.

Faraday's law of electromagnetic induction states that the induced e.m.f. in a circuit is proportional to the rate of change of flux linkage through the circuit.

Equation for Faraday's law; $E = -N\frac{\Delta\Phi}{\Delta t}$, where $N\frac{\Delta\Phi}{\Delta t}$ is the change in flux linkage per second.

For a moving conductor in a uniform magnetic field, the induced e.m.f. = Blv

For a changing magnetic field in a fixed coil,

induced e.m.f. = $NA\frac{\Delta B}{\Delta t}$

Units

The unit of magnetic flux density B is the **tesla** (T).

The unit of magnetic flux and of flux linkage is the **weber** (Wb), equal to 1 T m^2 or 1 V s.

The unit of rate of change of flux (or rate of change of flux linkage) is the weber per second (Wb s^{-1}), equal to 1 V.

17 Exam-style and Practice Questions

Launch additional digital resources for the chapter

1 In a ribbon microphone, a metal ribbon vibrates between the poles of a magnet when sound waves reach the microphone.

a Explain why an e.m.f. is induced across the ends of the ribbon when it vibrates.

b How would the e.m.f. be affected if:
i a stronger magnet was used,
ii a ribbon of greater mass was used?

2 **a** A bar magnet was positioned near a coil connected to a centre-reading meter. When the bar magnet was pushed into the coil, the meter pointer deflected briefly to the right.
i Explain why the pointer deflected briefly.
ii State and explain what is observed when the magnet is withdrawn from the coil.

b In **a**, the flux density of the magnet was 25 mT and the area of the 30 turn coil was $4.0 \times 10^{-4}\,m^2$. The magnet was pushed into the coil in 0.20 s. Calculate:
i the flux linkage through the coil,
ii the mean induced e.m.f.

3 A U-shaped magnet was placed at the centre of a horizontal stretched steel wire such that the magnetic field was vertical. When the wire was plucked at its centre, an alternating e.m.f. was induced between the ends of the wire.

a Explain why an alternating e.m.f. was induced between the ends of the wire.

b The length of wire between the poles was 28 mm and the magnetic flux density of the magnet was 78 mT. The peak voltage produced was 3.4 mV. Calculate the maximum speed of the wire between the poles.

4 **a** A straight conducting rod PQ of length 0.10 m moves through a uniform magnetic field at a speed of $20\,mm\,s^{-1}$. The induced e.m.f. was 0.60 mV when the conductor was moving perpendicular to the field. Calculate:
i the flux swept out in 5.0 s,
ii the magnetic flux density.

b In **a**, the field was vertically downwards and the rod was horizontal, as shown in Figure 4.1.
i State the polarity of each end of the rod.
ii If the rod was part of a complete circuit, state and explain the direction in which the induced current would pass through it.

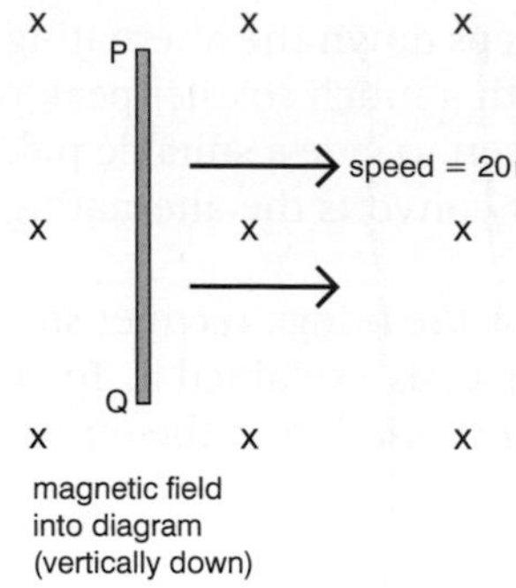

Figure 4.1

5 A rectangular coil of length 50 mm and width 20 mm has 25 turns. Figure 5.1 below shows the coil just before it was moved at a constant speed of $5\,mm\,s^{-1}$ into a uniform magnetic field of flux density 86 mT. The field lines are perpendicular to the plane of the coil.

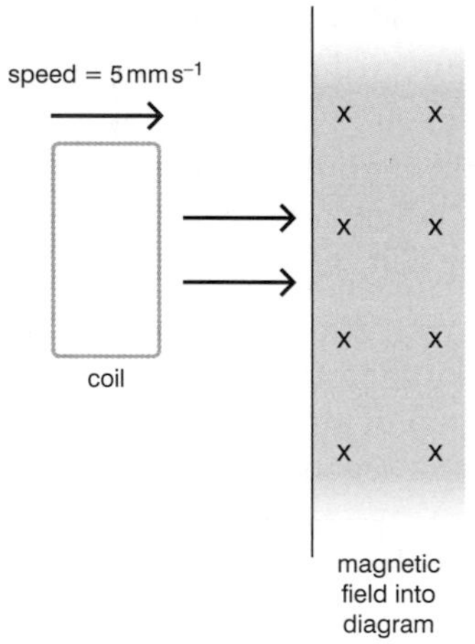

Figure 5.1

a Show that the coil takes 4.0 s to enter the field and calculate the flux linkage through the coil when it is completely in the field.

b Sketch a graph to show how:
i the flux linkage through the coil changes with time,
ii the induced e.m.f. varies with time t from $t = 0$ when the coil enters the field to $t = 5.0$ s.

6 An a.c. generator is used to provide electricity for a lighting circuit.

a Explain why the generator is easier to turn when the lamps are switched off.

b The generator coil has 120 turns on a rectangular coil of length 40 mm and width 30 mm. The coil spins in a magnetic field of flux density 220 mT.
i When the generator spins at a frequency of 20 Hz, show that the peak voltage is 4.0 V.
ii Calculate the peak voltage when the frequency is 25 Hz.

7 A bar magnet was held vertically in a horizontal coil connected to a data recorder, as shown in Figure 7.1 below. When the magnet was released, the data recorder was used to measure the voltage induced in the coil every 5 ms. The diagram shows how the voltage changed with time.

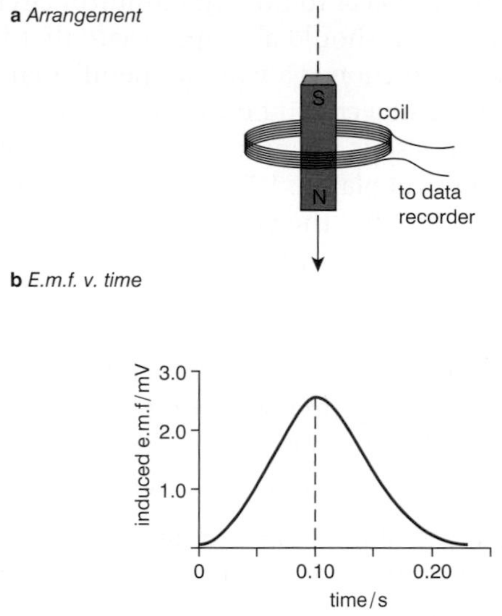

Figure 7.1

a **i** Without calculations, sketch a graph to show how the flux linkage through the coil changed with time.

ii Use your graph to explain the shape of the voltage v. time graph.

b Sketch the voltage v. time graph that would have been produced if the bar magnet had been released from a position above the coil so that it dropped through the coil.

8 A transformer is used to step down an alternating voltage of 230 V to 12 V. The transformer has a primary coil with 1000 turns.

a Calculate the number of turns on the secondary coil.

b The transformer is used to supply power to a 12 V, 60 W lamp. Calculate the current in:

i the secondary coil,

ii the primary coil when the lamp is on.

c The lamp is connected to the transformer by means of a cable of resistance 0.4 Ω.

i Estimate the power wasted due to the heating effect of the current in the cable.

ii Discuss whether or not it would be better to replace the lamp and transformer with a 230 V, 60 W lamp connected to the mains using the same cable.

9 **a** State Faraday's law of electromagnetic induction.

b The arrangement shown in Figure 9.1 can be used to test Faraday's law of electromagnetic induction. A cylindrical bar magnet is clamped horizontally on a wooden base which is designed to enable the wire to be rotated about the bar magnet when the handle is turned.

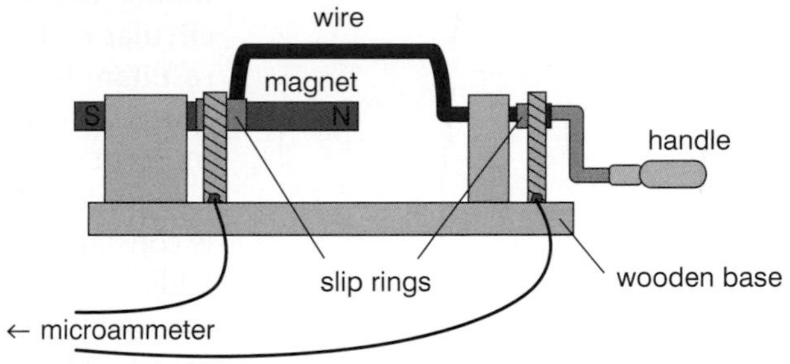

Figure 9.1

The wire is connected to a microammeter via two slip rings. When the handle is turned, the wire rotates about the magnet and cuts across the magnetic field lines, inducing a current in the circuit. The microammeter records the induced current in the circuit.

i Describe an experiment using the above apparatus and a stopwatch to investigate how the induced current varies with the frequency of rotation of the wire about the magnet.

ii Use your knowledge of Faraday's law to predict the relationship between the induced current and the frequency of rotation and how you would use your results to test your prediction.

10a **i** Describe the function of a transformer and explain how a step-up transformer works.

ii Explain why electrical power is transmitted more efficiently through cables at high voltage than at low voltage.

iii Cables of total resistance 5.0 Ω used to transfer electrical energy to a heating system are supplied with 5.0 kW of electrical power at a certain alternating voltage. Calculate and compare the output power and the efficiency of the transfer for root mean square input voltages of 100 V and 1000 V.

b **i** Describe, with the aid of a diagram, how a transformer and a bridge rectifier are used to convert alternating current to direct current.

ii Draw a graph to show how the output voltage and the input voltage vary over two cycles of the input voltage.

Key concepts: Fields

In studying Chapters 11 and 12, you will have gained a clear understanding of the force conditions necessary for an object to undergo uniform circular motion and simple harmonic motion. You should also appreciate that in uniform circular motion, because the velocity direction is always perpendicular to the resultant force, no work is done on the object and hence its kinetic energy and speed are constant. This consideration also is important where a satellite is in uniform circular motion about a spherical planet. When considering simple harmonic motion, you should appreciate that the energy of an oscillating system is constant provided dissipative forces such as air resistance are negligible. In addition, you will have developed your mathematical skills to a level where you can apply non-linear equations including sinusoidal equations and graphs to calculate the acceleration and the time period of the motion and the speed and position at any time. In addition, by investigating oscillating systems such as a loaded spring, you will have developed your practical and analytical skills by plotting straight-line graphs to test related non-linear theoretical equations.

In studying Chapters 13, 14 and 16, you should have gained awareness of important field concepts such as lines of force, uniform fields and field strength. Beyond these concepts, electric and gravitational fields share common features such as an inverse square law of force, similar equations for field strength and potential in radial fields and similar trajectories for objects moving across a uniform field. Magnetic fields differ from electric and gravitational fields in respect of these and other features. For example, charged objects moving across a uniform magnetic field follow a circular path because the magnetic force is perpendicular to the direction of the object's velocity (as well as being perpendicular to the field direction). The force on a moving charged object in a magnetic field is the reason why a current-carrying conductor in a magnetic field experiences a force when it is not aligned along the field lines, and it provides the basis of explanations of the motor effect and the Hall effect. Knowledge and understanding gained in Chapter 16 ought to have given you a sound basis for the study of electromagnetic induction in Chapter 17. Conservation of energy is also important in understanding and applying Lenz's law and Faraday's law to important applications such as the a.c. generator and the transformer. In all these chapters, you will have developed your mathematical skills by applying them to further non-linear equations and related graphs.

Your practical skills as well as your knowledge of electricity should have developed considerably in your studies of Chapter 15. Practical investigations into capacitors in Chapter 15 involve timing measurements as well as measurements of currents and potential differences. By making such measurements with and without a data logger, you should appreciate the benefits of using a data logger in these experiments. In addition, your mathematical knowledge and skills will have been enhanced by carrying out calculations and plotting graphs involving exponential decrease equations and graphs.

The following table provides an overview of the topics in Chapters 11 to 17, showing where key concepts have been developed.

Chapters 11–17 Key concepts

	Topic		Key concepts	
11.1–11.4	uniform motion in a circle	centripetal force and centripetal acceleration	forces and fields	vectors
		calculations involving angular speed, centripetal acceleration and Newton's laws of motion	use of maths	problem-solving
12.1–12.4	principles and applications of simple harmonic motion	conditions for SHM	forces and fields	restoring force
		calculations involving SHM equations and sinusoidal functions	use of maths	problem-solving
		simple pendulum to measure g	testing	practical skills including analysis and evaluation
		oscillations of a loaded spring		
12.5–12.6	energy and simple harmonic motion	kinetic and potential energy of an object	energy and matter	conservation of energy
		damping, forced oscillations and resonance		work done by dissipative forces
		SHM calculations of kinetic and potential energies	use of maths	problem-solving
13.1–13.4	gravitational forces and fields	Newton's law of gravitation	forces and fields	inverse-square law
		gravitational field strength, uniform and radial fields		field strength
		gravitational potential	energy and matter	conservation of energy
		calculations involving formulae for gravitational field strength and potential	use of maths	problem-solving
13.5	satellite motion	Newton's law of gravitation applied to uniform circular motion leading to speed and time period equations	forces and fields	use of inverse-square law
		calculations involving above equations	use of maths	problem-solving
14.1–14.4	electric fields	electric charge, line of force, Coulomb's law, electric field strength, uniform and radial fields	forces and fields	inverse-square law for radial fields, uniform fields
		electric potential, equipotentials	energy and matter	conservation of energy
		calculations involving formulae for electric field strength and potential	use of maths	problem-solving
14.5	comparison of electric and gravitational fields	field strength and potential	forces and fields energy and matter	inverse and inverse-square equations and graphs
15.1–15.4	capacitors	capacitance	forces and fields	electric charge
		capacitor combination rules		
		energy stored in a capacitor	energy and matter	conservation of energy
		capacitor charging and discharging	testing	practical skills including analysis
		calculations and graphs involving capacitor formulae, including capacitor discharge	use of maths	problem-solving

continued on p.268

Chapters 11–17 Key concepts *(continued)*

Topic			Key concepts	
16.1–16.4	magnetic fields	magnetic field lines, magnetic flux density, the motor effect, magnetic field patterns	forces and fields	force perpendicular to conductor and to the field lines
		calculations and graphs involving magnetic field formula	use of maths	problem-solving
	moving charges in a magnetic field	force on a moving charge in a magnetic field, Hall effect		
		use of a Hall probe		
16.5	charged particles in circular orbits	circular motion of charged particles in a magnetic field	forces and fields	force perpendicular to velocity
		calculations involving charged particles in circular orbits and use of the Hall voltage equation	testing	practical skills; measurement of magnetic flux density
			use of maths	problem-solving
17.1–17.4	electromagnetic induction	origin of induced e.m.f., magnetic flux	forces and fields	
		Lenz's law and Faraday's law	energy and matter	conservation of energy
		calculations involving magnetic flux, Faraday's law, the a.c. generator and the transformer	use of maths	problem-solving
				transformer efficiency

Question

This question is about the oscillations of a loaded spring. It tests how we use models to make predictions about a physical system and how we then test these predictions with an experimental investigation. It also tests practical skills involving timing the oscillations and measuring the extension of a loaded spring. Systematic errors can arise when making extension measurements, for example as a result of not using a plane mirror correctly when reading a pointer against a millimetre scale and also when timing oscillations, for example by miscounting the number of oscillations. Random errors can be minimised when timing oscillations by making repeat measurements of at least 20 oscillations.

As in the previous key concepts questions on p.92 and 165, physics theory and mathematics are used to predict the relationship between the static extension e of the spring and the time period $\boldsymbol{T}$ *of the oscillations. In this case, the predicted relationship is of the form* $\boldsymbol{T}^2 = \boldsymbol{ke}$ *(or* $\boldsymbol{T} = \boldsymbol{k}^{1/2}\boldsymbol{e}^{1/2}$*). So by plotting a graph of* $\boldsymbol{T}^2$ *against* $\boldsymbol{e}$ *(or* $\boldsymbol{T}$ *against* $\boldsymbol{e}^{1/2}$*), we can tell if the relationship is valid from whether or not the graph is linear (i.e. a straight-line graph) and passes through the origin. If the graph is linear, we can also determine the constant* $\boldsymbol{k}$*, as it is equal to the gradient of the graph. Also, the value of* $\boldsymbol{g}$ *can then be found from the result for* $\boldsymbol{k}$*.*

In some investigations, the relationship between two variables may not be easy to predict from the relevant theory or the theory might not give a straight-line graph. In these cases, assuming the relationship is of the form $\boldsymbol{y} = \boldsymbol{kx}^n$*, a graph of* $\log \boldsymbol{y}$ *against* $\log \boldsymbol{x}$ *may be plotted; a straight-line graph would confirm that the relationship is of the form* $\boldsymbol{y} = \boldsymbol{kx}^n$ *and the power n could then be found, as it is equal to the gradient of the graph.*

Before answering the question, students should be given the opportunity to do this or a similar investigation.

1 In an investigation, a small object was suspended from the lower end of a vertical steel spring, which was fixed at its upper end, as shown in Figure 1.

A horizontal marker pin P was attached to the lower end of the spring. The vertical position x of the pin was measured against the millimetre scale of a metre rule clamped vertically in a fixed position. The measurement was made three times without, then with, the small object suspended from the spring.

a The readings obtained are shown in Table 1.

Table 1

	x/mm			mean x/mm	extension e/mm
without the object on the spring	2	2	2	2.0	0
with the object on the spring	71	72	73		

i Copy and complete Table 1 by calculating the mean vertical position of P and the extension of the spring when the object was placed on it.

ii The readings were taken to a precision of 0.5 mm using a millimetre ruler. Estimate the percentage 'uncertainty' in the extension. *(2 marks)*

b The time period, T, of small vertical oscillations of the object on the spring was also measured by timing 20 oscillations three times. The timing readings for 20 oscillations were 10.98 s, 11.11 s and 10.97 s.

i Calculate the time period T.

ii Use the readings to estimate the percentage 'uncertainty' in T. *(2 marks)*

c i Give an expression for the extension e of the spring in terms of the mass m of the object and the spring constant k of the spring.

ii Hence show that $T = 2\pi\sqrt{\frac{e}{g}}$ *(3 marks)*

d The experiment was repeated with objects of different mass suspended from the spring. The measurements obtained are given in Table 2.

Table 2

object	e/mm	T/s
1	70	0.551
2	139	0.761
3	205	0.923
4	271	1.062
5	341	1.187
6	409	1.291

Plot a suitable graph using the above measurements to confirm the equation and to determine g. *(9 marks)*

e Discuss the accuracy of your determination of g. *(4 marks)*

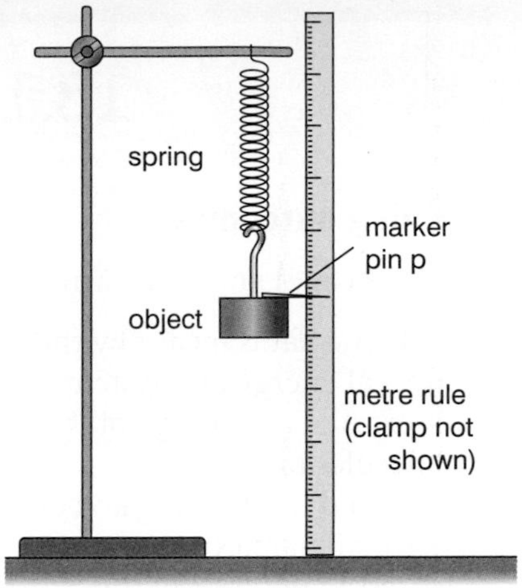

Figure 1

18 Thermal physics

18.1 Internal energy and temperature

Learning outcomes

On these pages you will learn to:

- explain what is meant by the internal energy of a system in terms of the energy of the molecules of a system
- recall and use the first law of thermodynamics $\Delta U = q + W$, expressed in terms of the increase in internal energy, the heating of the system and the work done on the system
- explain using a simple kinetic model for matter the structure of solids, liquids and gases

Figure 1 *Infrared image showing heat transfer from a house*

When you are outdoors in a cool climate in winter, you need to wrap up well otherwise heat transfer from your body to the surroundings takes place and you lose energy. In hot weather, if you are in a very hot room, you gain energy from the room due to heat transfer.

Energy transfer between two objects takes place if:

- one object exerts a force on the other one and makes it move. In other words, one object does work on the other (see Topic 2.4 for more about work).
- one object is hotter than the other so heat transfer takes place by means of conduction, convection or radiation. In other words, heat transfer is energy transfer due to a temperature difference.

Internal energy

The brake pads of a moving vehicle become hot if the brakes are applied for long enough. The work done by the frictional force between the brake pads and the wheel heats the brake pads. The brake pads gain energy from the kinetic energy of the vehicle. The temperature of the brake pads increases as a result and the internal energy of each brake pad increases.

As explained below, the internal energy of an object is the energy of its molecules due to their individual movements. The internal energy of an object due to its temperature is sometimes referred to as **thermal energy**. However, some of the internal energy of an object might be due to other causes. For example, an iron bar that is magnetised has more internal energy than if it is unmagnetised because of the magnetic interaction between its atoms.

The internal energy of an object changes as a result of:

- energy transfer to or from the object by heating or by radiation, or
- work done on or by the object, including work done by electricity.

If q represents the energy transfer to an object by heating and W represents the work done on it (leaving out work done to make an object move faster or to raise it), then

the object's change of internal energy, $\Delta U = q + W$

The equation follows from the **principle of conservation of energy**. The above equation is known as the **first law of thermodynamics**.

If the internal energy of an object is constant, either:

- there is no heat transfer and no work is done, or
- heat transfer and work done 'balance' each other out.

For example, the internal energy of a lamp filament increases when the lamp is switched on because work is done by the electricity supply pushing electrons through the filament. The filament becomes hot as a result. When it reaches its operating temperature, heat transfer to the surroundings takes place and it radiates light. Work done by the electricity supply pushing electrons through the filament is balanced by heat transfer including light radiated from the filament.

About molecules

A molecule is the smallest particle of a pure substance that is characteristic of the substance. For example, a water molecule consists of two hydrogen atoms joined to an oxygen atom.

An atom is the smallest particle of an element that is characteristic of the element. For example, a hydrogen atom consists of a proton and an electron.

- In a solid, the atoms and molecules are held to each other by forces due to the electrical charges of the protons and electrons in the atoms. The molecules in a solid vibrate randomly about fixed positions. The higher the temperature of the solid, the more the molecules vibrate. The energy supplied to raise the temperature of a solid increases the kinetic energy of the molecules. If the temperature is raised sufficiently, the solid melts. This happens because its molecules vibrate so much that they break free from each other and the substance loses its shape. The energy supplied to melt a solid raises the potential energy of the molecules because they break free from each other.
- In a liquid, the molecules move about at random in contact with each other. The forces between the molecules are not strong enough to hold the molecules in fixed positions. The higher the temperature of a liquid, the faster its molecules move. The energy supplied to a liquid to raise its temperature increases the kinetic energy of the liquid molecules. Heating the liquid more and more causes it to vaporise. The molecules have sufficient kinetic energy to break free and move away from each other.
- In a gas or vapour, the molecules also move about randomly but much further apart on average than in a liquid. Heating a gas or a vapour makes the molecules speed up and so gain kinetic energy.

The internal energy of an object is the sum of the random distribution of the kinetic and potential energies of its molecules.

Increasing the internal energy of a substance increases the kinetic and/or potential energy associated with the random motion and positions of its molecules. The kinetic and potential energy of each individual molecule differs between molecules and changes at random.

Summary test 18.1

1 Describe the energy transfers that occur when a low voltage heater connected to a battery is used to heat some water in a beaker.

2 a Explain why an electric motor becomes warm when it is used.

b A battery is connected to an electric motor which is used to raise a weight at a steady speed. When in operation, the electric motor is at a constant temperature which is above the temperature of its surroundings. Describe the energy transfers that take place.

3 a State what is meant by internal energy.

b Describe a situation in which the internal energy of an object is constant even though work is done on the object.

4 a State one difference between the motion of the molecules in a solid and the molecules in a liquid.

b Describe how the motion of the molecules in a solid changes when the solid is heated.

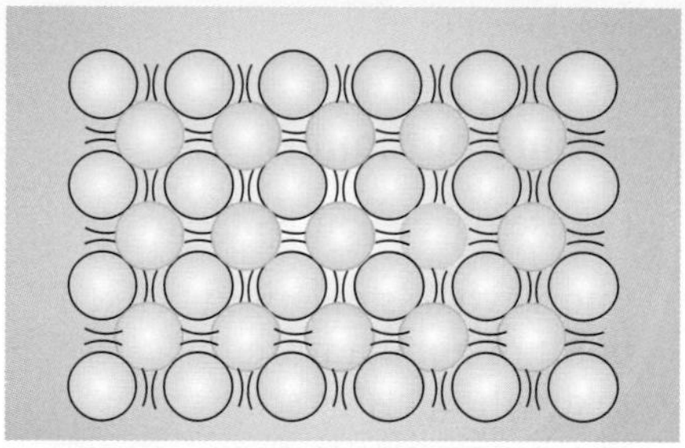

A solid is made up of particles arranged in a regular 3-dimensional structure. There are strong forces of attraction between the particles. Although the particles can vibrate, they cannot move out of their positions in the structure.

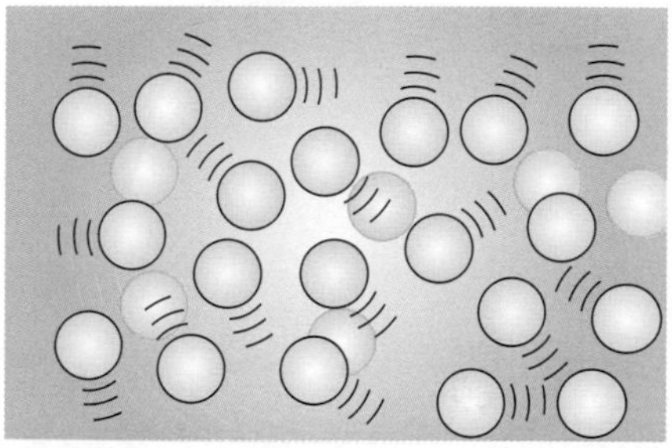

In a liquid the particles are free to move around. A liquid therefore flows easily and has no fixed shape. There are still forces of attraction between the particles.

When a liquid is heated, some of the particles gain enough energy to break away from the other particles. The particles which escape from the body of the liquid become a gas.

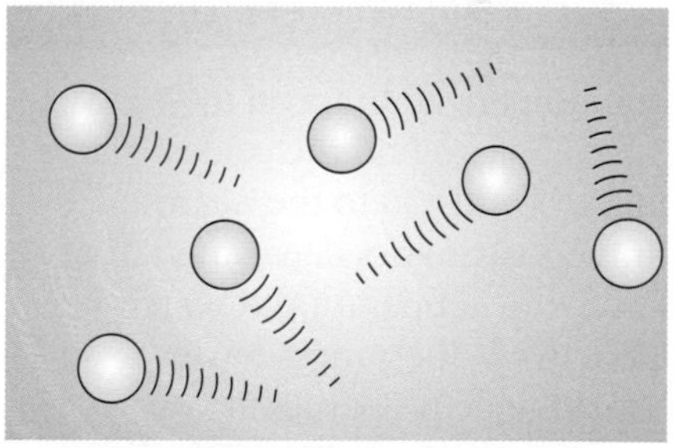

In a gas, the particles are far apart. There are almost no forces of attraction between them. The particles move about at high speed. Because the particles are so far apart, a gas occupies a very much larger volume than the same mass of liquid.

The molecules collide with the container. These collisions are responsible for the pressure which a gas exerts on its container.

Figure 2 *Particles in a solid, a liquid and a gas*

18.2 Temperature scales

Learning outcomes

On these pages you will learn to:

- know what is meant by a fixed point of temperature
- relate the Celsius scale of temperature to the thermodynamic (Kelvin) scale
- recognise that the gas thermometer is the standard against which other thermometers are calibrated

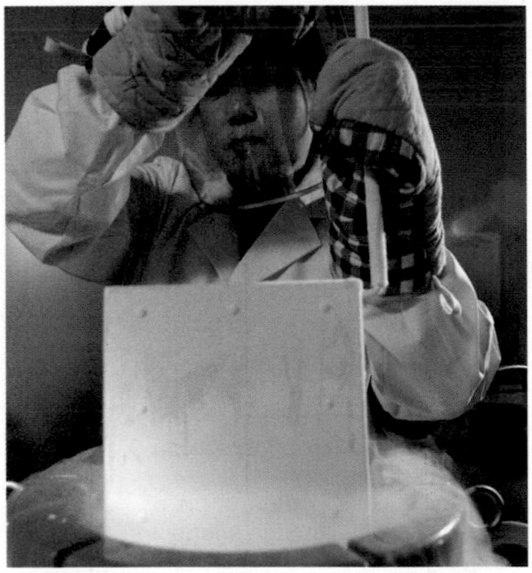

Figure 1 *A low temperature research laboratory*

The coldest places in the world

You don't need to travel to the South Pole to find the coldest places in the world. Go to the nearest university physics department that has a low temperature research laboratory. Substances have very strange properties at very low temperatures. For example, metals cooled to a few degrees within absolute zero become superconductors which means they have zero electrical resistance.

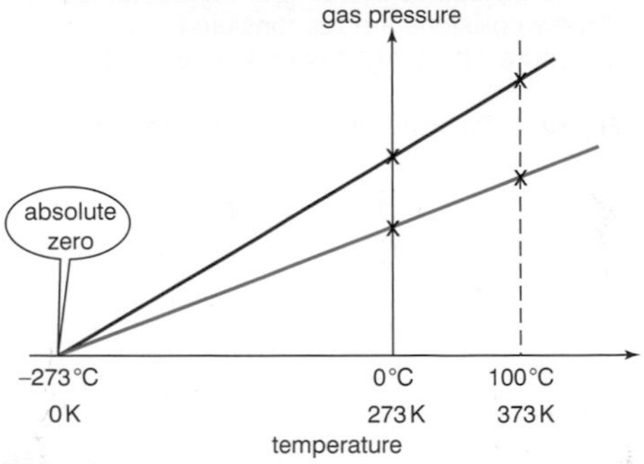

Figure 2 *Absolute zero*

The temperature of an object is a measure of the degree of hotness of the object. The hotter an object is, the more internal energy it has. Place your hand in cold water and it loses internal energy due to heat transfer. Place it in warm water and it gains internal energy due to heat transfer. If the water is at the same temperature as your hand, no overall heat transfer takes place. Your hand is in **thermal equilibrium** with the water. No overall heat transfer takes place between two objects at the same temperature.

The 'baby in the bath' rule

Before dipping the baby in the bath water, the parent tests the water by putting a hand (not the baby's hand!) in the water. If the baby is at the same temperature as the parent and the parent's hand is at the same temperature as the bath water, the baby will be at the same temperature as the bath water.

Practical temperature scales

A temperature scale is defined in terms of **fixed points** which are standard degrees of hotness which can be accurately reproduced.

- **The Celsius scale of temperature**, in °C, is defined in terms of:
  1. ice point, 0 °C, which is the temperature of pure melting ice,
  2. steam point, 100 °C, which is the temperature of steam at standard atmospheric pressure.
- **The thermodynamic (Kelvin) scale of temperature**, in kelvins (K) is defined in terms of:
  1. **absolute zero**, 0 K, which is the lowest possible temperature,
  2. the triple point of water, 273.16 K, which is the temperature at which ice, water and water vapour are in thermal equilibrium.

Because ice point on the thermodynamic scale is 273.15 K and steam point is 100 K higher,

temperature in °C = thermodynamic temperature in kelvins − 273.15

About absolute zero

The thermodynamic scale of temperature, also referred to as the absolute or kelvin scale, is based on absolute zero, the lowest possible temperature. No object can have a temperature below absolute zero. **An object at absolute zero has minimum internal energy**, regardless of the substances the object consists of.

As explained in Topic 19.1, the pressure of a fixed mass of gas in a sealed container of fixed volume decreases as the gas temperature is reduced. If the pressure measured at ice point and at steam point is plotted on a graph, as shown in Figure 2, the line between the two points always cuts the temperature axis at −273 °C, regardless of which gas is used or how much gas is used.

The thermodynamic scale of temperature starts at absolute zero. Its unit, the kelvin, is defined so that a temperature change of 1 K is the same as a temperature change of 1 °C. The Kelvin scale depends on a fundamental feature of nature, namely, the lowest possible temperature. In comparison, the Celsius scale depends on the properties of a substance, water, chosen for convenience rather than for any fundamental reason.

Thermometer	Thermometric property	Advantages	Disadvantages
Liquid-in-glass	the liquid's density (which changes with temperature causing the length of the liquid thread to change)	easy to use, portable	fragile, limited range
Thermocouple	e.m.f., E, across the junction of two different metal wires in contact with each other	fast response, wide range, remote readings	millivoltmeter needed
Resistance	resistance, R of a thermistor (or a metal wire)	wide range, accurate, measures small changes accurately	slow response
Gas	pressure, p of an ideal gas at constant volume	wide range, accurate, standard thermometer	bulky, slow response

Thermometers

The **thermometric property** of a thermometer is a physical property that varies smoothly with change of temperature and can therefore be used to measure temperature. Four different thermometers are compared in the table above and in Figure 3.

All thermometers are calibrated in terms of the temperature measured by a gas thermometer. This is a thermometer consisting of dry gas in a sealed container. The pressure of the gas is proportional to the thermodynamic temperature of the gas. In other words, equal increases of temperature cause equal increases of gas pressure.

By measuring the gas pressure, p_{Tr}, at the triple point of water (= 273.16 K by definition) and an unknown temperature T/K, the unknown temperature, in kelvins, can be calculated using

$$\frac{T}{273.16} = \frac{p}{p_{Tr}}$$

where p is the gas pressure at the unknown temperature.

To determine the temperature on the Celsius scale or the thermodynamic (i.e. Kelvin) scale using any other thermometer, it must be calibrated against a gas thermometer (or indirectly against a thermometer previously calibrated against a gas thermometer). This involves measuring the thermometric property at different temperatures, T, measured using the gas thermometer (or indirectly as above). A graph or 'calibration curve' can then be plotted of temperature T on the y-axis against the thermometric property on the x-axis.

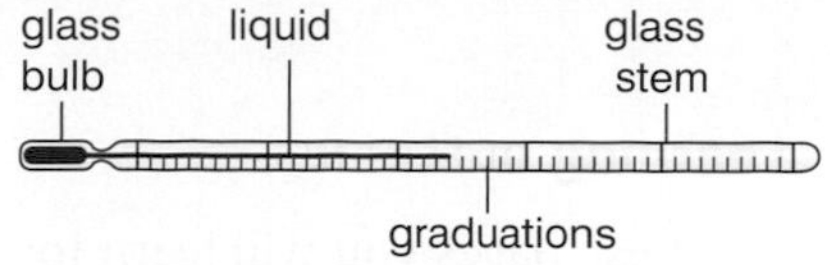

a *A liquid-in-glass thermometer*

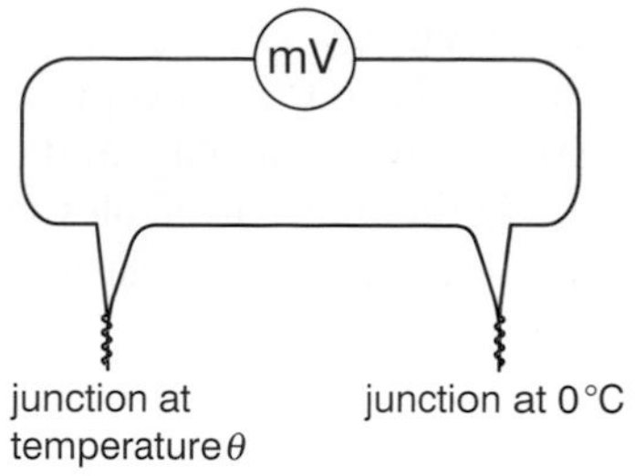

b *A thermocouple*

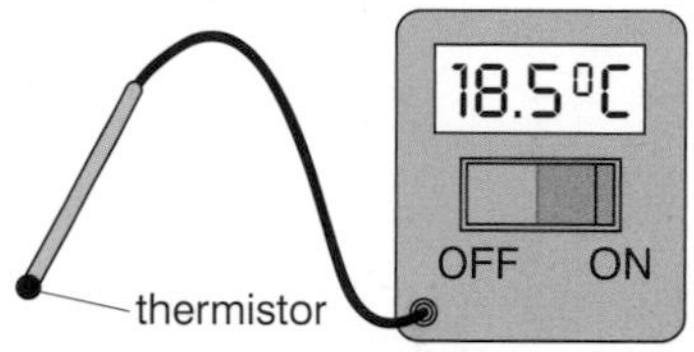

c *A resistance thermometer*

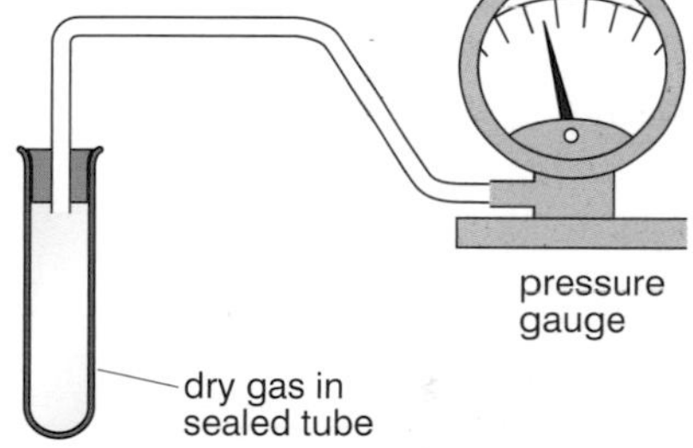

d *A gas thermometer*

Figure 3 Different types of thermometers

Note

If the thermometric property is X_0 at ice-point, X_{100} at steam point and X at an unknown temperature θ, the unknown temperature can be calculated on a centigrade scale (ie a scale of 100 equal intervals from ice point to steam point) using the equation $\theta = \frac{X - X_0}{X_{100} - X_0} \times 100$. This gives the values that would be obtained by dividing the interval between X_0 and X_{100} into 100 equal degrees. If a thermometer is graduated with such a scale, a calibration curve of the 'gas thermometer' temperature T against θ would need to be used to determine the temperature T from a thermometer reading, θ.

Summary test 18.2

1 a Define the *thermodynamic scale of temperature* and state its unit.

b State each of the following temperatures to the nearest degree on the thermodynamic scale:
i the temperature of pure melting ice,
ii 20 °C, **iii** −196 °C.

2 The pressure of a constant-volume gas thermometer was 100 kPa at a temperature of 273 K.

a Calculate the temperature of the gas when its pressure was 120 kPa.

b Calculate the pressure of the gas at 100 °C.

3 Explain why the 50 °C mark on the stem of a liquid-in-glass thermometer is not exactly mid-way between the 0 and 100 °C marks.

4 Explain why a gas thermometer is used to calibrate other types of thermometers.

18.3 Specific heat capacity

Learning outcomes

On these pages you will learn to:

- relate a rise in temperature of a body to an increase in its internal energy
- define and use the concept of specific heat capacity, and identify the main principles of its determination by electrical methods

Heating and cooling

Sunbathers on hot sandy beaches dive into the sea to cool off. Sand heats up much more readily than water does. Even when the sand is almost too hot to walk barefoot across, the sea water is refreshingly cool. The temperature rise of an object when it is heated depends on:

- the mass of the object,
- the amount of energy supplied to it,
- the substance or substances from which the object is made.

The specific heat capacity, *c*, of a substance is the energy needed to raise the temperature of unit mass of the substance by 1 K without change of state.

The unit of c is $\mathrm{J\,kg^{-1}\,K^{-1}}$.

Specific heat capacities of some common substances are shown in Table 1.

Table 1 *Some specific heat capacities*

Substance	Specific heat capacity / $\mathrm{J\,kg^{-1}\,K^{-1}}$
Aluminium	900
Concrete	850
Copper	390
Iron	490
Lead	130
Oil	2100
Water	4200

To raise the temperature of mass m of a substance from temperature T_1 to temperature T_2,

$$\text{the energy needed } \Delta Q = mc\,(T_2 - T_1)$$

For example, to calculate the heat that must be supplied to raise the temperature of 5.0 kg of water from 20 °C to 100 °C, using the above formula is $\Delta Q = 5.0 \times 4200 \times (100 - 20) = 1.7 \times 10^6$ J.

The heat capacity, *C*, of an object is the heat supplied to raise the temperature of the object by 1 K.

Therefore, for an object of mass m made of a single substance of specific heat capacity c, its heat capacity $C = mc$. For example, the heat capacity of 5.0 kg of water is $21\,000\,\mathrm{J\,K^{-1}} = 5.0\,\mathrm{kg} \times 4200\,\mathrm{J\,kg^{-1}\,K^{-1}}$.

The inversion tube experiment

In the inversion tube experiment, the gravitational potential energy of an object falling in a tube is converted into internal energy when it hits the bottom of a tube. Figure 1 shows the idea. The object is a collection of tiny lead spheres.

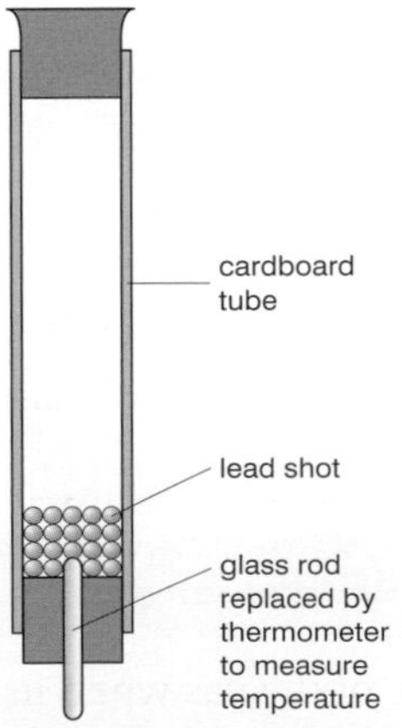

Figure 1 *The inversion tube experiment*

The tube is inverted each time the spheres hit the bottom of the tube. The temperature of the lead shot is measured initially and after a certain number of inversions.

Let m represent the mass of the lead shot.

For a tube of length L, the loss of gravitational potential energy for each inversion = mgL

Therefore, for n inversions, the loss of gravitational potential energy = $mgLn$

The gain of internal energy of the lead shot = $mc\Delta T$, where c is the specific heat capacity of lead and ΔT is the temperature rise of the lead shot.

Assuming all the gravitational potential energy lost is transferred to internal energy of the lead shot,

$$mc\Delta T = mgLn$$

$$\therefore \quad c = \frac{gLn}{\Delta T}$$

The experiment can therefore be used to measure the specific heat capacity of lead with no other measurements than the length of the tube, the temperature rise of the lead and the number of inversions.

Specific heat capacity measurements using electrical methods

1 Measurement of the specific heat capacity of a metal

A block of the metal of known mass m in an insulated container is used. A 12 V electrical heater is inserted into a hole drilled in the metal and used to heat the metal by supplying a measured amount of electrical energy. A thermometer inserted into a second hole drilled in the metal is used to measure the temperature rise ΔT (= its final temperature − its initial temperature).

The electrical energy supplied = heater current, I × heater p.d., V × heating time, t

∴ assuming no heat loss to the surroundings, $mc\Delta T = IVt$

$$\therefore \quad c = \frac{IVt}{m\Delta T}$$

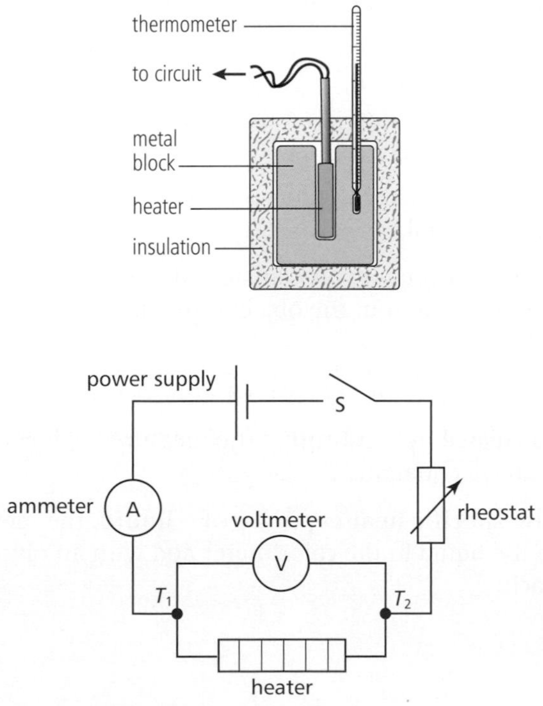

Figure 2 *Measuring c*

2 Measurement of the specific heat capacity of a liquid

A known mass of the liquid is used in an insulated calorimeter of known mass and known specific heat capacity. A 12 V electrical heater is placed in the liquid and used to heat it directly. A thermometer inserted into the liquid is used to measure the temperature rise, ΔT. Assuming no heat loss to the surroundings:

origin. The relationship between pressure p and temperature T, in kelvins, can therefore be written as

$$\frac{p}{T} = \text{constant}$$

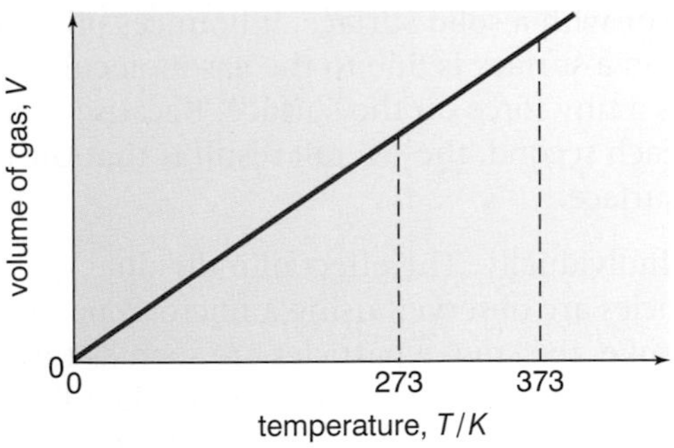

Figure 3 *Charles' law*

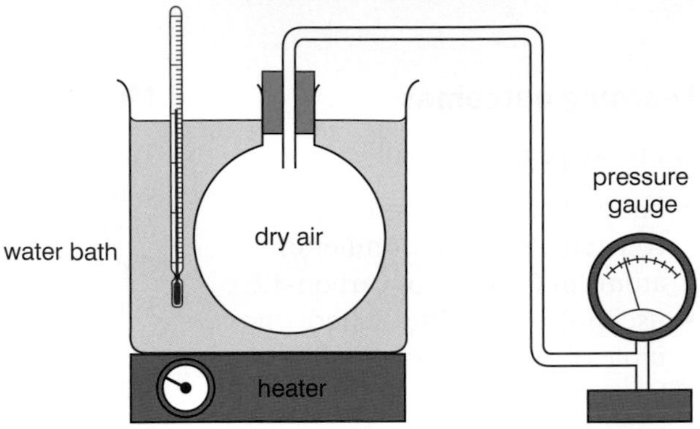

Figure 4 *The pressure law*

Extension

Measurement of the density of a powder using Boyle's law

A hand pump of volume V_P is used to compress the air in a sealed flask of known volume V_F. A pressure gauge is used to measure the pressure of the air in the flask before compression (p_0) and after compression. See Figure 5.

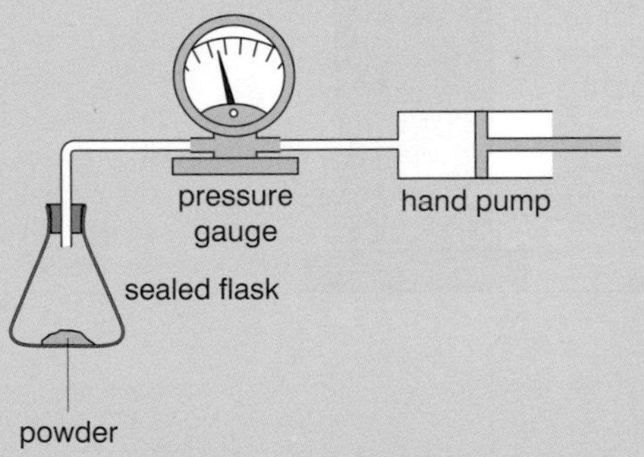

Figure 5 *Measuring the volume of a powder*

1 Without any powder in the flask; applying Boyle's law gives $p_1V_F = p_0\ (V_F + V_P)$, where p_1 is the measured pressure after compression.

$\therefore\ p_0V_F = p_1V_F - p_0V_P$

2 With a measured mass of powder in the flask; applying Boyle's law gives $p_2\ (V_F - v) = p_0\ (V_F - v + V_P)$, where p_2 is the measured pressure after compression and v is the powder volume.

$\therefore\ p_0V_F = p_2V_F - p_0V_P - (p_2 - p_0)v$

Subtracting the two equations gives

$0 = p_1V_F - p_2V_F + (p_2 - p_0)v$

$\therefore\quad (p_2 - p_0)\,v = p_2V_F - p_1V_F$

$$v = \frac{(p_2 - p_1)}{(p_2 - p_0)}V_F$$

Hence, the volume and the density of the powder can be determined.

Summary test 19.1

1 A hand pump of volume $2.0 \times 10^{-4}\,m^3$ is used to force air through a valve into a container of volume $8.0 \times 10^{-4}\,m^3$ which contains air at an initial pressure of 101 kPa. Calculate the pressure of the air in the container after one stroke of the pump, assuming the temperature is unchanged.

2 A sealed can contains air at a pressure of 101 kPa at 100 °C. The can is then cooled to a temperature of 20 °C. Calculate the pressure of the air in the can assuming the volume of the can is unchanged.

3 The volume of a fixed mass of gas at 15 °C was $0.085\,m^3$. The gas was then heated to 55 °C without change of pressure. Calculate the new volume of this gas.

4 In an experiment to measure the density of a powder, a hand pump was used to raise the pressure of the air in a flask of volume $1.20 \times 10^{-4}\,m^3$ first without, and then with the powder in the flask.

   1 Without the powder in the flask, the pressure increased from 110 kPa to 135 kPa.
   2 With 0.038 kg of powder in the flask, the pressure increased from 110 kPa to 141 kPa.

a Show that the volume of air in the hand pump initially was $2.7 \times 10^{-5}\,m^3$.

b Calculate the volume and the density of the powder.

19.2 The ideal gas law

Learning outcomes

On these pages you will learn to:

- know that the Avogadro constant N_A is the number of atoms in 0.012 kg of carbon-12
- use molar quantities where one mole of any substance is the amount containing a number of particles equal to the Avogadro constant N_A
- recall and solve problems using the ideal gas equation $pV = nRT$, where n = number of moles
- recognise the experimental basis of the ideal gas equation
- infer from a Brownian motion experiment the evidence for the movement of molecules

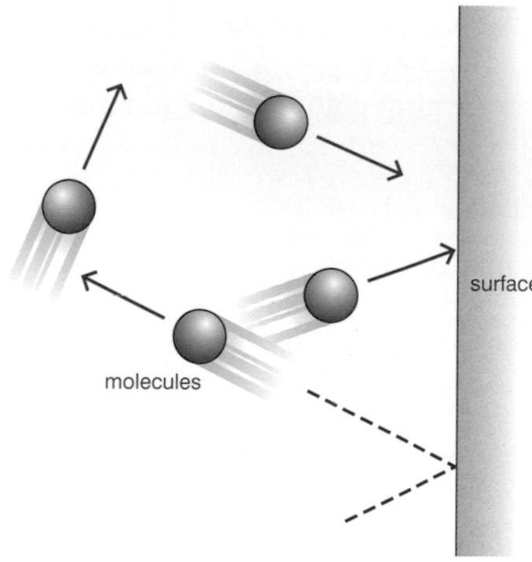

Figure 1 *Molecules in motion*

Molecules in a gas

The molecules of a gas move at random with different speeds. When a molecule collides with another molecule or with a solid surface, it bounces off without loss of speed. The pressure of a gas on a surface is due to the gas molecules hitting the surface. Each impact causes a tiny force on the surface. Because there are a very large number of impacts each second, the overall result is that the gas exerts a measurable pressure on the surface.

Molecules are too small to see individually. The effect of individual molecules in a gas can be seen if smoke particles are observed using a microscope. If a beam of light is directed through the smoke, the smoke particles are seen as tiny specks of light wriggling about unpredictably. This type of motion is called **Brownian motion** after Robert Brown who first observed it in 1827 with pollen grains in water. The motion of each particle is due to it being bombarded unevenly and at random by individual molecules. The particle is therefore subjected to a force due to the impacts which changes its magnitude and direction at random.

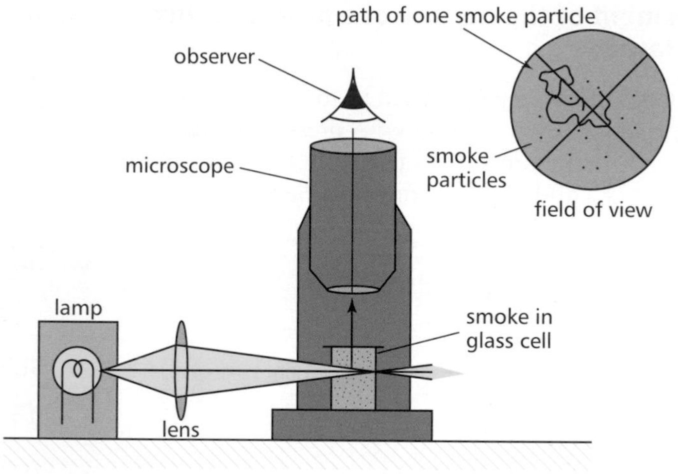

Figure 2 *Brownian motion*

The Avogadro number

The density of oxygen gas is 16 times that of hydrogen gas at the same temperature. Therefore, the mass of a certain volume of oxygen is 16 times that of the mass of the same volume of hydrogen at the same temperature. When such measurements were first made in the nineteenth century, Amadeo Avogadro put forward the hypothesis that equal volumes of gases at the same temperature and pressure contain an equal number of molecules.

How many molecules are in a certain amount of gas? Avogadro thought of the idea of counting atoms and molecules in terms of the number of atoms in 1 gram of hydrogen. Now we use 12 grams of the carbon isotope ${}^{12}_{6}C$ as the standard amount as hydrogen contains a small proportion of the isotope of hydrogen ${}^{2}_{1}H$ which cannot easily be removed.

The Avogadro constant, N_A, is defined as the number of atoms in exactly 12 grams of the carbon isotope ${}^{12}_{6}C$.

The value of N_A (to four significant figures) is 6.023×10^{23}. Therefore the mass of an atom of ${}^{12}_{6}C$ is 1.993×10^{-23} grams (= 12 grams/6.02×10^{23}).

One atomic mass unit (u) is $\frac{1}{12}$th of the mass of a ${}^{12}_{6}C$ atom.

The mass of a carbon atom is 1.993×10^{-26} kg, so 1 u = 1.661×10^{-27} kg.

Notes

1 The masses in atomic mass units of some different atoms (to 1 u) are: hydrogen, H = 1 u, carbon, C = 12 u, nitrogen, N = 14 u, oxygen, O = 16 u, copper, Cu = 64 u

2 The masses in atomic mass units of some different molecules (to 1 u) are: water H_2O = 18 u, carbon monoxide CO = 28 u, carbon dioxide CO_2 = 44 u, oxygen O_2 = 32 u

Molar mass

- **One mole** of a substance consisting of identical particles is defined as the quantity of substance that contains N_A particles. The number of moles in a certain quantity of a substance is its **molarity**. The unit of molarity is the mol.
- **The molar mass** of a substance is the mass of 1 mole of the substance. The unit of molar mass is kg mol^{-1}. For example, the molar mass of oxygen gas is 0.032 kg mol^{-1}. So 0.032 kg of oxygen gas contains N_A oxygen molecules.

Therefore:

1 the number of moles in mass m of a substance = $\frac{m}{M}$, where M is the molar mass of the substance,

2 the number of molecules in mass m of a substance = $N_A \frac{m}{M}$.

For example, because the molar mass of carbon dioxide is 44 grams:

2 moles of carbon dioxide has a mass of 88 grams and contains $2N_A$ molecules,

10 moles of carbon dioxide has a mass of 440 grams and contains $10N_A$ molecules,

n moles of carbon dioxide has a mass of $44n$ grams and contains nN_A molecules.

The ideal gas equation

An ideal gas is a gas that obeys the gas laws. The three experimental gas laws can be combined to give the equation

$$\frac{pV}{T} = \textbf{constant, for a fixed mass of ideal gas,}$$

where p is the pressure, V is the volume and T is the **thermodynamic** temperature. This equation takes in all situations where the pressure, volume and temperature of a fixed mass of gas change.

As explained above, equal volumes of ideal gases at the same temperature and pressure contain equal numbers of moles. Further measurements show that one mole of any ideal gas at 273 K and a pressure of 101 kPa has a volume of 0.0224 m^3.

Therefore, for 1 mole of any ideal gas, the value of $\frac{pV}{T}$ for 1 mole is equal to

$$8.31\,\text{J mol}^{-1}\,\text{K}^{-1} \left(= \frac{pV}{T} = \frac{101 \times 10^3\,\text{Pa} \times 0.0224\,\text{m}^3}{273\,\text{K}} \right).$$

This value is known as the **molar gas constant, *R*.**

Hence, the combined gas law may be written as

$$pV_m = RT$$

where V_m = volume of 1 mole of ideal gas at pressure p and temperature T.

Therefore, for n moles of ideal gas,

$$pV = nRT$$

where V = volume of the gas at pressure p and temperature T.

This equation is known as the **ideal gas equation**.

Note

The unit of R is the joule per mol per kelvin (J mol^{-1} K^{-1}) which is the same as the unit of $\frac{\text{pressure} \times \text{volume}}{\text{temperature}}$.

This is because the unit of pressure (the pascal = 1 N m^{-2}) × the unit of volume (m^3) is the joule (= 1 N m).

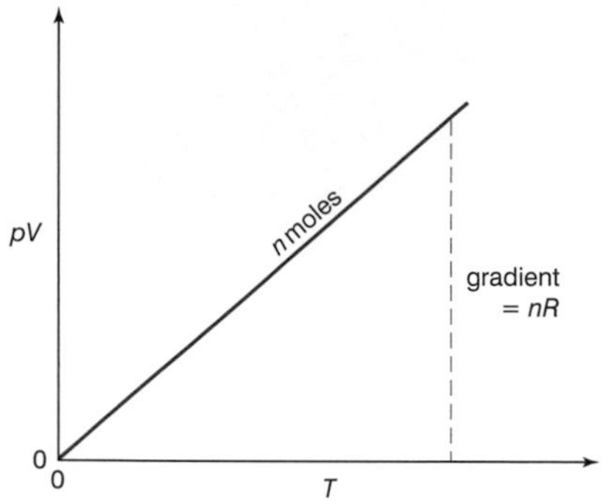

Figure 3 *A graph of pV against T for an ideal gas*

Notes

1. T is the temperature in kelvins.
2. The equation $pV = nRT$ can be written as $pV = NkT$ where N is the number of molecules in the gas, $n = \frac{N}{N_A}$ and $k = \frac{R}{N_A}$.
   k is referred to as the Boltzmann constant. See Topic 19.4.
3. A graph of pV against temperature T is a straight line through absolute zero and has a gradient equal to nR.
4. The density of an ideal gas of molar mass M, $\rho = \frac{\text{mass}}{\text{volume}} = \frac{nM}{V} = \frac{pM}{RT}$.
   Therefore, for an ideal gas at constant pressure, its density ρ is inversely proportional to its temperature T (as $\rho = \frac{pM}{RT} = \frac{\text{constant}}{T}$ for constant pressure).

Worked example

$R = 8.31\,\text{J mol}^{-1}\,\text{K}^{-1}$

Calculate the number of moles of air in a balloon when the air pressure in the balloon is 170 kPa, the volume of the balloon is $8.4 \times 10^{-4}\,\text{m}^3$ and the temperature of the air in the balloon is 17 °C.

Solution

$T = 273 + 17 = 290\,\text{K}$

Using $pV = nRT$ gives $n = \frac{pV}{RT} = \frac{170 \times 10^3 \times 8.4 \times 10^{-4}}{8.31 \times 290} = 5.9 \times 10^{-2}$ moles

Summary test 19.2

$N_A = 6.02 \times 10^{23}\,\text{mol}^{-1}$,
$R = 8.31\,\text{J mol}^{-1}\,\text{K}^{-1}$

1 A gas cylinder has a volume of $0.024\,\text{m}^3$ and is fitted with a valve designed to release the gas if the pressure of the gas reaches 125 kPa. Calculate:

a the maximum number of moles of gas that can be contained by this cylinder at 50 °C,

b the pressure in the cylinder of this amount of gas at 10 °C.

2 In an electrolysis experiment, $2.2 \times 10^{-5}\,\text{m}^3$ of a gas is collected at a pressure of 103 kPa and a temperature of 20 °C. Calculate:

a the number of moles of gas present,

b the volume of this gas at 0 °C and 101 kPa.

3 a Sketch a graph to show how the pressure of 2 moles of gas varies with temperature when the gas is heated from 20 °C to 100 °C in a sealed container of volume $0.050\,\text{m}^3$.

b The molar mass of the gas in **a** is 0.032 kg. Calculate the density of the gas.

4 The molar mass of air is 0.029 kg.

a Calculate the density of air at 20 °C and a pressure of 101 kPa.

b Calculate the number of molecules in $0.001\,\text{m}^3$ of air at 20 °C and a pressure of 101 kPa.

The kinetic theory of gases

Learning outcomes

On these pages you will learn to:

- explain how molecular movement causes the pressure exerted by a gas
- state the basic assumptions of the kinetic theory of gases and derive the relationship $pV = \frac{1}{3}Nm<c^2>$ where N = number of molecules
- solve problems using the above equation and the equation $p = \frac{1}{3}\rho c^2_{r.m.s.}$, where ρ is the density of the gas

The gas laws can be explained by assuming a gas consists of point molecules moving about at random, continually colliding with the container walls. Each impact causes a force on the container. The force of many impacts is the cause of the pressure of the gas on the container walls.

Explanation of Boyle's law: the pressure of a gas at constant temperature is increased by reducing its volume because the gas molecules travel a shorter distance between impacts at the walls due to the reduced volume. Hence there are more impacts per second so the pressure is greater.

Explanation of the pressure law: the pressure of a gas at constant volume is increased by raising its temperature. The average speed of the molecules is increased by raising the gas temperature so the impacts of the molecules on the container walls are harder and more frequent. Hence the pressure is raised as a result.

Molecular speeds

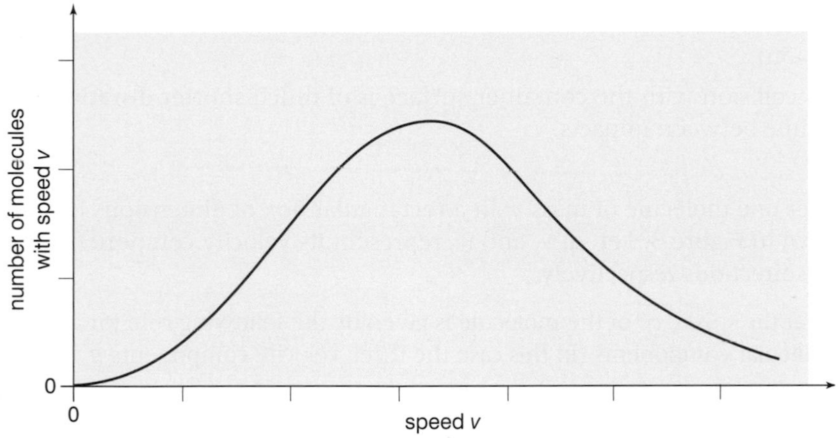

Figure 1 *Distribution of molecular speeds*

The molecules in an ideal gas have a continuous spread of speeds, as shown in Figure 1. The speed of an individual molecule changes when it collides with another gas molecule but the distribution stays the same, provided the temperature does not change.

The **root mean square speed** of the molecules, $c_{r.m.s.} = \frac{[(c_1^2 + c_2^2 + \ldots + c_N^2)]^{1/2}}{N}$

where $c_1, c_2, c_3, \ldots c_N$ represent the speeds of the individual molecules and N is the number of molecules in the gas.

Notes

1 The root mean square (r.m.s.) speed of the molecules of a gas is not the same as the mean speed which is the sum of the speeds divided by the number of molecules.

2 The symbols $<c^2>$ and $\bar{c^2}$ are sometimes used for the mean square speed, $c^2_{r.m.s.}$.

If the temperature of a gas is raised, its molecules move faster on average. The r.m.s. speed of the molecules increases. The distribution curve becomes flatter and broader as there are molecules at higher speeds (see Figure 2).

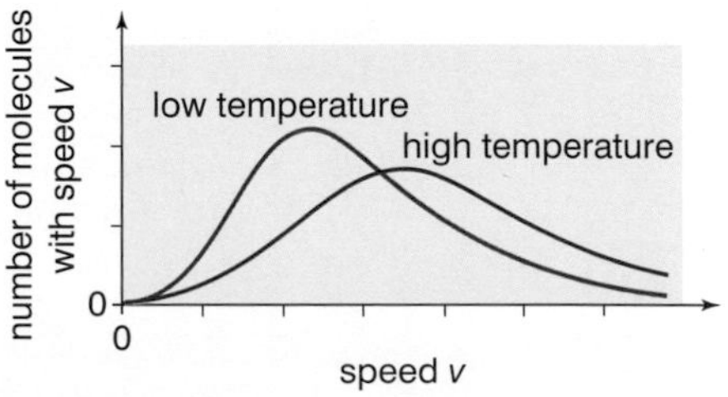

Figure 2 *The effect of temperature on the distribution of speeds*

The kinetic theory equation

For an ideal gas consisting of N identical molecules, each of mass m, in a container of volume V, the pressure p of the gas is given by

$$pV = \frac{1}{3}Nmc^2_{\text{r.m.s.}}$$

where $c_{\text{r.m.s.}}$ is the root mean square speed of the gas molecules.

We need to apply the laws of mechanics and statistics to the molecular model of a gas to derive the kinetic theory equation. In doing so, certain assumptions must be made about the molecules in a gas.

1. The molecules are point molecules. The volume of each molecule is negligible compared with the volume of the gas.
2. They do not attract each other. If they did, the effect would be to reduce the force of their impacts on the container surface.
3. They move about in continual random motion.
4. The collisions they undergo with each other and with the container surface are elastic collisions (i.e. there is no overall loss of kinetic energy in a collision).
5. Each collision with the container surface is of much shorter duration than the time between impacts.

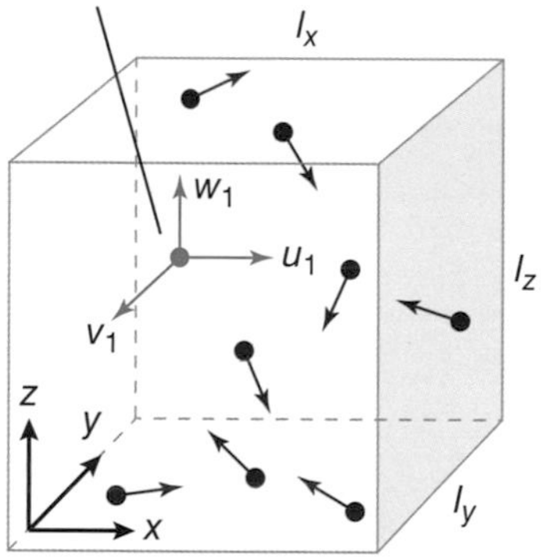

Figure 3 *Molecules in a box*

Part 1

Consider one molecule of mass m in a rectangular box of dimensions l_x, l_y and l_z as shown in Figure 3. Let u_1, v_1 and w_1 represent its velocity components in the x, y, and z directions respectively.

Note that the speed, c_1, of the molecule is given by the following rule for adding perpendicular components (in this case the three velocity components u_1, v_1 and w_1),

$$c_1^2 = u_1^2 + v_1^2 + w_1^2,$$

We will need to use this rule in Part 2.

Each impact of the molecule with the shaded face in Figure 3 reverses the x-component of velocity thus changing the x-component of its momentum from $+mu_1$ to $-mu_1$.

Therefore, the change of its momentum due to the impact

$$= \text{final momentum} - \text{initial momentum}$$

$$= (-mu_1) - (mu_1) = -2mu_1$$

The time, t, between successive impacts on this face is given by

$$t = \frac{\text{the total distance to the opposite face and back}}{x\text{-component of velocity}} = \frac{2l_x}{u_1}$$

Using Newton's second law therefore gives:

$$\text{the force on the molecule} = \frac{\text{change of momentum}}{\text{time taken}} = \frac{-2mu_1}{(2l_x/u_1)} = \frac{-mu_1^2}{l_x}$$

Since the force F_1 of the impact on the surface is equal and opposite to the force on the molecule in accordance with Newton's third law, then $F_1 = \frac{+mu_1^2}{l_x}$

As pressure = force/area, the pressure p_1 of the molecule on the surface is given by

$$p_1 = \frac{\text{force}}{\text{area of the shaded face } (l_y \times l_z)} = \frac{mu_1^2}{l_x \times l_y \times l_z} = \frac{mu_1^2}{V}$$

where V = the volume of the box = $l_x \times l_y \times l_z$

Part 2

For N molecules in the box moving at different velocities, the total pressure p is the sum of the individual pressures $p_1, p_2, p_3, \ldots p_N$ where each subscript refers to each molecule.

$$\text{Hence } p = \frac{mu_1^2}{V} + \frac{mu_2^2}{V} + \frac{mu_3^2}{V} + \ldots + \frac{mu_N^2}{V} = \frac{m}{V}(u_1^2 + u_2^2 + u_3^2 + \ldots + u_N^2)$$

$$= \frac{Nm\overline{u^2}}{V}$$

where $\overline{u^2} = (u_1^2 + u_2^2 + u_3^2 + \ldots + u_N^2)/N$

As the motion of the molecules is random, there is no preferred direction of motion. The equation above could equally well have been derived in terms of the y-components of velocity $v_1, v_2, v_3 \ldots v_N$ or the z-components of velocity $w_1, w_2, w_3 \ldots w_N$.

i.e. $$p = \frac{Nm\overline{v^2}}{V} \quad \text{where } \overline{v^2} = (v_1^2 + v_2^2 + v_3^2 + \ldots + v_N^2)/N$$

$$p = \frac{Nm\overline{w^2}}{V} \quad \text{where } \overline{w^2} = (w_1^2 + w_2^2 + w_3^2 + \ldots + w_N^2)/N$$

Therefore

$$p = \frac{Nm}{3V}(\overline{u^2} + \overline{v^2} + \overline{w^2})$$

The note below shows that, because the motion of the molecules is random, **the root mean square speed** of the gas molecules is given by the equation

$$c^2_{\text{r.m.s.}} = \overline{u^2} + \overline{v^2} + \overline{w^2}$$

Hence $$p = \frac{Nmc^2_{\text{r.m.s.}}}{3V} \quad \text{or} \quad pV = \frac{1}{3}Nmc^2_{\text{r.m.s.}}$$

Note

As explained in Part 1, the speed c of each molecule is related to its velocity components according to equations of the form:

$$c_1^2 = u_1^2 + v_1^2 + w_1^2, \quad c_2^2 = u_2^2 + v_2^2 + w_2^2, \quad c_3^2 = u_3^2 + v_3^2 + w_3^2, \ldots$$
$$\ldots c_N^2 = u_N^2 + v_N^2 + w_N^2$$

The root mean square speed of the molecules, $c_{\text{r.m.s.}}$ is defined by:

$$c^2_{\text{r.m.s.}} = (c_1^2 + c_2^2 + c_3^2 + \ldots + c_N^2)/N$$
$$= (u_1^2 + v_1^2 + w_1^2 + u_2^2 + v_2^2 + w_2^2 + u_3^2 + v_3^2 + w_3^2 + \ldots + u_N^2 + v_N^2 + w_N^2)/N$$
$$= \overline{u^2} + \overline{v^2} + \overline{w^2}$$

Pressure and density

In the kinetic theory equation $pV = \frac{1}{3}Nmc^2_{\text{r.m.s.}}$, Nm is the total mass of gas present because N is the number of molecules and m is the mass of each molecule.

Therefore the density of the gas, $\rho = \frac{\text{mass}}{\text{volume}} = \frac{Nm}{V}$

Rearranging the kinetic theory equation therefore gives $p = \frac{1}{3} \times \frac{Nm}{V} \times c^2_{\text{r.m.s.}}$

$$= \frac{1}{3}\rho c^2_{\text{r.m.s.}}$$

$$\boldsymbol{p = \frac{1}{3}\rho c^2_{\text{r.m.s.}}}$$

Summary test 19.3

$N_A = 6.02 \times 10^{23}\,\text{mol}^{-1}$,
$R = 8.31\,\text{J mol}^{-1}\,\text{K}^{-1}$

1 Explain in molecular terms why the pressure of a gas in a sealed container increases when its temperature is raised.

2 Calculate the r.m.s. speed of the molecules of a gas when its pressure is 1.10×10^5 Pa and its density is $1.20\,\text{kg m}^{-3}$.

3 A sealed flask of volume $1.50 \times 10^{-4}\,\text{m}^3$ contains oxygen gas at a pressure of 1.2×10^5 Pa at a temperature of 15 °C. Calculate:

a the number of moles of oxygen in the flask,

b the r.m.s. speed of the oxygen molecules.

Molar mass of oxygen = 0.032 kg

4 An ideal gas of molar mass 0.028 kg is in a container of volume $0.037\,\text{m}^3$ at a pressure of 100 kPa and a temperature of 300 K. Calculate

a the number of moles,

b the mass of gas present,

c the r.m.s. speed of the molecules of the gas.

19.4 Thermodynamics of ideal gases

Learning outcomes

On these pages you will learn to:

- recall that the Boltzmann constant $k = R/N_A$ and express the ideal gas equation as $pV = NkT$
- deduce that the average translational kinetic energy of a molecule is proportional to T using the equations $pV = \frac{1}{3}Nmc^2_{\text{r.m.s.}}$ and $pV = NkT$
- use the first law of thermodynamics, expressed in terms of the increase in internal energy, the heating of the system and the work done on the system ($\Delta U = q + W$), to solve problems

Molecules and kinetic energy

The mean **kinetic energy of a molecule of an ideal gas**

$$= \frac{\text{total kinetic energy of all the molecules}}{\text{total number of molecules}}$$

$$= (\tfrac{1}{2}mc_1{}^2 + \tfrac{1}{2}mc_2{}^2 + \tfrac{1}{2}mc_3{}^2 + \ldots + \tfrac{1}{2}mc_N{}^2)/N$$

$$= \tfrac{1}{2}m(c_1{}^2 + c_2{}^2 + c_3{}^2 + \ldots + c_N{}^2)/N = \tfrac{1}{2}mc^2_{\text{r.m.s.}}$$

The higher the temperature of a gas, the greater is the mean kinetic energy of a molecule of the gas.

For an ideal gas, by assuming the mean kinetic energy of a molecule $\frac{1}{2}mc^2_{\text{r.m.s.}} = \frac{3}{2}kT$, where $k = R/N_A$, then $3kT = mc^2_{\text{r.m.s.}}$.

Substituting $3kT$ for $mc^2_{\text{r.m.s.}}$ in the kinetic theory equation $pV = \frac{1}{3}Nmc^2_{\text{r.m.s.}}$ therefore gives $pV = \frac{1}{3}N \times 3kT = NkT$

As $Nk = NR/N_A = nR$, we then obtain the ideal gas equation $pV = nRT$. So we have derived the ideal gas equation (which is an experimental law) from the kinetic theory equation by assuming that the mean kinetic energy of an ideal gas molecule $= \frac{3}{2}kT$. Therefore we can say for an ideal gas at thermodynamic temperature T, **the mean kinetic energy of a molecule of the gas = $\frac{3}{2}kT$, where $k = \frac{R}{N_A}$.**

The constant k is called the Boltzmann constant.
Its value ($= \frac{R}{N_A}$) is $1.38 \times 10^{-23}\,\text{J K}^{-1}$

Note

- The total kinetic energy of 1 mole of an ideal gas $= N_A \times \frac{3}{2}kT = \frac{3}{2}RT$ (as $k = \frac{R}{N_A}$),
- The total kinetic energy of n moles of an ideal gas $= n \times \frac{3}{2}RT = \frac{3}{2}nRT$

The total kinetic energy of n moles of an ideal gas = $\frac{3}{2}nRT$

Note

The idea that the mean kinetic energy of a particle is proportional to the thermodynamic temperature can be applied in many other situations, for example to explain why conduction electrons can't escape from a metal at room temperature. See Topic 20.2.

Worked example

$N_A = 6.02 \times 10^{23}\,\text{mol}^{-1}$, $k = 1.38 \times 10^{-23}\,\text{J K}^{-1}$.

Calculate the r.m.s. speed of oxygen molecules at 0 °C.

The molar mass of oxygen = $0.032\,\text{kg mol}^{-1}$

Solution $T = 273\,\text{K}$

The mass of an oxygen molecule, $m = \dfrac{0.032}{6.02 \times 10^{23}}$

$= 5.3 \times 10^{-26}\,\text{kg}$

Rearranging $\frac{1}{2}mc^2_{\text{r.m.s.}} = \frac{3}{2}kT$ gives

$$\therefore \quad c^2_{\text{r.m.s.}} = \frac{3kT}{m} = \frac{3 \times 1.38 \times 10^{-23} \times 273}{5.3 \times 10^{-26}}$$

$= 2.13 \times 10^5\,\text{m}^2\,\text{s}^{-2}$

$\therefore$ root mean square speed, $c_{\text{r.m.s.}} = (2.13 \times 10^5)^{½} = 460\,\text{m s}^{-1}$

Using the first law of thermodynamics

As explained in Topic 18.1, the first law of thermodynamics tells us that the change of internal energy of a system is equal to the sum of the work done on or by it and the heat transfer to or from it.

The internal energy U of an ideal gas is due to the kinetic energy of its molecules. No internal forces of attraction exist between 'ideal gas' molecules so an ideal gas does not possess molecular potential energy. As explained above, for n moles of an ideal gas at temperature T, the total kinetic energy $= \frac{3}{2}nRT$. Therefore, its internal energy $= \frac{3}{2}nRT$.

To change the temperature of an ideal gas, its internal energy must be changed. To do this, work must be done on or by the gas or heat must be transferred to or from it.

Work W is done on a gas when its volume is reduced (i.e. it is compressed). At constant volume, no work is done because the forces due to pressure in the gas do not move the container walls.

- When a gas is **compressed**, work is done on it by the applied forces that push the pressure forces back.
- When a gas **expands**, the forces due to its pressure push the container walls back.

Consider when a gas in a piston expands at constant pressure as shown in Figure 1. When the piston moves a small distance Δs outwards:

- The work done W by the air in the cylinder is given by $W = F\Delta s$, where F is the force of the air on the piston.

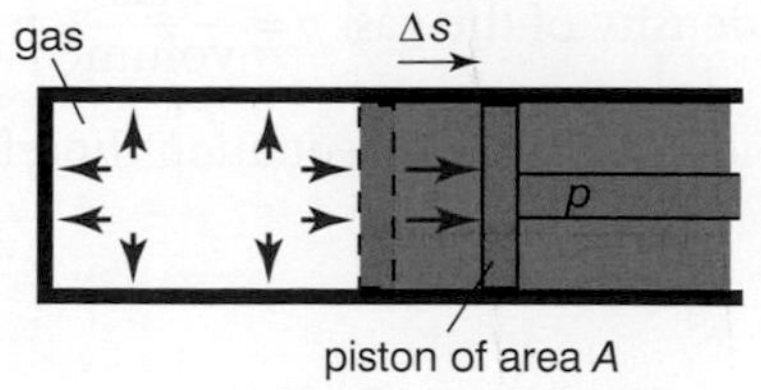

Figure 1

- The increase of volume of the air in the cylinder $V = A\Delta s$, where A is the cross-sectional area of the piston.
- Since pressure = force per unit area, then $F = pA$, where p is the air pressure (assumed constant as the movement of the piston is small). Hence the work done $W = F\Delta s = pA\ \Delta s = p\Delta V$.

More generally:

When a gas expands by a volume ΔV at constant pressure p, the work done by the gas, $W = p\Delta V$

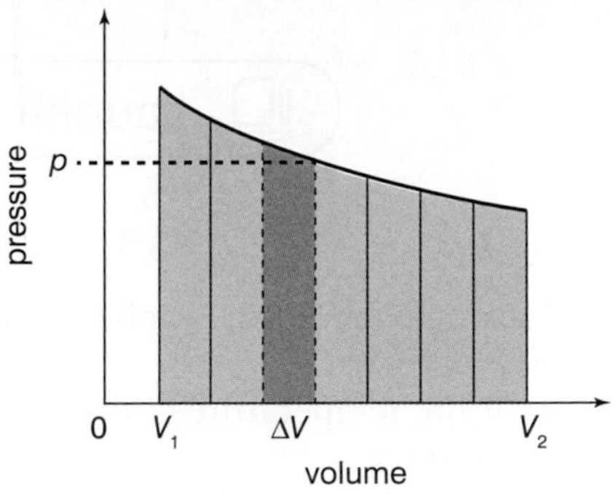

Figure 2 *Estimating work done*

Notes

1 The equation can be applied if the gas is compressed at constant pressure. In this case, energy is transferred to the gas by the force applied to it.

2 For an ideal gas, since $pV = nRT$, where n is the number of moles, then at constant pressure, $p\Delta V = nR\Delta T$. Therefore, the work done $= p\Delta V = nR\Delta T$.

3 **If the pressure changes,** the work done can be estimated from a graph of pressure against volume as shown in Figure 2. The work done is given by the area under the curve. This is because the work done in each small change of volume ΔV is represented by the area of the corresponding strip under the curve; so the total work done when the volume changes from V_1 to V_2 is represented by the total area under the curve from V_1 to V_2.

4 **Heat transfer** q to or from a gas may or may not cause a change of its temperature, depending on whether or not work is done on or by the gas. For example, if 1000 J of work is done on a gas by compressing it and 1200 J of heat energy is transferred from it by cooling it at the same time, its internal energy will decrease by 200 J and its temperature will decrease.

Summary test 19.4

$N_A = 6.02 \times 10^{23}\,\text{mol}^{-1}$, $R = 8.31\,\text{J K}^{-1}$, $k = 1.38 \times 10^{-23}\,\text{J K}^{-1}$.

1 The molar mass of oxygen is 0.032 kg. A cylinder of volume 0.025 m^3 contains oxygen gas at a pressure of 120 kPa and a temperature of 373 K. Calculate: **a** the number of moles of oxygen in the cylinder, **b** the mean kinetic energy of an oxygen molecule in the cylinder, **c** the total kinetic energy of all the gas molecules in the container.

2 Calculate: **a** the mean kinetic energy of a hydrogen molecule at 0 °C, **b** the root mean square speed of a hydrogen molecule at 0 °C. The molar mass of hydrogen gas = 0.002 kg.

3 Air consists mostly of nitrogen and oxygen in proportions 1 : 4 by mass.

a Explain why the mean kinetic energy of a nitrogen molecule in air is the same as that of an oxygen molecule in the same sample of air.

b Show that the r.m.s. speed of a nitrogen molecule in air is 1.07 × that of an oxygen molecule in the same sample of air.

Molar mass nitrogen = 0.028 kg; oxygen = 0.032 kg

4 An ideal gas in a cylinder at an initial temperature of 290 K is compressed and cooled from a volume of 0.025 m^3 to a volume of 0.019 m^3 at a constant pressure of 1.45×10^5 Pa. Calculate:

a **i** the number of moles of the gas,
ii the temperature of the gas after the change,

b the work done on the gas,

c the decrease of internal energy of the gas,

d the heat transferred from the gas.

Chapter Summary

The ideal gas law $pV = nRT$, where n is the number of moles of gas, T is the absolute temperature and R is the molar gas constant.

The Avogadro constant, N_A is defined as the number of atoms in 12 grams of the carbon isotope $^{12}_{6}C$

One mole of a substance consisting of identical particles is the quantity of substance that contains N_A particles of the substance.

The molar mass of a substance is the mass of one mole of that substance.

The kinetic theory of gases equation $pV = \frac{1}{3}Nmc^2_{\text{r.m.s.}}$, where $c_{\text{r.m.s.}}$ is the root mean square speed of the gas molecules

The mean kinetic energy of a molecule in a gas at absolute temperature $T = \frac{3}{2}kT$, where k is the Boltzmann constant ($= R/N_A$)

19 Exam-style and Practice Questions

Launch additional digital resources for the chapter

$N_A = 6.02 \times 10^{23}\,\text{mol}^{-1}$, $R = 8.3\,\text{J}\,\text{mol}^{-1}\,\text{K}^{-1}$

1 a In molecular terms, explain why the pressure of a gas increases:

i if the temperature of the gas is raised at constant volume,

ii if the amount of gas is increased at constant volume.

b Sketch a graph to show how the pressure of a gas in a sealed cylinder increases with the absolute temperature of the gas if the cylinder contains:

i 1 mole of gas,

ii 2 moles of gas.

2 a Assuming that air at atmospheric pressure consists of 80% nitrogen and 20% oxygen, show that the molar mass of air is 0.029 kg.

The molar mass of nitrogen = 0.028 kg,
The molar mass of oxygen = 0.032 kg.

b A rectangular room has a length of 5.0 m, a width of 4.0 m and a height of 2.5 m. The air in the room has a pressure of 100 kPa and a temperature of 290 K. Calculate:

i the number of moles of air in the room,

ii the number of gas molecules in the room.

3 A vehicle air bag inflates rapidly when an impact causes the production and release of a large quantity of nitrogen in a chemical reaction. In a test of an air bag, the bag inflates to a volume of $1.2\,\text{m}^3$ and a pressure of 103 kPa at a final temperature of 280 K. Calculate:

a the number of moles of gas in the bag,

b the initial pressure of the gas if it was released from a container of volume $5.6 \times 10^{-4}\,\text{m}^3$ at the same temperature.

4 A hot air balloon rises from the ground because it is partly filled with helium which is less dense than air.

a A certain hot air balloon contains $370\,\text{m}^3$ of helium gas at a temperature of 300 K. Calculate:

i the number of moles of helium gas in the balloon when the pressure of the gas is 110 kPa,

ii the density of the gas.

b The helium gas in the balloon fills most of the balloon from the top down. Heating the gas causes it to expand and displace some of the air in the lower part of the balloon.

i Explain why heating the gas causes its density to decrease.

ii Explain why the balloon ascends when the helium gas is heated.

The molar mass of helium = $4.0 \times 10^{-3}\,\text{kg}$

5 A hollow steel cylinder fitted with a pressure gauge is used to store nitrogen gas. The cylinder has an internal volume of $5.0 \times 10^{-3}\,\text{m}^3$.

The molar mass of nitrogen = 0.028 kg

a i Calculate the mass of nitrogen gas that must be stored in the cylinder at 300 K to give a gas pressure of 150 kPa.

ii Describe how the mass of gas remaining in the cylinder could be estimated from the pressure reading of the pressure gauge.

b The cylinder is fitted with a valve that releases gas from the cylinder if the gas pressure exceeds 160 kPa.

Calculate the maximum mass of nitrogen that could be stored in the cylinder at 350 K.

6 Argon is an inert gas. The gas is composed of single atoms that do not combine with each other. A certain light bulb contains $8.0 \times 10^{-5}\,\text{m}^3$ of argon gas. When the light bulb was at a temperature of 300 K, the pressure of the gas in the light bulb was 15 kPa.

a i Calculate the number of moles of argon gas in the light bulb.

ii Calculate the number of argon atoms in the light bulb.

iii The mean kinetic energy of an argon atom at 300 K is $6.2 \times 10^{-21}\,\text{J}$. Calculate the speed of an argon atom which has $6.2 \times 10^{-21}\,\text{J}$ of kinetic energy.

b When the light bulb was switched on, the temperature of the gas in the bulb increased to 350 K. Calculate:

i the pressure in the light bulb at this temperature,

ii the increase of kinetic energy of the gas molecules in the light bulb.

The molar mass of argon = 0.040 kg.

7 Helium gas released with oil from an oil well was collected and stored in a sealed underground cavern of volume $26\,000\,\text{m}^3$ at a temperature of 280 K. As a result of storing the helium gas, the pressure of the gas in the cavern increased from 100 kPa to 125 kPa.

a Calculate:

i the mass of helium gas stored in the cavern,

ii the mass of a helium molecule.

b The mean kinetic energy of a helium gas molecule at 280 K is $5.8 \times 10^{-21}\,\text{J}$. Calculate:

i the speed of a helium gas molecule which has $5.8 \times 10^{-21}\,\text{J}$ of kinetic energy,

ii the kinetic energy of all the helium gas molecules in the cavern at 280 K.

The molar mass of helium = 0.004 kg

8 In a vehicle braking system, pressure applied by the driver to the footbrake is transmitted from the master cylinder to the brake pads via sealed pipes filled with brake fluid.

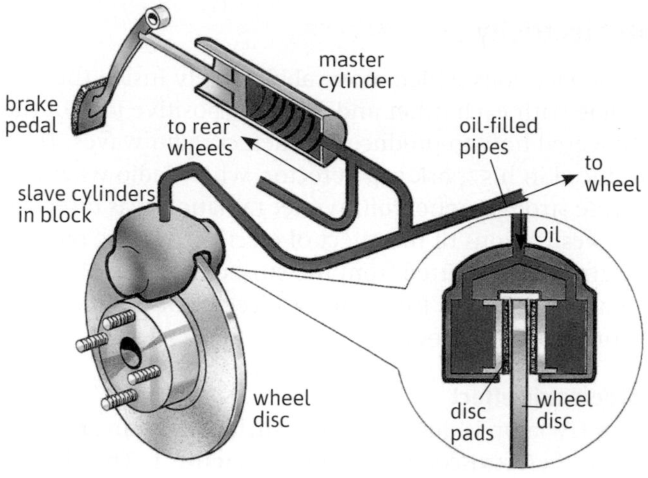

Figure 8.1

a Explain why a bubble of air in the brake fluid makes the brakes less effective than they should be.

b A vehicle brake system is filled with a volume of $1.2 \times 10^{-4}\,m^3$ of brake fluid. When a force of 20 N is applied to the piston in the master cylinder, the pressure in the brake system increases from 105 kPa to 155 kPa.

i Calculate the area of cross-section of the master cylinder.

ii Air leaks into the system when the brake fluid is replaced. When a force of 20 N is applied to the piston in the master cylinder, it travels 5 mm more than in **a** and the pressure in the system increases from 105 kPa to 145 kPa. Show that the volume of the air bubble before the brakes were applied was $7.3 \times 10^{-6}\,m^3$.

iii The temperature of the brake fluid was 290 K. Calculate the mass of air in the bubble.

The molar mass of air = 0.029 kg

9 An ideal gas in a sealed container of volume $2.71 \times 10^{-4}\,m^3$ is at a pressure of 125 kPa when the temperature is 20.0 °C.

a Calculate the number of moles of the gas.

b The temperature of the gas is reduced to 11.5 °C.

i Calculate the change of internal energy of the gas as a result of this change of temperature.

ii Calculate the pressure of the gas at 11.5 °C.

c i Draw a graph to show how the pressure of the gas in the sealed container changes between 0 °C and 20 °C.

ii On the same graph, show how the pressure of the gas would change if the number of moles in the container is half of the value you obtained in **a**.

10a i Define the *root mean square (r.m.s.) speed* of the molecules of a gas.

ii Calculate the r.m.s. speed of the molecules of an ideal gas which has a molar mass of 0.028 kg and is at a temperature of 290 K.

iii Show that the mean square speed of the molecules of an ideal gas is proportional to the absolute temperature of the gas.

b The graph in Figure 10.1 shows how the pressure and volume of the fuel in a cylinder of a petrol engine changes during part of an engine cycle.

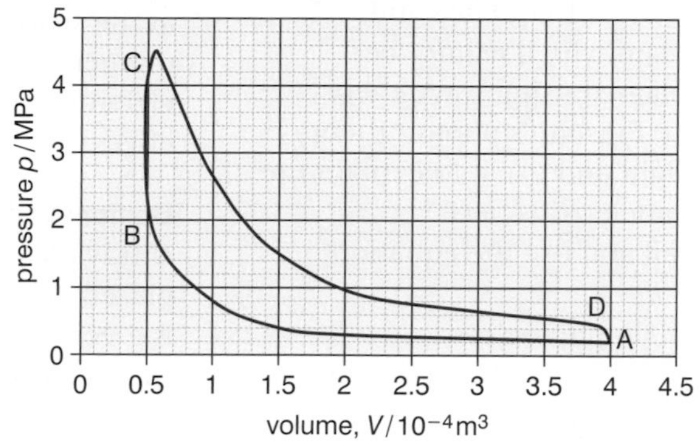

Figure 10.1

In simplified terms:

- From A to B, compression occurs as the fuel (petrol and air) in the cylinder is compressed by the piston in the cylinder.
- For B to C, combustion of the fuel occurs rapidly after the fuel is ignited at B by a spark from a spark plug. The pressure in the cylinder increases until no fuel remains in the cylinder.
- From C to D, expansion of the gas occurs as the piston is forced to reverse by the high pressure in the cylinder enabling the engine to do work until the exhaust valve opens at D.
- From D to A, the gaseous combustion products are expelled from the cylinder and the pressure drops back to its initial value.

i Explain why no work is done by the engine between B and C.

ii Explain why the internal energy of the gas in the cylinder decreases between C and D.

iii Show that the net work done by the engine in one cycle is approximately 330 J.

iv Combustion between B and C releases 790 J as thermal energy. Calculate the efficiency of the cycle.

v State two reasons why the efficiency is not higher than the value you calculated in **iv**.

20 Quantum theory

20.1 Photoelectricity

Learning outcomes

On these pages you will learn to:

- describe the photoelectric effect and its principal features
- recall the significance of threshold frequency and explain why it cannot be explained using wave theory
- state Einstein's photon theory of light and use it to explain the photoelectric effect
- recall and use the equation $E = hf$
- explain photoelectric phenomena in terms of photon energy and work function energy

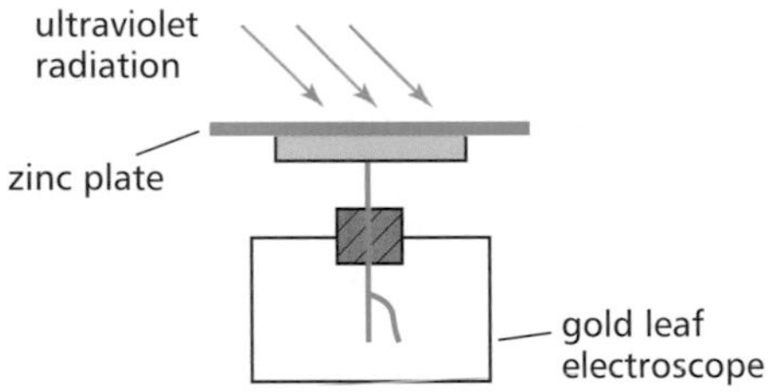

***Figure 1** Demonstrating photoelectricity*

Note

The **wavelength** of the incident light must be less than a **maximum** value equal to the speed of light/the threshold frequency. This value of the wavelength is called the **threshold wavelength** of the light.

The discovery of photoelectricity

A metal contains conduction electrons which move about freely inside the metal. These electrons collide with each other and with the positive ions of the metal. Heinrich Hertz discovered how to produce and detect radio waves. He found that the sparks produced in his spark gap detector, when radio waves were being transmitted, were stronger when ultraviolet radiation was directed at the spark gap. Further investigations of the effect of electromagnetic radiation on metals showed that electrons are emitted from the surface of a metal when electromagnetic radiation above a certain frequency is directed at the metal. This effect is known as the **photoelectric effect**.

Demonstration of the photoelectric effect

Ultraviolet radiation from a UV lamp is directed at the surface of a zinc plate placed on the cap of a gold leaf electroscope, as shown in Figure 1. This device is a very sensitive detector of charge. When it is charged, the thin gold leaf of the electroscope rises: it is repelled from the metal 'stem', because they are both charged with a like charge.

- If the electroscope is **charged negatively**, the leaf rises and stays in position. However, if ultraviolet light is directed at the zinc plate, the leaf gradually falls. The leaf falls because conduction electrons at the zinc surface leave the zinc surface when ultraviolet light is directed at it. The emitted electrons are referred to as **photoelectrons**.
- If the electroscope is **charged positively**, the leaf rises and stays in position, regardless of whether or not ultraviolet light is directed at the zinc plate. This is because the conduction electrons in the zinc plate are held on the zinc plate, as the plate is charged positive and electrons carry negative charge.

Puzzling problems

The following observations were made about **photoelectricity** after Hertz's discovery. These observations were a major problem, because they could not be explained using the wave theory of electromagnetic radiation.

- Photoelectric emission of electrons from a metal surface does not take place if the frequency of the incident electromagnetic radiation is below a certain value, known as the **threshold frequency**. This minimum frequency depends on the type of metal.
- The number of electrons emitted per second is proportional to the **intensity** of the incident radiation, provided the frequency is greater than the threshold frequency. If the frequency of the incident radiation is less than the threshold frequency, no photoelectric emission from that metal surface can take place, no matter how intense the incident radiation is.
- Photoelectric emission occurs without delay, as soon as the incident radiation is directed at the surface, provided the frequency of the radiation exceeds the threshold frequency. No matter how weak the intensity of the incident radiation is, electrons are emitted as soon as the source of radiation is switched on.

The wave theory of light cannot explain either the existence of a threshold frequency, or why photoelectric emission occurs without delay. According to wave theory, each conduction electron at the surface of a metal should gain some energy from the incoming waves, regardless of how many waves arrive each second. Wave theory therefore predicted that:

- Emission should take place with waves of any frequency.
- Emission would take longer using low intensity waves than using intensity waves.

The discovery of radio waves and X-rays confirmed the prediction by Maxwell that electromagnetic radiation exists beyond the known spectrum from ultraviolet to infrared radiation. Using Maxwell's theory of electromagnetic waves, physicists were very successful in predicting the properties of electromagnetic waves until the discovery of the photoelectric effect.

Einstein's explanation of photoelectricity

The photon theory of light was put forward by Einstein in 1905 to explain photoelectricity. Einstein assumed that light is composed of wave packets, or **photons**, each of energy equal to hf, where f is the frequency of the light and h is the Planck constant. The accepted value for h is 6.63×10^{-34} J s.

Energy of a photon = hf

For electromagnetic waves of wavelength λ, the energy of each photon $E = hf = \frac{hc}{\lambda}$ where c is the speed of the electromagnetic waves.

To explain photoelectricity, Einstein said that:

- When light is incident on a metal surface, an electron at the surface absorbs a **single** photon from the incident light and therefore gains energy equal to hf, where hf is the energy of a light photon.
- An electron can leave the metal surface, if the energy gained from a single photon exceeds **the work function,** ϕ, of the metal. This is the minimum energy needed by an electron to escape from the metal surface.

Hence the maximum kinetic energy of an emitted electron:

$$E_{K(max)} = hf - \phi$$

Emission can take place from a surface at zero potential, provided $E_{Kmax} > 0$, i.e. $hf > \phi$. Thus the threshold frequency of the metal:

$$f_{min} = \frac{\phi}{h}$$

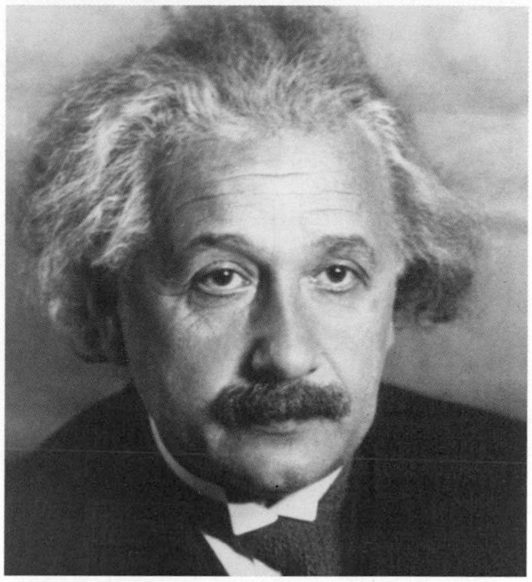

Figure 2 Albert Einstein 1879–1955

In 1905 Einstein published three papers, each of which changed existing ideas. His paper on Brownian motion showed that atoms must exist, and he established the photon theory of light in the second paper. Perhaps he is best remembered for his theories of relativity which he stated in the third paper.

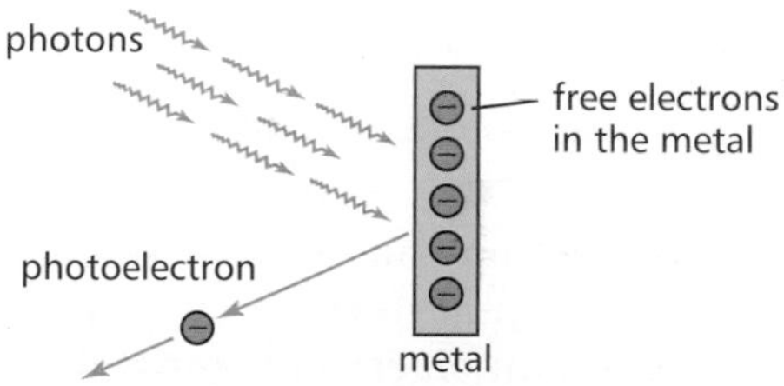

Figure 3 Explaining photoelectricity

Summary test 20.1

$h = 6.63 \times 10^{-34}$ J s, $c = 3.00 \times 10^{8}$ m s^{-1}

1 a What is meant by 'photoelectric emission' from a metal surface?

b Explain why photoelectric emission from a metal surface only takes place if the frequency of the incident radiation is greater than a certain value.

2 a Calculate the frequency and energy of a photon of wavelength:
i 450 nm, **ii** 1500 nm.

b A metal surface at zero potential emits electrons from its surface if light of wavelength 450 nm is directed at it, but not if light of wavelength 650 nm is used. Explain why photoelectric emission happens with light of wavelength 450 nm but not with light of wavelength 650 nm.

3 The work function of a certain metal plate is 1.1×10^{-19} J. Calculate:

a the threshold frequency of incident radiation,

b the maximum kinetic energy of photoelectrons emitted from this plate when light of wavelength 520 nm is directed at the metal surface.

4 Light of wavelength 635 nm is directed at a metal plate at zero potential. Electrons are emitted from the plate with a maximum kinetic energy of 1.5×10^{-19} J. Calculate:

a the energy of a photon of this wavelength,

b the work function of the metal,

c the threshold frequency of electromagnetic radiation incident on this metal.

20.2 More about photoelectricity

Learning outcomes

On these pages you will learn to:

- explain why the maximum photoelectric energy is independent of intensity, whereas the photoelectric current is proportional to intensity
- recall, use and explain the significance of $hf = \Phi + \frac{1}{2}mv_{max}^2$

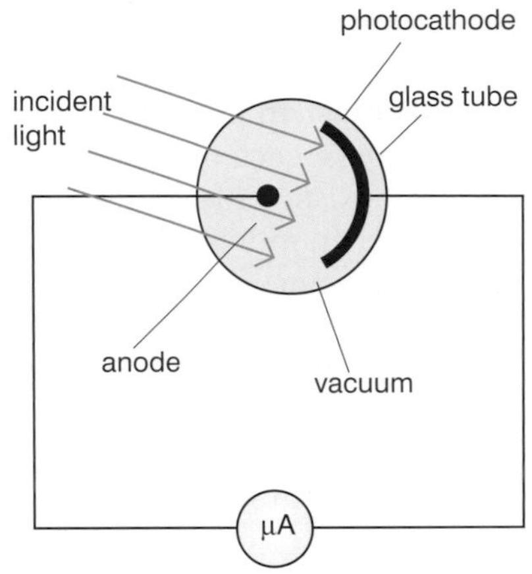

***Figure 1** Using a vacuum photocell*

The electron volt

The electron volt (eV) is a unit of energy equal to the work done when an electron is moved through a p.d. of 1 V.

For a charge q moved through a p.d., V:

$$\text{Work done} = qV$$

Therefore, the work done when an electron moves through a potential difference of 1 V is equal to **1.6×10^{-19} J** ($= 1.6 \times 10^{-19}$ C $\times$ 1 V). This amount of energy is defined as **1 electron volt.**

Examples

The work done on:

- an electron when it moves through a potential difference of 1000 V = 1000 eV,
- an ion of charge +2e when it moves through a potential difference of 10 V = 20 eV.

More about conduction electrons

The **average kinetic energy** of a conduction electron in a metal depends on the **temperature** of the metal. As the conduction electrons move about at random in the metal, they can be likened to the molecules of a gas. The average kinetic energy of a gas molecule is proportional to the absolute temperature of the gas. It can be shown that the average kinetic energy of an conduction electron in a metal at 300 K is therefore about 6×10^{-21} J.

- The **work function of a metal** is the **minimum** energy needed by a conduction electron to escape from the metal surface when the metal is at zero potential. The work function of a metal is of the order of 10^{-19} J, which is about 20 times greater than the average kinetic energy of a conduction electron in a metal at 300 K. In other words, a conduction electron in a metal at about 20 °C does not have sufficient kinetic energy to leave the metal.
- When a conduction electron **absorbs a photon**, its kinetic energy increases by an amount equal to the energy of the photon. Provided the energy of the photon exceeds the work function of the metal, the conduction electron can leave the metal. If the electron does not leave the metal, it collides repeatedly with other electrons and positive ions, and it quickly loses its extra kinetic energy.

Photoelectricity investigations

The vacuum photocell

A **vacuum photocell** is a glass tube that contains a metal plate, referred to as the **photocathode**, and a smaller metal electrode referred to as the **anode**. Figure 1 shows a vacuum photocell in a circuit. When light of frequency greater than the threshold frequency for the metal is directed at the photocathode, electrons emitted from the cathode transfer to the anode. The microammeter in the circuit can be used to measure the photoelectric current, which is proportional to the number of electrons per second that transfer from the cathode to the anode.

- For a photoelectric current I, the **number of photoelectrons per second** that transfer from the cathode to the anode is I/e, where e is the charge of the electron.
- The photoelectric current is proportional to the **intensity** of the light incident on the cathode. This is because the intensity of the incident light is a measure of the **energy per second** carried by the incident light, which is proportional to the number of photons per second incident on the cathode. Because each photoelectron must have absorbed one photon to escape from the metal surface, the number of photoelectrons emitted per second (i.e. the photoelectric current) is therefore proportional to the intensity of the incident light.
- The intensity of the incident light does **not** affect the maximum kinetic energy of a photoelectron. No matter how intense the incident light is, the energy gained by a photoelectron is due to the absorption of one photon only. Therefore, the maximum kinetic energy of a photoelectron is still given by $E_{Kmax} = hf - \phi$, as explained in Topic 20.1.

Measurement of the work function of a metal

Photoelectric emission can be stopped by making the photocathode sufficiently positive. Figure 2 shows how a potential divider can be used to make the photocathode increasingly positive, relative to the anode. As the potential difference is increased from zero, the microammeter reading decreases to zero. This happens because each photoelectron leaving the metal surface needs to do extra work, as the plate is at a positive potential. The kinetic energy of a photoelectron is therefore reduced, because each electron has to do work to overcome the attraction of the plate. When the microammeter reading is zero, photoelectric emission stops: this is because the kinetic energy of a photoelectron is reduced to zero before the electron can escape from the attraction of the plate.

- As explained earlier, the work done by a charged particle when it moves through a potential difference V is qV, where q is the charge of the particle. Therefore, to escape from a metal plate at positive potential V, the **extra work** needed to be done by a photoelectron is $\boldsymbol{eV}$, where e is the charge of the electron.
- At zero potential, the **maximum kinetic energy** of an emitted photoelectron is $\boldsymbol{hf - \phi}$, where f is the frequency of the incident radiation and ϕ is the work function of the metal.

Therefore, photoelectric emission is stopped when the potential V is such that

$$eV_S = hf - \phi$$

where V_S is the potential needed to stop emission.

In other words, the maximum kinetic energy of a photoelectron from a surface at zero potential:

$$E_{K(max)} = eV_S$$

By measuring V_S for different frequencies f, $E_{K(max)}$ can be calculated for each frequency; then a graph of $E_{K(max)}$ against frequency f can be plotted. Because $E_{K(max)} = hf - \phi$, the graph is a straight line with:

- a gradient h,
- a y-intercept equal to $-\phi$, and
- an x-intercept equal to the threshold frequency, $f_{min} = \dfrac{\phi}{h}$

The above equation can be written as $hf = \phi + \frac{1}{2}mv^2_{max}$ where v_{max} is the maximum speed of the emitted photoelectrons and $\frac{1}{2}mv^2_{max}$ is the maximum kinetic energy $E_{K(max)}$.

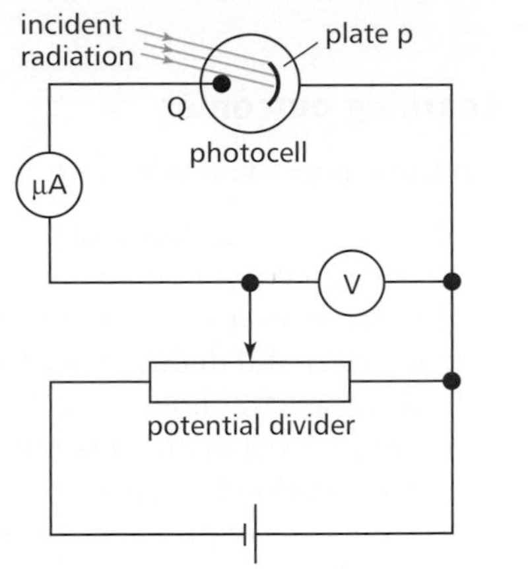

Figure 2 *Investigating photoelectricity*

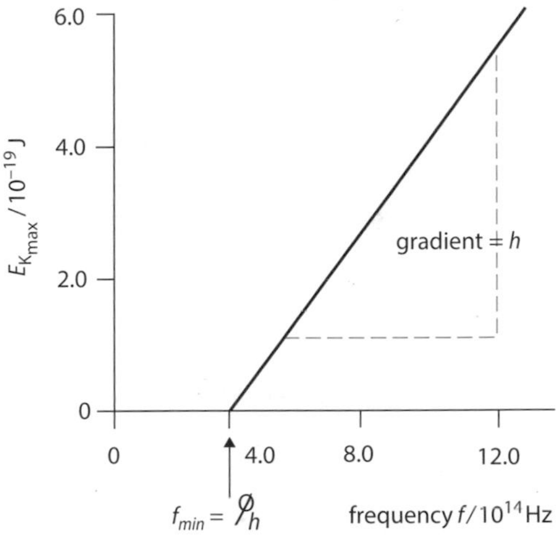

Figure 3 *A graph of $E_{K(max)}$ against frequency*

Summary test 20.2

$h = 6.63 \times 10^{-34}$ J s, $c = 3.00 \times 10^8$ m s^{-1}, $e = 1.6 \times 10^{-19}$ C

1 A vacuum photocell is connected to a microammeter. Explain the following observations:

  a When the cathode was illuminated with blue light of low intensity, the microammeter showed a non-zero reading.

  b When the cathode was illuminated with an intense red light, the microammeter reading was zero.

2 A vacuum photocell is connected to a microammeter. When light is directed at the photocell, the micro-ammeter reads 0.25 μA.

  a Calculate the number of photoelectrons emitted per second by the photocathode of the photocell.

  b Explain why the microammeter reading is doubled if the intensity of the incident light is doubled.

3 A narrow beam of light of wavelength 590 nm and of power 0.5 mW is directed at the photocathode of a vacuum photocell, which is connected to a micro-ammeter that reads 0.4 μA. Calculate:

  a the energy of a single light photon of this wavelength,

  b the number of photons per second incident on the photocathode,

  c the number of electrons emitted per second from the photocathode.

4 a Use Figure 3 to estimate:

  i the threshold frequency,

  ii the work function of the photocathode that gave the results used to plot the graph.

  b A metal surface has a work function of 1.9×10^{-19} J. Light of wavelength 435 nm is directed at the metal surface. Calculate the maximum kinetic energy of the photoelectrons emitted from this metal surface.

20.3 Energy levels and spectra

Learning outcomes

On these pages you will learn to:

- recognise the existence of discrete electron energy levels in isolated atoms (e.g. atomic hydrogen) and deduce how this leads to spectral lines
- distinguish between emission and absorption line spectra
- recall and solve problems using the relation $hf = E_1 - E_2$

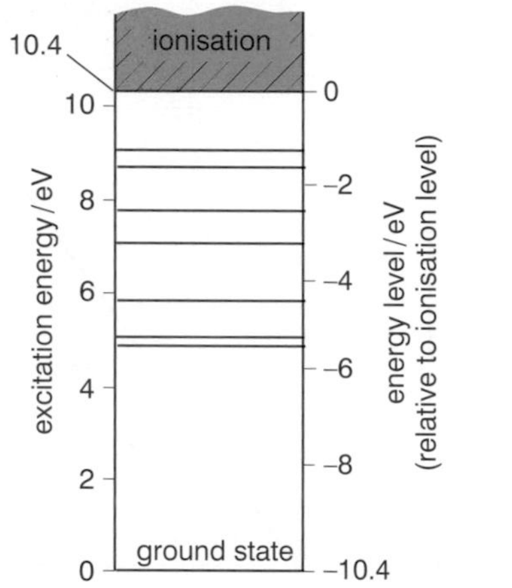

Figure 1 *The energy levels of the mercury atom*

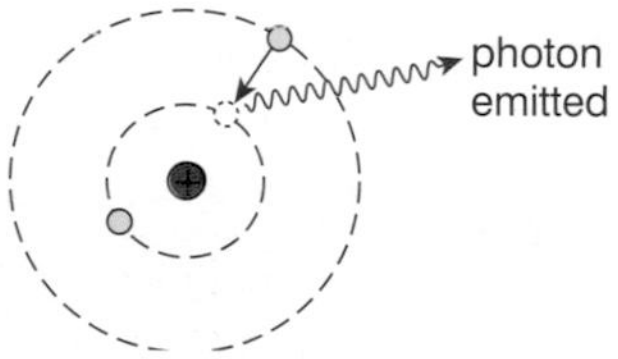

Figure 2 *De-excitation by photon emission*

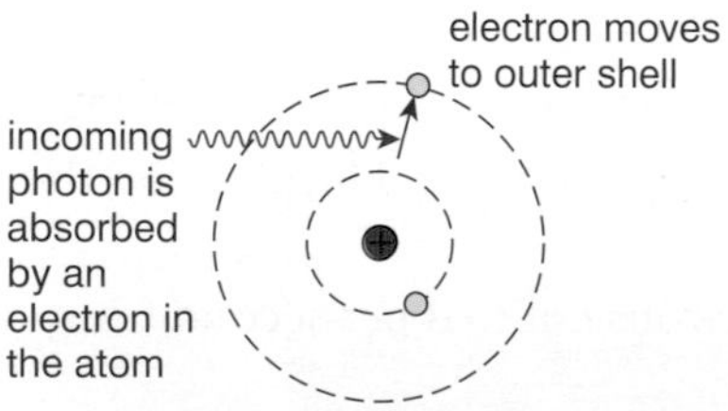

Figure 3 *Excitation by photon absorption*

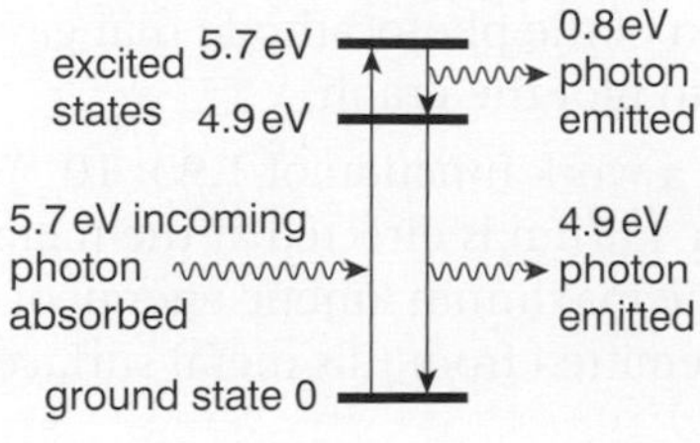

Figure 4 *Fluorescence*

Electrons in atoms

The electrons in an atom are trapped by the electrostatic force of attraction of the nucleus. They move about the nucleus in allowed orbits or 'shells' surrounding the nucleus. Thc energy of an electron in a shell is constant. An electron in a shell near the nucleus has less energy than an electron in a shell further away from the nucleus. Each shell can only hold a certain number of electrons. For example, the innermost shell (that is, the shell nearest to the nucleus) can only hold two electrons and the next nearest shell can only hold eight electrons.

Each type of atom has a certain number of electrons. For example, a helium atom has two electrons and a lithium atom has three electrons. Thus, in its lowest energy state:

- a helium atom has both electrons in the innermost shell,
- a lithium atom has two electrons in the innermost shell and one in the next shell.

The lowest energy state of an atom is called its **ground state**. When an atom in the ground state absorbs energy, one of its electrons moves to a shell at higher energy, so the atom is now in an **excited state**. We can use the excitation energy measurements to construct an energy level diagram for the atom, as shown in Figure 1. This shows the allowed energy values of the atom. Each allowed energy corresponds to a certain electron configuration in the atom. Note that the ionisation level may be considered as the 'zero' reference level for energy, instead of the ground state level.

De-excitation

Did you know that gases at low pressure emit light when they are made to conduct electricity? For example, a neon tube emits red-orange light when it conducts. The gas-filled tube used to measure excitation energies emits light when excitation occurs. This happens because the atoms absorb energy as a result of excitation by collision but they do not retain the absorbed energy permanently.

The electron configuration in an excited atom is unstable because an electron that moves to an outer shell leaves a vacancy in the shell it moves from. Sooner or later, the vacancy is filled by an electron from an outer shell transferring to it. When this happens, the electron emits a photon of energy equal to the energy lost by the electron. The atom therefore moves to a lower energy level. See Figure 2.

The energy of the photon is equal to the energy lost by the electron and therefore by the atom. For example, when a mercury atom at an excitation energy level of 4.9 eV de-excites to the ground state, it emits a photon of energy 4.9 eV.

An atom in an excited state may de-excite to the ground state indirectly. For example, a mercury atom at an excitation energy of 5.7 eV may de-excite:

- to the 4.9 eV level, emitting a 0.8 eV photon in the process, then
- to the ground state from the 4.9 eV level, emitting a 4.9 eV photon in this process.

In general, when an electron moves from energy level E_1 to a lower energy level E_2, **the energy of the emitted photon $hf = E_1 - E_2$**

Excitation using photons

An electron in an atom can absorb a photon and move to an outer shell where a vacancy exists – but only if the energy of the photon is exactly equal to the gain in the electron's energy (see Figure 3). In other words, the photon energy must be exactly equal to the difference between the final and initial energy levels of the atom. If the photon's energy is smaller or larger than the difference between the two energy levels, it will not be absorbed by the electron.

Fluorescence

An atom in an excited state can de-excite directly or indirectly to the ground state, regardless of how the excitation took place. Therefore, an atom can absorb photons of certain energies and then emit photons of the same or lesser energies. For example, a mercury atom in the ground state could be excited to its 5.7 eV energy level by absorbing a photon of energy 5.7 eV; the mercury could then de-excite to its 4.9 eV energy level by emitting a photon of energy 0.8 eV; then de-excite to the ground state by emitting a photon of energy 4.9 eV.

Figure 4 represents these changes on an energy level diagram.

This overall process explains why certain substances **fluoresce** or glow with visible light when they absorb ultraviolet radiation. Atoms in the substance absorb ultraviolet photons and become excited. When the atoms de-excite, they emit visible photons. When the source of ultraviolet radiation is removed, the substance stops glowing.

Line spectra

A rainbow is a natural display of the colours of the spectrum of sunlight. Raindrops split sunlight into a continuous spectrum of colours. Figure 5 shows the continuous spectrum produced by passing a beam of white light from a filament lamp through a glass prism. The wavelength of the light photons that produce the spectrum increases across the spectrum from deep violet at less than 400 nm to deep red at about 650 nm.

If we use a tube of glowing gas as the light source instead of a filament lamp, we see a spectrum of discrete lines of different colours, as shown in Figure 6. This type of spectrum is referred to as a **line emission spectrum**.

If the spectrum of light from a filament lamp is observed after passing it through a glowing gas, we see dark vertical lines in the continuous spectrum, as shown in Figure 7. The lines are due to absorption of photons of certain energies by gas atoms. This type of spectrum is referred to as a **line absorption spectrum**.

The wavelengths of the lines of a line spectrum of an element are characteristic of the atoms of that element. By measuring the wavelengths of a line spectrum, we can therefore identify the element that produced the light. No other element produces the same pattern of light wavelengths. This is because the energy levels of each type of atom are unique to that atom. So the photons emitted are characteristic of the atom.

- Each line in a line spectrum is due to light of a certain colour and therefore a certain wavelength.
- The photons that produce each line all have the same energy, which is different from the photons that produce any other line.
- Each photon is emitted when an atom de-excites due to one of its electrons moving from an inner shell.
- As explained above, if the electron moves from energy level E_1 to a lower energy level E_2, the energy of the emitted photon $hf = E_1 - E_2$.

For each wavelength λ, we can calculate the energy of a photon of that wavelength as its frequency $f = c / \lambda$, where c is the speed of light. Given the energy level diagram for the atom, we can therefore identify on the diagram the transition that causes a photon of that wavelength to be emitted.

Worked example

$c = 3.0 \times 10^8\,\mathrm{m\,s^{-1}}$, $e = 1.6 \times 10^{-19}\,\mathrm{C}$, $h = 6.63 \times 10^{-34}\,\mathrm{J\,s}$

A mercury atom de-excites from its 4.9 eV energy level to the ground state. Calculate the wavelength of the photon released.

Solution

$E_1 - E_2 = 4.9 - 0 = 4.9\,\mathrm{eV} = 4.9 \times 1.6 \times 10^{-19}\,\mathrm{J} = 7.84 \times 10^{-19}\,\mathrm{J}$

Therefore, $f = \dfrac{E_1 - E_2}{h} = \dfrac{7.84 \times 10^{-19}}{6.63 \times 10^{-34}} = 1.12 \times 10^{15}\,\mathrm{Hz}$

$\lambda = \dfrac{c}{f} = \dfrac{3.0 \times 10^8}{1.12 \times 10^{15}} = 2.52 \times 10^{-7}\,\mathrm{m}$

Figure 5 *Observing a continuous spectrum*

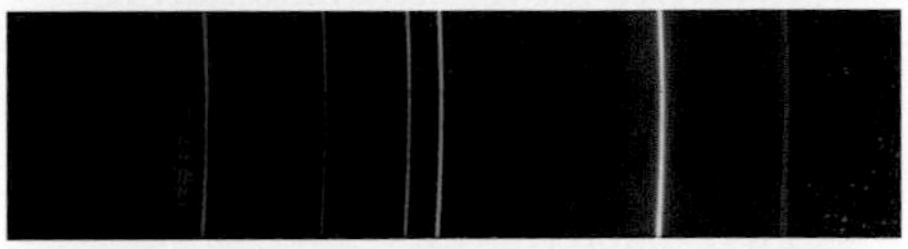

Figure 6 *A line emission spectrum*

Figure 7 *A line absorption spectrum*

Summary test 20.3

$e = 1.6 \times 10^{-19}\,\mathrm{C}$

1 Figure 1 shows some of the energy levels of the mercury atom.

  a Estimate the energy needed to excite the atom from the ground state to the highest excitation level shown in the diagram.

  b Mercury atoms in an excited state at 5.7 eV can de-excite directly or indirectly to the ground state. Show that the photons released could have six different energies.

2 a In terms of electrons, state two differences between excitation and de-excitation.

  b A certain type of atom has excitation energies of 1.8 eV and 4.6 eV.

    i Sketch an energy level diagram for the atom using these energy values.

    ii Calculate the possible photon energies from the atom when it de-excites from the 4.6 eV level, indicating on your diagram, by downward arrows, the energy change responsible for each photon energy.

3 Explain why the line spectrum of an element is unique to that element and can be used to identify it.

20.4 Wave–particle duality

Learning outcomes

On these pages you will learn to:

- show an understanding that the photoelectric effect provides evidence for a particulate nature of electromagnetic radiation while phenomena such as interference and diffraction provide evidence for a wave nature
- describe and interpret qualitatively the evidence provided by electron diffraction for the wave nature of particles
- recall and use the relation for the de Broglie wavelength $\lambda = h/p$

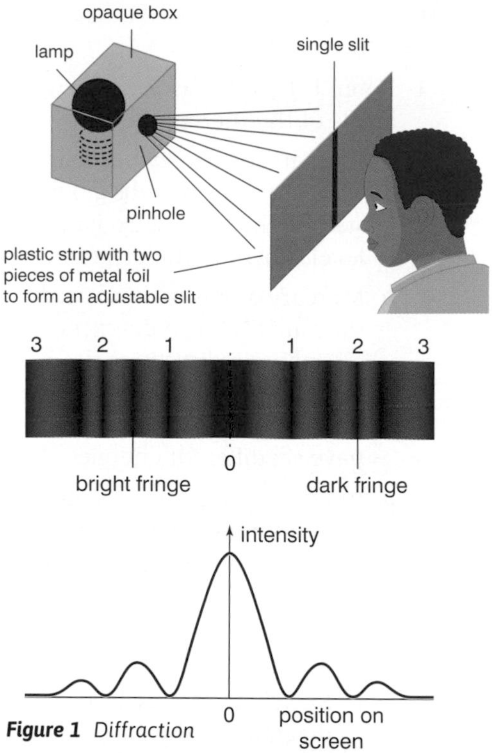

Figure 1 *Diffraction*

The dual nature of light

Light is part of the electromagnetic spectrum of waves. The theory of electromagnetic waves predicted the existence of electromagnetic waves beyond the visible spectrum. The subsequent discovery of X-rays and radio waves confirmed these predictions and seemed to show that the nature of light had been settled. Many scientists in the late nineteenth century reckoned that all aspects of physics could be explained using Newton's laws of motion, and the theory of electromagnetic waves. They thought that the few minor problem areas, such as photoelectricity, would be explained sooner or later using Newton's laws of motion and Maxwell's theory of electromagnetic waves. However, photoelectricity was not explained until Einstein put forward the radical theory that light consists of photons, which are 'particle-like' packets of electromagnetic waves. Light has a dual nature, in that it can behave as a wave or as a particle, according to circumstances.

- The **wave-like nature** is observed when **diffraction** of light takes place. This happens, for example, when light passes through a narrow slit. The light emerging from the slit spreads out in the same way as water waves spread out after passing through a gap. Although you were introduced to diffraction in Topic 8.3, you can see in Figure 1 how to observe diffraction of light. The light spreads out from the slit because the slit is very narrow. Notice that the pattern shows bright and dark fringes. This is because the wavelets from each wavefront that passes through the slit reinforce each other in certain directions only and cancel each other out in other directions.
- The **particle-like nature** is observed, for example, in the photoelectric effect. When light is directed at a metal surface and an electron at the surface absorbs a photon of frequency f, the kinetic energy of the electron is increased from a negligible value by hf. The electron can escape if the energy it gains from a photon exceeds the work function of the metal.

The particle-like nature of electromagnetic radiation was reinforced when it was discovered that X-rays can undergo collisions with electrons, causing the electrons to scatter by transferring momentum to them. Thus photons must possess momentum even though they are massless! From these experiments, it was deduced that a photon of energy E possesses momentum p equal to $\frac{E}{c}$, where c is the speed of light in a vacuum.

$$\textbf{Photon momentum } p = \frac{E}{c}$$

where E is the photon energy.

Matter waves

If light has a dual wave–particle nature, perhaps particles of matter also have a dual wave–particle nature. Electrons in a beam can be deflected by a magnetic field. This is evidence that electrons have a particle-like nature. The idea that matter particles also have a wave-like nature was first considered by de Broglie in 1923. By extending the ideas of duality from photons to matter particles, de Broglie put forward the hypothesis that:

- matter particles have a dual wave–particle nature,
- the wave-like behaviour of a matter particle is characterised by a wavelength, its **de Broglie wavelength**, λ, which is related to the momentum, p, of the particle by means of the equation

$$\lambda = \frac{h}{p}$$

Worked example

Calculate the momentum of an X-ray photon of wavelength 2.0×10^{-11} m.

Solution

For a photon of wavelength λ, its energy $E = hf$, where its frequency $f = \frac{c}{\lambda}$.

Therefore its momentum,

$p = \frac{E}{c} = \frac{hf}{c} = \frac{hc}{\lambda c} = \frac{h}{\lambda}$

$= 6.63 \times 10^{-34}\,\text{J s}/2.0 \times 10^{-11}\,\text{m}$

$= 3.3 \times 10^{-23}\,\text{kg m s}^{-1}$

Since the momentum of a particle is defined as its mass × its velocity, according to de Broglie's hypothesis, a particle of mass m moving at velocity v has a de Broglie wavelength given by

$$\lambda = \frac{h}{mv}$$

Note that the de Broglie wavelength of a particle can be altered by changing the velocity of the particle.

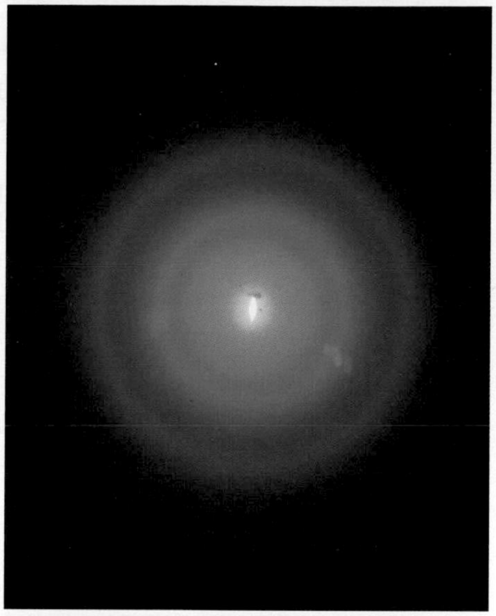

Figure 3 Electron diffraction

Evidence for de Broglie's hypothesis

The wave-like nature of electrons was discovered when, three years after de Broglie put forward his hypothesis, it was demonstrated that a beam of electrons can be diffracted. Figure 2 shows in outline how this is done. After this discovery, further experimental evidence, using other types of particles, confirmed the correctness of de Broglie's theory.

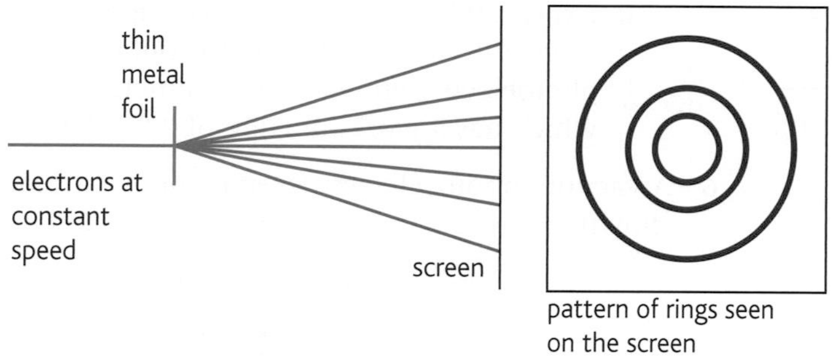

Figure 2 Diffraction of electrons

- A narrow beam of electrons in a vacuum tube is directed at a thin metal foil. A metal is composed of many tiny crystalline regions. Each region or 'grain' consists of positive ions arranged in fixed positions in rows in a regular pattern. The rows of atoms cause the electrons in the beam to be diffracted, just as a beam of light is diffracted by the slits.
- The electrons in the beam pass though the metal foil and are diffracted in certain directions only, as shown in Figure 2. They form a pattern of rings on a fluorescent screen at the end of the tube. Each ring is due to electrons diffracted by the same amount from grains of different orientations, at the same angle to the incident beam.
- The beam of electrons is produced by attracting electrons from a heated filament wire to a positively charged metal plate, which has a small hole at its centre. Electrons that pass through the hole form the beam. The speed of these electrons can be increased by increasing the potential difference between the filament and the metal plate. This makes the diffraction rings smaller, because the increase of speed makes the de Broglie wavelength smaller. So less diffraction occurs and the rings become smaller.

Energy levels explained

An electron in an atom has a fixed amount of energy that depends on the shell it occupies. Its de Broglie wavelength has to fit the shape and size of the shell. This is why its energy depends on the shell it occupies.

For example, an electron in a spherical shell moves round the nucleus in a circular orbit. The circumference of its orbit must be equal to a whole number of de Broglie wavelengths (circumference = $n\lambda$, where n = 1 or 2 or 3, etc.). You don't need to know this for your exam but this condition can be used to derive the energy level formula for the hydrogen atom – and it gives you a deeper insight into quantum physics.

Summary test 20.4

$h = 6.63 \times 10^{-34}$ J s, the mass of an electron = 9.11×10^{-31} kg, the mass of a proton = 1.67×10^{-27} kg

1 With the aid of an example in each case, explain what is meant by the dual wave–particle nature of:
 a light, **b** matter particles.

2 State whether each of the following experiments demonstrates the wave nature or the particle nature of matter or of light: (a) photoelectricity, (b) electron diffraction.

3 Calculate the de Broglie wavelength of:
 a an electron moving at a speed of 2.0×10^{7} m s^{-1}
 b a proton moving at the same speed.

4 Calculate the momentum and speed of:
 a an electron that has a de Broglie wavelength of 500 nm,
 b a proton that has the same de Broglie wavelength.

Launch additional digital resources for the chapter

1 When light at sufficiently high frequency, f, is incident on a metal surface, the maximum kinetic energy, E_{Kmax}, of a photoelectron emitted from the surface is given by:

$E_{Kmax} = hf - \phi$, where ϕ is the work function of the metal

a State what is meant by the *work function* ϕ.

b The following results were obtained using a certain metal X in an experiment to measure E_{Kmax} for different frequencies, f.

$f/10^{14}$ Hz	5.6	6.2	6.8	7.3	8.3	8.9
$E_{Kmax}/10^{-19}$ J	0.8	1.2	1.6	2.0	2.6	3.0

i Use these results to plot a graph of E_{Kmax} against f.

ii Determine the gradient of your graph.

iii Explain why your graph confirms the equation above.

iv Use your graph to determine the work function of the metal and hence calculate the threshold frequency of light for this metal.

b The metal was replaced by a different metal Y with a known work function of 3.3×10^{-19} J. Calculate the threshold frequency of this metal and draw and label a line on your graph to show the results you would expect to obtain for this metal.

2 a For a metal, state what is meant by:

i the *work function* of the metal,

ii the *threshold frequency* for photoelectric emission from the metal.

b State and explain the relationship between the work function, ϕ, and the threshold frequency, f, of a metal.

c i Explain why the existence of the threshold frequency for photoelectric emission from a metal supports the photon theory of light in favour of the wave theory of light.

ii State and explain one other experimental observation about photoelectric emission that supports the photon theory of light over the wave theory.

3 Electrons can escape from a certain metal surface when it is illuminated by blue light, but not when it is illuminated by red light.

a Explain why blue light causes emission of electrons from this metal, whereas red light does not.

b What difference would be made to the emission of electrons from this metal surface if:

i the intensity of the blue light were increased,

ii the surface were at a positive potential?

4 Light of wavelength 600 nm was used to illuminate a metal surface. The maximum kinetic energy of an electron emitted as a result was 1.3×10^{-19} J. Calculate:

a the energy of a photon of this wavelength,

b the work function of the metal surface.

5 a Explain the meaning of the term 'work function' of a metal surface.

b i Calculate the wavelength of a photon of energy 2.0×10^{-19} J.

ii Explain why photons of wavelength longer than the value calculated in part **i** could not cause photoelectric emission from a metal surface which has a work function of 2.0×10^{-19} J.

6 A vacuum photocell connected to a microammeter is shown below.

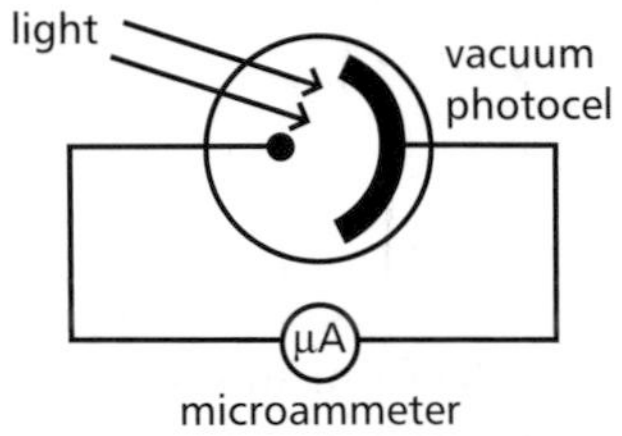

Figure 6.1

a Explain why:

i the microammeter registers a current when light of a certain wavelength is directed at the photocell,

ii the current increases when the light intensity is increased.

b When light of a longer wavelength is directed at the photocell, no current is registered.

i Explain why there is no current in this situation.

ii Explain why making the light more intense would not cause a current with light of this longer wavelength.

7 The spectrum of light from a sodium lamp has two prominent closely spaced lines at wavelengths of 589.0 nm and 589.6 nm. These lines are due to electron transitions from two closely spaced energy levels X and Y to the same lower energy level Z, as shown below (not to scale).

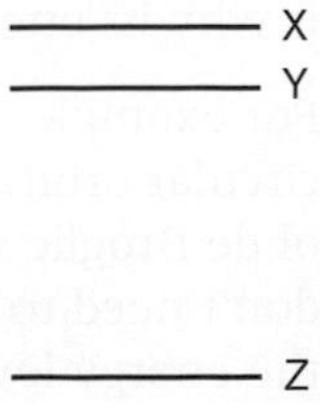

a Copy the diagram and show the electron transition that causes the emission of photons of wavelength:
i 589.0 nm, **ii** 589.6 nm.

b Calculate the energy difference between the two closely spaced energy levels.

8 The energy level diagram for a particular type of atom is shown in Figure 8.1.

a i Calculate the potential difference which electrons, initially at rest, must be accelerated through in order to ionise such an atom.

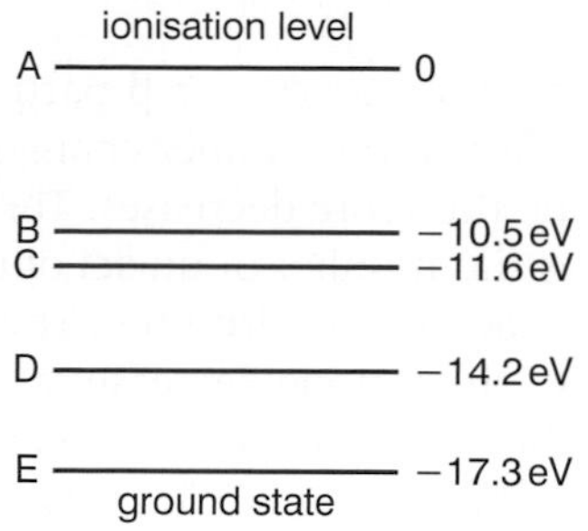

Figure 8.1

ii Calculate the wavelength of the electromagnetic radiation emitted by this type of atom when it de-excites from energy level D.

b i Determine the energy, in eV, of each photon that could be emitted from the atom as a result of de-exciting from level B, and state the electron transition that releases each photon.

ii Determine the maximum wavelength of a photon that could cause excitation from the ground state as a result of being absorbed by the atom in its ground state.

9 Protons from a cyclotron emerge at a maximum speed of $1.7 \times 10^7 \text{ m s}^{-1}$. Ignore relativistic effects in this question.

a For a proton moving at this speed, calculate:

i its kinetic energy,

ii its momentum and its de Broglie wavelength.

b Calculate the momentum of a photon with the same energy as the kinetic energy of a proton moving at a speed of $1.7 \times 10^7 \text{ m s}^{-1}$.

10 A narrow beam of electrons moving at a speed of $9.8 \times 10^7 \text{ m s}^{-1}$ is directed at a thin crystal. A fluorescent screen on the other side of the crystal shows a pattern of rings due to diffraction of electrons by the crystal, as shown in Figure 10.1.

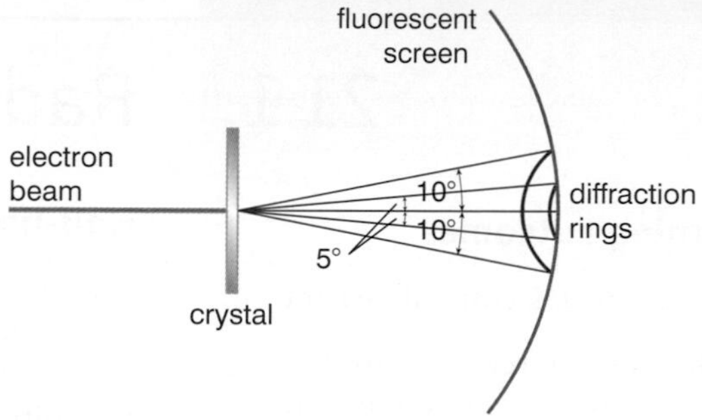

Figure 10.1

a Calculate the de Broglie wavelength, λ, of the electrons.

b Measurements show that the rings are produced by electrons diffracted through 5° and 10°. The angle of diffraction θ of each ring is given by the equation $2d\sin\theta = n\lambda$, where n is the order number of the ring and d is the spacing between the layers of atoms in the crystal. Use this information to calculate a value for d.

c State and explain how the diameter of the rings would change if the speed of the electrons was increased.

11 State whether each of the following experiments or observations using light demonstrates the wave-like or the particle-like nature of light.

i When light passes through a narrow gap, it diffracts.

ii When light is directed at a metal surface, electrons are emitted from the surface only if the light wavelength is less than a certain value.

12 a An electron in a beam has a speed of $1.5 \times 10^7 \text{ m s}^{-1}$. Calculate:

i the kinetic energy,

ii the momentum and de Broglie wavelength of this electron.

b Calculate the wavelength of a photon with the same energy as the electron in part **a**.

21 Nuclear physics

21.1 Radioactive decay

Learning outcomes

On these pages you will learn to:

- show an appreciation from the fluctuations in count rate that radioactive decay has a spontaneous and random nature
- define the terms activity and decay constant and recall and solve problems using $A = \lambda N$

Half-life

When a nucleus of a radioactive **isotope** emits an α- or a β-particle, it becomes a nucleus of a different isotope because its proton number changes. The number of nuclei of the initial radioactive isotope therefore decreases. The mass of the initial isotope decreases gradually as the number of nuclei of the isotope decreases. Figure 1 shows how the mass decreases with time. The curve is referred to as a decay curve. The mass of the isotope decreases at a slower and slower rate. Measurements show that the mass decreases exponentially which means that the mass drops by a constant factor (e.g. ×0.8) in equal intervals of time. For example, if the initial mass of the radioactive isotope is 100 g and the mass decreases by a factor of ×0.8 every 1000 seconds, then:

- after 1000 s, the mass remaining = 80 g (= 0.8 × 100 g),
- after 2000 s, the mass remaining = 64 g (= 0.8 × 0.8 × 100 g),
- after 3000 s, the mass remaining = 51 g (= 0.8 × 0.8 × 0.8 × 100 g).

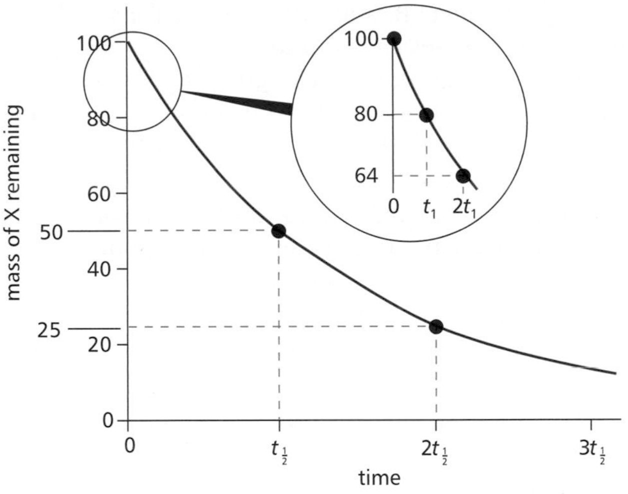

Figure 1 *A radioactive decay curve*

A convenient measure for the rate of decrease is the time taken for a decrease by half. This is the half-life of the process.

The half-life, $t_{\frac{1}{2}}$, of a radioactive isotope is the time taken for the mass of the isotope to decrease to half the initial mass.

This is the same as the time taken for the number of nuclei of the isotope to decrease to half the initial number.

Consider a sample of a radioactive isotope X which initially contains 100 g of the isotope.

- After 1 half-life, the mass of X remaining = 0.5 × 100 = 50 grams
- After 2 half-lives from the start, the mass of X remaining = $0.5^2 \times 100$ = 25 grams
- After 3 half-lives from the start, the mass of X remaining = $0.5^3 \times 100$ = 12.5 grams
- After n half-lives from the start, the mass of X remaining = $0.5^n\, m_0$, where m_0 = the initial mass

The mass of X decreases exponentially. This is because radioactive decay is a **random** process. We know this because the fluctuations in the count rate of a Geiger counter are random variations. So the number of nuclei that decay in a certain time is in proportion to the number of nuclei of X remaining.

Randomness

To understand the idea of randomness, consider a game of dice starting with 1000 dice, each representing a nucleus of X. The throw of a dice is a random process in which each face has a 1 in 6 chance of being uppermost.

After each throw, suppose all the dice showing '1' uppermost are removed. For example, after the first throw 167 dice (= 1000/6) would be removed and 833 (= 1000 − 167) would remain. The table below shows how many dice remain after each throw.

Throw	Number of dice		
	initially	removed	remaining
1st	1000	167	833
2nd	833	139	694
3rd	694	116	578
4th	578	96	482
5th	482	80	402

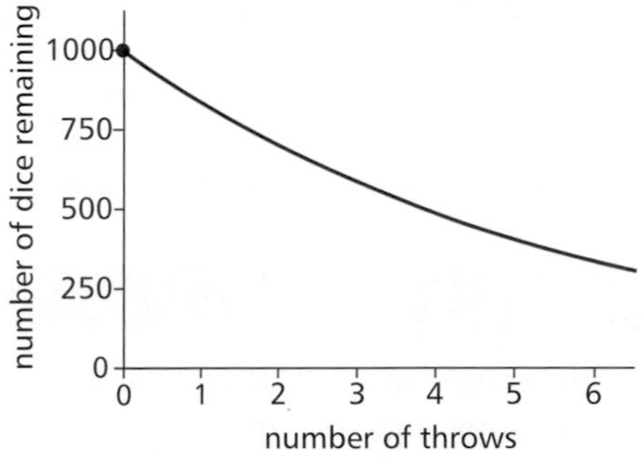

Figure 2 Exponential decrease

The analysis shows that 4 throws are needed to reduce the number of dice remaining to less than half the initial number. Prove for yourself that a further 4 throws would reduce the number of dice remaining to 25% of the initial number. Figure 2 shows how the number of dice remaining decreases with time. The curve has the same shape as Figure 1. The half-life of the process is 3.8 'throws'.

Activity

The activity *A* of a radioactive isotope is the number of nuclei of the isotope that disintegrate per second. In other words, it is the rate of change of the number of nuclei of the isotope. The unit of activity is the **becquerel (Bq),** where 1 Bq = 1 disintegration per second.

The activity of a radioactive isotope is proportional to the mass of the isotope. Because the mass of a radioactive isotope decreases with time due to radioactive decay, the activity decreases with time. Figure 3 shows an experiment in which the activity of a radioactive isotope of protoactinium $^{234}_{91}Pa$ is measured and recorded using a Geiger tube and a counter. This isotope is a β-emitter produced by the decay of the radioactive isotope of thorium $^{234}_{90}Th$. In this experiment, an organic solvent in a sealed bottle is used to separate protoactinium from thorium to enable the activity of the protoactinium to be monitored.

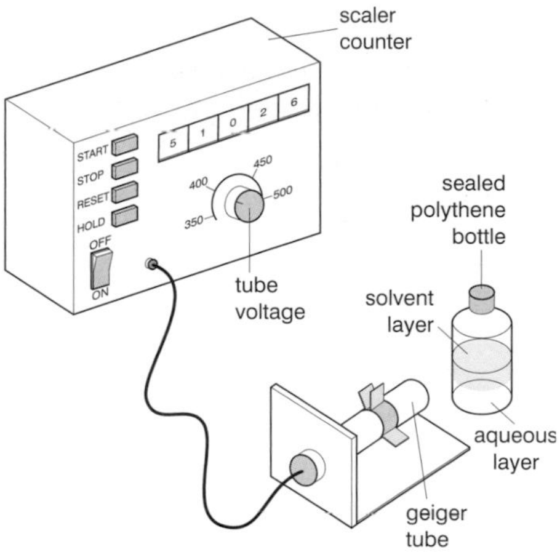

Figure 3 Measuring the activity of protoactinium

Before the experiment is carried out, the background count rate is measured without the bottle present. The bottle is then shaken to mix the aqueous and solvent layers and then placed near the end of the Geiger tube. The layers are allowed to separate as shown in Figure 3. The protoactinium is collected by the solvent and the thorium by the aqueous layer. The Geiger tube detects β-particles emitted by the decay of the protoactinium nuclei in the solvent layer.

The counter is used to measure the number of counts every 10 seconds. The count rate is the number of counts in each ten-second interval divided by 10 s. The background count rate is subtracted to give the corrected count rate. Since the activity is proportional to the corrected count rate, a graph of the corrected count rate against time, as in Figure 4, shows how the activity of the protoactinium decreases with time.

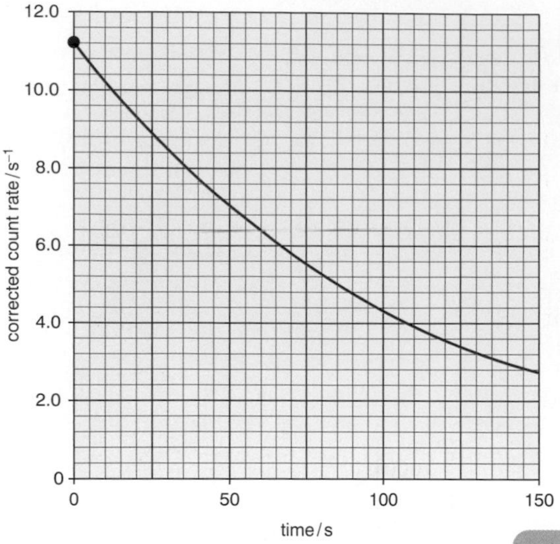

Figure 4 *A radioactive decay curve*

> **Note**
>
> **Reminder about molar mass and the Avogadro constant**
>
> As explained in Topic 19.2 for an element with a mass number A:
>
> - its molar mass is the mass of 1 mole,
> - one mole of the element contains N_A atoms where N_A is the Avogadro constant,
> - mass m of the element contains $\left(\frac{m}{M}\right) N_A$ atoms.

Activity and power

For a radioactive source of activity A that emits particles (or photons) of the same energy E, the energy per second released by radioactive decay in the source by the radiation is the product of its activity and the energy of each particle. In other words, the power of the source = AE.

The energy transfer per second from a radioactive source = AE

If the source is in a sealed container and emits only α-particles which are all absorbed by the container, the container gains thermal energy from the absorbed radiation equal to the energy transferred from the source. For example, for a source that has an activity of 30 MBq and emits particles of energy 2.5 MeV, the energy transfer per second from the source is given by

$$30 \times 10^6\,\text{Bq} \times 2.5\,\text{MeV} = 7.5 \times 10^7\,\text{MeV s}^{-1} = 1.2 \times 10^{-5}\,\text{J s}^{-1}.$$

Summary test 21.1

$N_A = 6.02 \times 10^{23}\,\text{mol}^{-1}$, $1\,\text{MeV} = 1.6 \times 10^{-13}\,\text{J}$

1 Figure 4 shows how the activity of protoactinium $^{234}_{91}\text{Pa}$ decreases with time.

a Use the graph to work out the half-life of this isotope,

b If the initial mass of the isotope was 48 g, calculate the mass of the isotope remaining after three half-lives.

2 A freshly prepared sample of a radioactive isotope X contains 1.8×10^{15} atoms of the isotope. The half-life of the isotope is 8.0 hours. Calculate:

a the number of atoms of this isotope remaining after:
- **i** 8 hours,
- **ii** 24 hours.

b the number of atoms of X that would have decayed after:
- **i** 8 hours,
- **ii** 24 hours.

c the energy transfer from the sample in 24 hours if the isotope emits α-particles of energy 5 MeV.

3 $^{131}_{53}\text{I}$ is a radioactive isotope of iodine which has a half-life of 8.0 days. A sample of this isotope has an initial activity of 38 kBq. Calculate the activity of this sample:

a 8.0 days later,

b 32 days later.

4 $^{137}_{55}\text{Cs}$ is a radioactive isotope of caesium which has a half-life of 35 years. A sample of this isotope has a mass of 1.0×10^{-3} kg.

a Calculate the number of atoms in 1.0×10^{-3} kg of this isotope.

b Calculate the number of atoms of the isotope remaining in the sample after 70 years.

The theory of radioactive decay

The random nature of radioactive decay

An unstable nucleus becomes stable by emitting an α- or a β-particle or a γ-photon. This is an unpredictable event. It happens at random and is spontaneous in the sense that it is a change without any external cause. Every nucleus of a radioactive isotope has an equal probability of becoming stable in any given time interval. Therefore, for a large number of nuclei of a radioactive isotope, the number of nuclei that disintegrate in a certain time interval depends only on the total number of nuclei present. The same idea was considered in the dice experiment. The greater the number of dice used, the more likely it is that 1 in every 6 dice show a particular number on the upper face.

Consider a sample of a radioactive isotope X that initially contains N_0 nuclei of the isotope.

Let N represent the number of nuclei of X remaining at time t after the start.

Suppose in time Δt, the number of nuclei that disintegrate is ΔN.

Because radioactive disintegration is a random process, ΔN is proportional to:

1 N, the number of nuclei of X remaining at time t,
2 the duration of the time interval Δt.

Therefore, $\Delta N = -\lambda N \Delta t$, where λ is a constant referred to as the **decay constant**. The minus sign is necessary because ΔN is a decrease.

So, the rate of disintegration, $\boldsymbol{\frac{\Delta N}{\Delta t} = -\lambda N}$

For a given radioactive isotope, its activity is the rate of disintegration $\frac{\Delta N}{\Delta t}$

Therefore, the activity A of N atoms of a radioactive isotope is given by:

$$\boldsymbol{A = \lambda N}$$

The solution of the equation $\frac{\Delta N}{\Delta t} = -\lambda N$ is $\boldsymbol{N = N_0 e^{-\lambda t}}$

where e^t is the exponential function.

Figure 1 shows that a graph of N against t gives a decay curve. The number of nuclei N decreases exponentially with time. In other words:

- in one half-life, the remaining number of nuclei $N_1 = 0.5\,N_0$
- in two half-lives, the remaining number of nuclei $N_2 = 0.25\,N_0$
- in n half-lives, the remaining number of nuclei $N = 0.5^n\,N_0$

The graph of the number of nuclei *N* against time *t* as represented by the equation $N = N_0 e^{-\lambda t}$ is shown in Figure 1 above. It is a curve with exactly the same shape as Figure 1 of Topic 21.1.

The mass, *m*, of a radioactive isotope decreases from initial mass, m_0, in accordance with the equation $\boldsymbol{m = m_0 e^{-\lambda t}}$ because the mass, m, is proportional to the number of nuclei, N, of the isotope.

The activity, *A*, of a sample of N nuclei of an isotope decays in accordance with the equation

$$\boldsymbol{A = A_0 e^{-\lambda t}}$$

where A_0 is the initial activity (or the activity at $t = 0$).

This is because the activity, A = the magnitude of the number of disintegrations per second = λN. Hence, $A = \lambda N_0 e^{-\lambda t} = A_0 e^{-\lambda t}$, where $A_0 = \lambda N_0$.

Learning outcomes

On these pages you will learn to:

- solve problems using the relationship $x = x_0 e^{-\lambda t}$, where x could represent activity, number of undecayed nuclei or count rate
- sketch a graph to show the exponential nature of radioactive decay
- define half-life and solve related problems using the equations above and the relationship $t_{½} = 0.693/\lambda$

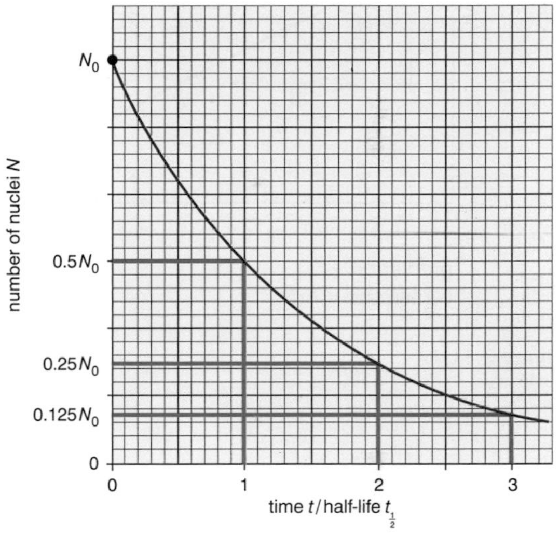

Figure 1 $N = N_0 e^{-\lambda t}$

The count rate, C, due to a sample of a radioactive isotope at a fixed distance from a Geiger tube is proportional to the activity of the source. Therefore, the count rate decreases with time in accordance with the equation $\boldsymbol{C = C_0 e^{-\lambda t}}$, where C_0 is the count rate at time $t = 0$.

The above equations for the number of nuclei, N, the activity, A, and the count rate, C, received from the source (i.e. the corrected count rate) are all of the same general form, namely $\boldsymbol{x = x_0 e^{-\lambda t}}$, where x represents N or A or C and x_0 represents the initial value.

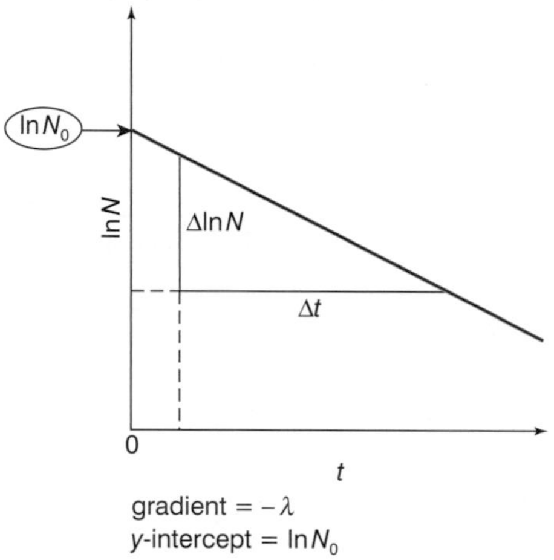

Figure 2 ln N v. t

1 The exponential function appears in any situation where the rate of change of a quantity is in proportion to the quantity itself. This is because the rate of change of each term in the function sequence is equal to the previous term in the sequence.

2 The exponential function, $e^x = 1 + x + \frac{x^2}{2!} + \frac{x^3}{3!} + \ldots$ (See Chapter 24, Mathematical skills)

Differentiating e^x with respect to x gives e^x $\left(\text{i.e. } \frac{d(e^x)}{dx} = e^x\right)$ because differentiating each term in the expression for e^x gives the previous term. The exponential function is indicated on a calculator as 'exp' or 'e^x' or 'inv ln'. (See Chapter 24, Mathematical skills)

3 The natural logarithm function, $\ln x$, is the inverse exponential function. In other words, if $y = e^x$, then $\ln y = x$. Therefore, $N = N_0 e^{-\lambda t}$ may be written $\ln N = \ln N_0 - \lambda t$.
The graph of $\ln N$ against t is therefore a straight line with:
- a gradient $= -\lambda$, and
- a y-intercept $= \ln N_0$.

4 The exponential decrease formula is also used in the theory of capacitor discharge. (See Topic 15.4)

Worked example

A sample of a radioactive isotope initially contains 1.2×10^{20} atoms of the isotope. The decay constant for the isotope is $3.6 \times 10^{-3}\,s^{-1}$. Calculate:

a the number of atoms of the isotope remaining after 1000 s,
b the activity of the sample after 1000 s.

Solution

a $N_0 = 1.2 \times 10^{20}$, $\lambda = 3.6 \times 10^{-3}\,s^{-1}$, $t = 1000\,s$,
$\lambda t = 3.6 \times 10^{-3}\,s^{-1} \times 1000\,s = 3.6$
$\therefore N = N_0 e^{-\lambda t} = 1.2 \times 10^{20} e^{-3.6} = 1.2 \times 10^{20} \times 2.7 \times 10^{-2} = 3.2 \times 10^{18}$
b Activity, $A = \lambda N = 3.6 \times 10^{-3} \times 3.2 \times 10^{18} = 1.2 \times 10^{16}\,Bq$

The decay constant

The decay constant λ is the probability of an individual nucleus decaying per second. If there are 10 000 nuclei present and 300 decay in 20 seconds, the decay constant is $0.0015\,s^{-1}$ $\left(= \frac{\left(\frac{300}{10\,000}\right)}{20\,s}\right)$.

In general, if the change of the number of nuclei ΔN in time Δt is given by $\Delta N = -\lambda N \Delta t$, then, the probability of decay, $\frac{\Delta N}{N} = \lambda \Delta t$ (the minus sign is not needed here as reference is made to decay).

So, the probability per unit time $= \frac{\frac{\Delta N}{N}}{\Delta t} = \lambda$

As explained on p.306, the **half-life**, $t_{1/2}$, of a radioactive isotope is the time taken for half the initial number of nuclei to decay. The longer the half-life, the smaller the decay constant because the probability of decay per second is smaller.

The half-life $t_{1/2}$ is related to the decay constant λ according to the equation

$$t_{1/2} = \frac{\ln 2}{\lambda}$$

As ln 2 = 0.693, this equation may be written as $t_{1/2} = \frac{0.693}{\lambda}$

Proof of $t_{1/2} = \frac{\ln 2}{\lambda}$

The proof of this equation is not part of this specification. It is provided below to help you develop a better understanding of the topic.

Let the number of nuclei $N = N_0$ at time $t = 0$, so at time $t = t_{1/2}$, $N = 0.5\,N_0$

Inserting $t = t_{1/2}$, $N = 0.5\,N_0$ into $N = N_0 e^{-\lambda t}$ gives $0.5\,N_0 = N_0 e^{-\lambda t_{1/2}}$

Cancelling N_0 and taking the natural logarithm (ln) of each side gives $\ln 0.5 = -\lambda t_{1/2}$

Because $\ln 0.5 = -\ln 2$, then $\ln 2 = \lambda t_{1/2}$.

Rearranging this equation gives $t_{1/2} = \frac{\ln 2}{\lambda}$

Note

To calculate N at time t, given values of N_0 and $t_{1/2}$,

- **either** calculate λ using $\lambda = \frac{\ln 2}{t_{1/2}}$ then use the equation $N = N_0 e^{-\lambda t}$,
- **or** calculate the number of half-lives, n, using $n = \frac{t}{t_{1/2}}$ then use $N = 0.5^n N_0$

Summary test 21.2

$N_A = 6.02 \times 10^{23}\,\text{mol}^{-1}$, $1\,\text{MeV} = 1.6 \times 10^{-13}\,\text{J}$

1 $^{131}_{53}\text{I}$ is a radioactive isotope of iodine which has a half-life of 8.0 days. A fresh sample of this isotope contains 4.2×10^{16} atoms of isotope. Calculate:

a the decay constant of this isotope,

b the number of atoms of this isotope remaining after 24 hours.

2 A radioactive isotope has a half-life of 35 years. A fresh sample of this isotope has an activity of 25 kBq. Calculate:

a the decay constant in s^{-1},

b the activity of the sample after 10 years.

3 **a** Calculate the number of atoms present in 1.0 kg of $^{226}_{88}\text{Ra}$.

b The isotope $^{226}_{88}\text{Ra}$ has a half-life of 1620 years. For an initial mass of 1.0 kg of this isotope, calculate:

i the mass of this isotope remaining after 1000 years,

ii how many atoms of the isotope will remain after 1000 years.

4 A fresh sample of a radioactive isotope has an initial activity of 40 kBq. After 48 hours, its activity has decreased to 32 kBq. Calculate:

a the decay constant of this isotope,

b its half-life.

21.3 Radioactive isotopes in use

Learning outcomes

On these pages you will learn to:

- show an understanding of the uses of radioactive isotopes in terms of the properties of the isotopes used and the radiation emitted
- use the equations in Topic 21.2 to solve problems in connection with the use of radioactive isotopes

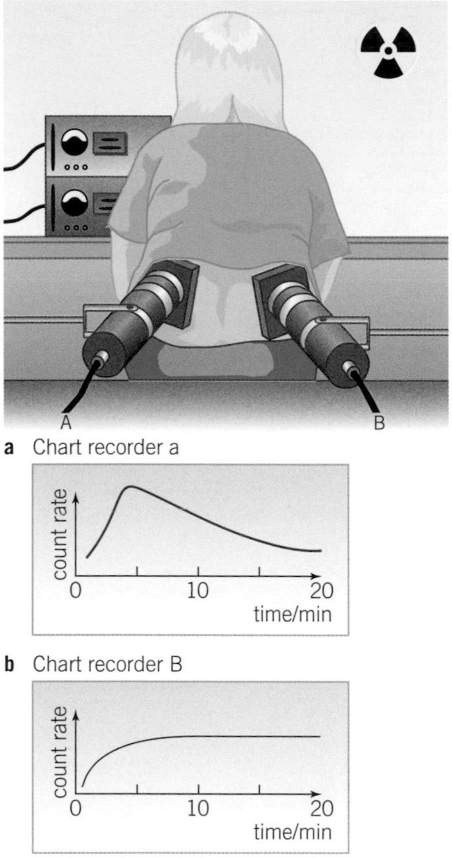

Figure 1 *Thyroid monitoring*

Radioactive isotopes are used for many purposes. The choice of an isotope for a particular purpose depends on its half-life and on the type of radiation it emits. For some uses, the choice also depends on how the isotope is obtained and on whether or not it produces a stable decay product. The following examples are intended to provide a wider awareness of important uses of radioactive substances and to set contexts in which knowledge and understanding of radioactivity is developed further. Read again Topics 10.3 and 10.4 about α, β and γ radiation if necessary.

Radioactive tracers

A radioactive tracer is used to follow the path of a substance through a system, as illustrated below.

- **Environmental uses**: for example, the detection of leaks in underground pipes that carry water or oil. Such a leak can be detected by injecting a radioactive tracer into the flow. Geiger tubes on the surface above the pipeline can then be used to detect leakage. The radioactive isotope used should have a half-life short enough so that it decays quickly after use and long enough so that the test can be completed before its activity becomes too low. In addition, it should be a β-emitter as α-radiation would be absorbed by the pipes and γ radiation would pass through the pipes without absorption.
- **Geological uses**: for example, to improve the recovery of oil from an underground reservoir. Water containing a radioactive tracer is injected into an oil reservoir at high pressure, forcing some of the oil out. Detectors at the production wells monitor breakthrough of the radioactive isotope. The results are used to build up a model of the reservoir to improve and control recovery. Because the time from injection to breakthrough can be many months, the tracer must have a suitably long half-life. A suitable tracer is 'tritiated' water ^3H_2O, a β-emitter with a half-life of 12 years.
- **Medical uses**: for example, to monitor the uptake of iodine by the thyroid gland (see Figure 1). The thyroid gland absorbs iodine to maintain its function of producing a hormone. The rate of uptake is measured by giving the patient a solution containing sodium iodide which includes a small quantity of the radioactive isotope of iodine, $^{131}_{53}I$, which emits beta and gamma radiation with a half-life of 8 days. The activity of the patient's thyroid and the activity of an identical sample prepared at the same time are measured 24 hours later. The percentage uptake by the patient is then calculated from (the corrected count rate of the thyroid / the corrected count rate of the identical solution) × 100%. A normal thyroid has a percentage uptake of 20–50% after 24 hours.
- **Agricultural research**: for example, to investigate the uptake of fertilisers by plants. This can be done by using a fertiliser which contains the radioactive isotope of phosphorus, $^{32}_{15}P$, which is a β-emitter with a half-life of 14 days. By measuring the radioactivity of the leaves, the amount of fertiliser reaching them can be determined.

Radioactive dating

- **Carbon dating**; living plants and trees contain a small percentage of the radioactive isotope of carbon, $^{14}_{6}C$, which is formed in the atmosphere as a result of cosmic rays knocking out neutrons from nuclei. These neutrons then collide with nitrogen nuclei to form carbon-14 nuclei.

$$^{1}_{0}n + ^{14}_{7}N \rightarrow ^{14}_{6}C + ^{1}_{1}p$$

Carbon dioxide from the atmosphere is taken up by living plants as a result of photosynthesis. So a small percentage of the carbon content of any plant is carbon-14. This isotope has a half-life of 5570 years so there is negligible decay during the life-time of a plant. Once a tree has died, no further carbon is taken in so the proportion of carbon-14 in the dead tree decreases as the carbon-14 nuclei decay. Because activity is proportional to the number of atoms still to decay, measuring the activity of the dead sample enables its age to be calculated, provided the activity of the same mass of living wood is known.

Worked example

A certain sample of dead wood is found to have an activity of 0.28 Bq. An equal mass of living wood is found to have an activity of 1.3 Bq. Calclulate the age of the sample.

The half-life of carbon-14 is 5570 years.

Solution

The half-life, $t_{½}$, in seconds = 5570 × 365 × 24 × 3600 s = 1.76×10^{11} s

$$\therefore \text{ the decay constant of carbon-14, } \lambda = \frac{0.693}{t_{½}} = \frac{0.693}{1.76 \times 10^{11}} = 3.95 \times 10^{-12}\,\text{s}^{-1}$$

Using activity $A = A_0 e^{-\lambda t}$, where $A = 0.28$ Bq and $A_0 = 1.30$ Bq gives

$$0.28 = 1.3e^{-\lambda t} \quad \text{so} \quad e^{-\lambda t} = \left(\frac{0.28}{1.30}\right) = 0.215$$

$$\therefore \quad \lambda t = 1.535$$

$$t = \frac{1.535}{\lambda} = \frac{1.535}{3.95 \times 10^{-12}\,\text{s}} = 3.88 \times 10^{11}\,\text{s} = 12\,300 \text{ years}$$

Note

A useful check is to estimate the number of half-lives needed for the activity to decrease from 1.30 Bq to 0.28 Bq. You should find that just over 2 half-lives are needed, corresponding to about 11 000 years.

- **Argon dating**; ancient rocks contain trapped argon gas as a result of the decay of the radioactive isotope of potassium, ${}^{40}_{19}\text{K}$ into the argon isotope ${}^{40}_{18}\text{Ar}$. This happens when its nucleus captures an inner shell electron. As a result, a proton in the nucleus changes into a neutron and a neutrino is emitted. The equation for the change is

$$ {}^{40}_{19}\text{K} + {}^{0}_{-1}\text{e} \rightarrow {}^{40}_{18}\text{Ar} + \nu $$

The potassium isotope ${}^{40}_{19}\text{K}$ also decays by β-emission to form the calcium isotope ${}^{40}_{20}\text{Ca}$. This process is 8 times more probable than electron capture.

$$ {}^{40}_{19}\text{K} \rightarrow {}^{0}_{-1}\beta + {}^{40}_{20}\text{Ca} + \bar{\nu} $$

The effective half-life of the decay of ${}^{40}_{19}\text{K}$ is 1250 million years. The age of the rock (i.e. the time from when it solidified) can be calculated by measuring the proportion of argon-40 to potassium-40. For every n potassium-40 atoms now present, if there is 1 argon-40 atom present, there must have originally been $n + 9$ potassium atoms. (i.e. 1 that decayed into argon-40 + 8 that decayed into calcium-40 + n remaining). The radioactive decay equation $N = N_0 e^{-\lambda t}$ can then be used to find the age of the sample. For example, suppose for every 4 potassium-40 atoms now present, a certain rock now has 1 argon-40 atom. Therefore, $N = 4$ and $N_0 = 13$. Substituting these values into the equation $N = N_0 e^{-\lambda t}$ gives $4 = 13e^{-\lambda t}$.

Therefore, $e^{-\lambda t} = \frac{4}{13} = 0.308$, which gives $t = \frac{-\ln 0.308}{\lambda}$. Substituting $\frac{0.693}{t_{½}}$ for λ into this equation gives $t = \left(\frac{-\ln 0.308}{0.693}\right) t_{½} = 1.70\, t_{½}$. The age of the sample is therefore 2120 million years.

Note

A useful check is to estimate the number of half-lives needed for N to decrease from 13 to 4. You should find that between 1 and 2 half-lives are needed, corresponding to an age of between 1250 and 2500 million years.

Figure 2 *Pistons (and piston rings) fit into this engine block*

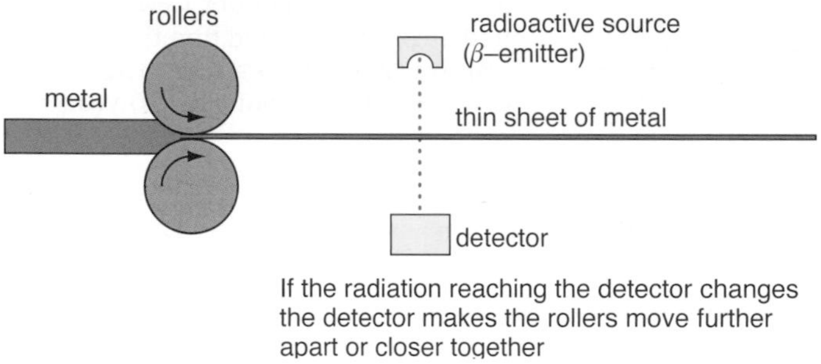

Figure 3 *The manufacture of metal foil*

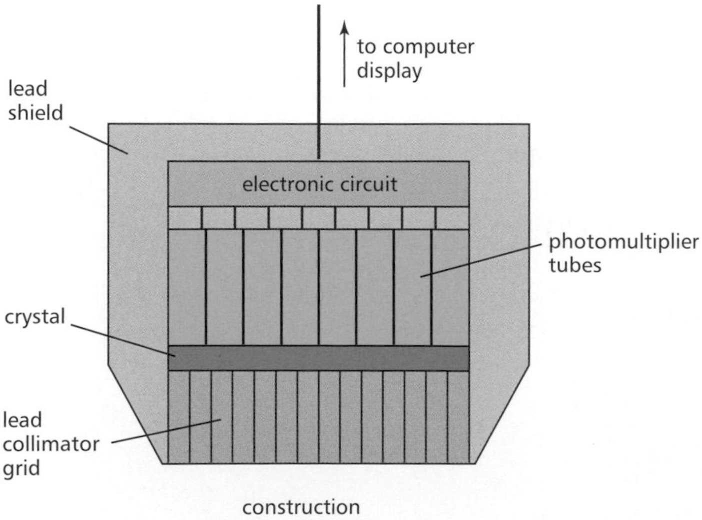

Figure 4 *The gamma camera*

Industrial uses

The examples below are just two of a wide range of applications of radioactivity in industry.

- **Engine wear**
  The rate of wear of a piston ring in an engine can be measured by fitting a ring that is radioactive. As the ring slides along the piston compartment, radioactive atoms transfer from the ring to the engine oil. By measuring the radioactivity of the oil, the mass of radioactive metal transferred from the ring can be determined and the rate of wear calculated. A metal ring can be made radioactive by exposing it to neutron radiation in a nuclear reactor. Each nucleus that absorbs a neutron becomes unstable and disintegrates by β-emission.
- **Thickness monitoring**
  Metal foil is manufactured by using rollers to squeeze plate metal on a continuous production line, as shown in Figure 3. A detector measures the amount of radiation passing through the foil. If the foil is too thick, the detector reading drops. A signal from the detector is fed back to the control system to make the rollers move closer together and so make the foil thinner. The source used is a β-emitter with a long half-life. α radiation would be absorbed completely by the foil and γ radiation would pass straight through without absorption.
- **Power for remote devices** such as satellites and weather sensors can be obtained using a radioactive isotope in a thermally insulated sealed container which absorbs all the radiation emitted by the isotope.
  A thermocouple attached to the container produces electricity as a result of the container becoming warm through absorbing radiation. For mass, m, of the isotope, its activity, $A = \lambda N$, where N is the number of radioactive atoms present in mass m. If each disintegration of a nucleus releases energy E, the energy transfer per second from the source = λNE. The source needs to have a reasonably long half-life so it does not need to be replaced frequently but a very long half-life would require too much mass to generate the necessary power.

Medical and health uses

As explained earlier, radioactive isotopes are used as medical tracers. Further uses of radioactive isotopes in medicine include the gamma camera used to form images of joints and organs, the PET scanner and gamma therapy to destroy cancerous tissues.

- **The gamma camera** is designed to detect γ radiation from sites inside the body where a γ-emitting isotope is located. For example, bone deposits can be located using a phosphate tracer containing the radioactive isotope of technetium, $^{99}_{52}Te$ which is a γ-emitter that has a half-life of 6 hours.
  The γ-photons from inside the body are absorbed by a lead collimator grid unless they travel parallel to narrow

channels through the collimator. Each γ-photon that passes through the grid strikes a large sodium iodide crystal, causing a flash of light which is detected by a photomultiplier tube in an array of tubes. The tubes are connected to a computer which displays an image of the γ-emitting sources in the body.

- **Gamma therapy** is used to destroy tumours inside the body. A narrow beam of γ radiation from the radioactive isotope of cobalt, $^{60}_{27}Co$, is directed at the tumour from different directions by moving the source or by moving the patient. This movement is necessary to ensure healthy tissue in the path of the beam is exposed much less than the target tissue. The cobalt-60 source has a half-life of 5.3 years and emits γ-photons of energies 1.17 MeV and 1.33 MeV. The source is enclosed in a thick lead container. When the source is to be used, it is rotated to the inner end of an exit channel so that a beam of γ radiation emerges after passing along the exit channel. When the source is not in use, it is rotated away from the inner end of the exit channel so that no γ radiation can emerge from the container.
- **Food preservation** can be achieved by irradiating food with γ radiation. About 30% of the world's food is lost through spoilage. The major cause is bacteria, moulds and yeast which grow on food. Some bacteria produce toxic waste products that cause food poisoning. Irradiation of food with γ radiation kills 99% of the disease-carrying organisms in the food, such as *Salmonella* which infects poultry and *Clostridium*, the cause of botulism. The treatment is not suitable for all foods. Red meat turns brown and develops an unpleasant taste, eggs develop a smell and tomatoes go soft.

Summary test 21.3

$N_A = 6.02 \times 10^{23}\,mol^{-1}$, $1\,MeV = 1.6 \times 10^{-13}\,J$

1 a Explain why living wood is slightly radioactive.

b A sample of ancient wood of mass 0.5 g is found to have an activity of 0.11 Bq. A sample of living wood of the same mass has an activity of 0.13 Bq. Calculate the age of the sample of wood.

The half-life of radioactive carbon $^{14}_{6}C$ is 5570 years.

2 The radioactive isotope of iodine, $^{131}_{53}I$, is used for medical diagnosis of the kidneys. The isotope has a half-life of 8 days. A sample of the isotope is given to a patient in a glass of water. The passage of the isotope through each kidney is then monitored using two detectors outside the body. The isotope is required to have an activity of 800 kBq at the time it is given to the patient.

a Calculate:

i the activity of the sample 24 hours after it was given to the patient,

ii the activity of the sample when it was prepared 24 hours earlier,

iii the mass of $^{131}_{53}I$ in the sample when it was prepared.

b The reading from the detector near one of the patient's kidneys rises then falls. The reading from the other detector which is near the other kidney rises and does not fall. Discuss the conclusions that can be drawn from these observations.

3 a i In the manufacture of metal foil, describe how the thickness of the foil is monitored using a radioactive source and a detector.

ii Explain why the source needs to be a β-emitter, not an α-emitter or a γ-emitter.

b i Explain why a cobalt-60 source used for γ-therapy is enclosed in a thick lead-lined container.

ii Explain why a beam of γ radiation used to destroy a tumour inside a patient is directed at the tumour from different directions during treatment.

4 a A cardiac pacemaker is a device used to ensure that a faulty heart beats at a suitable rate. The required electrical energy in one type of pacemaker is obtained from the energy released by a radioactive isotope. The radiation is absorbed inside the pacemaker. As a result, the absorbing material gains thermal energy and heats a thermocouple attached to the absorbing material. The voltage from the thermocouple provides the source of electrical energy for the pacemaker.

i Discuss whether the radioactive source should be an α-emitter, a β-emitter or a γ-emitter.

ii The radioactive source needs to have a reasonably long half-life, otherwise it would need to be replaced frequently. Discuss the disadvantages of using a radioactive source with a very long half-life.

b The energy source for a remote weather station is the radioactive isotope of strontium $^{90}_{38}Sr$, which has a half-life of 28 years. It emits β-particles of energy 0.40 MeV. For a mass of 10 g of this isotope, calculate:

i its activity,

ii the energy released per second.

21.4 Energy from the nucleus

Learning outcomes

On these pages you will learn to:

- show an appreciation of the association between energy and mass as represented by $E = mc^2$ and recall and use this relationship
- show an understanding of the energy released in radioactive decay
- solve problems relating the energy released in radioactive decay to the change of total mass

Energy and mass

In 1905, Einstein published the theory of special relativity in which he showed that moving clocks run more slowly than stationary clocks, fast-moving objects appear shorter than when stationary, the mass of a moving object changes with its speed and no material object can travel as fast as light. He also showed that the mass of an object increases (or decreases) when it gains (or loses) energy, E, in accordance with the equation

$$E = mc^2$$

where m is the change of its mass and c is the speed of light in free space which is $3.0 \times 10^8\,\mathrm{m\,s^{-1}}$.

For example:

- A sealed torch that radiates 10 W of light for 10 h (= 36 000 s) would lose 0.36 MJ of energy (= 10 W × 36 000 s). Its mass would therefore decrease by 4.0×10^{-12} kg (= $0.36\,\mathrm{MJ}/c^2$), an insignificant amount compared with the mass of the torch.
- A mass of a 1000 kg car that speeds up from a standstill to $30\,\mathrm{m\,s^{-1}}$ would gain 450 kJ of kinetic energy so its mass when moving at $30\,\mathrm{m\,s^{-1}}$ would be 5.0×10^{-12} kg (= $450\,\mathrm{kJ}/c^2$) more than when it is rest.
- An unstable nucleus that releases a 5 MeV γ photon would lose 8.0×10^{-13} J of energy. Its mass would therefore decrease by 8.9×10^{-30} kg (= $8.0 \times 10^{-13}\,\mathrm{J}/c^2$), which is not an insignificant amount compared with the mass of a nucleus.

The equation applies to all energy changes of any object. These two examples show that such changes are important in nuclear reactions but are not usually significant otherwise. A century after Einstein published his theory, the reason why the mass of an object changes when energy is transferred to or from it is still not clearly understood. However, it is known for every type of particle, there is a corresponding antiparticle with the same mass and opposite charge (if charged). We also know that:

- When a particle and its corresponding antiparticle meet, they **annihilate** each other and two gamma (γ) photons are produced, each of energy mc^2 where m is the mass of the particle or antiparticle.
- A single γ photon of energy in excess of $2mc^2$ can produce a particle and an antiparticle, each of mass m, in a process known as pair production.

Energy changes in reactions

Reactions on a nuclear or sub-nuclear scale do involve significant changes of mass. For example, in radioactive decay, if we know the exact rest mass of each particle involved, we can calculate the energy released (Q) from the difference Δm in the total mass before and after the reaction. In general, for a spontaneous reaction in which no energy is supplied,

the energy released, $Q = \Delta mc^2$

In any change where energy is released such as radioactive decay, the total mass after the change is always less than the total mass before the change. This is because, in the change, some of the mass is converted to energy which is released.

1 In α decay, the nucleus recoils when the α-particle is emitted so the energy released is shared between the α-particle and the nucleus. Applying conservation of momentum to the recoil, you should be able to show that the energy released is shared between the α-particle and the nucleus in inverse proportion to their masses.

2 In β decay, the energy released is shared in variable proportions between the β-particle and the neutrino or antineutrino released in the decay. When the β-particle has maximum kinetic energy, the neutrino or antineutrino has negligible kinetic energy in comparison. The maximum kinetic energy of the β-particle is very slightly less than the energy released in the decay because of recoil of the nucleus.

Note

1 To calculate the energy corresponding to a mass difference of 1 atomic mass unit ($1\,\mathrm{u} = 1.6605 \times 10^{-27}$ kg), using $E = mc^2$ gives
$E = 1.6605 \times 10^{-27}\,\mathrm{kg} \times (2.9979 \times 10^8\,\mathrm{m\,s^{-1}})^2$
$= 1.4923 \times 10^{-10}\,\mathrm{J} = 931.5\,\mathrm{MeV}$ using exact values.

2 When calculating Q in beta decay, assume that the mass of the neutrino is negligible.

3 If the mass of each atom is given instead of the mass of its nucleus, calculate the mass of each nucleus by subtracting the mass of the electrons (= Z_m) in the atom from the mass of each atom.

Worked example

The polonium isotope $^{210}_{84}Po$ emits α-particles and decays to form the stable isotope of lead $^{206}_{82}Pb$. Write down an equation to represent this process and calculate the energy released when a $^{210}_{84}Po$ nucleus emits an α-particle.

mass of $^{210}_{84}Po$ nucleus = 209.93667 u

mass of $^{206}_{82}Pb$ nucleus = 205.92936 u

mass of α-particle = 4.00150 u

1 u is equivalent to 931.5 MeV

Solution

$$^{210}_{84}Po \longrightarrow {}^{4}_{2}\alpha + {}^{206}_{82}Pb \text{ (+ energy released } Q)$$

mass difference = total initial mass − total final mass

= 209.93667 − (205.92936 + 4.00150)

= 5.81×10^{-3} u

energy released Q = mass difference in u × 931.5

= 5.41 MeV

More about the strong nuclear force

The fact that most nuclei are stable tells us there must be an attractive force, the **strong nuclear force**, between any two protons or neutrons in the nucleus.

- The strength of the strong nuclear force can be estimated by working out the force of repulsion between two protons at a separation of 1 fm ($= 10^{-15}$ m), the approximate size of the nucleus. The strong nuclear force must be greater in magnitude that this force of repulsion.

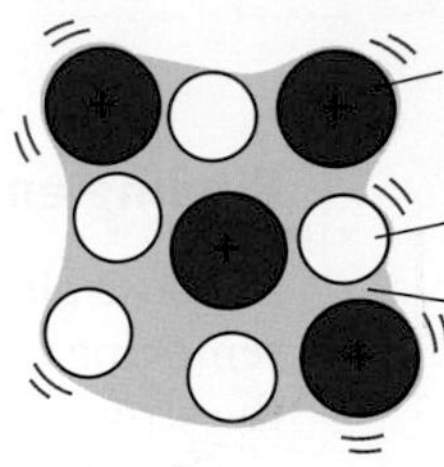

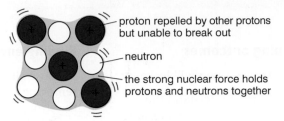

Figure 1 *The strong nuclear force*

Prove for yourself, using Coulomb's law of force (see Topic 14.1), that the force of repulsion between two protons at a separation of 10^{-15} m is of the order of 200 N. So the strong nuclear force is at least 200 N.

- The range of the strong nuclear force is no more than about 2 to 3 × 10^{-15} m. The diameter of a nucleus can be measured from high-energy electron scattering experiments. The results show that nucleons are evenly spaced at about 10^{-15} m in the nucleus and therefore the strong nuclear force acts only between nearest neighbour nucleons.
- The energy needed to pull a nucleon out of the nucleus is of the order of millions of electron volts (MeV). This can be deduced because the strong nuclear force is at least about 200 N and it acts over a distance of about 2 to 3 × 10^{-15} m. The work done by the strong nuclear force over this distance is therefore about 5×10^{-13} J (= 200 N × 2.5 × 10^{-15} m) which is about 3 MeV as 1 MeV = 1.6×10^{-13} J.
- The strong nuclear force between two nucleons must become repulsive at separations of about 0.5 fm or less, otherwise nucleons would pull each other closer and closer together and be much smaller than it is.

Summary test 21.4

Magnitude of the charge of the electron = 1.60×10^{-19} C, rest mass of an electron = 9.11×10^{-31} kg, 1 u = 931.5 MeV, $g = 9.81\,\text{m s}^{-2}$

1 Calculate the increase of mass of:

a a 10 kg object when it is raised through a height of 2.0 m,

b an electron when it is accelerated from rest through a p.d. of: **i** 5000 V, **ii** 5 MV.

2 The bismuth isotope $^{212}_{83}Bi$ emits α-particles and decays to form the stable isotope of thallium $^{208}_{81}Tl$.

a Write down an equation to represent this process and calculate the energy released.
Mass of $^{212}_{83}Bi$ nucleus = 211.94562 u,
Mass of $^{208}_{81}Tl$ nucleus = 207.93746 u,
Mass of α-particle = 4.00150 u

b Explain without calculation why the thallium nucleus in the above decay gains a small proportion of the energy released.

3 The strontium isotope $^{90}_{38}Sr$ emits β^- particles and decays to form the stable isotope of yttrium $^{90}_{39}Y$.

a Write down an equation to represent this process and calculate the energy released.
Mass of $^{90}_{38}Sr$ nucleus = 89.88640 u,
Mass of $^{90}_{39}Y$ nucleus = 89.88525 u,
Mass of β^- particle = 0.00055 u

b Explain without calculation why the kinetic energy of the β-particle released when the strontium nucleus decays varies from zero up to a maximum.

4 The sodium isotope $^{25}_{11}Na$ emits β^- particles and decays to form the stable isotope of magnesium $^{25}_{12}Mg$.

Calculate the Q-value of this decay.

Mass of $^{25}_{11}Na$ nucleus = 24.98931 u,

Mass of $^{25}_{12}Mg$ nucleus = 24.98528 u

21.5 Binding energy

Learning outcomes

On these pages you will learn to:

- define and understand the terms mass defect and binding energy
- calculate the binding energy per nucleon of a nuclide
- sketch the variation of binding energy per nucleon with nucleon number
- explain the relevance of binding energy per nucleon to nuclear fusion and to nuclear fission

Binding energy and mass defect

Suppose all the nucleons in a nucleus were separated from one another, removing each one from the nucleus in turn. Work must be done to overcome the strong nuclear force and separate each nucleon from the others. The potential energy of each nucleon is therefore increased when it is removed from the nucleus.

The binding energy of the nucleus is the work that must be done to separate a nucleus into its constituent neutrons and protons.

When a nucleus forms from separate neutrons and protons, energy is released as the strong nuclear force does work pulling the nucleons together. The energy released is equal to the binding energy of the nucleus. Because energy is released when a nucleus forms from separate neutrons and protons, the mass of a nucleus is less than the mass of the separated nucleons.

The mass defect Δm of a nucleus is defined as the difference between the mass of the separated nucleons and the combined mass of the nucleus.

- Calculation of the mass defect of a nucleus of known mass: a nucleus of an isotope ${}^{A}_{Z}X$ is composed of Z protons and $(A - Z)$ neutrons. Therefore, for a nucleus ${}^{A}_{Z}X$ of mass M_N, **its mass defect $\Delta m = Zm_p + (A - Z)m_n - M_N$.**
- Calculation of the binding energy of a nucleus: the mass defect Δm is due to energy released when the nucleus formed from separate neutrons and protons. The energy released in this process is equal to the binding energy of the nucleus. Therefore, **the binding energy of a nucleus** = $c^2(\Delta m)$.

Worked example

The mass of a nucleus of the bismuth isotope ${}^{212}_{83}Bi$ is 211.80012 u. Calculate the binding energy of this nucleus in MeV.

The mass of a proton, m_p = 1.00728 u; the mass of a neutron, m_n = 1.00866 u

1 u is equivalent to = 931.5 MeV

Solution

Mass defect $\Delta m = 83m_p + (212 - 83)m_n - M_N$ = 1.92126 u

$\therefore$ binding energy = 1.92126 u × 931.5 MeV/u = 1790 MeV

Note

1 The mass of an atom of an isotope ${}^{A}_{Z}X$ is measured using a mass spectrometer. The mass of a nucleus can then be calculated by subtracting the mass of Z electrons from the atomic mass.
2 The atomic mass unit, 1 u = 1.661×10^{-27} kg. This is defined as $\frac{1}{12}$th of the mass of an atom of the carbon isotope ${}^{12}_{6}C$.
3 The energy corresponding to a mass of 1 u = 931.5 MeV. See Topic 23.4 if necessary.

Nuclear stability

The binding energy of each nuclide is different. The **binding energy per nucleon** of a nucleus is the work done to remove a **nucleon** from a nucleus; it is therefore a measure of the stability of a nucleus. For example, the binding energy per nucleon of the ${}^{212}_{83}Bi$ nucleus is 8.4 MeV per nucleon (= 1790 MeV/212 nucleons).

If the binding energy per nucleon of two different nuclides are compared, the nucleus with more binding energy per nucleon is the more stable of the two nuclei. Figure 1 shows a graph of the binding energy per nucleon v. mass number A for all the known nuclides. This graph is a curve which has a maximum value of 8.7 MeV per nucleon between $A = 50$ and $A = 60$. Nuclei with mass numbers in this range are the most stable nuclei. As explained below, energy is released in:

- **nuclear fission**, the process in which a large unstable nucleus splits into two fragments which are more stable than the original nucleus. The binding energy per nucleon increases in this process, as shown in Figure 1.
- **nuclear fusion**, the process of making small nuclei fuse together to form a larger nucleus. The product nucleus has more binding energy per nucleon than the smaller nuclei. So the binding energy per nucleon also increases in this process, provided the nucleon number of the product nucleus is no greater than about 50.

Note The change of binding energy per nucleon is about 0.5 MeV in a fission reaction and can be more than 20 times as much in a fusion reaction.

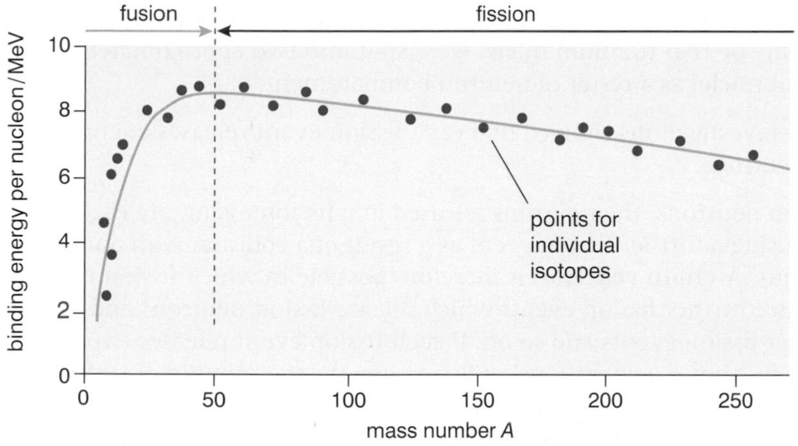

***Figure 1** Binding energy per nucleon for all known nuclides*

Summary test 21.5

mass of a proton, $m_p = 1.00728\,u$; mass of a neutron, $m_n = 1.00866\,u$;
1 u is equivalent to = 931.5 MeV

1 a Explain what is meant by the *binding energy* of a nucleus.

b Sketch a curve to show how the binding energy per nucleon of a nucleus varies with its mass number A, showing the approximate scale on each axis.

2 Calculate the binding energy per nucleon, in MeV per nucleon, of:

a a $^{12}_{6}C$ nucleus (mass = 12 u by definition),

b a $^{56}_{26}Fe$ nucleus (mass = 55.92067 u).

3 a Calculate the binding energy per nucleon in MeV per nucleon, of:
i an α-particle, **ii** a $^{3}_{2}He$ nucleus.
Mass of an α-particle = 4.00150 u; mass of a $^{3}_{2}He$ nucleus = 3.01493 u

b Use the results of your calculations in **a** to explain why an α-particle rather than a $^{3}_{2}He$ nucleus is emitted by a large unstable nucleus.

4 Calculate the binding energy per nucleon, in MeV, of the $^{2}_{1}H$ nucleus.
Mass of $^{2}_{1}H$ nucleus = 2.01355 u

21.6 Fission and fusion

Learning outcomes

On these pages you will learn to:

- explain what is meant by nuclear fission and by a chain reaction
- calculate the energy released in a fission or fusion event from the masses of the nuclei and other particles involved
- explain what is meant by nuclear fusion and explain in terms of forces the necessary condition to fuse two nuclei

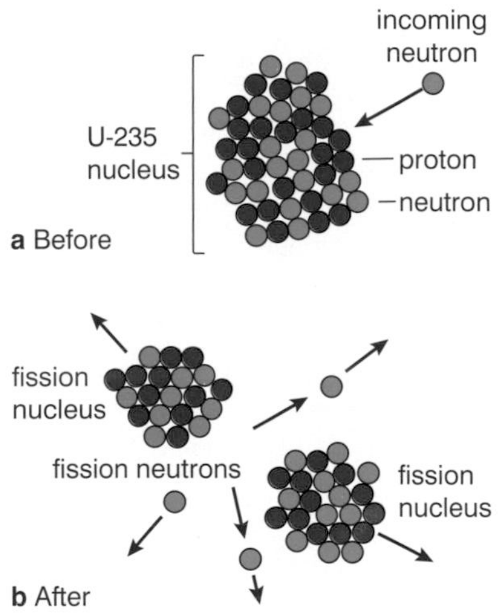

***Figure 1** Induced fission*

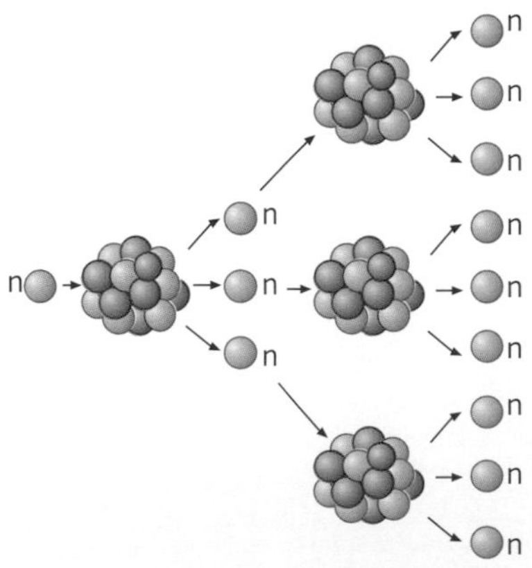

***Figure 2** A chain reaction in a nuclear reactor*

Note

The fragment nuclei have excess neutrons compared with stable nuclei of the same proton number. They become stable by emitting β^- particles. See Topic 10.4.

Induced fission

Fission of a nucleus occurs when a nucleus splits into two approximately equal fragments. This happens when the uranium isotope $^{235}_{92}U$ is bombarded with neutrons, a discovery made by Hahn and Strassmann in 1938. The process is known as induced fission. The plutonium isotope, $^{239}_{94}Pu$, is the only other isotope that is fissionable. This isotope is an artificial isotope formed by bombarding nuclei of the uranium isotope $^{238}_{92}U$ with neutrons.

Hahn and Strassmann knew that bombarding different elements with neutrons produces radioactive isotopes. Uranium is the heaviest of all the naturally occurring elements; scientists thought that neutron bombardment could turn uranium nuclei into even heavier nuclei. Hahn and Strassmann undertook the difficult work of analysing chemically the products of uranium after neutron bombardment to try to discover any new elements heavier than uranium. Instead, they discovered that many lighter elements such as barium were present after bombardment, even though the uranium was pure before. The conclusion could only be that uranium nuclei were split into two approximately equal fragment nuclei as a result of neutron bombardment.

Further investigations showed that each fission event releases energy and two or three neutrons.

- Fission neutrons, the neutrons released in a fission event, are each capable of causing a further fission event as a result of a collision with another $^{235}_{92}U$ nucleus. A **chain reaction** is therefore possible in which fission neutrons produce further fission events which release fission neutrons and cause further fission events and so on. If each fission event releases two neutrons on average, after n 'generations' of fission events, the number of fission neutrons would be 2^n. Prove for yourself that fission of 6×10^{23} $^{235}_{92}U$ nuclei (i.e. 235 g of the isotope) would happen in 79 generations. As explained below, each fission event releases about 200 MeV of energy. Because each event takes no more than a fraction of a second, a huge amount of energy is released in a very short time. Using the above figures, complete fission of 235 g of $^{235}_{92}U$ would release about 10^{13} J (= $6 \times 10^{23} \times 200$ MeV). This is about a million times more than the energy released as a result of burning a similar mass of fossil fuel.
- Energy is released when a fission event occurs because the fragments repel each other (as they are both positively charged) with sufficient force to overcome the strong nuclear force trying to hold them together. The fragment nuclei and the fission neutrons therefore gain kinetic energy. The two fragment nuclei are smaller and therefore more tightly bound than the original $^{235}_{92}U$ nucleus. In other words, they have more binding energy so they are more stable than the original nucleus. The energy released is equal to the change of binding energy. The binding energy of each nucleon increases from about 7.5 MeV to about 8.5 MeV as a result of the fission event. As there are about 240 nucleons in the original nucleus, the energy released in a fission event is of the order of 200 MeV (= 240 × about 1 MeV).
- Many fission products are possible when a fission event occurs. For example, the equation below shows a fission event in which a $^{235}_{92}U$ nucleus is split into a barium $^{144}_{56}Ba$ nucleus and a krypton $^{90}_{36}Kr$ nucleus and two neutrons are released.

$$^{235}_{92}U + ^{1}_{0}n \longrightarrow ^{144}_{56}Ba + ^{90}_{36}Kr + 2\,^{1}_{0}n + \text{energy released, } Q$$

- The energy released, Q, can be calculated using $E = mc^2$ in the form $Q = c^2(\Delta m)$, where Δm is the difference between the total mass before and after the event.
- In the above equation, the mass difference

$$\Delta m = M_{U\text{-}235} - M_{Ba\text{-}144} - M_{Kr\text{-}90} - m_n$$

where M represents the appropriate nuclear mass and m_n is the mass of the neutron.

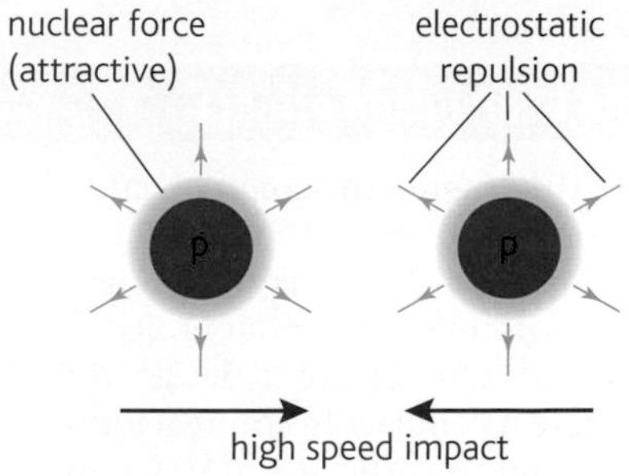

Figure 3 *Fusion of two protons*

Nuclear fusion

Fusion takes place when two nuclei combine to form a bigger nucleus. The binding energy curve Topic 21.5 shows that if two light nuclei are combined, the individual nucleons become more tightly bound together. The binding energy per nucleon of the product nucleus is greater than of the initial nuclei. In other words, the nucleons become even more trapped in the nucleus when fusion occurs. As a result, energy is released equal to the increase of binding energy.

Nuclear fusion can only take place if the two nuclei that are to be combined collide at high speed. This is necessary to overcome the electrostatic repulsion between the two nuclei so that they can become close enough to interact through the strong nuclear force. Some examples of nuclear fusion reactions are shown in Figure 4 and described below.

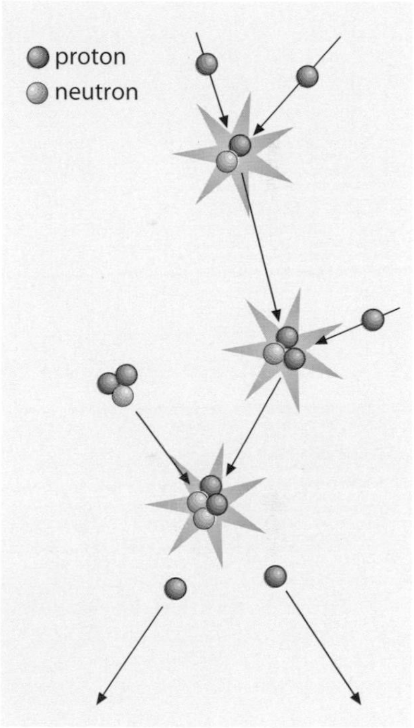

Figure 4 *Fusion reactions inside the Sun*

1 The fusion of two protons produces a nucleus of deuterium (the hydrogen isotope ${}^2_1H$), a β^+ particle and a neutrino.

$${}^1_1p + {}^1_1p \longrightarrow {}^2_1H + {}^{\ 0}_{+1}\beta + \nu$$

2 The fusion of a proton and a deuterium nucleus ${}^2_1H$ produces a nucleus of the helium isotope ${}^3_2He$ and 5.5 MeV of energy.

$${}^2_1H + {}^1_1p \longrightarrow {}^3_2He$$

3 The fusion of two nuclei of helium isotope, ${}^3_2He$, produces a nucleus of the helium isotope ${}^4_2He$, two protons and 12.9 MeV of energy.

$${}^3_2He + {}^3_2He \longrightarrow {}^4_2He + 2{}^1_1p$$

In each case, the energy released in the reaction may calculated using $E = mc^2$ in the form $Q = c^2(\Delta m)$, where Δm is the difference between the total mass before and after the event.

Solar energy is produced as a result of fusion reactions inside the Sun. The temperature at the centre of the Sun is thought to be 10^8 K or more. At such temperatures, atoms are stripped of their electrons. Matter in this state is referred to as 'plasma'. The nuclei of the plasma move at very high speeds because of the enormous temperature. When two nuclei collide, they fuse together because they overcome the electrostatic repulsion due to their charge and approach each other closely enough to interact through the strong nuclear force. Protons (i.e. hydrogen nuclei) inside the Sun's core fuse together in stages (corresponding to equations 1, 2 and 3 above) to form helium ${}^4_2He$ nuclei. For each helium nucleus formed, 25 MeV of energy is released. This corresponds to 7 MeV per proton, considerably more than the energy released per nucleon in a fission event.

Fusion power

Fusion reactors are still at the prototype stage even though scientific teams in several countries have been working on fusion research for more than 50 years. Prototype fusion reactors such as JET, the Joint European Torus, in the United Kingdom have produced large amounts of power but only for short periods of time. JET produces less power than it uses but the less powerful International Thermonuclear Experimental Reactor (ITER) due to start up in 2025 is designed to produce several times more power than it uses.

Energy is released in JET by fusing nuclei of deuterium ${}^{2}_{1}H$ and tritium ${}^{3}_{1}H$ to produce nuclei of the helium isotope ${}^{4}_{2}He$ and neutrons, as below.

$${}^{2}_{1}H + {}^{3}_{1}H \longrightarrow {}^{4}_{2}He + {}^{1}_{0}n + 17.6\,MeV$$

The neutrons are absorbed by a 'blanket' of lithium surrounding the reactor vessel. The reaction between the neutrons and the lithium nuclei, as shown below, produces tritium which is then used in the main reaction. Deuterium occurs naturally in water as it forms 0.01% of naturally occurring hydrogen.

$${}^{6}_{3}Li + {}^{1}_{0}n \longrightarrow {}^{4}_{2}He + {}^{3}_{1}H + 4.8\,MeV$$

The plasma is contained in a doughnut-shaped steel container and is heated by passing a very large current through it. A magnetic field is used to confine the plasma so it does not touch the sides of its steel container, otherwise it would lose its energy. In theory, the energy released per second should be more than is needed to heat the plasma so the reactor ought to give a continuous output of power. However, at the present time, power can only be produced for a short time as the plasma becomes unstable at such high temperatures.

***Figure 5** The JET fusion reactor*

The STEP fusion project

The UK entered the commercial fusion reactor race in 2019, with the Spherical Tokamak for Energy Production (STEP) fusion project. The aims of STEP are to design and operate a compact fusion reactor by 2024 and construct a 100 MW power plant by 2040.

Summary test 21.6

1 u is equivalent to 931.5 MeV

1 a Explain why the protons in a nucleus do not leave the nucleus even though they repel each other.

b Explain why the mass of a nucleus is less than the mass of the separated protons and neutrons from which the nucleus is composed.

2 a What is meant by *nuclear fission*?

b i The incomplete equation below represents a reaction that takes place when a neutron collides with a nucleus of the uranium isotope ${}^{235}_{92}U$. Determine the values of *a* and *b* in this equation.

$${}^{235}_{92}U + {}^{1}_{0}n \longrightarrow {}^{136}_{a}Xe + {}^{b}_{36}Kr + 2{}^{1}_{0}n + \text{energy released, } Q$$

ii Calculate the energy, in MeV, released in this fission reaction.

Masses:
${}^{235}_{92}U$ nucleus 234.993 u, ${}^{136}_{a}Xe$ nucleus 135.877 u,
${}^{b}_{36}Kr$ nucleus 97.886 u, neutron 1.00866 u

3 a What is meant by *nuclear fusion*?

b Hydrogen nuclei fuse together to form helium nuclei in the Sun. Two stages in this process are represented by the following equations:

$${}^{1}_{1}p + {}^{1}_{1}p \longrightarrow {}^{2}_{1}H + {}^{0}_{+1}\beta$$

$${}^{2}_{1}H + {}^{1}_{1}p \longrightarrow {}^{3}_{2}He$$

i Describe the reactions that these equations represent.

ii Calculate the energy released in each reaction.

Masses: β-particle 0.00055 u, proton 1.00728 u,
${}^{2}_{1}H$ nucleus 2.01355 u, ${}^{3}_{2}He$ nucleus 3.01493 u

4 a Explain why light nuclei do not fuse when they collide unless they are moving at a sufficiently high speed.

b Calculate the energy released in the following fusion reaction:

$${}^{3}_{2}He + {}^{3}_{2}He \longrightarrow {}^{4}_{2}He + 2{}^{1}_{1}p$$

Masses: proton 1.00728 u, ${}^{3}_{2}He$ nucleus 3.01493 u, α-particle 4.00150

c Show that about 25 MeV of energy is released when a ${}^{4}_{2}He$ nucleus is formed from 4 protons.

Chapter Summary

The **half-life**, $t_{\frac{1}{2}}$, of a radioactive isotope is the time taken for the mass of the isotope to decrease to half the initial mass. This is the same as the time taken for the number of nuclei of the isotope to decrease to half the initial number.

The decay constant, λ, is the probability of an individual nucleus decaying per second.

The **activity**, *A*, of a radioactive isotope is the number of nuclei of the isotope that disintegrate per second. The unit of activity is the becquerel (Bq), equal to one disintegration per second.

Binding energy of a nucleus is the work that must be done to separate a nucleus into its constituent neutrons and protons.

Binding energy = mass defect × c^2

Binding energy/nucleon is greatest for nuclei of mass number 57.

Fission is the splitting of a ${}^{235}_{92}U$ nucleus or a ${}^{239}_{94}Pu$ nucleus into two approximately equal fragments. Induced fission is fission caused by an incoming neutron colliding with a ${}^{235}_{92}U$ nucleus or a ${}^{239}_{94}Pu$ nucleus.

Fusion is the fusing together of light nuclei to form a heavier nucleus.

Equations

1 $A = \lambda N$

2 $N = N_0 e^{-\lambda t}$, $A = A_0 e^{-\lambda t}$, $C = C_0 e^{-\lambda t}$

3 Binding energy = mass defect × c^2

21 Exam-style and Practice Questions

Launch additional digital resources for the chapter

1 u is equivalent to 931.5 MeV

1 a A radioactive isotope of polonium, $^{210}_{84}Po$, has a half-life of 140 days. Calculate:

i the decay constant in s^{-1},

ii the mass of a sample of this isotope which has an activity of 1.0×10^{12} Bq.

b The isotope emits α-particles and forms a stable isotope of lead. Each disintegration in the isotope releases 5.3 MeV of energy. The isotope is to be used to supply power for a remote weather station for at least 1 year.

i Show that 0.72 g of this isotope will release 100 W of power.

ii Calculate the power supplied after 1 year by an initial mass of 0.72 g of this isotope.

2 a Calculate the number of atoms present in 1.0 kg of $^{226}_{88}Ra$.

b The radioactive isotope $^{226}_{88}Ra$ has a half-life of 1620 years. Calculate the activity of 1 milligram of this isotope.

3 a Radioactive disintegration is a random process yet it is possible to calculate reasonably accurately the number of atoms in a radioactive source of known activity and half-life. Explain why.

b A radioactive isotope X with a half-life of 44 hours disintegrates to form a stable product. A pure sample of X is prepared with an activity of 60 kBq. Calculate:

i the activity of the sample after 24 hours,

ii the time taken for the activity of the sample to decrease to 10% of its initial activity.

4 The radioactive isotope of cobalt, $^{60}_{27}Co$, has a half-life of 5.3 years. It emits γ-photons of energy 1.3 MeV.

a Calculate:

i the activity of a sample of mass of 10 g of this isotope,

ii the energy transfer per second from this sample.

b The sample is in a thick lead container of mass 920 kg. Calculate the temperature rise of the container in 24 hours, assuming no energy transfer occurs from the container.

Specific heat capacity of lead = $130\,J\,kg^{-1}\,K^{-1}$.

c $^{60}_{27}Co$ is used for gamma therapy to destroy tumours.

i What property of γ radiation is made use of in this application?

ii What precautions are taken to ensure the patient is not exposed to the beam unnecessarily?

5 a Explain what is meant by the binding energy of a nucleus.

b Calculate the binding energy per nucleon, in MeV, of:

i an α-particle,

ii a deuterium $^{2}_{1}H$ nucleus.

Masses: neutron = 1.00866 u, proton = 1.00728 u, $^{2}_{1}H$ nucleus = 2.01355 u, α-particle = 4.00150 u

c Discuss why an α-particle does not break up into two $^{2}_{1}H$ nuclei.

6 a In a nuclear reaction, a neutron collided with a nucleus of the lithium isotope $^{6}_{3}Li$. A tritium nucleus $^{3}_{1}H$ and another nucleus X was formed as a result.

Write down an equation that represents this reaction and identify the nucleus X.

b The mass loss in the above reaction was 0.00514 u.

i Calculate the energy released, in J, in this reaction.

ii Calculate the mass of X, given the masses of the other nuclei are as follows: $^{3}_{1}H$ 3.0155 u, $^{6}_{3}Li$ 6.01348 u, neutron 1.00866 u

7 a i State two properties of the strong nuclear force.

ii Explain why energy is released when a $^{235}_{92}U$ nucleus undergoes induced fission.

b i Copy and complete the induced fission equation below:

$$^{235}_{92}U + ^{1}_{0}n \rightarrow ^{140}_{?}Xe + ^{93}_{38}Sr + ?\,^{1}_{0}n$$

ii Calculate the energy released, in J, in the above reaction.

Masses: $^{235}_{92}U$ 235.0439 u, $^{1}_{0}n$ 1.00866 u, $^{140}_{?}Xe$ 139.9216 u, $^{93}_{38}Sr$ 92.9140 u

8 a State one difference and one similarity between a fusion reaction and a fission reaction.

b i Explain why fusion in a plasma containing light nuclei only takes place if the temperature of the plasma is of the order of 10^{8} K.

ii Two deuterium ($^{2}_{1}H$) nuclei fuse together to form a tritium ($^{3}_{1}H$) nucleus and a proton.
Write down the equation which represents this reaction and use the information below to calculate the energy released, in MeV, in this reaction.
Masses $^{2}_{1}H$ 2.0136 u, $^{3}_{1}H$ 3.0155 u, proton 1.0073 u

9 a Explain why the spent fuel rods from a nuclear reactor are more radioactive after removal from the reactor than they were before they were used in the reactor.

b The radioactive isotope, $^{90}_{38}Sr$, is a β^--emitter which has a half-life of 28 years. It is produced as a fission product in a fuel rod in a nuclear reactor. If it escapes into the environment, it can be absorbed by the body in place of calcium. Calculate:

i the activity of 1 mg of this isotope,

ii the activity in 100 years of a sample of this isotope that has a mass of 1 mg at the present time.

10 a State what is meant by the binding energy of a nucleus.

b The mass of the hydrogen $^{2}_{1}H$ nucleus is 2.01355 u.

i State how many protons and how many neutrons are in this nucleus.

ii The mass of a proton is 1.00728 u and the mass of a neutron is 1.00867 u. Calculate the binding energy per nucleon of the hydrogen $^{2}_{1}H$ nucleus.

c The binding energy per nucleon of the hydrogen $^{3}_{1}H$ nucleus is 2.832 MeV.

i Calculate the ratio of the binding energy of the hydrogen $^{3}_{1}H$ nucleus to that of the hydrogen $^{2}_{1}H$ nucleus.

ii Student A suggests the ratio should be 1.5 on the grounds that the ratio of the number of nucleons in the two nuclei is 3 : 2. Student B suggests the ratio should be 3 on the grounds that there are three force bonds between the nucleons in the $^{3}_{1}H$ nucleus and only one in the $^{2}_{1}H$ nucleus. State which suggestion is closer to the binding energy ratio calculated in **ci** and discuss how this suggestion could be applied to other known light nuclei such as the helium $^{4}_{2}He$ nucleus which has a binding energy of 28 MeV.

d A helium $^{4}_{2}He$ nucleus is formed when a hydrogen $^{3}_{1}H$ nucleus and a hydrogen $^{2}_{1}H$ nucleus are fused. In this process, a particle Y is released.

i Identify particle Y and write an equation to represent this reaction.

ii The binding energy of a hydrogen $^{3}_{1}H$ nucleus is about 9 MeV. Estimate the energy released in the above reaction.

11 a A thermal nuclear reactor is designed to release energy at a steady rate as a result of induced nuclear fission.

i Explain what is meant by induced *nuclear fission*.

ii State an isotope that undergoes fission in a thermal nuclear reactor.

iii Explain what is meant by a *chain reaction* in nuclear fission.

iv Explain why the mass of fissile material in a nuclear reactor must be greater than a minimum mass in order for a chain reaction to occur.

b The fuel rods in the core of a nuclear reactor are surrounded by a moderator. Control rods in the core are used to control the rate of release of energy from the fuel rods. A coolant is pumped through the core to transfer energy from the reactor core to a heat exchanger.

i State the purpose of the moderator.

ii Name a substance which control rods are made from and describe how the control rods are used to keep the rate of release of energy constant.

iii State two physical properties of the coolant.

22 Medical imaging

22.1 Ultrasonic imaging

Learning outcomes

On these pages you will learn to:

- explain the principles of the generation and detection of ultrasonic waves using piezo-electric transducers
- explain the main principles behind the use of ultrasound to obtain diagnostic information about internal structures
- define specific acoustic impedance and explain the importance in relation to the intensity reflection coefficient at a boundary
- recall and solve problems by using the equation $I = I_0 e^{-\mu x}$ for the attenuation of ultrasound in matter

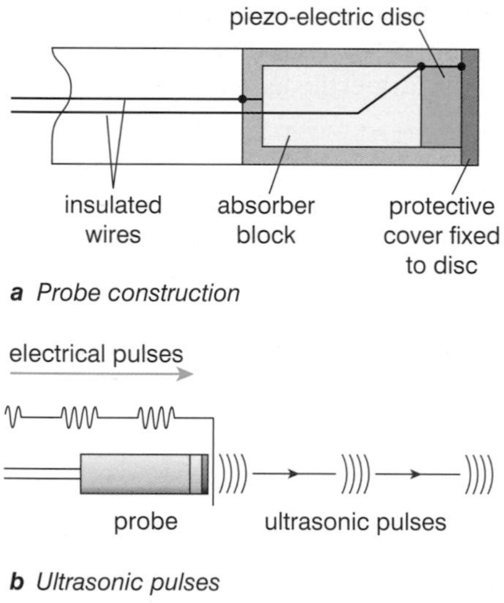

Figure 1 *An ultrasonic probe*

Producing ultrasonic waves

Ultrasonics are sound waves at frequencies of more than about 18 kHz above the range of the human ear. Unlike X-rays, ultrasonic radiation is non-ionising radiation and therefore does not damage living tissue. For medical imaging, ultrasonics at frequencies between about 1 and 10 MHz are used as diffraction would be significant at lower frequencies and reduced intensity due to absorption would be significant at higher frequencies.

An **ultrasonic probe** used in ultrasonic scanning is a hand-held device placed in contact with the body surface to direct pulses of ultrasound into the body. Each emitted pulse is partially reflected by internal boundaries in the body. The reflected pulses are then detected by the probe before the next pulse is emitted.

Figure 1 shows the construction of an ultrasonic probe.

- The probe contains a piezo-electric **transducer** in the shape of a disc. Piezo-electricity is a property of certain solids whereby a p.d. applied between opposite faces causes a change of distance between the two faces. When an alternating p.d. is applied between the faces of the disc, the disc vibrates due to its changing thickness.
- By applying an alternating p.d. of frequency equal to the resonant frequency of vibration of the disc, the disc vibrates at resonance and creates ultrasonic waves in the surrounding medium. The thickness of the disc determines its resonant frequency.
- An absorber pad of 'backing material' behind the disc prevents ultrasonic waves created at the two surfaces of the disc from cancelling each other out. The pad also damps the vibrations of the disc rapidly after each pulse is emitted.

More about piezo-electricity

A **transducer** is any device that is designed to convert energy from one form to another. A piezo-electric transducer generates a p.d. when it is squeezed. The piezo-electric material contains positive and negative ions, which are held together by the electrostatic forces they exert on each other. The centre of the negative charge of each molecule is in the same position as the centre of positive charge. When pressure is applied to opposite surfaces of the material, the centres of charge of the positive and negative ions are displaced slightly in opposite directions, causing a potential difference between the two surfaces. The effect is known as the **piezo-electric effect** and is displayed by crystals such as quartz.

To apply a p.d. across a quartz crystal, opposite surfaces to which the pressure is applied must be coated with metal so that an electrical connection can be made to each surface.

The piezo-electric effect is reversible in that the application of a p.d. across a piezo-electric material causes the distance across the material to increase or decrease according to the polarity of the p.d.

Absorption and reflection of ultrasonic waves

Ultrasonic waves can be reflected and refracted just like sound waves. When ultrasonic waves reach a boundary between two substances, some of the wave energy is reflected and some is transmitted as shown in Figure 3.

Considering the energy reaching the boundary in 1 second, as energy cannot be created or destroyed, the sum of the reflected energy and the transmitted energy is equal to the incident energy. Hence,
incident intensity I = the reflected intensity I_R + the transmitted intensity I_T.

The fraction of the incident energy that is reflected or transmitted depends on:

- the angle of incidence θ of the incident waves,
- the densities ρ_1 and ρ_2 of the two substances,
- the wave speeds c_1 and c_2 of ultrasonic waves in the two substances.

The **acoustic impedance** of a substance, Z is defined by the equation $\mathbf{Z = \rho c}$

The unit of Z is given by the product of the unit of density (i.e. $kg\,m^{-3}$) and the unit of speed (i.e. $m\,s^{-1}$). Hence the unit of Z is $kg\,m^{-2}\,s^{-1}$.

Specific values for different substances are given in Table 1. Notice the very small value for air and the large values for quartz and bone compared with the other substances listed in the table.

Table 1

Substance	Acoustic impedance $Z/kg\,m^{-2}\,s^{-1}$
air	430
blood	1.59×10^6
bone	6.80×10^6
fat	1.38×10^6
muscle	1.70×10^6
quartz	1.52×10^7
soft tissue	1.63×10^6
water	1.50×10^6

When ultrasonic waves are incident on a boundary between two substances with acoustic impedances Z_1 and Z_2, the ratio of the reflected intensity to the incident intensity $\frac{I_R}{I}$, the reflection coefficient, is given by the equation

$$\frac{I_R}{I} = \frac{(Z_2 - Z_1)^2}{(Z_2 + Z_1)^2}$$

- If Z_1 and Z_2 are almost equal, the ratio is close to zero and the reflected intensity is very small compared with the incident intensity. In other words, most of the wave energy is transmitted.
- If Z_1 and Z_2 are very different, the ratio is close to 1 so most of the wave energy is reflected. The transmitted intensity is very small compared with the incident intensity. In other words, most of the wave energy is reflected.

The intensity of the reflected pulses from the ultrasonic probe when they return to it depends on:

1 the ratio of the reflected intensity to the incident intensity at each boundary, and
2 the absorption of the ultrasonic waves by each substance they pass through.

When ultrasound is directed from a probe into the body, any air trapped between the probe and the body will cause most of the ultrasound energy to be reflected from the body. This is because the acoustic impedances of air and soft tissue are very different so the ratio I_R/I is close to 1. To eliminate such trapped air, a **coupling medium** such as a gel is applied between the probe and the body surface. Such substances have similar acoustic impedances to soft tissue so the ratio I_R/I is close to zero. In other words, most of the wave energy is transmitted into the body.

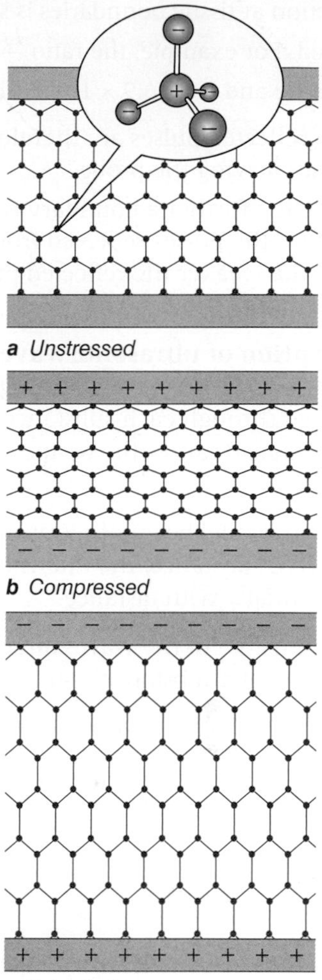

Figure 2 *Piezo-electricity*

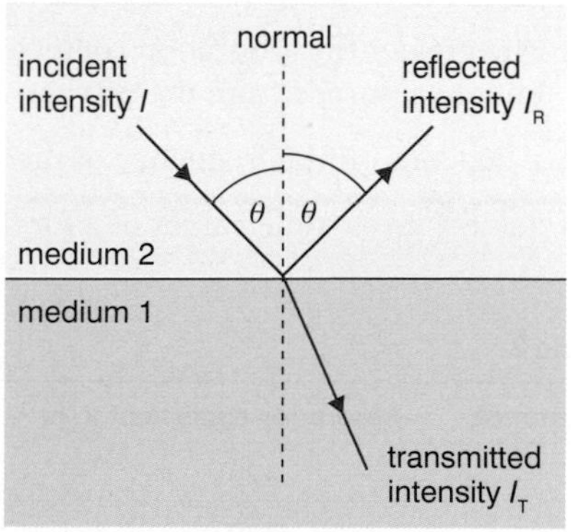

Figure 3 *Reflection and transmission at a boundary*

Reflection at tissue boundaries is significant and can't be avoided. For example, the ratio $\frac{I_R}{I}$ for a boundary between soft tissue and fat is 6.9×10^{-3} (using values from Table 1).

- The reflected pulses are therefore much weaker than the pulses leaving the probe.
- Also, the further a boundary is from the probe, the further the pulses reflected from that boundary travel and the weaker the reflected pulses from it will be due to absorption (see below).

Absorption of ultrasonic waves depends on the substances the waves pass through and the distance travelled through each substance. The energy absorbed by the substance causes the temperature of the substance to increase.

When a parallel beam of ultrasonic waves travels through a substance, the intensity of the waves decreases exponentially with distance.

The intensity I of the waves after travelling through distance x of a substance is given by the equation

$$I = I_0 e^{-\mu x}$$

where I_0 is the incident intensity and μ is the absorption coefficient of the substance. The unit of μ is m^{-1}.

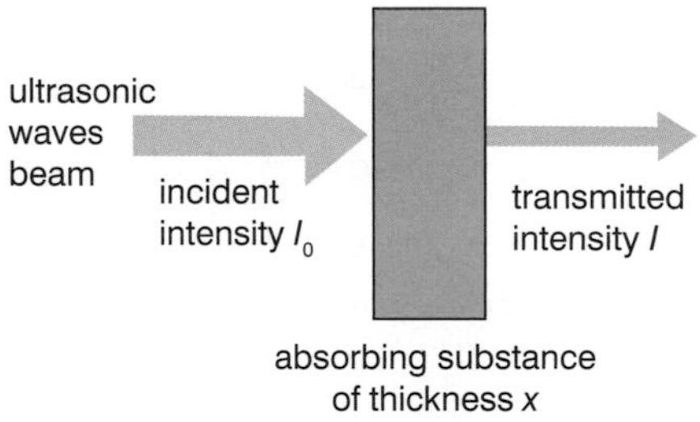

Figure 4 *Absorption*

Note

1 The greater the absorption coefficient μ, the greater the energy absorbed from the waves over any given distance.

2 μ depends on the frequency of the ultrasonic waves.

3 Table 2 gives some values of μ for different substances.

Table 2

Substance	Absorption coefficient μ / m^{-1}
air	120
bone	130
muscle	23
water	0.02

Ultrasonic scans

An **ultrasonic scanner** consists of an ultrasonic probe connected to a control unit and a visual display unit. In the simplest scan system, referred to as the **A-scan system**, a pulse generator is used to supply electrical pulses to the probe and to trigger the oscilloscope time base each time a pulse is generated. Figure 5 shows an A-scan of an eye.

- In each scan, a pulse is generated by the probe and, before the next pulse is generated, the probe detects reflected pulses from the boundaries in the path of the pulse. The reflected pulses or 'echoes' detected by the probe are amplified and displayed on the oscilloscope.
- Each time the time base is triggered, the oscilloscope beam sweeps across the screen from left to right. The time base of the oscilloscope is adjusted to display on the screen all the reflected pulses for each transmitted pulse. As a result, each pulse on the screen will be very narrow as the duration of each pulse is much shorter than the time for each 'sweep' of the oscilloscope screen.

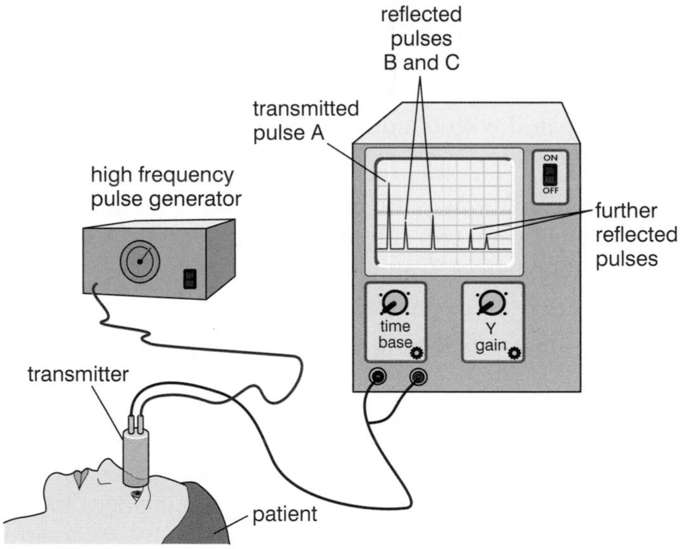

Figure 5 *The A-scan system*

1 The position of each reflected pulse on the screen depends on the transit time of the ultrasonic pulse (i.e. the time taken by the ultrasonic pulse to travel from the probe to the internal boundary that reflected the pulse and back).
2 The further a reflected pulse on the screen appears from the transmitted pulse:
   - the longer the transit time of the pulse in the body and the further away the boundary is from the probe,
   - the smaller the pulse height will be because the ultrasonic pulse is partially reflected by the boundaries it passes through and the substances it passes through absorb some of its energy.

3 The transit time is proportional to the distance from the probe to the boundary. Therefore, the greater the distance from the probe to the boundary, the further the pulse appears across the screen. The oscilloscope can be used to measure the transit time, t, of a pulse. The distance travelled by the pulse $s = vt$, where v is the speed of ultrasonic waves in the body. Therefore, the distance from the probe to the internal boundary causing the pulse $= \frac{1}{2}vt$. Using this equation, the screen could be calibrated in terms of distance from the probe so the distance between a boundary and any other boundary or the probe can be measured directly.

In the **B-scan system**, the probe has a number of ultrasonic transducers side by side, each one sending out ultrasonic pulses in a slightly different direction to the others. The signals from the transducers due to the reflected pulses are processed by a computer such that each reflected pulse is displayed as a bright spot on the screen in the correct direction and at the correct distance from the probe. As the probe is moved over the body surface, the bright spots on the screen build up a two-dimensional image of the reflecting boundaries scanned. The image may be enhanced and stored electronically.

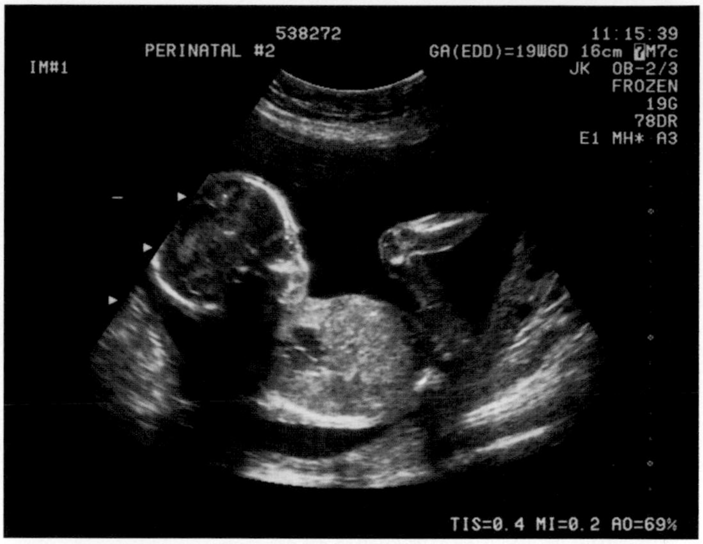

Figure 6 *A B-scan image of an unborn baby*

Comparison with X-rays

- B-scans are used for pre-natal scans (i.e. to observe unborn babies in the womb) rather than X-ray CT scans. This is because ultrasonic waves are non-ionising and, at the intensities used in scanning, do not damage human tissue.
- Ultrasound reflects at bone/tissue boundaries as well as at internal boundaries between soft substances such as fat, muscle and tissue because the acoustic impedance of such substances differs. Therefore, ultrasonic images show such 'soft' boundaries whereas X-ray images do not (because X-rays are not reflected at such boundaries).

Summary test 22.1

Use the data in Table 1, p.327, where necessary.

1 An ultrasonic probe generates ultrasonic waves at a frequency of 2.5 MHz. The speed of ultrasound in air $= 350\,\text{m}\,\text{s}^{-1}$ and in soft tissue $= 1550\,\text{m}\,\text{s}^{-1}$.

a Calculate the wavelength of the ultrasonic waves from this probe: **i** in air, **ii** in soft tissue.

b Explain why ultrasonic waves of much lower frequency are unsuitable for medical imaging.

2 a i With the aid of a diagram, describe the construction of an ultrasonic probe and how it produces ultrasonic waves.

ii Explain the function of the backing block in an ultrasonic probe.

b In the A-scan arrangement shown in Figure 5, on the previous page, the furthest boundary from the probe is the retina.

i Explain the presence of each pulse on the screen in terms of the cross-section of the patient's eye.

ii Calculate the distance between the boundary responsible for pulse B and the retina in Figure 5, if the distance from the probe to the furthest boundary is 24 mm.

3 a Use the data in Table 1 to calculate the reflection coefficient of the boundary between: **i** air and skin, **ii** water and skin. Assume skin is soft tissue.

b Use the results of your calculation to explain why a gel must be applied between an ultrasonic probe and the skin when the probe is used.

c A body organ has a density of $1040\,\text{kg}\,\text{m}^{-3}$ and the speed of sound through it is $1580\,\text{m}\,\text{s}^{-1}$.

i Calculate the acoustic impedance of the organ tissue.

ii Use the data in Table 1 to calculate the reflection coefficient of the boundary between the organ and the surrounding soft tissue.

4 a State the main differences between an A-scan and a B-scan.

b Ultrasonic waves and X-rays are both used for medical imaging. Explain why an ultrasonic scan rather than an X-ray scan is used for scanning a baby in the womb.

22.2 X-rays

Learning outcomes

On these pages you will learn to:

- explain the principles of the production of X-rays by electron bombardment of a metal target
- describe the main features of a modern X-ray tube, including control of the intensity and hardness of the X-ray beam
- show an understanding of the use of X-rays in imaging internal body structures, including a simple analysis of the causes of sharpness and contrast in X-ray imaging
- recall and solve problems by using the equation $I = I_0 e^{-\mu x}$ for the attenuation of X-rays in matter
- show an understanding of the purpose and principles of computed tomography (CT scanning)
- show an understanding of how the image of an 8-voxel cube can be developed using CT scanning

The production and properties of X-rays

X-ray imaging in medicine is an example of a diagnostic technique that is non-invasive. X-rays are electromagnetic waves of wavelength of the order of 0.1 nm or less. Figure 1 shows how a diagnostic X-ray tube works. The current through the filament wire heats the wire, which causes electrons to be emitted from the wire. These electrons are attracted from the filament or 'cathode' to the **anode** when the anode is positive relative to the filament, typically 20–100 kV for X-ray imaging. The electrons are stopped when they collide with the anode and they emit X-rays in the process.

For an anode potential V, the maximum energy of an X-ray photon = eV so the maximum frequency $f_{max} = eV/h$
Hence the minimum wavelength λ_{min} is given by:

$$\lambda_{min} = \frac{hc}{eV}$$

Note

X-ray tubes used for therapy to destroy tumours are designed differently because they need to produce photons at higher energies. Such tubes operate at voltages from 250 kV to an upper limit (due to insulation breakdown) of 300 kV. X-ray photons from such tubes can destroy tumours no deeper than 5 cm beneath the skin. For deeper tumours, gamma photons with energies of the order of 1 MeV from radioactive isotopes are used.

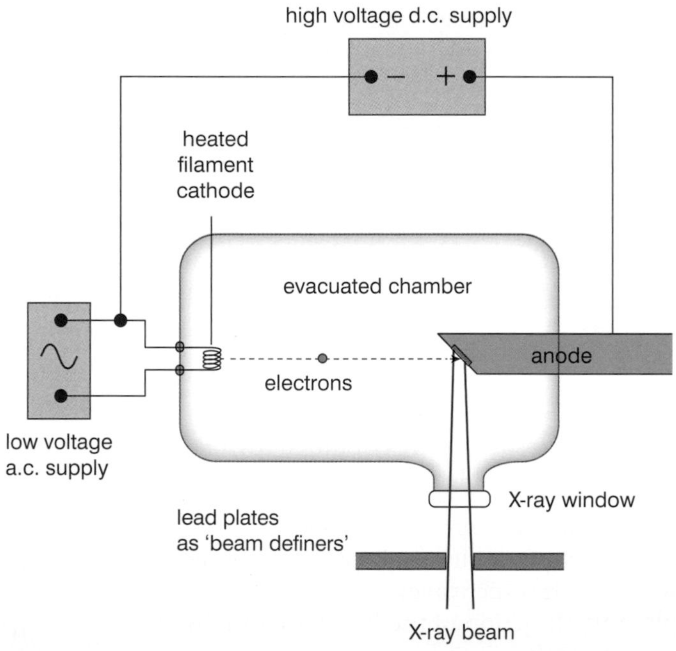

Figure 1 *The X-ray tube*

The spectrum of photon energies from an X-ray tube

An X-ray tube produces a continuous spectrum of photon energies up to the maximum value of eV, where V is the maximum tube voltage, as shown in Figure 2. Raising the tube voltage increases the intensity at all photon energies up to the maximum photon energy as well as increasing the maximum photon energy.

In addition, intensity 'spikes' are produced, which are characteristic of the atoms of the anode and do not change position when the tube voltage is altered. However, if the tube voltage is reduced sufficiently, each spike will disappear when the maximum photon energy is less than the energy of the photons at the spike.

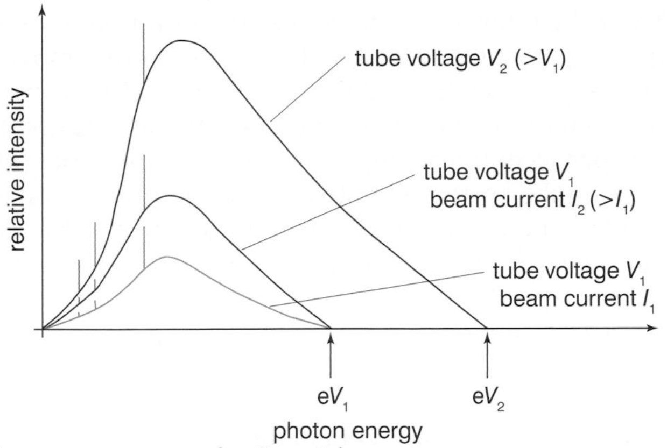

Figure 2 *The energy spectrum of an X-ray tube*

The spikes are caused by the excitation of atoms in the anode when electrons from the filament collide with them. As a result, electrons in the atoms move temporarily from the innermost shells of the atom to higher energy levels. When these electrons return to their original levels, they emit X-ray photons at energies which are characteristic of the anode atoms. These emitted X-rays form patterns of line spectra, each pattern corresponding to electrons returning from the outer energy levels to a particular electron shell. The pattern for electrons returning to:

- the innermost energy shell ($n = 1$) is referred to as the K-series,
- the second energy shell ($n = 2$) is referred to as the L-series,
- the third shell ($n = 3$) is referred to as the M-series.

Notes

1 The energy of an X-ray photon is often expressed in electronvolts (eV) where $1\,\text{eV} = 1.6 \times 10^{-19}\,\text{J}$.

2 The power supplied to an X-ray tube = IV, where I is the beam current. The % efficiency of an X-ray tube is the percentage of the power supplied emitted as X-radiation. A typical X-ray tube has an efficiency of about 1%. The wasted energy is dissipated as heat at the anode.

X-ray imaging

When an X-ray picture is made, X-rays from the X-ray tube are directed for a specified time at the relevant area of the body with a film cassette on the other side of the body. Bones, teeth and other dense matter in the path of the X-rays absorb X-rays much more than muscle and body tissue does. When the film is developed, the areas of the film exposed to X-rays are darker than the unexposed areas so a negative image of the bones, teeth, etc. is formed on the developed film.

For any given application, the following factors must be taken into account before the X-ray tube is used.

1 **The penetrating power or 'hardness' of the X-ray beam** is increased by increasing the tube voltage. The higher the energy of the X-ray photon, the further it can travel through matter. The X-rays used to give an image of the bones of a broken arm do not need to penetrate as far as the X-rays used to give an image of an organ in the body. Increasing the tube voltage increases the maximum energy of the photons emitted by the tube so the beam is more penetrating. Low energy photons are easily absorbed. An **aluminium filter** placed between the X-ray tube and the patient is used to absorb

such photons which would otherwise be absorbed by the body and cause unnecessary exposure of the body to X-rays.

2 **The intensity of the X-ray beam** is the radiation energy per second passing through unit area at right angles to the area. The darkening of an X-ray film depends on the intensity of the X-radiation as well as on the duration of exposure. The greater the intensity or the longer the duration of exposure, the darker the exposed parts of the film will be. An organ that moves would need a shorter duration of exposure and therefore greater intensity than a bone in the arm which can be held still. The intensity depends on the number of electrons per second reaching the anode and therefore on the tube current (since the tube current is a measure of the number of electrons per second reaching the anode). The tube current is controlled by the current through the filament wire. If the filament current is increased, the intensity of the X-ray beam is increased because:
   - the filament becomes hotter and emits more electrons per second,
   - more electrons per second hit the anode so more X-ray photons are released each second.

3 **The width of the beam** is set using 'beam definer' lead plates to ensure that only the part of the patient to be X-rayed is exposed to X-rays. Lead plates surrounding the X-ray tube are used to prevent people other than the patient being exposed to X-rays (see Figure 3).

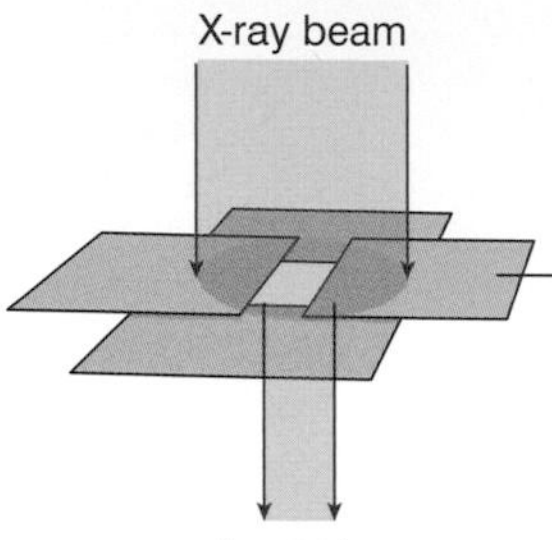

Figure 3 *Beam definers*

Image quality

Image sharpness

The sharpness of an X-ray image is determined by how clearly the edges of structures in the image are defined. A sharp image is one in which such edges can clearly be seen.

To form a sharp image on the film, the X-rays need to originate from a small area of the anode and X-rays scattered by body organs and tissues need to be stopped from reaching the film.

If the area of the anode is too large, the images will be blurred at the edges by large partial shadows as shown in Figure 4. However if the area is too small, the intense concentration of electrons in this area of the target area will damage the anode. To prevent overheating of the anode, it usually consists of a tungsten metal block set in a copper cylinder which is kept cool by pumping water or oil through it. Tungsten is chosen as it has a high melting point and copper is chosen as it is an excellent conductor of heat.

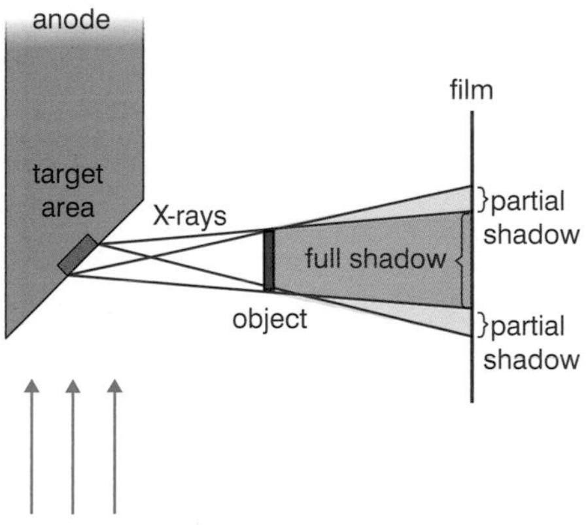

Figure 4 *Sharpness*

- X-ray photons may be scattered by atoms in the body tissues which they pass through. Some scattered X-ray photons may be scattered into the shadow areas on the film of bones or body organs. This would lessen the contrast between the images on the film of bones and body organs and the surrounding tissues. To eliminate scattered X-rays, a lead grid is placed between the patient and the film, as shown in Figure 5. Lead is used because it is a very effective absorber of X-rays.

The grid holes are aligned with the direction in which the unscattered X-rays are travelling so unscattered X-rays that enter the holes of the grid pass straight through it. However, scattered X-rays are absorbed by the grid because they travel mostly through lead after reaching the grid.

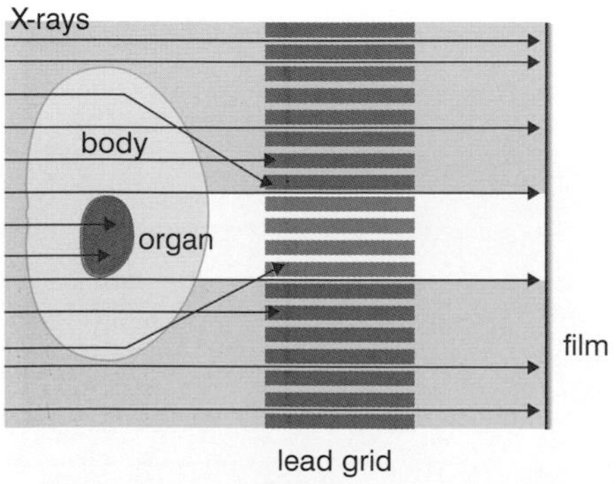

Figure 5 *Using a lead grid*

Contrast

An X-ray image with good contrast has areas where the film is very dark due to exposure to X-rays and other areas that are hardly darkened by X-rays. Bones and teeth are good absorbers of X-rays so they give images that stand out in good contrast with the surrounding tissue.

Body organs such as the stomach are not as effective at absorbing X-rays as bones and teeth. In order to obtain good X-ray images of an organ, a **contrast medium** is used. For example, a patient about to undergo a stomach X-ray is given a drink containing barium sulphate. Because barium is a good absorber of X-rays, X-rays that would otherwise pass through the stomach are absorbed so the contrast between the image of the stomach and its surroundings is vastly improved. A contrast medium is also used to obtain X-ray images of blood vessels where the contrast medium is injected into the bloodstream.

Contrast is lessened if:

- **The duration of exposure is too long**. The light areas and the dark areas of the film both become darker but the increase of darkness is greater in the light areas of the film. So the difference in darkness between the light and dark areas is reduced.
- **The X-rays are too penetrating**. Increasing the energy of the photons by increasing the tube voltage would increase their penetrating power. So more X-rays would pass through the organ and reach the shadow area of the film.
- **Too much scattering of X-ray photons** occurs when they pass through the tissue surrounding the organ.

Contrast can be improved if the film in its cassette is covered with a sheet of fluorescent substance. X-rays directed at a fluorescent substance cause the atoms of the substance to emit light photons. Each X-ray photon might cause many light photons to be emitted, darkening the film more in the areas which are exposed to X-rays. In addition to improving the contrast, the exposure of the patient to X-rays can be reduced as the film is more sensitive to darkening.

Absorption of X-rays by matter

When a beam of X-rays spreads out in a vacuum from a source, the intensity decreases as the distance from the source increases. If the beam spreads out in all directions, the intensity at distance r from the source is proportional to $1/r^2$. This is the same rule as for gamma photons from a point source. See Topic 10.4 for an explanation.

When a beam of X-rays passes through matter, some X-ray photons are absorbed and some are scattered. The transmitted beam is said to be **attenuated** because it is less intense than the incident beam.

For a parallel beam of intensity I_0 directed at normal incidence at an absorber of thickness x, the transmitted intensity I is given by:

$$I = I_0 e^{-\mu x}$$

where μ is the **attenuation** (or absorption) **coefficient** of the absorber.

The variation of intensity with thickness is shown in Figure 7. The curve decreases exponentially with increase of thickness.

The half thickness, $X_{1/2}$, of an absorber is the thickness required to reduce the intensity of the beam to half its initial value.

For thickness $x = X_{1/2}$, $\quad I = \frac{1}{2}I_0$.

Substituting these values into the above equation gives

$$\tfrac{1}{2}I_0 = I_0 e^{-\mu X_{1/2}}$$

Cancelling I_0, rearranging and taking natural logs of both sides of this equation gives:

$$\mu X_{1/2} = \ln 2$$

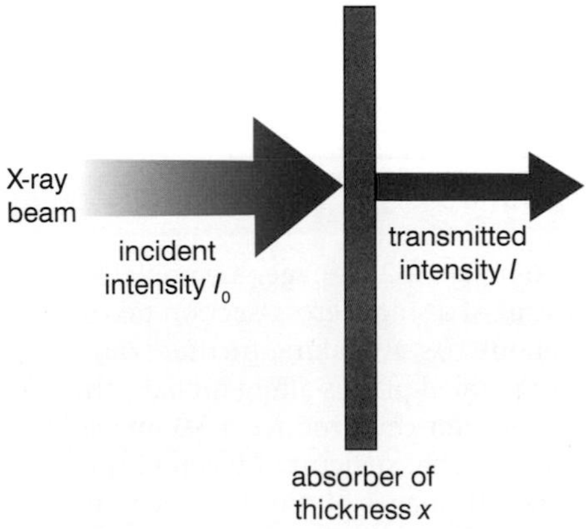

Figure 6 Attenuation

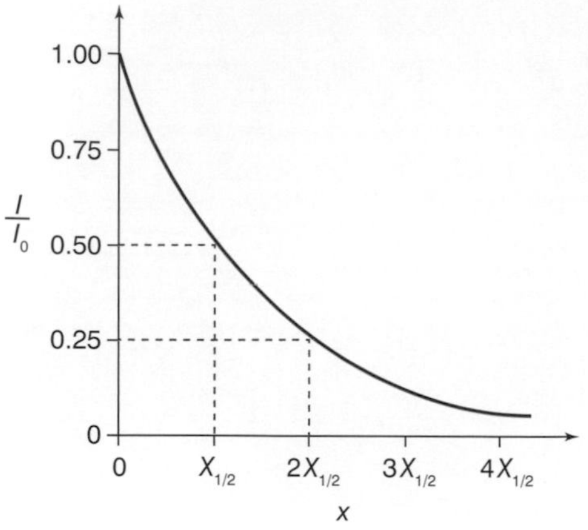

Figure 7 *Transmitted intensity against thickness*

Notes

1 The unit of μ is m^{-1} (or cm^{-1} or mm^{-1}).

2 μ is proportional to the density of the absorber. Bones absorb X-rays more than body tissue does because bones are more dense.

3 Absorption of X-rays due to knocking electrons out of atoms is proportional to Z^3, where Z is the atomic number of the atom. This is why a suitable element with a high value of Z (e.g. barium) is used as a contrast medium when an X-ray picture is made of a soft organ such as the stomach.

Computed tomography (CT)

In a computed tomography (CT) scan, a narrow collimated beam of X-rays is directed through the patient at an array of detectors. The X-ray tube and the detectors are rotated in small steps around an axis along the length of the patient. The detectors are connected to a computer and the signals from the **detectors** are processed to form an image of the cross-section of the body which is exposed to the beam.

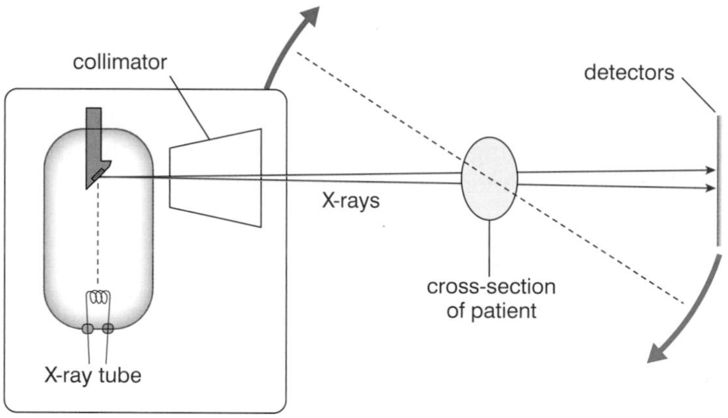

Figure 8 *CT scanning*

- The tube and the detectors are mounted in a gantry which is rotated through 360° about the patient lying in a stationary position on a flat couch between the tube and the detector.
- Each detector receives X-rays that have travelled along a straight path between the tube and the detector.

The intensity detected by each detector depends on the absorption of X-rays along the path from the tube to the detector. This absorption depends on the different densities and thicknesses of the tissue along the path.

By considering the body divided into small volume elements, referred to as 'voxels', each voxel would contribute a certain 'pixel' value to the intensity reduction along any given path.

- Each detector records the sum of the pixel values along each path,
- Each voxel contributes its pixel value to the detector reading (i.e. the detected intensity) many times as the tube and detectors are rotated round the patient.

The signals from the detectors can be processed by a computer to work out each pixel value for each voxel and hence display a 2-D 'mosaic' image of the cross-section. The contrast and sharpness of the pixels can be adjusted to give the best possible image. In addition, the image can be stored and transmitted electronically.

Note

To obtain a 2D image, a complete scan of a single cross-section takes about 10 s. By taking further scans in parallel planes along an axis, the computer can produce a 3D image of an internal structure by combining the 2D images of multiple sections. In addition, the computer can produce an image of a cross-section through the patient in any plane, not just in the scanning planes.

A model scan

Figure 9 shows a simplified version of the process where four voxels are scanned from four different directions which are horizontal, diagonal, vertical and the opposite diagonal.

The pixel value of each voxel is worked out by reconstructing an image consisting of four image voxels and assigning the detector readings to the image voxels according to the path the X-rays pass through in each direction. In this 2×2 example:

- Each image voxel receives a contribution four times (once for each direction) from its corresponding object voxel and once from each of the other three object voxels. The total value of all the image voxels after the complete scan is therefore seven times the sum of the object voxel values.
- The sum of the object voxel values is therefore the total value of all the image voxels ÷ 7. This sum is effectively a background value for each of the four image voxels.

Subtracting the background value from the final value of each image voxel leaves a value equal to three times the contribution from the corresponding object voxel. The value of each object voxel can then be calculated.

Suppose the image voxel values shown in Figure 9 after a complete scan are 22, 16, 25 and 28.

- Their total value = 22 + 16 + 25 + 28 = 91.
- The sum of the object voxel values = 91 ÷ 7 = 13.
- Subtracting this background value from the values of the image voxels gives 9, 3, 12 and 15.
- Dividing these 'corrected' values by 3 gives 3, 1, 4 and 5 for the object voxel values.

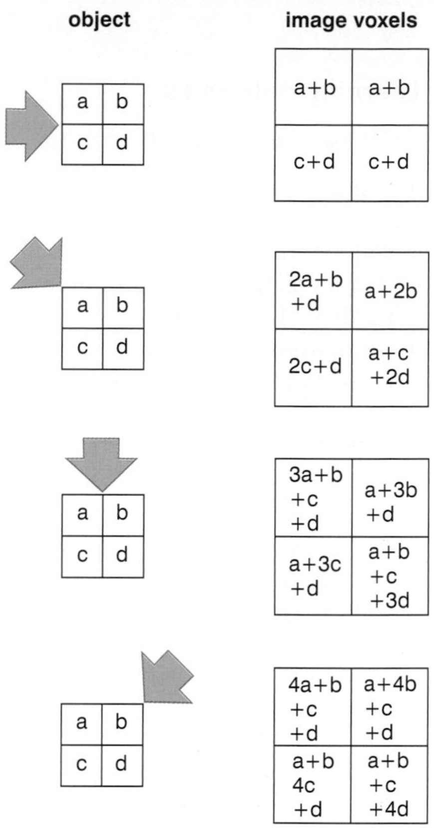

Figure 9 *A simplified scan*

Summary test 22.2

1 An X-ray tube operates at an anode potential of 50 kV.

a i Show that the minimum wavelength of the X-rays from this tube is 0.025 nm.

ii When the beam current at 50 kV is 0.2 mA, the tube operates at an efficiency of 1.5%. Calculate the radiation energy per second produced by the tube.

b State and explain what change in the X-ray beam occurs if:

i the tube current is increased,

ii the anode voltage is increased.

2 a What is meant by: **i** the sharpness, **ii** the contrast of an X-ray image.

b State the function of a scattering grid and explain why it is necessary when an X-ray image is obtained.

3 a A metal plate of thickness 1.3 mm placed in the path of a collimated X-ray beam reduces the beam intensity by 70%. Show that the absorption coefficient of the metal plate is $9.3 \times 10^{-2}\ m^{-1}$.

b Explain why a contrast medium is used when an X-ray picture of the stomach is made.

4 a With the aid of a diagram, outline the principle of operation of a CT scanner.

b In the simplified model in Figure 9, the values of the image voxels after a complete scan were 21, 15, 21 and 27. Calculate: **i** the background value of the image voxels, **ii** the pixel value of each object voxel.

22.3 PET scanning

Learning outcomes

On these pages you will learn to:

- explain what is meant by a positron emission from a suitable radioactive nucleus and that a positron is the antimatter particle of the electron
- recall that a positron and an electron annihilate each other when they collide in a process that causes two gamma-ray photons to be released from a single point
- understand why two gamma rays from a positron emission travel in opposite directions and that if the radioactive nucleus is inside the body, the gamma-ray photons can be detected outside the body
- understand that a medical tracer is a substance that contains radioactive nuclei that is introduced into the body and absorbed by the tissue under investigation
- recall that the point of origin of each pair of gamma-ray photons can be determined by using signals from a ring of gamma-ray detectors around the body and the information from such signals can be used to trace an image of the radioactive nuclei in the body

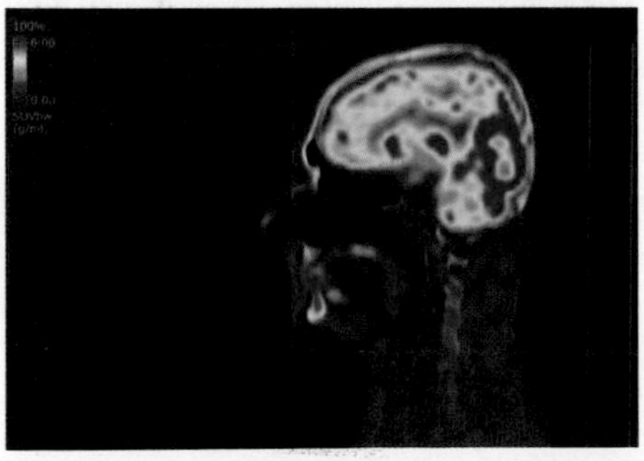

Figure 1 A PET image

About annihilation

The positron is the antimatter counterpart or 'antiparticle' of the electron. It has the same rest mass as an electron, but it has an equal and opposite charge to that of the electron. If an antiparticle meets its particle counterpart, they **annihilate** each other and radiation is released in the form of two gamma-ray photons. Because momentum and energy are conserved in the process, the two photons must travel in opposite directions each with the same energy. The creation of a single photon only would violate the principle of conservation of momentum because it would possess far more momentum than the initial momentum of the electron and positron would have possessed. The two photons possess equal and opposite momentum and, since their energy is proportional to their momentum, they move in opposite directions with equal energy.

The rest mass of a positron or an electron is 9.11×10^{-31} kg. When an electron and a positron with negligible kinetic energy annihilate each other, radiation energy is created due to the complete loss of mass of $2 \times 9.11 \times 10^{-31}$ kg in accordance with Einstein's equation $E = mc^2$. Using this equation gives energy equal to 1.64×10^{-13} J or 1.02 MeV. Each of the two gamma-ray photons is therefore released with 0.51 MeV of energy, assuming that the electron and positron lose most their kinetic energy before they annihilate each other.

A positron emitted from an unstable 'proton-rich' nucleus in body tissue travels about 1–2 mm before it collides with an electron and both the electron and the positron are annihilated in the interaction, producing two gamma photons in the process. Positrons referred to as β^+ particles are emitted by unstable nuclei that have a greater proton to neutron ratio than stable nuclei of the same element. In comparison, β^- particles are electrons emitted by unstable nuclei that have a smaller proton to neutron ratio than stable nuclei of the same element. See Topic 10.4.

Note

Positrons are emitted with an initial kinetic energy characteristic of the positron-emitting isotope. For example, the positrons from the fluorine isotope $^{18}_{9}F$ are emitted with initial kinetic energies of 0.6 MeV. If the isotope is inside the body, the positrons lose most of their kinetic energy within a distance of a few millimetres before they are annihilated. So each annihilation event from this isotope causes two 0.51 MeV gamma photons to be released in opposite directions.

Positron-emitting isotopes as tracers in medicine

Certain radioactive isotopes are suitable for use as tracers in medicine because they can be attached to substances that are absorbed by specific organs or types of tissue such as cancer cells in the body. Positron-emitting tracers differ from other gamma-emitting tracers in that each positron releases two gamma photons in opposite directions from the decay of each unstable nucleus, whereas other gamma-emitting tracers release a single photon from the decay of each unstable nucleus. As explained later, this key difference enables the tracer to be scanned and mapped more precisely than, for example, with a CT or a gamma camera scan.

The positron-emitting fluorine isotope $^{18}_{9}F$ is used to detect cancer cells because such cells have a higher uptake of glucose than normal cells. Glucose molecules containing an atom of the positron-emitting fluorine isotope $^{18}_{9}F$ instead of an oxygen atom attach themselves preferentially to cancer cells. So when a nucleus of this isotope decays, two gamma photons are released close to the cancer cell. The gamma photons released in this way can be detected outside the body and the location of the tracer nuclei can be traced as described on the next page.

Note

The positron-emitting isotope ${}^{18}_{9}F$ is chosen because:

- It decays directly to a stable product (i.e. a non-radioactive product).
- Its half-life is such that it decays to a stable product within a day or so; in terms of radiation risk it becomes harmless relatively quickly.

Production of positron-emitting isotopes

Positron-emitting isotopes suitable for medical use are produced by exposing stable isotopes to high-energy protons from, for example, a cyclotron. For instance, if nuclei of the stable oxygen isotope ${}^{18}_{8}O$ are exposed to high energy protons, some of them may have a neutron knocked out and replaced by a proton to become nuclei of the fluorine isotope ${}^{18}_{9}F$ which is a positron emitter with a half-life of 110 minutes. Using chemical techniques including ion exchange, atoms of this particular isotope can be attached to glucose molecules by replacing oxygen atoms to form 18-FDG (fluorodeoxyglucose) molecules. In the body, these molecules preferentially attach themselves to cancer cells due to the higher uptake of glucose by cancer cells. With a half-life of just under 2 hours, 18-FDG needs to be prepared just a few hours before being administered to a patient in a PET scanner.

The PET scanner

A small quantity of the tracer such as the fluorine isotope ${}^{18}_{9}F$ is injected into the patient via a saline drip. To allow the tracer to be distributed in the body, the PET scan starts about 40 minutes later.

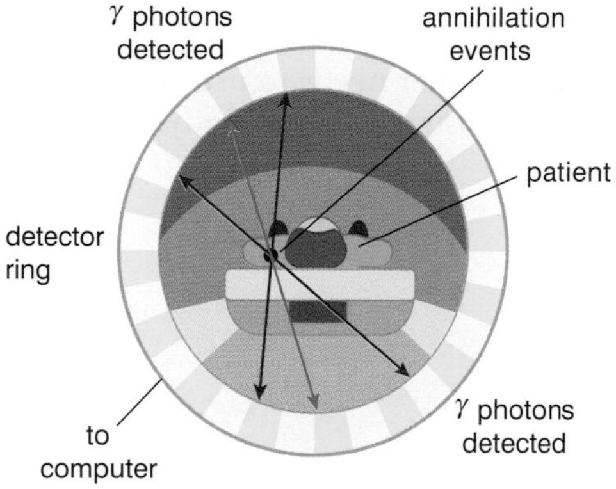

Figure 2 *A PET scanner*

Each positron travels less than a millimetre in the patient before it meets an electron and they annihilate each other to produce two γ photons travelling in opposite directions.

- A ring of detectors connected to a computer registers a positron emission when two of its detectors each detect a γ photon within a very short time interval (called the 'coincidence time' window) of about 5–10 ns.
- Such an emission must be from a point very close to, or along a straight line between, the two detectors as the γ photons travel at the same speed in opposite directions and are created at the same time. The position along the line, called a 'line of response' or LOR, can be found by mapping this and other LORs (from other nearby ${}^{18}_{9}F$ decays) that cross each other at the same point.
- Although each detector is narrow, there is a small degree of uncertainty about the LOR between two detectors and therefore of the two γ photons. In addition, there is some uncertainty about the point of origin in relation to the position of the positron-emitting nucleus.

- Using a computer to process the arrival times of the γ photons (see Figure 2) enables an image of the tracer in the tissue to be created. PET scans can take 20 minutes or more to gather sufficient data to map out the location of the positron-emitting isotope in the body.

Summary test 22.3

1 The fluorine isotope $^{18}_{9}F$ emits positrons and becomes stable. The isotope is prepared by irradiating the oxygen isotope $^{18}_{8}O$ with protons.

a Write down an equation to represent the production of a $^{18}_{9}F$ nucleus from an $^{18}_{8}O$ nucleus.

b Write down an equation to represent the decay of a nucleus of $^{18}_{9}F$.

2 A sample of the fluorine isotope $^{18}_{9}F$ in a solution has an activity of 220 MBq. This isotope has a half-life of 110 minutes.

a Calculate the mass of the isotope in the sample when its activity is 220 MBq.

b Estimate the activity of this sample after 24 hours.

3 When a positron-emitting nucleus in a substance decays, two gamma-ray photons are emitted from the substance.

a Explain why gamma photons are emitted each time a positron-emitting nucleus in a substance decays.

b Explain why two gamma photons are always emitted in opposite directions from each decay.

4 a State what is meant by a *radioactive tracer*.

b Explain why a PET scanner needs to use positron-emitting isotopes to trace radioactive substances in the body rather than other gamma-emitting isotopes.

Launch additional digital resources for the chapter

1 An ultrasound probe generates ultrasound waves at a frequency of 2.5 MHz.

a Calculate the wavelength of the ultrasound waves from this probe:
i in air, **ii** in soft tissue.

b i Explain why ultrasound waves of much lower frequency are unsuitable for medical imaging.
ii With the aid of a diagram, describe the construction of an ultrasound probe and how it produces ultrasound waves.

c i Define the *specific acoustic impedance* of a substance.
ii Use the data in Table 2, p.328, to calculate the reflection coefficient of the boundary between: (1) air and skin and (2) water and skin. Assume that skin is soft tissue.

d Use the results of your calculation in **cii** to explain why a gel is applied between an ultrasound probe and the skin when the probe is used.

2 a An X-ray tube operates at a potential difference of 40 kV.
i Calculate the minimum wavelength of the X-rays from this X-ray tube.
ii Sketch a graph to show how the intensity of the X-rays from this tube varies with the energy of the photons.

b i What is meant by a *contrast medium* as used in X-ray imaging?
ii Give an example of a contrast medium, stating an organ it is used for.

c Describe a CT scanner and explain how it works.

3 a i State the type of radiation used to form an image in a PET scanner.
ii State the type of substance used to produce the above radiation and describe how the above radiation is produced.

b In a PET scanner, the detectors are arranged radially and equally spaced along a ring. The detectors are linked to a computer which registers any two detectors that are each triggered by radiation in the same short time interval. Figure 1 represents the ring of detectors when two detectors X and Y have been triggered in the same time interval.

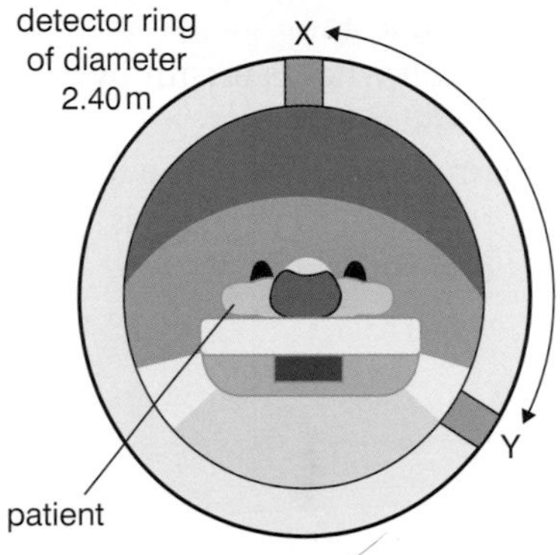

Figure 3.1

i Describe how the radiation from the substance in a patient undergoing a PET scan could have triggered the two detectors within the same short time.
ii The ring has a diameter of 2.40 m. The arc XY subtends an angle of 135° at the centre of the ring. Use this information to determine the distance along a straight line from X to Y.
iii Explain why the source of the radiation must be at a location on or close to the straight line from X to Y.
iv Discuss what further information could be used to locate the source between X and Y.

23 Astrophysics and cosmology

23.1 Astronomical distances

Learning outcomes

On these pages you will learn to:

- explain what is meant by the luminosity of a star
- use the inverse square law to relate the luminosity of a star to the intensity of the radiation from a star
- recall what is meant by a standard candle
- understand how the distances to stars and galaxies is determined

Stars and galaxies

On a clear night, the stars we see are pinpoints of light that may or may not differ in brightness. Each one is a massive glowing ball of gas, mostly hydrogen, with a core at its centre where nuclear fusion takes place. The radiation from the core heats the star's outer layers and causes them to emit electromagnetic radiation into space in all directions. The Sun is our nearest star. Light from the Sun takes about 500 seconds to reach the Earth. In comparison, light takes 4.3 years to reach us from Proxima Centauri, the next nearest star. All the stars except the Sun appear as pinpoints of light in the night sky because they are so far away even though each one is a glowing ball of gas emitting radiation into space in all directions.

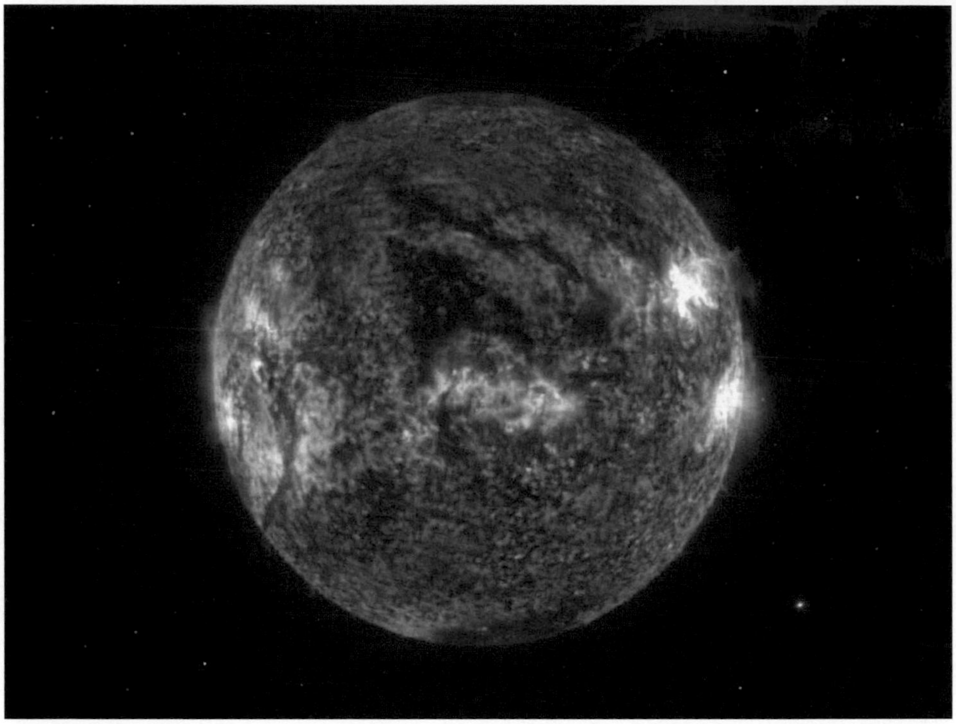

Figure 1 *The Sun*

Galaxies are vast collections of stars held together by their own gravity. A galaxy is like an 'island' of stars. The Sun is one of about 100 000 million stars in the Milky Way galaxy, which is a spiral galaxy about 100 000 light years across. The Milky Way galaxy is the second largest of a cluster of 'local' nearby galaxies, the largest being the Andromeda galaxy twice as large as the Milky Way galaxy and about 2.5 million light years away. There are thought to be over 100 000 million galaxies in the Universe, the furthest being over 13 000 million light years away. In this chapter, you will discover how astronomers used the laws of physics to measure these enormous distances.

Luminosity and intensity

The luminosity L of a star is the total power of the radiation emitted by a star.

Luminosity is measured in watts. The luminosity of the Sun is about 4×10^{26} W. In other words, the Sun emits radiant energy at a rate of about 4×10^{26} J/s. The luminosity of Proxima Centauri is about 10 000 times greater than that of the Sun. The Earth is over a quarter of a million times further from Proxima Centauri than it is from the Sun. So even though Proxima Centauri emits 10 000 times more radiant energy than the Sun, because it is much further away than the Sun, the intensity of the radiation from it at the Earth is much less than the intensity of sunlight at the Earth.

The intensity of the radiation from a star in the night sky depends on the **radiant flux intensity, *F*,** of its radiation at the Earth. This is the radiation energy per second per unit surface area received from the star at normal incidence on a surface.

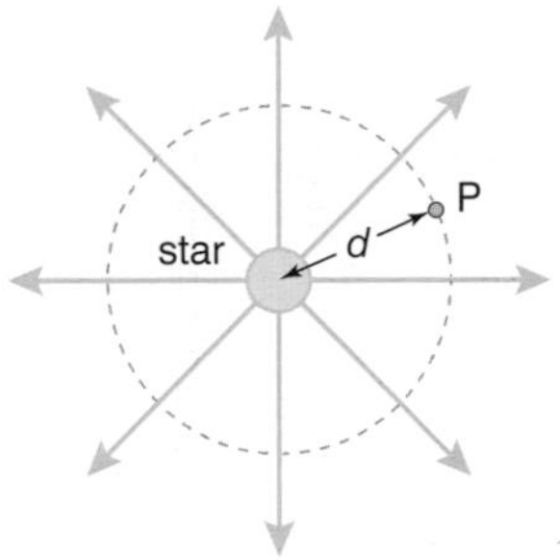

Figure 2 *Radiation from a star*

Consider a star of luminosity L at the centre of an imaginary sphere of radius d. The radiation from the star spreads out equally in all directions. The radiant flux intensity F of the radiation at the sphere is given by the equation

$$F = \frac{\text{the energy per second emitted by the star}}{\text{the area of the sphere}}$$

Note the unit of F is the watt per square metre ($\mathrm{W\,m^{-2}}$) or joules per second per square metre ($\mathrm{J\,s^{-1}\,m^{-2}}$). Since the energy per second emitted by the star is its luminosity L and the surface area of the sphere is $4\pi d^2$:

$$F = \frac{L}{4\pi d^2}$$

The equation shows that:

the radiant flux intensity of the radiation from a star at a distance d from the centre of a star is proportional to:

- **its luminosity L**
- **$\frac{1}{d^2}$, the inverse of the square of the distance d from the sphere to the centre of the star.**

Worked example

The radiant flux intensity of solar radiation at the Earth's surface is about 1400 W $\mathrm{m^{-2}}$. The mean distance from the Earth to the Sun is 1 AU which is 1.50×10^{11} m. Estimate the luminosity of the Sun.

Solution

Rearranging the equation $F = \frac{L}{4\pi d^2}$ gives $L = F \times 4\pi d^2$

Hence: $L = 1400\,\mathrm{W\,m^{-2}} \times 4\pi \times (1.5 \times 10^{11}\,\mathrm{m})^2 = 4.0 \times 10^{26}\,\mathrm{W}$

About parallax

To calculate the distance d to a nearby star, consider Figure 3 which shows the 'six month' angular shift of a nearby star's position relative to stars much further away.

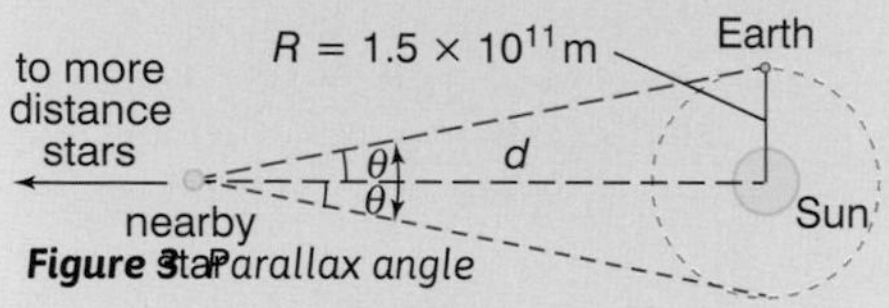

Figure 3 *Parallax angle*

The parallax angle, θ, is defined as the angle subtended by the star to the line between the Sun and the Earth, as shown in Figure 3. This angle is half the angular shift of the star's line of sight over six months. From the triangle consisting of the three lines between the Sun, the star and the Earth as shown in Figure 1.3,

$\tan\theta = R/d.$

and $d = R/\tan\theta$ where R is the mean distance from the centre of the Sun to the centre of the Earth.

Standard candles

The stars in the night sky might seem to be in fixed positions relative to each other, but accurate measurements of their positions show that some of them change their relative position slightly against the background of the other stars. This 'parallax' effect is due to the Earth's orbital motion round the Sun, and is because the direction of line of sight from the Earth to a nearby star changes slightly as the Earth orbits the Sun, as shown in Figure 3. The effect can be measured and used to calculate the distance to a nearby star. By measuring the radiant flux intensity of a star, its luminosity can then be calculated if its distance from the Earth is known.

An object of known luminosity is known as a **standard candle**, because it can be used to determine the distance to more distant stars and to galaxies. For example, a supernova event is a star that explodes and becomes much brighter before fading away. Type 1a supernovae are known to reach a peak luminosity of about 10^{36} W, about ten thousand million times more than the Sun's luminosity. By measuring the maximum radiant flux intensity of such a supernova and knowing its peak luminosity, its distance from the Earth can be calculate.

Summary test 23.1

$c = 3.00 \times 10^8\,\text{m}\,\text{s}^{-1}$

1 The luminosity of Proxima Centauri is about 10 000 times greater than the luminosity of the Sun. Proxima Centauri is about 250 000 times further from the Sun than the Earth is. Use this information to estimate the radiant flux intensity at the Earth of the radiation from Proxima Centauri. Assume that the radiant flux intensity of solar radiation at the Earth is $1400\,\text{W}\,\text{m}^{-2}$.

2 Sirius is the brightest star in the sky. Its luminosity is about 25 times that of the Sun. It is 8.1×10^{16} m from the Earth.

The luminosity of the Sun = 4.0×10^{26} W

a Calculate how long light takes to travel from Sirius to the Earth.

b Estimate the radiant flux intensity at the Earth of the radiation from Sirius.

3 A supernova is a star that explodes and outshines its host galaxy before it fades away. In 2014, a supernova event occurred in a distant galaxy. Its luminosity peaked at about 4.4×10^{36} W. The maximum radiant flux intensity at the Earth from this supernova was $0.21\,\text{pW}\,\text{m}^{-2}$. Estimate the distance from the Earth to this supernova.

4 The Earth's orbit around the Sun is slightly elliptical. The mean distance from the Earth to the Sun is 1.5×10^{11} m. The percentage change of the radiant flux intensity of solar radiation incident on the Earth is about 1.7% when it moves from its least distance to its greatest distance from the Sun. Estimate the change of the distance from the Earth to the Sun when the Earth moves from its greatest distance to its least distance from the Sun.

Note

Since the parallax angle θ is always less than 10°, using the small angle approximation:

$\tan\theta \approx \theta$ gives

$$\theta = \frac{R}{d},$$

where θ is in radians and $d = \frac{R}{\theta}$

Stellar radii

Learning outcomes

On these pages you will learn to:

- recall and use Wien's law of radiation to estimate the peak surface temperature of a star
- use the Stefan–Boltzmann's law of radiation to relate the luminosity of a star to its radius and its surface temperature
- use the above laws of radiation to estimate the radius of a star

Starlight

Stars differ in colour as well as brightness. Viewed through a telescope, stars that appear to be white to the unaided eye appear in their true colours. This is because a telescope collects much more light than the unaided eye, thus activating the colour-sensitive cells in the retina. **Charge coupled devices** (CCDs) with filters and colour-sensitive photographic film show that stars vary in colour from red to orange and yellow to white, to bluish-white.

Like any glowing object, a star emits thermal radiation which includes visible light and infrared radiation. For example, if the current through a torch bulb is increased from zero to its working value, the filament glows dull red then red then orange-yellow, as the current increases and the filament becomes hotter. The spectrum of the light emitted shows that there is a continuous spread of colours which change their relative intensities as the temperature is increased. This example shows that:

- **The thermal radiation from a hot object at constant temperature consists of a continuous range of wavelengths.**
- **The distribution of intensity with wavelength changes as the temperature of the hot object is increased.**

Figure 1 shows how the intensity distribution of such radiation varies with wavelength for different temperatures, as indicated.

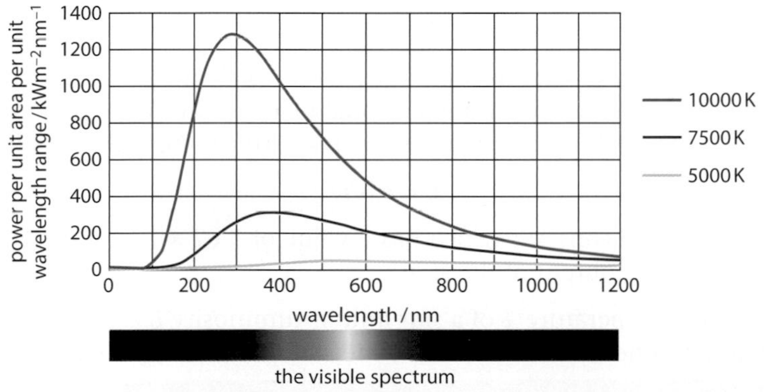

Figure 1 *Black body radiation curves*

Note

The curves are referred to as **black body** radiation curves, a black body being defined as a body that is a perfect absorber of radiation (i.e. absorbs 100% of radiation incident on it at all wavelengths) and therefore emits a continuous spectrum of wavelengths. A small hole in the door of a furnace is an example of a black body, because any thermal radiation that enters the hole from outside would be completely absorbed by the inside walls. We can assume that a star is a black body because any radiation incident on it would be absorbed, and none would be reflected or transmitted by the star. In addition, the spectrum of thermal radiation from a star is a continuous spectrum with an intensity distribution that matches the shape of a black body radiation curve.

The laws of thermal radiation

The radiation curves in Figure 1 are obtained by measuring the intensity of the thermal radiation from a black body at different constant temperatures. Each curve has a peak which is higher and at shorter wavelength than the curves at lower temperatures. The following two laws of thermal radiation were obtained by analysing the radiation curves:

1 Wien's displacement law: The wavelength at peak intensity, λ_P, is inversely proportional to the thermodynamic temperature T of the object, in accordance with the following equation known as **Wien's law**:

$$\lambda_{max} T = 0.0029\,\mathrm{m\,K}$$

Therefore, if λ_{max} for a given star is measured from its spectrum, the above equation can be used to calculate the temperature T of the light-emitting outer layer, the **photosphere**, of the star. The photosphere is sometimes referred to as the **surface** of a star.

Notice that the unit symbol 'm K' stands for 'metre kelvin' not milli kelvin!

Worked example

The peak intensity of thermal radiation from the Sun is at a wavelength of 500 nm. Calculate the surface temperature of the Sun.

Solution

Rearranging $\lambda_{max} T = 0.0029\,\mathrm{m\,K}$ gives $T = \dfrac{0.0029\,\mathrm{m\,K}}{500 \times 10^{-9}\,\mathrm{m}} = 5800\,\mathrm{K}$

2 Stefan–Boltzmann's law: The luminosity L (i.e. total energy per second emitted) of a black body at temperature T is proportional to its surface area A and to T^4, in accordance with the following equation known as **Stefan–Boltzmann's law**:

$$L = \sigma A T^4$$

where σ is the Stefan constant which has a value of $5.67 \times 10^{-8}\,\mathrm{W\,m^{-2}\,K^{-4}}$. In effect, L is the power output of the star.

Therefore, if the temperature T of a star and its luminosity L are known, the surface area A and hence the radius R of the star can be calculated.

Worked example

$\sigma = 5.67 \times 10^{-8}\,\mathrm{W\,m^{-2}\,K^{-4}}$

A star has a luminosity of 6.0×10^{28} W and a surface temperature of 3400 K.

a Show that its surface area is $7.8 \times 10^{21}\,\mathrm{m^2}$.

b Calculate: **i** its radius, **ii** the ratio of its radius to the radius of the Sun. (Radius of Sun = 7.0×10^{8} m)

Solution

a Rearranging $L = \sigma A T^4$ gives: $A = \dfrac{L}{\sigma T^4}$

Hence $A = \dfrac{6.0 \times 10^{28}}{5.67 \times 10^{-8} \times (3400)^4} = 7.9 \times 10^{21}\,\mathrm{m^2}$

b i For a sphere of radius R, its surface area $A = 4\pi R^2$

Rearranging this equation gives: $R^2 = \dfrac{A}{4\pi} = \dfrac{7.9 \times 10^{21}}{4\pi} = 6.3 \times 10^{20}\,\mathrm{m^2}$

Hence $R = 2.5 \times 10^{10}$ m

ii Ratio of radius to Sun's radius $= \dfrac{2.5 \times 10^{10}\,\mathrm{m}}{7.0 \times 10^{8}\,\mathrm{m}} = 36$

Extension

The radiation curves in Figure 1 show how the power per unit area of the electromagnetic radiation from a star, in each nanometre of its wavelength range, varies with the wavelength of the radiation and with the surface temperature of the star.

The area under each curve gives the power radiated per unit surface area of a star at the indicated temperature. This can be estimated from the graph by counting the number of complete squares under the curve and multiplying this number by $10\,000\,\mathrm{kW\,m^{-2}}$, which is the power per unit area represented by a single square (ie $100\,\mathrm{KW\,m^{-2}\,nm^{-1}} \times 100\,\mathrm{nm}$). This estimate can be compared with the power per unit area calculated using the Stefan–Boltzmann law as explained opposite.

Notes

1 For two stars X and Y that have the same luminosity:

- Luminosity of X, $L_X = \sigma A_X T_X^4$, where A_X = surface area of X and T_X = surface temperature of X.
- Luminosity of Y, $L_Y = \sigma A_Y T_Y^4$, where A_Y = surface area of Y and T_Y = surface temperature of Y.

Therefore $\dfrac{L_X}{L_Y} = \dfrac{\sigma A_X T_X^4}{\sigma A_Y T_Y^4} = \dfrac{A_X T_X^4}{A_Y T_Y^4}$

2 For equal luminosity: $\sigma A_X T_X^4 = \sigma A_Y T_Y^4$ hence $\dfrac{A_X}{A_Y} = \dfrac{T_X^4}{T_Y^4}$

Therefore, if X and Y have equal surface temperatures, they must have the same radius. If their surface temperatures are unequal, the cooler star must have a bigger radius than the hotter star.

Summary test 23.2

1 The spectrum of light from a star has its peak intensity at a wavelength of 620 nm. Calculate the temperature of the star's light-emitting surface.

2 A star has a power output of 6.0×10^{28} W and a surface temperature of 3400 K.

a Show that it surface area is $7.9 \times 10^{21}\,\text{m}^2$.

b Calculate:

i its radius,

ii the ratio of its radius to the radius of the Sun. (Radius of Sun = 7.0×10^8 m)

3 A star has a surface temperature which is twice that of the Sun and a diameter that is four times as large as the Sun's diameter. Show that it emits approximately 250 times as much energy per second as the Sun.

4 Two stars, X and Y, have the same surface temperature of 5400 K. Star X emits 100 times more power that star Y.

a State and explain which star, X or Y, has the bigger diameter.

b X has a diameter of 2.0×10^9 m. Calculate the luminosity of star X.

c Calculate the diameter of star Y.

23.3 Hubble's law and the Big Bang theory

Learning outcomes

On these pages you will learn to:

- explain why the emission spectra from distant galaxies show an increase in wavelength compared with the spectra from laboratory and other sources
- understand what is meant by a red shift and calculate its value using an appropriate equation from suitable data
- explain why red shift leads to the idea that the Universe is expanding
- recall and use Hubble's law and explain how Hubble's law leads to the Big Bang theory

Red shift

The wavelengths of the electromagnetic radiation from a star or galaxy moving towards Earth are shorter than they would be if the star or galaxy was stationary. If the star or galaxy had been moving away from the Earth, the wavelengths would be longer than if the star or galaxy was stationary. As explained in Topic 8.9, this effect applies to all waves and is known as the **Doppler effect**. For electromagnetic waves, the effect can be seen by comparing the line emission spectrum of the light from a laboratory sources with that from a galaxy moving away from us, as shown in Figure 1.

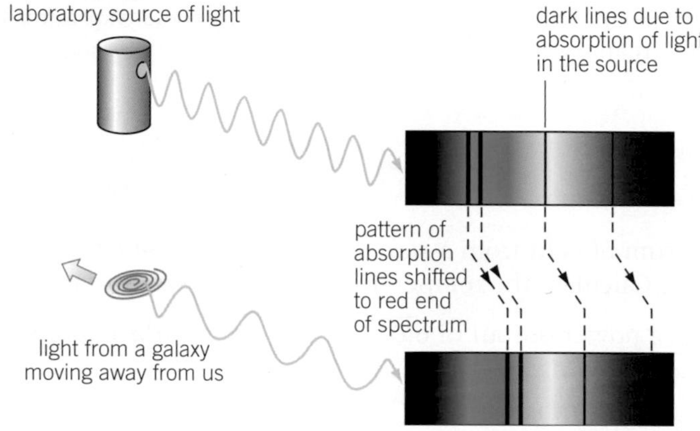

Figure 1 *Red shift*

The lines of the line emission spectrum of the light from the galaxy are shifted towards the red part of the visible spectrum because the light is lengthened in wavelength. This shift in the wavelength due to the light source moving away (i.e. receding) from the Earth is called a **red shift**. As explained in Topic 8.9, for light of wavelength, λ, from a star or galaxy receding at speed v, the change of the wavelength $\Delta\lambda = (v/c)\,\lambda$. Hence:

$$\frac{\Delta\lambda}{\lambda} = \frac{v}{c} \qquad \text{where } c = \text{the speed of light in a vacuum}$$

Notes

1 The equation can also be written in terms of frequency f as: $\frac{\Delta f}{f} = \frac{v}{c}$

2 The formulae above can only be applied to electromagnetic waves emitted by stars travelling at speeds much less than the speed of light c.

3 By measuring the shift in wavelength of a line of a star or galaxy's line spectrum, the speed v of the star or galaxy relative to Earth can be found using the rearranged equation: $v = c\left(\frac{\Delta\lambda}{\lambda}\right)$

Worked example

$c = 3.0 \times 10^8\,\text{m s}^{-1}$

A spectral line of a star is found to be displaced from its laboratory value of 434 nm by + 0.087 nm. State whether the star is moving towards or away from the Earth and calculate its speed relative to the Earth.

Solution

The star is moving away from the Earth because the wavelength of its light is increased.

Rearranging $\Delta\lambda = \frac{v\lambda}{c}$ gives $v = \frac{c\Delta\lambda}{\lambda} = \frac{3.0 \times 10^8 \times 0.087 \times 10^{-9}}{434 \times 10^{-9}} = 6.0 \times 10^4\,\text{m s}^{-1}$

Hubble's law

The Universe consists of galaxies, each containing millions of stars, separated by vast empty spaces. Edwin Hubble and other astronomers studied the light spectra of many galaxies and were able to identify prominent spectral lines as in the spectra of individual stars but 'red-shifted' to longer wavelengths. Hubble studied galaxies which were close enough to be resolved into individual stars; for each galaxy, he measured:

- **its red shift and then calculated its *speed of recession* (i.e. the speed at which it was moving away),**
- **its distance from Earth by observing individual stars of known luminosity as standard candles.**

Hubble's results showed that galaxies are receding from us, each moving at speed v, that is directly proportional to the distance, d, to it. This discovery, referred to as **Hubble's law**, is usually expressed as the following equation:

$$v = Hd$$

where H, the constant of proportionality, is referred to as the **Hubble constant**. H represents the speed of recession per unit distance from Earth. The value of H is $2.2 \times 10^{-18}\,\text{s}^{-1}$ or $21\,\text{km}\,\text{s}^{-1}$ per million light years.

Figure 2 shows the pattern of typical measurements of the speed of recession v and distance d plotted on a graph is a straight line through the origin. Note that the gradient of the graph is equal to the Hubble constant H.

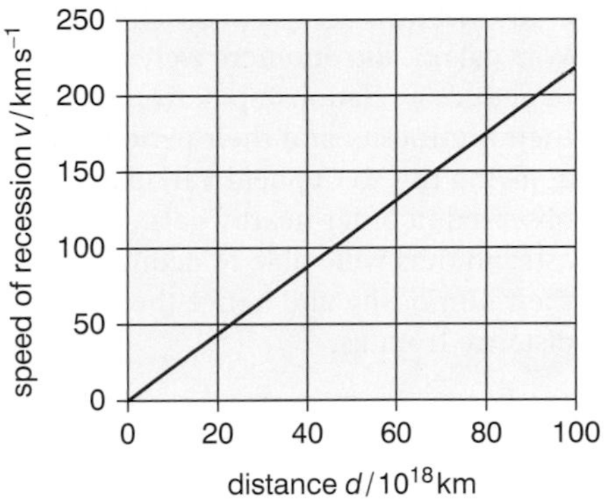

Figure 2 *Speed of recession v. distance*

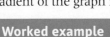

Worked example

$c = 3.0 \times 10^8\,\text{m}\,\text{s}^{-1}$, $H = 2.2 \times 10^{-18}\,\text{s}^{-1}$

The wavelength of a spectral line in the spectrum of light from a distant galaxy was measured at 398.6 nm. The same line measured in the laboratory has a wavelength of 393.3 nm. Calculate:

a the speed of recession of the galaxy,

b the distance to the galaxy.

Solution

a $\Delta\lambda = 398.6 - 393.3 = 5.3\,\text{nm}$

Rearranging, $\Delta\lambda = \dfrac{v\lambda}{c}$ gives $v = \dfrac{c\Delta\lambda}{\lambda} = \dfrac{3.0 \times 10^8\,\text{m}\,\text{s}^{-1} \times 5.3 \times 10^{-9}\,\text{m}}{393.3 \times 10^{-9}\,\text{m}}$

$= 4.0 \times 10^6\,\text{m}\,\text{s}^{-1}$

b Rearranging, $v = Hd$ gives $d = \dfrac{v}{H} = \dfrac{4.0 \times 10^6\,\text{m}\,\text{s}^{-1}}{2.2 \times 10^{-18}\,\text{s}^{-1}} = 1.8 \times 10^{24}\,\text{m}$

The Big Bang theory

Hubble's law tells us that the distant galaxies are receding from us. The conclusion we must draw from this discovery is that the galaxies are all moving away from each other and **the Universe must therefore be expanding**. At first, some astronomers thought this expansion is because the Universe was created in a massive 'primordial' explosion and has been expanding ever since. This theory was referred to by its opponents as the **Big Bang theory**.

With no evidence for a primordial explosion other than an explanation of Hubble's law, many astronomers supported an alternative theory that the Universe is unchanging, the same now as it ever was. This theory, known as the **Steady State theory**, explained the expansion of the Universe by supposing matter entering the Universe at 'white holes' pushes the galaxies apart as it

> **Note**
>
> Since Hubble's discovery, astronomers have used other methods to measure the distances to distant stars and galaxies. One such method used 'Cepheid variable' stars which are stars that vary periodically in luminosity with a period of variability that increases with luminosity. So, by timing the variability of such stars at known distances in the Milky Way galaxy, astronomers were able to deduce a relationship between their luminosity and their period. By applying this to Cepheid variables observed in other nearby galaxies, astronomers were able to deduce their luminosity and hence their distance from us.

enters. The Big Bang theory was accepted in 1965 when radio astronomers discovered microwave radiation from all directions in space. Steady state theory could not explain the existence of this **cosmic microwave background radiation (CMBR)**, but the Big Bang theory could.

Evidence for the Big Bang theory

The spectrum of microwave radiation from space matched the theoretical spectrum of thermal radiation from an object at a temperature of 2.7 K. Because the radiation was detected from all directions in space with little variation in intensity, it was realised it must be universal or 'cosmic' in origin.

This background cosmic microwave radiation is explained readily by the Big Bang theory as radiation that was created in the Big Bang and has been travelling through the Universe ever since the Universe became transparent. As the Universe expanded after the Big Bang, its mean temperature has decreased and is now about 2.7 K. The expansion of the Universe has gradually increased the background cosmic microwave radiation to its present range of wavelengths.

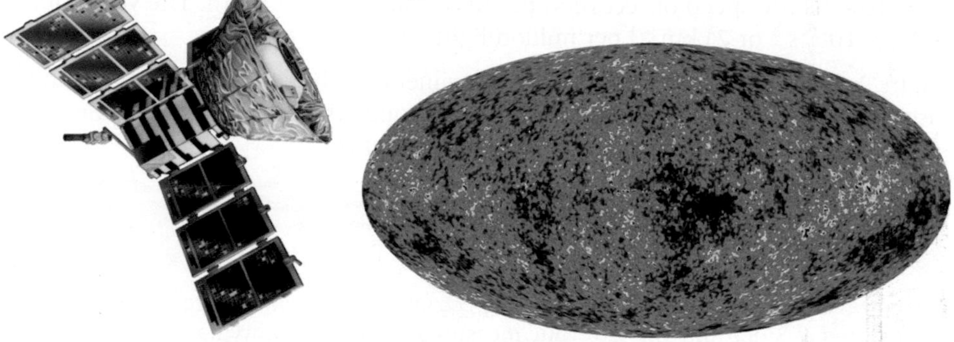

Figure 3 *Cosmic background microwave radiation*
a *The Cosmic Background Explorer (COBE)* ***b*** *Microwave map of the Universe*

Estimate of the age of the Universe

The speed of light in free space, c, is 300 000 km s^{-1}. No material object can travel as fast as light. Therefore, even though the speed, v, of a galaxy increases with its distance d, no galaxy can travel as fast as light. Therefore the furthermost galaxy could not be any further than a distance equal to cT where T is the age of the Universe.

Applying this distance to Hubble's law, $v = Hd$, therefore gives $c = H(cT)$.

Hence, $T = 1/H = (2.2 \times 10^{-18}\,\text{s}^{-1})^{-1} = 4.5 \times 10^{17}\,\text{s}$. Prove for yourself that this gives about 14 000 million years (i.e.
14 billion years) as an estimate of the age of the Universe.

> **Note**
>
> Measurements of cosmic microwave background radiation using different satellites and other methods give values of H between 2.1 and $2.3 \times 10^{-18}\,\text{s}^{-1}$.

Dark energy

Astronomers in 1998 studying type Ia supernova were astounded when they discovered very distant supernovas much further away than expected. To reach such distances, they must have been accelerating. The astronomers concluded that the expansion of the Universe is accelerating and has been for about the past 5000 million years. Before this discovery, most astronomers expected that the Universe is decelerating, as very distant objects would be slowed down by the force of gravity from other galaxies. Many more observations since then have confirmed the Universe is accelerating. Scientists think that no known force could cause an acceleration of the expansion of the Universe and that a hitherto unknown type of force must be releasing hidden energy referred to as **dark energy**.

Evidence for accelerated expansion of the Universe

Evidence for accelerated expansion of the Universe is based on differing distance measurements to type Ia supernova by two different methods:

1 The red shift method: Measurement of the red shift of each of these distant type Ia supernova and use of Hubble's law gives the distance to each one.

2 The luminosity method: Type Ia supernova at peak intensity are known to be 10^9 times more luminous that the Sun. The distance to such a supernova can be calculated from its luminosity L and its radiant flux intensity F. See Topic 23.1.

The two methods give results that are different and indicate that the distant type Ia supernova are dimmer and therefore further away than their red shift indicates.

The nature of dark energy

The nature of dark energy is unclear. It is thought to be a form of background energy present throughout space and time. It is more prominent than gravity at very large distances, because gravity becomes weaker and weaker with increased distance whereas the force associated with dark energy is thought to be constant. Current theories suggest it makes up about 70% of the total energy of the Universe. The search for further evidence of dark energy will continue with observations using larger telescopes and more sensitive microwave detectors on satellites.

Summary test 23.3

$c = 3.0 \times 10^8\,\mathrm{m\,s^{-1}}$, $H = 2.2 \times 10^{-18}\,\mathrm{s^{-1}}$

1 **a** State Hubble's law.

b What conclusion was drawn by astronomers when Hubble's law was discovered?

2 The wavelength of a spectral line in the spectrum of light from a distant galaxy was measured at 597.2 nm. The same line measured in the laboratory has a wavelength of 589.6 nm. Calculate:

a the speed of recession of the galaxy,

b the distance to the galaxy.

3 A supernova in a distant galaxy was observed to have a wavelength of 525 nm for a certain spectral line in the spectrum of its light. The same line measured in the laboratory has a wavelength of 486 nm.

a **i** Calculate the speed of recession of this galaxy.

ii Estimate the distance to this galaxy.

b The peak radiant flux intensity of the supernova at the Earth was measured at $2.7 \times 10^{-15}\,\mathrm{W\,m^{-2}}$. Estimate the peak luminosity of the supernova.

4 **a** State what is meant by *cosmic microwave background radiation* (CMBR).

b What is thought to be the origin of CMBR?

23 Exam-style and Practice Questions

Launch additional digital resources for the chapter

$\sigma = 5.67 \times 10^{-8}\,W\,m^{-2}\,K^{-4}$

1 a i State Wien's displacement law.

ii The black body radiation curve for a certain star has its peak intensity at a wavelength of 740 nm. Calculate the surface temperature of this star.

b The star is 6.0×10^{17} m from the Earth. The radiation flux intensity of the radiation from this star at the Earth is $3.5 \times 10^{-8}\,W\,m^{2}$. Calculate the luminosity of the star.

c i Show that the radiant energy per second per unit surface area emitted from the star is equal to σT^4, where T is the surface temperature of the star.

ii Hence calculate the surface area of the star and determine its diameter.

2 a Define the *radiant flux intensity* at the Earth of the electromagnetic radiation from a star.

b The table below gives the luminosity of three stars A, B and C in terms of the Sun's luminosity, their distances from the Earth and the wavelength of their peak intensity.

Star	Luminosity of star / Luminosity of the Sun	Distance from Earth/10^{17} m	Peak intensity wavelength/nm
A	10.6	1.58	380
B	130	7.39	230
C	0.0035	0.57	940

i State and explain which star is hottest and which star is coolest, giving a reason for your answer.

ii List the stars in order of increasing radiant flux intensity at the Earth.

c The luminosity of the Sun is 4.0×10^{26} W. For star A, calculate:

i its luminosity,

ii its surface temperature,

iii the radius of its surface.

d i The surface temperature of the Sun is 5800 K. Show that the radius R of a star is proportional to $L\lambda_{max}{}^4$, where L is its luminosity and λ_{max} is its peak intensity wavelength.

ii List the three stars in order of increasing radius, giving reasons for their order.

3 A supernova is an event that occurs when a large star explodes. In 1987, a supernova was observed in a galaxy 1.5×10^{23} m from the Earth.

a When the supernova was at its brightest, the radiant flux intensity at the Earth from it was $1.4 \times 10^{-13}\,W\,m^{-2}$. Calculate the luminosity of the supernova when it was at its brightest.

b The star that exploded was identified from past records as a blue star.

i Assuming the peak intensity wavelength of the radiation from the blue star was about 300 nm, estimate the surface temperature of this star.

ii The luminosity of the blue star was about 4×10^{29} W. Show that its diameter was about ten times larger than that of the Sun. The diameter of the Sun is 1.4×10^{9} m.

4 a i Explain what is meant by a *red shift* in the light from a star or a galaxy.

ii Describe how a red shift is used to determine the speed of recession of a star or galaxy moving away from the Earth.

b The table below shows some speeds of recession of six distant galaxies at different distances from the Earth.

Galaxy	Distance, d / 10^{19} km	Speed, v / km s^{-1}
A	4.6	63
B	10.5	328
C	20.9	437
D	25.3	719
E	32.0	609
F	38.5	1030

i Plot a graph of the speed of recession v on the y-axis against the distance d on the x-axis.

ii Use your graph to determine a value of the Hubble constant, H.

iii Estimate the uncertainty in the value of H you obtained from your graph.

c Hubble discovered that the speed of recession of galaxies at known distances is proportional to their distance from Earth. The measurements he used gave a value of H which was seven times greater than the currently accepted value. Later distance measurements using different methods since then have led to the present accepted value. Discuss the continuing *validity* of Hubble's law, even though the value of the Hubble constant is very different to the value estimated by Hubble.

Key concepts: Thermal and nuclear physics, medical imaging and astrophysics and cosmology

In studying Chapter 18, you will have gained a sound understanding of internal energy and absolute temperature and a good grasp of the practical skills involved in measuring specific heat capacities and specific latent heats. In addition, by considering conservation of energy, you will have learned how to analyse problems involving specific heat and latent heat and to carry out the necessary calculations in order to solve such problems.

After studying Chapter 19 on gases, you should appreciate the experimental basis of the ideal gas equation and how the kinetic theory of gases provides a full explanation of the properties of an ideal gas. The random motion of the molecules of an ideal gas is a key concept here and you should recognise that the observation of Brownian motion provides direct evidence of this random motion. In studying gases, you will have developed your mathematical skills in carrying out calculations using the ideal gas equation and the kinetic theory of gases equation. In addition, you should be able to combine these two equations to show that the mean kinetic energy of the molecules of an ideal gas is proportional to the absolute temperature of the gas.

In Chapter 20, you should be aware that the results of investigations in photoelectricity could not be explained using the wave theory of light and consequently led to the establishment by Einstein of the photon theory of light. The success of the photon theory should be evident to you from your studies of energy levels and spectra as well as from the successful explanation of photoelectricity and from the discovery that photons scattered by electrons lose momentum to the electrons and therefore possess momentum even though they have no rest mass. In contrast to the properties described above, where experiments and investigations preceded important theoretical work, the wave nature of matter was predicted by de Broglie and only accepted when experimental evidence of electron diffraction was discovered. Throughout Chapter 20, in addition to furthering your knowledge and understanding of the structure and properties of matter, you should have further developed your problem-solving skills as well as your analytical and mathematical skills.

Your studies of nuclear physics in Chapter 21 build on previous studies on radioactivity in Chapter 10. After studying Chapter 21, you should be aware that radioactive decay is a random process that gives a full explanation of the exponential decrease in the activity of a radioactive isotope. In addition, you should be able to appreciate how the concept of binding energy and mass defect enables us to explain why energy is released in nuclear changes including fission and fusion. In studying radioactive decay, you will have further developed your mathematical skills in solving problems and carrying out calculations involving exponential decreases and nuclear reactions.

Chapter 22, on medical imaging, provides awareness of the important role of physics in medicine. The three topics, ultrasonic imaging, X-ray imaging and positron emission tomography (PET scanning) are not the only imaging techniques used, but they do involve techniques applicable to different situations.

Chapter 23 Astrophysics and cosmology builds on previous studies on waves in order to develop understanding and awareness of how astronomers can use data from their observations to work out the size of a star and the distances to stars and galaxies.

The following table provides an overview of the topics in Chapters 18 to 23, showing where key concepts have been developed.

Chapters 18–23: Key concepts

Topic			Key concepts	
18.1–18.2	internal energy and temperature	first law of thermodynamics	energy and matter	conservation of energy
		absolute temperature		internal energy
18.3–18.4	specific heat capacity and specific latent heat	measuring specific heat capacity and specific latent heat	testing	practical skills
		heat capacity, latent heat	energy and matter	conservation of energy
		calculations involving specific heat capacity and specific latent heat	use of maths	problem-solving
19.1–19.2	ideal gases	testing the gas laws	testing	practical skills
		Brownian motion molar mass	energy and matter	random motion molecular theory
		calculations involving the ideal gas equation	use of maths	problem-solving
19.3–19.4	kinetic theory of gases	derivation of the kinetic theory of gases equation	models	kinetic theory of gases model
		random motion, root mean square speed, internal energy of an ideal gas	energy and matter	random motion
		calculations involving the kinetic theory of gases equation and the ideal gas equation	use of maths	problem-solving
20.1–20.3	electrons and photons	electromagnetic waves, photoelectricity, threshold frequency, work function energy levels and spectra	energy, matter and waves	nature of light
		the photon model used to explain photoelectricity, Einstein's photoelectric equation	models	photon model
		calculations on photoelectricity and energy levels	use of maths	problem-solving
20.4	wave particle duality	wave properties of electrons, electron diffraction	energy, matter and waves	dual nature of matter
		de Broglie's hypothesis and equation, photon momentum		
		calculations involving de Broglie's equation photon momentum	use of maths	problem-solving

continued on p.353

Chapters 18–23: Key concepts *(continued)*

Topic			Key concepts	
21.1–21.3	radioactive decay	activity, half-life	energy, matter and waves	
		random nature of radioactive decay	models	randomness
		theory of radioactive decay		exponential decrease
		calculations using the radioactive decay equation	use of maths	problem-solving
21.4–21.6	energy from the nucleus	binding energy, mass defect	energy and matter	conservation of energy
		binding energy curve		
		fission and fusion	models	chain reaction
		induced fission		
		calculations of binding energy per nucleon and of energy released in nuclear reactions	use of maths	problem-solving
22.1–22.3	medical imaging	ultrasonic imaging	energy, matter and waves	use of waves to probe matter
		X-ray imaging in medicine		
		PET scanning		ionising and non-ionising radiation
		CT scans	models	3D images
		calculations involving acoustic impedance, reflection coefficients, absorption using $I = I_0e^{-\mu x}$ and minimum photon wavelength from an X-ray tube	use of maths	problem-solving
23.1–23.3	astrophysics and cosmology	luminosity and radiation, intensity,	energy and waves	intensity of radiation from a point source
		astronomical distances	energy and waves	
		stellar radii	energy and waves	radiation laws applied to observations
		Hubble's law and expansion of the Universe	energy and waves	proportionality (of red shift and distance) and prediction of expansion of Universe
		the Big Bang	models	explanation of expansion of the Universe
		Calculations using the inverse square law, Wien's displacement law and the Stefan–Boltzmann's law	use of maths	problem-solving

Question

The first part of the question is about the nuclear model of the atom and why scientists predicted that the nucleus is composed of neutrons as well as protons before there was any direct evidence of the existence of neutrons. It also gives some insight into how such direct evidence was obtained using alpha radiation. The question also tests the application of the principles of conservation of energy and of momentum to the emission of α-particles from an α-emitting isotope to explain why the α-particles are emitted with well-defined energy.

*This well-defined energy is the reason why α-particles from an α-emitting isotope have a well-defined range in air, and the final part of the question also tests the relationship between the range **R** in air of α-particles and their initial kinetic energy **E** using experimental data from several α-emitting isotopes. A graph of log **R** against log **E** is plotted on the assumption that the relationship is of the form $\mathbf{R} = \mathbf{k}\mathbf{E}^n$, where n is an unknown constant. A straight-line graph would confirm that the relationship is of this form, enabling the power n to be found as it is equal to the gradient of the graph. Knowledge of this value of n provided physicists with a further challenge beyond the question to develop a theoretical model that 'confirms' the value of n found using the graphical method. The question thus shows how we can use data from experimental investigations to find relationships that can then be investigated by devising theoretical models that explain the relationships. In addition, knowing the value of n enables **E** to be calculated for other α-emitting isotopes by measuring the range **R** and using the now-known relationship between **E** and **R**.*

Before answering the question, students should be given the opportunity to do this or a similar investigation.

1 The neutron–proton model of the nucleus was first put forward by Rutherford to explain the general composition of the nucleus. The existence of the neutron was not proved experimentally until some years later.

a Rutherford knew from his alpha-scattering experiments that the charge of a nucleus is equal to $+Ze$, where Z is the atomic number of the nucleus and e is the magnitude of the charge of the electron. He also knew that the mass of a nucleus is approximately equal to $A\,m_u$, where A is the mass number of the nucleus and m_u is the mass of a hydrogen nucleus.

i By reference to the isotopes ${}^{1}_{1}H$ and ${}^{4}_{2}He$, give **two** reasons why Rutherford's neutron–proton model was considered more than an untested hypothesis when it was first put forward. *(2 marks)*

ii Explain why Rutherford concluded that the hydrogen ${}^{1}_{1}H$ nucleus is a single proton. *(1 mark)*

b Evidence of the existence of neutrons was found when it was discovered that α-particles knocked uncharged particles out of beryllium foil. Investigations into the nature of the uncharged radiation showed it consisted of particles each uncharged and of about the same mass as a proton.

i Complete the nuclear reaction below to show the result of an inelastic collision between an α-particle and a beryllium $^{9}_{4}Be$ nucleus.

$$^{4}_{2}\alpha + ^{9}_{\_}Be \longrightarrow ^{\_}_{6}C + ^{1}_{0}n$$ *(1 mark)*

ii Explain why the above change only takes place if the α-particle has sufficient kinetic energy. *(3 marks)*

c The α-particles from any α-emitting isotope have the same initial kinetic energy and a well-defined range in air at atmospheric pressure.

i The isotope $^{210}_{84}Po$ emits α-particles with initial kinetic energy of 5.3 MeV. Show that the nucleus recoils with an initial kinetic energy of 0.1 MeV and hence calculate the total energy released by the nucleus when it emits an α-particle. *(4 marks)*

ii The table below shows the range *R* in air and the initial kinetic energy *E* of α-particles from several α-emitting isotopes.

R/mm	39	48	53	57	66	78
E/MeV	5.3	6.0	6.5	6.8	7.4	8.3

Plot a suitable graph to find out if the relationship between *R* and *E* is of the form $R = kE^n$, where *k* and *n* are constants, and determine a value for *n*. Explain your choice of graph. *(9 marks)*

24 Mathematical skills

24.1 Data handling

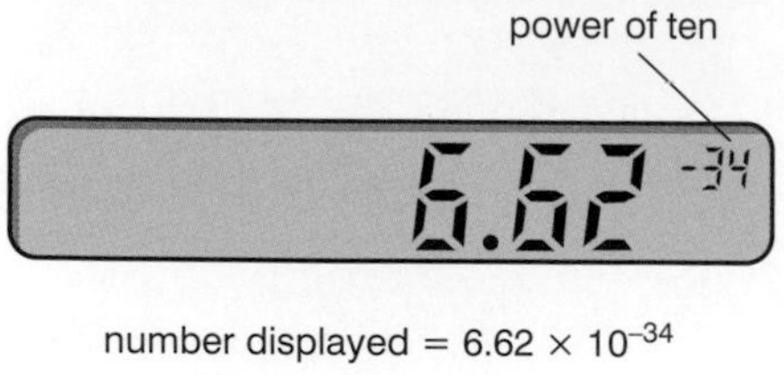

Figure 1 *Displaying powers of ten*

Using a calculator

(EXP) (or (EE) on some calculators)
This is the calculator button you press to key in a **power of ten**. To key in a number in standard form (e.g. 3.0×10^8), the steps are as follows:
Step 1 Key in the number between 1 and 10 (e.g. ③ ⊙ ⓪).
Step 2 Press the calculator button (EXP) (or (EE) on some calculators).
Step 3 Key in the power of ten (e.g. ⑧).

The display should now read 3.0 08 which should be read as 3.0×10^8 (not 3.0^8 which means 3.0 multiplied by itself 8 times). If the power of ten is a negative number (e.g. 10^{-8} not 10^8), press the calculator button (+/_) after Step 3 to change the sign of the power of ten.

(Inv)
This is the button you press if you want the calculator to give the value of the **inverse of a function**. For example, if you want to find out which angle has a sine of 0.5, you key in ⓪⊙⑤, then (Inv), then (sin), to obtain the answer of 30°. Some calculators have a 'second function' button that you press instead of the 'inv' button.

(log) or (lg)
This is the button you press to find out what a number is as a power of ten. For example, press (log), then key in ① ⓪ ⓪ and the display will show ②, because $100 = 10^2$. **Logarithmic scales** have equal intervals for each power of ten.

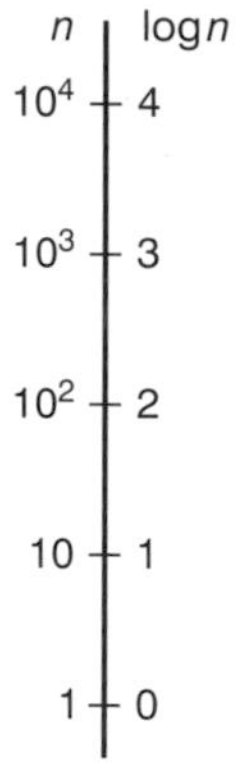

Figure 2 *A logarithmic scale*

(deg/rad)
Angles can be expressed in degrees or **radians** where 2π radians = 360 degrees. A scientific calculator has a button you can press to use either degrees or radians. Make sure you know how to switch your calculator from one of these two modes to the other. Many marks have been lost in examinations as a result of forgetting to use the correct mode. For example:

- $\sin 30° = 0.50$ whereas $\sin 30$ rad $= -0.99$,
- inv $\sin 10° = 0.17$ whereas inv $\sin 10$ rad $= -0.54$

Also, take care when you calculate the sine, cosine or tangent of the product of a number and an angle. For example, if you forget about the brackets in a calculation of $\sin(2 \times 30°)$, the answer 1.046 is obtained instead of the correct answer of 0.866. The reason for the error is that, unless you insert the brackets, the calculator is programmed to work out $\sin 2°$ then multiply the answer by 30.

Note

The label for each function on your calculator may differ from the label shown here. If you are unsure which button to press, check with your teacher.

(y^x)
This is used to raise any number to any power, for example, if you want to work out the value of 2^8, key in ② onto the display, then press the (y^x) button, then ⑧ and press ⊜. The display should then show ② ⑤ ⑥ as the decimal value of 2^8.

The (y^x) button can be used to find roots. For example, given the equation $T^4 = 5200$, you can find T by keying in ⑤ ② ⓪ ⓪ onto the display, then pressing the y^x button, followed by $(1 \div 4)$ which will give the answer 8.49.

Worked example

Calculate the cube root of 2.9×10^6.

Solution

Step 1 Key in 2.9×10^6 as explained earlier.

Step 2 Press the (y^x) button.

Step 3 Key in (( ) (1) (÷) (3) ( )).

Step 4 Press (=).

The display should show 1.426 02, so the answer is 142.6.

Significant figures

In general, always write a numerical answer to the same number of significant figures as the data used for the calculation. Because a calculator display shows a large number of digits, you should always round up, or round down, the answer displayed on a calculator to achieve the appropriate number of significant figures. For example, a calculator will show 9.063077870 × 10⁻¹ for the sine of 65°. Because the data used (65°) is given to two significant figures, the sine of 65° should then be rounded off to 9.1×10^{-1}, and written as 0.91 because 9.06 to two significant figures is rounded up to 9.1.

In calculations where there is more than one stage, carry forward the results of intermediate calculations without rounding off the data (or rounding it off to at least one more significant figure than the data has) and round off the final calculation only. Rounding off the data at the intermediate stage may result in an incorrect 'rounded-off' final answer.

For example, consider the calculation of the density ρ of the material in a solid cylinder of radius r, length l and mass m using the equations volume $V = \pi r^2 l$ and density $\rho = m \div V$, where $r = 2.30 \times 10^{-3} m$, $l = 0.152$ m and $m = 0.223$ kg.

- Rounding off to three significant figures at the final stage only gives $\rho = 8830\,\text{kg}\,\text{m}^{-3}$. The same answer is obtained if the value of the volume is rounded off to four significant figures before the density is calculated and rounded off to three significant figures.
- Rounding off the calculation of V to three significant figures before the density is calculated and rounded off to three significant figures gives $8810\,\text{kg}\,\text{m}^{-3}$.

Summary test 24.1

Write your answers to each of the following questions in standard form where appropriate, and to the same number of significant figures as the data.

1 Copy and complete the following conversions:

a **i** 500 mm = … m
ii 3.2 m = … cm
iii 9560 cm = … m

b **i** 0.45 kg = … g
ii 1997 g = … kg
iii 54 000 kg = …g

c **i** $20\,\text{cm}^2 = \ldots\text{m}^2$
ii $55\,\text{mm}^2 = \ldots\text{m}^2$
iii $0.05\,\text{cm}^2 = \ldots\text{m}^2$

2 **a** Write the following values in standard form:
i 150 million km in metres
ii 365 days in seconds
iii 630 nm in metres
iv 25.7 mg in kilograms
v 150 m in millimetres
vi 1.245 μm in metres

b Write the following values with a prefix instead of in standard form:
i 3.5×10^4 m = … km
ii 6.5×10^{-7} m = … nm
iii 3.4×10^6 g = …kg
iv 8.7×10^8 W = … MW = … GW

3 **a** Use the equation 'average speed = $\dfrac{\text{distance}}{\text{time}}$', to calculate the average speed in $\text{m}\,\text{s}^{-1}$ of:
i a vehicle that travels a distance of 9000 m in 450 s,
ii a vehicle that travels a distance of 144 km in 2 h,
iii a particle that travels a distance of 0.30 nm in a time of 2.0×10^{-18} s,
iv the Earth on its orbit of radius 1.5×10^{11} m, given the time taken per orbit is 365.25 days.

b Use the equation resistance = $\dfrac{\text{potential difference}}{\text{current}}$, to calculate the resistance of a component for the following values of current I and p.d. V:
i $V = 15$ V, $I = 2.5$ mA
ii $V = 80$ mV, $I = 16$ mA
iii $V = 5.2$ kV, $I = 3.0$ mA
iv $V = 250$ V, $I = 0.51$ μA
v $V = 160$ mV, $I = 53$ mA

4 **a** Calculate each of the following:
i 6.7^3
ii $(5.3 \times 10^4)^2$
iii $(2.1 \times 10^{-6})^4$
iv $(0.035)^2$
v $(4.2 \times 10^8)^{1/2}$
vi $(3.8 \times 10^{-5})^{1/4}$

b Calculate each of the following:
i $\dfrac{2.4^2}{3.5 \times 10^3}$
ii $\dfrac{3.6 \times 10^{-3}}{6.2 \times 10^2}$
iii $\dfrac{8.1 \times 10^4 + 6.5 \times 10^3}{5.3 \times 10^4}$
iv $7.2 \times 10^{-3} + \dfrac{6.2 \times 10^4}{2.6 \times 10^6}$

24.2 Trigonometry

More about the rules of trigonometry

Angles and arcs

- Angles are measured in **degrees or radians**. The scale for conversion is $360° = 2\pi$ radians. The symbol for the radian is rad so $1 \text{ rad} = \dfrac{360}{2\pi} = 57.3°$ (to three significant figures).
- The circumference of a circle of radius $r = 2\pi r$. So the circumference can be written as the angle in radians (2π) round the circle $\times\, r$.
- For a segment of a circle, the length of the arc of the segment is in proportion to the angle θ which the arc makes at the centre of the circle. This is shown in Figure 1. Because the arc length is $2\pi r$ (i.e. the circumference) for an angle of $360°$ ($= 2\pi$ radians):

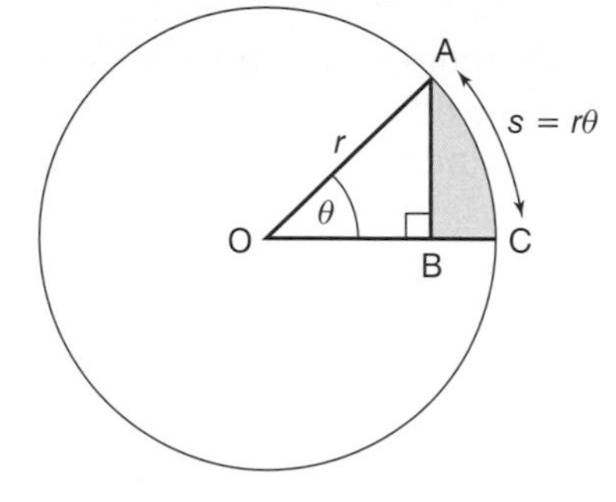

Figure 1 *Arcs and segments*

$$\frac{\text{arc length, } s}{2\pi r} = \frac{\theta \text{ (in degrees)}}{360} = \frac{\theta \text{ (in radians)}}{2\pi}$$

Equating $\dfrac{\text{arc length, } s}{2\pi r}$ to $\dfrac{\theta \text{ in radians}}{2\pi}$ gives

$$\textbf{arc length } s = r\theta$$

where θ is the angle subtended in radians.
Note that for $s = r$, $\theta = 1$ rad ($= 360/2\pi = 57.3°$)

The small angle approximation

For angle θ less than about $10°$,

$\sin\theta \approx \tan\theta \approx \theta$ in radians, and **$\cos\theta \approx 1$**

To explain these approximations, consider Figure 1 again. If angle θ is sufficiently small, then the segment OAC will be almost the same as triangle OAB, as shown in Figure 2.

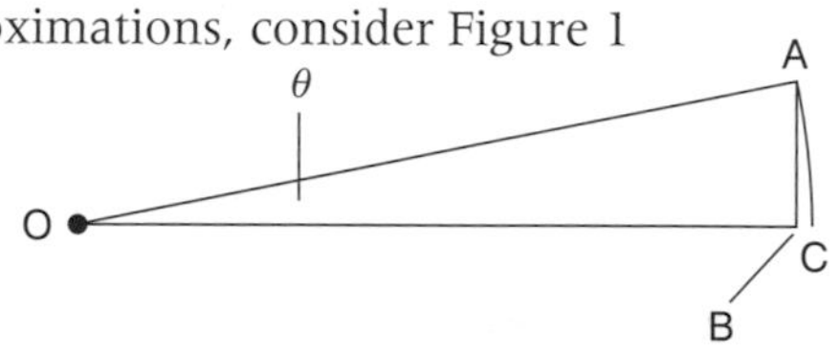

Figure 2 *The small angle approximation*

- AB ≈ arc length s so $\sin\theta = \dfrac{AB}{OA} \approx \dfrac{s}{r} = \theta$ in radians.
  $\therefore \sin\theta \approx \theta$ in radians.

- OB ≈ radius r, so $\tan\theta = \dfrac{AB}{OB} \approx \dfrac{s}{r} = \theta$ in radians.
  $\therefore \tan\theta \approx \theta$ in radians.
  and $\cos\theta = \dfrac{OB}{OA} \approx \dfrac{r}{r} = 1$
  $\therefore \cos\theta \approx 1$

Use a calculator to prove to yourself that for $\sin 10° = 0.1736$, $\tan 10° = 0.1763$ and $10° = 0.1745$ rad. Also, $\cos 10° = 0.9848$. So the small angle approximation is almost 99% accurate up to $10°$. Figure 3 shows how $\sin\theta$ and $\cos\theta$ change as θ increases.

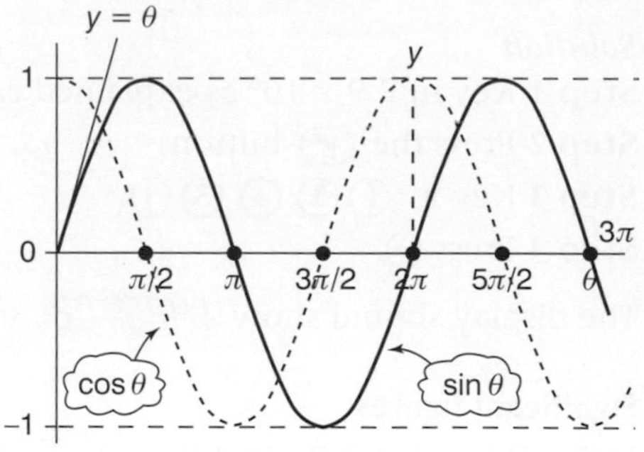

Figure 3 *Sine and cosine curves*

More triangle rules

In addition to the 'right-angle triangle' rules described on page x, the following rules apply to any triangle:

- Area of a triangle $= \frac{1}{2} \times$ its height $(h) \times$ its base (b). See Figure 4a.
- For a triangle with sides of lengths a, b and c and angles A, B and C opposite sides a, b and c, as shown in Figure 4b.

$$\frac{a}{\sin A} = \frac{b}{\sin B} = \frac{c}{\sin C} \qquad \text{(the sine rule)}$$

$$a^2 = b^2 + c^2 - 2bc\cos A \qquad \text{(the cosine rule)}$$

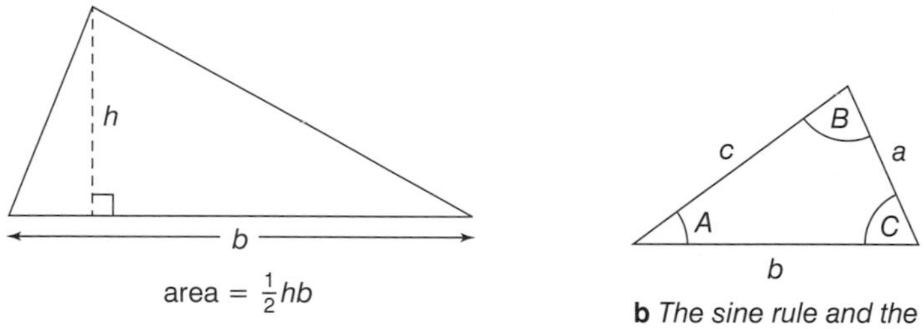

a *The area of a triangle*

b *The sine rule and the cosine rule*

Figure 4

Pythagoras' theorem and trigonometry

Pythagoras' theorem states that for any right-angled triangle:

The square of the hypotenuse = the sum of the squares of the other two sides

Applying Pythagoras' theorem to the right-angled triangle in Figure 5 gives:

$$h^2 = o^2 + a^2$$

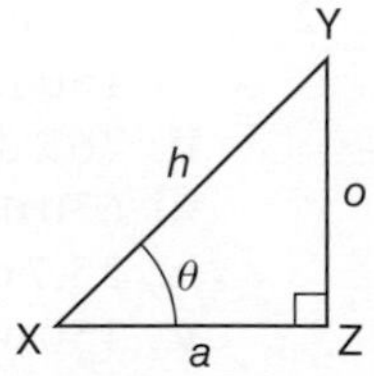

Figure 5 *A right-angled triangle*

Since $o = h\sin\theta$ and $a = h\cos\theta$, the above equation may be written:
$h^2 = h^2\sin^2\theta + h^2\cos^2\theta$

Cancelling h^2 therefore gives the following useful link between $\sin\theta$ and $\cos\theta$:

$$\mathbf{1 = \sin^2\theta + \cos^2\theta}$$

Vector rules

Resolving a vector

As explained in Topic 1.1, any vector can be resolved into two perpendicular components in the same plane as the vector, as shown by Figure 6. The force vector F is resolved into a horizontal component $F \cos \theta$ and a vertical component $F \sin \theta$, where θ is the angle between the line of action of the force and the horizontal line.

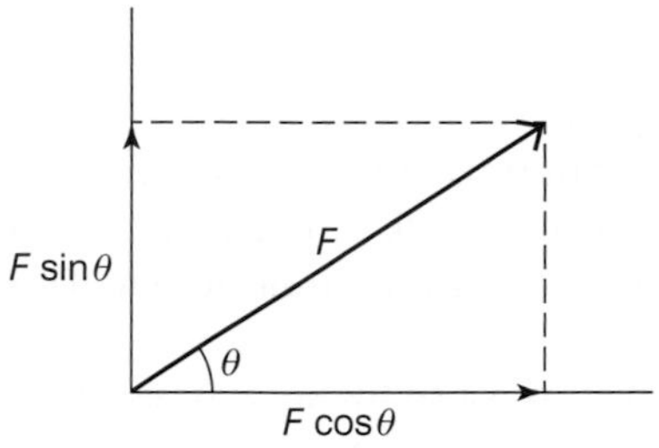

Figure 6 *Resolving a vector*

Note The component at angle θ to the direction of the vector is always $\cos \theta \times$ the magnitude of the vector.

Adding two vectors at angle θ to each other

Consider an object, O, acted on by forces F_1 and F_2 at angle θ to each other, as shown in Figure 7a. The magnitude and direction of the resultant force F_R can be found by resolving one of the forces into components that are parallel and perpendicular to the other force, as explained in Topic 1.1.

The method described in Topic 1.1 gives

$$F_R{}^2 = [(F_1 \cos\theta + F_2)^2 + (F_1 \sin\theta)^2].$$

Squaring the first term on the right-hand side gives

$$F_R{}^2 = F_1{}^2 \cos^2\theta + 2F_1F_2 \cos\theta + F_2{}^2 + F_1{}^2 \sin^2\theta$$

Because $\cos^2\theta + \sin^2\theta = 1$, then $F_1{}^2 \cos^2\theta + F_1{}^2 \sin^2\theta = F_1{}^2$

Hence

$$\mathbf{F_R{}^2 = F_1{}^2 + 2F_1F_2 \cos\theta + F_2{}^2}$$

This equation above can also be obtained by applying the cosine rule to the force triangle shown in Figure 7b with $a = F_R$, $b = F_1$ and $c = F_2$ and $A = 180 - \theta$.

This gives $F_R{}^2 = F_1{}^2 + F_2{}^2 - 2F_1F_2 \cos(180 - \theta)$. The equation above then follows since $\cos(180 - \theta) = -\cos\theta$.

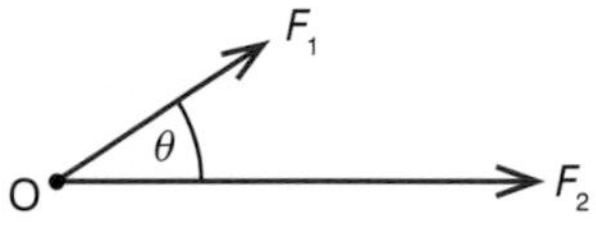

Figure 7a *Vector diagram for F_1 and F_2*

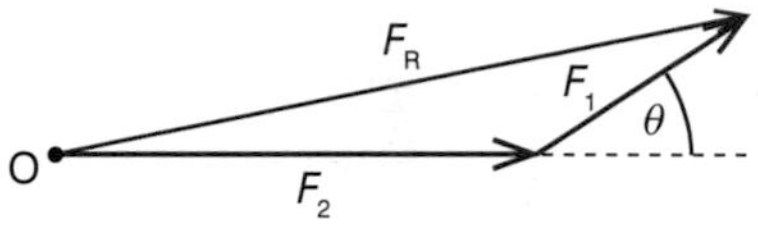

Figure 7b *Using the components to find the resultant force F_R*

Summary test 24.2

1 a Calculate the circumference of a circle of radius 0.25 m.

b Calculate the length of the arc of a circle of radius 0.25 m for the following angles, between the arc and the centre of the circle:
i 360°, **ii** 240°, **iii** $\pi/3$ rad, **iv** 0.6π rad.

c An aircraft travels a distance of 30 km due north from an airport P to an airport Q. It then travels due east for a distance of 18 km, to an airport R.

Calculate:
i the distance from P to R, **ii** the angle QPR.

2 a Measure the diameter of a coin to the nearest mm. Calculate the angle subtended at your eye, in degrees, by the coin held at a distance of 50 cm from your eye.

b i Estimate the angular width of the Moon, in degrees, at your eye by holding a millimetre scale at 50 cm from your eye and measuring the distance on the scale covered by the lunar disc.

ii The diameter of the Moon is 3500 km. The average distance to the Moon from the Earth is 380 000 km. Calculate the angular width of the Moon as seen from the Earth and compare the calculated value with your estimate in **b i**.

3 a Use the small angle approximation to calculate $\sin\theta$ for: **i** $\theta = 2.0°$, **ii** $\theta = 8.0°$.

b A triangle has sides of lengths a, b, and c and internal angles A, B and C opposite sides a, b and c respectively. Sketch each of the following triangles and use the cosine rule to calculate a for:
i $b = 80$ mm, $c = 60$ mm and $A = 60°$,
ii $b = 75$ mm, $c = 40$ mm and $A = 70°$,
iii $b = 120$ mm, $c = 45$ mm and $A = 120°$.

c For each triangle in **b**, use the sine rule as appropriate to calculate B and C.

4 a Calculate the perpendicular components A and B of each of the vectors in Figure 8.

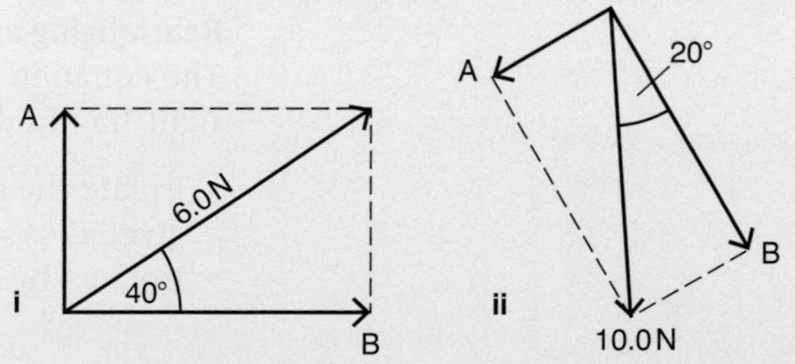

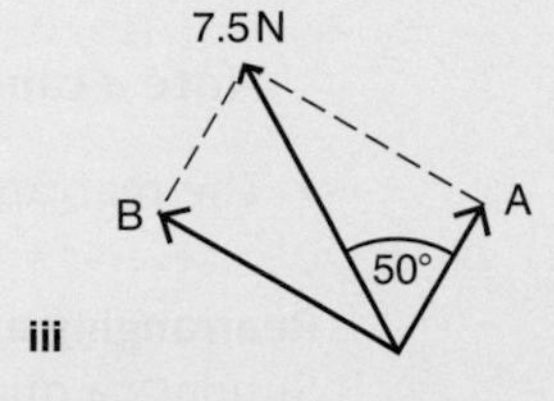

Figure 8

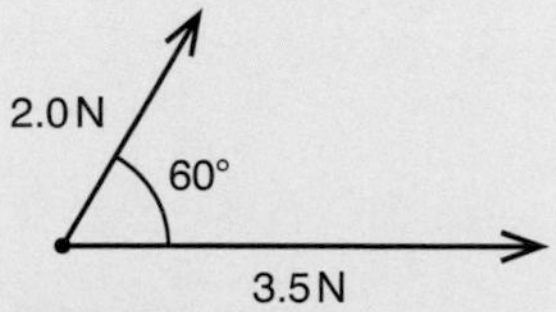

Figure 9

b Calculate the magnitude and direction of the resultant of the two vectors shown in Figure 9.

24.3 More about algebra

Signs and symbols

Table 1 Signs

Sign	Meaning
>	greater than
<	less than
⩾	greater than or equal to
⩽	less than or equal to
≫	much greater than
≪	much less than
≈	approximately equals
$<x>$	mean value
$<x^2>$	mean square value
∝	is proportional to
Δ	change of
√	square root

Signs you need to recognise

- **Inequality signs** are often used in physics. You need to be able to recognise the meaning of the signs in Table 1. For example, the inequality $I \geqslant 3$ A means that the current is greater than or equal to 3 A. This is the same as saying that the current is not less than 3 A.
- The **approximation sign** is used where an estimate or an order-of-magnitude calculation is made, rather than an accurate calculation. For an order-of-magnitude calculation, the final value is written with one significant figure only, or even rounded up or down to the nearest power of ten. Order-of-magnitude calculations are useful as a quick check after using a calculator. For example, if you are asked to calculate the density of a 1.0 kg metal cylinder of height 0.100 m and diameter 0.071 m, you ought to obtain a value of $2530\,\text{kg}\,\text{m}^{-3}$ using a calculator. Now let's check the value:

$$\text{Volume} = \pi\,(\text{radius})^2 \times \text{height} \approx 3 \times (0.04)^2 \times 0.1 \approx 48 \times 10^{-5}\,\text{m}^3$$

$$\text{Density} = \frac{\text{mass}}{\text{volume}} \approx \frac{1.0}{50 \times 10^{-5}} \approx 2000\,\text{kg}\,\text{m}^{-3}$$

  This confirms our 'accurate' calculation.

- **Proportionality** is represented by the ∝ sign. A simple example of its use in physics is for Hooke's law: the tension in a spring is proportional to its extension.

$$\text{Tension, } T \propto \text{extension, } e$$

By introducing a constant of proportionality k, the link above can be made into an equation:

$$T = ke$$

where k is defined as the spring constant. See Topic 5.4. With any proportionality relationship, if one of the variables is increased by a given factor (e.g. × 3), the other variable is increased by the same factor. So in the above example, if T is trebled, then extension e is also trebled. A graph of tension T on the y-axis against extension e on the x-axis would give a straight line through the origin.

More about equations and formulas

Rearranging an equation with several terms

The equation $v = u + at$ is an example of an equation with two terms on the right-hand side. These terms are u and at. To make t the subject of the equation:

- Isolate the term containing t on one side, by subtracting u from both sides, to give: $v - u = at$
- Isolate t by dividing both sides of the equation $v - u = at$ by a to give:

$$\frac{v-u}{a} = \frac{at}{a} = t$$

  Note a cancels out in the expression $\frac{at}{a}$.

- The rearranged equation may now be written: $t = \frac{v-u}{a}$

Rearranging an equation containing powers

Suppose a quantity is raised to a power in a term in an equation and that quantity is to be made the subject of the equation. For example, consider the equation $V = \frac{4}{3}\pi r^3$, where r is to be made the subject of the equation:

- Isolate r^3 from the other factors in the equation by dividing both sides by 4π then multiplying both sides by 3 to give: $\frac{3V}{4\pi} = r^3$
- Take the cube root of both sides to give: $\left(\frac{3V}{4\pi}\right)^{1/3} = r$
- Rewrite the equation with r on the left-hand side if necessary.

More about powers

- Powers add for identical quantities when two terms are multiplied together. For example, if $y = ax^m$ and $z = bx^n$, then: $yz = ax^m bx^n = abx^{m+n}$
- An equation of the form $y = \frac{k}{z^x}$ may be written in the form $y = kz^{-n}$
- The nth root of an expression is written as the power $\frac{1}{n}$.

  For example, the square root of x is $x^{½}$. Therefore, rearranging $y = x^n$ to make x the subject gives $x = y^{1/n}$

Quadratic equations

Quadratic equations occur in physics where a formula contains the square of a variable. The equation $s = ut + \frac{1}{2}at^2$ for displacement at constant acceleration is an example. See Topic 1.4.

Any quadratic equation can be written in the form $ax^2 + bx + c = 0$, where a, b and c are constants. The general solution of this equation is

$$x = \frac{-b \pm \sqrt{b^2 - 4ac}}{2a}$$

Note that every quadratic equation has two solutions, one given by the + sign before the square root sign in the above expression, and the other given by the – sign.

For example, consider the solution of the equation $2x^2 + 5x - 3 = 0$.

As $a = 2$, $b = 5$ and $c = -3$, the solution is

$$x = \frac{-5 \pm \sqrt{5^2 - (4 \times 2 \times -3)}}{2 \times 2} = \frac{-5 \pm \sqrt{49}}{4}$$

$$= \frac{-5 \pm 7}{4} = +0.50 \text{ or } -3$$

A graph of $y = 2x^2 + 5x - 3$ is shown in Figure 1. Note that the two solutions above are the values of the x-intercepts, which is where $y = 0$.

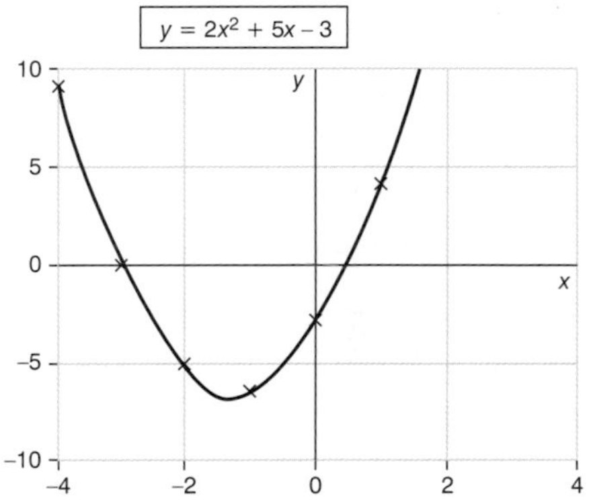

Figure 1 $y = 2x^2 + 5x - 3$

How to check a physics equation

Derived units written in terms of their **base units** can be used to check equations. The physical quantities on each side of an equation must match in terms of base units. If they don't match, the equation cannot be correct. For example, consider:

- the equation $v = \sqrt{(2gR)}$, which is used to calculate the escape speed v of an object from the surface of a planet of radius R and surface gravitational field strength g:

  left-hand side base units = $\mathrm{m\,s^{-1}}$
  right-hand side base units = $\sqrt{\mathrm{m\,s^{-2} \times m}} = \mathrm{m\,s^{-1}}$

  The equation has the same combination of base units on each side, so it is correct; we say it is **homogeneous** in terms of the base units.

- the equation $W = QV$ is used to calculate the work done W to move a charge Q through a potential difference V:

  left-hand side base units (see Table 2) = $\mathrm{kg\,m^2\,s^{-2}}$
  right-hand side base units = $(\mathrm{A\,s}) \times (\mathrm{kg\,m^2\,s^{-3}\,A^{-1}}) = \mathrm{kg\,m^2\,s^{-2}}$

  The equation has the same combination of base units on each side, so it is homogeneous.

Note

For simple equations such as this, homogeneity can sometimes be checked faster by recalling basic relationships between physical quantities. In the second example, one volt is one joule per coulomb, so the unit of QV is the joule per coulomb × the coulomb, which is the joule.

Table 2 *Links between units*

Quantity	Symbol	Unit	Unit symbol	Other forms of unit	Base unit	Chapter reference
acceleration	a	metre per second^2	$m\,s^{-2}$		$m\,s^{-2}$	1
activity of radioactive source	A	becquerel	Bq	s^{-1}	s^{-1}	23
angle	θ	radian or degree	rad or °			1,11
angular displacement	θ	radian or degree	rad or °			11
angular frequency	ω	radian per second	$rad\,s^{-1}$			11
angular speed or velocity	ω	radian per second	$rad\,s^{-1}$			11
capacitance	C	farad	F	$C\,V^{-1}$	$kg^{-1}\,m^{-2}\,s^4\,A^2$	15
charge	q, Q	coulomb	C		A s	6
decay constant	λ	second $^{-1}$	s^{-1}		s^{-1}	23
density	ρ	kilogram per cubic metre	$kg\,m^{-3}$		$kg\,m^{-3}$	5
electric field strength	E	newton per coulomb	$N\,C^{-1}$	$V\,m^{-1}$	$kg\,m\,s^{-3}\,A^{-1}$	6
electric potential	V	volt	V	$J\,C^{-1}$	$kg\,m^2\,s^{-3}\,A^{-1}$	14
energy, work	E, U	joule	J		$kg\,m^2\,s^{-2}$	2
force	F	newton	N		$kg\,m\,s^{-2}$	2
frequency	f	hertz	Hz	s^{-1}	s^{-1}	8
gravitational field strength	g	newton per kilogram	$N\,kg^{-1}$		$m\,s^{-2}$	1,2
gravitational potential	ϕ	joule per kilogram	$J\,kg^{-1}$		$m^2\,s^{-2}$	13
magnetic flux density	B	tesla	T	$N\,A^{-1}\,m^{-1}$ $Wb\,m^{-2}$	$kg\,s^{-2}\,A^{-1}$	18
magnetic flux	Φ	weber	Wb	$T\,m^2$, $V\,s$	$kg\,m^2\,s^{-2}\,A^{-1}$	18
momentum	p	kilogram metre per second	$kg\,m\,s^{-1}$	N s	$kg\,m\,s^{-1}$	4
permeability of free space	μ_0	henry per metre	$H\,m^{-1}$		$kg\,m\,s^{-2}\,A^{-2}$	18
permittivity of free space	ε_0	farad per metre	$F\,m^{-1}$		$kg^{-1}\,m^{-3}\,s^4\,A^2$	14
phase difference	ϕ	radian or degree	rad or °			8
potential difference, e.m.f.	V, E	volt	V	$J\,C^{-1}$	$kg\,m^2\,s^{-3}\,A^{-1}$	6, 7
power	P	watt	W	$J\,s^{-1}$	$kg\,m^2\,s^{-3}$	2
pressure	p	pascal	Pa	$N\,m^{-2}$	$kg\,m^{-1}\,s^{-2}$	5
resistance	R	ohm	Ω	$V\,A^{-1}$	$kg\,m^2\,s^{-3}\,A^{-2}$	6
resistivity	ρ	ohm metre	$\Omega\,m$		$kg\,m^3\,s^{-3}\,A^{-2}$	6
spring constant	k	newton per metre	$N\,m^{-1}$		$kg\,s^{-2}$	5
stress, Young modulus	σ, E	pascal	Pa	$N\,m^{-2}$	$kg\,m^{-1}\,s^{-2}$	5
velocity, speed	v	metre per second	$m\,s^{-1}$		$m\,s^{-1}$	1
wavelength	λ	metre	m		m	8

Summary test 24.3

1 a Complete each of the following statements:

i If $x > 5$, then $\frac{1}{x} < \ldots$

ii If $4 < x < 10$, then $\ldots < \frac{1}{x} < \ldots$

iii If $x^2 > 100$ then $\frac{1}{x} \ldots$

b Make t the subject of each of the following equations:

i $v = u + at$ **ii** $s = \frac{1}{2}at^2$

iii $y = k(t - t_0)$ **iv** $F = \frac{mv}{t}$

c Solve each of the following equations:

i $2z + 6 = 10$ **ii** $2(z + 6) = 10$

iii $\frac{2}{z - 4} = 8$ **iv** $\frac{4}{z^2} = 36$

2 a Make x the subject of each of the following equations:

i $y = 2x^{½}$ **ii** $2y = x^{-½}$

iii $yx^{1/3} = 1$ **iv** $y = \frac{k}{x^2}$

b Solve each of the following equations:

i $x^{-½} = 2$ **ii** $3x^2 = 24$

iii $\frac{8}{x^2} = 32$ **iv** $2(x^{½} + 4) = 12$

3 a Use the data given with each equation below to calculate:

i The surface area of cross-section A, of a wire of radius $r = 0.34$ mm and length $L = 0.840$ m, using the equation $A = \pi r^2 L$.

ii The radius r of a sphere, of volume $V = 1.00 \times 10^{-6}\,\text{m}^3$, using the formula $V = \frac{4}{3}\pi r^3$.

iii The time period T of a simple pendulum of length $L = 1.50$ m, using the formula $T = 2\pi(L/g)^{0.5}$, where $g = 9.81\,\text{m s}^{-2}$.

iv The speed v of an object, of mass $m = 0.20$ kg and kinetic energy $E_k = 28$ J, using the formula:

$$E_k = \frac{1}{2}mv^2$$

b Express each of the following combinations of physical quantities in SI base units:

i pressure × volume,

ii momentum^2 ÷ energy,

iii resistance × capacitance.

4 a Solve each of the following quadratic equations.

i $2x^2 + 5x - 3 = 0$

ii $x^2 - 7x + 8 = 0$

iii $3x^2 + 2x - 5 = 0$

b Use the data and the given equation to write down a quadratic equation and so determine the unknown quantity in each case:

i $s = ut + \frac{1}{2}at^2$, where $s = 20$ m, $u = 4\,\text{m s}^{-1}$ and $a = 6\,\text{m s}^{-2}$; find t.

ii $P = \frac{V^2 R}{(R + r)^2}$ where $P = 16$ W, $V = 12$ V, $r = 2.0\,\Omega$; find R.

The general equation for a straight-line graph

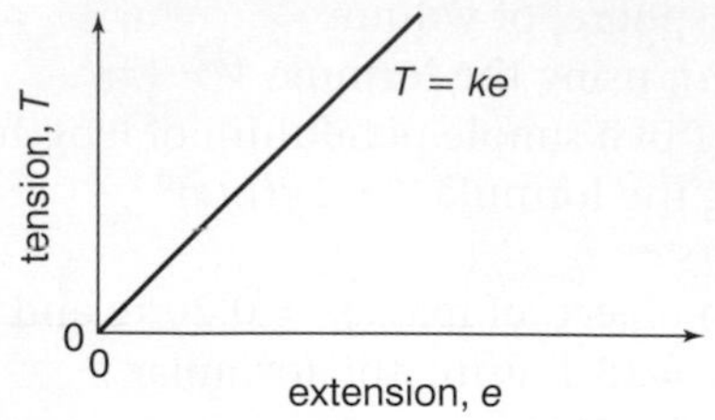

Figure 1

Links between two physical quantities can be established most easily by plotting a graph. One of the physical quantities is represented by the vertical scale (the 'ordinate', often called the y-axis) and the other quantity by the horizontal scale (the 'abscissa', often called the x-axis). The **coordinates** of a point on a graph are the x- and y-values, usually written (x, y) of the point.

The simplest link between two physical variables is where the plotted points define a straight line. For example, Figure 1 shows the link between the tension in a spring and the extension of the spring; the gradient of the line is constant and the line passes through the origin. Any situation where the y-variable is proportional to the x-variable gives a straight line through the origin. For Figure 1, the gradient of the line is the spring contant k. The relationship between the tension T and the extension e may therefore be written as $T = ke$.

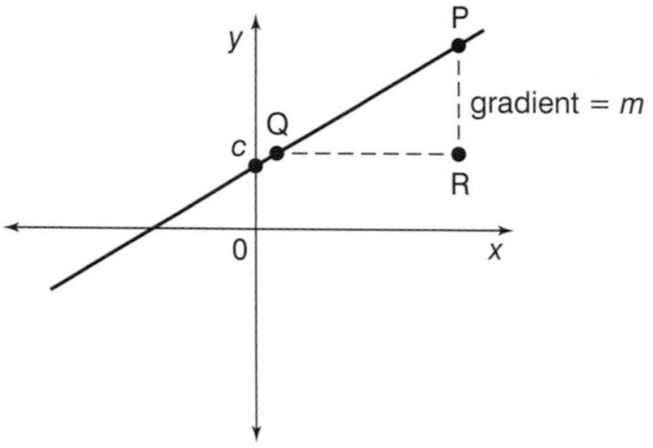

Figure 2 $y = mx + c$

The general equation for a straight-line graph is usually written in the form:

$$y = mx + c$$

where m = the gradient of the line, and c = the y-intercept.

- The **gradient** m can be measured by marking two points, P and Q, as far apart as possible on the line. The triangle PQR as shown in Figure 2 is then used to find the gradient. If (x_P, y_P) and (x_Q, y_Q) represent the x- and y-coordinates of points P and Q respectively, then

$$\text{gradient } m = \frac{y_P - y_Q}{x_P - x_Q}$$

- The **y-intercept**, c, is the point at $x = 0$, where the line crosses the y-axis. To find the y-intercept of a line on a graph that does not show $x = 0$, measure the gradient as above then use the coordinates of any point on the line with the equation $y = mx + c$ to calculate c. For example, rearranging $y = mx + c$ gives $c = y - mx$. Therefore, using the coordinates of point Q in Figure 2, the y-intercept $c = y_Q - mx_Q$.

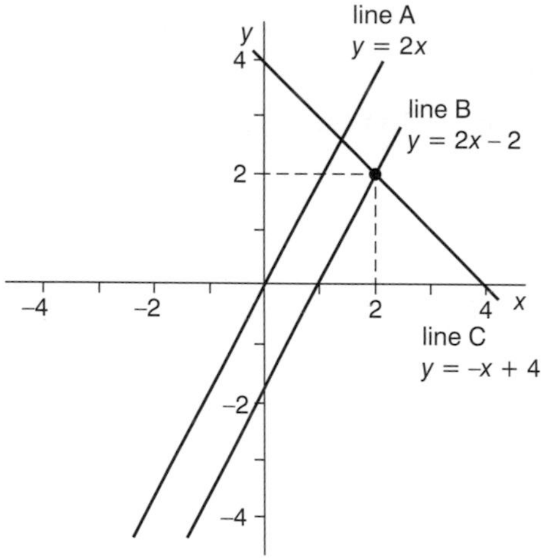

Figure 3 *Straight-line graphs*

Examples of straight-line graphs

In Figure 3

- **Line A**: $c = 0$, so the line passes through the origin; its equation is $y = 2x$.
- **Line B**: $m > 0$, so the line has a positive gradient; its equation is $y = 2x - 2$.
- **Line C**: $m < 0$, so the line has a negative gradient; its equation is $y = -x + 4$.

Straight-line graphs and physics equations

You need to be able to work out gradients and intercepts for equations you meet in physics that generate straight-line graphs. Some further examples are described below:

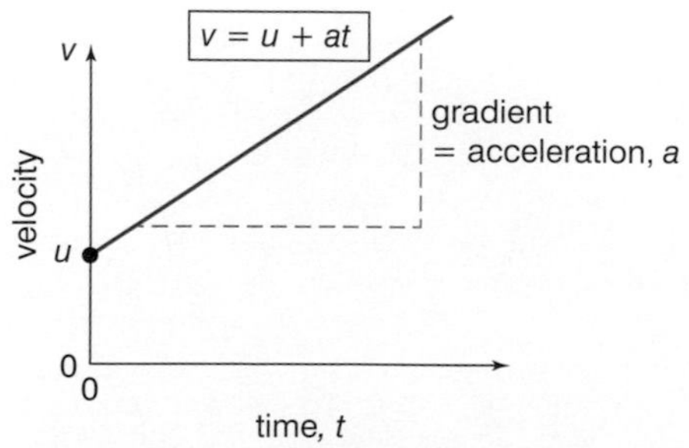

Figure 4 *Motion at constant acceleration*

Motion at constant acceleration

The velocity v of an object moving at constant acceleration a at time t is given by the equation $v = u + at$, where u is its speed at time t. Figure 4 shows the corresponding graph of velocity v on the y-axis against time t on the x-axis.

Rearranging the equation as $v = at + u$ and comparing this with $y = mx + c$ shows that:

- the gradient, m = acceleration a, and
- the y-intercept, c = the initial velocity u.

P.d. and current for a battery

The p.d. V across the terminals of a battery, of e.m.f. E and internal resistance r, varies with current in accordance with the equation $V = E - Ir$ (see Topic 7.3). Figure 5 shows the corresponding graph of p.d. V on the y-axis against current I on the x-axis.

Rearranging the equation as $V = -rI + E$ and comparing this with $y = mx + c$ shows that:

- the gradient, $m = -r$, and
- the y-intercept, $c = E$ so the intercept on the y-axis gives the e.m.f. E of the battery.

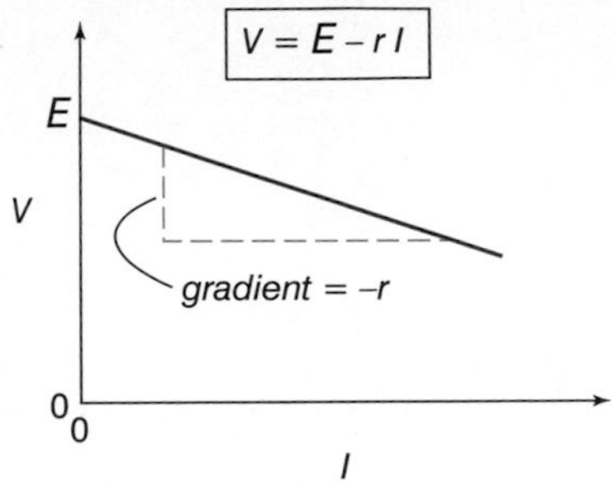

Figure 5 P.d. v. current for a battery

Photoelectric emission

The maximum kinetic energy $E_{K(max)}$ of a photoelectron, emitted from a metal surface of work function ϕ, varies with frequency f of the incident radiation in accordance with the equation $E_{K(max)} = hf - \phi$ (see Topic 20.1). Figure 6 shows the corresponding graph of $E_{K(max)}$ on the y-axis against f on the x-axis.

Comparing the equation $E_{K(max)} = hf - \phi$ with $y = mx + c$ shows that:

- the gradient, $m = h$, and
- the y-intercept, $c = -\phi$.

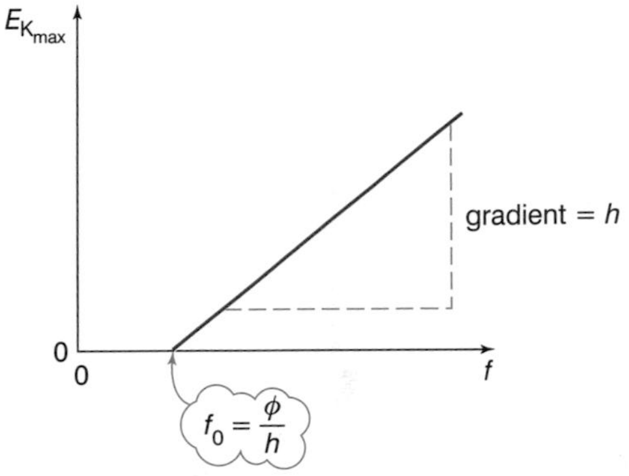

Figure 6 Photoelectric emission

Note

The x-intercept is where $y = 0$ on the line. Let the coordinates of the x-intercept be $(x_0, 0)$. Therefore $mx_0 + c = 0$ so $x_0 = \frac{-c}{m}$. In Figure 6, the x-intercept is therefore $\frac{\phi}{h}$. This is the threshold frequency f_0, $= \frac{\phi}{h}$.

Simultaneous equations

Two equations with two variable quantities, x and y, in each can be solved to find the values of x and y. Such a pair of equations is referred to as **simultaneous equations** because they have the same solution. They are described as **linear** because they contain terms in x and y and do not contain any higher order terms such as x^2 or y^2.

The general equation for a straight-line graph is $y = mx + c$, as explained in the previous section. Two straight lines on a graph can be represented by two such equations. Provided the two lines are not parallel to one another, they cross each other at a single point. The coordinates of this point are the values of x and y that fit both equations. In other words, these coordinates are the solution of a pair of simultaneous equations representing the two straight lines. For example, the two straight lines in Figure 7 $y = 2x - 2$ and $y = -x + 4$ meet at the point (2, 2) so $x = 2$, $y = 2$ are the only values of x and y that fit both equations.

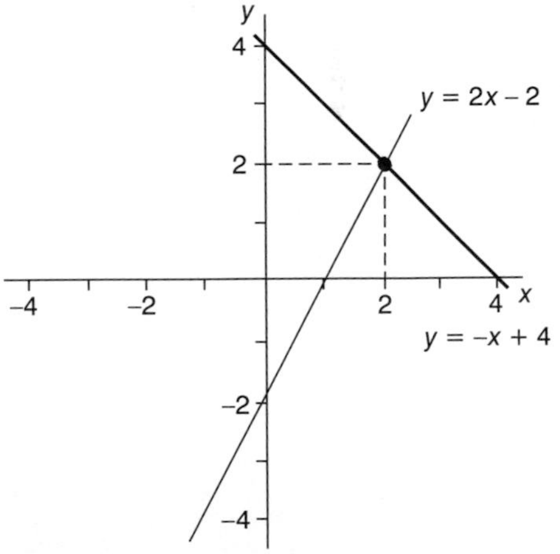

Figure 7 A graphical solution

The graphical approach to finding the solution of a pair of simultaneous equations takes time and is not as accurate as a systematic algebraic method. This method can best be explained by considering an example, as follows

$$\mathbf{2x - y = 2} \quad \text{(equation 1)}$$
$$\mathbf{x + y = 4} \quad \text{(equation 2)}$$

Make the coefficient of x the same in both equations by multiplying one or both equations by a suitable number. In the above equation, this is most easily achieved by multiplying equation 2 throughout by 2 to give $2x + 2y = 8$.

The two equations to be solved are now

$$2x - y = 2 \quad \text{(equation 1)}$$
$$2x + 2y = 8 \quad \text{(modified equation 2)}$$

Subtracting modified equation 2 from equation 1 gives

$$(2x - y) - (2x + 2y) = 2 - 8$$
$$\therefore -y - 2y = -6$$
$$-3y = -6$$
$$y = \frac{-6}{-3} = 2$$

Substituting this value into equation 1 or equation 2 enables the value of x to be determined. Using equation 2 for this purpose gives $x + 2 = 4$ and hence $x = 4 - 2 = 2$.

The solution of the two equations is, therefore, $x = 2$, $y = 2$.

Simultaneous equations with two unknown quantities can occur in several parts of the A level physics course, for example:

- $\boldsymbol{v = u + at}$ in kinematics (see Topic 1.4)
- $\boldsymbol{V = E - Ir}$ in electricity (see Topic 7.3)
- $\boldsymbol{E_{K(max)} = hf - \phi}$ (see Topic 20.1)

Summary test 24.4

1 For each of the following equations that represent straight-line graphs, write down:

i the gradient,
ii the y-intercept,
iii the x-intercept:

a $y = 3x - 3$

b $y = -4x + 8$

c $y + x = 5$

d $2y + 3x = 6$

2 a A straight line on a graph has a gradient $m = 2$ and passes through the point (2, −4). Work out:

i the equation for this line,
ii its y-intercept.

b The velocity v (in $\mathrm{m\,s^{-1}}$) of an object varies with time t (in s) in accordance with the equation
$v = 5 + 3t$.

Determine:

i the acceleration of the object,
ii the initial velocity of the object.

3 a Plot the equations $y = x + 3$ and $y = -2x + 6$ over the range from $x = -3$ to $x = +3$. Write down the coordinates of the point P where the two lines cross.

b Write down the equation for the line OP, where O is the origin of the graph.

4 Solve the following pairs of simultaneous equations, after making y the subject of each equation if necessary:

a $y = 2x - 4$, $y = -x + 2$

b $y = 3x - 4$, $x + y = 8$

c $2x + 3y = 4$, $x + 2y = 2$

Gradients

The gradient of a straight line

The gradient of a straight line $= \frac{\Delta y}{\Delta x}$, where Δy is the change of the quantity plotted on the y-axis and Δx is the change of the quantity plotted on the x-axis. The gradient of a straight line is obtained by drawing as large a gradient triangle as possible and measuring the height Δy and the base Δx of this triangle, using the scale on each axis (Figure 1).

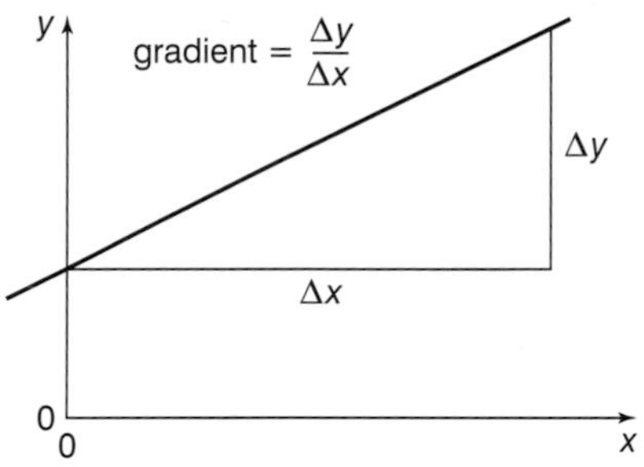

Figure 1 *Constant gradient*

Note

As a rule, when you plot a straight-line graph, always choose a scale for each axis that covers at least half the length of each axis. This will enable you to draw the line of best fit as accurately as possible, as explained on Practical skills p.xxiv. The measurement of the gradient of the line will therefore be more accurate. If the y-intercept is required and it cannot be read directly from the graph, it can be calculated by substituting the value of the gradient and the coordinates of a point on the line into the equation $y = mx + c$.

The gradient at a point on a curve

The gradient of a point on a curve is equal to the gradient of the tangent to the curve at that point.

- The **tangent to the curve at a point** is a straight line that touches the curve at that point without cutting across it. To see why, mark any two points on a curve and join them by a straight line. The gradient of the line is $\frac{\Delta y}{\Delta x}$, where Δy is the vertical separation of the two points and Δx is the horizontal separation. Now repeat with one of the points closer to the other; the straight line is now closer in direction to the curve. If the points are very close, the straight line between them is almost along the curve. The gradient of the line is then virtually the same as the gradient of the curve at that position (Figure 2). In other words, the gradient of the straight line $\frac{\Delta y}{\Delta x}$ becomes equal to the gradient of the curve as $\Delta x \to 0$. The curve gradient is written as:

$$\frac{\mathbf{d}y}{\mathbf{d}x}$$

  where $\frac{\mathrm{d}}{\mathrm{d}x}$ means 'rate of change'.

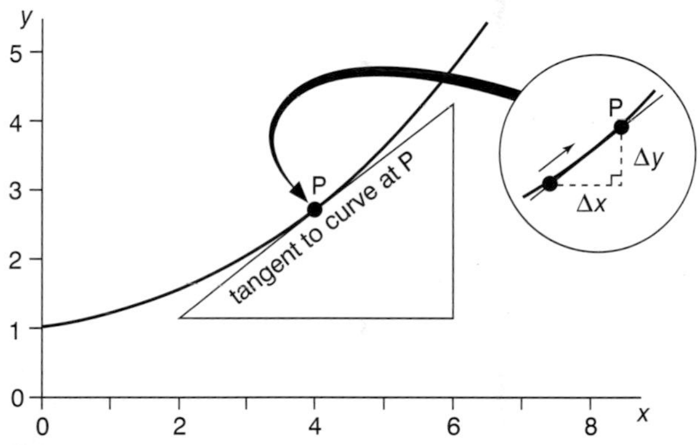

Figure 2 *Tangents and curves*

- The **gradient of the tangent** is a straight line and is obtained as explained above. Drawing the tangent to a curve requires practice. This skill is often needed in practical work. The **normal** at the point where the tangent touches the curve is the straight line perpendicular to the tangent at that point. An accurate technique for drawing the normal to a curve using a plane mirror is shown in Figure 3. At the point where the normal intersects the curve, the curve and its mirror image should join smoothly without an abrupt change of gradient where they join. After positioning the mirror surface correctly, the normal can be drawn and then used to draw the tangent to the curve.

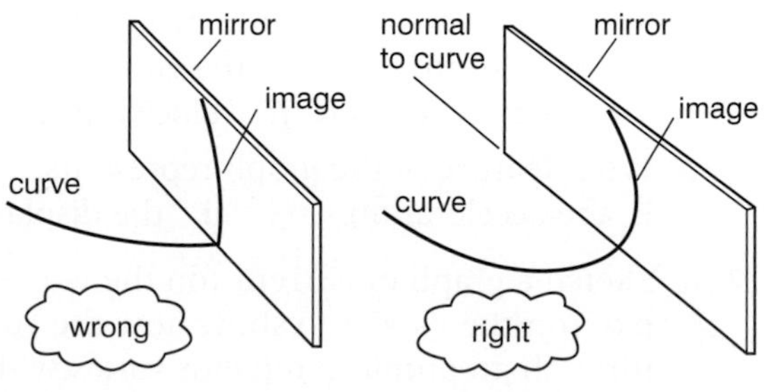

Figure 3 *Drawing the normal to a curve*

Turning points

A **turning point on a curve** is where the gradient of the curve is zero. This happens where a curve reaches a **peak** with a fall either side (i.e. a **maximum**) or where it reaches a **trough** with a rise either side (i.e. a **minimum**).

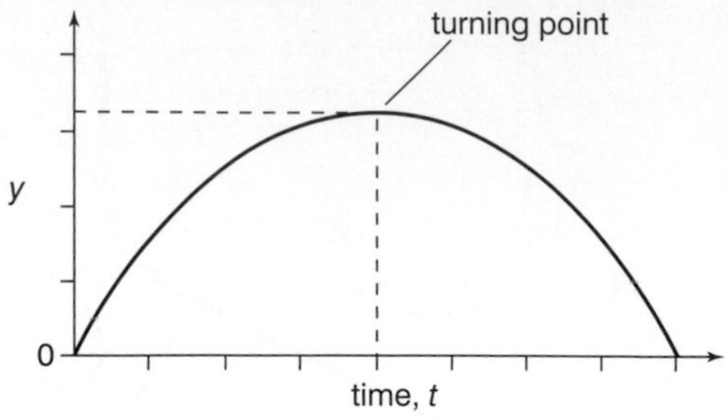

Figure 4 *Turning points*

Where the gradient represents a physical quantity, a turning point is where that physical quantity is zero. Figure 4 shows an example of a curve with a turning point. This is a graph of the vertical height against time for a projectile that reaches a maximum height, then descends as it travels horizontally. The gradient represents the vertical component of velocity. At maximum height, the gradient of the curve is zero so the vertical component of velocity is zero at that point.

Areas under graphs

The **area** under a line on a graph can give useful information. For example, consider Figure 5a which is a graph of the tension in a spring against its extension. Since 'tension × extension' is 'force × distance', which equals work done, then the area under the line represents the work done to stretch the spring.

Figure 5b shows a tension v. extension graph for a rubber band. Unlike Figure 5a, the area under the curve is not a triangle; but it still represents work done, in this case the work done to stretch the rubber band. (See Topic 5.6.)

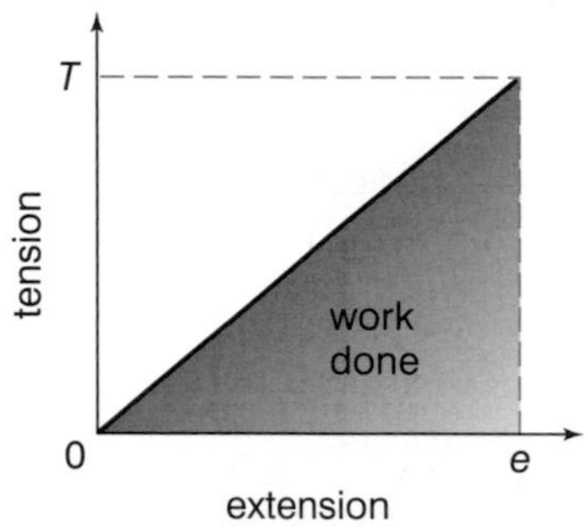

a *Spring*

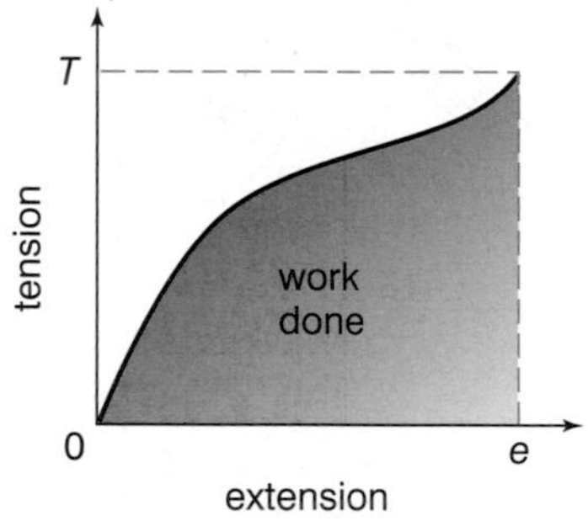

b *Rubber band*

Figure 5 *Tension v. extension*

Note

The **product** of the *y*-variable and the *x*-variable must represent a physical variable with a physical meaning if the area is to be of use.

Even where the area does represent a physical variable, it may not have any physical meaning. For example, for a graph of p.d. against current, the product of p.d. and current represents power – but this physical quantity has no meaning in this situation.

Examples

More examples of curves where the area is useful include:

- Velocity against time, where the area under the line and the time axis represents **displacement**.
- Acceleration against time, where the area under the line and the time axis represents **change of velocity**.
- Power against time, where the area between the curve and the time axis represents **energy**.
- Charge and potential difference, where the area between the curve and the p.d. axis represents **energy**.

Summary test 24.5

1 a Sketch a velocity against time graph (with time on the *x*-scale) to represent the equation $v = u + at$, where v is the velocity at time t.

b What feature of the graph represents:
i the acceleration, **ii** the displacement?

2 a Sketch a graph of current (on the *y*-axis) against p.d. (on the *x*-axis) to show how the current through an ohmic conductor varies with p.d.

b How can the resistance of the conductor be determined from the graph?

3 An electric motor is supplied with energy at a constant rate:

a Sketch a graph to show how the energy supplied to the motor increases with time.

b Explain how the power supplied to the motor can be determined from the graph.

4 A steel ball bearing was released in a tube of oil and it fell to the bottom of the tube.

a Sketch graphs to show how the following changed with time for the ball bearing, from the instant of release to the point of impact at the bottom of the tube:
i the velocity, **ii** the acceleration.

b i What is represented on graph **ai** by 1 the gradient, 2 the area under the line?

ii What is represented on graph **aii** by the area under the line?

Logarithms

Any number can be expressed as any other number raised to a particular power. You can use the $x^{\square}$ key on a calculator to show, for example that $8 = 2^3$ and $9 = 2^{3.17}$ In these examples, 2 is referred to as the base number and is raised to a different power in each case to generate 8 or 9. The power is defined as the **logarithm** of the number generated.

In general, for a number $\boldsymbol{n = b^p}$ where b is the base number, then $\boldsymbol{p = log_b\, n}$ where $\log_b$ means a logarithm using b as the base number.

Note that $\log_b (b^p) = p$ as $b^p = n$ and $\log^b n = p$.

Applying the general definition above gives the following rules to remember when working with logs:

1 For any two numbers m and n,
$\log (nm) = \log n + \log m$
Let $p = \log n$ and let $q = \log m$ so $n = b^p$ and $m = b^q$.
$\therefore nm = b^p\, b^q = b^{p+q}$ so $\log (nm) = p + q = \log m + \log n$

2 For any two numbers m and n,
$\log \left(\frac{n}{m}\right) = \log n - \log m$
Let $p = \log n$ and let $q = \log m$ so $n = b^p$ and $m = b^q$. Therefore $\frac{1}{m} = \frac{1}{b^q} = b^{-q}$
$\therefore \frac{n}{m} = b^p\, b^{-q} = b^{p-q}$ so $\log \left(\frac{n}{m}\right) = p - q = \log n - \log m$.

3 For any number m raised to a power p,
$\log (m^p) = p \log m$

This is because $m^p = m$ multiplied by itself p times.

$$\text{Therefore, } \log m^p = \underbrace{\{\log m + \log m + \ldots + \log m\}}_{\longleftarrow\, p \text{ terms}\, \longrightarrow} = p \log m$$

The following particular bases are used extensively in physics.

1 Base 10 logs, written as $\log_{10}$ or lg
For example,

- $100 = 10^2$ so $\log_{10} 100 = 2$,
- $50 = 10^{1.699}$ so $\log_{10} 50 = 1.699$,
- $10 = 10^1$ so $\log_{10} 10 = 1$,
- $5 = 10^{0.699}$ so $\log_{10} 5 = 0.699$

This illustrates the product rule for logs (i.e. $\log (nm) = \log n + \log m$)) since $\log_{10} 50 = \log_{10} 5 + \log_{10} 10 = 0.699 + 1 = 1.699$.

Uses of base 10 logs
In graphs where a logarithmic scale is necessary to show the full range of a variable that covers a very wide range, as shown in Figure 1. Notice in Figure 1 that the frequency increases by ×10 in equal intervals along the horizontal axis.

In data analysis where a relationship between two variables is of the form $y = kx^n$ and k and n are unknown constants. Applying the above rules to an equation of the form $y = kx^n$,

$$\log_{10} y = \log_{10} k + \log_{10} x^n = \log_{10} k + n \log_{10} x$$

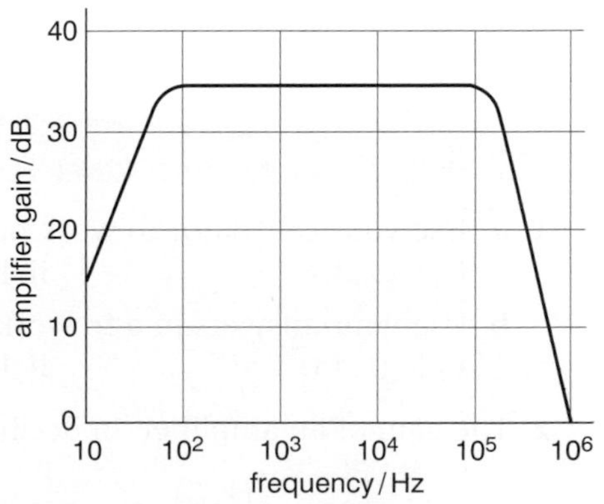

Figure 1 *Logarithmic scale*

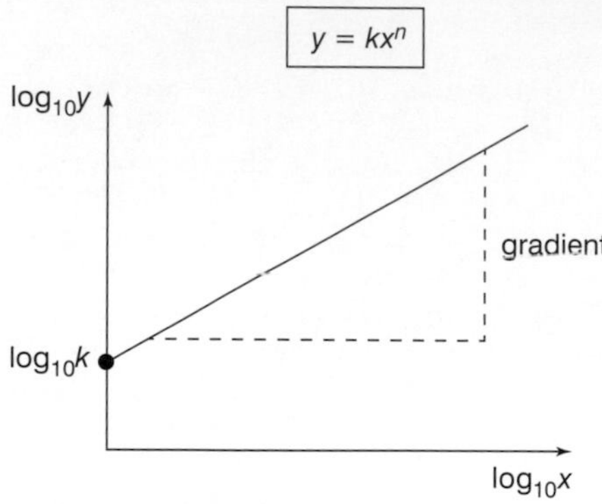

Figure 2 *Using logs to test* $y = kx^n$

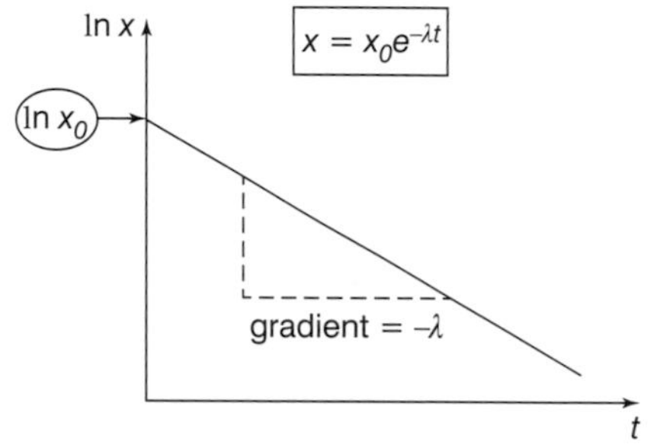

Figure 3 *Using logs to test* $x = x_0e^{-lt}$

The graph of $\log_{10}y$ (on the vertical axis) against $\log_{10}x$ is, therefore, a straight line of gradient n with an intercept equal to $\log k$ (see Figure 2).

In certain formulae where a ×10 scale is used. For example, the gain of an amplifier in decibels (dB) is a ×10 scale defined by the formula:

$$\text{voltage gain/dB} = 10\log_{10}\left(\frac{V_{out}}{V_{in}}\right),$$

where V_{out} and V_{in} are the output and input voltages respectively.

If $V_{out} = 50\,V_{in}$, the gain of the amplifier is 17 dB (= $10\log_{10}50$).

2 Natural logs, written as $\log_e$ or ln

Here, e is the exponential number used as the base of natural logarithms and is equal to 2.718. For example,

- $2.718 = e^1$ so $\ln 2.718 = 1$
- $7.389 = e^2$ so $\ln 7.389 = 2$
- $20.009 = e^3$ so $\ln 20.009 = 3$
- In general, for any number n, if p is such that $n = e^p$, then $\ln n = p$.

Uses of natural logarithms

Natural logs are used in the equations for radioactive decay (Topic 21.2) and capacitor discharge (Topic 15.4) or any other process where the rate of change of a quantity is proportional to the quantity itself. For example, the rate of decrease of p.d. across a capacitor discharging through a resistor is proportional to the p.d. across the capacitor. This type of change is described as an exponential decrease because the quantity decreases by the same factor in equal intervals of time.

Applying the general rule (that if p is such that $n = e^p$, then $p = \ln n$) to the equation $x = x_0e^{-\lambda t}$ gives $\ln x = \ln x_0 - \lambda t$.

Therefore, a graph of $\ln x$ (on the vertical axis) against t is a straight line with a gradient equal to $-\lambda$ and a y-intercept equal to $\ln x_0$. See Figure 3.

Comparing the equation for capacitor discharge $Q = Q_0e^{-t/RC}$ with the equation for radioactive decay $N = N_0e^{-\lambda t}$:

- for capacitor discharge, $\ln Q = \ln Q_0 - \frac{t}{RC}$ so a graph of $\ln Q$ (on the vertical axis) against t is a straight line which has a gradient $\frac{-1}{RC}$ and $\ln Q_0$ as its y-intercept,
- for radioactive decay, $\ln N = \ln N_0 - \lambda t$ so a graph of $\ln N$ (on the vertical axis) against t is a straight line which has a gradient $-\lambda$ and $\ln N_0$ as its y-intercept.

Summary test 24.6

1 a Use your calculator to work out:
i $\log_{10}3$, **ii** $\log_{10}15$.

b Use your answers in **a** to work out:
i $\log_{10}45$, **ii** $\log_{10}5$.

2 The gain of an amplifier, in decibels, is given by the formula $10\log_{10}\left(\frac{V_{out}}{V_{in}}\right)$.

a Calculate the gain, in decibels (dB), for:
i $V_{out} = 12\,V_{in}$, **ii** $V_{out} = 5\,V_{in}$.

b Show that the gain, in decibels, of an amplifier for which $V_{out} = 60\,V_{in}$ is equal to the sum of the gain in **i**, and the gain in **ii** above.

3 Write down the gradient and the y-intercept of a line on a graph representing the equation $\log_{10}y = n\log_{10}x + \log_{10}k$ for:
a $y = 3x^5$,
b $y = \frac{1}{2}x^3$,
c $y = x^2$.

4 a Use your calculator to work out:
i $\ln 3$,
ii $\ln 15$.

b Use your answers in **a** to work out:
i $\ln 45$,
ii $\ln 5$.

Exponential decrease

Rates of change

Consider a variable quantity y that changes with respect to a second quantity x as shown in Figure 1. The gradient of the curve at any point is the rate of change of y with respect to x at that point. This can be worked out from the graph by drawing a tangent to the curve at that point and measuring the gradient of the tangent. Figure 1 shows the idea. The rate of change of y with respect to x at point P is equal to the gradient of the tangent to the curve at P which is $\frac{\Delta y}{\Delta x}$.

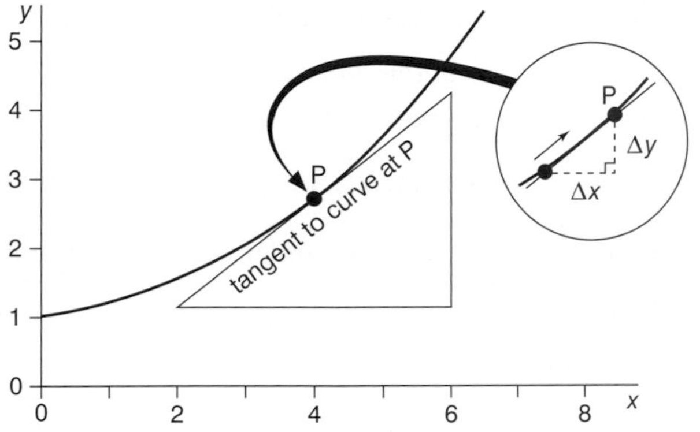

Figure 1 *Tangents and curves*

The rate of change of y with respect to x can be worked out algebraically if the equation relating y and x is known. This process is known as differentiation. For example:

- For $y = x^2$, then increasing x to $x + \Delta x$ increases y to $y + \Delta y$ where $y + \Delta y = (x + \Delta x)^2$.

  Multiplying out $(x + \Delta x)^2$ gives $y + \Delta y = x^2 + 2x\Delta x + \Delta x^2$

  Subtracting $y = x^2$ from this equation gives $\Delta y = 2x\Delta x + \Delta x^2$

  Dividing by Δx therefore gives $\frac{\Delta y}{\Delta x} = \frac{2x\Delta x + \Delta x^2}{\Delta x} = 2x + \Delta x$

  Therefore, as $\Delta x \to 0$, $\frac{\Delta y}{\Delta x} \to 2x$ which is therefore the formula for the gradient at x.

  This is written $\frac{dy}{dx} = 2x$, where $\frac{dy}{dx}$ is the mathematical expression for the rate of change of y with respect to x.

- For the general expression $y = x^n$, it can be shown that $\frac{dy}{dx} = nx^{n-1}$

  For example, if $y = 3x^5$, then $\frac{dy}{dx} = 15x^4$.

Note that differentiation of functions is not required in the A level specification. The information on differentiation is provided to help you develop your understanding of exponential change in the next section.

Exponential change happens when the rate of change of a quantity is proportional to the quantity itself. Such a change can be an increase (i.e. exponential growth) or a decrease (i.e. exponential decay). In both cases, the quantity changes by a fixed proportion in equal intervals of time. The A level specification requires knowledge and understanding of exponential decrease but not of exponential growth. The notes below will, therefore, concentrate on exponential decrease.

Summary test 24.7

1 a For each exponential decrease equation, write down the initial value at $t = 0$ and the decay constant:
   i $x = 2e^{-3t}$,
   ii $x = 12e^{-t/5}$,
   iii $x = 4e^{-0.02t}$.
  b For each exponential decrease equation above, work out the half-life.

2 A radioactive isotope has a half-life of 720 s and it decays to form a stable product. A sample of the isotope is prepared with an initial activity of 12.0 kBq. Calculate the activity of the sample after:
  a 1 minute,
  b 5 minutes,
  c 1 hour.

3 A capacitor of capacitance 22 μF discharged from a p.d. of 12.0 V through a 100 kΩ resistor.
  a Calculate:
   i the time constant of the discharge circuit,
   ii the half-life of the exponential decrease.
  b Calculate the capacitor p.d.:
   i 2.0 s,
   ii 5.0 s after the discharge started.

4 A certain exponential decrease process is represented by the equation $x = 1000e^{-5t}$.
  a i Calculate the half-life of the process.
   ii Calculate N when t = 0.5 s.
  b Show that the above equation can be rearranged as an equation of the form $\ln x = a + bt$ and determine the values of a and b.

In your studies of capacitor discharge (Topic 15.4) and of radioactive decay (Topic 21.2), you will have met and used the equation $\frac{dx}{dt} = -\lambda x$ and the solution of this equation $x = x_0 e^{-\lambda t}$.

- Let's consider why the equation $\frac{dx}{dt} = -\lambda x$ represents an exponential decrease which is a change where the variable quantity x decreases with time at a rate in proportion to the quantity.
  If x decreases by Δx in time Δt, the rate of change is $\frac{\Delta x}{\Delta t}$. This is written as $\frac{dx}{dt}$ in the limit $\Delta t \to 0$.
  For an exponential decrease, the rate of change is negative and is proportional to x, therefore, $\frac{dx}{dt} = -\lambda x$, where λ is referred to as the decay constant.
- Now consider why the solution of this equation is $x = x_0 e^{-\lambda t}$, where x_0 is a constant.
  Applying the rules of differentiation to the function
  $x = x_0(1 + t + \frac{t^2}{2 \times 1} + \frac{t^3}{3 \times 2 \times 1} + \frac{t^4}{4 \times 3 \times 2 \times 1}$ + similar higher order terms) gives
  $\frac{dx}{dt} = x_0\,(0 + 1 + t + \frac{t^2}{2 \times 1} + \frac{t^3}{3 \times 2 \times 1}$ + similar higher order terms)
  which is the same as x.
  So, $\frac{dx}{dt} = x$ if x is the above function.
  It can be shown that the function in brackets may be written as e^t, where e is referred to as the exponential number.

  Therefore,
  $e^t = 1 + t + \frac{t^2}{2 \times 1} + \frac{t^3}{3 \times 2 \times 1} + \frac{t^4}{4 \times 3 \times 2 \times 1}$ + similar higher order terms …
  The value of e, the exponential number, can be worked out by substituting $t = 1$ in the above expression for e^t, giving $e = 1 + 1 + \frac{1}{2} + \frac{1}{6}$ + etc. = 2.718 to four significant figures.
- To show that the solution of the equation $\frac{dx}{dt} = -\lambda x$ is $x = x_0 e^{-\lambda t}$, divide both sides of the equation by $-\lambda$ to give $\frac{1}{-\lambda}\frac{dx}{dt} = x$.
  Substituting z for $-\lambda t$ therefore gives $\frac{dx}{dz} = x$ which has the solution $x = x_0 e^{z} = x_0 e^{-\lambda t}$

The half-life, $t_{½}$ of an exponential decrease is the time taken for x to decrease from x_0 to $\frac{x_0}{2}$.
Substituting $x = \frac{x_0}{2}$ and $t = t_{½}$ into $x = x_0 e^{-\lambda t}$ gives $\frac{x_0}{2} = x_0 e^{-\lambda t_{½}}$.
Applying logs to both sides gives $\ln x_0 - \ln 2 = \ln x_0 - \lambda t_{½}$ which simplifies to $\lambda t_{½} = \ln 2$
$\therefore t_{½} = \frac{\ln 2}{\lambda} = \frac{0.693}{\lambda}$

The time constant, τ of an exponential decrease is the time taken for x to decrease from x_0 to $\frac{x_0}{e}$ (= 0.368 x_0 as $\frac{1}{e} = 0.368$).
Substituting $x = \frac{x_0}{e}$ and $t = \tau$ into $x = x_0 e^{-\lambda t}$ gives $\frac{x_0}{e} = x_0 e^{-\lambda \tau}$.
Applying logs to both sides gives $\ln x_0 - \ln e = \ln x_0 - \lambda \tau$ which simplifies to $\tau = \frac{1}{\lambda}$ (as $\ln e = 1$)

For capacitor discharge, $\lambda = \frac{1}{CR}$; therefore $\tau = \frac{1}{\lambda} = CR$.

Note

As explained on the previous page, $\ln(e^{-\lambda t}) = -\lambda t$.

Therefore, $\ln x = \ln(x_0 e^{-\lambda t}) = \ln x_0 + \ln(e^{-\lambda t}) = \ln x_0 - \lambda t$.

Chapter Summary

Trigonometry

- 1 radian $= \dfrac{360}{2\pi}$ degrees.
- Arc length $s = r\theta$ where θ is the angle in radians.
- For small angles (i.e. $\theta <$ about 10°) $\sin\theta \approx \tan\theta \approx \theta$ in radians, and $\cos\theta \approx 1$.
- The cosine rule; $a^2 = b^2 + c^2 - 2bc\cos A$
- The sine rule: $\dfrac{a}{\sin A} = \dfrac{b}{\sin B} = \dfrac{c}{\sin C}$.

Algebra

- Linear simultaneous equations: to solve a pair of simultaneous equations with two unknown variables, x and y,
  1 make the coefficient of x the same in both equations by multiplying one or both equations by a suitable number, then
  2 combine the two equations to eliminate x and so find y, then
  3 substitute the value of y into either equation to find x.
- The general solution of the quadratic equation $ax^2 + bx + c = 0$
  is $x = \dfrac{-b \pm \sqrt{(b^2 - 4ac)}}{2a}$

Logarithms

- For a number $n = b^p$, where b is the base number, $p = \log_b n$.
- $\log(nm) = \log n + \log m$.
- $\log(n/m) = \log n - \log m$.
- $\log(m^p) = p\log m$.
- For $y = kx^n$, $\log_{10} y = \log_{10} k + n\log_{10} x$:
  the graph of $\log_{10} y$ (on the vertical axis) against $\log_{10} x$ is a straight line of gradient n with an intercept equal to $\log k$.
- For $n = e^p$, $\ln n = p$.
- For $x = x_0 e^{-\lambda t}$, $\ln x = \ln x_0 - \lambda t$:
  the graph of $\ln x$ (on the vertical axis) against t is a straight line with a gradient equal to $-\lambda$ and a y-intercept equal to $\ln x_0$.

Exponential decrease

- Exponential change happens when the rate of change of a quantity is proportional to the quantity itself.
- For an exponential decrease, the rate of change is negative and is proportional to x, therefore, $\dfrac{dx}{dt} = -\lambda x$, where λ is referred to as the decay constant.
- The solution of this equation is $x = x_0 e^{-\lambda t}$, where x_0 is a constant.
- Half-life, $t_{1/2} = \dfrac{0.693}{\lambda}$ (= time for x to decrease from x_0 to $\dfrac{x_0}{2}$).
- Time constant, $\tau = \dfrac{1}{\lambda}$ (= time for x to decrease from x_0 to $\dfrac{x_0}{e}$).

Glossary

A-scan system ultrasound scan in which pulses from an ultrasonic transducer are detected by the probe after being reflected by internal boundaries and then displayed on an oscilloscope

absolute temperature, T, in kelvins = temperature in °C + 273(.15)

absolute zero is the temperature at which an object has minimum internal energy

acceleration change of velocity per unit time

acceleration of free fall acceleration of an object acted on only by the force of gravity

acoustic impedance the product of the density of a substance and the speed of the ultrasonic or sound waves through it

activity, A, of a radioactive isotope is the number of nuclei of the isotope that disintegrate per second. The unit of activity is the becquerel (Bq), equal to 1 disintegration per second

air resistance the force of the air opposing the motion of an object moving through the air (see drag force)

alpha radiation consists of particles that are each composed of two protons and two neutrons. An alpha (α) particle is emitted by a heavy unstable nucleus which is then less unstable as a result. Alpha radiation is easily absorbed by paper, has a range in air of no more than a few centimetres and is more ionising than beta (β) or gamma (γ) radiation

aluminium filter aluminium plate used in X-radiography to absorb 'soft' (i.e. low energy) X-rays

amplifier output p.d. that is proportional to the input p.d. supplied to it

amplitude maximum displacement of a vibrating particle; for a transverse wave, it is the distance from the middle to the peak of the wave

angular displacement, in radians, in time $t = 2\pi ft = \dfrac{2\pi t}{T}$ for an object in uniform circular motion

angular frequency 2π × frequency of oscillating motion. The unit of ω is the radian per second (rad s^{-1})

annihilation process whereby a particle and its corresponding antiparticle collide, annihilate each other and produce photons

anode positive terminal of an electrical device

antineutrino an uncharged antiparticle with a very low rest mass compared with an electron and which is emitted from a nucleus when it emits a β^- particle

antinode fixed point in a stationary wave pattern where the amplitude is a maximum

antiquark the antiparticle of a quark

Archimedes' principle The upthrust on a body wholly or partially immersed in a fluid is equal to the weight of liquid displaced by the cylinder

atomic mass unit, u, (correctly referred to as the unified atomic mass constant) is equal to 1.66×10^{-27} kg. It is defined as $\frac{1}{12}$ of the mass of an atom of the carbon isotope $^{12}_{6}\text{C}$

atomic number number of protons in the nucleus of the atom. It is also the order number of the element in the Periodic Table

attenuation reduction of intensity of a beam due to absorption or scattering

attenuation coefficient measure of the reduction of intensity per unit distance of a beam travelling through a substance and is equal to ln 2/half thickness of the substance

audio wave electrical waves produced from sound waves for example by a microphone

Avogadro constant, N_A, is defined as the number of atoms in 12 grams of the carbon isotope $^{12}_{6}\text{C}$

background radioactivity is radioactivity due to radioactive substances which may be in the ground or in building materials or elsewhere in the environment. Background radioactivity is also caused by cosmic radiation

baryons matter particles and antiparticles that each consist of three quarks or three antiquarks

base units the five units that define the SI system (the metre, the kilogram, the second, the ampere and the kelvin)

beta minus (β^-) particle an electron created and emitted by an unstable neutron-rich nucleus when a neutron in the nucleus changes into a proton – an antineutrino is created and emitted at the same time

beta plus (β^+) particle a positron created and emitted by an unstable proton-rich nucleus when a proton in the nucleus changes into a neutron – a neutrino is created and emitted at the same time

beta radiation consists of beta-particles (β) which are electrons emitted by unstable nuclei with too many neutrons compared to protons. Beta radiation has a range in air of about a metre and is less ionising than alpha (α) radiation and more ionising than gamma (γ) radiation

Big Bang theory The Universe was created in a massive 'primordial' explosion and has been expanding ever since

binding energy of a nucleus:

- the work that must be done to separate a nucleus into its constituent neutrons and protons,
- binding energy = mass defect × c^2,
- binding energy/nucleon is greatest for nuclei of mass number 57

biofuel fuel obtained from biomass; see biomass

biomass biological material from living or recently living organisms used as fuel (eg animal waste or woodchip)

black body an object that absorbs all the radiation incident on it; the radiation emitted by a black body is called *black body radiation*

bonds forces that hold atoms or molecules together

brittle snaps without stretching or bending when subject to stress

Brownian motion is the random and unpredictable motion of a particle such as a smoke particle and is caused by molecules of the surrounding substance (which are all much smaller than smoke particles) colliding at random with the particle

capacitance of a capacitor is defined as the charge stored per unit p.d. The unit of capacitance is the farad (F), equal to 1 coulomb per volt

For a capacitor of capacitance C at p.d. V, the charge stored, $Q = CV$

capacitor combination rules

1 **capacitors in parallel**; combined capacitance $C = C_1 + C_2 + C_3 + \ldots$

2 **capacitors in series**; combined capacitance is given by w $\frac{1}{C} = \frac{1}{C_1} + \frac{1}{C_2} + \frac{1}{C_3} + \ldots$

capacitor energy; energy stored by the capacitor, $W = \frac{1}{2}QV = \frac{1}{2}CV^2$

capacitor discharge through a fixed resistor R;

1 time constant = RC

2 exponential decrease equation for current or charge or p.d.; $x = x_0 e^{-t/RC}$

carrier wave electromagnetic waves used to carry a signal

centre of gravity point where the weight of a body may be considered to act

centripetal acceleration:

1 for an object moving at speed v in uniform circular motion, its centripetal acceleration $a = \frac{v^2}{r}$ towards the centre of the circle

2 for a satellite in a circular orbit, its centripetal acceleration $\frac{v^2}{r} = g$

centripetal force is the resultant force on an object that moves along a circular path. For an object of mass m moving at speed v along a circular path of radius r, the centripetal force = $\frac{mv^2}{r}$

chain reaction a series of reactions in which each reaction causes a further reaction. In a nuclear reactor, each fission event is due to a neutron colliding with a $^{235}_{92}U$ nucleus which splits and releases two or three further neutrons which can go on to produce further fission. A steady chain reaction occurs when one fission neutron on average from each fission event produces a further fission event

charge carriers charged particles that move through a substance when a p.d. is applied across it

coherence two sources of waves are coherent if they emit waves with a constant phase difference

collisions an **elastic** collision is one in which the total kinetic energy after the collision is the same as before the collision. A **totally inelastic** collision is where the colliding objects stick together

contrast medium substance passed through a body organ or blood vessel to enhance the X-ray image of the body part by increasing the absorption of X-rays

cooling by evaporation the decrease in the temperature of a liquid due to evaporation from its open surface

cosmic microwave background radiation (CMBR) radiation that was created by the Big Bang and has been travelling through the Universe ever since the Universe became transparent

Coulomb's law of force between two point charges states that the force is proportional to the product of the charges and inversely proportional to the square of the distance between the charges. For two point charges Q_1 and Q_2 at distance apart r, the force F between the two charges is given by the equation $F = \frac{Q_1 Q_2}{4\pi\varepsilon_0 r^2}$, where ε_0 is the absolute permittivity of free space

couple pair of equal and opposite forces acting on a body but not along the same line

coupling medium gel applied between the body surface and the surface of an ultrasonic probe to exclude air at the interface to ensure the ultrasonic waves are not reflected at the body surface

critical mass is the minimum mass of the fissile isotope (e.g. the uranium isotope $^{235}_{92}U$) in a nuclear reactor necessary to produce a chain reaction. If the mass of the fissile isotope in the reactor is less than the critical mass, a chain reaction does not occur because too many fission neutrons escape from the reactor or are absorbed without fission

cycle interval for a vibrating particle (or a wave) from a certain displacement and velocity to the next time the particle (or the next particle) has the same displacement and velocity

damped oscillations of an oscillating system are due to the presence of resistive forces due to friction and drag. For a lightly damped system, the amplitude of oscillations decreases gradually. For a heavily damped system displaced from equilibrium then released, the system slowly returns to equilibrium without oscillating. For a critically damped system, the system returns to equilibrium in the least possible time without oscillating

Dark energy unknown force releasing hidden energy thought to be causing the expansion of the Universe to accelerate

de Broglie wavelength A particle of matter has a wave-like nature because it can behave as a wave. For example, electrons directed at a thin crystal are diffracted by the crystal. The de Broglie wavelength, λ, of a matter particle depends on its momentum, p, in accordance with de Broglie's equation

$$\lambda = \frac{h}{p} = \frac{h}{mv}$$

decay constant, λ is the probability of an individual nucleus decaying per second

density of a substance mass per unit volume of the substance

diffraction is the spreading of waves when they pass through a gap or round an obstacle. X-ray diffraction is used to determine the structure of crystals, metals and long molecules. Electron diffraction and neutron diffraction are also used to probe the structure of materials. High-energy electron scattering is used to determine the diameter of the nucleus

diffraction grating plate with many close equally-spaced parallel slits that diffracts light at normal incidence into a direction that depends on its wavelength

diode an electrical device that conducts in one direction only

dispersion splitting of a beam of white light by a glass prism into the colours of a spectrum

displacement distance in a given direction

Doppler effect the effect of relative motion between a source of waves and an observer on the observed frequency, causing it to differ from the emitted frequency

Doppler shift the difference between the observed frequency and the emitted frequency of the waves from a source due to relative motion between the source and the observer

drag force the force of fluid resistance on an object moving through the fluid

ductile stretches easily without breaking

efficiency $= \dfrac{\text{useful energy transferred}}{\text{total energy supplied}}$

effort the force applied to a machine to make it move

elastic deformation regain of shape of an object after it has been deformed

elastic limit point beyond which a wire is permanently stretched

electric potential (at a point) work done in bringing unit positive charge from infinity to the point

electrolysis process of electrical conduction in a solution or molten compound due to ions moving to the oppositely charged electrode

electrolyte a solution or molten compound that conducts electricity

electromotive force (e.m.f.) the amount of electrical energy per unit charge produced inside a source of electrical energy

electron fundamental particle with a fixed negative charge equal and opposite to that of a proton and a mass approximately $\dfrac{1}{1840}$th of the mass of a proton

electron volt amount of energy equal to 1.6×10^{-19} J, defined as the work done when an electron is moved through a p.d. of 1 V

engine force the force that drives a vehicle

equilibrium state of an object when at rest or in uniform motion

equipotential a line or a surface of constant gravitational potential

error bar representation of an uncertainty on a graph

error of measurement uncertainty of a measurement

evaporation the process by which a liquid turns to vapour below its boiling point

excited state an energy state of an atom with more energy than the lowest energy state

explosion In an explosion where two objects fly apart, the two objects carry away equal and opposite momentum

exponential decrease Exponential change happens when the rate of change of a quantity is proportional to the quantity itself. For an exponential decrease of a quantity x, $\dfrac{dx}{dt} = -\lambda x$, where λ is referred to as the decay constant
The solution of this equation is $x = x_0 e^{-\lambda t}$, where x_0 is a constant

Faraday's law of electromagnetic induction states that the induced e.m.f. in a circuit is proportional to the rate of change of magnetic flux linkage through the circuit

For a changing magnetic field in a fixed coil, induced e.m.f. $= -NA\dfrac{\Delta B}{\Delta t}$

fission is the splitting of a $^{235}_{92}$U nucleus or a $^{239}_{94}$Pu nucleus into two approximately equal fragments. Induced fission is fission caused by an incoming neutron colliding with a $^{235}_{92}$U nucleus or a $^{239}_{94}$Pu nucleus

fluoresce light emitted from a substance as a result of high energy radiation or particles being directed at the substance

force resultant force on an object

= rate of change of its momentum

$= \dfrac{\text{change of momentum}}{\text{time taken}}$

(= mass × acceleration for fixed mass)

forward biased the direction of a diode in a circuit in order for the diode to conduct

forward voltage the potential difference necessary across a diode to make it conduct in its forward direction

free and forced oscillations

- **free oscillations** are oscillations where there is no damping and no periodic force acting on the system so the amplitude of the oscillations is constant
- **forced oscillations** are oscillations of a system that is subjected to an external periodic force

frequency the number of cycles of a wave that pass a point per second

fusion is the fusing together of light nuclei to form a heavier nucleus

galvanometer a centre-reading electrical meter used to detect and measure an electric current.

geothermal energy energy transferred towards the Earth's surface from underground rocks heated by radioative substances deep within the Earth

gravitational field strength, *g*, is the force per unit mass on a small test mass placed in the field

- $g = \frac{F}{m}$, where F is the gravitational force on a small mass m
- at distance r from a point mass M, $g = \frac{GM}{r^2}$
- at or beyond the surface of a sphere of mass M, $g = \frac{GM}{r^2}$ where r is the distance to the centre
- at the surface of a sphere of mass M and radius R, $g_S = \frac{GM}{r^2}$

gravitational potential, ϕ, (at a point) the work done per unit mass to move a small object from infinity to the point

gravitational potential energy energy due to position in a gravitational field; and is equal to the work done to move a small object from infinity to a point in the field

ground heat heat flow from underground

ground state lowest energy state of an atom

hadrons matter particles and antiparticles that can interact through the strong interaction. Hadrons are subdivided into baryons (which consist of three quarks or three antiquarks) and mesons (which consist of a quark and an antiquark). Protons and neutrons are baryon hadrons

half-life, $t_{½}$, of a radioactive isotope is the time taken for the mass of the isotope to decrease to half the initial mass. This is the same as the time taken for the number of nuclei of the isotope to decrease to half the initial number

Hall effect the effect of causing a potential difference between the opposite sides of a conductor or semiconductor when a magnetic field is used to deflect charge carriers passing though the material

Hall voltage the potential difference between the opposite sides of a conductor or semiconductor when a magnetic field is used to deflect charge carriers passing though the material

Hooke's law the extension of a spring is proportional to the force needed to extend it

Hubble's law the galaxies are receding from us, each moving at speed v, that is directly proportional to the distance, d, to it

Hubble constant, H, the speed of recession of a galaxy per unit distance from Earth

ideal gas law, $pV = nRT$, where n is the number of moles of gas, T is the absolute temperature and R is the molar gas constant

intensity of radiation at a surface is the radiation energy per second per unit area at normal incidence to the surface. The unit of intensity is $J\,s^{-1}\,m^{-2}$ or $W\,m^{-2}$

intensity of waves the power per unit area that waves transfer through an area perpendicular to the direction of the waves

interference formation of points of cancellation and reinforcement where two coherent waves pass through each other

internal energy of an object is the sum of the random distribution of the kinetic and potential energies of its molecules

internal resistance resistance inside a source of electrical energy

inverse-square laws

- **force** Newton's law of gravitation and Coulomb's law of force between electric charges are described as inverse-square laws because the force between two point objects (masses in the case of gravitation and charge in the case of charges) is inversely proportional to the square of the distance between the two objects. Because the two laws above are inverse-square laws, the field strength due to a point mass or a point charge varies with distance according to the inverse of the square of the distance to the point object
- **intensity** the intensity of γ radiation from a point source varies with the inverse of the square of the distance from the source. The same rule applies to radiation from any point source that spreads out equally in all directions and is not absorbed

ion a charged atom

ionising radiation produces ions in substances it passes through. It destroys cell membranes and damages vital molecules such as DNA directly or indirectly by creating 'free radical' ions which react with vital molecules

isotopes forms of the same element with different numbers of neutrons and the same number of protons in their nuclei

kaon a meson that contains a strange quark and an antiquark or a strange antiquark and a quark

kinetic energy is the energy of a moving object due to its motion. For an object of mass m moving at speed v, its kinetic energy $E_K = \frac{1}{2}mv^2$, provided $v \ll c$ (the speed of light in free space)

kinetic energy of a molecule of an ideal gas The mean kinetic energy of a molecule of an ideal gas $= \frac{3}{2}kT$,

where the Boltzmann constant $k = \frac{R}{N_A}$

kinetic theory of a gas
- Assumptions: a gas consists of identical point molecules which do not attract one another. The molecules are in continual random motion colliding elastically with each other and with the container
- It can be shown that the pressure p of N molecules of such a gas in a container of volume V is given by the equation $pV = \frac{1}{3} Nm <c^2>$, where m is the mass of each molecule and $<c^2>$ is the mean square speed of the gas molecules
- Assuming that the mean kinetic energy of a gas molecule $\frac{1}{2} m <c^2> = \frac{3}{2} kT$, where $k = \frac{R}{N_A}$, it can be shown from $pV = \frac{1}{3} N m <c^2>$ that $pV = nRT$ which is the ideal gas law

Kirchhoff's first law at a junction, the total current in = the total current out

Kirchhoff's second law the sum of the e.m.fs round a complete loop in a circuit = the sum of the p.ds round the loop

latent heat of fusion (energy to melt a solid) is used to break the bonds that lock the molecules of the solid into fixed positions

latent heat of vaporisation (energy to boil a liquid) is used to break the bonds that prevent molecules moving away from each other

Lenz's law states that the direction of the induced current is always such as to oppose the change that causes the current

lepton matter particles and antiparticles that can interact through the weak interaction. They cannot interact through the strong interaction. Leptons are thought to be elementary. Electrons, positrons, neutrinos and antineutrinos are examples of leptons

light-dependent resistor (LDR) resistor which is designed to have a resistance that changes with change of intensity

light-emitting diode (LED) a diode that emits light when it conducts

line of force or a field line of a gravitational field (or electrical field) A line followed by a small mass (or small charge) acted on by no other forces than the force of the field.
Line of force or a field line of magnetic field is the line along which a free north pole would move

linear two quantities are said to have a linear relationship if the change of one quantity is proportional to the change of the other

load the force to be overcome by a machine when it shifts or raises an object

logarithmic scale This is a scale such that equal intervals correspond to a change by a constant factor or multiple (e.g. ×10)

logarithms For a number $n = b^p$, where b is the base number, $p = \log_b n$
- $\log(nm) = \log n + \log m$ and $\log(n/m) = \log n - \log m$
- $\log(m^p) = p \log m$
- natural logs; for $n = e^p$, then $\ln n = p$
- base 10 logs; for $n = 10^p$, then $\log_{10} n = p$

log graphs
1. For $y = k x^n$, $\log_{10} y = \log_{10} k + n \log_{10} x$; the graph of $\log_{10} y$ (on the vertical axis) against $\log_{10} x$ is therefore a straight line of gradient n with an intercept equal to log k
2. For $x = x_0 e^{-\lambda t}$, $\ln x = \ln x_0 - \lambda t$; the graph of $\ln x$ (on the vertical axis) against t is a straight line with a gradient equal to $-\lambda$ and a y-intercept equal to $\ln x_0$

longitudinal waves waves with a direction of vibration of the particles parallel to the direction of energy transfer by the waves

luminosity the total power of the radiation emitted by a star

magnetic field region near a magnet or a current-carrying wire in which another magnet or current-carrying wire experiences a force

magnetic flux, Φ $= BA$ for a uniform magnetic field that is perpendicular to an area A

magnetic flux density is defined as the force per unit length per unit current on a current-carrying conductor at right angles to the field lines. The unit of magnetic flux density B is the tesla (T). B is sometimes referred to as the magnetic field strength

magnetic flux linkage through a coil of N turns $= N\Phi = NBA$, where B is the magnetic flux density perpendicular to area A. The unit of magnetic flux and of flux linkage is the **weber** (Wb), equal to $1\,\mathrm{T\,m^2}$ or $1\,\mathrm{V\,s}$

magnetic force
- $\boldsymbol{F = BIl \sin\theta}$ gives the force F on a current-carrying wire of length l in a uniform magnetic field B at angle θ to the field lines, where I is the current. The direction of the force is given by Fleming's left-hand rule where the field direction is the direction of the field component perpendicular to the wire
- $\boldsymbol{F = BQv \sin\theta}$ gives the force F on a particle of charge Q moving through a uniform magnetic field B at speed v in a direction at angle θ to the field. If the velocity of the charged particle is perpendicular to the field, $F = BQv$. The direction of the force is given by Fleming's left-hand rule, provided the current is in the direction in which positive charge would flow
- $\boldsymbol{BQv = \frac{mv^2}{r}}$ gives the radius of the orbit of a charge moving in a direction at right angles to the lines of a magnetic field

Malus's law For polarised light passed through a polariser, its transmitted intensity $\boldsymbol{I = I_0 \cos^2\theta}$ where θ is the angle the polariser is rotated through from its orientation at the maximum transmitted intensity I_0

mass defect the difference between the mass of the separated nucleons (ie protons and neutrons from which the nucleus is composed) and the nucleus

mass measure of the inertia or resistance to change of motion of an object

matter waves the wave-like behaviour of particles of matter

mean kinetic energy of a molecule in a gas at absolute temperature $T = \frac{3}{2} kT$, where k is the Boltzmann constant $\left(= \frac{R}{N_A}\right)$

meson matter particles and antiparticles that each consist of a quark and an antiquark

microphone an electrical device that converts sound waves into electrical waves

mole One mole of a substance consisting of identical particles is the quantity of substance that contains N_A particles of the substance. The **molar mass** of a substance is the mass of one mole

molecule the smallest particle in a substance that can be identified as belonging to the substance

moment of a force about a point force × perpendicular distance from line of action of force to the point

momentum for an object, its momentum is defined as its mass × its velocity. For a photon, its momentum is equal to its energy / c (the speed of light in a vacuum). The unit of momentum is $kg\,m\,s^{-1}$

muon (symbol μ) a negatively charged particle with a rest mass over 200 times the rest mass of the electron. Muons and antimuons decay into electrons and antineutrinos or positrons and neutrinos, respectively

neutrino an uncharged particle with a very low rest mass compared with an electron and which is emitted from a nucleus when it emits a β^+ particle

neutron a particle with no charge and a mass approximately the same as a proton. One or more neutrons are in every atomic nucleus except that of the smallest hydrogen nucleus

Newton's law of gravitation the gravitational force F between two point masses m_1 and m_2 at distance r apart is given by $\boldsymbol{F = G m_1 m_2 / r^2}$

Newton's laws of motion

- **first law**: an object continues at rest or in uniform motion unless it is acted on by a resultant force
- **second law**: the rate of change of momentum of an object is proportional to the resultant force on it
- **third law**: when two objects interact, they exert equal and opposite forces on one another

Newton's second law may be written as $F = \frac{\Delta p}{\Delta t}$, where p is the momentum (= mv) of the object and F is the force in newtons. For constant mass, $\Delta p = m\Delta v$ so $F = \frac{m\Delta v}{\Delta t} = ma$

node fixed point of no displacement

nuclear fission splitting of certain large nuclei such as a $^{235}_{92}U$ nucleus or a $^{239}_{94}Pu$ nucleus into two approximately equal fragments. Induced fission is fission caused by an incoming neutron colliding with a $^{235}_{92}U$ nucleus or a $^{239}_{94}Pu$ nucleus

nuclear fusion fusing together of light nuclei to form a heavier nucleus

nucleon a neutron or a proton in the nucleus

nucleon number the number of neutrons and protons in a nucleus

nuclide a nucleus of a certain isotope

nuclide of an isotope $^{A}_{Z}\mathbf{X}$ is a nucleus composed of Z protons and $(A - Z)$ neutrons, where Z is the proton number (and also the atomic number of element X) and A is the number of protons and neutrons in a nucleus

Ohm's law the p.d. across a metallic conductor is proportional to the current through it provided the physical conditions do not change

pair production process whereby a photon of sufficient energy produces a particle and its corresponding antiparticle

particle-like nature properties that are characteristic of particles such as momentum or deflection by electric or magnetic fields

path difference the difference in distances from two coherent sources to an interference fringe

period of a wave time for one complete cycle of a wave to pass a point

phase difference the fraction of a cycle between the vibrations of two vibrating particles, measured either in radians or degrees

phase difference, in radians, $= \frac{2\pi\Delta t}{T_P}$, for two objects oscillating with the same time period, T_P, where Δt is the time between successive instants when the two objects are at maximum displacement in the same direction

photoconduction electrical conduction due to light

photoelectricity emission of electrons from a metal surface when the surface is illuminated by light of frequency greater than a minimum value, known as the threshold frequency

photon electromagnetic radiation consists of photons. Each photon is a wave packet of electromagnetic radiation. The energy of a photon, $E_{ph} = hf$, where f is the frequency of the radiation and h is the Planck constant. The momentum of a photon $= \frac{E_{ph}}{c}$

piezo-electric effect property of certain solids whereby a p.d. applied between opposite faces causes a change of distance between the two faces and where applied stress causes a potential difference

piezo-electric transducer the component in an ultrasonic probe that produces ultrasonic waves when an alternating p.d. is applied to it

pion a meson that contains an up or down quark and an up or down antiquark

plastic deformation deformation of a solid beyond its elastic limit

polarised waves transverse waves that vibrate in one plane only

positron the antimatter counterpart of the electron. It is positively charged and its rest mass is the same as that of the electron

potential difference the energy transferred per unit charge to a component when electric charge passes through it

potential divider two or more resistors in series connected to a source of fixed potential difference. The source p.d. is divided between the resistors as they are in series with each other

potential energy the energy of an object due to its position

potential gradient change of potential per unit change of distance in a given direction

power = rate of transfer of energy = $\frac{\text{energy transferred}}{\text{time taken}}$

pressure force per unit area acting on a surface perpendicular to the surface

principle of conservation of energy states that in any change, the total amount of energy after the change is always equal to the total amount of energy before the change

principle of conservation of momentum states that when two or more bodies interact, the total momentum is unchanged, provided no external forces act on the bodies

probable error estimate of the uncertainty of a measurement

projectile a projected object in motion acted on only by the force of gravity

proton a particle with a fixed positive charge of $+1.60 \times 10^{-19}$ C and a mass of 1.663×10^{-27} kg . One or more protons are in every atomic nucleus

quark protons and neutrons and all other hadrons consist of quarks and/or antiquarks. There are six types of quarks (up, down, strange, charm, top, bottom) referred to as quark flavours.

quark model (or standard model) a quark can join with an antiquark to form a meson or with two other quarks to form a baryon. An antiquark can join with two other antiquarks to form an antibaryon

radian measure of an angle defined such that 2π radians = 360°

radiant flux intensity, *F* radiation energy per second per unit surface area received from the star at normal incidence on a surface. The unit of radiant flux intensity is W m^{-2}

random error error of measurement with no obvious cause

red shift increase of wavelength of electromagnetic radiation from a receding star or galaxy due to its receding motion

relative speed the difference between the speeds of two objects moving along the same straight line in the same direction (or the sum of their speeds if they are moving in opposite directions).

renewable energy This is energy from a source that is continually renewed. Examples include hydroelectricity, tidal power, geothermal power, solar power, wave power and wind power

resistance $\frac{\text{p.d.}}{\text{current}}$

resistivity resistance per unit length × area of cross-section

resonance large-amplitude oscillations that occur in a lightly-damped system when the frequency of the applied force is equal to the natural frequency of oscillation of the system

reverse biased the direction of a diode in a circuit in order for the diode not to conduct

Rutherford's α-particle scattering experiment demonstrated that every atom contains a positively charged nucleus which is much smaller than the atom and where all the positive charge and most of the mass of the atom is located

satellite motion a satellite is a small object in orbit round a larger object. For a satellite moving at speed v in a circular orbit of radius r round a planet, its centripetal acceleration, $\frac{v^2}{r} = g$. Substituting $v = \frac{2\pi r}{T}$, where T is its time period, and $g = \frac{GM}{r^2}$, where M is the mass of the planet, it can be shown that $T^2 = \left(\frac{4\pi^2}{GM}\right) r^3$. A satellite in a geostationary orbit is always directly above the same point on the Equator. This is because it is in a circular orbit in the same plane as the Equator and it has a time period of exactly 24 hours

scalar a physical quantity with magnitude only

semiconductor a substance in which the number of charge carriers increases when its temperature is raised

SI system the scientific system of units

simple harmonic motion an object oscillates in simple harmonic motion if its acceleration is proportional to the displacement of the object from equilibrium and is always directed towards equilibrium:

- the acceleration, a, of an object oscillating in simple harmonic motion is given by the equation $a = -(2\pi f)^2 x$, where x = displacement from equilibrium, and f = frequency of oscillations
- the solution of this equation depends on the initial conditions. If $x = 0$ and the object is moving in the + direction at time $t = 0$, then $x = A \sin(2\pi ft)$. If the object is at maximum displacement, $+A$, at time $t = 0$, then $x = A\cos(2\pi ft)$

sinusoidal wave a wave that has the same shape as a sine wave

specific heat capacity, *c*, of a substance is the energy needed to raise the temperature of 1 kg of the substance by 1 K without change of state. The energy needed to raise the temperature of mass m of a substance from T_1 to $T_2 = mc(T_2 - T_1)$, where c is the specific heat capacity of the substance

spectrometer optical instrument used to measure the wavelengths of the lines in a spectrum

speed change of distance per unit time

standard candle an object of known luminosity

stationary waves wave pattern with nodes and antinodes formed when two or more progressive waves of the same frequency and amplitude pass through each other

strain extension per unit length of a solid when deformed

stress force per unit area of cross-section in a solid perpendicular to the cross-section

strong interaction/strong nuclear force the force in a nucleus responsible for holding the protons and neutrons in the nucleus together and for interactions between hadrons

superposition the effect of two waves adding together when they meet

systematic error error of measurement with a known cause

terminal velocity the maximum speed reached by an object when the drag force on it is equal and opposite to the force causing the motion of the object

thermistor resistor which is designed to have a resistance that changes with temperature

thermodynamic temperature, T, in kelvins = temperature in °C + 273(.15)

threshold frequency minimum frequency of light that can cause photoelectric emission

threshold wavelength maximum wavelength of light that can cause photoelectric emission

time period (or period) is the time taken for one complete cycle of oscillations

torque of a couple force × perpendicular distance between the lines of action of the forces

total internal reflection a light ray travelling in a substance is totally internally reflected at a boundary with a substance of lower refractive index, if the angle of incidence is greater than a certain value known as the critical angle

transducer any device designed to convert energy from one form to another. A piezo-electric transducer generates a p.d. when it is squeezed

transformer a transformer converts the amplitude of an alternating p.d. to a different value. It consists of two insulated coils, the primary coil and the secondary coil, wound round a soft iron laminated core

- The transformer rule states that the ratio of the secondary voltage to the primary voltage is equal to the ratio of the number of secondary turns to the number of primary turns
- For a transformer that is 100% efficient, the output power (= secondary voltage × secondary current) = the input power (= primary voltage × primary current)

transverse waves waves with a direction of vibration of the particles perpendicular to the direction of energy transfer by the waves

ultrasonic probe hand-held device used in medicine to direct ultrasonic pulses into the body

ultrasonics sound waves at frequencies above the range of the human ear, which is about 18 kHz

uniform field a region where the field strength is the same in magnitude and direction at every point in the field

- the electric field between two oppositely charged parallel plates is uniform. The electric field strength $E = \frac{V}{d}$, where V is the p.d. between the plates and d is the perpendicular distance between the plates
- the gravitational field of the Earth is uniform over a region which is small compared to the scale of the Earth
- the magnetic field inside a solenoid carrying a constant current is uniform

upthrust the upward force on a body in a fluid due to the pressure in the fluid

vector a physical quantity with magnitude and direction

velocity change of displacement per unit time

velocity selector arrangement with perpendicular electric and magnetic fields that only allow charged particles at a particular speed through

voltage gain the ratio of the output p.d. from an amplifier to its input p.d.

W-boson a charged particle that is created in a nucleus when it undergoes beta decay. The W^- boson then decays into an electron and an antineutrino. The W^+ boson then decays into a positron and a neutrino

wave-like nature properties that are characteristic of waves such as interference or diffraction

wave particle duality

- matter particles have a wave-like nature as well as a particle-like nature. For example, electrons directed at a thin crystal are diffracted by the crystal. This is wave-like behaviour in contrast with the particle-like behaviour of electrons in a beam which is deflected by a magnetic field. See de Broglie wavelength
- photons have a particle-like nature, as shown in the photoelectric effect, as well as a wave-like nature as shown in diffraction experiments

wavelength distance between two adjacent wave peaks

weak nuclear force/weak interaction the force in a nucleus responsible for beta decay and for interactions between leptons

weight the force of gravity acting on an object

work force × distance moved in the direction of the force

work done work is energy transferred when a force moves its point of application in the direction of the force. The work done W by a force F when its point of application moves through displacement s at angle θ to the direction of the force is given by $\boldsymbol{W = Fs\cos\theta}$

work function of a metal minimum amount of energy needed by an electron to escape from a metal surface

X-rays electromagnetic radiation of wavelength less than about 1 nm. X-rays are emitted from an X-ray tube as a result of fast-moving electrons from a heated filament (as the cathode) being stopped on impact with the metal anode. X-rays are ionising and they penetrate matter. Thick lead plates are needed to absorb a beam of X-rays

Answers to summary tests

Chapter 1

1.1

1 a 63 km b 18°
2 a 40 m s^{-1} East, 69 m s^{-1} North b 21 km
3 6.1 kN vertically up, 2.2 kN horizontal
4 a i 10.4 km ii 6.0 km
b i 20.4 km ii 10.0 km
c 22.7 km
5 a 3.7 N at 33° to 3.1 N
b 17.1 N at 21° to 16 N
c 1.4 N at 45° to 3 N and to 1 N
6 a 14.0 N b 6.0 N c 10.8 N at 22° to the 10 N force
d 13.5 N at 10° to the 10 N force

1.2

1 a 80 km h^{-1} b 22 m s^{-1}
2 a 2.5×10^4 km h^{-1} b 7.0×10^3 m s^{-1}
3 a 45 000 m b 28.3 m s^{-1}
4 b i 4.0 km
ii 30 m s^{-1} then, 25 m s^{-1} in the opposite direction

1.3

1 a 1.5 m s^{-2} b −2.5 m s^{-2}
2 a 0.45 m s^{-2} b 7.9 m s^{-1}
3 a 0–20 s: straight line from the origin to 12 m s^{-1}; 20–60 s: flat line; 60–90 s straight line from 12 m s^{-1} to zero speed at 90 s.
b 0.60 m s^{-2}, 0, −0.40 m s^{-2}
4 a The velocity increases with time at a decreasing rate and reaches a constant value.
b The acceleration decreases with time (at a decreasing rate) to zero.

1.4

1 a 2.0 m s^{-2} b 221 m
2 a 43 s b −0.93 m s^{-2}
3 a i 0.2 m s^{-2} ii 90 m
b i −0.75 m s^{-2} ii 8.0 s d 3.0 m s^{-1}
4 a 5.0 m s^{-2} b 7.5 m c 18 m d 6.4 m s^{-1}

1.5

1 a 4.0 m b 8.8 m s^{-1}
2 a 3.2 s b 31 m s^{-1}
3 a i 3.9 s ii 38 m s^{-1}
4 a 1.6 m s^{-2} b 3.6 m s^{-1} c 0.64 m

1.6

1 a i 83 s ii 127 s
b i The graph should show a straight line from 0 to 100 m at 83 s then a straight line down to zero displacement at 210 s.
ii The graph should show a straight line from 0 to 100 m at 83 s then a straight line for 127 m to a distance of 200 m at 210 s.
c The graph should be a flat line at 1.20 m s^{-1} from 0 to 83 s then a flat line at −0.79 m s^{-1} (= 100 m / 127 s) from 83 s to 210 s.
2 a 600 s
b i The line should be flat at 8.8 m s^{-1} from 0 to 200 s the at zero velocity from 200 to 800 s.
ii The line should be flat at 2.2 m s^{-1} from 0 to 800 s
3 a i The displacement increases from O to A at an increasing rate for 1200 s then at a decreasing rate over 300 s to a constant displacement at B then at this value for 600 s to C before decreasing at an increasing rate for a further 900 s when it suddenly stops at D.
ii See graph below

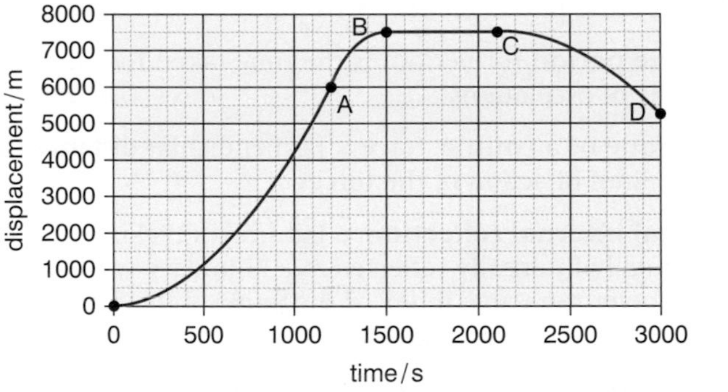

b i 5250 m ii 9750 m
4 a i 0.61 s ii 5.9 m s^{-1} iii 0.43 s iv 4.2 m s^{-1}

1.7

1 a i 52 s ii 0.49 m s^{-2} b i 406 m ii −1.04 m s^{-2}
2 a 15 m b −0.13 m s^{-2}
c 0.67 m s^{-1} downwards, 13.4 m from the start
3 a i 80 m ii 8.0 m s^{-1} b i 65 s ii −0.12 m s^{-2}
4 a i 180 m s^{-1} ii 2.7 km
b 4.4 km
c 290 m s^{-1}

1.8

1 a 32 m s^{-1} b 2.8 s c 39 m
2 a 3.0 s b 49 m c 34 m s^{-1} (at 62° to the horizontal)
3 a 0.20 s b 11.7 m s^{-1}
4 a 354 m b i 1020 m ii 1020 m c 146 m s^{-1}

1.9

1 a 470 mm b 3.0 m s^{-2}
c 2.7 m s^{-1} (at 79° to the vertical)
2 a 25.8 m b 2.3 s c 4.6 s d 179 m
3 a 3.5 m s^{-1}, 3.0 m s^{-1} b i 150 m ii 20 m
4 a 21 m s^{-1}, 10 m s^{-1} b 8.7 m c 27 m s^{-1}

d The range would have been less because it would not reach the same maximum height so would not be in the air for as long and the horizontal component of velocity would progressively decrease from its initial value.

Chapter 2

2.1

1 **a** $0.24\,m\,s^{-2}$ **b** 190 N **c** 0.024
2 **a** $-2.4\,m\,s^{-2}$ **b** 12 000 N
3 **a** 360 N **b** 23 s
4 **a** $-1.3 \times 10^5\,m\,s^{-2}$ **b** 260 N

2.2

1 **a** 5400 N **b** 7700 N
2 **a** 60 N **b** 270 N
3 **a** 11.8 kN **b** 11.8 kN **c** 12.2 kN **d** 12.2 kN
4 **a** $1.0\,m\,s^{-2}$ **b** 12.5 N

2.3

1 **a i** $0.04\,m\,s^{-1}$ **ii** 1.5 N
b The drag force would be smaller but its weight would be even smaller so its rate of descent would be less.
2 Crouching reduces the drag force so the cyclist with the same 'driving force' could reach a higher speed before the drag force equals the driving force.
3 **a** $0.14\,m\,s^{-2}$ **b** 520 m
4 The component of weight down the slope increases the resultant force on the vehicle but the drag force at any given speed is unchanged. So the speed at which the resultant force is zero is higher.

2.4

1 **a** 200 J **b** 4.5 J
2 **a** 48 J **b** 24 J **c** 0
3 **a** 1000 J **b** 600 J **c** 400 J
4 **a** 2.4 N **b** 0.12 J

2.5

1 **a** 9.0 J **b** 9.0 J **c** 1.8 m
2 **a i** 2.9 J **ii** 2.4 J **b** 0.5 J
3 **a i** 16 kJ **ii** 5.8 kJ **b i** 10.2 kJ **ii** 20 N
4 **a** 590 kJ **b** 2.4 kJ **c** 470 kJ **d** 122 kJ **e** 1.6 kN

2.6

1 **a** 1.13 kJ **b** $62.5\,J\,s^{-1}$
2 500 MW
3 **a** 156 MJ **b** 140 MJ **c** 12 MW
4 122 m

2.7

1 **a** 450 J **b** $1800\,J\,s^{-1}$
2 **a** 480 J **b** 50 J **c** 10%
3 **a** $570\,MJ\,s^{-1}$ **b** $6.2 \times 10^5\,kg$
4 **a** 600 s **b** 3.7 MJ **c** 8%

2.8

1 $1.0 \times 10^7\,m^2$
2 4 MW
3 $125\,m^3\,s^{-1}$
4 **a** $6.3 \times 10^{11}\,kg$ **b** 430 MW

Chapter 3

3.1

1 **a** 7.3 N **b** 7.3 N at 31.5° to vertical
2 **a** See 3.1 Fig 3 with $\theta = 30°$
b i 2.7 N **ii** 4.7 N
3 **a** 139 N **b** 95 N
4 **a** 73° **b** 7.0 N

3.2

1 300 N
2 **a** See 3.2 Fig 2 with an extra weight shown and all the force arrows for the weights labelled and at the correct distances from the pivot.
b 6.2 N
3 27 cm
4 6.75 N

3.3

1 0.51 N at 100 mm mark, 0.69 N at 800 mm mark
2 **a** 122 N at 1.0 m end and 108 N from the other end, both vertically upwards
b 122 N at 1.0 m end and 108 N at the other end, both vertically downwards
3 620 kN, 640 kN
4 **a** 100 N, 50 N **b** 150 N

3.4

1 The centre of gravity of the bookcase and books on it is higher than if the books were on the bottom shelf. It would be more unstable because the line of action of its weight would reach the outside its base more readily if it was tilted too far at the top.
2 89 N
3 **a** 48°
b Yes, they will raise the overall centre of mass so it will topple on a less steep slope
4 The centre of gravity of a fully loaded lorry is higher than when it is unloaded. It would be more unstable because the line of action of its weight would reach the outside its base more readily when a side wind acts on it.

3.5

1 **a** 50 N **b** 250 N
2 **a** 1800 N m **b** 1800 N
3 6 kN
4 10.8 kN

3.6

1 **a** 15 N **b** 3.0 N **c** 10.8 N
2 7 N
3 **a** 6.8 N **b** 52°
4 18.0 N
5 **a** 17 kN **b** 17 kN
6 **a** 6.2 N **b** 11.2 N
7 50 mm away from pivot

8 **a** 6.8 N **b** 9.8 N
9 **a** 2200 N **b** 3100 N
10 **a** The diagram should show a uniform horizontal beam XY acted on by a labelled vertical force at each end and two labelled vertical downward forces at the correct positions to represent the weights of the beam and the person.
b 950 N at X, 750 N at Y
11 **a** 2820 kN **b** 1660 kN and 1540 kN
12 **a** 8.0 N and 16 N **b** 38 N and 76 N
13 **a** The angle of the beam to the horizontal should be labelled 8.1°. Three labelled vertical force arrows should be shown acting on the beam representing its weight, the tension in the cable and the support force on it from the ground.
b 11 kN, 11 kN
14 **a** The diagram should show a labelled tension force acting on each top corner of the picture and a force vector for its weight acting vertically downwards at the centre of the picture.
b 28.4 N

Chapter 4

4.1

1 **a** **i** 1.2×10^{-18} kg m s^{-1} **ii** 0.050 kg m s^{-1}
iii 14 kg m s^{-1}
b **i** 6.0 kg **ii** 20 m s^{-1}
2 **a** 3.6×10^{5} kg m s^{-1} **b** 60 s
3 **a** 5.4×10^{6} kg m s^{-1} **b** 45 s
4 **a** 4.1 Fig 3 with F and t replaced by 400 N and 20 s and an extra step of 20 N for a further 20 s.
b **i** 9.0×10^{3} kg m s^{-1} **ii** -8.4×10^{3} kg m s^{-1}
iii 1.0 m s^{-1}

4.2

1 **a** 1600 kg m s^{-1} **b** 3200 N
2 **a** 3000 kg m s^{-1} **b** 7.5 kN
3 **a** 4.2×10^{-23} kg m s^{-1} **b** 1.9×10^{-13} N
4 **a** 2.1×10^{-23} kg m s^{-1} **b** 9.5×10^{-14} N

4.3

1 0.72 m s^{-1},
2 0.70 m s^{-1} in the same direction
3 0.050 m s^{-1} in the direction the 1.0 kg trolley was moving in
4 0.63 m s^{-1} in the opposite direction to its initial direction.

4.4

1 **a** **i** Its total energy is the same at the end as at the start as its energy changes from potential to kinetic to elastic energy in the descent and from elastic to kinetic to potential during the ascent.
ii Its total energy is less at the end than at the start as some of its elastic energy is transferred to thermal energy of the ball in the impact and then dissipated to the surroundings. So it has less energy in its ascent than in its descent.
b Assuming air resistance is negligible, its KE just before impact = its loss of PE on the descent from height = $mg \times 1.2$ m and its KE just after impact = its gain of PE on the ascent = $mg \times 0.9$ m. So its loss of KE due to the impact = $mg \times 0.3$ m. Therefore its % loss of KE = ($mg \times 0.3$ m / $mg \times 1.2$ m) × 100% = 25%

2 **a** 9.0 m s^{-1} in the same direction **b** 24 kJ
3 **a** 1.1 m s^{-1} in the reverse direction **b** 20 J
4 **a** **i** 1.0 m s^{-1}
b **i** 0.9 m s^{-1} **ii** 0.9 m s^{-1}

4.5

1 0.35 m s^{-1}
2 **a** 0.25 m s^{-1}; the mass of A and X was greater than the mass of B, so B moved away faster.
b After they separated, they have equal and opposite momentum therefore (0.50 kg + m_X) × 0.25 m s^{-1} = 0.50 kg × 0.30 m s^{-1} so $0.25m_X$ = 0.150 − 0.125 which gives m_X = (0.150 − 0.125) × 4 = 0.10(0) kg.
3 **a** **i** 0.10 m s^{-1} **ii** 15 mJ **b** 0.19 m s^{-1}
4 **a** 8.9 m s^{-1} **b** **i** 1.1 J **ii** 79 J

Chapter 5

5.1

1 **a** 8.0×10^{-4} m^3 **b** 3.1×10^{3} kg m^{-3}
2 **a** 6.3 kg **b** 2.0×10^{-3} m^3 **c** 3.1×10^{3} kg m^{-3}
3 **a** 9.6×10^{-6} m^3 **b** 7.5×10^{-2} kg
4 **a** **i** 0.29 kg **ii** 0.12 kg **b** 2.3×10^{3} kg m^{-3}

5.2

1 **a** The pressure of the water on the lower half of the ball is greater than the water pressure on the top half. So the ball experiences an upthrust.
The volume of liquid displaced by the ball is equal to the volume of the ball. Because water is much denser than air, the weight of the liquid displaced is much greater than the weight of the ball.
So the upthrust on the ball is much greater than its weight.
b Cork is used because it doesn't absorb water and it has a much smaller density than water has. When a cork object is in water, the upthrust on it doesn't need to be very large to support its weight so it doesn't need to be fully immersed.
2 **a** When it is in water, there is an upthrust acting on it. The reading on the newton-meter is therefore less than when it is in air as the upthrust helps to support it in water.
b **i** The upthrust = 5.2 N − 3.3 N = 1.9 N. The weight of water displaced = upthrust = 1.9 N. The mass of water displaced = weight of water displaced/g = 1.9 N/9.81 N kg^{-1} = 0.194 kg. The volume of the object = the volume of water displaced = mass of water/density of water = 0.194 kg/1000 kg m^{-3} = 1.94×10^{-4} m^3

ii The mass of the object = its weight/g = $5.2\,N/9.81\,N\,kg^{-1} = 0.530\,kg$. Therefore the density of the object = mass/volume = $0.531\,kg/1.94 \times 10^{-4}\,m^{-3} = 2740\,kg\,m^{-3}$ = $2700\,kg\,m^{-3}$ to 2 significant figures.

3 a i A has the greatest density because it sinks and B and C both float. A must be more dense than water whereas B and C must be less dense than water.
ii Both B and C have a density less than that of water. Because B floats higher in the water than C, the difference between its density and the density of water must be greater than the corresponding difference for C. So the density of B must be less than the density of C. Therefore B has the lowest density.

4 a The extra weight causes the tube to float lower in the water. Therefore the length of the tube above the water decreases (*linearly*) as the total weight is increased.
b Remove the tube from the water. (Place a soft pad at the bottom of the tube.) Add a suitable metal object (e.g. a steel nail) of known weight to the tube. Use a millimetre ruler to measure the length L of the tube (and cork) above the water. Repeat the procedure five more times, each time adding another object of known weight. Record the measurements in a table including a column for the total weight added, W. Plot a graph of $y = L$ against $x = W$. If the (*linear*) prediction is correct, the graph should have a negative gradient (*and be a straight line with a constant gradient*).

5.3

1 1.2 kPa
2 120 kN
3 13 kN
4 a 10.3 m
b i 1.47 kPa ii This gas pressure is 1.5% ((= 1.47 kPa/101 kPa) × 100%) above atmospheric pressure so it is not normal.

5.4

1 a 0.40 m b 12.5 N
2 a 20 N b 100 mm c $200\,N\,m^{-1}$
3 a 40 N b 200 mm
4 a $12.3\,N\,m^{-1}$ b $8.8 \times 10^{-2}\,J$ c 2.2 N

5.5

1 $1.0 \times 10^{9}\,Pa$
2 $1.3 \times 10^{11}\,Pa$
3 a $9.4 \times 10^{8}\,Pa$ b $1.2 \times 10^{-2}\,m$
4 a No, the UTS for glass is greater than for copper.
b No, the initial gradient for steel is greater than for glass.
c Yes, the copper curve extends more than the glass curve.

5.6

1 a $3.3 \times 10^{6}\,Pa$ b $2.8 \times 10^{-4}\,m$ c 0.21 J
2 a 2.3 mm b $1.7 \times 10^{-2}\,J$
3 a 470 kN b 47 J
4 a 10.5 J b 3.5 J

Chapter 6

6.1

1 a Electrons are negatively charged so free electrons in the can are attracted towards the positively charged rod. Some of the free electrons transfer through the point of contact from the can onto the rod so the can is left with a positive charge.
b Electrons transfer through the thread from the sphere to the ground.

2 a Free electrons on the conductor transfer to the ground through the wire when the conductor is earthed.
b The conductor therefore loses negative change and is left with a positive charge when the wire is removed.

3 a The positive charge of the object attracts free electrons in the metal plate to the surface of the plate.
b There is a force of attraction between the object and the electrons at the surface because the object is positively charged and the electrons are negatively charged. There is also a force of repulsion between the object and the positive ions in the plate but this force is less because the ions are further from the object than the electrons at the surface are.

4 **a and b** 6.1 Fig 4 with full lines not dashed and with the field direction arrows from + to −.

6.2

1 a i 3.5 C ii 210 C b i 3.0 A ii 0.15 A
2 a 3.8×10^{14} b 1.9×10^{21}
3 a 800 C b i 1600 s ii 8000 s
4 $1.0\,mm\,s^{-1}$

6.3

1 a 29 kJ b 720 J
2 a 2 A b 22 kJ
3 a i 48 kJ ii 3.5 A b 5 A
4 a 12 kJ b 4.5 W c 2700 s

6.4

1 a 6.0 Ω, 9.9 V, 0.125 mA, 160 Ω, 2.5 mA b 7.5 Ω
2 31 Ω
3 0.11 mΩ
4 a $1.8 \times 10^{-6}\,\Omega\,m$ b 33 mm

6.5

1 a 0.25 A, 12 Ω
b The filament would become brighter and hotter until it melts and breaks as a result.
2 a 0.03 mA b 0.38 mA
4 a 30.4 Ω b 46 °C

Chapter 7

7.1

1 a 1.0 A, 4.0 A b 5.0 A c 30 W
2 a 7.1 Fig 5 with the battery pd changed and without the other pd values shown.
  b i 2.0 V ii 0.20 A
3 a 7.1 Fig 3 with a resistor changed to an ammeter and the values / labels changed.
  b i 4.0 V ii 2.0 V iii 10 Ω
4 a 3.6 V b 30 Ω

7.2

1 a 16 Ω b 3.0 Ω c 4 Ω
2 a 2 Ω b 6 Ω c 1.0 A d 4.0 W
3 a 3.6 Ω b 0.83 A
  c 2 Ω: 0.5 W; 4 Ω: 1.0 W; 9 Ω: 1.0 W, d 2.5 W
4 a 14.4 W b 2.4 Ω

7.3

1 a 6.0 Ω b 2.0 A c 3.0 V d 9.0 V
2 a 0.5 A b 1.25 V c 0.63 W d 0.13 W
3 a 2.0 Ω b 1.5 V
4 12 V, 2 Ω

7.4

1 a 12.0 Ω b 0.25 A
  c 4 Ω: 0.25 A, 1.0 V; 24 Ω: 0.08 A, 2.0 V; 12 Ω: 0.17 A, 2.0 V
2 a i 20.0 Ω ii 1.05 A iii 1.05 A, 15.8 V
  b i 2.49 A ii 3.17 A
3 a i 2.0 W ii 2.0 W b i 2.0 W ii 8.0 W
4 a Q: 0.6 V, 0.06 mA; P: 2.4 V, 0.48 mA
  b P: 0.6 V, 0.12 mA; Q: 2.4 V, 0.24 mA

7.5

1 a 7.5 Fig 1 with a battery instead of a single cell and the values/labels changed; 1 kΩ: 0.75 V; 5.0 kΩ: 3.75 V
  b 1 kΩ: 1.3 V; 5.0 kΩ: 3.2 V
2 Sketch: 7.5 Fig 3c with a light bulb connected between C and B and the labels / values changed; as the contact is moved up from B, the light bulb filament begins to glow and becomes increasingly bright until it reaches its normal brightness when the contact is at A.
3 a i 0.5 A ii 8.0 Ω: 4.0 V; 4.0 Ω: 2.0 V
  b i 3.0 V ii 4.0 V
4 a i 2.8 V ii 6.4 kΩ
  b The voltmeter reading increases because the LDR resistance decreases so the pd across the LDR drops as its share of the 5.0 V cell pd decreases.

Chapter 8

8.1

1 Sound waves are longitudinal; the other 3 are transverse.
2 See 8.1 Fig 2.
3 As 8.1 Fig 4a, with the unnecessary labels removed and arrows to show the reversal of the direction of motion at the peaks and troughs.
4 Q moves from zero displacement to the right then back to zero after half a cycle, as in Fig 7.5b, then it moves to the left of zero displacement then back to zero in the next half-cycle.

8.2

1 a 0.10 m b 1.9×10^{-2} m
2 a 10 GHz b 5.0×10^{14} Hz
3 1.0 V, 1.0 kHz
4 a i amplitude = 9 mm ii 180° iii 270°
  b +9 mm

8.3

1 The reflected wavefront is at 30° to the reflector with its top end near the right hand side of the reflector and its direction arrows normal to the wavefront and pointing away from the reflector.
2 Label the wavefronts below the boundary A, B and C from left to right. The 2 missing sections of B and C above the boundary should be both parallel to the section of A above the boundary and the lower end of each missing section should join the corresponding wavefront below the boundary. The direction arrows of the wavefront sections above the boundary should be perpendicular to the wavefronts and pointing away from the boundary.
3 a decreased diffraction
  b increased diffraction
  c increased diffraction compared with b
  d little change
4 a Waves from the transmitter are diffracted at the gap and they spread out beyond the gap so they are detected by the detector.
  b If the detector was made a little wider, there would be less diffraction so the detector signal would be reduced. Making the gap too wide would enable waves to reach the detector directly so the signal would increase.

8.4

1 a An electromagnetic wave is a transverse wave that consists of an oscillating electric field perpendicular to and in phase with an oscillating magnetic field of the same frequency.
  b i In a polarised electromagnetic wave, the oscillations of each field are always in the same perpendicular plane. The plane of polarisation is defined by the plane of the electric field oscillations.
    ii In an unpolarised electromagnetic wave, the plane of polarisation is not constant and it changes at random.
2 The intensity of the observed light decreases and is a minimum when the angle of rotation is 90°. The intensity then increases as the angle of rotation is increased and reaches maximum intensity at 180°.

3 The radio waves are polarised. The signal becomes weaker as the aerial is rotated away from the plane of polarisation of the radio waves and is weakest when it is perpendicular to the plane of polarisation. Further rotation of the aerial causes the signal to become stronger as it is rotated nearer to the plane of polarisation.

4 a Malus's law states that when polarised light is passed through a polariser, the intensity of the transmitted light is given by the equation $\boldsymbol{I = I_0 \cos^2\theta}$ where θ is the angle the polariser is rotated through from its orientation at the maximum transmitted intensity I_0.
 b For $\theta = 10°$, $I = I_0\cos^2 10° = 0.970\ I_0$ therefore $I = 0.970 I_0 = 97\% \times I_0$. Therefore the % reduction in the transmitted intensity is 3%.

8.5

1 a When the peak is opposite the trough, the rope would be momentarily flat.
 b The peak would be near the right end of the rope and the trough would be at the other end, both travelling away from the centre.

2 a The lines of the gaps would be closer together.
 b The lines of the gaps would be further apart (because the wavelength would be greater).

3 a The signal would decrease gradually (as the intensity of the waves would decrease).
 b The signal may increase due to increased diffraction but a considerable reduction in the gap width would cause the signal to decrease due to the reduced intensity of the diffracted waves.

4 a The signal decreases midway between A and B because the detector is then at a point of cancellation where the intensity is a minimum. The signal then increases as the detector reaches a point of reinforcement at B.
 b The intensity is a minimum midway between A and B and a maximum at B. So the signal increases as the detector moves towards B from the midway position.

8.6

1 a 8.6 Fig 1 a and c superimposed with an appropriate arrows to indicate the vibration direction of the node over 1 cycle.
 b 8.0 m

2 a 2.0 m b i 180° ii 225° iii 0

3 Progressive wave; all particles vibrate with the same amplitude and a phase difference (over each wavelength) that increases with distance apart. Stationary wave; all particles between adjacent nodes vibrate in phase with amplitudes that increase from zero at the nodes to a maximum midway between the nodes.

4 a The reflected waves and the incident waves form a stationary wave pattern with nodes as the zero signal positions which are 15 mm apart.
 b 30 mm

8.7

1 a 1.6 m b 410 m s^{-1}

2 a 0.4 m b 0.53 m

3 a The frequency of each harmonic would be greater (because the wavelength of each harmonic on the wire would be shorter) so the pitch of each note (or frequencies of the sound waves in each notes created) would be higher.
 b The frequency of each harmonic would be greater due to the increase of tension so the pitch of each note (or frequencies of the sound waves in each notes created) would be higher.

4 The notes are different in pitch because the mass per unit length of the two wires differs so the frequency of corresponding harmonics differs. The steel wire has a greater mass per unit length so the speed and therefore each of its harmonic frequencies is lower than that of the corresponding harmonic of the nylon wire.

8.8

1 a 2.40 m, 140 Hz b 425 Hz

2 a 2.0 m b 57 Hz

3 a 71 Hz b 142 Hz

4 68–680 Hz

8.9

1 a 1200 Hz
 b When the car is directly under the bridge, its direction of motion is perpendicular to the straight line between an overhead observer and the car. So at that instant, the car is neither approaching or moving away from the bridge. So the observed frequency is equal to the emitted frequency, which is 1100 Hz.

2 a Towards Y, because Y observes a higher frequency than the frequency emitted by the horn
 b 1060 Hz

3 a 2040 m s^{-1} b 3.4 GHz

4 a The reflected pulses would be too weak to be detected, as they would diffract too much on reflection. Also, the variable speed of the air would affect the speed of the waves.
 b The small particles in the air would be very poor reflectors of microwaves because they are so small compared with the wavelength of the microwaves.

Chapter 9

9.1

1 a The fringes would be closer together because the fringe spacing would be smaller.
 b The interference fringes would disappear although a diffraction pattern from the remaining slit would be seen.

2 550 nm

3 0.9 mm

4 0.75 m

9.2

1 **a** See 9.1 Fig 2
b The fringe pattern would consist of alternate bright and dark fringes. The bright fringes would be yellow-orange, the same colour as the sodium lamp.

2 **a** See 9.1 Q1b answer
b The fringe spacing would be unchanged but the bright fringes would be wider and the dark fringes narrower.

3 1.1 mm

4 The central fringe would be white and the other inner fringes would be blue at the edges nearest the central fringe, and red at the edges furthest from the central fringe. The edges of the outer fringes would overlap so these fringes would be less distinct.

9.3

1 **a** 10.9°, 22.2° **b** 5

2 **a** 2 **b** 0.58° (= 35′)

3 **a** 1090 **b** 69.9°

4 **a** $599\,\text{mm}^{-1}$ **b** **i** 3 **ii** 50° 40′

Chapter 10

10.1

1 **a** Most of an atom is empty space.
b There is a positively charged nucleus at the centre of the atom. The nucleus is much smaller than the atom and is where most of the atom's mass is located.

2 **a** Alpha particles from the source would collide with air molecules and be stopped. So the chamber needs to be evacuated.
b If the foil is not thin enough, each alpha particle would be scattered by the nuclei of the atoms in the foil more than once.
c Alpha particles of different speeds moving along the same initial path would be scattered differently so the scattered alpha particles at each angle of deflection would not have followed the same path and the measurements would not confirm Rutherford's theory.
d If the beam was too wide, the scattering at nearby angles of deflection would overlap and the measurements may not confirm Rutherford's theory.

3 **a** See 10.1 Fig 5; the force on the particle at X should be shown as an arrow acting on the particle pointing away from the centre of the nucleus.
b **i** The KE decreases as the particle approaches X where its KE is a minimum and then increases as it moves away from X.
ii Its PE increases from zero as the particle approaches X where its PE is a maximum and then decreases to zero at infinity as it moves away from X.

4 **a** The activity of the source decreases because more and more nuclei become stable and the number of radioactive nuclei in the source decreases. If the source did not have a long half-life, later readings would be significantly lower than earlier readings due to the decrease of the source activity.
b Alpha particles with different kinetic energies approaching a nucleus along the same path would be deflected by different amounts so the reading at each angle of deflection would not be due to alpha particles of the same kinetic energy.

10.2

1 **a** 6p, 6n **b** 8p, 8n **c** 92p, 143n
d 11p, 13n **e** 29p, 34n

2 **a** neutron **b** electron **c** neutron

3 **a** **i** $+3.2 \times 10^{-19}\,\text{C}$ **ii** 63
b 8 neutrons and 10 electrons

4 Most of the hydrogen atoms are ${}^{1}_{1}\text{H}$ atoms and therefore have a single proton as a nucleus. Less energy is needed for a neutron to knock a ${}^{1}_{1}\text{H}$ nucleus (i.e. a proton) out compared with knocking a carbon nucleus out. This is because a neutron hitting a carbon nucleus is likely to recoil and retain some kinetic energy, whereas a neutron hitting a ${}^{1}_{1}\text{H}$ nucleus is likely to be stopped and use most of its kinetic energy to eject the ${}^{1}_{1}\text{H}$ nucleus.

10.3

1 **a** Beta particles
b Use a magnet to deflect a narrow beam of them; they should deflect as in 10.3 Fig 1 if they are beta particles.

2 **a** Alpha radiation
b The radiation affects the film and causes it to blacken when the film is developed. The key prevented the radiation reaching the film underneath the key so the image of the key was seen on the film when it was developed.

3 **a** **i** gamma **ii** alpha
b Alpha radiation in air has a certain range. When the alpha source was moved beyond the range of the source , the current dropped to zero because the alpha particles could not reach the ionisation chamber.

4 **a** 42% **b** 58%

10.4

1 **a** ${}^{238}_{92}\text{U} \rightarrow {}^{234}_{90}\text{Th} + {}^{4}_{2}\alpha$ **b** ${}^{228}_{90}\text{Th} \rightarrow {}^{224}_{88}\text{Ra} + {}^{4}_{2}\alpha$

2 **a** ${}^{64}_{29}\text{Cu} \rightarrow {}^{64}_{30}\text{Zn} + {}^{0}_{-1}\beta\ (+\ \bar{\nu})$
b ${}^{32}_{15}\text{P} \rightarrow {}^{32}_{16}\text{S} + {}^{0}_{-1}\beta\ (+\ \bar{\nu})$

3 **a** ${}^{213}_{84}\text{Po}$, ${}^{209}_{82}\text{Pb}$, ${}^{209}_{83}\text{Bi}$
b **i** 83 p + 130 n **ii** 83 p + 126 n

4 **a** **i** ${}^{205}_{84}\text{Th} \rightarrow {}^{201}_{82}\text{Pb} + {}^{4}_{2}\alpha$
ii ${}^{201}_{82}\text{Pb} \rightarrow {}^{201}_{81}\text{Th} + {}^{0}_{+1}\beta + \nu$
b **1** They have equal and opposite charge.
2 The positron is an antimatter particle whereas the β^- particle is a matter particle (or the positron is the antiparticle of the β^- particle (or electron)).

10.5

1 a The creation of charged atoms (ions) by adding or removing electrons to or from uncharged atoms.
b Alpha radiation from outside the body is absorbed by the layer of dead skin at the body suface whereas beta radiation can penetrate the skin.

2 a Ionising radiation affects living cells by damaging or destroying cell membranes or damaging DNA molecules in the cells, causing the cells to divide uncontrollably.
b i A film badge provides a record of how much ionising radiation of each type its wearer has been exposed to.
ii See 10.5 Fig 3 and related text.

3 a Lead is very dense and a thick lead plate absorbs α, β and γ radiation. A lead-lined storage box absorbs the radiation from the radioactive sources inside it and so prevents the radiation from leaving the box.
b Long handles ensure the user is as far away as possible from the radioactive source while it is being moved.

4 Solid sources should only be moved by robots or using long-handles tongs and should be out of its storage box for as little time as possible. Liquid or gas or powdered sources should be in sealed containers and only moved as above. No sources should be allowed to make contact with the skin. The eyes should never be exposed to ionising radiation from a radioactive source.

10.6

1 a A W boson is charged; a photon is uncharged. A W boson has a very short range; a photon has an unlimited range.
b lifetime $\cong$ range/speed of light = 3×10^{-15} m/3×10^{8} s = 10^{-23} s

2 a Leptons interact through the weak interaction; hadrons interact through the strong interaction.
b i a proton or a neutron ii a π meson
iii a K meson

3 a i uud ii udd
b One of the up quarks in the proton changes into a down quark and emits a W^+ boson, which then decays into a β^+ particle and a neutrino.

4 a −1
b $\pi^- = \bar{u}d$, X = ssu. X must contain a u quark rather than a d quark to ensure conservation of charge.

Chapter 11

11.1

1 a 1.75×10^{-3} rad b 0.105 rad c 6.28 rad
2 a 20 ms b i 0.31 rad ii 310 rad
3 a 470 m s^{-1} b i 0.0042° ii 7.3×10^{-5} rad
4 a 7.0 km s^{-1} b i 0.050° ii 8.7×10^{-4} rad

11.2

1 a 0.23 m s^{-1} b i 7.9×10^{-4} m s^{-2} ii 5.1×10^{-2} N
2 a 0.53 m s^{-1}, 0.66 m s^{-2} b 9.9×10^{-2} N
3 a i 3.0×10^{4} m s^{-1} ii 5.9×10^{-3} m s^{-2}
b i 7.9×10^{3} m s^{-1} ii 5.1×10^{3} s
4 a 8.4 m s^{-1} b 88 m s^{-2} c 175 N

11.3

1 a 6.7 m s^{-2} b 3.8 kN
2 a 4.1 m s^{-2} b 3.0 kN
3 Sprinters run much faster than marathon runners so on a banked rather than a flat circular track, sprinters would be less likely to slip as there would be a component of their weight acting parallel to the track to maintain their circular motion.
4 a 40 m s^{-1}

11.4

1 a 30 m s^{-1} b i 11.3 m s^{-2} ii 690 N
2 a 25 m s^{-1} b 20 m s^{-2} c 2000 N
3 a 13 m s^{-1} b 13 m s^{-2} c 240 N
4 −0.04 N

Chapter 12

12.1

1 Her/his velocity would increase until the bungee rope is vertical and starts to stretch, causing the velocity to decrease gradually to zero at the lowest point of the descent. The stretched rope would then pull the jumper upwards, increasing their velocity until the rope becomes slack after which the jumper's velocity would decrease to zero at or below the platform.

2 a No frictional forces are present and the oscillations are at constant amplitude.
b Depress the free end of the ruler and record its initial position against a vertical millimetre scale then release it and record its lowest position after every five cycles. The readings should be unchanged if it oscillates freely.

3 a 0.48 s b 2.1 Hz
4 a $\frac{\pi}{2}$ radians b π radians

12.2

1 a +25 mm, changing direction from up to down
b 0, moving down
c −25 mm, changing direction from down to up
d 0, moving up
2 a 0.5 Hz b i −0.25 m s^{-2} ii 0 iii 0.25 m s^{-2}
3 a 0.5 Hz b −0.32 m s^{-2}
4 a −32 mm, 0.32 m s^{-2} b 0, 0

12.3

1 a 0.33 Hz b 0.25 m s^{-2}
2 a i 12 mm ii 0.63 s b 10.1 mm
3 a 2.1 Hz b 0.057 m
4 a 3.7 Hz
b i 8.7 mm, −191 mm s^{-1}
ii −12 mm, −16 mm s^{-1}

12.4

1 **a** **i** 0.33 s **ii** 3.1 Hz
b **i** 0 **ii** $-3.7\,m\,s^{-2}$ **iii** $-7.5\,m\,s^{-2}$
2 **a** **i** 3.0 Hz **ii** 0.33 s **b** $f_2 < f_1 \therefore m_2 > m_1$
3 **a** **i** 70 mm **ii** $21\,N\,m^{-1}$
b **i** Calculate ω as $\omega^2 = k / m$ then calculate the frequency f using $f = \omega / 2\pi$
ii 0.53 s
4 **a** **i** 1.25 N **ii** $2.5\,m\,s^{-2}$
b **i** $\omega^2 = k/m = 25\,N\,m^{-1} / 0.50\,kg = 50\,rad^2\,s^{-2}$ therefore $a = -\omega^2 x = -50x$
ii 1.1 Hz, +46 mm
5 **a** **i** 2.0 s **ii** 1.0 s **b** 5.0 s
6 The mass–spring system would have the same time period on the Moon as it has on the Earth. The simple pendulum would have larger time period on the Moon than it has on the Earth.

12.5

1 **a** In the first quarter cycle, its KE decreases to zero and its PE increases to maximum. In the next quarter cycle, its KE increases to a maximum and its PE decreases to a minimum. The above sequence is repeated in the next 2 quarter cycles.
b See 12.5 Fig 2
2 **a** $60\,N\,m^{-1}$ **b** **i** $7.5 \times 10^{-2}\,J$ **ii** $7.5 \times 10^{-2}\,J$
3 **a** **i** In each half-cycle starting at maximum displacement, its PE decreases and its KE increases from zero then its PE increases and its KE decreases to zero. Air resistance causes the total energy to decrease gradually.
ii The water in the tube oscillates with a decreasing amplitude between the two sides of the U tube. In each half-cycle starting at maximum displacement, its PE decreases and its KE increases from zero then its PE increases and its KE decreases to zero. Fluid friction (viscosity) causes the kinetic energy of the water and the amplitude of the oscillations to decrease rapidly to zero.
b The suspension would be slow to respond and therefore less effective giving an less comfortable ride.
4 **a** 82 mm **b** 44 mm

12.6

1 **a** Resonance is when the amplitude of an oscillating system becomes very large as a result of being subjected to a periodic force of the same frequency as the natural frequency of the system.
b The periodic force is then in phase with the displacement of the oscillating system so the amplitude becomes very large.
2 **a** It would be lower because the increased mass makes the time period longer so the frequency is lower.
b It would be higher because the spring constant of stiffer springs is greater and so the time period is shorter and the frequency is higher.
3 The rotation of the drum causes a periodic force to act on the panel. At a certain frequency of rotation of the drum, the panel resonates because the frequency of rotation is equal to the panel's natural frequency of vibration.
4 **a** When the vehicle passes over the speed bumps, it experiences an upward force at each speed bump. If the frequency of this periodic force is equal to the natural frequency of the suspension system, resonance occurs and the chassis moves up and down violently because amplitude of the oscillations becomes very large.
b **i** Resonance would not occur because the periodic force frequency would no longer be equal to the natural frequency.
ii The mass of the system would be greater so the natural frequency of the system would be lower and resonance would occur at a lower speed than before.

Chapter 13

13.1

1 **a** A line along which a small mass would move if no other forces acted on it.
b See 13.1 Fig 2 and related text.
2 **a** **i** 33 N **ii** 160 N **b** **i** $16\,N\,kg^{-1}$ **ii** $4.0\,N\,kg^{-1}$
3 See 13.1 page 1.
4 **a** The field should be radial as in 13.5 Fig 2.
b The field should be mostly radial except near the surface above the mass where the line would be closer together.

13.2

1 **a** 235 J
b $\Delta V = 235\,J/12\,kg = 19.6\,J\,kg^{-1}$
2 **a** $2.0\,MJ\,kg^{-1}$ **b** **i** $-61\,MJ\,kg^{-1}$ **ii** $2.1 \times 10^9\,J$
3 **a** **i** −250 J **ii** −200 J **iii** −200 J
b **i** 50 J **ii** 0
4 **a** $\Delta\phi$ between X and the equipotential 1 km away $= 5\,kJ\,kg^{-1}$ so $\Delta\phi$ between X and an equipotential 10 m away $= 50\,J\,kg^{-1}$ $(= (5000\,J\,kg^{-1} \times 10\,m / 1000\,m)$ so $W = m\Delta\phi = 1\,kg \times 50\,J\,kg^{-1} = 50\,J$
b $5\,N\,kg^{-1}$ **c** 25 MJ

13.3

1 **a** $1.3 \times 10^{-6}\,N$ **b** 5.4 mm
2 **a** 780 N **b** $6.0 \times 10^{24}\,kg$
3 **a** 54 N **b** 0.24 N
4 **a** **i** 16.6 N **ii** 0.2 N
b 16.4 N towards the centre of the Earth

13.4

1 **a** $7.35 \times 10^{22}\,kg$
b At the Earth, $g_{Moon} = GM_{Moon} / d^2$ (where d = Earth–Moon distance) $= 6.67 \times 10^{-11}\,N\,m^2\,kg^{-2} \times 7.35 \times 10^{22}\,kg / (3.8 \times 10^8\,m)^2 = 3.4 \times 10^{-5}\,N\,kg^{-1}$. Therefore, $g_{Moon} / g_s = 3.4 \times 10^{-5}\,N\,kg^{-1} / 9.81\,N\,kg^{-1} = 3 \times 10^{-6}$.

2 **a** **i** $68\,N\,kg^{-1}$ **ii** $5.9 \times 10^{-3}\,N\,kg^{-1}$
b Use $g = GM / r^2$ to calculate the Earth's gravitational field strength at 260 000 km from the Earth's centre and to calculate the Sun's gravitational field strength at 1.5×10^{11} m (-2.6×10^8 m) from the Sun.
3 **a** $0.028\,N\,kg^{-1}$
b At height h above the Earth, $g = g_s R^2 / (R + h)^2$. For h = 10 km, $h << R$ (the Earth's radius) so at 10 km above the surface, $g = g_s$.
c 7.1×10^6 J
4 $-2.8\,MJ\,kg^{-1}$, 1410 MJ

13.5

1 **a** X moves faster across the sky so its time period of less and therefore its radius of orbit is less.
b Satellite TV dishes need to point to a geostationary satellite which is a satellite that stays in the same position directly above a point on the surface at the equator. If the dish is not aligned correctly, the signal it receiver from the satellite will be too weak to detect.
2 **a** 3.4×10^6 m **b** $3.0\,N\,kg^{-1}$ **c** 5.2×10^{23} kg
3 **a** **i** v^2 / r
ii centripetal force mv^2 / r = force of gravity on the satellite mg where m is the mass of the satellite therefore $v^2 = gr$.
b **i** $9.5\,N\,kg^{-1}$ **ii** $7.9\,km\,s^{-1}$ **iii** 5200 s
4 **a** See 13.5 page 1
b **i** Insert appropriate values into the equation in part a and calculate v.
ii 7100 s

Chapter 14

14.1

1 **a** 1.4×10^{-3} N **b** $4.0 \times 10^4\,V\,m^{-1}$
2 **a** **i** negative **ii** 1.3×10^{-7} C
b **i** 7.3×10^{-3} N **ii** towards the metal surface
3 **a** **i** $9.0 \times 10^4\,V\,m^{-1}$ **ii** 7.2×10^{-14} N **b** 80 mm
4 The acceleration of each electron towards the positive plate, $a = F/m = eE/m$. The time taken, t, by each electron to cross the field $= x/v$. Since $y = \frac{1}{2}at^2$, $y = \frac{1}{2}(eE/m) \times (x/v)^2 = \frac{1}{2}kx^2$, where $k = eE/mv^2$

14.2

1 **a** 3.7×10^{-11} N **b** 2.6×10^{-10} N
2 **a** **i** 69 mm **ii** 3.6×10^{-6} N **b** 2.5×10^{-5} N repulsion
3 **a** 6.1 nC, negative **b** 2.2×10^{-2} N
4 **a** 2.7 nC, attract **b** 6.2×10^{-2} m, repel

14.3

1 **a** **i** -8.0×10^{-18} J **ii** $+7.2 \times 10^{-17}$ J
b $+8.0 \times 10^{-17}$ J
2 **a** -1.8×10^{-3} J **b** $+1.2 \times 10^{-3}$ J
3 **a** **i** $250\,V\,m^{-1}$
ii 8.0×10^{-17} N (towards the negative plate)
b -8.0×10^{-19} J
4 **a** See Topic 14.3
b **i** $3000\,V\,m^{-1}$
ii As 14.3 Figure 4 with the x-axis labelled 'distance h / mm' and the y-axis labelled 'potential / V', and with the values 60 and 20 in place of ΔV and Δd respectively.

14.4

1 **a** $5.3 \times 10^6\,V\,m^{-1}$ **b** 10 mm
2 **a** **i** $3.7 \times 10^8\,V\,m^{-1}$ **ii** 5.6×10^{-3} N towards Q_1
b Insert appropriate values into the equation for the electric field strength near a point charge to calculate the electric field strength of each charge at the midpoint. The resultant value of E should be zero because the two calculated field strength values should be of equal strength and they are in opposite directions.
3 **a** **i** $4.5 \times 10^8\,V\,m^{-1}$ towards Q_2
ii $2.6 \times 10^8\,V\,m^{-1}$ away from Q_1
iii $4.90 \times 10^8\,V\,m^{-1}$ at 83.4° to the line between the two charges
b **i** Q_1 and Q_2 are both positive charge so a test charge at any point on the line between them would experience two forces in opposite directions. At some point along the line nearer Q_2 than Q_1 the two forces would be equal in magnitude and opposite in direction so the resultant electric field strength at that point would be zero.
ii 11 mm from Q_1, 9 mm from Q_2
4 **a** -9.0×10^6 V
b **i** The electric potential near a point charge Q is proportional to Q / r where r is the distance from Q. For zero electric potential at a point P due to two point charges, $Q_1 / r_1 + Q_2 / r_2 = 0$ therefore $r_2 / r_1 = -Q_2 / Q_1 = -(-30\,mC) / (+15\,mC) = 2.0$ which is in agreement with the given distances of 20 mm for r_2 and 10 mm for r_1.
ii $2.0 \times 10^9\,V\,m^{-1}$ directly towards Q_2

Chapter 15

15.1

1 **a** 5.0 μF **b** 2.2 V **c** 9.9 mC
d 1.4 μF **e** 11 V **f** 3.4 mC
2 **a** 264 μC **b** 106 s
3 **a** 27.5 μC **b** 5.5 μF
4 **a** 910 μC **b** 220 μF **c** 700 μC **d** 7.4 V

15.2

1 **a** 5.0 μF
b 2.0 μF; 6.0 μC, 3.0 V; 3.0 μF; 9.0 μC, 3.0 V
2 **a** **i** 2.4 μF
ii 6.0 μF; 11.0 μC, 1.8 V, 4.0 μF; 11.0 μC, 2.7 V
b **i** 3.2 μF
ii 4 μF; 14.4 μC, 3.6 V, 10 μF; 9.0 μC, 0.90 V, 6.0 μF; 5.4 μC, 0.90 V

3 1. All in series 6.0 μF 2. All in parallel 79 μF
3. Two in series with the third in parallel, 25 μF, 30 μF, 54 μF
4. Two in parallel in series with the third, 9 μF, 16 μF, 19 μF

4 **a** 4.9 μF
b 2 μF; 12 μC, 6.0 V, 4 μF; 17 μC, 4.3 V, 10 μF; 17 μC, 1.7 V

15.3

1 **a** 30 μC, 45 μJ **b** 60 μC, 180 μJ
2 **a** 0.45 C, 2.0 J **b** 10 W
3 2.2 μF; 5.4 μC, 6.6 μJ, 10 μF; 5.4 μC, 1.5 μJ
4 **a** 56 μC, 340 μJ **b** 6.9 μF
c 8.2 V **d** 4.7 μF; 160 μJ, 2.2 μF; 73 μJ

15.4

1 **a** **i** 300 μC **ii** 5.0 s **b** **i** 5 s approx **ii** 20 kΩ
2 **a** **i** 0.61 mC **ii** 0.45 mA **b** 0.23 V, 11 μA
3 **a** 13 μC, 40 μJ **b** 0.62 V **c** 0.42 μJ
4 **a** 0.34 mJ **b** 1.4 s **c** 0.32 mJ

Chapter 16

16.1

1 **a** S **b** NW
2 **a** S **b** sudden switch to N
3 **a** clockwise **b** reversed
4 **a** field due to coil along axis
b 90° if coil field >> Earth's field

16.2

1 **a** S **b** W **c** N
2 **a** N → S **b** vertical up
3 **a** The current in the wires down the two long sides of the coil is in opposite directions. The two long sides of the coil experience a force due to the magnetic field which is equal in magnitude and opposite in direction. These two forces act as a couple because they are in opposite directions and they do not act along the same line (except when the plane of the coil is perpendicular to the field lines).
b When the coil is perpendicular to the field lines, the force on each long side acts along the same line as the force on the other long side so they have no turning effect.
4 **a** It reverses the direction of the current in the coil when the plane of the spinning coil moves through the position where it is perpendicular to the field lines. As a result, the coil continues to spin in the same direction.
b **i** It would spin faster.
ii It would spin in the opposite direction.
iii It would oscillate about the position where its plane is perpendicular to the field lines.

16.3

1 **a** 0.14 T **b** **i** 0 **ii** 11 mN **iii** 22 mN
2 **a** 2.4×10^{-2} N west
b 4.5 A west
c 0.20 T vertically down
d South, 8.0×10^{-3} N
3 0.10 N due south at 20° below the horizontal
4 Long sides: 2.7 N on each side perpendicular to the plane of the coil and in opposite directions; Short sides: 0

16.4

1 **i** The force would reverse in direction.
ii The force would be reduced in magnitude.
iii The force would increase.
2 **i** 1.9×10^{-13} N **ii** 9.6×10^{-14} N
3 **a** The velocity of each electron is parallel to the field so the field does not exert a force on the electrons.
b The initial velocity has a component parallel to the field and a component perpendicular to the field. Therefore the parallel component of velocity is constant and the perpendicular component continually changes direction without changing its magnitude. So the electrons move along a helical path (ie spiral) around the field lines.
4 **a** The magnetic field exerts a force on the conduction electrons that pushes them towards one edge of the slice, making that edge negative which causes the opposite edge to become positive and thus creating a potential difference between the negative edge and the opposite edge.
b **i** 4.4 m s^{-1} **ii** 8.5×10^{-20} N

16.5

1 **a** **i** Each electron in the beam is acted on by a magnetic force perpendicular to their direction of motion. The force causes each electron to move on a circular path because it is always perpendicular to the direction of motion of the electron, so it acts as a centripetal force.
ii 21 mm **b** 2.8 mT
2 **a** 4.7 mT **b** 17.5 mm
3 **a** Use the equation on 16.5 page 1 with the given data including the values of e and m.
b 1.1 MeV
4 **a** 8.0×10^{6} C kg^{-1} **b** 1.4×10^{7} C kg^{-1}

Chapter 17

17.1

1 **a** The motion of the magnet into the coil causes an induced e.m.f in the coil that creates a current in the circuit, which is detected by the meter.
b Any two of the following: Move the magnet faster or use a stronger magnet or wind more turns of wire on the coil.
2 **a** The coil in the motor spins between the poles of the magnets in the motor and so an e.m.f is induced in the coil which creates a current in the circuit that passes through the lamp and lights it.

b The lamp would be brighter because the induced e.m.f would be greater so the current in the lamp would be greater.

3 a A dynamo contains a magnet and a coil. In 14.5 Fig 14.5.3, when the dynamo turns, the magnet spins and the coil is fixed. As a result an e.m.f. is induced in the coil. Because a lamp is connected to the coil, the induced e.m.f. creates a current in the circuit which passes through the lamp and lights it.

b When the lamp lights, the forces needed to turn the dynamo transfer energy to the dynamo to generate the electric current and to overcome friction in the dynamo. Less force is needed when the lamp is disconnected because less energy is transferred to the dynamo as no current is generated.

4 a i Vertically downwards

ii The eastern end: each conduction electron in the rod is pushed by the magnetic field towards the westward end so the western end becomes negative and the eastern end positive.

b The rod does not cut across the field lines because it is parallel to them so no e.m.f. is induced in it.

17.2

1 a i 1.1 mWb **ii** 2.0 s **iii** 0.54 mV

b i The graph should have its *y*-axis labelled 'flux linkage / mWb' and its horizontal axis labelled 'time / s'. The line should be a sloped straight line for the 2 s from the origin to 1.08 mWb then flat for the next 4 s at 1.08 mWb then for the next 2 s, it is a sloped straight line down to the *x*-axis.

ii The graph should show the negative as well as the positive parts of the *y*-axis with its *y*-axis labelled 'emf / V' and its horizontal axis labelled 'time / s'. The line is flat at 0.54 mV for the first 2 s then zero for the next 4 s then −0.54 mV for the next 2 s then zero from 8 s after the start.

2 a 1.4 mWb **b** 23 mV

3 a i $4.5 \times 10^{-4}\,\text{m}^2$ **ii** 1.5(4) mWb

b i 3.1 mWb **ii** 32 mV

4 a 8.0 μWb **b** 40 μV

17.3

1 a As 17.3 Fig 3 with appropriate values and labels shown on both axes and the other labels removed. Note the time period is 50 ms.

b As above with a peak emf of 12 V (because the induced emf is proportional to the frequency) and with a time period of 33 ms.

2 a i 4.3A (=1000 W/230 V = 4.347 A),

ii 6.1 A ($= 4.347\,\text{A} \times \sqrt{2} = 6.148\,\text{A}$)

b Peak power = 2 × mean power = 2000 W

c Power *P* is proportional to the square of the pd (assuming the resistance is unchanged) so a 5% drop in the pd *V* causes an approximate 10% drop in the power supplied.

3 a 26 mWb

b i The speed of rotation of each side = the angular speed of the coil × the radius of rotation *r* (which is half its width) $= 2\pi f / r = 2\pi \times 50\,\text{Hz} / 0.019\,\text{m} = 6.0\,\text{m s}^{-1}$.

ii $E_0 = 2NBlv = 2 \times 80 \times 0.13\,\text{T} \times 0.065\,\text{m} \times 6.0\,\text{m s}^{-1} = 8.1\,\text{V}$

4 a A back emf is induced in the motor when it spins. The faster the motor spins, the greater the back e.m.f so the motor current is small because the back emf acts against the battery e.m.f.

b When the load is increased, the motor spins slower so the back emf is smaller. The current increases because the back emf acting against the battery e.m.f is less.

17.4

1 a The alternating current in the primary coil creates an alternating magnetic field (via the core) in the secondary coil which induces an alternating e.m.f. in the secondary coil.

b The current in the secondary coil increases when a device is connected across it and this reduces the magnetic flux in the core and therefore reduces the back emf in the primary coil so the primary current increases.

2 a The induced e.m.f. in the secondary coil is proportional to the rate of change of magnetic flux through it. If the magnetic flux in it is less than in the primary coil due to the transformer design, the peak secondary e.m.f is less than it could be.

b Direct current in the primary coil of a transformer would not create changing magnetic flux so there would not be an induced emf in the secondary coil. Alternating current does create changing magnetic flux in a transformer so does cause an induced emf in the secondary coil.

3 a 11.5 V

b i 0.26 A **ii** 5.2 A

4 a Electrical power = current × voltage so the current needed to deliver a certain amount of power is much reduced if the voltage applied to the transmission cables is increased much more. The power dissipated in the cables due to the resistance heating effect of the current is therefore much reduced

b i 17 A (16.7 A to 3 sig. fig.) **ii** 56 kW

Chapter 18

18.1

1 The electric current in the circuit transfers energy from the battery to the heater which increases its store of thermal energy so it becomes hot. The heater transfers energy by heating the water which increases its store of thermal energy. The current also heats the circuit wires due to their resistance, causing energy transfer to the wires and the surroundings.

2 a Friction in the motor bearings and resistance heating due to the current in the wires cause the motor to become warm.
  b The current in the circuit transfers energy to the motor coil from the battery. The current in the coil does work on the coil to make it turn. In the process, energy is transferred from the coil to the weight which gains potential energy.
3 a Internal energy is the energy of its molecules due to their individual movements.
  b An electric lamp at constant brightness has energy transferred to it by the electric current in it and transfers energy at the same rate to the surroundings as it heats the surroundings and radiates light. So its internal energy is constant.
4 a The molecules in a solid are held to each other in fixed positions and they vibrate about these fixed positions whereas the molecules in a liquid move about at random but still in contact with each other.
  b The molecules vibrate more and more about their fixed positions as the solid's temperature increases to its melting point. At this temperature, substance changes from a solid to a liquid as the molecules break free from each other and move about at random state. Above this temperature, the molecules move about more and more as the temperature is increased.

18.2

1 a See 18.2 page 1
  b i 273 K ii 293 K iii 77 K
2 a 328 K b 137 kPa
3 The liquid thread does not expand by equal lengths for equal increases of temperature because the volume of the thermometer bulb changes with temperature.
4 Any ideal gas in a gas thermometer always gives the same temperature reading at any temperature not just at the fixed points.

18.3

1 a 23 kJ b 536 kJ
2 a 270 s b 10.3 MJ
3 a 320 J b 130 J kg^{-1} K^{-1}
4 3.2 kW

18.4

1 a The vibrating molecules in the solid need energy in order to increase their kinetic energies so they can break free from each other.
  b The internal energy of the water molecules needs to be reduced, so their kinetic energy is reduced so that they move slower and form strong bonds in fixed positions with each other.
2 0.16 kg
3 a 4.2 J s^{-1} b 6500 s
4 a 22 J s^{-1} b 6.5 kJ

Chapter 19

19.1

1 126 kPa
2 79 kPa
3 0.097 m^3
4 a Apply Boyle's law to the data without the powder present and with the final volume as $(1.20 \times 10^{-4}\,\text{m}^3 - V_P)$ where V_P is the volume of the pump.
  b $2.33 \times 10^{-5}\,\text{m}^3$, 1600 kg m^{-3}

19.2

1 a 1.1 moles b 109 kPa,
2 a 9.3×10^{-4} moles b $2.1 \times 10^{-5}\,\text{m}^3$
3 a Use the ideal gas equation to calculate the pressure at each temperature in kelvins. The temperature axis should be labelled in °C. A straight line should be drawn between the two plotted points.
  b 1.3 kg m^{-3},
4 a 1.2 kg m^{-3} b 2.5×10^{22}

19.3

1 Increasing the temperature of the gas causes the root mean square speed of the molecules to increase. Therefore the molecules have more momentum on average so their change of momentum on impact is greater. Also, because the container volume is constant, the molecules move faster on average so they collide with the container walls more frequently. Therefore, the mean force exerted by each molecule is greater.
2 524 m s^{-1}
3 a 7.52×10^{-3} b 474 m s^{-1}
4 a 1.48 moles b 4.2×10^{-2} kg c 5.2×10^2 m s^{-1}

19.4

1 a 0.97 moles b 7.7×10^{-21} J c 4.5 kJ
2 a 5.7×10^{-21} J b 1.8×10^3 m s^{-1}
3 a The mean kinetic energy of a molecule of an ideal gas depends only on the temperature of the gas. The gas temperature is the same throughout the gas so the mean kinetic energy of any gas molecule is the same as that of any other molecule.
  b In any sample of air, the mean kinetic energy of a nitrogen molecule is equal to that of an oxygen molecule in the same sample. The kinetic energy of a gas molecule is equal to $\frac{1}{2} m\, c^2_{\text{r.m.s.}}$, where m is its mass and $c^2_{\text{r.m.s.}}$ is its mean square speed. Therefore $\frac{1}{2} m_n (c_{\text{r.m.s.}})^2_n = \frac{1}{2} m_o (c_{\text{r.m.s.}})^2_o$ where the subscripts n and o are used respectively to denote the masses and mean square speeds of a nitrogen molecule and an oxygen molecule.
  Rearranging this equation gives

$$\frac{\text{the mean square speed of a nitrogen molecule, } (c_{\text{r.m.s.}})^2_n}{\text{the mean square speed of an oxygen molecule, } (c_{\text{r.m.s.}})^2_o} = \frac{\text{mass of an oxygen molecule, } m_o}{\text{mass of a nitrogen molecule, } m_n}$$

The ratio of molecular masses m_o/m_n = the ratio of the molar masses $= \frac{0.032}{0.028} = 1.14$
Therefore the ratio of mean square speeds = 1.14
Hence the r.m.s. speed of a nitrogen molecule = $1.14^{0.5}$ = 1.07 × the r.m.s. speed of an oxygen molecule.

4 a i 1.50 **ii** 220 K **b** 870 J **c** 1310 J **d** 2180 J

Chapter 20

20.1

1 a The emission of electrons from the surface of a metal when light above a certain frequency is directed at the surface.
b The work function of a metal is the minimum amount of energy an electron at the surface of a metal needs to leave the surface. Light consists of photons each of energy equal to hf where f is the frequency of the light. When light is directed at the surface of a metal, an electron at the surface could absorb a photon and leave the metal surface if the energy it gains from the photon is greater than the work function of the metal.

2 a i 6.7×10^{14} Hz, 4.4×10^{-19} J
ii 2.0×10^{14} Hz, 1.3×10^{-19} J
b The energy of a 450 nm photon is greater than the work function of the metal, so if an electron at the metal surface absorbs a 450 nm photon, the electron can leave the surface. An electron at the metal surface that absorbs a 650 nm photon could not leave the surface as a 650 nm photon has less energy than the work function.

3 a 1.7×10^{14} Hz **b** 2.7×10^{-19} J

4 a 3.1×10^{-19} J **b** 1.6×10^{-19} J **c** 2.5×10^{14} Hz

20.2

1 a The blue light had a frequency greater than the threshold frequency of the metal so the energy of each of its photons was greater than the work function of the metal. Hence an electron at the metal surface could leave the surface if it absorbed a blue photon.
b The red light had a frequency below the threshold frequency of the metal so the energy of each of its photons was less than the work function of the metal . Hence an electron at the metal surface that absorbed a red photon would not have had sufficient energy to leave the surface and it would have quickly lost the energy it gained in collisions with other electrons no matter how many photons were incident on the surface.

2 a 1.6×10^{12}
b The number of electrons per second leaving the surface (ie the photoelectric current) is proportional to the number of photons per second incident on the surface. So if the intensity is doubled, the number of photons per second incident on the surface is doubled and hence the photoelectric current is doubled.

3 a 3.4×10^{-19} J
b 1.5×10^{15}
c 2.5×10^{12}

4 a i 4.0×10^{14} Hz **ii** 2.7×10^{-19} J,
b 2.7×10^{-19} J

20.3

1 a 9.0 eV
b There are three energy levels below the 5.7 eV level. There are three possible transitions to the ground state, two to the first excited state and one to the second excited state making six transitions in total.

2 a 1. The energy of an electron in an atom increases in an excitation and decreases in a de-excitation.
2. Excitation can occur through photon absorption or electron collision. De-excitation only occurs through photon emission.
b i The diagram should show the ground state and two excited levels at 1.8 eV and 4.6 eV
ii photon energies 1.8 eV, 2.8 eV, 4.6 eV

3 The energy levels of each type of atom are unique to that atom. So the photons emitted are characteristic of that atom.

20.4

1 a Light has wave-like properties, such as diffraction. It consists of wave packets called photons, which have particle-like properties such as in the photoelectric effect, in which a single photon is absorbed by a single electron.
b Matter particles transfer momentum when they collide hence have particle-like properties and they also have wave-like properties, for example they can be diffracted by thin crystals.

2 a The particle-like nature of light.
b The wave-like nature of particles.

3 a 3.6×10^{-11} m **b** 1.9×10^{-14} m

4 a 1.3×10^{-27} kg m s^{-1}, 1.5×10^{3} m s^{-1}
b 1.3×10^{-27} kg m s^{-1}, 0.78 m s^{-1}

Chapter 21

21.1

1 a 73 s **b** 6 g

2 a i 9.0×10^{14} **ii** $2.2(5) \times 10^{14}$
b i 9.0×10^{14} **ii** 15.8×10^{14} **c** 1.3×10^{3} J

3 a 19 kBq **b** 2.4 kBq

4 a 4.4×10^{21} **b** 1.1×10^{21} atoms

21.2

1 a 1.0×10^{-6} s^{-1} **b** 3.9×10^{16}

2 a 6.3×10^{-10} s^{-1} **b** 20.5 kBq

3 a 2.7×10^{24} **b i** 0.65 kg **ii** 1.7×10^{24}

4 a 1.3×10^{-6} s^{-1} **b** 149 hours

21.3

1 a The carbon dioxide content of atmosphere contains a small percentage of radioactive carbon isotope $^{14}_{6}C$. Living wood contains this isotope because trees and plants absorb carbon dioxide from the atmosphere.
b 1340 years

2 a i 730 kBq ii 870 kBq iii 1.9×10^{-13} kg
b The flow through the first kidney is normal because the radioactive isotope flows through it. The second kidney is blocked because the radioactive isotope enters it but does not leave it.

3 a i The foil is made by using rollers either side of a metal plate to squeeze the plate as it passes between the rollers. The foil then passes between a source of beta radiation and a detector which supplies a feedback signal to the rollers. If the foil becomes too thin, the amplitude of the signal increases and the pressure of the rollers on the foil is reduced. The opposite happens if the foil becomes too thick.
ii The intensity of the beta radiation passing through the foil decreases with increasing thickness of the foil. Alpha radiation would be absorbed by the foil. Gamma radiation would almost all pass through the foil.
b i Gamma radiation is absorbed by thick lead. The thick lead lining in the container prevents gamma radiation from the source harming people near the container.
ii A narrow hole in the lead container allows a narrow beam of gamma radiation to be directed at the tumour in the patient in order to destroy it. Directing the beam at the tumour from different directions reduces the damage caused by gamma radiation to the normal tissues outside the tumour.

4 a i Human tissues are damaged by alpha, beta, and gamma radiation. An alpha source would be most suitable as alpha radiation from the source inside the pacemaker would all be absorbed inside the pacemaker and would not reach the tissues outside the pacemaker. Radiation from a beta or gamma source would reach the tissues outside the pacemaker and damage them.
ii The longer the half-life of a radioactive isotope is, the less its activity per mole is. A greater amount of the radioactive isotope in the pacemaker would therefore be needed to provide sufficient power if a source with a much longer half-life was used.
b i 5.2×10^{13} Bq ii $3.3\,J\,s^{-1}$

21.4

1 a 2.18×10^{-15} kg
b i 8.89×10^{-33} kg ii 8.89×10^{-30} kg

2 a $^{212}_{83}Bi \rightarrow {}^{208}_{81}Tl + {}^{4}_{2}\alpha$; 6.2 MeV
b When the alpha particle is emitted, the nucleus recoils with equal and opposite momentum. The energy released in the process is shared between the nucleus and the alpha particle in inverse proportion to their masses. So the nucleus gains a small proportion of the energy released.

3 a $^{90}_{38}Sr \rightarrow {}^{90}_{39}Y + {}^{0}_{-1}\beta + \bar{\nu}$; 0.56 MeV
b When the beta particle is emitted, a neutrino or an antineutrino is emitted so the energy released is shared between the nucleus and the two emitted particles. The nucleus takes a small proportion of the energy released and the rest is shared between the two emitted particles so the beta particle's kinetic energy can be any value from zero to a maximum.

4 3.2 MeV

21.5

1 a The binding energy is the work that must be done to separate a nucleus into its constituent neutrons and protons.
b See 21.5 Figure 1, without the individual points and unnecessary labels.

2 a 7.4 MeV b 8.8 MeV

3 a i 7.1 MeV ii 2.6 MeV
b The binding energy per nucleon of an alpha particle is greater than that of the other particle. Both particles need kinetic energy to escape from the nucleus. An alpha particle has more kinetic energy after it forms than the other particle would have because more binding energy is released when the alpha particle forms

4 1.1 MeV

21.6

1 a They do not have sufficient kinetic energy to overcome the strong nuclear force holding them in the nucleus.
b When a nucleus is formed from separate proton and neutrons, the strong nuclear force between them binds them together and their potential energy decreases so their total mass decreases in accordance with $E = mc^2$.

2 a Nuclear fission is the process that occurs when a large unstable nucleus splits into two approximately equal fragments and energy is released. This can happen spontaneously or can be induced in certain nuclei when a free neutron collides with such a nucleus.
b i $a = 56$, $b = 98$ ii 206 MeV

3 a Nuclear fusion is the process that occurs when two nuclei combine to form a larger nucleus and energy is released. This can only happen if the nucleon number of the nucleus formed is no greater than about 50.
b i In the first interaction, one of the two protons changes into a neutron and a positron is released. The neutron and the other proton form a hydrogen $^{2}_{1}H$ nucleus. In the second interaction, a third proton collides with the hydrogen $^{2}_{1}H$ nucleus to form a $^{3}_{2}He$ nucleus.
ii 0.43 MeV, 5.5 MeV

4 **a** The nuclei need to have sufficient kinetic energy to overcome the mutual coulomb repulsion force between them so that they can approach each other closely enough for the strong nuclear force between them to attract them together.

b 12.9 MeV

c The mass difference between 4 separate protons and an alpha particle and 2 released positrons $= (4 \times 1.00728\,u) - (4.00150\,u + 0.00110\,u)$ $= 2.65 \times 10^{-2}\,u$. Therefore the energy released $= 2.65 \times 10^{-2}\,u \times 931.5\,\text{MeV/u} \approx 25\,\text{MeV}$

Chapter 22

22.1

1 **a** **i** 0.14 mm **ii** 0.62 mm

b At this frequency, their wavelength in the body is significant compared with the transducer and with body structures so diffraction is significant. As a result, they spread out and weaken too much so they do not form distinct images.

2 **a** **i** See Topic 22.1, Figure 1 and related notes

ii The backing block prevents ultrasonic waves created at the two surfaces of the disc from cancelling each other out. The pad also damps the vibrations of the disc rapidly after each pulse is emitted.

b **i** A is due to reflection at the cornea; B is due to reflection at the front surface of the eye lens; C is due to reflection at the back surface of the eye lens; the furthest pulse (at the right side of the screen) is due to reflection at the retina.

ii about 13 mm

3 **a** **i** 0.999 **ii** 1.73×10^{-3}

b Gel and skin have similar acoustic impedances so a gel–skin interface hardly reflects any ultrasonic waves from the probe. If the gel was not used, trapped air between the probe and the body would reflect most of the ultrasonic waves because air and skin have very different acoustic impedances.

c **i** $1.64 \times 10^{6}\,\text{kg}\,\text{m}^{-2}\,\text{s}^{-1}$ **ii** 1.6×10^{-5}

4 **a** A B-scan gives a two-dimensional image whereas an A-scan only gives information about the distances from the probe to reflecting boundaries in one direction. A B-scan requires a multi-transducer probe whereas an A-scan uses a probe with a single transducer.

b Ultrasonic waves are non-ionising unlike X-rays so they would not harm the baby whereas X-rays might.

22.2

1 **a** **ii** 0.15 W

b **i** The intensity increases because more electrons strike the anode each second so more X-ray photons are produced.

ii The beam becomes more penetrating because the maximum energy of the X-ray photons in the beam is increased. This happens because the electrons strike the anode with more kinetic energy when the tube voltage is increased.

2 **a** **i** Sharpness is to do with how clearly the edges of structures in an X-ray image can be seen.

ii Contrast is to do with the relative darkening of an X-ray film in different areas of the film. Poor contrast occurs if the lightest areas are not much lighter than the darkest areas.

b A scattering grid placed between the patient and the film prevents X-rays scattered in the body of the patient from reaching the film. Without the grid, such X-rays would darken the shadow areas of the film which would reduce the contrast and sharpness of the image.

3 **b** A contrast medium is necessary when an X-ray image of a soft body organ such as the stomach is being created. If the organ is filled with the contrast medium, it absorbs X-rays more effectively. This improves the contrast of the image so the edges and structures of the image can be seen more clearly.

4 **a** See Topic 22.2, Figure 8 for a suitable diagram. The X-ray tube produces a narrow beam which is detected by electronic detectors on the other side of the patient. The tube and the detectors are on a gantry which is rotated in steps about the patient so that the beam always passes through the same cross-section of the patient. The detector signals are recorded at each step and used to construct a cross-sectional image of the patient.

b **i** 12 **ii** 3, 1, 3, 5

22.3

1 **a** ${}^{18}_{8}\text{O} + {}^{1}_{1}\text{p} \rightarrow {}^{18}_{9}\text{F} + {}^{1}_{0}\text{n}$

b ${}^{18}_{9}\text{F} \rightarrow {}^{18}_{8}\text{O} + {}^{0}_{1}\beta + \nu$

2 **a** No. of nuclei N = activity A/decay constant λ $= A \times \text{half-life}/\ln 2 = 220 \times 10^{6} \times 110 \times 60\,\text{s}/\ln 2$ $= 2.09 \times 10^{12}$ nuclei. Mass of each nucleus $= 0.018\,\text{kg}/6.02 \times 10^{23} = 3.00 \times 10^{-26}\,\text{kg}$. Mass of N nuclei $= 2.09 \times 10^{12} \times 3.00 \times 10^{-26}\,\text{kg} = 6.2 \times 10^{-14}\,\text{kg}$

b $A = 220 \times 10^{6} \times e^{-(\ln 2/110 \times 60) \times (24 \times 60)} = 25\,\text{kBq}$ (Alternative method: No. of half-lives = 24×60 mins/110 mins = 13.1 half-lives. Therefore activity $= 220\,\text{MBq}/2^{13.1} = 25\,\text{kBq}$

3 **a** The positron is the antiparticle of the electron. When a positron-emitting nucleus decays, it emits a positron which is annihilated when it collides with an electron. The loss of mass and energy is converted into radiation energy in the form of gamma photons.

b Two gamma photons are produced in each annihilation event. Because momentum is conserved and the total momentum before the event is negligible, the two photons are created with equal and opposite momentum. Because energy is conserved and they have equal and opposite momentum, they each have the same energy after the event.

4 a A radioactive tracer is a radioactive emitter of gamma (or beta) radiation attached to the molecules of a fluid introduced into a system to obtain information about the system by using external detectors to detect the emissions.

b A PET scanner is designed to locate positron emissions from positron-emitting substances by using a ring of detectors to detect two gamma photons emitted in opposite directions within a very short time of each other. Other gamma-emitting isotopes emit gamma photons in random directions so are unlikely to trigger the PET detectors.

Chapter 23

23.1

1 For Proxima Centauri, $F = L/4\pi d^2$
$= 10\,000\,L_{SUN}/4\pi\ (250\,000\ d_{SUN})^2$
$= \{10\,000/(250\,000)^2\} \times (L_{SUN}/4\pi\, d_{SUN}^2)$
$= 1.6 \times 10^{-7} F_{SUN}$. Therefore, for Proxima Centauri, $F = 1.6 \times 10^{-7} \times 1400\,\text{W m}^{-2} = 0.22\,\text{mW m}^{-2}$.

2 a 2.7×10^8 s

b For Sirius, $F = L/4\pi d^2$
$= 4.0 \times 10^{26}\,\text{W}/(4\,\pi \times (8.1 \times 10^{16}\,\text{m})^2$
$= 4.9 \times 10^{-9}\,\text{W m}^{-2}$

3 Rearranging $F = L/4\,\pi\, d^2$ hence $d^2 = L/4\pi F$
$= 4.4 \times 10^{36}\,\text{W}/(4\,\pi \times (2.1 \times 10^{-13}\,\text{W m}^{-2}))$
$= 1.7 \times 10^{48}\,\text{m}^2$. Therefore $d = 1.3 \times 10^{24}$ m.

4 Since F is inversely proportional to the square of the distance from the Earth to the Sun, a change of F by 1.7% is due to a change of distance of about 3.4% (= 2 × 1.7%). Therefore the distance changes by about 5.1×10^9 m (= 3.4% × 1.5×10^{11} m)

23.2

1 Rearranging $\lambda_{max}\, T = 0.0029$ m K gives
$T = 0.0029\,\text{m K}/620 \times 10^{-9}\,\text{m} = 4700\,\text{K}$

2 a Rearranging $L = \sigma A T^4$ gives $A = L/\sigma T^4$

Hence $A = \dfrac{6.0 \times 10^{28}}{5.67 \times 10^{-8} \times (3400)^4} = 7.9 \times 10^{21}\,\text{m}^2$

b i For a sphere of radius R, its surface area $A = 4\pi R^2$ Rearranging this equation gives

$R^2 = \dfrac{A}{4\pi} = \dfrac{7.9 \times 10^{21}}{4\pi} = 6.3 \times 10^{20}\,\text{m}^2$.

Hence $R = 2.5 \times 10^{10}$ m

ii Ratio of radius to Sun's radius $= \dfrac{2.5 \times 10^{10}\,\text{m}}{7.0 \times 10^{8}\,\text{m}} = 36$

3 Since the surface area is proportional to the diameter squared, the surface area of the star is 16 × the surface area of the Sun. Therefore

$\dfrac{L_{star}}{L_{Sun}} = \dfrac{A_{star}\, T_{star}^4}{A_{Sun}\, T_{Sun}^4} = \dfrac{16 A_{Sun}\,(2T_{Sun})^4}{A_{Sun}\, T_{Sun}^4} = 16 \times 2^4 = 256 \approx 250.$

4 a Since they have the same surface temperature and X has greater luminosity than Y, the surface area of X and therefore its diameter must be greater than that of Y.

b $L_X = \sigma A_X T_X^4 = 5.67 \times 10^{-8}\,\text{W m}^{-2}\,\text{K}^{-4} \times 4\pi \times (0.5 \times 2.0 \times 10^9\,\text{m})^2 \times (5400\,\text{K})^4 = 6.0 \times 10^{26}\,\text{W}$

c Since $T_X = T_Y$, $\dfrac{L_X}{L_Y} = \dfrac{\sigma\, A_X\, T_X^4}{\sigma\, A_Y T_Y^4} = \dfrac{A_X}{A_Y}$, therefore $A_X / A_Y = L_X / L_Y = 100$.
Hence the ratio of the square of their diameters is 100 so the diameter of X is 10 × the diameter of Y. Hence the diameter of Y is 2.0×10^8 m.

23.3

1 a Hubble's law states that the distant galaxies are receding and that their speed of recession is proportional to their distance from Earth.

b The Universe is expanding.

2 a $v = c\,(\Delta\lambda / \lambda) = 300\,000\,\text{km s}^{-1} \times 7.6\,\text{nm}/589.6\,\text{nm}$
$= 3900\,\text{km s}^{-1}$

b Rearranging $v = Hd$ gives $d = \dfrac{v}{H} = \dfrac{3.9 \times 10^6\,\text{ms}^{-1}}{2.2 \times 10^{-18}\,\text{s}^{-1}}$
$= 1.8 \times 10^{24}$ m

3 a i $v = c(\Delta\lambda / \lambda) = 300\,000\,\text{km s}^{-1} \times 39\,\text{nm}/486\,\text{nm}$
$= 24\,000\,\text{km s}^{-1}$

ii Rearranging $v = Hd$ gives $d = \dfrac{v}{H} = \dfrac{2.4 \times 10^7\,\text{ms}^{-1}}{2.2 \times 10^{-18}\,\text{s}^{-1}}$
$= 1.1 \times 10^{25}$ m

b $L = F \times 4\pi d^2$
$= 2.7 \times 10^{-15}\,\text{W m}^{-2} \times 4\pi \times (1.1 \times 10^{25}\,\text{m})^2$
$= 4.1 \times 10^{36}$ W

4 a CMBR is background microwave radiation that is detected from all directions in space.

b CMBR is thought to be electromagnetic radiation that has been travelling through space after being created in the Big Bang when the Universe was created in a massive explosion.

Chapter 24

24.1

1 a i 0.500 m ii 320 cm iii 95.6 m
b i 450 g ii 1.997 kg iii 5.4×10^7 g
c i $2.0 \times 10^{-3}\,\text{m}^2$ ii $5.5 \times 10^{-5}\,\text{m}^2$ iii $5 \times 10^{-6}\,\text{m}^2$

2 a i 1.5×10^{11} m ii 3.15×10^7 s
iii 6.3×10^{-7} m iv 2.57×10^{-8} kg
v 1.50×10^5 mm vi 1.245×10^{-6} m
b i 35 km ii 650 nm iii 3.4×10^3 kg
iv 870 MW (= 0.87 GW)

3 a i 20 m s^{-1} ii 20 m s^{-1} iii $1.5 \times 10^8\,\text{m s}^{-1}$
iv $3.0 \times 10^4\,\text{m s}^{-1}$
b i $6.0 \times 10^3\,\Omega$ ii 5.0 Ω iii $1.7 \times 10^6\,\Omega$
iv $4.9 \times 10^8\,\Omega$ v 3.0 Ω

4 a i 301 ii 2.8×10^9 iii 1.9×10^{-23}
iv 1.2×10^{-3} v 2.0×10^4 vi 7.9×10^{-2}
b i 1.6×10^{-3} ii 5.8×10^{-6} iii 1.7 iv 3.1×10^{-2}

24.2

1 a 1.57 m
b i 1.57 m ii 1.05 m iii 0.26 m iv 0.47 m
c i 35 km ii 31°

2 a The angle subtended in degrees should be equal to 0.115 × the diameter in mm. b ii 0.5°

3 **a** **i** 0.035 **ii** 0.140
b **i** 72 mm **ii** 72 mm **iii** 148 mm
c **i** $B = 74°$, $C = 46°$ **ii** $B = 78°$, $C = 32°$
iii $B = 45°$, $C = 15°$

4 **a** **i** 3.9 N, 4.6 N, **ii** 3.4 N, 9.4 N **iii** 4.8 N, 5.7 N
b 4.8 N at 21° to the 3.5 N

24.3

1 **a** **i** 0.2 **ii** 0.1, 0.25 **iii** <0.1
b **i** $t = \frac{v - u}{a}$ **ii** $t = \sqrt{\frac{2s}{a}}$
iii $t = \frac{y}{k} + t_0$ **iv** $t = \frac{mv}{F}$
c **i** 2 **ii** −1 **iii** 4.25 **iv** $\frac{1}{3}$

2 **a** **i** $x = \frac{y^2}{4}$ **ii** $x = \frac{1}{4y^2}$
iii $x = \frac{1}{y^3}$ **iv** $x = \sqrt{\frac{k}{y}}$
b **i** 0.25 **ii** ±2.8 **iii** ±0.5 **iv** 4

3 **a** **i** $3.1 \times 10^{-7}\,m^2$ **ii** $6.2 \times 10^{-3}\,m$
iii 2.5 s **iv** $17\,m\,s^{-1}$
b **i** $kg\,m^2\,s^{-2}$ **ii** kg **iii** s

4 **a** **i** ½ or −3 **ii** 1.44 or 5.56 **iii** 1 or −1.67
b **i** $t = -\frac{20}{6}$ s or 2 s **ii** $R = 1.0\,\Omega$ or $4.0\,\Omega$

24.4

1 **a** **i** 3 **ii** −3 **iii** 1
b **i** −4 **ii** 8 **iii** 2
c **i** −1 **ii** 5 **iii** 5
d **i** −1.5 **ii** 3 **iii** 2

2 **a** **i** $y = 2x - 8$ **ii** −8
b **i** $3\,m\,s^{-2}$ **ii** $5\,m\,s^{-1}$

3 **a** (1, 4) **b** $y = 4x$

4 **a** $x = 2$, $y = 0$
b $x = 3$, $y = 5$
c $x = 2$, $y = 0$

24.5

1 **a** See 1.3 Fig 2
b **i** gradient **ii** area under line

2 **a** See 6.5 Fig 3a
b Resistance $= \frac{1}{\text{gradient}}$

3 **a** y = energy, x = time; the graph should be a straight line with a positive constant gradient from the origin.
b The power is constant and is represented by the gradient of the line.

4 **a** **i** As end of chapter 2, Q11 Fig 11.1, with the line becoming flat and the y-axis labelled 'velocity'.
ii The graph is an exponential decrease curve similar to 22.2 Figure 7 with the *y*-axis labelled acceleration and the *x*-axis labelled time.
b **i** **1** acceleration **2** distance fallen
ii velocity

24.6

1 **a** **i** 0.477 **ii** 1.176 **b** **i** 1.653 **ii** 0.699

2 **a** **i** 10.8 dB **ii** 7.0 dB **b** 17.8 dB

3 **a** $n = 5$, $k = 3$ **b** $n = 3$, $k = \frac{1}{2}$ **c** $n = 2$, $k = 1$

4 **a** **i** 1.10 **ii** 2.71 **b** **i** 3.81 **ii** 1.61

24.7

1 **a** **i** 2, 3 **ii** 12, 0.2 **iii** 4, 0.02
b **i** 0.23 s **ii** 3.5 s **iii** 35 s

2 **a** 11.3 kBq **b** 9.0 kBq **c** 0.38 kBq

3 **a** **i** 2.20 s **ii** 1.52 s **b** **i** 4.83 V **ii** 1.24 V

4 **a** **i** 0.14 s **ii** 82 **b** $a = 6.9$, $b = -5$

Chapter 1

1 **a i** A vector has magnitude and direction. A scalar has magnitude only.
ii velocity (or acceleration) **iii** mass or volume
b 7.8 N

2 **a** 1.48×10^3 s **c** 15.8 m s^{-1}

3 **a** 40 m s^{-1} **b** 80 m s^{-1} **c** 2.0 m s^{-2}

4 **b** 0.375 m s^{-2}, 0, −0.75 m s^{-2}; 300 m, 900 m, 150 m
c 11.25 m s^{-1}

5 **a** 24 m **b** 22 m s^{-1}

6 **a i** 10 m **ii** −0.20 m s^{-2} **b iii** 0.80 m s^{-1} downhill

7 **a i** 14.7 m s^{-1} **ii** 12.6 m **iii** 15.7 m s^{-1} downwards

8 **a** 20 m s^{-1} **c** 5.2 m

9 **a i** 150 m s^{-1}, 1.88 km **b i** 25 s **ii** 243 m s^{-1}

10 **a** 1.31 m **b i** 3.62 m **ii** 23.9 m s^{-1}, 16° below the horizontal
c i The speed on impact would have been the same because its loss of potential energy (from start to finish) would have been unchanged so its gain of kinetic energy would be the same.
ii The angle (A) of its direction to the horizontal would be greater because the ratio of its vertical component of velocity to its horizontal component (which is equal to tan A) would be greater as the sum of the squares of these components is unchanged (equal to its speed squared) and its horizontal component is slightly less so its vertical component is greater than before. (OR Because it would be in the air for a longer time and its vertical displacement from maximum height would be greater. Therefore, its vertical component of velocity would be greater at impact and its horizontal component of velocity would be slightly less.)

Chapter 2

1 **a** Friction acts on the object as it slides on the floor. The force pushing the object must be equal and opposite to the frictional force to maintain steady speed in a constant direction.
b Friction is absent on ice so no applied force is needed to maintain constant velocity.

2 **a** 0.25 m s^{-2} **b** 110 N **c** 0.026

3 **a** 1.5 ms **b** -7.8×10^4 m s^{-2} **c** 190 N

4 **a i** 3.3×10^4 kg **ii** 0.18 m s^{-2} **iii** 110 s **iv** 1.1 km
b i 0.15 ms^{-2}, **ii** 129 s

5 **a i** the gradient of the line,
ii The speed increased from zero and gradually became constant.
iii

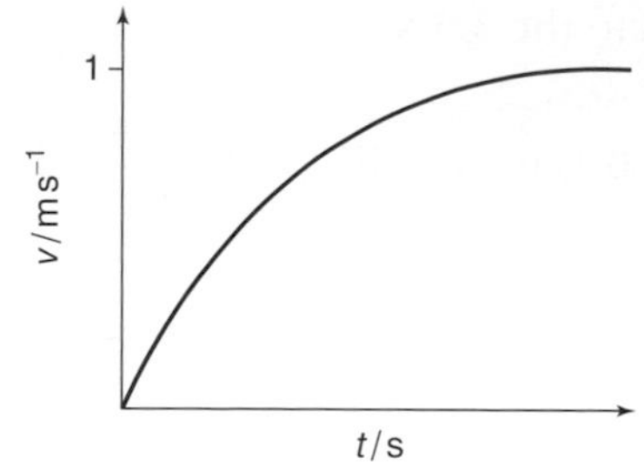

Figure 1

b As its speed increased, the drag force opposing its motion gradually increased. Therefore the resultant force which is the difference between its weight and the drag force gradually decreased until it became zero and the velocity is then constant.

6 **a** 1.7 kJ **b** 38 m s^{-1} **c** 73 m

7 **a** 61 J **b** 61 J **c** 7.0 m s^{-1}

8 **a i** 4.7 J **ii** 4.7 J **iii** 4.8 ms^{-1}
b i 3.1 J **ii** 3.1 J **iii** 4.0 ms^{-1}

9 **a i** 6.9 MJ **ii** 7.9 MJ **b i** 740 kW **ii** 0.99 kg

10 **a** 1.8 kN **b** 2.3 kN **c i** 16 kW **ii** 3.5 kW
d KE gain each second

11 **a i** Resultant force = weight (or mg) – resistive force but resistive force = 0 initially (as initial speed = 0). So resultant force initially = mg hence initial acceleration = resultant force/mass = $mg/m = g$ = 9.81 m/s^2.
ii Graph of acceleration (a) on the y-axis and time (t) on the x-axis; curve with a negative decreasing gradient starting at a = 9.8 m s^{-2} (or g) at t = 0; non-zero gradient at t = 5.6 s (or object does not reach terminal (or constant speed).
iii Resistive force of air on object increases as object's speed increases so the resultant force on the object (= weight – resistive force) decreases with increasing speed. Since acceleration = resultant force / mass, acceleration decreases as the object's speed increases so the object's rate of increase of speed decreases.
b i Weight = 8.2(4) N
ii Acceleration, a, at midpoint = gradient of speed–time graph at t = 2.8 s gives a = 4.0 to 4.5 m s^{-2}, so resultant force = 3.4 to 3.8 N.
c i Kinetic energy gained = 306 J
ii Distance fallen estimated from area under the curve = 100 to 105 m; Loss of potential energy = 845 J [830 to 860 J acceptable]
iii Work done against air resistance = loss of P.E. – gain of K.E. = average resistive force × distance fallen so average resistive force = (loss of P.E. – gain of K.E.)/distance fallen = 5.0 N to 5.5 N (using values above)

Chapter 3

1 a 2.5 kN b 4.5 kN
2 a i 270 N ii 225 N
b The moment of the wind force about the direction of motion tends to overturn the boat. By leaning as shown, the weight of the crew provides an equal and opposite moment to prevent overturning.
3 a i 570 N ii 100 N b i 540 N ii 6180 N
c 30 N down theslope
4 7.5 kN, 6.5 kN
5 a i The point where the weight of the body may be considered to act.
ii force × perpendicular distance from the line of action of the force to the point about which moments are being considered.
b i 60 N
6 b 12.7 N
7 a 152 N, 106 N
b The paraglider is acted on by three forces which are the tension T in the cable, the weight W of the paraglider and the force F of the parachute. F cannot be in the opposite direction to T because the three force vectors form a triangle with a vertical side representing W. See Topic 3.5 Figure 1.
8 a 130 kN b 147 kN, 113 kN
9 a i A *couple* is a pair of equal and opposite forces acting on a body but not along the same line.
ii The *torque* of a couple is the magnitude of one force multiplied by the perpendicular distance between the lines of action of the forces.
b i Torque = force × distance therefore force = torque/perpendicular distance = 84 N m/0.44 m = 190 N
ii Perpendicular distance (from the axis of rotation of the nut) to the line of action of the force would be less than 0.44 m so the force would need to be greater to give the same torque. OR Force has a component parallel to the handle and a component perpendicular to the handle. Perpendicular component would need to be 190 N to achieve the same torque so magnitude of the force would need to be greater than 190 N.

Chapter 4

1 a 140 N s b 20 m s^{-1}
2 a 9000 kg m s^{-1} b 60 s
3 b 19.2 m s^{-1} c 4.8 N
4 a 7.5×10^{-23} kg m s^{-1} b 1.5×10^{-19} N
5 a 2.6×10^{-23} kg m s^{-1} b 1.3×10^{-20} N
6 a 1.5 m s^{-1} b i 2.7 kJ
b ii kinetic energy of the wagons is changed to sound energy and heat energy of the colliding parts.
7 a 24 m s^{-1} b i 163 kJ ii 91 kJ iii 72 kJ
8 a 2.5×10^5 m s^{-1} b i 1.3×10^{-14} J ii 7.5×10^{-13} J
9 a Duration of impact = 0.68 s (±0.2 s)

$$\text{Impact force} = \frac{\text{change of momentum}}{\text{time taken}} = \frac{25\,000\,\text{kg} \times 1.2\,\text{m s}^{-1}}{0.68\,\text{s}} = 37\,\text{kN}$$

b i For a system of interacting objects, the total momentum remains constant provided no external resultant force acts on the system.
ii The total momentum is conserved, the total final momentum = the total initial momentum
Hence (25 000 kg × 1.2 m s^{-1}) + (15 000 kg × v) = 27 000 kg m s^{-1}, where v = velocity of P after the impact
$15\,000v = 27\,000 - (25\,000 \times 1.2)$
$= 27\,000 - 30\,000 = -3000$

$$\text{so } v = \frac{-3000\,\text{kg m s}^{-1}}{15\,000} = -0.20\,\text{m s}^{-1}$$

c i A *perfectly elastic collision* is one in which there is no loss of kinetic energy.
ii For P, its loss of K.E. = initial K.E. – final K.E. = 0.5 × 15 000 kg × (1.8 m s^{-1})^2 – 0.5 × 15 000 kg × (–0.2 m s^{-1})^2 = 24.3 kJ – 0.3 kJ = 24 kJ
For Q, its gain of K.E. = 0.5 × 25 000 kg × (1.2 m s^{-1})^2 = 18 kJ
iii The collision is not elastic because P has lost more kinetic energy than Q has gained, so the total kinetic energy has decreased.
10 a i Before emission, the nucleus was stationary so it had no momentum. No external forces acted on the nucleus so, in accordance with the principle of conservation of momentum, the total momentum was also zero immediately after the emission. Therefore the momentum of the nucleus after emission must be equal in magnitude and opposite in direction to the momentum of the α-particle.
ii The nucleus and the α-particle have different masses and, since momentum is mass × velocity, their velocities must differ in magnitude and hence their speeds must differ.
b i 1.6×10^7 m s^{-1} ii speed = 3.1×10^5 m s^{-1}; K.E. = 1.6×10^{-14} J
c i 2.4 ns ii 3.3×10^{15} m^2 s^{-2}
11 a i The force of the rubber band on each trolley changes but at any instant the rubber band pulls on each trolley with the same amount of force (but in opposite directions) even though the force changes. The duration of the force on each trolley is the same. Since the force on each trolley is equal to its rate of change of momentum, the two trolleys gain an equal magnitude of momentum but in opposite directions. (OR The total initial momentum is zero so the total final momentum at any instant is zero. Therefore the two trolleys gain equal and opposite momentum.)

ii The two trolleys gain equal and opposite momentum and, since momentum is mass × velocity, the velocity of the trolley with lower mass must be greater in magnitude than that of the other trolley. So, at any instant, the trolley with the lower mass must have a different (or greater) speed than that of the other trolley.

b KE = $\frac{1}{2}mv^2 = \frac{1}{2}(mv)^2/m = \frac{1}{2}$ (momentum)^2/mass. At any instant, their momentum is equal in magnitude so the trolley with the lower mass must have more kinetic energy than the other trolley.

Chapter 5

1 a $2.0 \times 10^{-7}\,m^3$ **b** $1.8 \times 10^{-3}\,kg$ **c** $7.2 \times 10^{-4}\,kg\,m^{-1}$

2 a 28 kN **b i** 52 kJ **ii** 88 kJ **iii** P.E. of the platform

3 a i $33\,N\,m^{-1}$ **ii** 480 mm **b** 1.5 N

4 a See Topic 5.6. **b** 58 N

5 a i The limit to which it can be stretched and regain its original length when the applied force is removed,

ii Plastic behaviour occurs when the strip is stretched beyond its elastic limit so it does not regain its original length when the stretching force is removed.

b i See Topic 5.6 Figure 2 **ii** As each part of the tyre rolls over the road surface, some of the work done when that part of the tyre is squashed and stretched increases the internal energy of the rubber molecules so the tyre becomes warm.

6 a 5200 N **b i** $1.0 \times 10^7\,Pa$

ii $e = \frac{Fl}{AE}$, where F = 4500 N and the values of A, E and l are given

iii 15 J

7 a $9.4 \times 10^{10}\,Pa$ **b** $1.0 \times 10^{-2}\,J$

8 a i The extension beyond which the spring would not return to its original length when it is unloaded.

ii The tension (or force) per unit extension needed to stretch the spring, provided its limit of proportionality is not exceeded.

iii $\frac{1}{2}W^2/k$

b P **i** $\frac{3e}{2}$ **ii** $\frac{2k}{3}$ Q **i** $\frac{e}{3}$ **ii** $3k$

9 a 10.0 J (± 0.1 J) **b** stiff at first then much less stiff, then stiffer but not as stiff as at the start

10 a i The water exerted an upthrust on the cylinder so the support force on the cylinder from the newton-meter was reduced from 1.95 N. There was an equal and opposite force to the upthrust on the water from the cylinder. The weight of water in the beaker was unchanged. So the force on the top-pan balance increased.

ii 0.15 N

b i

Upthrust, *U*/N	0	0.03	0.07	0.10	0.13	0.15

ii See Figure 2.

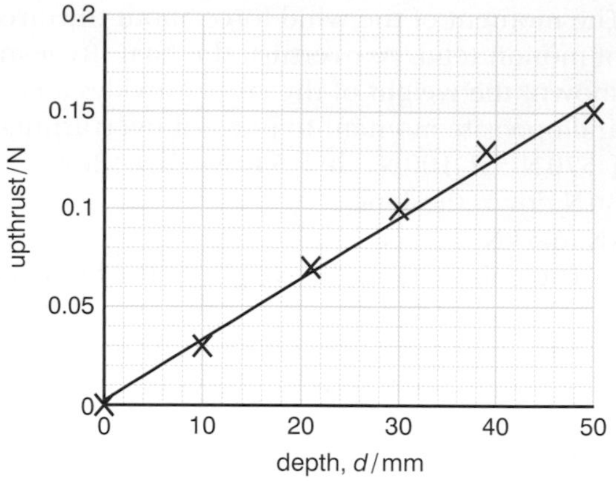

Figure 2

iii Gradient = 0.157 N/50 mm = $3.14 \times 10^{-3}\,N\,mm^{-1}$

c i At a depth d of 80 mm, the upthrust would be equal to the graph gradient × 0.080 m. Hence the upthrust at this depth = 80 mm × $3.14 \times 10^{-3}\,N\,mm^{-1}$ = 0.25 N (to 2 significant figures)

ii The weight of the cylinder is 1.95 N and its length is 80 mm. The weight of water displaced at d = 80 mm is equal to the upthrust which is 0.25 N. The ratio of the cylinder's weight to the weight of water displaced at d = 80 mm is 7.80 (= 1.95 N/0.25 N) and this is equal to their density ratio as they have the same volume. So the density of the cylinder is $7.8 \times 1000\,kg\,m^{-3}$ = $7800\,kg\,m^{-3}$.

Chapter 6

1 a 450 C **b** 675 J

2 a 0.33 A **b** 150 J

3 a See Topic 6.5 Figure 2a or b **b i** See Topic 6.5 Figure 3b **ii** The filament resistance increases as the bulb becomes brighter **iii** As the current in the bulb increases, the bulb becomes brighter and hotter. The filament is a metal wire and the resistance of a metal wire increases as its temperature increases. This is because the metal atoms in the filament vibrate more and they make it harder for conduction electrons to pass through the filament.

4 b i 2.4 V **ii** 0.60 V **c** 0.048 A

5 a Set up the circuit shown (as Topic 6.4 Figure 1) and measure the resistance R of different lengths L of the wire (as explained in Topic 6.4). Use a metre ruler to measure the length of wire connected in the circuit. Plot a graph of resistance R on the y-axis against length L on the x-axis. Determine the gradient of the line which is the resistance per unit length of the wire. The resistivity is calculated by multiplying the resistance per unit length by the area of cross-section of the wire.

b **i** $5.5 \times 10^{-7}\,\Omega\,\text{m}$ **ii** $4.4\,\Omega$

6 **a**

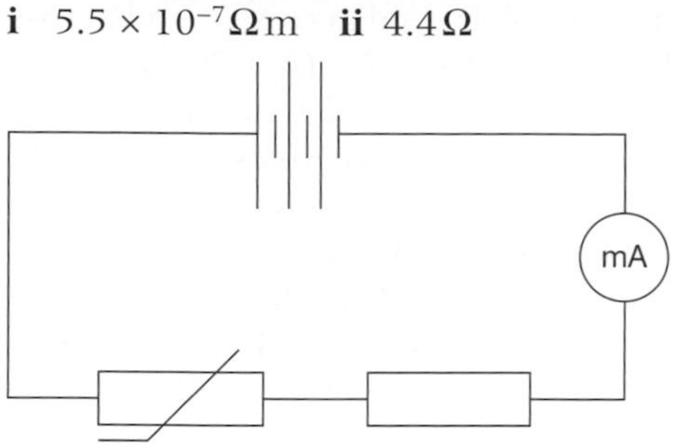

Figure 4

b **i** The ammeter reading would increase as the resistance of the thermistor would decrease so the current in the circuit would increase.

ii Connect an extra $270\,\Omega$ resistor in series

7 **a** **i** $0.21\,\Omega$ **ii** 2.8 V **iii** 36 W **b** 28A

8 **a** $530\,\Omega$ **b** 18.5 m

9 **a** 3.25 V **ii** 1.25 V **iii** $250\,\Omega$

b The milliammeter reading would increase as the resistance of the LDR would decrease so the current in the circuit would increase.

10 **b** $5.1 \times 10^{-7}\,\Omega\,\text{m}$

11 **a** As Topic 7.1, Figure 5, without the voltage values under the variable resistor and X and with the battery labelled 18 V

b $I = P/V = 24\,\text{W}/12\,\text{V} = 2.0\,\text{A}$ **c** Power supplied by the battery = $I\,V = 2.0\,\text{A} \times 18\,\text{V} = 36\,\text{W}$, power dissipated = power supplied by battery – power supplied to X = 36 W – 24 W = 12 W

12 **a** **i** Resistance per unit length for 1 aluminium wire

$$= \frac{R}{l} = \frac{\rho}{A} = \frac{2.5 \times 10^{-8}\,\Omega\,\text{m}}{\pi \times (0.5 \times 4.0 \times 10^{-3}\,\text{m})^2}$$

$= 1.99 \times 10^{-3}\,\Omega\,\text{m}^{-1}$

Since the 50 wires are identical and in parallel, their effective cross-sectional are is $50A$ where A is the area of cross-section of one wire. The resistance per metre of all 50 wires = one-fiftieth of the resistance per unit length of one wire = $1.99 \times 10^{-3}\,\Omega\,\text{m}^{-1}/50 = 3.98 \times 10^{-5}\,\Omega\,\text{m}^{-1}$

ii Resistance per unit length for 1 steel strand

$$= \frac{R}{l} = \frac{\rho}{A} = \frac{1.6 \times 10^{-7}\,\Omega\,\text{m}}{\pi \times (0.5 \times 3.0 \times 10^{-3}\,\text{m})^2}$$

$= 2.26 \times 10^{-2}\,\Omega\,\text{m}^{-1}$

Since the 7 strands are identical and in parallel, their effective cross-sectional is $7A$ where A is the area of cross-section of one strand. The resistance per metre of all 7 strands = one-seventh of the resistance per unit length of one strand $= 2.26 \times 10^{-2}\,\Omega\,\text{m}^{-1}/7 = 3.23 \times 10^{-3}\,\Omega\,\text{m}^{-1}$

b **i** Current in each component = p.d./resistance of 1000 m. The p.d. across each component is the same so the ratio

$$\frac{\text{current in the aluminium wires}}{\text{the current in the steel strands}} = \frac{\text{resistance of the steel strands}}{\text{resistance of the aluminium wires}}$$

$$= \frac{3.23 \times 10^{-3}\,\Omega\,\text{m}^{-1} \times 1000\,\text{m}}{3.98 \times 10^{-5}\,\Omega\,\text{m}^{-1} \times 1000\,\text{m}} = 81$$ so the steel strands conduct $\frac{1}{82}$ of the total current.

Therefore the current in the steel strands = 100 A/82 = 1.22 A and the current in the aluminium strands = 100 A – 1.22 A = 98.78 A

ii p.d. across the cable = current × resistance of either component = $98.78\,\text{A} \times 3.98 \times 10^{-5}\,\Omega\,\text{m}^{-1} \times 1000\,\text{m}$(for the aluminium wires) = 3.93 V

For the aluminium wires, the power dissipated = current × p.d. = 98.78 A × 3.93 V = 388.2 W

For the steel strands, the power dissipated = current × pd = 1.22 A × 3.93 V = 4.8 W

13 **a** See Figure 3.

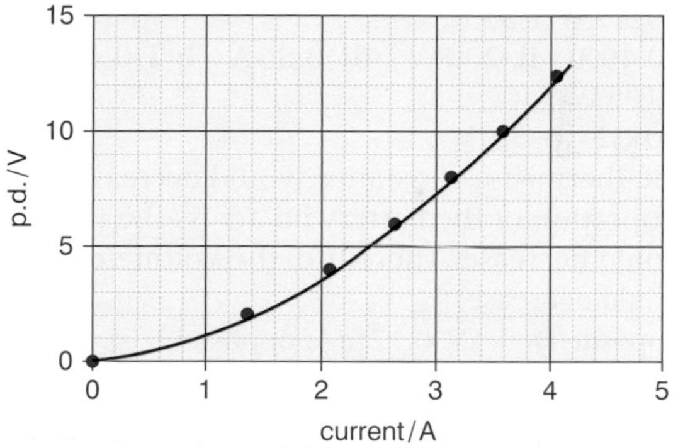

Figure 3

b **i** $1.50\,\Omega$ **ii** $1.70\,\Omega$ **iii** $3.06\,\Omega$

c % uncertainty for 2.00 A measurement = (±0.02/2.00) × 100% = 1.0%

% uncertainty in the corresponding p.d. of 3.4 V = (±0.02/3.4) × 100% = 0.6%

% uncertainty in the resistance = 1.6%, so the uncertainty in the resistance = 1.6% of $1.70\,\Omega$ = $0.03\,\Omega$ rounded up from $0.027\,\Omega$.

d **i** As the current increases, the wire becomes hotter so the resistance of the filament increases. Because the length and area of cross-section of the wire do not change (significantly), the resistivity therefore increases as the filament becomes hotter.

ii The drift velocity $v = I/Anq$, where A is the area of cross-section of the filament, n is the number of charge carriers per unit volume in the filament and q is the charge of each charge carrier. Assuming A, n and q do not change significantly when the wire becomes hotter, v increases when the current increases.

Chapter 7

1 **a** 6.0 Ω **b** 2.0 Ω: **i** 1.0 A **ii** 2.0 V **iii** 2.0 W; 6.0 Ω: **i** 0.67 A **ii** 4.0 V **iii** 2.7 W; 12.0 Ω: **i** 0.33 A **ii** 4.0 V **iii** 1.3 W

2 **a** The brightness increases from zero as the resistance is increased from zero.
b When the variable resistor has zero resistance, no current passes through the light bulb as the variable resistor 'short-circuits' the bulb and there is no pd across the bulb.

3 **a** **i** 20 Ω **ii** 15 Ω, 2.25 V, 3 Ω, 0.45 V **iii** 2.7 V
b **i** 15 Ω, 0.34 W; 3 Ω, 0.068 W **ii** 0.045 W

4 **a** **i** 0.75 A **ii** 4.5 V **b** **i** 3.4 W **ii** 4.5 W
c The power supplied to each heating element and to the internal resistance would be greater as the current in each would be greater. The battery would 'run down' faster as the energy stored in the battery would be used at a faster rate.

5 **a** 2.0 Ω **b** 1.5 A

6 **a** **i** 0.40 A **ii** 200 W **iii** 0.86 A **b** **i** 0.40 A **ii** 105 W

7 **a** 5.0 kΩ **b** 3.0 V
c The thermistor resistance would decrease so the proportion of the battery pd across the thermistor would decrease. Therefore, the voltmeter reading would decrease.

8 **a** Opposite D_1 0.6 V, 0.12 mA; opposite D_2 4.4 V, 0.88 mA
b Opposite D_1 0.12 mA, opposite D_2 1.7 mA

9 **a** 1.08 V **b** **i** ±0.67% **ii** ±0.41%, ±0.57% **c** ±0.02V

10 **a** **i** With S across Y, the emf of Y, $E_Y = I(R + r)$ and the p.d. across cell Y , $V_Y = IR$ where I is the current through S. So $\frac{E_Y}{V_Y} = \frac{(R+r)}{R}$. Since $V_Y = kl_2$ and $E_Y = kl_1$ where k is the p.d. per unit length along the potentiometer slide wire, substituting these into the equation above gives $\frac{l_1}{l_2} = \frac{(R+r)}{R}$.
ii $\frac{l_1}{l_2} = \frac{(R+r)}{R}$
Substituting in the values gives: $\frac{741}{625} = \frac{5.00 + r}{5.00}$
Rearranging in terms of r gives: $r = \frac{741 \times 5.00}{625} - 5 = 0.928\,\Omega$
b **(i)** The driver cell's e.m.f. decreased during the measurements **(ii)** The measurements all need to be repeated.

11 **a** **i** 2.75R **ii** $E_X/11R$
b **i** The p.d. across resistor A and across resistor B are both equal to I_X R. The current through resistor C is $(I_X + I_Y)$ in accordance with Kirchhoff's 1st law so the p.d. across C is $(I_X + I_Y)R$. For the loop consisting of E_X and resistors *A*, *C* and *B*, applying Kirchhoff's 2nd law gives E_X = p.d. across A + p.d. across C + p.d. across B = $I_X R + (I_X + I_Y)R + I_X R = (3I_X + I_Y)R$.
ii For the loop consisting of E_Y and resistors *D*, *C* and *E*, applying Kirchhoff's 2nd law gives E_Y = p.d. across D + pd across C + pd across E $= I_Y R + (I_X + I_Y)R + I_Y R = (I_X + 3\,I_Y)R$.
iii Substituting $E_X = E_Y = 6.0\,V$ and $R = 2.0\,\Omega$ into the two equations above gives $6.0\,V = (3\,I_X + I_Y) \times 2.0\,\Omega$ and $6.0\,V = (I_X + 3\,I_Y) \times 2.0\,\Omega$. Subtracting the two equations gives $0 = (2\,I_X - 2\,I_Y) \times 2.0\,\Omega$ which means $I_X = I_Y$. Therefore $4\,I_X = 6.0\,V/2.0\,\Omega = 3.0\,A$. So $I_X = I_Y = 0.75\,A$. The p.d. across C $= (I_X + I_Y)R = 1.5\,A \times 2.0\,\Omega = 3.0\,V$.

Chapter 8

1 **a** **i** The vibrations of a transverse wave are perpendicular to the direction of propagation of the waves whereas in a longitudinal wave, the vibrations are parallel to the direction of propagation.
ii sound **iii** electromagnetic waves
b **i** Diffraction is the spreading of a wave after it passes through a gap or round an obstacle.
ii Refraction is the change of direction of a wave when it crosses a boundary at which its speed changes.

2 **a** Polarised transverse waves vibrate in one plane only. As with any transverse wave, the vibrations are perpendicular to the direction of propagation.
b Sound waves are longitudinal and longitudinal waves vibrate parallel to the direction of propagation and therefore can never be polarised.

3 **a** 250 Hz
b Two complete wave cycles of the same amplitude should be drawn from one side to the other side of the screen

4 **a** See Topic 8.5.
b **i** At maximum loudness, the observer is at a position where the sound waves from each loudspeaker reinforce each other. At minimum loudness, the observer has a position where the waves cancel each other. The maxima and minima are equally spaced along XY.
ii Reducing the frequency increases the wavelength of the waves. As a result the path difference to each maximum and minimum increases except for the maximum at the centre. Successive minima are therefore further apart.

5 **a** **i** Reinforcement occurs at both P and Q because the path difference in both cases is a whole number of wavelengths, different in each case.
ii Cancellation occurs at the midpoint of P and Q because the path difference at this position is a whole number + half a wavelength so the microwaves from the two gaps are out of phase by 180°.

b The detector signal would increase as the waves from one gap would not reach the transmitter to cancel the waves from the other gap.

6 a i The amplitude is the same as at the midpoint. The frequency is the same. There is a phase difference of 180°.

ii The amplitude is the same. The frequency is the same. The vibrations are in phase.

b i 0.40 m

ii The pattern should show 4 equally spaced loops.

7 a At each resonant length, sound waves directly from the speaker reinforce sound waves from the speaker that have travelled down the tube and reflected at the water surface then partially reflected again at the open end. This happens when the length of the air column in the tube is an odd multiple of a quarter of the sound wavelength.

b i See Topic 8.7 Figure 3 **ii** 330 ms^{-1}

8 a i 1.65 m **ii** See Topic 8.8 Figure 4a.

b i 400 Hz

ii At each resonant length, sound waves directly from the speaker reinforce sound waves from the speaker that have travelled along the tube and reflected at the far end then partially reflected again at the open end where the speaker is. This happens when the length of the air column in the tube is a multiple of a half of the sound wavelength.

9 a i Every point along a progressive wave vibrates with the same amplitude, whereas the amplitude of a stationary wave differs along the wave and is zero at equally spaced points called *nodes*.

ii For a progressive wave, the phase difference between any two points increases with their separation. For a stationary wave, any two points between adjacent nodes (or separated by an even number of nodes) vibrate in phase (or any two points separated by an odd number of nodes vibrate out of phase by half a cycle).

b i Both points vibrate with the same amplitude, but their displacement at any given time is different. When one of the particles is at zero displacement, the other particle is at maximum displacement (either positive or negative). They vibrate with a phase difference of 270° (or three-quarters of a cycle or $3\pi/2$).

ii The two progressive waves must pass through each other.

1. in opposite directions with the same amplitude.
2. and at the same frequency and speed.

Chapter 9

1 a i Light from the two slits reinforce at a bright fringe

ii Light from the two slits cancels at a dark fringe.

b i The two slits emit waves with a constant phase difference,

ii Two separate light sources emit waves at random so they are not coherent sources. Therefore, the points of cancellation and reinforcement continually move about so no dark fringes can be observed.

c 530 nm

2 a i See Figure 6; the diffracted wavefronts should both spread out and be at the same spacing as the incident wavefronts.

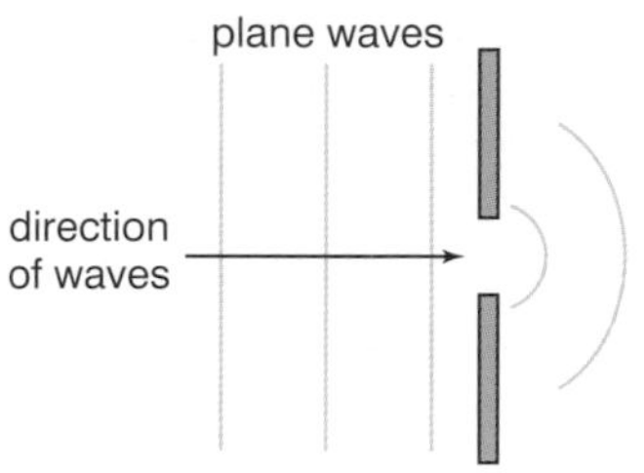

Figure 6

ii The two wavefronts should spread out less and the central part of each should be flatter.

b i Number of orders each side of the zero order = 2

ii 49.1°

3 a i fringe spacing, $x = 4.8\,mm/4 = 1.2\,mm$
Therefore $\lambda = ax/D = 0.40\,mm \times 1.2\,mm/0.810\,m = 5.9 \times 10^{-9}\,m$

ii % uncertainty in each measurement:
$\Delta D = 5\,mm/810\,mm \times 100\% = 0.6\%$;
$\Delta a = 0.02\,mm/0.40\,mm \times 100\% = 5.0\%$:
$\Delta x = 0.5\,mm/\ 4.8\,mm \times 100\% = 10.4\%$. The measurement of the slit spacing is the least precise because it has the greatest uncertainty.

b If white light had been used, the centre of each fringe at and near the middle, M, of fringe pattern would be white but their edges would be tinged with red at the edge furthest from M and with blue at the edge nearest M. Further from the centre, the edges of the fringes would overlap and be less distinct.

4 a White light consists of light of all wavelengths in the visible spectrum. Each colour of the visible spectrum is due to a band of wavelengths.

b Grating line spacing $d = 1/600\,000\,m^{-1} = 1.667 \times 10^{-6}\,m$; order number $n = 2$

i For red light: $n\lambda = d \sin\theta$ gives
$\lambda = (d \sin\theta)/2 = (1.667 \times 10^{-6}\,m \times \sin 48°)/2 = 619\,nm$

ii For blue light: $n\lambda = d \sin\theta$ gives
$\lambda = (d \sin\theta)/2 = (1.667 \times 10^{-6}\,m \times \sin 31°)/2 = 429\,nm$

c Applying the condition $\theta \leq 90^{\circ}$ and the equation $n\lambda = d \sin\theta$ to third order diffraction gives $3\lambda \leq d$. Therefore wavelengths greater than $3d$ (= 500 nm as $d = 1/600\,000$ m) cannot undergo 3rd order diffraction with this grating. Therefore only wavelengths below 500 nm are seen in the third order spectrum.

d i For red light in glass: $n\lambda' = d \sin\theta$ gives $\lambda' = (d \sin\theta)/2 = (1.667 \times 10^{-6}\text{m} \times \sin 30^{\circ})/2 = 417\,\text{nm}$
For blue light in glass: $n\lambda' = d \sin\theta$ gives $\lambda' = (d \sin\theta)/2 = (1.667 \times 10^{-6}\text{m} \times \sin 20^{\circ})/2 = 285\,\text{nm}$

ii Ratio for red light = 619 nm/417 nm = 1.48; Ratio for blue light = 429 nm/285 nm = 1.51

iii A 1° difference in the above calculations would mean for blue light, for example, using values of sin 32 not sin 31 for the wavelength in air and sin 19 not sin 20 for the wavelength in glass to calculate an upper limit for the ratio. In this case, the ratio would therefore be 1.63 not 1.51. The uncertainty in the ratio therefore is significantly greater than the difference between the calculated ratio values in **di** for blue light, so at this level of uncertainty it cannot be concluded that the ratios for red and blue light are different.

5 a Slit spacing, $a = 0.12$ mm.
Therefore fringe spacing, $x = \lambda D/a = 560 \times 10^{-9}\text{m} \times 1.800\,\text{m}/1.2 \times 10^{-4}\text{m} = 8.4 \times 10^{-3}\text{m}$

b The slit spacing with the 2 outer slits only would be 3×0.12 mm so the fringe spacing would be one-third of the fringe spacing in **a** (i.e. $8.4 \times 10^{-3}\text{m}/3$ or $2.8 \times 10^{-3}\text{m}$). Therefore instead of a wave with a central peak and two peaks either side, the sketch should show 15 peaks (i.e. a central one with 7 equally spaced peaks either side). The peaks would be lower, as more fringes would be created per unit distance so the light would be less intense at each bright fringe.

c Further fringes patterns due to different pairs of slits would be superimposed on the pattern described in **a** or **b**. Suppose the slits are labelled A, B ,C and D from left to right, the two slits on each side (AB and CD) would give fringes with the same spacing as the two inner slits BC but their fringe patterns would be displaced slightly in opposite directions from the inner slits fringe pattern (OR A further fringe pattern with twice the fringe spacing would be caused due to interference of light from AC and BD and these would also be displaced from the fringes due to BC.)

6 a Grating line spacing, $d = 1/500\,000\,\text{m}^{-1} = 2.00 \times 10^{-6}$ m; order number $n = 4$

i For the blue line: $n\lambda = d \sin\theta$ gives $\lambda = (d \sin 59^{\circ}25')/4 = (2.000 \times 10^{-6}\text{m} \times \sin 59.42^{\circ})/4 = 430\,\text{nm}$

ii For the orange line: Assume $n = 3$. Therefore $3\lambda = (d \sin\theta)$ gives $\lambda = (2.000 \times 10^{-6}\text{m} \times \sin 62.25^{\circ})/3 = 590\,\text{nm}$. Assuming $n = 2$ would give a wavelength of 885 nm which is beyond the visible spectrum. Assuming $n = 4$ would give a wavelength of 442 nm which would not be orange. Therefore, the wavelength of the orange line is 590 nm and it is a third order line.

b A white light spectrum consists of a continuous distribution of colours of the spectrum from deep red through to deep blue, corresponding to a continuous range of wavelengths which increase across the spectrum from blue to red. A line emission spectrum consists of narrow discrete lines of different colours against a black background. Each line is due to light of a particular wavelength corresponding to the colour of the line.

Chapter 10

1 a i The nucleus is very small in diameter compared with the diameter of anatom. Most α particles do not pass close enough to the nucleus to be affected by the electrostatic force of repulsion of the nucleus.

ii α particles that pass close to the nucleus experience strong repulsion from the nucleus which causes a significant deflection. If an *a* particle approaches a nucleus very closely, the force of repulsion from the nucleus deflects the α particle by more than 90°.

b See Topic 10.1 Figure 3.

2 a As the particle approaches, the kinetic energy decreases to zero at the closest approach then increases as the particle moves away becoming equal to the initial kinetic energy at infinity.

b i Factors include its initial speed or kinetic energy, its direction and the size of the charge on the nucleus. See Topic 10.1

ii It could penetrate the nucleus and cause a rearrangement of the nucleons with a different particle (e.g. a neutron being emitted)

3 ${}^{6}_{3}\text{Li} + {}^{1}_{0}\text{n} \rightarrow {}^{3}_{1}\text{H} + {}^{4}_{2}\alpha$

4 a γ

b i The reading would be unchanged as the radiation would pass through the second plate with negligible absorption.

ii The reading would decrease significantly as lead absorbs γ radiation much more effectively than aluminium does.

5 a i $7.7(3)\,\text{s}^{-1}$ **ii** 710

b See Topic 10.5.

6 a i 90 p, 144 n **ii** 91 p, 143 n

b X = ${}^{64}_{29}\text{Cu}$, ${}^{64}_{30}\text{Zn}$

7 a Proton number increased by 1, nucleon number unchanged.

b γ radiation is much more penetrating than β radiation
c By the inverse-square law $\frac{I_4}{I_1} = \frac{1.0^2}{4.0^2} = \frac{1}{16}$ (or 0.0625)

8 a X experiences a greater force of repulsion by the nucleus than Y does because its initial direction takes it closer to the nucleus. So X's rate of change of momentum is greater and so it is deflected more than Y.
b Most of the incident α-particles pass through atoms with little or no deflection so most of the atom is empty space. The nucleus is very small in size and most of the mass of the atom is located there. The nucleus must be positively charged because it repels α-particles which are positively charged.
c i If the foil was too thick, individual α-particles would be scattered several times so the deflections would be random. The foil needs to be thin enough so that individual α-particles are scattered only once.
ii If their kinetic energies differed, α-particles with differing kinetic energies moving along the same initial path would be deflected by different amounts. The measured count rate at any particular angle would be due to α-particles with different kinetic energies moving along different paths.

9 a i 83 **ii** 126
b $2.3 \times 10^{17}\,\mathrm{kg\,m^{-3}}$
c In bismuth metal, the atoms are in contact with each other with little space between them. Almost all the mass of a bismuth atom is in its nucleus which is about 10^5 times smaller in diameter than that of an atom. So the volume of the nucleus is about 10^{15} times smaller than the volume of the atom. Since density = mass/volume, the density of the nucleus is about 10^{15} times larger than that of the atom.

10 a $a = 2$, $b = 228$, $c = 88$, $d = -1$, $e = 89$
b 3

Chapter 11

1 a Its direction of motion keeps changing so its velocity keeps changing as velocity is speed in a certain direction.
b It has an acceleration towards the centre of its circular path because its rate of change of velocity is towards the centre.

2 a $1.0\,\mathrm{km\,s^{-1}}$ **b** $2.7 \times 10^{-3}\,\mathrm{m\,s^{-2}}$

3 a i $2.3\,\mathrm{m\,s^{-1}}$ **ii** $430\,\mathrm{m\,s^{-2}}$ **b i** 12 Hz **ii** $170\,\mathrm{m\,s^{-2}}$

4 a i 5.6 Hz **ii** $530\,\mathrm{ms^{-2}}$
b i $34\,\mathrm{m\,s^{-2}}$
ii The chain will come off the gear wheel if it moves too fast round the gear wheel and the tension in the chain is too small to provide the necessary centripetal force on the chain at this speed.

5 a Friction between the tyres and the road is not sufficient to provide the necessary centripetal force to the car at this speed.
b i $3.2\,\mathrm{m\,s^{-2}}$ **ii** $11\,\mathrm{m\,s^{-1}}$

6 a i $29\,\mathrm{ms^{-1}}$
b ii The force of gravity provides the centripetal force because $v^2 / r = 9.81\,\mathrm{ms^{-2}}$. Therefore $r = 23^2 / 9.81 = 54\,\mathrm{m}$

7 a The resultant of the train weight and the normal reaction force of the track on the train provides the centripetal force needed to maintain the circular motion along the horizontal curve of the track.
b The resultant force would not be large enough to maintain the circular motion. The train would tilt outwards as its centre of mass would move away from the centre of curvature of the track.

8 a The tension in the springs pulling the brake pads onto the shaft would not be large enough to keep the pads on the shaft. The pads would move away from the shaft until they press on the collar.
b 19 Hz

9 a Angular velocity of an object about a fixed point is its angular displacement per second.
b i $15.1\,\mathrm{rad\,s^{-1}}$
ii $10.3\,\mathrm{m\,s^{-1}}$ (Note the radius of rotation = length of the string + the radius of ball)
iii The centripetal force for constant frequency is constant. At the lowest point, the tension T acts upwards on the ball and the force of gravity of the ball acts downwards. So, at this point, the tension in the string supports the weight of the ball as well as providing the centripetal force so the tension is greatest at this point. For tension T in the string, when the ball is at the lowest point $T - mg = mv^2/r$ therefore $T = mv^2/r + mg = (0.16\,\mathrm{kg} \times (10.3\,\mathrm{m\,s^{-1}})^2/0.682 + (0.25\,\mathrm{kg} \times 9.81\,\mathrm{m\,s^{-2}}) = 26.5\,\mathrm{N}$
iv 2.86 Hz

Chapter 12

1 a 0.71 Hz **ii** $-1.6\,\mathrm{ms^{-2}}$
b The rate of decrease of the amplitude would increase.

2 a See Topic 12.5.
b i 0.57 Hz ii $-0.64\,\mathrm{m\,s^{-2}}$

3 a $\pi/2$ radians
b The graphs should show 2.25 cycles of two sinusoidal waves with the wave for X starting at maximum displacement of +60 mm and the wave for Y starting at zero displacement and reaching +60 mm displacement in the first quarter cycle.

4 **a** **i** The inertia of the ring causes it to stay momentarily at rest so it is displaced relative to its initial equilibrium position. This causes the trailing spring to be compressed and the other one stretched. The springs exert a resultant force on the ring and accelerate it.

ii The greater the acceleration, the greater the displacement of the ring.

b The inertia of the ring causes it to continue moving forward when the supports stop moving. As a result the ring oscillates about the position where the springs exert zero resultant force on it.

c **i** For a given acceleration, the displacement would be greater.

ii For a given acceleration, the displacement would be greater.

5 **a** 24 N m^{-1} **b** **i** 2.0 Hz **ii** 0.50 s

6 **a** amplitude, angular frequency

b 36 mm, 0.29 Hz **c** −15 mm **d** 0.12 m s^{-2}

7 **a** The engine vibrations subject the panel to a periodic force at the resonant frequency of the panel.

b The panel would be stiffer so its resonant frequency would be higher. This would solve the problem only if the resonant frequency is above the highest engine vibration frequency.

8 **a** **i** The vibrations cause the spring to oscillate and the oscillations are damped by the frictional force of the damper pads.

ii The frictional force on the dampers transfers energy from the vibrations to the dampers which gain internal energy so become warm.

b The vibrations caused large oscillations of the spring at or near the resonant frequency. The increase of the damping force reduces the amplitude of the oscillations.

9 **a** **i** 2.21 s, 0.453 Hz **ii** 0.325 mJ **iii** Measure the maximum horizontal or vertical displacement every 10 cycles for 50 or more cycles. If the displacement decreases and the average decrease per cycle is more than the uncertainty in the initial displacement, the decrease is significant.

b If P stops, then Q must gain all the momentum that P had. Since they are identical, Q must therefore have the same initial speed as P had immediately before the impact so Q's initial kinetic energy must the same as the kinetic energy P had immediately before the impact. Therefore the collision must be elastic.

10 **a** **i** Suitable scales and axes labelled correctly; points plotted correctly; correct graph shape, see Figure 7.

ii 18 mm, 1.8 Hz **iii** 0.20 m s^{-1}, 4.1 mJ

b **i** Curve should start and end at same points as first curve; curve should be below 1st curve by an decreasing amount from the peak to each end; peak should be slightly to the left of peak of first curve.

ii Energy transfers to the oscillating system from the oscillator and energy transfers from the system to the surroundings (at the same rate). (Note: The energy of the oscillating system is constant as long as the oscillator is on.)

c 1.3 Hz (= 1.8 Hz/2$^{0.5}$ as f is inversely proportional to $m^{0.5}$.)

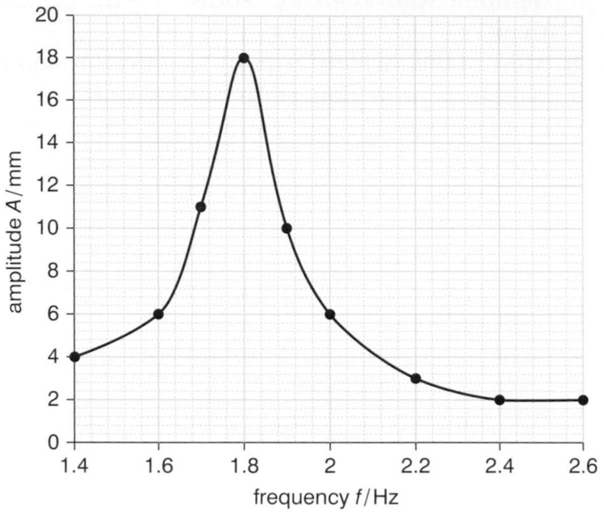

Figure 7

Chapter 13

1 **a** See Topic 13.1 Figure 2.

b **i** X and Y should be diametrically opposite at the same distance from the centre at or above the surface.

ii Z should be twice as far from the centre as X.

2 **a** 1.9 × 10^{18} N **b** 3.2 × 10^{-7} N kg^{-1}

3 **b** 3.1 N kg^{-1}

4 **a** 2.8 × 10^{-3} N kg^{-1}

b **i** 2.8 × 10^{-3} m s^{-2} **ii** 1.0 km s^{-1} **iii** 2.4 × 10^{6} s

5 **a** 2.3 × 1010 m

b An object would experience equal and opposite gravitational forces from Jupiter and the Sun at a certain position between Jupiter and the Sun. These two forces would tend to stretch the object and may eventually pull it apart.

6 **a** 670 N

b The astronaut is acted on by the force of gravity due to the Earth so is not weightless. However, as this force provides the centripetal force needed to maintain circular motion, no other force acts on the astronaut so the astronaut is unsupported.

7 **a** 8.9 N kg^{-1} **b** **ii** 5400 s

8 **a** **i** See Topic 13.5.

ii Such a satellite remains at the same relative position above the equator as the Earth rotates. Satellite dishes on the ground pointed at the satellite do not need to be moved to continue to point at the satellite.

b i $8.4\,\mathrm{N\,kg^{-1}}$
iii Adjacent satellites are separated by 30° along their orbit. Each satellite takes about 8 minutes to move through 30°. As a satellite moves away from a mobile phone with which it is contact, the signal is switched to the nearest satellite moving towards it to maintain the signal. This happens about once every 8 minutes.

9 a Gravitational field strength at a point in a gravitational field is the force per unit mass on a small test mass placed at that point in the field.
b i $17.3\,\mathrm{km\,s^{-1}}$ **ii** Let m represent the mass of Io. The gravitational force F on Io $= GMm/r^2$ where M is the mass of Jupiter and r is the (mean) radius of the orbit. The centripetal acceleration of Io, $a = v^2/r$. Using $F = ma$ gives $mv^2/r = GMm/r^2$ which gives $v^2 = GM/r$ and hence $v = \sqrt{\frac{GM}{r}}$.
Rearranging this equation gives

$$M = \frac{rv^2}{G} = \frac{4.22 \times 10^8 \times (17.3 \times 10^3)^2}{6.67 \times 10^{-11}}$$

$$= 1.89 \times 10^{27}\,\mathrm{kg}$$

iii $1320\,\mathrm{kg\,m^{-3}}$
c The least distance between Io and Europa is $2.49 \times 10^5\,\mathrm{km}$. The force on Io due to Jupiter (or Europa) $\propto m/d^2$ where m = Jupiter (or Europa) and d is the relevant distance from Io to Jupiter (or Europa). The ratio of (m/d^2) for Europa to that of Jupiter gives a measure of the required significance. For Europa $m/d^2 = (4.92 \times 10^{22}\,\mathrm{kg}/(2.49 \times 10^8\,\mathrm{m})^2 = 7.94 \times 10^5\,\mathrm{kg\,m^{-2}}$. For Jupiter $m/d^2 = (1.89 \times 10^{27}\,\mathrm{kg}/(4.22 \times 10^8\,\mathrm{m})^2 = 1.06 \times 10^{10}\,\mathrm{kg\,m^{-2}}$. The ratio of the m/d^2 values is 7.2×10^{-5}. Although this may seem insignificant, the fact that the three bodies are in line every few days stretches Io periodically creating resonant disturbances in Io which heat Io internally and cause volcanic activity.

10 a i Newton's law of gravitation states that there is a gravitational force of attraction between any two point objects and that this force is proportional to the mass of each object and inversely proportional to the square of their distance apart.
ii At height h above the surface of the Earth, the force on a small object of mass $m = \frac{GMm}{(R+h)^2}$, where the mass of the Earth $= M$ and its radius $= R$.

$$\text{Therefore } g = \frac{F}{m} = \frac{GM}{(R+h)^2} = \frac{GM}{R^2\left(1 + \frac{h}{R}\right)^2}$$

$$= \frac{g_S}{\left(1 + \frac{h}{R}\right)^2} \text{ since } g_S = GM/R^2.$$

iii The graph should be a smooth curve from the point ($h/R = 0$, $g = g_S$) through the following points ($h/R = 1$, $g = g_S/4$), ($h/R = 2$, $g = g_S/9$), ($h/R = 3$, $g/g_S = 1/16$), ($h/R = 4$, $g = g_S/25$), ($h/R = 5$, $g = g_S/36$) in accordance with the inverse square law. No line should be shown between the origin and the first point.
b i Centripetal acceleration $= \omega^2 r = (2\pi\,\mathrm{rad}/(24 \times 3600\,\mathrm{s}))^2 \times (0.5 \times 12756 \times 10^3\,\mathrm{m}) = 0.034\,\mathrm{m\,s^2}$.
Since the difference in g at the Equator and the poles is $0.05\,\mathrm{m\,s^{-2}}$, centripetal acceleration is significant.
ii Using $g = GM/R^2$ with appropriate values for the radii and including $G = 6.67 \times 10^{-11}\,\mathrm{N\,m^2\,kg^{-2}}$ gives $9.80\,\mathrm{m\,s^{-2}}$ for g at the poles and $9.86\,\mathrm{m\,s^{-2}}$ for g at the Equator. This is also a significant factor although, since the two effects taken together exceed the observed difference of $0.05\,\mathrm{m\,s^{-2}}$, other factors (e.g. geological, solar, lunar) need to be taken into account.

11 a Gravitational potential at a point is defined as the work done per unit mass to move a small object from infinity to that point.
b i $\phi = -\frac{GM}{r}$, $\phi_S = -\frac{GM}{R}$
ii $\phi/\phi_S = R/r$ therefore ϕ/ϕ_S is inversely proportional to r. The graph should be a curve starting at $\phi/\phi_S = 1$, $r/R = 1$ then decreasing smoothly, passing through the following points:

r/R	1.5	2.0	2.5	3.0	3.5	4.0	4.5	5.0
ϕ/ϕ_S	0.667	0.500	0.400	0.333	0.286	0.250	0.220	0.200

iii $\phi = -\frac{GM}{r}$ and $g = -\frac{GM}{r^2}$ therefore multiplying both sides of $g = -\frac{GM}{r^2}$ by r gives $gr = -\frac{GM}{r^2} \times r = -\frac{GM}{r}$; hence $\phi = gr$.
c i $-2.82\,\mathrm{J\,kg^{-1}}$ **ii** $-2.88\,\mathrm{J\,kg^{-1}}$
iii The difference is due in part to the gravitational potential of the Earth at the Moon's surface.

Chapter 14

1 a The electric field strength has the same magnitude and direction at all positions; the lines of force are parallel lines.
b i The electron carries a negative charge and therefore the direction of the electric force on it is in the opposite direction to the field direction.
ii $1.1 \times 10^6\,\mathrm{V\,m^{-1}}$

2 a 33 nC
b The electric field of the spheres at the midpoint is equal in magnitude and opposite in direction. A positively charged particle at the midpoint therefore experiences equal and opposite electric forces of attraction to the spheres.

3 **a** **i** The sketch should show lines from the + to the – charge, straight across between the centres and curved above and below.
ii $2.9 \times 10^5\,V\,m^{-1}$ towards Y
b **i** The position needs to be 75 mm further from X than from Y in order that the electric field strength of X is equal and opposite to that of Y.
ii 75 mm to Y, 150 mm to X

4 **a** A sufficiently strong electric field pulls an electron attached to an atom away from the atom and so ionises the atom.
b **i** 100 nC **ii** $680\,V\,m^{-1}$

5 **b** 49 nC

6 **a** See Topic 14.5.

7 **a** Electric field strength at a point in an electric field is the force per unit charge on a positive test charge at that point.
b **i** $3.56 \times 10^{-7}\,C$, $2.83 \times 10^{-6}\,C\,m^{-2}$ **ii** $80\,kV\,m^{-1}$
c **i** The electric field of the sphere attracts conduction electrons in the rod towards X, so X becomes positive and Y becomes negative. The electrical force of attraction on end X due to the sphere is greater than the electrical force of repulsion on end Y due to the sphere. So there is an overall force of attraction on the rod towards the sphere.
ii See Figure 8. (Note: The rod is at constant potential with the potential at its midpoint approximately unchanged. The potentials at X and Y are at the same potential as the midpoint of XY. The surface potential of the sphere is unchanged. So the potential should decrease smoothly from the sphere to X then it is constant from X to Y then it decreases smoothly from Y.)

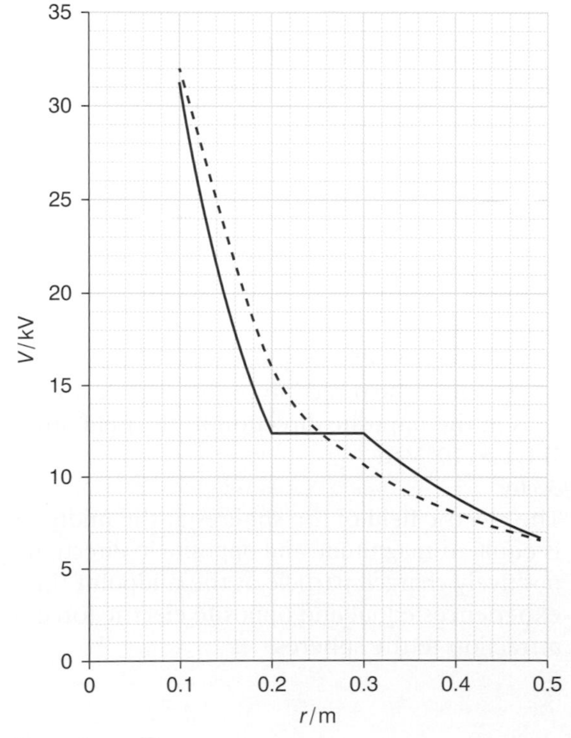

Figure 8

iii The electric field strength is stronger than before and it decreases between the sphere and X. From X to Y, the field strength is zero, whereas it decreased over the same distance before the rod was present. Beyond Y, the field is stronger and decreases gradually to zero at infinity.

8 **a** Electric field strength at a point in an electric field is the force per unit charge on a positive test charge placed at that point.
b **i** The arrow on each line should be from plate B to plate A.
ii U should be anywhere between the parallel lines between A and B.
iii The magnitude of the field strength decreases from A to B. The direction of the field changes along the line as it is always tangential to the line.
b **i** The arrow at P should be horizontal pointing in the same direction as the field lines. The arrow at Q should be horizontal and pointing in the opposite direction to the field lines.
ii The perpendicular distance d between the two force arrows = PN sin 60 = 3.5×10^{-9} sin 60 = $3.0(3) \times 10^{-9}\,m$
Torque = $Fd = qEd = 1.6 \times 10^{-19}\,C \times 50\,000\,V\,m^{-1} \times 3.03 \times 10^{-9}\,m = 2.4 \times 10^{-24}\,N\,m$

9 **a** The electric potential at a point in an electric field is the work done per unit charge on a positive test charge when it is moved from infinity to that point.
b **i** $8.20 \times 10^{-8}\,N$
ii $-4.35 \times 10^{-18}\,J$
iii $\frac{mv^2}{r} = \frac{e^2}{4\pi\varepsilon_0 r^2}$ where m is the mass of the electron.
Therefore K.E. = $\frac{1}{2}mv^2 = \frac{1}{2} \times \frac{e^2}{4\pi\varepsilon_0 r}$ which is 0.5 × the magnitude of the potential energy of the atom.
iv P.E. = –27.2 eV so the total energy = K.E. + P.E. = (0.5 × 27.2 eV) – 27.2 eV = –13.6 eV

Chapter 15

1 **a** Charge stored per unit p.d.
b **i** See Topic15.1.
ii 320 μC, 40 μF

2 **a** **i** 1.33 μF **ii** 8.0 μC, 4.0 V, 16 μJ **b** 6.0 μF

3 **a** 36 μC, 160 μJ **b** **i** 10 μF **iii** 65 μJ

4 **a** **i** 0.95 J **ii** 19 W
b The combined capacitance is 0.5 × 47000 μF compared with 2 × 47000 μF in parallel. The battery p.d. is the same as before. Since charge stored = capacitance × p.d., the charge stored is less than when they are in parallel.
The energy stored is $\frac{1}{2}CV^2$, so the energy stored when the capacitors are connected in series is $\frac{1}{4}$ of the energy stored when they are connected in parallel.

5 **a** 5.6 mC, 34 mJ **b i** 47 s **iii** 31 mJ

6 **a** 480 mC, 2.88 J **b ii** 1.1 s

7 **a** The potential V of X is given by the equation $V = \frac{Q}{4\pi\varepsilon_0 R}$. Since its capacitance $C = Q/V$, rearranging $V = \frac{Q}{4\pi\varepsilon_0 R}$ gives $C = \frac{Q}{V} = 4\pi\varepsilon_0 R$

b i Conduction electrons in Y are attracted to X because X is positively charged so they transfer to X via the conductor.

ii X gains the negative charge of the electrons transferred to it so its overall charge becomes less positive and its potential decreases. Y loses negative charge so it becomes positively charged (as it was initially uncharged). Therefore the potential of Y increases. The flow of electrons from Y to X continues until X and Y are at the same potential.

iii Let V_f be the potential of X and Y when the flow of electrons ceases and let q represent the positive charge on Y. Therefore the potential of Y = $\frac{q}{4\pi\varepsilon_0 \times 0.2R}$. The charge now on X = $Q - q$ so the potential of X = $\frac{Q-q}{4\pi\varepsilon_0 R}$. Since the two potentials are equal, $Q - q = \frac{q}{0.2} = 5q$ so $q = Q/6$. Therefore the potential of X = $\frac{Q - \frac{1}{6}Q}{4\pi\varepsilon_0 R} = \frac{\frac{5}{6}Q}{4\pi\varepsilon_0 R} = \frac{5}{6}$ or 0.83 × the original potential of X.

c i The energy stored on a capacitor = $\frac{1}{2}QV$. All the charge on both spheres is at a potential of 0.83 × the original potential. The total charge of both spheres is unchanged at Q. So the total energy stored is 0.83 x the original energy stored on X.

ii Energy is dissipated due to the resistance heating of the current in the conductor.

8 **a** Taking natural logs on both sides of $V = V_0 e^{-t/CR}$ gives $\ln V = \ln V_0 + \ln(e^{-t/CR})$

As $\ln(e^{-t/CR}) = -\frac{t}{CR}$, then $\ln V = \ln V_0 - \frac{t}{CR} = a - bt$

Hence $a = \ln V_0$ and $b = \frac{1}{CR}$

b i

t/s	210	240	270	300
mean V/V	1.427	1.233	1.033	0.887
ln V	0.356	0.209	0.032	−0.120

ii For correct labels on each axis, for suitable scales, for correctly plotted points, see Figure 9.

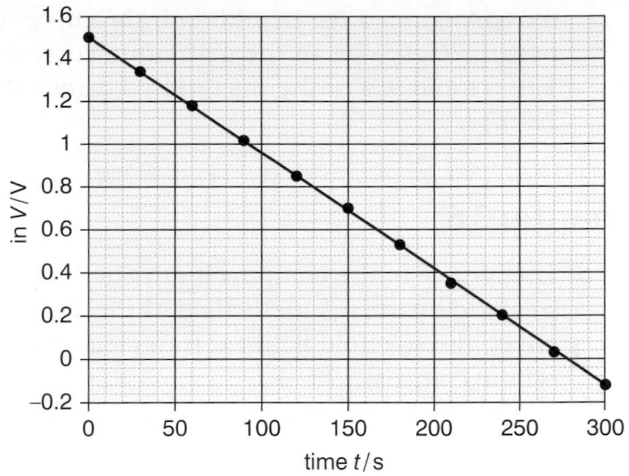

Figure 9

iii time constant (= RC) = $\frac{1}{\text{gradient of graph}}$

gradient of graph = $5.40 \times 10^{-3}\,\text{s}^{-1}$

time constant = $\frac{1}{5.40 \times 10^{-3}} = 185\,\text{s}$

iv $C = \frac{\text{time constant}}{R} = \frac{185}{6.8 \times 10^4}$

$= 2.72 \times 10^{-3}\,\text{F} = 2720\,\mu\text{F}$

c i The range of each set of readings is no more than 0.03 V, except for the reading at t = 150 s which is 0.12 V. This exception is due to an anomalous reading.

The readings are therefore reliable because the range of each set of readings is very small compared with the mean value.

ii Apart from the exception at t = 150 s, the precision of the readings is therefore no more than ±0.02 V.

Using the smallest mean value (i.e. 0.887 V), for 0.887 ± 0.02 V, ln V = −0.143 for V = 0.867 V and −0.098 for V = 0.907 V. The range of this value of ln V is therefore 0.045.

Thus the % uncertainty in this value of ln V is about 3% (= $\frac{1}{2}$ × 0.045/0.887 × 100%).

Given R is accurate to 1%, the value of C is accurate to within 4% (= 3% + 1%)

Note: If time permits, you could estimate the random error in V and hence in ln$V$ for every point and represent them as error bars on the graph. This would allow you to draw lines of maximum and minimum gradient and so determine maximum and minimum values for the time constant to give an uncertainty value for C.

9 **a** 4.7 μA **ii** As the charge on the capacitor decreases, the capacitor p.d. decreases. Since this p.d. acts directly across the resistor, the current through the resistor decreases.

b i 1.0(3) s **ii** $I = I_0 e^{-(t/RC)}$ and t/RC = 5.0 s/1.03 s = 4.85

Therefore $I = I_0 e^{-(t/RC)} = 4.7\,\mu\text{A} \times e^{-4.85} = 0.037\,\mu\text{A}$

c ii The graph should be an exponential decrease curve from 4.7 μA at t = 0 to just above zero (approx

$0.04\,\mu A$) at $t = 5.0\,s$. In addition, the curve should pass through 37% of the maximum current at 1.0(3) s and 9% of the maximum current at 2.0(6) s.

Chapter 16

1 a The N-pole of the compass is attracted to the solenoid by the magnetic field of the solenoid. In effect, the solenoid field exerts a clockwise couple on the compass.

b i The solenoid field becomes weaker so the couple it exerts on the compass becomes smaller and the compass turns back towards the bar magnet.

ii The compass would turn clockwise.

2 a 0.23 T **b** 0.052 N

3 a ii West

b A horizontal force due to the $18\,\mu T$ component of the Earth's field acts on the cable as well as the force due to the vertical component.

4 a i 1.7 N

ii The force reverses direction every half-turn. The magnitude of the force is the same before and after reversal but the turning effect of the force decreases to zero as the coil turns perpendicular to the field.

b The perpendicular distance between the lines of action of the force on each side is greatest when the coil is parallel to the field so the couple on the coil due to the forces is a maximum.

5 a 90 mT **b** 18°

6 b Horizontal due west

7 a See Topic 16.3 Figure 7. **b** 0.52 N

8 a $2.6 \times 10^{-3}\,N\,m$

b The weight exerts a moment of $2.5 \times 10^{-3}\,N\,m$ (= weight × spindle radius). The magnetic couple is $2.6 \times 10^{-3}\,N\,m$, which is greater than the couple exerted by the weight ($2.5 \times 10^{-3}\,N\,m$), so the motor should just be able to lift the weight.

9 a See Topic 16.5.

b $3.5 \times 10^{7}\,m\,s^{-1}$

10 a i See Topic 16.5. **ii** 18 mm

b The beam radius increases until the beam is straight when the field is zero. As the reversed field increases from zero, the beam curves in the opposite direction on a circular path with a decreasing radius of curvature.

11 b i 0.73 T

12 a The radius of curvature r of the path of an ion is given by $r = mv/BQ$. Since the speed of the ions is the same and they are in the same uniform field, radius r is proportional to m/Q. Hence ions with different specific charges move along different paths.

b i 0.84 T **ii** $1.4 \times 10^{7}\,C\,kg^{-1}$

13 a Magnetic flux density is the force per unit length per unit current on a current-carrying wire perpendicular to the field lines.

b i Multiplying each balance reading in grams by $9.81 \times 10^{-3}\,N/g$ gives 0, 3.63, 6.87, 10.00, 14.42, 17.85 mN to two decimal places.

ii See Figure 10.

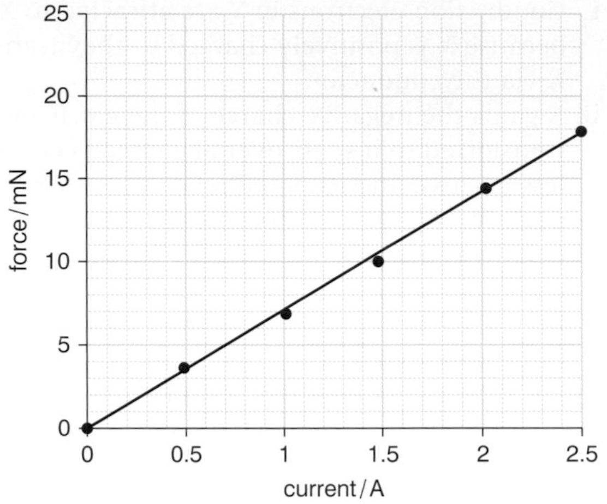

Figure 10

iii Force per unit current = the gradient of the line: answer should be between 7.0 and $7.1 \times 10^{-3}\,N\,m^{-1}$

iv B = force per unit length/wire length = 0.17 T (= $7.05 \times 10^{-3}\,N\,m^{-1}/0.042\,m$ to 2 s.f.)

c i An estimate of the maximum and minimum gradient values gives about $\pm 0.2\,mN\,m^{-1}$ which gives a % uncertainty in the force per unit length of 2.8%

ii The % uncertainty in the wire length = 4.8% (= 2 mm/42 m × 100%)
The % uncertainty in B = 7.6% therefore the uncertainty in $B = \pm 0.01\,T$

Chapter 17

1 a The ribbon is a conductor that cuts across the field lines when it moves backwards and forwards. As a result an alternating e.m.f. is induced in the ribbon.

b i The amplitude of the e.m.f. would be greater,

ii The ribbon would vibrate with a smaller amplitude.

2 a i The movement of the magnet into the coil induced an emf in the coil (when the coil was in motion) due to the changing magnetic flux in the coil. The induced emf caused a current briefly in the coil and the galvanometer.

ii The pointer deflects briefly to the left

b i 0.30 mWb **ii** 1.5 mV

3 a The wire is a conductor that cuts across the field lines when it vibrates. As a result an alternating emf is induced in the wire.

b $1.6\,m\,s^{-1}$ ($1.56\,m\,s^{-1}$ to three significant figures)

4 a i 3.0 mWb **ii** 0.30 T
b i P +, Q – **ii** from Q to P in the rod (and from P to Q round the rest of the circuit)
5 a 2.2 mWb (2.15 mWb to three significant figures)
b i The graph should show a steady increase in the flux linkage from zero to 2.2 mWb in 4.0 s then constant flux linkage.
ii The graph should show the induced e.m.f. is 0.54 mV in the first 4 seconds then it is zero.
6 a When the lamps are on, the current through the circuit causes reaction forces on the generator coil which create a couple opposing the applied torque. When the lamps are off, there is no current in the circuit so there is no opposing couple.
b ii 5.0 V
7 a i The graph should show the flux linkage falling from an initial value and decreasing to zero.
ii The induced e.m.f. = – the gradient of the graph in **ai**. The gradient changes from zero and becomes increasingly negative before becoming less and less negative until it is zero.
b The graph should have a positive peak as in the graph shown then becoming negative to give a negative peak then becoming zero.
8 a 52 **b i** 5.0 A **ii** 0.26 A
c i 10 W,
ii The current through the cable would be much smaller so less power would be dissipated in the cable.
However, the cable might not be intended for use at 230 V so would be unsafe.
9 a Faraday's law of electromagnetic induction states that the induced e.m.f. in a circuit is proportional to the rate of change of magnetic flux linkage through the circuit.

b i The induced current is measured when the handle is turned at different constant frequencies. For each frequency, a steady rate of rotation is maintained and (with the aid of another student) the time taken for twenty turns is measured at that frequency. The measurements should be recorded in a table and each frequency calculated (= $20/t$ where t is the time for 20 turns). A graph of the induced current (on the y-axis) against the frequency (on the x-axis) should be plotted.
ii The induced current is proportional to the induced e.m.f. and the rate of change of the flux linkage is proportional to the frequency of rotation. The induced current should be proportional to the frequency. A graph of the induced current (on the y-axis) against the frequency (on the x-axis) should give a straight line through the origin.
10 a i A transformer increases or decreases the peak voltage of an alternating voltage. The magnetic flux produced by the primary coil passes through the secondary coil. The ratio of the e.m.f. induced in the secondary coil to the alternating voltage applied to the primary coil is equal to the ratio of number of turns of the secondary coil to the number of turns on the primary coil. In a step-up transformer, there are more turns on the secondary coil than on the primary coil so the peak e.m.f. induced in the secondary coil is greater than the peak voltage applied to the primary coil.
ii For transmission of a certain amount of power, the greater the voltage of the cables, the smaller the current through the cables is. Therefore the greater the voltage, the smaller the power wasted due to the heating effect of the resistance of the cables.
iii For 5.0 kW of power at 100 V, the current in the cables would be 50 A (= 5000 W/100 V). Therefore the power wasted in the cables due to resistance heating would be 12 500 W = 12.5 kW (= $50^2 \times 5.0\,\Omega$). For the same amount of power at 1000 V, the current would be 5.0 A and the power wasted would be 125 W (= $5.0^2 \times 5.0\,\Omega$) which is 1/100 th of the power wasted at 100 V.

Chapter 18

1 a See Topic 18.1 Figure 2. **b** See Topic 18.1.
2 a The sum of the random distribution of kinetic and potential energies of the molecules,
b Increasing the internal energy at the melting or boiling point enables the molecules to break free from each other and therefore increases their potential energies. As the temperature is constant, the sum of the kinetic energies of the molecules is unchanged.
3 a i 540 kJ **ii** 430 kJ
4 a i Some of the electrical energy is wasted as heat due to friction at the bearings and due to resistance heating by the electric current passing through the motor coil.
ii Electric energy from the power supply is transferred to the load as work is done on the load and also transferred to the surroundings as heat as explained above and as sound.
b i 8.8% **ii** 87 J, 1.7%
5 a $1.9 \times 10^{-2}\,K\,s^{-1}$ **b** $900\,J\,kg^{-1}\,K^{-1}$
6 a i 780 kJ **ii** 620 kJ
b i 89 kJ **ii** Heat loss to surroundings, energy used to vaporise water below boiling point
7 a 7.1 kW **b** $0.045\,kg\,s^{-1}$

8 **a** *Specific latent heat of vaporisation* is the energy needed to change the state of unit mass of the substance from liquid to vapour without change of temperature.

b **i** See Figure 11.

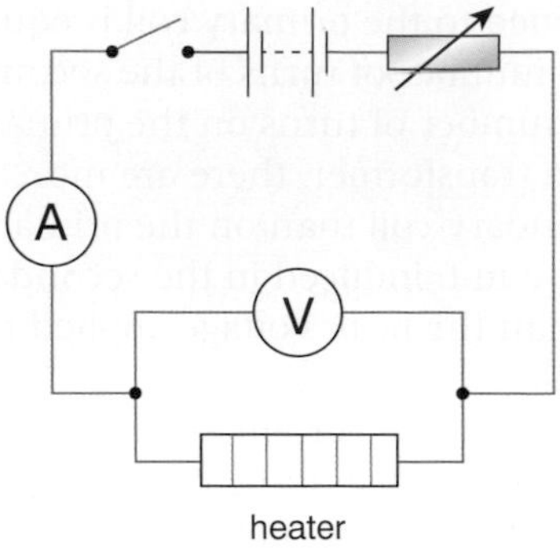

a Heater circuit
Figure 11

ii $850\,\text{J}\,\text{kg}^{-1}$

9 **a** The *internal energy* of an object is the sum of the random distribution of the kinetic and potential energies of its molecules.

b **i** $25.0\,^{\circ}\text{C}$ **ii** $2350\,\text{J}\,\text{kg}^{-1}\,^{\circ}\text{C}^{-1}$

c % uncertainties: mass of water 0.3%; mass of rock 1.3%; temperature change of water 0.7%; temperature change of rock 5.0%;
total % uncertainty = 7.3%
Uncertainty in specific heat capacity of the rock $= \pm 170\,\text{J}\,\text{kg}^{-1}\,^{\circ}\text{C}^{-1}$

Chapter 19

1 **a** **i and ii** See Topic 19.3.

b See Figure 12 below.

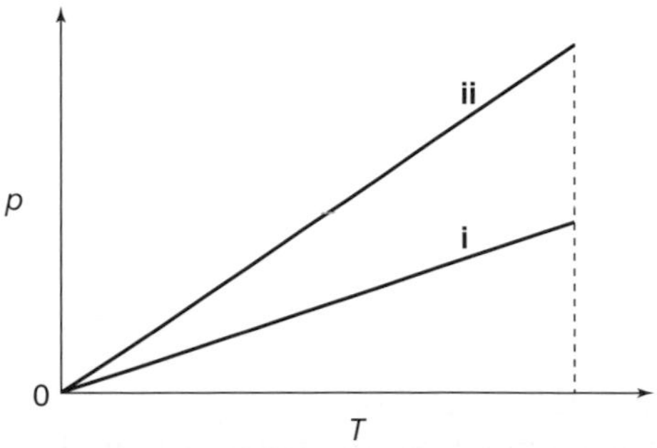

Figure 12

2 **b** **i** 2.1×10^3 moles

a **ii** 1.3×10^{27} (1.25×10^{27} to three significant figures)

3 **a** 53 moles **b** 220 MPa

4 **a** **i** 1.6×10^4 moles **ii** $0.18\,\text{kg}\,\text{m}^{-3}$

b **i** The mass of helium is constant and the volume increases so the density decreases as it is equal to the mass per unit volume.

ii The helium gas in the balloon displaces air. As air is more dense than helium, an upthrust acts on the balloon which is greater than the weight of the balloon.

5 **a** **i** $8.4 \times 10^{-3}\,\text{kg}$

ii The pressure is proportional to the mass of gas provided the temperature is constant. Therefore at pressure p, the mass remaining = $(p/\text{kPa})/150 \times$ the mass at 150 Pa (assuming the temperature is 300 K when the reading is taken).

b $7.7 \times 10^{-3}\,\text{kg}$

6 **a** **i** 4.8×10^{-4} moles **ii** 2.9×10^{20} **iii** $430\,\text{m}\,\text{s}^{-1}$

b **i** 17.5 kPa **ii** 0.30 J

7 **a** **i** $1.1 \times 10^3\,\text{kg}$ **ii** $6.7 \times 10^{-27}\,\text{kg}$

b **i** $1.3 \times 10^3\,\text{m}\,\text{s}^{-1}$ **ii** $9.8 \times 10^8\,\text{J}$

8 **a** When the brakes are applied, the pressure is not transmitted through the oil as effectively as it compresses the bubble. So the braking force is reduced.

b **i** $4.0 \times 10^{-4}\,\text{m}^2$ **iii** 9.2 mg

9 **a** 0.0139 mole **b** **i** 1.47 J **ii** 121 kPa

c **i** The graph should be a straight line passing through (20 °C, 125 kPa), (11.5 °C, 121 kPa) and (0 °C, 116 kPa).

ii The graph should be a straight line passing through (20 °C, 62.5 kPa), (11.5 °C, 60.5 kPa) and (0 °C, 58 kPa).

10 **a** The *r.m.s. speed* is the square root of the mean value of the square of the speeds of the molecules.

b $508\,\text{m}\,\text{s}^{-1}$ **ii** Combining the ideal gas equation and the kinetic theory equation for 1 mole of ideal gas gives $RT = \frac{1}{3}N_A m c_{\text{r.m.s.}}^2$ ($= pV$) and, since the molar mass $M = N_A m$, then $c_{\text{r.m.s.}}^2 = \left(\frac{3R}{M}\right)T$.
Therefore the mean square speed $c_{\text{r.m.s.}}^2$ is proportional to the absolute temperature T because $\left(\frac{3R}{M}\right)$ is constant.

c **i** The volume of the gas does not change, so no work is done.

ii No energy is transferred by heating to or from the gas between C and D and the gas does work since it expands from C to D so its internal energy decreases.

iii The area of the loop represents the net work done in a cycle. There are about 165 small squares in the loop and each small square represents energy transfer of 2 J ($= 0.2 \times 10^6\,\text{Pa} \times 0.1 \times 10^{-4}\,\text{m}^3$) so the work done is about 330 J.

iv 42%

v Energy is transferred to the surroundings by the hot exhaust gases. The engine block becomes hot and transfers energy by heating to the cooling system. Sound waves created by the vibrations of the engine transfer energy to the surroundings.

Chapter 20

1 a The *work function* of a metal is the minimum energy needed by an electron to escape from the metal's surface.

b i See Figure 13.

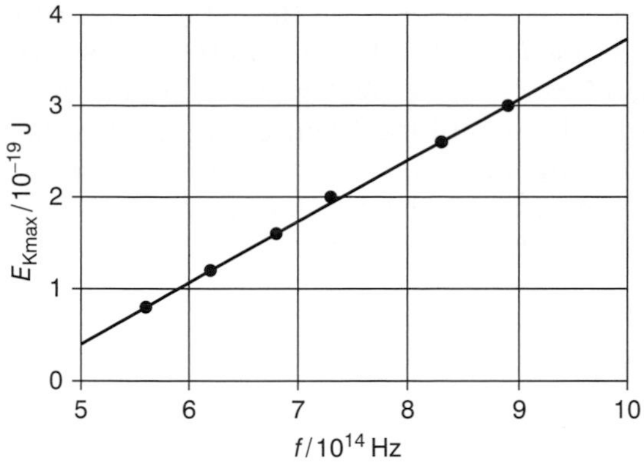

Figure 13

ii Gradient = (2.40 – 0.40/8.0 – 5.0) = 6.67 × 10^{-34} J Hz^{-1}

iii The graph is a straight line so it fits the straight line equation $y = mx + c$ equation where $y = E_{Kmax}$, $x = f$, m = the Planck constant h and $c = -\phi$, where ϕ is the work function of the metal. The graph gradient is very close to the accepted value of h.

iv The work function = c as it is the y-coordinate at $x = 0$. Since $E_{Kmax} = 0.40 \times 10^{-19}$ J at $f = 5.0 \times 10^{14}$ Hz, then substituting these values and the gradient value above into $y = mx + c$ gives $0.40 \times 10^{-19}\,J = (6.67 \times 10^{-34}\,J\,Hz^{-1} \times 5.0 \times 10^{14}\,Hz) + c$. Hence $c = 0.40 \times 10^{-19}\,J - (6.67 \times 10^{-34}\,J\,Hz^{-1} \times 5.0 \times 10^{14}\,Hz) = -2.94 \times 10^{-19}$ J. Therefore the work function is equal to 2.94×10^{-19} J.
(Note: a non-graphical value can be obtained by recognising the line rises from the y-intercept by 3.4 squares over 5 squares along the x-axis. Therefore extrapolating backwards to zero frequency would mean a descent of 3.4 squares to reach $x = 0$ which would be 3.0 below the x-axis, corresponding to a work function of 3.0×10^{-19} J.

b The threshold frequency = ϕ/h
$= 3.3 \times 10^{-19}\,J/6.63 \times 10^{-34}\,J\,s = 5.0 \times 10^{14}$ Hz
The line should be parallel to the existing line and pass through the point on the x-axis, where $f = 5.0 \times 10^{14}$ Hz

2 a i The work function of a metal is the minimum amount of energy needed by a conduction electron to escape from the surface of a metal when the metal is at zero potential.

ii The threshold frequency f_o of a metal is the minimum frequency of incident light that will cause photoelectric emission from the surface of the metal.

b $\phi = hf_o$

c i According to wave theory, photoelectric emission from a metal surface should take place at *any* frequency of light as a conduction electron at the surface of the metal will repeatedly absorb energy from the incident waves until it has sufficient energy to leave the surface. The fact that there is a threshold frequency of light below which photoelectric emission does not take place contradicts the wave theory. The photon theory of light assumes light consists of photons, each with energy in proportion to the frequency of light according to Einstein's photon equation $E = hf$. The photon theory therefore explains the existence of the threshold frequency because photoelectric emission can only take place if a conduction electron at the surface of the metal absorbs of photon of energy hf greater than or equal to the work function of the metal (i.e. $hf \geq \phi$).

ii Photoelectric emission takes place instantly provided the frequency of light is greater than or equal to the threshold frequency of the metal. This supports the photon theory because conduction electrons at the surface are emitted as soon as the light is incident on the surface. In comparison, according to wave theory, conduction electrons at the surface would need to wait to accumulate sufficient energy from the incident light waves so emission would not be instant.

3 a Light consists of photons of energy proportional to the light frequency. A photon of blue light therefore has more energy than a photon of red light. Each photon directed at the metal surface that is absorbed by a single electron at or near the metal surface gives all its energy to the electron. The energy gained by an electron from a photon of blue light is sufficient to enable it to leave the metal surface. The energy gained by an electron from a photon of red light is not sufficient to enable it to leave the metal surface.

b i Increasing the intensity of blue light causes more photons of blue light to be directed at the surface each second so more electrons are released from the surface each second.

ii If the metal surface is at a positive potential, the maximum kinetic energy of the electrons emitted from the surface would be less as each electron leaving the surface would need to do work to move away from the positively charged surface. If the positive potential is too large, no electrons would have sufficient energy to leave the surface so photoelectric emission would be stopped.

4 **a** 3.3×10^{-19} J **b** 2.0×10^{-19} J

5 **a** The work function of a metal surface is the minimum energy needed by an electron to escape from the surface.

b **i** 9.9×10^{-7} m

ii A photon of longer wavelength would have less energy than 2.0×10^{-19} J and would not therefore be able to give an electron at the metal surface enough energy to escape from the surface.

6 **a** **i** Light consists of photons of energy proportional to the light frequency. Each photon in the beam directed at the metal surface that is absorbed by a single electron at or near the metal surface gives all its energy to the electron. At this light frequency, the energy gained by such an electron is sufficient to enable it to leave the metal surface. The electrons that do leave the metal surface are attracted to the central anode and they then move round the circuit. The microammeter registers a current due to the flow of electrons from the anode round the circuit back to the metal.

ii More photons are directed at the metal surface so more electrons are emitted from its surface. The current therefore increases.

b **i** The light photons have less energy than in a) and each one does not have enough energy to enable an electron to escape from the metal. The energy of a light photon is less than the work function which is the minimum energy needed by an electron to escape from the metal surface.

ii Making the incident light more intense increases the number of photons directed at the metal each second. However, no electrons can escape from the metal because no photon has enough energy to enable an electron to escape.

7 **a** **i** X to Z **ii** Y to Z **b** 3.4×10^{-22} J

8 **a** **i** 17.3 V

ii Energy E of photon released = (−14.2 eV) − (−17.3 eV) = 3.1 eV. Photon frequency = E/h = (3.1 eV × 1.6×10^{-19} J/eV)/6.63×10^{-34} J = 7.48×10^{14} Hz. Photon wavelength = c/f = 3.00×10^{8} m s^{-1} /7.48×10^{14} Hz = 4.0×10^{-7} m.

b **i** B→E 6.8 eV B→D 3.7 eV B→C 1.1 eV
C→E 5.7 eV C→D 2.6 eV D→E 3.1 eV

ii Energy of photon that causes excitation from E to D = 3.1 eV, so its wavelength = 4.0×10^{-7} m (as calculated in **aii**)

9 **a** **i** Kinetic energy of proton = $\frac{1}{2}mv^2$ = 0.5 × 1.67 × 10^{-27} kg × $(1.7 \times 10^{7}\,\text{m s}^{-1})^2$ = 2.41×10^{-13} J

ii Momentum, p = 1.67×10^{-27} kg × 1.7×10^{7} m s^{-1} = 2.84×10^{-20} kg m s^{-1};
de Broglie wavelength = h/p = 6.63×10^{-34} J s/ 2.84×10^{-20} kg m s^{-1} = 2.33×10^{-14} m

b Momentum of photon = E/c = 2.41×10^{-13} J/3.00 × 10^{8} m s^{-1} = 8.03×10^{-22} kg m s^{-1}

10 **a** de Broglie wavelength = h/mv = 6.63×10^{-34} J s/ (9.11×10^{-31} kg × 9.8×10^{7} m s^{-1}) = 7.43×10^{-12} m

b 5° ring; $n = 1$ gives $2d \sin 5°$ = 7.43×10^{-12} m, therefore d = 7.43×10^{-12} m/2 sin 5° = 4.26×10^{-11} m
10° ring; $n = 2$ gives $2d \sin 10°$ = 2 × 7.43×10^{-12} m, therefore d = 7.43×10^{-12} m/sin 10° = 4.28×10^{-11} m
Therefore d = 4.3×10^{-11} m (Note: Data does not justify three significant figures in the answer.)

c Increasing the speed of the electrons reduces their de Broglie wavelength and, since d does not change, the angle of diffraction of each ring decreases.

Chapter 21

1 **a** **i** 5.7×10^{-8} s^{-1} **ii** 6.1×10^{-3} g

b **ii** 16 W

2 **a** 2.7×10^{24} **b** 36 MBq

3 **a** Because radioactive decay is random, the probability of decay is the same for all the nuclei of a given isotope. Therefore, for a large number of nuclei, the number of nuclei that decay per second (i.e. the activity) depends only on the number of radioactive nuclei present. As a result, the number of radioactive nuclei decreases exponentially with time. The rate of decay is characterised by the half life of the isotope which is defined as the time taken for half the initial number of nuclei to decay. If the activity at a certain time is known, the number of atoms present at that time is given by the half life × the activity ÷ ln 2.

b **i** 41 kBq **ii** 146 hours

4 **a** **i** 4.1×10^{14} Bq **ii** 86 J s^{-1}

b 62 K

c **i** Gamma radiation is weakly ionising and therefore can penetrate body tissue and kill cancer cells.

ii The cobalt source is at the centre of a solid lead container which stops all the gamma radiation from the source except radiation that passes along a narrow channel from the source to the outside. A lead cover at the end of the channel used to stop the beam is removed only for a limited time when the beam is directed accurately at the part of the patient to be 'treated'. In this way, the patient is exposed to the beam for the least time necessary.

5 **a** The binding energy of a nucleus is the work that must be done to separate a nucleus into its constituent neutrons and protons.

b **i** 28 MeV **ii** 2.2 MeV

c The binding energy per nucleon of an alpha particle is about 7 MeV which is much greater than that of a deuterium nucleus which is about 1 MeV so the protons and neutrons in an alpha particle are bound much more strongly together than in a deuterium nucleus. In an alpha particle, the strong nuclear force holding the neutrons and protons together is much stronger than the electrostatic force of repulsion between the two protons. This force is independent of whether a nucleon is a neutron or a proton.

6 a ${}^{6}_{3}\text{Li} + {}^{1}_{0}\text{n} \rightarrow {}^{3}_{1}\text{H} + {}^{4}_{2}\alpha$
X is an alpha particle

b i 7.7×10^{-13} J **ii** 4.00150 u

7 a i It has a range of $2–3 \times 10^{-15}$ m; it is an attractive force beyond 0.5×10^{-15} m and is repulsive at less than 0.5×10^{-15} m; it is charge-independent (i.e. the same between protons and neutrons).

ii The binding energy per nucleon of the two fission nuclei is greater than that of the uranium −235 nucleus. The nucleons in the fission nuclei are more tightly bound than in the uranium nucleus. Therefore energy must be released when the uranium nucleus undergoes fission.

b i ${}^{235}_{92}\text{U} + {}^{1}_{0}\text{n} \rightarrow {}^{140}_{54}\text{Xe} + {}^{93}_{38}\text{Sr} + 3\,{}^{1}_{0}\text{n}$

ii 2.8×10^{-11} J

8 a Similarity: the binding energy per nucleon increases in both cases. Difference: a large nucleus splits into two smaller nuclei in a fission reaction; in a fusion reaction, two small nuclei fuse to form a larger nucleus.

b i Two light nuclei need to approach each other within 2–3 fm in order that a strong nuclear force can fuse them together. At 10^{8} K, the nuclei have sufficient kinetic energy to overcome the electrostatic repulsion that would otherwise stop them fusing.

ii ${}^{2}_{1}\text{H} + {}^{2}_{1}\text{H} \rightarrow {}^{3}_{1}\text{H} + {}^{1}_{1}\text{p}$ Energy released 4.1 MeV

9 a They contain many different fission products which are neutron-rich and therefore β^{-} and γ emitters which are highly radioactive. They also contain radioactive isotopes such as plutonium 239 which is highly radioactive.

b i 5.2×10^{9} Bq, **ii** 4.4×10^{8} Bq

10 a The binding energy of a nucleus is the work that must be done to separate a nucleus into its constituent neutrons and protons.

b i 1 proton and 1 neutron **ii** 1.118 MeV

c i 3.8 **ii** B: For the nucleus ${}^{A}_{Z}\text{X}$, there are $A(A - 1)/2$ force bonds which gives a BE ratio of $A(A - 1)/2$ in relation to the ${}^{2}_{1}\text{H}$ nucleus. Applied to the helium ${}^{4}_{2}\text{He}$ nucleus, the ratio in relation to the ${}^{2}_{1}\text{H}$ nucleus is therefore 6, which differs greatly from the actual ratio which is about 12. Student A's suggestion doesn't work either as it predicts a ratio of 2 (= 4 for ${}^{4}_{2}\text{He}$/2 for ${}^{2}_{1}\text{H}$).

d i Y is a proton: ${}^{3}_{1}\text{H} + {}^{2}_{1}\text{H} \rightarrow {}^{4}_{2}\text{He} + {}^{1}_{1}\text{p}$

ii Energy released ≈ 17 MeV

11 a i Nuclear fission is the splitting of a large unstable nucleus into two smaller nuclei as a result of the original nucleus being struck by an incident neutron.

ii Uranium ${}^{235}_{92}\text{U}$

iii When nuclear fission of a fissionable nucleus occurs, the nucleus splits in two and several neutrons referred to as fission neutrons are released. In a nuclear reactor, these fission neutrons collide with other fissionable nuclei causing a chain reaction in which further fission events occur and more neutrons are released which then cause more fission events and so on.

iv In a nuclear reactor, the number of fission neutrons produced per second depends on the mass of the fissile material in the reactor core. However, the number of fission neutrons per second that are lost from the reactor core depends on the surface area of the fuel rods. For a sustainable chain reaction, the mass of fissile material must therefore be greater than a critical mass or else more neutrons per second are lost than are produced and chain reactions cease.

b i In a thermal nuclear reactor, the fission neutrons are released with too much kinetic energy to induce fission events. The moderator is necessary in a thermal nuclear reactor because the moderator atoms reduce the kinetic energy of the fission neutrons that collide with them to a level at which further fission events occur.

ii Boron or cadmium: The fission neutrons are absorbed by the control rod nuclei without causing fission. The rate of production of fission neutrons can therefore be reduced by inserting the control rods further into the reactor core or increased by partially withdrawing the control rods from the reactor core. Automatic sensors are used to monitor the rate of fission events in the reactor core and to control the position of the control rods in the core to maintain a constant fission rate in the core.

iii The coolant must be fluid and non-corrosive.

Chapter 22

1 a 0.14 mm, **ii** 0.62 mm

b i Ultrasonic waves of much lower frequency would diffract more due to their longer wavelength. Increased diffraction would reduce the resolution (i.e. detail) of the images.

ii An ultrasonic probe contains a piezo-electric transducer in the shape of a disc is applied inside and at the end of a tube which contains a backing block that keeps the disc in place. When an alternating p.d. of frequency equal to the natural frequency of vibration of the disc is applied, the disc vibrates in resonance and creates ultrasound waves in the surrounding medium at the same frequency as the alternating p.d.

c i *Specific acoustic impedance* of a substance is the product of its density and the speed of ultrasound in it. **ii 1** 0.999 **2** 1.66×10^{-3}

d The probe is applied to the body via a suitable gel so that most of the ultrasound energy enters the body. This is because an air-skin boundary reflects almost 100% of the incident ultrasonic energy, whereas a gel-skin boundary reflects very little of the incident ultrasonic energy.

2 a i $\lambda_{min} = \frac{hc}{eV} = \frac{6.63 \times 10^{-34}\,\text{Js} \times 3.00 \times 10^{8}\,\text{ms}^{-1}}{1.60 \times 10^{-19}\,\text{C} \times 40000\,\text{V}}$
$= 3.1 \times 10^{-11}\,\text{m}$

ii See Topic 22.2, Figure 2. The graph should show a continuous distribution between zero intensity at zero energy and at maximum energy with peak intensity between. X-ray spikes should be shown at specific energies.

b i A contrast medium is an X-ray absorbing substance that is used to fill a soft organ or blood vessel to make its image stand out from the surrounding tissues.

ii A barium meal containing barium sulphate is a contrast medium. It is given to a patient before an X-ray image of the stomach is taken.

c A CT scanner consists of an X-ray tube and a ring of thousands of small solid-state detectors linked to a computer. The patient lies stationary on a bed along the axis of the ring. The X-ray tube automatically moves round the inside of the ring, turning as it moves so the X-ray beam is always directed at the centre of the ring. The detector signals are simultaneously recorded by the computer each time the X-ray tube moves round the ring through a fraction of a degree until the tube has moved through 180°. For each position of the X-ray tube, the X-rays from the tube travel through the cross-section of the patient under investigation and reach the detectors.
The signal from each detector depends on the different types of tissue along the path and how far the X-ray beam passes through each type of tissue. By considering the patient as a collection of small volume elements (voxels), the signal from each detector is considered to have been attenuated (i.e. reduced in intensity) by each voxel along the path of the X-rays to the detector. The total attenuation along a given path is the sum of the attenuation by each voxel along the path. The computer is programmed to process all the detector signals to determine the attenuation due to each voxel, and then to display a digital image consisting of corresponding pixels, each of greyness or colour according to the attenuation caused by the voxel.

3 a i gamma radiation **ii** A positron-emitting substance is used. Nuclei of the substance are unstable and each one that decays emits a positron which is the antiparticle of the electron. When the emitted positron collides with an electron, they annihilate each other and create two gamma ray photons which travel away from each other in opposite directions.

b i The two gamma ray photons are released simultaneously and because they travel in opposite directions, each one reaches the detector which is along its line of travel which is in the opposite direction to the other one.

ii The straight line XY is a chord of the circular ring which subtends an angle θ of 135° to the centre of the chord. The centre of the circle O, the midpoint M of the chord and either X or Y form a right-angle triangle with a hypotenuse of length 2.40 m. Since the angle MOX is 67.5° (= 0.5 × 135°), the distance MX = 2.40 m × sin 67.5° = 2.217 m. Therefore XY = 4.43 m.

iii The two gamma photons travel in opposite directions from a point at or near the line XY. The positron -emitting nucleus must therefore have been on or near the line XY because the positron emitted by it would have travelled no further than 2 or 3 mm from the nucleus.

iv More simultaneous or near-simultaneous detections of two gamma photons by separate detectors need to be recorded. For each such detection, straight line drawn on a scale diagram of the detector ring between the two relevant detectors would cross the XY line. Where all the lines or groups of lines cross at or near each other, the crossing points must correspond to where the positron-emitting substance is located.
OR The position of the positron-emitting nucleus along XY can be deduced if the time interval Δt between the arrival of the two gamma photons is measured. The time taken T for a photon to travel from X to Y can be calculated by dividing the length of XY by the speed of light. The distance along XY from the first detector triggered is equal to $(\Delta t / T) \times$ XY.

c 1. The PET scanner uses gamma radiation from a positron-emitting substance that has been taken in by the body. The CT scanner uses gamma radiation from an X-ray tube.

2. The PET scanner gives an image of the location of the positron-emitting substance in the body. The CT scanner gives an image of an internal organ in the body.

3. PET scanner detectors are designed to detect gamma photons in very short time intervals. CT scanner detectors measure the intensity of the gamma radiation from the X-ray tube after it has passed through the patient.

Chapter 23

1 a i $\lambda_{max} T = 0.0029\,\text{m K}$ where T is the surface temperature of the star and λ_{max} is the wavelength at which the electromagnetic radiation emitted from the star is at maximum intensity.

ii $T = 0.0029\,\text{m K}/\lambda_{max} = 0.0029\,\text{m K}/740 \times 10^{-9}\,\text{m} = 3900\,\text{K}$

b $L = F \times 4\pi d^2 = 3.5 \times 10^{-8}\,\text{W m}^2 \times 4\pi\,(6.0 \times 10^{17}\,\text{m})^2 = 1.6 \times 10^{29}\,\text{W}$

c i Stefan's law states that the luminosity of a star $L = \sigma A T^4$ where A is the surface area of the star and T is its surface temperature. Since luminosity is the radiant energy per second emitted from the surface, then L/A is the radiant energy per second emitted per unit surface area. Hence $L/A = \sigma T^4$.

ii Rearranging the above equation gives $A = L/\sigma T^4 = 6.2 \times 10^{26}\,\text{W}/[5.67 \times 10^{-8}\,\text{W m}^{-2}\,\text{K}^{-4} \times (9700\,\text{K})^4] = 1.2 \times 10^{22}\,\text{m}^2$. Since $A = 4\pi R^2$ where R is the surface radius, then $R = (A/4\pi)^{1/2} = 3.1 \times 10^{10}\,\text{m}$. Therefore its diameter $= 6.2 \times 10^{10}\,\text{m}$.

2 a i The surface temperature of a star is inversely proportional to its peak intensity wavelength. B is hottest because it has the shortest peak intensity wavelength and C is the coolest because it has the longest peak intensity wavelength.

ii The radiant flux intensity from a star is proportional to its luminosity $L \div$ its distance d squared. A calculation of L/d^2 using the above values gives 4.25 L_{SUN} for A, 1.83 L_{SUN} for B and 0.011 L_{SUN} for C. Therefore, in terms of increasing radiant flux intensity, the order is C, B, A.

b i For star A, $L = 10.6 \times 4.0 \times 10^{26}\,\text{W} = 4.2 \times 10^{27}\,\text{W}$ **ii** Surface temperature $T = 0.0029\,\text{m K}/380 \times 10^{-9}\,\text{m} = 7630\,\text{K}$ **iii** Rearranging $L = \sigma A T^4$ gives $A = L/\sigma T^4 = 4.2 \times 10^{27}\,\text{W}/(5.76 \times 10^{-8}\,\text{W m}^{-2}\,\text{K}^{-4} \times (7630\,\text{K})^4 = 2.15 \times 10^{19}\,\text{m}$. Therefore $R = (A/4\pi)^{1/2} = 1.3 \times 10^9\,\text{m}$.

c i Inserting $A = 4\pi R^2$ and $T = 0.0029\,\text{m K}/\lambda_{max}$ into $L = \sigma A T^4$ gives $L = \sigma(4\pi R^2) \times (0.0029\,\text{m K})^4/\lambda_{max}{}^4$ which simplifies after rearrangement to $R^2 = z\lambda_{max}{}^4 L$ where $z = 4\pi\sigma\,(0.0029\,\text{m K})^4$. Hence R^2 is proportional to $\lambda_{max}{}^4 L$.

ii Values of $L\lambda_{max}{}^4$ for stars B and C relative to A are $(130/10.6) \times (230\,\text{nm}/380\,\text{nm})^4 = 1.6$ for B and $(0.0035/10.6) \times (940\,\text{nm}/380\,\text{nm})^4 = 0.012$ for C. In terms of increasing radius, the order of the three stars is C, A, B.

3 a $L = 4\pi d^2 F = 4\pi\,(1.5 \times 10^{23}\,\text{m})^2 \times 1.4 \times 10^{-13}\,\text{W m}^{-2} = 4.0 \times 10^{34}\,\text{W}$

b i $\lambda_{max} \approx 300\,\text{nm}$; $T = 0.0029\,\text{m K}/300 \times 10^{-9}\,\text{m} = 10\,000\,\text{K}$

ii Rearranging $L = \sigma A T^4$ gives $A = L/\sigma T^4 = 4.0 \times 10^{29}\,\text{W}/[5.67 \times 10^{-8}\,\text{W m}^{-2}\,\text{K}^{-4} \times (10\,000\,\text{K})^4] = 7.1 \times 10^{10}\,\text{m}^2$. Since $A = 4\pi R^2$ where R is the surface radius, $R = (A/4\pi)^{1/2} = 7.5 \times 10^9\,\text{m}$. Therefore its diameter $= 15 \times 10^9\,\text{m} \approx 10 \times$ diameter of the Sun.

4 a *Red shift* is the increase in the wavelength of the light from a light source that is moving away from the observer.

b The increase of wavelength $\Delta\lambda$ in the light from a receding star or galaxy is proportional to its speed of recession v in accordance with the Doppler shift equation $\Delta\lambda = (v/c)\,\lambda$. The increase of wavelength can be determined by comparing the wavelength of the lines of the line spectrum of the light from the star or galaxy with the wavelength of the same lines in the light from a laboratory source. The speed of recession is then calculated by dividing the increase of wavelength for each line by its wavelength and multiplying the result by the speed of light.

c i See Figure 14. **ii** H = gradient $= 2.4 \times 10^{-18}\,\text{s}^{-1}$ ($= 1000\,\text{km s}^{-1}/41 \times 10^{22}\,\text{m}$)

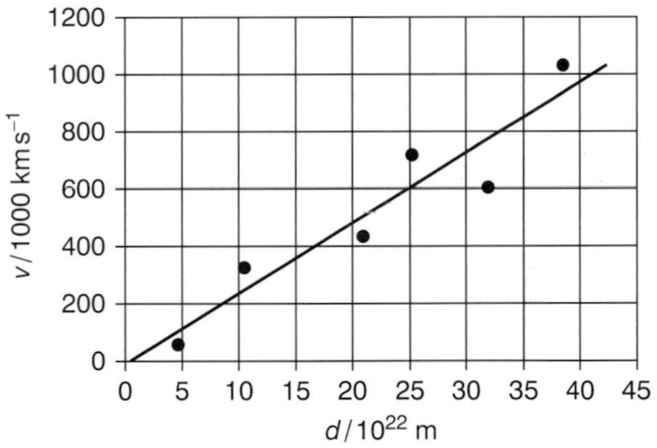

Figure 14

iii Uncertainty estimate $\pm\,0.2 \times 10^{-18}\,\text{s}^{-1}$

c Hubble showed that the speed of recession is proportional to the distance from the Earth. All subsequent measurements confirm this relationship. The changed value of H does not affect the conclusion that the results confirm the proportionality relationship. The changed value of H is due to more accurate methods of measuring the distances.

Answers to questions for Key concepts: Forces (p.90–93)

	Answers	Marks	Comments
1 a i	to ensure its speed of projection was the same every time the test was carried out	**1**	
ii	It might dislodge the bar on impact.	**1**	
	It might bend the tube as it rolled down inside it.	**1**	
b i	mean value of y/mm: 20.3, 71.7, 159, 278.3, 433, 637.7	**2**	2 marks for all correct
		2	1 mark for 3 correct
	x^2/m^2: 0, 0.0420, 0.162, 0.333, 0.643, 0.990, 1.452		as above
ii	$t = x/U$ (from $x = Ut$ rearranged)	**1**	
	substituting in $y = \frac{1}{2}gt^2$ gives		
	$y = \frac{1}{2}g(x/U)^2 = (\frac{1}{2}g/U^2)x^2 = kx^2$		
	where $k = \frac{1}{2}g/U^2$	**1**	
iii	graph: 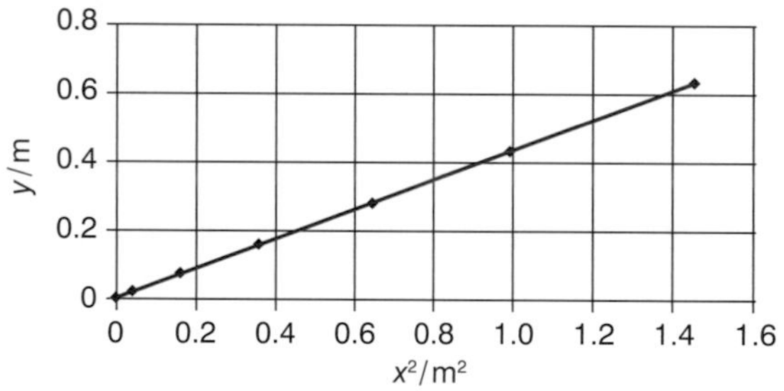		Three sets of measurements are given of the vertical distance fallen for different horizontal distances. In part b, the hypothesis is given that vertical distance is directly proportional to the horizontal distance squared. The mean value of each vertical distance has to be calculated then used to plot a graph of the vertical distance against the horizontal distance squared in order to test the hypothesis.
	labelled correctly	**1**	
	correct units shown on axes	**1**	
	points plotted correctly	**1**	
	best-fit line drawn	**1**	
iv	k = the gradient of the straight line	**1**	Accept answer for U in the range 3.1–3.5 m s^{-1}.
	$= 0.61/1.4 = 0.44\,m^{-1}$	**1**	
	so $U = \sqrt{\frac{g}{2k}} = \sqrt{\frac{9.8}{2 \times 0.44}} = 3.3\,m\,s^{-1}$	**1**	
c	The y values for each distance measurement have a range of more than 2 mm. (For example, for $x = 1.205$ m, the range of y is 12 mm.)	**1**	Part c asks about the precision and accuracy of the measurements. You need to realise here that the uncertainty of each mean value is not given by the precision of a mm scale, but by estimating the spread of each set of measurements (e.g. half the range of the vertical distance measurements).
	So, the y values have a measurement error of more than 2 mm.	**1**	
	For the x-values, although the readings using a mm rule can be made to within 1 mm, the exact position of impact on the bar is uncertain and introduces a further measurement error equal to the diameter of the bar.	**1** **1**	
	Total mark	**20**	

Answers to questions for Key concepts: Electricity, waves and radioactivity (p.162–165)

	Answers	Marks	Comments
1 a i	four out of five components shown correctly for all components shown correctly 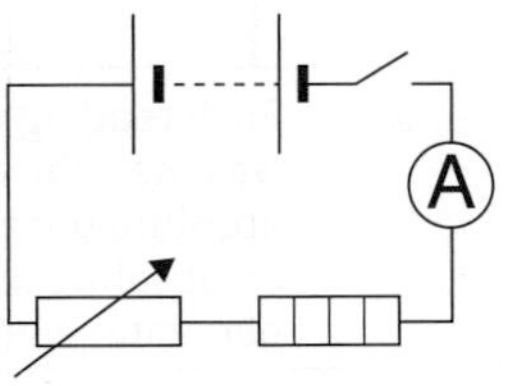 	1 1	The question is about an investigation into the heating effects of an electric current. Part a tests your knowledge of electric circuits.
ii	by adjusting the variable resistor if the current (or the ammeter reading) changed to keep the ammeter reading the same in each test	1 1	Correct reference to ammeter and variable resistor needed for both marks.
b i	(Use the thermometer to) measure the water temperature at intervals after the current is switched off. The insulation is effective if the water temperature does not decrease (or change) over at least 600 s.	1 1	Reference to time scale of at least 600 s necessary for the second mark.
ii	The temperature rise depends on the heating time and the volume (or mass of water). If either of these quantities differs in a test, the temperature rise will be unreliable.	1 1	
c i	the energy supplied to the heater in the time $t = I^2Rt$ The temperature rise is proportional to the energy supplied.	1 1	Here you have to use your knowledge of electrical power and resistance to justify a given prediction and then use your data analysis skills to plot a graph and use it to test the prediction.
ii	I^2 / A^2: 1.44, 4.00, 9.61, 13.7, 20.3	1	
iii	graph 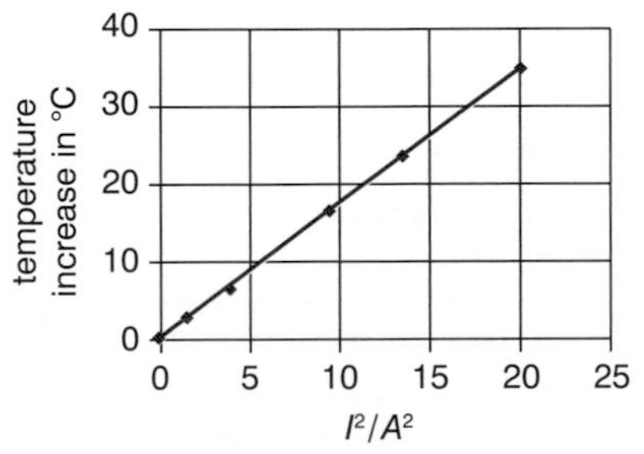 		
	for correctly labelled axes	1	
	for suitable scales	1	
	for correctly plotted points	1	
	for best-fit line	1	
	The graph is a straight line through the origin, so the temperature increase is directly proportional to the current squared, as predicted.	1 1	
d	The resistance of the heater element increased as the water temperature increased. The battery provided a constant e.m.f. So the resistance of the variable resistor had to be reduced to keep the total circuit resistance constant and hence keep the current constant.	1 1 1	You are expected to use your knowledge and understanding about the effect of temperature on the resistance of a metal to explain the observation.
	Total mark	**18**	

Answers to questions for Key concepts: Fields (p.266–269)

	Answers	Marks	Comments
1 a i	With the object on the spring: mean value of $x = 72\,mm$ $e = 70\,mm$	1	
ii	1.4%	1	Each reading was ±0.5 mm. As the extension was the subtraction of two readings, the absolute uncertainties are added to give an absolute uncertainty of ±1.0 mm and a percentage uncertainty of $\frac{1}{70} \times 100 = \pm 1.4\%$.
b i	0.551 s	1	$T_{av} = 11.02\,s$
ii	0.6%	1	The absolute uncertainty can be taken as half the range of the values, so the uncertainty in T_{av} is: $(11.11 - 10.97)/2 = \pm 0.07\,s$
c i	$mg = ke$, therefore $e = mg/k$	1	
ii	Using the above equation gives $\frac{m}{k} = \frac{e}{g}$ Substituting this expression for $\frac{m}{k}$ into the mass–spring time period equation $T = 2\pi\sqrt{\frac{m}{k}}$ gives the required equation.	1 1	
d	Plot either T^2 against e or T against $\sqrt{e}$: correct labels and units suitable scales all points plotted correctly best-fit line According to the equation, the line should pass through the origin and the gradient is equal to $4\pi^2/g$ for the T^2 against e line *or* $2\pi/\sqrt{g}$ for the T against $\sqrt{e}$ line. To determine g, the gradient of the line should be measured. Gradient given the correct unit ($s^2\,m^{-1}$ or $s^2\,mm^{-1}$). A large triangle used correctly to determine the gradient and used with the appropriate gradient formula above to find g.	4 1 1 1 1 1	Draw a triangle or use points that cover over half of the line you have drawn. It improves accuracy. Make sure the points used are on your line and not just two of the plotted points. For example, the graph for T^2 against e is shown below. Its gradient = $4.074\,s^2\,m^{-1}$. Hence $g = 4\pi^2/4.074$ $= 9.69\,m\,s^{-2}$ 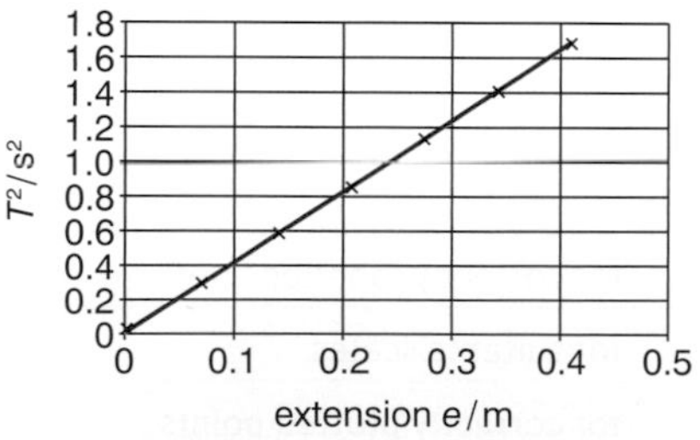
e	The line is straight through the origin. The uncertainty in the gradient can be estimated by considering the uncertainty in each measurement and put error bars on each plotted point, as explained on p.xxi. Draw best-fit lines through the error bars with maximum and minimum gradients, and then give an estimate of the uncertainty in the gradient value.		An alternative method is to estimate the uncertainty in the y-coordinate of the point where the extension is 0.400 m, which is between $1.62\,s^2$ and $1.65\,s^2$, giving a gradient of between $1.62\,s^2/0.400\,m$ (= $4.05\,s^2\,m^{-1}$) and $1.65\,s^2/0.400\,m$ (= $4.13\,s^2\,m^{-1}$) or $4.09 \pm 0.04\,s^2\,m^{-1}$. Prove for yourself that these estimates give $g = 9.65 \pm 0.09\,m\,s^{-2}$.
	Total mark	**20**	

Answers to questions for Key concepts: Thermal and nuclear physics, medical imaging and astrophysics and cosmology (p.351–355)

	Answers	Marks	Comments
1 a i	The $^{4}_{2}$He nucleus has twice the charge of a $^{1}_{1}$H nucleus or proton but it cannot be just two protons because its mass is four times greater than the mass of proton. • The extra mass of 2 units cannot be due to a single particle as mass numbers are integer units.	1	
	Alternative: Rutherford used the neutron–proton model to explain why the mass number of any nucleus heavier than the $^{1}_{1}$H nucleus is greater than its atomic number. The extra mass must be due to 'unit mass' particles, as mass numbers are integer units.	1	
ii	The $^{1}_{1}$H nucleus is the smallest known nucleus and has the least charge / is a single particle / is a proton.	1	
b i	$^{9}_{4}$Be, $^{12}_{6}$C	1	
b ii	any 3 from: The α-particle has to have sufficient kinetic energy: • to overcome the electrostatic repulsion of the nucleus • to reach the nucleus closely enough • to experience the strong nuclear force • which pulls the α-particle into the nucleus • as the strong nuclear force is stronger than the electrostatic force at close range.	3	You may recognise that low-energy alpha particles are scattered elastically.
c i	$E_K = \frac{1}{2}mv^2 = \frac{p^2}{2m}$, where p = momentum = mv From conservation of momentum, the momentum of the emitted α-particle $p_\alpha = -p_{nuc}$, where p_{nuc} is the momentum of the recoil nucleus.	1	Don't forget that the recoil nucleus has a mass of 206 m_u because the a-particle has left the Po-210 nucleus.
	Therefore $\frac{E_k \text{ for the recoil nucleus}}{E_k \text{ for the } \alpha \text{ particle}} = \frac{\left(\frac{p^2}{2m}\right)_{nuc}}{\left(\frac{p^2}{2m}\right)_\alpha} = \frac{m_\alpha}{m_{nuc}} = \frac{4}{206}$	1	
	E_K for the recoil nucleus = $\frac{4}{206} \times 5.3\,\text{MeV} = 0.1\,\text{MeV}$	1	
	So the total energy released = 5.3 + 0.1 = 5.4 MeV.	1	
c ii	If $R = kE^n$, a graph of ln R on the y-axis against ln E on the x-axis should give a straight line with a gradient equal to n.	1	This is an excellent example of the use of a log–log graph to determine the numerical power of an equation of the form $y = kx^n$. If this method does not generate a straight-line graph, the equation cannot be of the form above.

R/mm	39	48	53	57	66	78
E/MeV	5.3	6.0	6.5	6.8	7.4	8.3
ln R	3.66	3.87	3.97	4.04	4.19	4.36
ln E	1.67	1.79	1.87	1.92	2.00	2.12

	Answers	Marks	Comments
	Correct values to 2 or more significant figures for: • ln R • ln E	 1 1	
	suitable scales	1	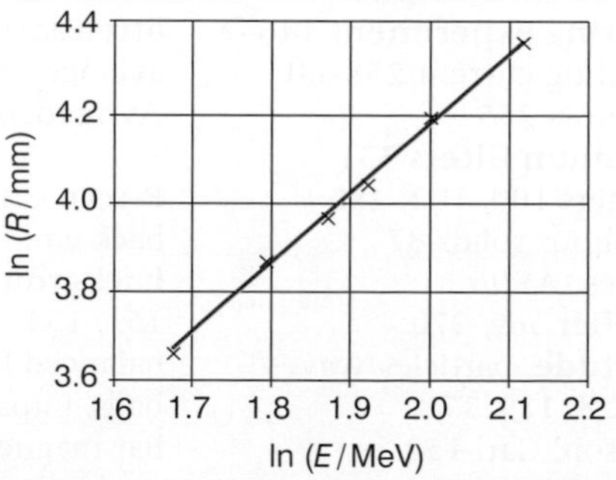
	correctly labelled scales	1	
	all points plotted correctly	1	
	best-fit line	1	
	correct calculation of gradient e.g. gradient = $\frac{4.34 - 3.72}{2.10 - 1.70} = 1.55$ $n = 1.55$	1	
	n in the range = 1.5 to 1.6	1	
	Total mark	20	

Index

Headings in **bold** indicate glossary terms. Page numbers in *italic* refer to figures; page numbers in **bold** refer to tables.